Summary of Project Analysis Methods

Analysis Method	Description	Single Project Evaluation	Mutually Exclusive Projects	
			Revenue Projects	**Service Projects**
Payback period PP	A method for determining when in a project's history it breaks even. Management sets the benchmark PP°.	$PP < PP^\circ$	Select the one with the shortest PP.	
Discounted payback period $PP(i)$	A variation of payback period that factors in the time value of money. Management sets the benchmark PP^*.	$PP(i) < PP^*$	Select the one with the shortest $PP(i)$.	
Present worth $PW(i)$	An equivalent method that translates a project's cash flows into a net present value.	$PW(i) > 0$	Select the one with the largest PW.	Select the one with the least negative PW.
Future worth $FW(i)$	An equivalence method variation of the PW: a project's cash flows are translated into a net future value.	$FW(i) > 0$	Select the one with the largest FW.	Select the one with the least negative FW.
Capitalized equivalent $CE(i)$	An equivalence method variation of the PW of a perpetual or very long-lived project that generates a constant annual net cash flow.	$CE(i) > 0$	Select the one with the largest CE.	Select the one with the least negative CE.
Annual equivalence $AE(i)$	An equivalence method and variation of the PW: a project's cash flows are translated into an annual equivalent sum.	$AE(i) > 0$	Select the one with the largest AE.	Select the one with the least negative AE.
Internal rate of return IRR	A relative percentage method that measures the yield as a percentage of investment over the life of a project: the IRR must exceed the minimum required rate of return (MARR).	$IRR > MARR$	Incremental analysis: If $IRR_{A2-A1} > MARR$, select the higher cost investment project, A2.	
Benefit–cost ratio $BC(i)$	An equivalence method to evaluate public projects by finding the ratio of the equivalent benefit over the equivalent cost.	$BC(i) > 1$	Incremental analysis: If $BC(i)_{A2-A1} > 1$, select the higher cost investment project, A2.	

CONTEMPORARY ENGINEERING ECONOMICS

CHAN S. PARK • MING J. ZUO • RONALD PELOT

CONTEMPORARY ENGINEERING ECONOMICS

A CANADIAN PERSPECTIVE   THIRD CANADIAN EDITION

Pearson Canada
Toronto

Library and Archives Canada Cataloguing in Publication

Park, Chan S.
Contemporary engineering economics : a Canadian
perspective / Chan S. Park,
Ming J. Zuo, Ronald Pelot. — 3rd Canadian ed.

Includes index.
ISBN 978-0-321-53876-5

1. Engineering economy. 2. Engineering economy—
Canada. I. Zuo, Ming J. II. Pelot, Ron, 1957- III. Title.

TA177.4.P37 2011 658.15024'62 C2010-904795-8

ISBN 978-0-321-53876-5

Vice-President, Editorial Director: Gary Bennett
Executive Editor: Cathleen Sullivan
Marketing Manager: Michelle Bish
Developmental Editor: Darryl Kamo
Project Manager: Sarah Lukaweski
Production Editor: Heather Sangster, Strong Finish
Copy Editor: Meghan Newton, Strong Finish
Proofreader: Aspasia Bissas, Strong Finish
Compositor: MPS Limited, a Macmillan Company
Photo Researcher: MRM Associates
Permissions Researcher: MRM Associates
Art Director: Julia Hall
Interior and Cover Designer: Anthony Leung
Cover Image: Veer Inc. (Electric Power Collage); GettyImages (Rooftop Garden, Centre Left); ShutterStock
(Calgary, Alberta, Canada, Bottom Left).

1 2 3 4 5   15 14 13 12 11

Printed and bound in the United States of America.

In memory of my mother, Hon Yong Hee, whose wisdom as a parent has inspired those she influenced.

Chan S. Park

To my wife, Ning; and my children, Kevin and Lillian.

Ming J. Zuo

To my parents, Vi and Gerry, my family, and my friends.

Ronald Pelot

CONTENTS

Preface   xix

PART 1 BASICS OF FINANCIAL DECISIONS   1

Chapter 1   Engineering Economic Decisions   2

1.1	Role of Engineers in Business	5
	1.1.1 Types of Business Organization	5
	1.1.2 Engineering Economic Decisions	6
	1.1.3 Personal Economic Decisions	7
1.2	What Makes the Engineering Economic Decision Difficult?	7
1.3	Economic Decisions versus Design Decisions	9
1.4	Large-Scale Engineering Projects	9
	1.4.1 How a Typical Project Idea Evolves	9
	1.4.2 Impact of Engineering Projects on Financial Statements	13
	1.4.3 A Look Back in 2009: Did Toyota Make the Right Decision?	13
1.5	Common Types of Strategic Engineering Economic Decisions	14
1.6	Fundamental Principles of Engineering Economics	15
	Summary	17

Chapter 2   Understanding Financial Statements   19

2.1	Accounting: The Basis of Decision Making	22
2.2	Financial Status for Businesses	24
	2.2.1 The Balance Sheet	26
	2.2.2 The Income Statement	30
	2.2.3 The Cash Flow Statement	34
2.3	Using Ratios to Make Business Decisions	38
	2.3.1 Debt Management Analysis	39
	2.3.2 Liquidity Analysis	42
	2.3.3 Asset Management Analysis	43
	2.3.4 Profitability Analysis	45
	2.3.5 Market Value Analysis	47
	2.3.6 Limitations of Financial Ratios in Business Decisions	48
	Summary	49
	Problems	50
	Short Case Studies	56

Chapter 3   Time Value of Money and Economic Equivalence   58

3.1	**Interest: The Cost of Money**	**60**
	3.1.1 The Time Value of Money	61
	3.1.2 Elements of Transactions Involving Interest	62
	3.1.3 Methods of Calculating Interest	65
	3.1.4 Simple Interest versus Compound Interest	68
3.2	**Economic Equivalence**	**69**
	3.2.1 Definition and Simple Calculations	69
	3.2.2 Equivalence Calculations: General Principles	72
	3.2.3 Looking Ahead	77
3.3	**Development of Interest Formulas**	**77**
	3.3.1 The Five Types of Cash Flows	78
	3.3.2 Single-Cash-Flow Formulas	79
	3.3.3 Uneven Payment Series	86
	3.3.4 Equal Payment Series	90
	3.3.5 Linear Gradient Series	102
	3.3.6 Geometric Gradient Series	108
	3.3.7 Limiting Forms of Interest Formulas	114
3.4	**Unconventional Equivalence Calculations**	**114**
	3.4.1 Composite Cash Flows	115
	3.4.2 Determining an Interest Rate to Establish Economic Equivalence	120
	Summary	**125**
	Problems	**125**
	Short Case Studies	**135**

Chapter 4   Understanding Money and Its Management   139

4.1	**Nominal and Effective Interest Rates**	**141**
	4.1.1 Nominal Interest Rates	141
	4.1.2 Effective Annual Interest Rates	142
	4.1.3 Effective Interest Rates per Payment Period	145
	4.1.4 Continuous Compounding	146
4.2	**Equivalence Calculations With Effective Interest Rates**	**148**
	4.2.1 When Payment Period Is Equal to Compounding Period	149
	4.2.2 Compounding Occurs at a Different Rate Than That at Which Payments Are Made	150
4.3	**Equivalence Calculations With Continuous Payments**	**157**
	4.3.1 Single-Payment Transactions	157
	4.3.2 Continuous-Funds Flow	157

4.4   **Changing Interest Rates**   **161**
       4.4.1   Single Sums of Money   161
       4.4.2   Series of Cash Flows   163
4.5   **Debt Management**   **164**
       4.5.1   Commercial Loans   164
       4.5.2   Loan versus Lease Financing   172
       4.5.3   Mortgages   176
4.6   **Investment in Bonds**   **183**
       4.6.1   Bond Terminology   183
       4.6.2   Types of Bonds   184
       4.6.3   Bond Valuation   185
       Summary   **189**
       Problems   **190**
       Short Case Studies   **201**

PART 2 EVALUATION OF BUSINESS AND ENGINEERING ASSETS 205

Chapter 5 Analysis of Independent Projects 206

5.1   **Describing Project Cash Flows**   **209**
       5.1.1   Loan versus Project Cash Flows   209
       5.1.2   Independent versus Mutually Exclusive Investment Projects   212
5.2   **Initial Project Screening Method**   **212**
       5.2.1   Payback Period: The Time It Takes to Pay Back   213
       5.2.2   Benefits and Flaws of Payback Screening   215
       5.2.3   Discounted Payback Period   216
       5.2.4   Where Do We Go From Here?   217
5.3   **Present-Worth Analysis**   **217**
       5.3.1   Net-Present-Worth Criterion   218
       5.3.2   Meaning of Net Present Worth   222
       5.3.3   Basis for Selecting the MARR   224
5.4   **Variations of Present-Worth Analysis**   **225**
       5.4.1   Future-Worth Analysis   225
       5.4.2   Capitalized Equivalent Method   229
5.5   **Annual Equivalent-Worth Criterion**   **234**
       5.5.1   Fundamental Decision Rule   234
       5.5.2   Annual-Worth Calculation With Repeating Cash Flow Cycles   237
       5.5.3   Capital Costs versus Operating Costs   238
       5.5.4   Benefits of AE Analysis   241

5.5.5	Unit Profit or Cost Calculation	242
5.5.6	Pricing the Use of an Asset	243
5.6	**Rate of Return Analysis**	**245**
5.6.1	Return on Investment	245
5.6.2	Return on Invested Capital	247
5.6.3	Simple versus Nonsimple Investments	248
5.6.4	Methods of Finding i^*	249
5.7	**Internal-Rate-of-Return Criterion**	**256**
5.7.1	Relationship to PW Analysis	256
5.7.2	Decision Rule for Simple Investments	256
5.7.3	Decision Rule for Nonsimple Investments	259
	Summary	261
	Problems	264
	Short Case Studies	285
	Appendix 5A: Computing IRR for Nonsimple (Mixed) Investments	287
5A.1	**Predicting Multiple i^*s**	**287**
5A.2	**Net-Investment Test: Pure versus Mixed Investments**	**289**
5A.3	**External Interest Rate for Mixed Investments**	**292**
5A.4	**Calculation of Return on Invested Capital for Mixed Investments**	**293**
5A.5	**Trial-and-Error Method for Computing IRR for Mixed Investments**	**297**
	Summary	298
	Problems	299

Chapter 6   Comparing Mutually Exclusive Alternatives   303

6.1	**Basic Elements of Mutually Exclusive Analysis**	**305**
6.1.1	The "Do Nothing" Decision Option	305
6.1.2	Service Projects versus Revenue Projects	306
6.1.3	Scale of Investment	306
6.2	**Total-Investment Approach**	**308**
6.2.1	Present-Worth Comparison	308
6.2.2	Annual Equivalent Comparison	310
6.3	**Incremental Investment Analysis**	**312**
6.3.1	Flaws in Project Ranking by IRR	312
6.3.2	Incremental Investment Analysis	312
6.4	**Analysis Period**	**320**
6.4.1	Analysis Period Differs From Project Lives	321
6.4.2	Analysis Period Is Not Specified	329
6.5	**Make-or-Buy Decision**	**336**

6.6	Life-Cycle Cost Analysis	339
6.7	Design Economics	346
	Summary	354
	Problems	356
	Short Case Studies	374

PART 3 ANALYSIS OF PROJECT CASH FLOWS 383

Chapter 7 Cost Concepts Relevant to Decision Making 384

7.1	General Cost Terms	386
	7.1.1 Manufacturing Costs	386
	7.1.2 Nonmanufacturing Costs	388
7.2	Classifying Costs for Financial Statements	388
	7.2.1 Period Costs	389
	7.2.2 Product Costs	389
7.3	Cost Classification for Predicting Cost Behaviour	392
	7.3.1 Volume Index	392
	7.3.2 Cost Behaviours	392
7.4	Future Costs for Business Decisions	398
	7.4.1 Differential Cost and Revenue	398
	7.4.2 Opportunity Cost	402
	7.4.3 Sunk Costs	404
	7.4.4 Marginal Cost	404
7.5	Estimating Profit From Production	409
	7.5.1 Calculation of Operating Income	410
	7.5.2 Sales Budget for a Manufacturing Business	410
	7.5.3 Preparing the Production Budget	411
	7.5.4 Preparing the Cost-of-Goods-Sold Budget	413
	7.5.5 Preparing the Nonmanufacturing Cost Budget	414
	7.5.6 Putting It All Together: The Budgeted Income Statement	416
	7.5.7 Looking Ahead	417
	Summary	418
	Problems	419
	Short Case Study	425

Chapter 8 Depreciation 426

8.1	Asset Depreciation	428
	8.1.1 Economic Depreciation	429
	8.1.2 Accounting Depreciation	430

8.2	**Factors Inherent in Asset Depreciation**	**430**
	8.2.1 Depreciable Property	430
	8.2.2 Cost Basis	431
	8.2.3 Useful Life and Salvage Value	433
	8.2.4 Depreciation Methods: Book and Tax Depreciation	433
8.3	**Book Depreciation Methods**	**434**
	8.3.1 Straight-Line (SL) Method	434
	8.3.2 Accelerated Methods	436
	8.3.3 Units-of-Production (UP) Method	442
8.4	**Capital Cost Allowance for Income Tax**	**442**
	8.4.1 CCA System	443
	8.4.2 Available-for-Use Rule	445
	8.4.3 Calculating the Capital Cost Allowance	447
	8.4.4 The 50% Rule	450
	8.4.5 CCA for Individual Projects	454
8.5	**Additions or Alterations to Depreciable Assets**	**456**
	8.5.1 Revision of Book Depreciation	456
	8.5.2 Revision of Tax Depreciation	456
8.6	**Tax Depreciation in the U.S. (Optional)**	**459**
	8.6.1 MACRS Depreciation	459
	8.6.2 MACRS Depreciation Rules	460
	Summary	**464**
	Problems	**466**
	Short Case Studies	**473**

Chapter 9 Corporate Income Taxes 475

9.1	**Income Tax Fundamentals**	**477**
	9.1.1 Tax Rate Definitions	477
	9.1.2 Capital Gains (Losses)	478
9.2	**Net Income**	**480**
	9.2.1 Calculation of Net Income	480
	9.2.2 Treatment of Depreciation Expenses	481
	9.2.3 Taxable Income and Income Taxes	481
	9.2.4 Net Income versus Cash Flow	483
9.3	**Corporate Income Tax Calculation**	**486**
	9.3.1 Combined Corporate Tax Rates on Operating Income	486
	9.3.2 Income Taxes on Operating Income	488
	9.3.3 Corporate Operating Losses	490
	9.3.4 Income Taxes on Non-operating Income	490
	9.3.5 Investment Tax Credits	491

9.4   Practical Issues Around Corporate Income Tax   493
    9.4.1   Incremental Corporate Tax Rate Specified   493
    9.4.2   Claiming Maximum Deductions   494
    9.4.3   Timing of Corporate Income Tax Payments   494
9.5   Depreciable Assets: Disposal Tax Effects   495
    9.5.1   Calculation of Disposal Tax Effects   496
    Summary   499
    Problems   500
    Short Case Studies   505

Chapter 10   Developing Project Cash Flows   507

10.1   Cost–Benefit Estimation for Engineering Projects   510
    10.1.1 Simple Projects   511
    10.1.2 Complex Projects   511
10.2   Incremental Cash Flows   512
    10.2.1 Elements of Cash Outflows   513
    10.2.2 Elements of Cash Inflows   514
    10.2.3 Classification of Cash Flow Elements   514
10.3   Developing Cash Flow Statements   516
    10.3.1 When Projects Require Only Operating and Investing Activities   518
    10.3.2 When Projects Require Working-Capital Investments   521
    10.3.3 When Projects Are Financed With Borrowed Funds   526
    10.3.4 When Projects Result in Negative Taxable Income   528
    10.3.5 When Projects Require Multiple Assets   531
    10.3.6 When Investment Tax Credits Are Allowed   534
10.4   Generalized Cash Flow Approach   537
    10.4.1 Setting Up Net Cash Flow Equations   537
    10.4.2 Presenting Cash Flows in Compact Tabular Formats   538
    10.4.3 Lease-or-Buy Decision   540
10.5   The After-Tax Cash Flow Diagram Approach   545
    10.5.1 When There Is No Debt Financing   546
    10.5.2 When There Is Debt Financing in the Form of an Amortized Loan   547
    10.5.3 When There Is Debt Financing in the Form of Bond-Type Loan   547
    10.5.4 Other Considerations   549
    Summary   550
    Problems   551
    Short Case Studies   563

PART 4 SPECIAL TOPICS IN ENGINEERING
ECONOMICS 567

Chapter 11 Replacement Decisions 568

11.1	**Replacement Analysis Fundamentals**	**570**
	11.1.1 Basic Concepts and Terminology	570
	11.1.2 Opportunity Cost Approach to Comparing Defender and Challenger	573
11.2	**Economic Service Life**	**575**
11.3	**Replacement Analysis When the Required Service Is Long**	**580**
	11.3.1 Required Assumptions and Decision Frameworks	581
	11.3.2 Replacement Strategies Under the Infinite Planning Horizon	582
	11.3.3 Replacement Strategies Under the Finite Planning Horizon	587
	11.3.4 Consideration of Technological Change	590
11.4	**Replacement Analysis With Tax Considerations**	**591**
	Summary	**607**
	Problems	**608**
	Short Case Studies	**619**

Chapter 12 Capital-Budgeting Decisions 626

12.1	**Methods of Financing**	**628**
	12.1.1 Equity Financing	629
	12.1.2 Debt Financing	631
	12.1.3 Capital Structure	632
12.2	**Cost of Capital**	**637**
	12.2.1 Cost of Equity	637
	12.2.2 Cost of Debt	640
	12.2.3 Calculating the Cost of Capital	642
12.3	**Choice of Minimum Attractive Rate of Return**	**643**
	12.3.1 Choice of MARR When Project Financing Is Known	643
	12.3.2 Choice of MARR When Project Financing Is Unknown	645
	12.3.3 Choice of MARR Under Capital Rationing	647
12.4	**Capital Budgeting**	**651**
	12.4.1 Evaluation of Multiple Investment Alternatives	651
	12.4.2 Formulation of Mutually Exclusive Alternatives	651
	12.4.3 Capital-Budgeting Decisions With Limited Budgets	653
	Summary	**657**
	Problems	**658**
	Short Case Studies	**664**

Chapter 13   Economic Analysis in the Public Sector   670

13.1   **What Is the Service Sector?**   672
        13.1.1 Characteristics of the Service Sector   673
        13.1.2 How to Price Service   673

13.2   **Economic Analysis in Health-Care Service**   675
        13.2.1 Economic Evaluation Tools   675
        13.2.2 Cost-Effectiveness Analysis   676
        13.2.3 How to Use a CEA   677

13.3   **Economic Analysis in the Public Sector**   680
        13.3.1 What Is Benefit–Cost Analysis?   681
        13.3.2 Framework of Benefit–Cost Analysis   681
        13.3.3 Valuation of Benefits and Costs   682
        13.3.4 Quantifying Benefits and Costs   684
        13.3.5 Difficulties Inherent in Public-Project Analysis   688

13.4   **Benefit–Cost Ratios**   689
        13.4.1 Definition of Benefit–Cost Ratio   689
        13.4.2 Relationship Between B/C Ratio and NPW   692
        13.4.3 Comparing Mutually Exclusive Alternatives: Incremental Analysis   692

13.5   **Analysis of Public Projects Based on Cost-Effectiveness**   695
        13.5.1 Cost-Effectiveness Studies in the Public Sector   695
        13.5.2 A Cost-Effectiveness Case Study   696

        Summary   704
        Problems   705
        Short Case Studies   710

PART 5   HANDLING RISK AND UNCERTAINTY   717

Chapter 14   Inflation and Its Impact on Project Cash Flows   718

14.1   **Meaning and Measure of Inflation**   720
        14.1.1 Measuring Inflation   720
        14.1.2 Actual versus Constant Dollars   726

14.2   **Equivalence Calculations Under Inflation**   729
        14.2.1 Market and Inflation-Free Interest Rates   729
        14.2.2 Constant-Dollar Analysis   730
        14.2.3 Actual-Dollar Analysis   731
        14.2.4 Mixed-Dollar Analysis   734

14.3   **Effects of Inflation on Project Cash Flows**   735

	14.3.1 Multiple Inflation Rates	738
	14.3.2 Effects of Borrowed Funds Under Inflation	740
14.4	**Rate-of-Return Analysis Under Inflation**	**743**
	14.4.1 Effects of Inflation on Return on Investment	743
	14.4.2 Effects of Inflation on Working Capital	746
	Summary	**749**
	Problems	**751**
	Short Case Studies	**759**

Chapter 15   Project Risk and Uncertainty   762

15.1	**Origins of Project Risk**	**764**
15.2	**Methods of Describing Project Risk**	**765**
	15.2.1 Sensitivity Analysis	765
	15.2.2 Sensitivity Analysis for Mutually Exclusive Alternatives	769
	15.2.3 Break-Even Analysis	771
	15.2.4 Scenario Analysis	774
15.3	**Probability Concepts for Investment Decisions**	**776**
	15.3.1 Assessment of Probabilities	776
	15.3.2 Summary of Probabilistic Information	781
	15.3.3 Joint and Conditional Probabilities	783
	15.3.4 Covariance and Coefficient of Correlation	785
15.4	**Probability Distribution of NPW**	**785**
	15.4.1 Procedure for Developing an NPW Distribution	785
	15.4.2 Aggregating Risk Over Time	791
	15.4.3 Decision Rules for Comparing Mutually Exclusive Risky Alternatives	796
15.5	**Risk Simulation**	**798**
	15.5.1 Computer Simulation	799
	15.5.2 Model Building	800
	15.5.3 Monte Carlo Sampling	803
	15.5.4 Simulation Output Analysis	808
	15.5.5 Risk Simulation With @RISK	810
15.6	**Decision Trees and Sequential Investment Decisions**	**813**
	15.6.1 Structuring a Decision-Tree Diagram	814
	15.6.2 Worth of Obtaining Additional Information	819
	15.6.3 Decision Making After Having Imperfect Information	822
	Summary	**827**
	Problems	**828**
	Short Case Studies	**837**

PART 6 PERSONAL INVESTMENT ANALYSIS AND STRATEGY 843

Chapter 16 Principles of Investing and Personal Income Tax 844

16.1	**Investing in Financial Assets**	**846**
	16.1.1 Investment Basics	846
	16.1.2 Cash Investments	847
	16.1.3 Bond Investments	849
	16.1.4 Stock Investments	852
16.2	**Investment Strategies**	**855**
	16.2.1 How to Determine Your Expected Return	856
	16.2.2 Investment Risks	859
	16.2.3 Investment Timing Issues	862
	16.2.4 Diversification Reduces Risk	863
	16.2.5 Dollar Cost Averaging	866
16.3	**Personal Income Tax**	**867**
	16.3.1 Calculation of Taxable Income	868
	16.3.2 Income Taxes on Ordinary Income	869
	16.3.3 Timing of Personal Income Tax Payments	872
	16.3.4 Professional/Business Income	873
16.4	**Effects of Income Taxes on Investment Returns**	**874**
	16.4.1 Bond Investments After Tax	875
	16.4.2 Stock Investments After Tax	878
	16.4.3 Registered Retirement Savings Plans (RRSPs)	880
	16.4.4 Registered Education Savings Plans (RESPs)	882
	Summary	**883**
	Problems	**884**
	Short Case Studies	**889**

English/French Concordance 891

Appendix Interest Factors for Discrete Compounding 901

Index 931

PREFACE

What Is "Contemporary" About Engineering Economics?

Decisions made during the engineering design phase of product development determine the majority of the costs associated with the manufacturing of that product (some say that this value may be as high as 85%). As design and manufacturing processes become more complex, engineers make decisions that involve money more often than ever before. With more than 75% of the total gross domestic product (GDP) in Canada provided by the service sector, engineers work on various economic decision problems in the service sector as well. The competent and successful twenty-first-century engineer must have an improved understanding of the principles of science, engineering, and economics, coupled with relevant design experience. Increasingly, in the new world economy, successful businesses will rely on engineers with such expertise.

Economic and design issues are inextricably linked in the product/service life cycle. Therefore, one of our strongest motivations for writing this text was to bring the realities of economics and engineering design into the classroom and to help students integrate these issues when contemplating many engineering decisions. Of course, our underlying motivation for writing this book was not simply to address contemporary needs but also to address the ageless goal of all educators: to help students learn. Thus, thoroughness, clarity, and accuracy of presentation of essential engineering economics represent our aims at every stage in the development of the text.

Although the major focus of this book is the economic evaluation of engineering projects, the methodology is directly applicable to any financial situation. Hence, a secondary theme in the text is personal financial management and the analysis of personal investments.

The Third Canadian Edition

Much of the content in the third Canadian edition of *Contemporary Engineering Economics: A Canadian Perspective* has been streamlined to provide materials in depth and to reflect the challenges in contemporary engineering economics. Some of the highlighted changes are as follows:

- Coverage of Basic Financial Reports has been expanded to include the analysis of reports. The idea is to emphasize the link between the financial viability of the enterprise with economic decisions around individual projects. This material now appears as new Chapter 2.
- Various cost elements and types of costs are relevant to the evaluation of a project's attractiveness. Given the importance of understanding costs from various perspectives, this material has been brought together in a new Chapter 7 dedicated to this topic and ties back to the expanded discussion on financial reporting in new Chapter 2.
- Our belief that a secondary benefit of this course is to equip students to make better personal investment decisions may not be shared by all instructors. The third Canadian edition therefore gathers all of the personal investment material, including personal income tax considerations, into a new standalone Chapter 16 at the end of the book.

- The new edition includes revisions to all of the corporate and personal income tax information, consistent with current Canada Revenue Agency rules.
- Replacement decisions were dealt with in the previous edition in two different chapters (Chapters 6 and 11). In the third Canadian edition, this topic is addressed in Chapter 11 only.
- The chapter on the economic analysis in the public sector is expanded and appears as Chapter 13: "Economic Analysis in the Public Sector." This revised chapter now provides economic analysis unique to service sectors beyond the government sector. Increasingly, engineers seek their career in the service sector, such as in health care, financial institutions, transportation, and logistics. In this chapter, we present some unique features that must be considered when evaluating investment projects in the service sector.
- The opening vignettes at the beginning of each chapter, as well as the examples included throughout the book, have been updated.
- Short case studies have been added to the end-of-chapter material, and these can be used as the basis for in-class discussions or group projects.
- Finally, the third Canadian edition of this text is printed in two colours, making it easier to read and to see highlighted material.

Overview of the Text

Although it contains little advanced math and few truly difficult concepts, the introductory engineering economics course is often a curiously challenging one for engineering students of all levels. There are several likely explanations for this difficulty.

1. The course is the student's first analytical consideration of money (a resource with which he or she may have had little direct contact beyond paying for tuition, housing, food, and textbooks).
2. The emphasis on theory may obscure for the student the fact that the course aims, among other things, to develop a practical set of analytical tools for measuring project worth. This is unfortunate since, at one time or another, virtually every engineer—not to mention every individual—is responsible for the wise allocation of limited financial resources.
3. The mixture of industrial, civil, mechanical, electrical, chemical, and manufacturing engineering students, as well as other undergraduates who take the course, often fail to "see themselves" using the skills the course and text are intended to foster. This is perhaps less true for industrial engineering students, whom many texts view as their primary audience, but other disciplines are often motivationally shortchanged by a text's lack of applications that appeal directly to them.

Goal of the Text

This text aims not only to provide sound and comprehensive coverage of the concepts of engineering economics but also to address the difficulties of students outlined above, all of which have their basis in inattentiveness to the practical concerns of engineering economics. More specifically, this text has the following chief goals:

1. To build a thorough understanding of the theoretical and conceptual basis upon which the practice of financial project analysis is built.

2. To satisfy the practical needs of the engineer toward making informed financial decisions when acting as a team member or project manager for an engineering project.

3. To incorporate all critical decision-making tools—including the most contemporary, computer-oriented ones that engineers bring to the task of making informed financial decisions.

4. To appeal to the full range of engineering disciplines for which this course is often required: industrial, civil, mechanical, electrical, computer, mining, chemical, and manufacturing engineering, as well as engineering technology.

Prerequisites

The text is intended for undergraduate engineering students in second year or higher. The only mathematical background required is elementary calculus. For Chapter 15, a first course in probability or statistics is helpful but not necessary, since the treatment of basic topics there is essentially self-contained.

Taking Advantage of the Internet

The integration of computer use is another important feature of *Contemporary Engineering Economics: A Canadian Perspective*. Students have greater access to and familiarity with the various spreadsheet tools, and instructors have greater inclination either to treat these topics explicitly in the course or to encourage students to experiment independently.

A remaining concern is that the use of computers will undermine true understanding of course concepts. This text does not promote the use of trivial spreadsheet applications as a replacement for genuine understanding of and skill in applying traditional solution methods. Rather, it focuses on the computer's productivity-enhancing benefits for complex project cash flow development and analysis.

Additionally, spreadsheets are introduced via Microsoft® Excel examples. For spreadsheet coverage, the emphasis is on demonstrating a chapter concept that embodies some complexity that can be much more efficiently resolved on a computer than by traditional longhand solutions.

Supplements

- Instructor's Solutions Manual
- Instructor PowerPoint Slides
- Image Library
- Companion Website

On the Companion Website

Together with the publication of the textbook, we have developed a Companion Website. The website will provide instant, online access to materials that complement the textbook. Specifically, the following materials will be made available on the Companion Website:

- **Student Resources:** Sample test questions are provided so that students can test their understanding of the materials covered in the textbook and know the answers right away. Excel files and computer notes are provided to explain the equations used in the Excel files so that students can learn how to use Excel to solve engineering economic analysis problems.

- **Instructor Resources:** The Instructor's Solutions Manual provides classification of end-of-chapter problems, solutions to all end-of-chapter problems, and lecture notes. Instructor PowerPoint Slides and an Image Library are also provided.
- **Other Resources:** These include capital tax factors, advanced topics on disposal tax effects, and updated Canadian tax information.
- **Online Tools:** Provided on the Companion Website are interest tables, Cash Flow Analyzer, loan analysis, and depreciation analysis tools. Cash Flow Analyzer is an integrated online Java program that is menu-driven for convenience and flexibility. It provides (1) a flexible and easy-to-use cash flow editor for data input and modifications and (2) an extensive array of computational modules and user-selected graphic outputs.

CourseSmart for Students and Instructors

CourseSmart goes beyond traditional expectations—providing instant, online access to the textbooks and course materials you need at a lower cost for students. And even as students save money, you can save time and hassle with a digital eTextbook that allows you to search for the most relevant content at the very moment it's needed. Whether it's evaluating textbooks or creating lecture notes to help students with difficult concepts, CourseSmart can make life a little easier. See how when you visit **www.coursesmart.com**.

Acknowledgments

A number of individuals provided assistance in the preparation of the earlier editions of this textbook. Their contributions have carried over directly or indirectly into the third Canadian edition. We again acknowledge the following people: John R. McDougall, Alberta Research Council; Douglas A. Hackbarth, Hackbarth Environmental Consultants; Ron G. Bryant, RBC Dominion Securities; Barry J. Walker, Peterson Walker Chartered Accountants; Carl F. Hunter, Dalcor Consultants Ltd.; Tony D'Andrea, Toronto Dominion Bank; Glen A. Murney, University of Alberta; Scott Dunbar, University of British Columbia; Ken Rose, Queen's University; Gervais Soucy, University of Sherbrooke; Sastry VanKamamidi, Saint Mary's University; and Jerry Ward, University of New Brunswick.

We sought help from a number of people during the preparation of the third Canadian edition. Their assistance included clerical help with the manuscript, specialized advice on financial or income tax matters, and verification of the solutions manual. The contributions of the following individuals were greatly appreciated: Val Arychuk, Nelson Epp, Jill Stanton, Cara Wood, Lina Xu, Mayank Pandey, and Ajit Sahoo from the University of Alberta; Sandy Wishloff of KPMG; and Navin Chari of Dalhousie University.

The possibility of a third Canadian edition was first presented to us by Michelle Sartor at Pearson Canada. Jeff Chamberlin maintained communication between the authors and Pearson Canada during development of the third Canadian edition. When the project began, we worked with Pamela Voves, Senior Development Editor. Cathleen

Sullivan, Executive Editor, assumed responsibility for the overall project and, along with Darryl Kamo, Developmental Editor, saw it through to completion. The production staff at Pearson Canada then published the book on a tight schedule. We thank all of these people for their belief in, and commitment to, this project.

Finally, the preparation of this manuscript would not have been possible without the support and understanding of our families.

Chan S. Park, *Auburn University, Auburn, Alabama*
Ming J. Zuo, *University of Alberta, Edmonton, Alberta*
Ronald Pelot, *Dalhousie University, Halifax, Nova Scotia*

Basics of Financial Decisions

ONE

Engineering Economic Decisions

Google Cofounder Sergey Brin Comes to Class at Berkeley[1] Sergey Brin, cofounder of Google, showed up for class as a guest speaker at the University of California, Berkeley, on October 3, 2005. Casual and relaxed, Brin talked about how Google came to be, answered students' questions, and showed that someone worth US$11 billion (give or take a billion) still can be comfortable in an old pair of blue jeans. Indistinguishable in dress, age, and demeanour from many of the students in the class, Brin covered a lot of ground in his remarks, but ultimately it was his unspoken message that was most powerful: To those with focus and passion, all things are possible. In his remarks to the class, Brin stressed simplicity. Simple ideas sometimes can change the world, he said. Likewise, Google started out with the simplest of ideas, with a global audience in mind. In the mid-1990s, Brin and Larry Page were Stanford students pursuing doctorates in computer science. Brin recalled that at that time there were some five major Internet search engines, the importance of searching was being de-emphasized, and the owners of these major search sites were focusing on creating portals with increased content offerings. "We believed we could build a better search. We had a simple idea—that not all pages are created equal. Some are more important," related Brin. Eventually, they developed a unique approach to solving one of computing's biggest challenges: retrieving relevant information from a massive set of data.

According to Google lore,[2] by January 1996 Larry and Sergey had begun collaboration on a search engine called BackRub, named for its unique ability to analyze the "back links" pointing to a given website.

---

[1] *UC Berkeley News*, October 4, 2005, UC Regent.
[2] Google Corporate Information: www. google.com/corporate/history.html.

How Google Works

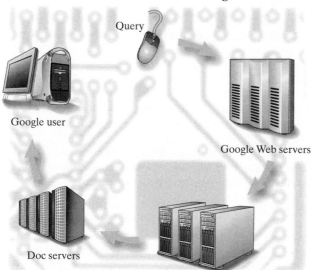

Query

Google user

Google Web servers

Doc servers

Index servers

1. The Web server sends the query to the index servers—it tells which pages contain the words that match the query.

2. The query travels to the Doc servers (which retrieve the stored documents) and snippets are generated to describe each search result.

3. The search results are returned to the user in a fraction of a second.

Larry, who had always enjoyed tinkering with machinery and had gained some "notoriety" for building a working printer out of Lego bricks, took on the task of creating a new kind of server environment that used low-end PCs instead of big expensive machines. Afflicted by the perennial shortage of cash common to graduate students everywhere, the pair took to haunting the department's loading docks in hopes of tracking down newly arrived computers that they could borrow for their network. A year later, their unique approach to link analysis was earning BackRub a growing reputation among those who had seen it. Buzz about the new search technology began to build around campus. Eventually, in 1998 they decided to create a company named Google by raising US$25 million from venture capital firms Kleiner Perkins Caufield & Byers and Sequoia Capital. Since taking their Internet search engine public in August 2004, the dynamic duo behind Google has seen their combined fortune soar to US$24 billion. At a recent price of over US$500, Google trades at 29 times trailing earnings, after starting out at US$85. The success has vaulted both Larry and Sergey into *Forbes* magazine's list of the World's Billionaries 2009.[3] The net worth of the pair is estimated at US$12 billion each.

---

[3] "The World's Billionaires," *Forbes*, March 11, 2009.

A Little Google History

- **1995**
  - Developed in dorm room of Larry Page and Sergey Brin, graduate students at Stanford University
  - Nicknamed BackRub
- **1998**
  - Raised US$25 million to set up Google, Inc.
  - Ran 100,000 queries a day out of a garage in Menlo Park
- **2005**
  - Over 4000 employees worldwide
  - Over 8 billion pages indexed
- **2009**
  - The latest version of Google Earth makes a splash with Ocean, a new feature that provides a three-dimensional look at the ocean floor and information about one of the world's greatest natural resources.
  - After adding Turkish, Thai, Hungarian, Estonian, Albanian, Maltese, and Galician, Google Translate is capable of automatic translation between 41 languages, covering 98% of the languages read by Internet users.

The story of how the Google founders got motivated to invent a search engine and eventually transformed their invention to a multibillion-dollar business is a typical one. Companies such as Dell, Microsoft, and Yahoo! all produce computer-related products and have market values of several billion dollars. These companies were all started by highly motivated young university students just like Brin. One thing that is also common to all these successful businesses is that they have capable and imaginative engineers who constantly generate good ideas for capital investment, execute them well, and obtain good results. You might wonder about what kind of role these engineers play in making such business decisions. In other words, what specific tasks are assigned to these engineers, and what tools and techniques are available to them for making such capital investment decisions? We answer these questions and explore related issues throughout this book.

CHAPTER LEARNING OBJECTIVES

After completing this chapter, you should understand the following concepts:

- The role of engineers in business.
- Types of business organization.
- The nature and types of engineering economic decisions.
- What makes the engineering economic decisions difficult.
- How a typical engineering project idea evolves in business.
- Fundamental principles of engineering economics.

1.1 Role of Engineers in Business

Yahoo!, Apple Computer, Microsoft Corporation, and Sun Microsystems produce computer products and have a market value of several billion dollars each. These companies were all started by young university students with technical backgrounds. When they went into the computer business, these students initially organized their companies as proprietorships. As the businesses grew, they became partnerships and were eventually converted to corporations. This chapter begins by introducing the three primary forms of business organization and briefly discusses the role of engineers in business.

1.1.1 Types of Business Organization

As an engineer, you should understand the nature of the business organization with which you are associated. This section will present some basic information about the type of organization you should choose should you decide to go into business for yourself.

The three legal forms of business, each having certain advantages and disadvantages, are proprietorships, partnerships, and corporations.

Proprietorships

A **proprietorship** is a business owned by one individual. This person is responsible for the firm's policies, owns all its assets, and is personally liable for its debts. A proprietorship has two major advantages. First, it can be formed easily and inexpensively. No legal and organizational requirements are associated with setting up a proprietorship, and organizational costs are therefore virtually nil. Second, the earnings of a proprietorship are taxed at the owner's personal tax rate, which may be lower than the rate at which corporate income is taxed. Apart from personal liability considerations, the major disadvantage of a proprietorship is that it cannot issue stocks and bonds, making it difficult to raise capital for any business expansion.

Partnerships

A **partnership** is similar to a proprietorship, except that it has more than one owner. Most partnerships are established by a written contract between the partners. The contract normally specifies salaries, contributions to capital, and the distribution of profits and losses. A partnership has many advantages, among which are its low cost and ease of formation. Because more than one person makes contributions, a partnership typically has a larger amount of capital available for business use. Since the personal assets of all the partners stand behind the business, a partnership can borrow money more easily from a bank. Each partner pays only personal income tax on his or her share of a partnership's taxable income.

On the negative side, under partnership law each partner is liable for a business's debts. This means that the partners must risk all their personal assets—even those not invested in the business. And while each partner is responsible for his or her portion of the debts in the event of bankruptcy, if any partners cannot meet their pro rata claims, the remaining partners must take over the unresolved claims. Finally, a partnership has a limited life, insofar as it must be dissolved and reorganized if one of the partners quits.

Corporations

A **corporation** is a legal entity created under provincial or federal law. It is separate from its owners and managers. This separation gives the corporation four major advantages: (1) It can raise capital from a large number of investors by issuing stocks and bonds; (2) it permits easy transfer of ownership interest by trading shares of stock; (3) it allows limited liability—personal liability is limited to the amount of the individual's investment in the business; and (4) it is taxed differently than proprietorships and partnerships, and under certain conditions, the tax laws favour corporations. On the negative side, it is expensive to establish a corporation. Furthermore, a corporation is subject to numerous governmental requirements and regulations.

As a firm grows, it may need to change its legal form because the form of a business affects the extent to which it has control of its own operations and its ability to acquire funds. The legal form of an organization also affects the risk borne by its owners in case of bankruptcy and the manner in which the firm is taxed. Apple Computer, for example, started out as a two-person garage operation. As the business grew, the owners felt constricted by this form of organization: It was difficult to raise capital for business expansion; they felt that the risk of bankruptcy was too high to bear; and as their business income grew, their tax burden grew as well. Eventually, they found it necessary to convert the partnership into a corporation.

In Canada, the overwhelming majority of business firms are proprietorships, followed by corporations and partnerships. However, in terms of total business volume (dollars of sales), the quantity of business transacted by proprietorships and partnerships is several times less than that of corporations. Since most business is conducted by corporations, this text will generally address economic decisions encountered in corporations.

1.1.2 Engineering Economic Decisions

What role do engineers play within a firm? What specific tasks are assigned to the engineering staff, and what tools and techniques are available to it to improve a firm's profits? Engineers are called upon to participate in a variety of decisions, ranging from manufacturing, through marketing, to financing decisions. We will restrict our focus, however, to various economic decisions related to engineering projects. We refer to these decisions as **engineering economic decisions**.

In manufacturing, engineering is involved in every detail of a product's production, from conceptual design to shipping. In fact, engineering decisions account for the majority (some say 85%) of product costs. Engineers must consider the effective use of capital assets such as buildings and machinery. One of the engineer's primary tasks is to plan for the acquisition of equipment (**capital expenditure**) that will enable the firm to design and produce products economically.

With the purchase of any fixed asset—equipment, for instance—we need to estimate the profits (more precisely, cash flows) that the asset will generate during its period of service. In other words, we have to make capital expenditure decisions based on predictions about the future. Suppose, for example, you are considering the purchase of a deburring machine to meet the anticipated demand for hubs and sleeves used in the production of gear couplings. You expect the machine to last 10 years. This decision thus involves an implicit 10-year sales forecast for the gear couplings, which means that a long waiting period will be required before you will know whether the purchase was justified.

An inaccurate estimate of the need for assets can have serious consequences. If you invest too much in assets, you incur unnecessarily heavy expenses. Spending too little on fixed assets, however, is also harmful, for then the firm's equipment may be too obsolete to produce products competitively, and without an adequate capacity, you may lose a portion of your market share to rival firms. Regaining lost customers involves heavy marketing expenses and may even require price reductions or product improvements, both of which are costly.

1.1.3 Personal Economic Decisions

In the same way that an engineer can play a role in the effective utilization of corporate financial assets, each of us is responsible for managing our personal financial affairs. After we have paid for nondiscretionary or essential needs, such as housing, food, clothing, and transportation, any remaining money is available for discretionary expenditures on items such as entertainment, travel, and investment. For money we choose to invest, we want to maximize the economic benefit at some acceptable risk. The investment choices are unlimited and include savings accounts, guaranteed investment certificates, stocks, bonds, mutual funds, registered retirement savings plans, rental properties, land, business ownership, and more.

How do you choose? The analysis of one's personal investment opportunities utilizes the same techniques that are used for engineering economic decisions. Again, the challenge is predicting the performance of an investment into the future. Choosing wisely can be very rewarding, while choosing poorly can be disastrous. Some investors in the energy stock Enron who sold prior to the fraud investigation became millionaires. Others, who did not sell, lost everything.

A wise investment strategy is a strategy that manages risk by diversifying investments. With such an approach, you have a number of different investments ranging from very low to very high risk in various business sectors. Since you do not have all your money in one place, the risk of losing everything is significantly reduced. (We discuss some of these important issues in Chapter 15.)

1.2 What Makes the Engineering Economic Decision Difficult?

The economic decisions that engineers make in business differ very little from the financial decisions made by individuals, except for the scale of the concern. Suppose, for example, that a firm is using a lathe that was purchased 12 years ago to produce pump shafts. As the production engineer in charge of this product, you expect demand to continue into the foreseeable future. However, the lathe has begun to show its age: It has broken frequently during the last two years and has finally stopped operating altogether. Now you have to decide whether to replace or repair it. If you expect a more efficient lathe to be available in the next one or two years, you might repair the old lathe instead of replacing it. The major issue is whether you should make the considerable investment in a new lathe now or later. As an added complication, if demand for your product begins to decline, you may have to conduct an economic analysis to determine whether declining profits from the project offset the cost of a new lathe.

Let us consider a real-world engineering decision problem on a much larger scale, as taken from an article from *The Washington Post*.[4] Ever since Hurricane Katrina hit the city of New Orleans in August 2005, the U.S. federal government has been under pressure to show strong support for rebuilding levees in order to encourage homeowners and businesses to return to neighbourhoods that were flooded when the city's levees crumbled under Katrina's storm surge. Many evacuees have expressed reluctance to rebuild without assurances that New Orleans will be made safe from future hurricanes, including Category 5 storms, the most severe. Some U.S. Army Corps of Engineers officers estimated that it would cost more than US$1.6 billion to restore the levee system merely to its design strength—that is, to withstand a Category 3 storm. New design features would include floodgates on several key canals, as well as stone-and-concrete fortification of some of the city's earthen levees, as illustrated in Figure 1.1. Donald E. Powell, the administration's coordinator of post-Katrina recovery, insisted that the improvements would make the levee system much stronger than it had been in the past. But he declined to say whether the administration would support further upgrades of the system to Category 5 protection, which would require substantial reengineering of existing levees at a cost that could, by many estimates, exceed US$30 billion.

Obviously, this level of engineering decision is far more complex and more significant than a business decision about when to introduce a new product. Projects of this nature involve large sums of money over long periods of time, and it is difficult to justify the cost of the system purely on the basis of economic reasoning, since we do not know when another Katrina-strength storm will be on the horizon. Even if they decide to rebuild the levee systems, should they build just enough to withstand a Category 3 storm, or should they build to withstand a Category 5 storm? Any engineering economic decision pertaining to this type of extreme event will be extremely difficult to make.

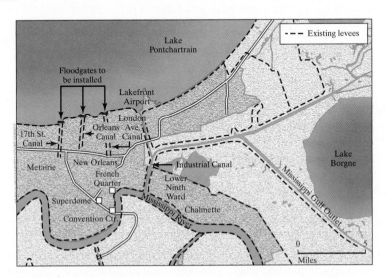

Figure 1.1 The White House pledged US$1.5 billion to armour existing New Orleans levees with concrete and stone, build floodgates on three canals, and upgrade the city's pumping system.

[4] Joby Warrick and Peter Baker, "Bush Pledges US$1.5 Billion for New Orleans—Proposal Would Double Aid From U.S. for Flood Protection," *The Washington Post*, December 16, 2005, p. A03.

In this book, we will consider many types of investments—personal investments as well as business investments. The focus, however, will be on evaluating engineering projects on the basis of their economic desirability and on dealing with investment situations that a typical firm faces.

1.3 Economic Decisions versus Design Decisions

Economic decisions differ in a fundamental way from the types of decisions typically encountered in engineering design. In a design situation, the engineer utilizes known physical properties, the principles of chemistry and physics, engineering design correlations, and engineering judgment to arrive at a workable and optimal design. If the judgment is sound, the calculations are done correctly, and we ignore technological advances, the design is time invariant. In other words, if the engineering design to meet a particular need is done today, next year, or in five years' time, the final design would not change significantly.

In considering economic decisions, the measurement of investment attractiveness, which is the subject of this book, is relatively straightforward. However, the information required in such evaluations always involves predicting or forecasting product sales, product selling prices, and various costs over some future time frame—5 years, 10 years, 25 years, and so on.

All such forecasts have two things in common. First, they are never completely accurate compared with the actual values realized at future times. Second, a prediction or forecast made today is likely to be different from one made at some point in the future. It is this ever-changing view of the future that can make it necessary to revisit and even change previous economic decisions. Thus, unlike engineering design, the conclusions reached through economic evaluation are not necessarily time invariant. Economic decisions have to be based on the best information available at the time of the decision and a thorough understanding of the uncertainties in the forecasted data.

1.4 Large-Scale Engineering Projects

In the development of any product, a company's engineers are called upon to translate an idea into reality. A firm's growth and development depend largely upon a constant flow of ideas for new products, and for the firm to remain competitive, it has to make existing products better or produce them at a lower cost. In the next section, we present an example of how a design engineer's idea eventually turned into an innovative automotive product.

1.4.1 How a Typical Project Idea Evolves

The Toyota Motor Corporation introduced the world's first mass-produced car powered by a combination of gasoline and electricity. Known as the Prius, this vehicle is the first of a new generation of Toyota cars whose engines cut air pollution dramatically and boost fuel efficiency to significant levels. Toyota, in short, wants to launch and dominate a new "green" era for automobiles—and is spending US$1.5 billion to do it. Developed for the Japanese market initially, the Prius uses both a gasoline engine and an electric motor as its motive power source. The Prius emits less pollution than ordinary cars, and it gets more mileage, which means less output of carbon dioxide. Moreover, the Prius gives a highly responsive performance and smooth acceleration. The following information from *BusinessWeek*[5] illustrates how a typical strategic business decision is made by the engineering staff of a large company. Additional information has been provided by Toyota Motor Corporation.

Why Go for a Greener Car?

Toyota first started to develop the Prius in 1995. Four engineers were assigned to figure out what types of cars people would drive in the 21st century. After a year of research, a chief engineer concluded that Toyota would have to sell cars better suited to a world with scarce natural resources. He considered electric motors. But an electric car can travel only 215 kilometres before it must be recharged. Another option was fuel-cell cars that run on hydrogen. But he suspected that mass production of this type of car might not be possible for another 15 years. So the engineer finally settled on a hybrid system powered by an electric motor, a nickel–metal hydride battery, and a gasoline engine. From Toyota's perspective, it was a big bet, as oil and gasoline prices were bumping along at record lows at that time. Many green cars remain expensive and require trade-offs in terms of performance. No carmaker has ever succeeded in selling consumers en masse something they have never wanted to buy: *cleaner air*. Even in Japan, where a litre of regular gas can cost as much as US$1, carmakers have trouble pushing higher fuel economy and lower carbon emissions. Toyota has several reasons for going green. In the next century, as millions of new car owners in China, India, and elsewhere take to the road, Toyota predicts that gasoline prices will rise worldwide. At the same time, Japan's carmakers expect pollution and global warming to become such threats that governments will enact tough measures to clean the air.

What Is So Unique in Design?

It took Toyota engineers two years to develop the current power system in the Prius. The car comes with a dual engine powered by both an electric motor and a newly developed 1.5-litre gasoline engine. When the engine is in use, a special "power split device" sends some of the power to the driveshaft to move the car's wheels. The device also sends some of the power to a generator, which in turn creates electricity, to either drive the motor or recharge the battery. Thanks to this variable transmission, the Prius can switch back and forth between motor and engine, or employ both, without creating any jerking motion. The battery and electric motor propel the car at slow speeds. At normal speeds, the electric motor and gasoline engine work together to power the wheels. At higher speeds, the battery kicks in extra power if the driver must pass another car or zoom up a hill.

When the car decelerates and brakes, the motor converts the vehicle's kinetic energy into electricity and sends it through an inverter to be stored in the battery. (See Figure 1.2.)

- When engine efficiency is low, such as during start-up and with midrange speeds, motive power is provided by the motor alone, using energy stored in the battery.
- Under normal driving conditions, overall efficiency is optimized by controlling the power allocation so that some of the engine power is used for turning the generator to supply electricity for the motor while the remaining power is used for turning the wheels.

---

[5] Emily Thornton (Tokyo), Keith Naughton (Detroit), and David Woodruff, "Japan's Hybrid Cars—Toyota and Rivals Are Betting on Pollution Fighters—Will They Succeed?" *BusinessWeek*, December 4, 1997.

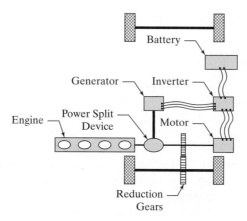

Figure 1.2   Arrangement of components of the Toyota Hybrid System II.
(*Source: Evaluation of 2004 Toyota Prius Hybrid Electric Drive System*, Oak Ridge National Laboratory, ONR/TM2004/247, U.S. Department of Energy.)

- During periods of acceleration, when extra power is needed, the generator supplements the electricity being drawn from the battery, so the motor is supplied with the required level of electrical energy.
- While decelerating and braking, the motor acts as a generator that is driven by the wheels, thus allowing the recovery of kinetic energy. The recovered kinetic energy is converted to electrical energy that is stored in the battery.
- When necessary, the generator recharges the battery to maintain sufficient reserves.
- When the vehicle is not moving and when the engine moves outside of certain speed and load conditions, the engine stops automatically.

So the car's own movement, as well as the power from the gasoline engine, provides the electricity needed. The energy created and stored by deceleration boosts the car's efficiency. So does the automatic shutdown of the engine when the car stops at a light. At higher speeds and during acceleration, the companion electric motor works harder, allowing the gas engine to run at peak efficiency. Toyota claims that, in tests, its hybrid car has boosted fuel economy by 100% and engine efficiency by 80%. The Prius emits about half as much carbon dioxide, and about one-tenth as much carbon monoxide, hydrocarbons, and nitrous oxide, as conventional cars.

Is It Safe to Drive on a Rainy Day?

Yet, major hurdles remain to be overcome in creating a mass market for green vehicles. Car buyers are not anxious enough about global warming to justify a "sea-level change" in automakers' marketing. Many of Japan's innovations run the risk of becoming impressive technologies meant for the masses but bought only by the elite. The unfamiliarity of green technology can also frighten consumers. The Japanese government sponsors festivals at which people can test drive alternative-fuel vehicles. But some turned down that chance on a rainy weekend in May because they feared that riding in an electric car might electrocute them. An engineer points out, "It took 20 years for the automatic transmission to become popular in Japan." It certainly would take that long for many of these technologies to catch on.

How Much Would It Cost?

The biggest question remaining about the mass production of the vehicle concerns its production cost. Costs will need to come down for Toyota's hybrid to be competitive around the world, where gasoline prices are lower than in Japan. Economies of scale will help as production volumes increase, but further advances in engineering also will be essential. With its current design, Prius's monthly operating cost would be roughly twice that of a conventional automobile. Still, Toyota believes that it will sell more than 12,000 Prius models to Japanese drivers during the first year. To do that, it is charging just US$17,000 a car. The company insists that it will not lose any money on the Prius, but rivals and analysts estimate that, at current production levels, the cost of building a Prius could be as much as US$32,000. If so, Toyota's low-price strategy will generate annual losses of more than US$100 million on the new compact.

Will There Be Enough Demand?

Why buy a US$17,000 Prius when a US$13,000 Corolla is more affordable? It does not help Toyota that it launched the Prius in the middle of a sagging domestic car market. Nobody really knows how big the final market for hybrids will be. Toyota forecasted that hybrids would account for a third of the world's auto market as early as 2005, but Japan's Ministry of International Trade and Industry expected 2.4 million alternative-fuel vehicles, including hybrids, to roam Japan's backstreets by 2010. Nonetheless, the Prius has set a new standard for Toyota's competitors. There seems to be no turning back for the Japanese. The Japanese government requested that carmakers slash carbon dioxide emissions by 20% and cut nitrous oxide, hydrocarbon, and carbon monoxide emissions by 80% by 2010. The government may also soon give tax breaks to consumers who buy green cars.

The Prius first went on sale in Japan in December 1997. By June 1998, total sales for the Prius reached over 7,700 units. With this encouraging sales trend, Toyota finally announced that the Prius would be introduced in the North American and European markets by the year 2000 with an estimated total sales volume of approximately 20,000 units per year in the North American and European market combined. As with the 2000 North American and European introduction, Toyota also planned to use the subsequent two years to develop a hybrid vehicle optimized to the usage conditions of each market.

What Is the Business Risk in Marketing the Prius?

Engineers at Toyota Motors have stated that California would be the primary market for the Prius outside Japan, but they added that an annual demand of 50,000 cars would be necessary to justify production. Despite Toyota management's decision to introduce the hybrid electric car into the world market, the Toyota engineers were still uncertain whether there would be enough demand. Furthermore other automakers just do not see how Toyota can achieve the economies of scale needed to produce green cars at a profit. The primary advantage of the design, however, is that the Prius can cut auto emissions in half. This is a feature that could be very appealing at a time when government air-quality standards are becoming more rigorous and consumer interest in the environment is strong. However, in the case of the Prius, if a significant reduction in production cost never materializes, demand may remain insufficient to justify the investment in the green car.

1.4.2 Impact of Engineering Projects on Financial Statements

Engineers must understand the business environment in which a company's major business decisions are made. It is important for an engineering project to generate profits, but it also must strengthen the firm's overall financial position. How do we measure Toyota's success in the Prius project? Will enough Prius models be sold, for example, to keep the green-engineering business as Toyota's major source of profits? While the Prius project will provide comfortable, reliable, low-cost driving for the company's customers, the bottom line is its financial performance over the long run.

Regardless of a business's form, each company has to produce basic financial statements at the end of each operating cycle (typically a year). These financial statements provide the basis for future investment analysis. In practice, we seldom make investment decisions solely on the basis of an estimate of a project's profitability, because we must also consider the overall impact of the investment on the financial strength and position of the company.

Suppose that you were the president of the Toyota Corporation. Suppose further that you even hold some shares in the company, making you one of the company's many owners. What objectives would you set for the company? While all firms are in business in hopes of making a profit, what determines the market value of a company are not profits per se, but cash flow. It is, after all, available cash that determines the future investments and growth of the firm. Therefore, one of your objectives should be to increase the company's value to its owners (including yourself) as much as possible. To some extent, the market price of your company's stock represents the value of your company. Many factors affect your company's market value: present and expected future earnings, the timing and duration of those earnings, and the risks associated with them. Certainly, any successful investment decision will increase a company's market value. Stock price can be a good indicator of your company's financial health and may also reflect the market's attitude about how well your company is managed for the benefit of its owners.

1.4.3 A Look Back in 2009: Did Toyota Make the Right Decision?

Clearly, there were many doubts and uncertainties about the market for hybrids in 1998. Even Toyota engineers were not sure that there would be enough demand in the world market to justify the production of the vehicle. Eleven years after the decision to launch the Prius globally, it turns out that the decision was right. The Prius was subsequently introduced worldwide in 2000. The larger, more powerful, and more fuel-efficient second-generation Prius was launched in 2003. On June 7, 2007, Toyota announced that the cumulative worldwide sales of Prius had reached 757,600. It had passed the 1 million mark by May 2008 and the 1.2 million mark by January 2009. The third-generation Prius introduced in 2009 features a larger 1.8-litre engine and three different driving modes offering more performance, fuel efficiency, and environmental friendliness than ever before.[6]

The Prius, which is sold today in more than 40 countries and regions, has its largest markets in Japan and North America. Nearly 60% of all Prius sales have been in North America, where 183,800 vehicles were sold in 2007. More than 14,000 Canadians have embraced the performance of the Prius. Indeed, the new Prius proved that Toyota has achieved its goal: to create an eco-car with high-level environmental performance but with the conventional draw of a modern car. These features, combined with its efficient handling and attractive design, are making the Prius a popular choice of individuals and companies alike.

[6] Toyota Motor Corporation (www.toyota.co.jp/en/), News Releases, 2009.

Toyota made its investors happy, as the public liked the new hybrid car, resulting in an increased demand for the product. This, in turn, caused stock prices, and hence shareholder wealth, to increase. In fact, the new, heavily promoted green car turned out to be a market leader in its class and contributed to sending Toyota's stock to an all-time high in late 2007. Any successful investment decision on Prius's scale will tend to increase a firm's stock prices in the marketplace and promote long-term success. Thus, in making a large-scale engineering project decision, we must consider its possible effect on the firm's market value. (In Chapter 2, we discuss the financial statements in detail and show how to use them in our investment decision making.)

1.5 Common Types of Strategic Engineering Economic Decisions

The story of how the Toyota Corporation successfully introduced a new product and became the market leader in the hybrid electric car market is typical: Someone had a good idea, executed it well, and obtained good results. Project ideas such as the Prius can originate from many different levels in an organization. Since some ideas will be good, while others will not, we need to establish procedures for screening projects. Many large companies have a specialized project analysis division that actively searches for new ideas, projects, and ventures. Once project ideas are identified, they are typically classified as (1) equipment or process selection, (2) equipment replacement, (3) new product or product expansion, (4) cost reduction, or (5) improvement in service or quality. This classification scheme allows management to address key questions: Can the existing plant, for example, be used to attain the new production levels? Does the firm have the knowledge and skill to undertake the new investment? Does the new proposal warrant the recruitment of new technical personnel? The answers to these questions help firms screen out proposals that are not feasible, given a company's resources.

The Prius project represents a fairly complex engineering decision that required the approval of top executives and the board of directors. Virtually all big businesses face investment decisions of this magnitude at some time. In general, the larger the investment, the more detailed is the analysis required to support the expenditure. For example, expenditures aimed at increasing the output of existing products or at manufacturing a new product would invariably require a very detailed economic justification. Final decisions on new products, as well as marketing decisions, are generally made at a high level within the company. By contrast, a decision to repair damaged equipment can be made at a lower level. The five classifications of project ideas are as follows:

- **Equipment or Process Selection.** This class of engineering decision problems involves selecting the best course of action out of several that meet a project's requirements. For example, which of several proposed items of equipment shall we purchase for a given purpose? The choice often hinges on which item is expected to generate the largest savings (or the largest return on the investment). For example, the choice of material will dictate the manufacturing process for the body panels in the automobile. Many factors will affect the ultimate choice of the material, and engineers should consider all major cost elements, such as the cost of machinery and equipment, tooling, labour, and material. Other factors may include press and assembly, production and engineered scrap, the number of dies and tools, and the cycle times for various processes.

- **Equipment Replacement.** This category of investment decisions involves considering the expenditure necessary to replace worn-out or obsolete equipment. For example, a company may purchase 10 large presses, expecting them to produce stamped metal parts for 10 years. After five years, however, it may become necessary to produce the parts in plastic, which would require retiring the presses early and purchasing plastic moulding machines. Similarly, a company may find that, for competitive reasons, larger and more accurate parts are required, making the purchased machines obsolete earlier than expected.

- **New Product or Product Expansion.** Investments in this category increase company revenues if output is increased. One common type of expansion decision includes decisions about expenditures aimed at increasing the output of existing production or distribution facilities. In these situations, we are basically asking, "Shall we build or otherwise acquire a new facility?" The expected future cash inflows in this investment category are the profits from the goods and services produced in the new facility. A second type of expenditure decision includes considering expenditures necessary to produce a new product or to expand into a new geographic area. These projects normally require large sums of money over long periods.

- **Cost Reduction.** A cost-reduction project is a project that attempts to lower a firm's operating costs. Typically, we need to consider whether a company should buy equipment to perform an operation currently done manually or spend money now in order to save more money later. The expected future cash inflows on this investment are savings resulting from lower operating costs.

- **Improvement of Product Quality.** Continuous improvement of the quality of a successful product and elimination of faulty designs may also be worthy projects to pursue. In 2009–10, Toyota had to recall millions of vehicles, including the Prius, to fix problems in their brakes and gas pedals. These recalls and their impact on sales of new vehicles cost Toyota billions of US dollars. If projects had been implemented in time to eliminate such design problems, these costs would have been saved.

Most of the examples in the previous sections were related to economic decisions in the manufacturing sector. The decision techniques we develop in this book are also applicable to various economic decisions in the service sector.

1.6 Fundamental Principles of Engineering Economics

This book is focused on the principles and procedures engineers use to make sound economic decisions. To the first-time student of engineering economics, anything related to money matters may seem quite strange when compared to other engineering subjects. However, the decision logic involved in solving problems in this domain is quite similar to that employed in any other engineering subject. There are fundamental principles to follow in engineering economics that unite the concepts and techniques presented in this text, thereby allowing us to focus on the logic underlying the practice of engineering economics.

- **Principle 1: A nearby penny is worth a distant dollar.** A fundamental concept in engineering economics is that money has a time value associated with it. Because we can earn interest on money received today, it is better to receive money earlier than later. This concept will be the basic foundation for all engineering project evaluation.

- **Principle 2: All that counts are the differences among alternatives.** An economic decision should be based on the *differences* among the alternatives considered. All that is common is irrelevant to the decision. Certainly, any economic decision is no better than the alternatives being considered. Thus, an economic decision should be based on the objective of making the best use of limited resources. Whenever a choice is made, something is given up. The opportunity cost of a choice is the value of the best alternative given up.

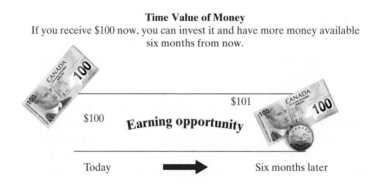

Time Value of Money
If you receive $100 now, you can invest it and have more money available six months from now.

$100   Earning opportunity $101

Today   →   Six months later

Comparing Buy versus Lease
Whatever you decide, you need to spend the same amount of money on fuel and maintenance.

Option	Monthly Fuel Cost	Monthly Maintenance	Cash Outlay at Signing	Monthly Payment	Salvage Value at End of Year 3
Buy	$960	$550	$6,500	$350	$9,000
Lease	$960	$550	$2,400	$550	0

Irrelevant items in decision making

- **Principle 3: Marginal revenue must exceed marginal cost.** Effective decision making requires comparing the additional costs of alternatives with the additional benefits. Each decision alternative must be justified on its own economic merits before being compared with other alternatives. Any increased economic activity must be justified on the basis of the fundamental economic principle that marginal revenue must exceed marginal cost. Here, *marginal revenue* means the additional revenue made possible by increasing the activity by one unit (or small unit). *Marginal cost* has an analogous definition. Productive resources—the natural resources, human resources, and capital assets available to make goods and services—are limited. Therefore, people cannot have all the goods and services they want; as a result, they must choose some things and give up others.

Marginal cost: The cost associated with one additional unit of production, also called the incremental cost.

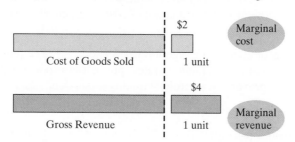

Marginal Analysis
To justify your action, marginal revenue must exceed marginal cost.

- **Principle 4: Additional risk is not taken without the expected additional return.** For delaying consumption, investors demand a minimum return that must be greater than the anticipated rate of inflation or any perceived risk. If they didn't receive enough to compensate for anticipated inflation and the perceived investment risk, investors would purchase whatever goods they desired ahead of time or invest in assets that would provide a sufficient return to compensate for any loss from inflation or potential risk.

Risk and Return Trade-Off
Expected returns from bonds and stocks are normally higher than the expected return from a savings account.

Investment Class	Potential Risk	Expected Return
Savings account (cash)	Low/None	1.5%
Bond (debt)	Moderate	4.8%
Stock (equity)	High	11.5%

Risk-return trade-off: Invested money can render higher profits only if it is subject to the possibility of being lost.

The preceding four principles are as much statements of common sense as they are theoretical precepts. These principles provide the logic behind what is to follow. We build on them and attempt to draw out their implications for decision making. As we continue, keep in mind that, while the topics being treated may change from chapter to chapter, the logic driving our treatment of them is constant and rooted in the four fundamental principles.

SUMMARY

■ This chapter has given us an overview of a variety of engineering economic problems that are commonly found in the business world. We examined the role, and the increasing importance, of engineers in the firm, as evidenced by the development at Toyota of the Prius, a hybrid vehicle. Commonly, engineers are called upon to participate in a variety of strategic business decisions ranging from product design to marketing.

■ The term **engineering economic decision** refers to any investment decision related to an engineering project. The facet of an economic decision that is of most interest from an engineer's point of view is the evaluation of costs and benefits associated with making a capital investment.

■ The five main types of engineering economic decisions are (1) equipment or process selection, (2) equipment replacement, (3) new product or product expansion, (4) cost reduction, and (5) improvement in product quality or service.

■ The factors of **time** and **uncertainty** are the defining aspects of any investment project.

■ The four fundamental principles that must be applied in all engineering economic decisions are (1) the time value of money, (2) differential (incremental) cost and revenue, (3) marginal cost and revenue, and (4) the trade-off between risk and reward.

On the Companion Website that accompanies this text, you will find Excel templates and exercises, as well as the following analysis tools: Cash Flow Analyzer, Depreciation Analysis, Loan Analysis, and Interest Tables.

TWO

Understanding Financial Statements

RIM Keeps Hiring Despite Mobile Slump; 600 New Positions[1] Facing global economic recession, many cellphone and portable-data device makers are cutting their workforces for their own survival. In January 2009, Nokia, the number-one phone maker globally, slashed 1700 jobs while Motorola, the biggest U.S. phone maker, eliminated 4000 jobs in its latest round of cuts, and Microsoft, which makes software for phones, said it planned to reduce up to 5000 positions. Google Inc., whose software runs on the G1 smartphone, said in May 2009 it would cut 200 sales and marketing jobs. In contrast, Research In Motion Ltd. (RIM), the Waterloo, Ontario–based BlackBerry maker, is expanding its workforce to keep rolling out new BlackBerry models. As of April 16, 2009, RIM had 469 job openings in the Americas; 88 in Europe, the Middle East, and Africa; and 51 in Asia-Pacific. This is signalling that RIM is withstanding a recession that has forced its rivals to trim payrolls.

RIM was formed in 1984 by Mike Lazaridis and Doug Fregin. Both were engineering students then, one at the University of Waterloo and the other at the University of Windsor. The starting capital of the company was a small government grant and a family loan of $15,000. RIM's first contract was with General Motors (GM): For $600,000, RIM delivered a networked display board that scrolled messages across LED signs at GM's Oshawa plant. In the late 1980s, RIM signed a contract with Rogers Cantel Communications to design connectivity software for a data network called Mobitex. In the early 1990s, RIM developed its first DigiSync Film KeyKode Reader for film editing and won an

---

[1] Hugo Miller, "RIM Keeps Hiring Despite Mobile Slump: 600 New Positions," Bloomberg News, April 17, 2009.

Academy Award (1998) for scientific and technical achievement. Eventually, RIM focused its business on developing wireless data management products. These included wireless point-of-sale terminals, wireless modems, RIMgate, and two-way data pagers, among others. These products and the knowledge gained in the process of developing them are the building blocks of the hugely successful BlackBerry brand smartphones, first launched in January 1999. Mike Lazaridis, co-CEO of RIM, summarized the process: "You have to understand, the BlackBerry didn't happen overnight; it happened over a decade. It's not like one day we woke up and said, 'Eureka!'"

1984	Mike Lazaridis and Doug Fregin founded Research In Motion Ltd.
1985	**$600,000** contract to develop networked display terminals for General Motors
1997	Listed on the Toronto Stock Exchange (TSE:RIM) and raised more than **$115 million** from investors
1999	Revenues of **US$47.5 million**; net income of **US$6.8 million**; listed on the NASDAQ Stock Exchange (NASDAQ:RIMM) and raised **US$250** million from investors; launched the BlackBerry line of products
2002	Revenues of **US$294.1 million**; net loss of **US$28.3 million**
2004	Revenues of **US$595 million**; net income of **US$52 million**; more than 2 million BlackBerry subscribers worldwide
2006	Revenues of **US$1,526 million**; net income of **US$374 million**; more than 5 million BlackBerry subscribers worldwide

Today, BlackBerry devices are RIM's sole focus. The financial results released April 2, 2009, demonstrate the excellent performance of RIM in the global economic recession. All figures were given in U.S. dollars and followed U.S. Generally Accepted Accounting Principles (GAAP).

Revenue for the fiscal-year ended February 28, 2009, was US$11.07 billion, up 84% from US$6.01 billion the year before. RIM shipped approximately 26 million devices during fiscal-year 2009. At the end of fiscal 2009, the total BlackBerry subscriber account base was approximately 25 million, and net income was US$1.89 billion, or US$3.30 per share diluted, up 46.3% over fiscal 2008.

"We are very pleased to report another record quarter with standout subscriber growth that speaks volumes about the early success and momentum of our new BlackBerry products," said Jim Balsillie, co-CEO at RIM. "RIM experienced an extraordinary year in fiscal-year 2009, shipping our 50 millionth BlackBerry smartphone and generating US$11 billion in revenue. Looking ahead into fiscal 2010, we see exceptional opportunities for RIM and its partners to leverage the investments and success of the past year to continue growing market share and profitability."

RIM's revenue for the first quarter of fiscal 2010, ending May 30, 2009, was US$3.42 billion. Gross margin for the first quarter was 43.6%. Net subscriber account additions in the first quarter were 3.8 million. Earnings per share for the first quarter were US$0.98 per share diluted.

RIM's stock price at the Toronto Stock Exchange started in 1997 at about $1.50 per share. In October 2010, it was traded at about $48 per share. It has been declining from $71 since January 1, 2010. No cash dividends have been paid to shareholders. Stock splits have been done by issuing more shares to shareholders in the form of dividends. Its return on equity has been steadily increasing over the past few years and reaching more than 40% per year.

If you want to explore investing in RIM stock, what information would you go by? You would certainly prefer that RIM have a record of accomplishment of profitable operations, earning a profit (net income) year after year. The company would need a steady stream of cash coming in and a manageable level of debt. How would you determine whether the company met these criteria? Investors commonly use the financial statements contained in the annual report as a starting point in forming expectations about future levels of earnings and about the firm's riskiness.

Before making any financial decision, it is good to understand an elementary aspect of your financial situation—one that you'll also need for retirement planning, estate planning, and, more generally, to get an answer to the question, "How am I doing?" It is called your **net worth**. If you decided to invest $10,000 in RIM stocks, how would that affect your net worth? You may need this information for your own financial planning, but it is routinely required whenever you have to borrow a large sum of money from a financial institution. For example, when you are buying a home, you need to apply for a mortgage. Invariably, the bank will ask you to submit your net-worth statement as a part of loan processing. Your net-worth statement is a snapshot of where you stand financially at a given point in time. The bank will determine how creditworthy you are by examining your net worth. In a similar way, a corporation prepares the same kind of information for its financial planning or to report its financial health to stockholders or investors. The reporting document is known as the financial statements. In this chapter, we will discuss basic accounting concepts, financial statements, and how information contained in such statements may be used in ratio analysis for business decision making. The materials covered in this chapter will also be useful in Chapter 10 when we introduce the income statement approach for after-tax cash flow analysis.

Net worth is the amount by which a *company's* or individual's assets exceed the company's or individual's liabilities.

CHAPTER LEARNING OBJECTIVES

After completing this chapter, you should understand the following concepts:

- The role of accounting in economic decisions.
- The types of financial statements prepared for investors and regulators.
- How to use financial statements to manage a business.
- How to conduct the ratio analysis and what the numbers really mean.

2.1 Accounting: The Basis of Decision Making

We need financial information when we are making business decisions. Virtually all businesses and most individuals keep accounting records to aid in making decisions. As illustrated in Figure 2.1, accounting is the information system that measures business activities, processes the resulting information into reports, and communicates the results to decision makers. For this reason, we call accounting "the language of business." The better you understand this language, the better you can manage your financial well-being, and the better your financial decisions will be.

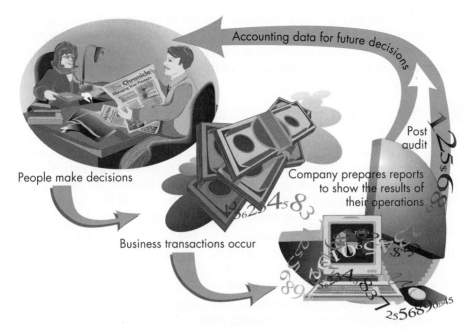

People make decisions

Business transactions occur

Accounting data for future decisions

Post audit

Company prepares reports to show the results of their operations

Figure 2.1   The accounting system, which illustrates the flow of information.

Personal financial planning, education expenses, loans, car payments, income taxes, and investments are all based on the information system we call accounting. The uses of accounting information are many and varied:

- **Individual people** use accounting information in their day-to-day affairs to manage bank accounts, to evaluate job prospects, to make investments, and to decide whether to rent an apartment or buy a house.
- **Business managers** use accounting information to set goals for their organizations, to evaluate progress toward those goals, and to take corrective actions if necessary. Decisions based on accounting information may include which building or equipment to purchase, how much merchandise to keep on hand as inventory, and how much cash to borrow.
- **Investors and creditors** provide the money a business needs to begin operations. To decide whether to help start a new venture, potential investors evaluate what return they can expect on their investment. Such an evaluation involves analyzing the financial statements of the business. Before making a loan, banks determine the borrower's ability to meet scheduled payments. This kind of evaluation includes a projection of future operations and revenue, based on accounting information.

An essential product of an accounting information system is a series of financial statements that allows people to make informed decisions. For personal use, the net-worth statement is a snapshot of where you stand financially at a given point in time. You do that by adding your assets—such as cash, investments, and pension plans—in one column and your liabilities—or debts—in the other. Then subtract your liabilities from your assets to find your net worth. In other words, your net worth is what you would be left with if you sold everything and paid off all you owe. For business use, financial statements are the documents that report financial information about a business entity to decision makers. They tell us how a business is performing and where it stands financially. The purpose of this chapter is not to present the bookkeeping aspects of accounting, but to acquaint you with financial statements and to give you the basic information you need to make sound engineering economic decisions through the remainder of the book.

2.2 Financial Status for Businesses

Just like figuring out your personal wealth, all businesses must prepare their financial status reports. Of the various reports corporations issue to their stockholders, the annual report is by far the most important, containing basic financial statements as well as management's opinion of the past year's operations and the firm's future prospects. What would managers and investors want to know about a company at the end of the fiscal-year? Following are four basic questions that managers or investors are likely to ask:

- What is the company's financial position at the end of the fiscal period?
- How well did the company operate during the fiscal period?
- On what did the company decide to use its profits?
- How much cash did the company generate and spend during the fiscal period?

As illustrated in Figure 2.2, the answer to each of these questions is provided by one of the following financial statements: the balance sheet statement, the income statement, the statement of retained earnings, and the cash flow statement. The fiscal-year (or operating cycle) can be any 12-month term, but is usually January 1 through December 31 of a calendar year.

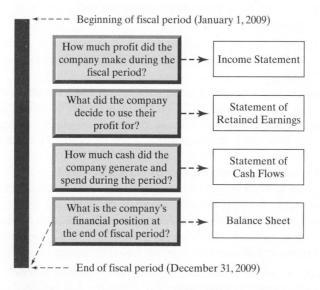

Figure 2.2 Information reported in the financial statements.

As mentioned in Section 1.1.2, one of the primary responsibilities of engineers in business is to plan for the acquisition of equipment (capital expenditure) that will enable the firm to design and produce products economically. This type of planning will require an estimation of the savings and costs associated with the acquisition of equipment and the degree of risk associated with execution of the project. Such an estimation will affect the business' **bottom line** (profitability), which will eventually affect the firm's stock price in the marketplace. Therefore, engineers should understand the various financial statements in order to communicate with upper management regarding the status of a project's profitability. The situation is summarized in Figure 2.3.

Bottom line is slang for net income or accounting profit.

For illustration purposes, we will use data from Research In Motion, manufacturer of the BlackBerry products, to discuss the basic financial statements. Since going public in 1997, RIM has been posting its annual reports at its website (www.rim.com). Its fiscal-year-end date is the 52nd or the 53rd week ending on the last Saturday of February or the first Saturday of March. For example, its 2007 fiscal-year ended March 3, 2007, and its 2009 fiscal-year ended February 28, 2009. In addition to the financial statements for the past fiscal-year, these annual reports also include the management's projection of the company's performance for the coming fiscal-year.

What can individual investors make of all this information contained in a company's annual report? Actually, we can make quite a bit. As you will see, investors use the information contained in a company's annual report to form expectations about future earnings and dividends. Therefore, the annual report is certainly of great interest to investors.

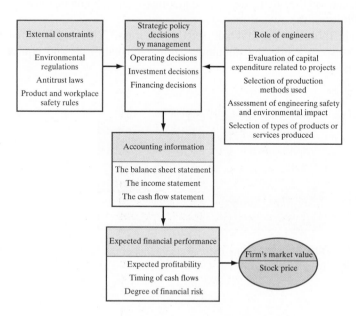

Figure 2.3 Summary of major factors affecting stock prices.

2.2.1 The Balance Sheet

What is the company's financial position at the end of the reporting period? We find the answer to this question in the company's **balance sheet statement**. A company's balance sheet, sometimes called its **statement of financial position**, reports three main categories of items: assets, liabilities, and stockholders' equity. Assets are arranged in order of liquidity. Liquidity is the ease with which an asset can be converted into an economy's medium of exchange. The most liquid assets appear at the top of the page, the least liquid assets at the bottom of the page. (See Figure 2.4.) Because cash is the most liquid of all assets, it is always listed first. Current assets are so critical that they are separately broken out and totalled. They are what will hold the business afloat for the next year.

Liabilities are arranged in order of payment, the most pressing at the top of the list, the least pressing at the bottom. Like current assets, current liabilities are so critical that they are separately broken out and totalled. They are what will be paid out during the next year.

A company's financial statements are based on the most fundamental tool of accounting: the accounting equation. The **accounting equation** shows the relationship among assets, liabilities, and owners' equity:

$$\text{Assets} = \text{Liabilities} + \text{Owners' Equity}$$

Every business transaction, no matter how simple or complex, can be expressed in terms of its effect on the accounting equation. Regardless of whether a business grows or contracts, the equality between the assets and the claims against the assets is always maintained. In other words, any change in the amount of total assets is necessarily accompanied by an equal change on the other side of the equation—that is, by an increase or decrease in either the liabilities or the owners' equity.

As shown in Table 2.1, the first half of RIM's year-end 2009 and 2008 balance sheets lists the firm's assets, while the remainder shows the liabilities and equity, or claims against those assets.

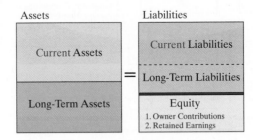

Figure 2.4   Using the four quadrants of the balance sheet.

> **Current assets account** represents the value of all assets that are reasonably expected to be converted into cash within one year.

Assets

The dollar amount shown in the assets portion of the balance sheet represents how much the company owns at the time it issues the report. We list the asset items in the order of their "liquidity," or the length of time it takes to convert them to cash.

- **Current assets** can be converted to cash or its equivalent in less than one year. Current assets generally include three major accounts.

TABLE 2.1 Consolidated Balance Sheets for Research In Motion Limited, Incorporated under the Laws of Ontario (United States dollars, in thousands except per share data) (unaudited)

	February 28, 2009	March 1, 2008
Assets		
Current		
Cash and cash equivalents	$ 835,546	$ 1,184,398
Short-term investments	682,666	420,709
Trade receivables	2,112,117	1,174,692
Other receivables	157,728	74,689
Inventory	682,400	396,267
Other current assets	187,257	135,849
Deferred income tax asset	183,872	90,750
	4,841,586	3,477,354
Long-Term Investments	720,635	738,889
Capital assets	1,334,648	705,955
Intangible assets	1,066,527	469,988
Goodwill	137,572	114,455
Deferred income tax asset	404	4,546
	$ 8,101,372	$ 5,511,187
Liabilities		
Current		
Accounts payable	$ 448,339	$ 271,076
Accrued liabilities	1,238,602	690,442
Income taxes payable	361,460	475,328
Deferred revenue	53,834	37,236
Deferred income tax liability	13,116	–
Current portion of long-term debt	–	349
	2,115,351	1,474,431
Deferred income tax liability	87,917	65,058
Income taxes payable	23,976	30,873
Long-term debt	–	7,259
	2,227,244	1,577,621
Shareholders' equity		
Capital stock	2,208,235	2,169,856
Retained earnings	3,545,710	1,653,094
Paid-in capital	119,726	80,333
Accumulated other comprehensive income	457	30,283
	5,874,128	3,933,566
	$ 8,101,372	$ 5,511,187

Source: Press Release, Research In Motion Ltd., April 2, 2009. Reprinted with permission.

1. The first is *cash and cash equivalents*. A firm typically has a cash account at a bank to provide the funds needed to conduct day-to-day business. Although all assets are stated in terms of dollars, only cash represents actual money. If a company's short-term cash position is negative, it is called short-term borrowing or credit line and is shown as a liability. Cash equivalent items such as marketable securities are often listed together with cash. Short-term investments may be included in cash equivalents or listed separately. As shown in Table 2.1, RIM's cash position at the end of fiscal-year 2009 was US$836 million in cash and cash equivalents and US$683 million in short-term investments.

2. The second type of current assets is *accounts receivables*, which represent money that is owed to the firm by its customers but has not yet been paid. For example, when RIM receives an order from a customer, it sends an invoice along with the shipment to the buyer. The unpaid invoice falls into the category called "accounts receivable." As this bill is paid, it will be deducted from the accounts receivable account and placed into the cash account. A typical firm will have a 30- to 60-day accounts receivable, depending on the frequency of its bills and the payment terms for its customers. As shown in Table 2.1, RIM listed two lines of receivables, trade receivables in the amount of US$2,112 million and other receivables in the amount of US$158 million at the end of its fiscal-year 2009.

3. The third major type of current assets is *inventories*, which constitute the dollar amount of materials, supplies, work-in-process, and finished goods available for sale. RIM had US$682 million in inventory at the end of fiscal-year 2009.

Other current assets may also be listed separately in a company's balance sheet. As shown in Table 2.1, RIM listed US$187 million in other current assets and US$184 million in deferred income tax asset at the end of fiscal-year 2009.

- **Fixed assets** or **capital assets** are relatively permanent and would take usually longer than a year to convert into cash. The most common capital assets include the physical investment in the business, such as land, buildings,[2] factory machinery, office equipment, furniture and fixtures, and vehicles. With the exception of land, most fixed assets have a finite useful life. For example, buildings and equipment are used up over a period of years. Each year, a portion of the usefulness of these assets expires, and a portion of their total cost should be recognized as a depreciation expense. The term *depreciation* denotes the accounting process for this gradual conversion of capital assets into expenses. Thus capital assets in the amount of US$1,335 million shown in Table 2.1 represent the current book value of these assets of RIM at the end of fiscal-year 2009 after deducting depreciation expenses. Detailed discussions on capital assets and various depreciation methods will be given in Chapter 8.

- Finally the balance sheet shows **other assets**. Typical assets in this category include investments made in other companies and intangible assets such as licences, patents, goodwill, copyrights, franchises, and so forth. Intangible assets may also have lives longer than one year. The amount of decrease in the value of an intangible asset is often called *amortization* or *depreciation*. How the amortization is accounted for depends on the company. Goodwill appears on the balance sheet only when the company has purchased another business in its entirety. Goodwill includes the excess of the purchase price of business acquisitions over the fair value of identifiable net assets acquired in such acquisitions. The fair market value is defined as the price that a

---

[2] Land and buildings are commonly called real assets to distinguish them from equipment and machinery.

buyer is willing to pay when the business is offered for sale. The fair market value is often larger than the book value of a business. As shown in Table 2.1, RIM listed long-term investments of US$721 million, intangible assets of US$1,067 million, goodwill of US$138 million, and deferred income tax asset of US$0.4 million.

Liabilities and Shareholders' Equity (Owners' Net Worth)

The claims against assets are of two types: liabilities and shareholders' equity. Shareholders or stockholders are the owners of the company. The liabilities of a company indicate where the company obtained the funds to acquire its assets and to operate the business. Liability is the money the company owes. We usually divide liabilities into current liabilities and other liabilities. Shareholders' equity is that portion of the assets of the company that is provided by the investors (owners). Therefore, shareholders' equity is actually the liability of the company to its owners.

- **Current liabilities** are those debts that must be paid in the near future (normally, within one year). Accounts payables are money owed to suppliers. Accrued expenses or liabilities are the opposite of prepaid expenses. Firms will typically incur periodic expenses such as wages, rent, interest, and taxes. Income taxes payable may be listed as a separate item. RIM sometimes bills the customer prior to performing the service for the customer; the prebilling is recorded as deferred revenue. Advance payments and deposits from customers for services are also included in the deferred revenue item. Another item of the current liabilities is the "current portion" of any long-term debt, i.e., the principal payments that are due within one year. In Table 2.1, the current liabilities of RIM at the end of fiscal-year 2009 included US$448 million in accounts payable, US$1,239 million in accrued expenses, US$361 million in income taxes payable, US$54 million in deferred revenue, and US$13 million in deferred income tax liability.

> **Current liabilities** are bills that are due to creditors and suppliers within a short period of time.

- **Other liabilities** include long-term liabilities, such as bonds, mortgages, and long-term notes, which are due and payable more than one year in the future. RIM reported US$88 million in deferred income tax liability, US$24 million in income taxes payable, and zero long-term debt.

- **Common stock** is the aggregate par value of the company's stock issued. Companies rarely issue stocks at a discount (i.e., at an amount below the stated par). Normally, corporations set the par value low enough so that, in practice, stock is usually sold at a premium. The total common stock at RIM was US$2,208 million at the end of fiscal-year 2009.

- **Preferred stock** is a hybrid between common stock and debt. In case the company goes bankrupt, it must pay its preferred stockholders after its creditors, but before its common stockholders. Preferred dividend is fixed, so preferred stockholders do not benefit if the company's earnings grow. Many companies, for example, RIM as of the end of fiscal-year 2009, do not use preferred stocks. The term **capital stocks** used by RIM may include both common stocks and preferred stocks.

- Outstanding stock is the number of shares issued that actually are held by the public. The corporation can buy back part of its issued stock and hold it as **treasury stock**.

> **Treasury stock** is not taken into consideration when calculating earnings per share or dividends.

- **Paid-in capital** (**capital surplus**) is the amount of money received from the sale of stock in excess of the par (stated) value of the stock. RIM had a paid-in capital of US$120 million as of the end of fiscal-year 2009.

Retained earnings refers to earnings not paid out as dividends but instead are reinvested in the core business or used to pay off debt.

The amount of **retained earnings** represents the cumulative net income of the firm since its beginning, less the total dividends that have been paid to stockholders. In other words, retained earnings indicate the amount of assets that have been financed by plowing profits back into the business. These retained earnings belong to the common stockholders. The retained earnings at RIM were US$3,546 million at the end of fiscal-year 2009.

- **Shareholders' equity** or **stockholders' equity** represents the amount that is available to the owners after all other debts have been paid. It is the portion of the assets of the company that are provided by the investors (owners). It generally consists of common stock, preferred stock, treasury stock, paid-in capital (capital surplus), and retained earnings. The stockholders' equity, or net worth, is equal to assets minus liabilities (long and short terms). In the case of RIM, based on Table 2.1, we get the following net worth of the owners:

$$US\$8,101 \text{ million} - US\$2,227 \text{ million} = US\$5,874 \text{ million}.$$

2.2.2 The Income Statement

The second financial report is the **income statement**, which indicates whether the company is making or losing money during a stated *period*, usually a year. Most businesses prepare quarterly and some monthly income statements as well. The company's accounting period refers to the period covered by an income statement.

Basic Income Statement Equation

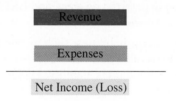

For RIM, the accounting period covers 52 or 53 weeks ending on the last Saturday of February or the first Saturday of March. Table 2.2 gives the 2009 and 2008 income statements for RIM.

Reporting Format

Typical items that are listed in the income statement are as follows:

- **Revenue** is the money received from goods sold and services rendered during a given accounting period. As given in Table 2.2, RIM's revenue during fiscal-year 2009 was US$11,065 million.
- **Net revenue** represents the revenue less any sales returns and allowances. When these adjustments are minor, as in RIM's case, revenue is used in the place of net revenue.
- **Cost of sales** (also called **cost of goods sold** or **cost of revenue**) represents the expenses and costs of doing business as deductions from the revenues. The largest expense for a typical manufacturing firm is its production costs, including such items as labour, materials, depreciation, and overhead, known as cost of goods sold. These costs are also called variable costs as they are directly affected by the volume of the

TABLE 2.2 Consolidated Statements of Operations for Research In Motion Limited, Incorporated under the Laws of Ontario (United States dollars, in thousands except per share data) (unaudited)

	Fiscal-Year Ended	
	February 28, 2009	March 1, 2008
Revenue	$ 11,065,186	$ 6,009,395
Cost of sales	5,967,888	2,928,814
Gross margin	5,097,298	3,080,581
Gross Margin %	46.1%	51.3%
Expenses		
Research and development	684,702	359,828
Selling, marketing, and administration	1,495,697	881,482
Amortization	194,803	108,112
	2,375,202	1,349,422
Income from operations	2,722,096	1,731,159
Investment income	78,267	79,361
Income before income taxes	2, 800,363	1,810,520
Provision for income taxes		
Current	948,536	587,845
Deferred	(40,789)	(71,192)
	907,747	516,653
Net Income	$ 1,892,616	$ 1,293,667
Earnings per share		
Basic	3.35	2.31
Diluted	3.30	2.26
Weighted average number of common shares outstanding		
Basic	565,059	559,778
Diluted	574,156	572,830

Source: Press Release, Research In Motion Ltd., April 2, 2009. Reprinted with permission.

company's business. RIM's cost of sales during fiscal-year 2009 was US$5,968 million. The depreciation or amortization expenses of certain capital assets and licences (part of intangible assets) were also charged to cost of sales at RIM. RIM believes that these amortization expenses are directly related to the number of BlackBerry units it produces and services.

- Net revenue less cost of sales gives the **gross margin**, also called net operating profit. RIM's gross margin was US$5,097 million, or 46.1% of the revenue (or sales) in 2009.

- **Operating expenses** (also called **expenses**, **overhead**, **fixed costs**, or **general and administrative costs**) include cost items that are not directly related to manufacturing the products or delivering the services. They may include research and development expenses, rental expenses, administrative expenses, marketing expenses, property taxes, interest charges on debts, and depreciation or amortization of capital and intangible assets. Each year, a portion of the usefulness of the capital assets used by the company is recognized as a depreciation expense. This non-cash expense of depreciation or amortization is a part of the operating expenses. During 2009, RIM incurred US$2,375 million in operating expenses, including US$685 million in research and development, US$1,496 million in selling, marketing, and administration, and US$195 million in amortization of certain capital and all intangible assets other than licences.

- **Income from operations** or **operating income** is calculated by subtracting operating expenses from the gross margin. RIM's income from operations during fiscal-year 2009 was US$2,722 million.

- **Investment income** is income generated from investment activities that are not directly related to the manufacturing and/or service activities of the company. RIM's investment income during fiscal-year 2009 was US$78 million.

- **Income before income taxes** or **taxable income** is the sum of income from operations and investment income. RIM's taxable income during fiscal-year 2009 was US$2,800 million.

- Finally, we determine **net income** (or **net profit**) by subtracting income taxes from the taxable income. Net income is also known as accounting income. RIM's total taxes for fiscal-year 2009 were US$908 million. Its net income for 2009 was US$1,893 million.

Earnings per Share

EPS is generally considered to be the single most important variable in determining a share's price.

Another important piece of financial information provided in the income statement is the **earnings per share (EPS)**.[3] In simple situations, we compute this by dividing the "available earnings to common stockholders" by the number of shares of common stock outstanding. Stockholders and potential investors want to know what their share of profits is, not just the total dollar amount. The presentation of profits on a per-share basis allows the stockholders to relate earnings to what they paid for a share of stock. Naturally, companies want to report a higher EPS to their investors as a means of summarizing how well they managed their businesses for the benefit of the owners. Interestingly, RIM earned US$3.35 per share in 2009, up from US$2.31 in 2008, but it paid no dividends.

Retained Earnings

As a supplement to the income statement, many corporations also report their retained earnings during the accounting period. When a corporation earns profits, it must decide

---

[3] In reporting EPS, the firm is required to distinguish between "basic EPS" and "diluted EPS." Basic EPS is the net income of the firm, divided by the number of shares of common stock outstanding. By contrast, the diluted EPS includes all common stock equivalents (convertible bonds, preferred stock, warrants, and rights), along with common stock. Therefore, diluted EPS will usually be less than basic EPS.

what to do with these profits. The corporation may decide to pay out the profits as dividends to its stockholders. Or it may retain the profits in the business to finance expansion or support other business activities.

When a corporation declares dividends, preferred stock has priority over common stock. Preferred stock pays a stated dividend, much like the interest payment on bonds. The dividend is not a legal liability until the board of directors has declared it. However, many corporations view the dividend payment to preferred stockholders as a liability. Therefore, "available for common stockholders" reflects the net earnings of the corporation less the preferred stock dividends. When preferred and common stock dividends are subtracted from net income, the remainder is retained earnings (profits) for the year. As mentioned previously, these retained earnings are reinvested into the business.

Even when a separate statement of retained earnings is not available, the information can be derived from the income statement and the balance sheet. We will illustrate this in the following example.

EXAMPLE 2.1 Understanding RIM's Balance Sheet and Income Statement

Tables 2.1 and 2.2 are the balance sheet and the income statement of RIM released April 2, 2009. They show how RIM generated its net income during the fiscal-year.

The Balance Sheet. RIM's US$8,101 million of total assets shown in Table 2.1 were necessary to support its total revenue of US$11,065 million shown in Table 2.2.

- RIM obtained the bulk of the funds it used to buy assets (see liabilities in Table 2.1)

  1. By buying on credit from its suppliers (accounts payable), US$448 million.

  2. By using other accrued liabilities, US$1,239 million.

  3. By issuing common stock to investors, US$2,208 million.

  4. By plowing earnings into the business, as reflected in the retained earnings account, US$3,546 million.

- The net increase in the book value of certain capital assets was US$629 million (US$1,335 million in 2009 minus US$706 million in 2008; Table 2.1). The net increase in intangible assets other than licences was US$597 million (US$1,067 million in 2009 minus US$470 million in 2008; Table 2.1). The net increase in goodwill acquired was US$23 million (US$137 million in 2009 minus US$114 million in 2008; Table 2.1). Remember that the depreciation of certain other capital assets and licences have been charged to the cost of sales. To find the total new acquisitions of all fixed assets including all capital assets, all intangible assets, and goodwill, we need to add the total non-cash amortization expenses of US$328 million listed in the company's cash flow statement shown in Table 2.3 (to be further discussed after this example). Thus, the acquisition of all fixed assets in the year equals $1,577 million. This total investment in new assets (capital, intangible, and business) can also be obtained from the cash flows from investing activities given in Table 2.3.

- RIM had no long-term debt at the end of fiscal 2009 (Table 2.1). The long-term debt at the end of fiscal 2008 was only US$7 million. All long-term debt was paid off in fiscal 2009.

- RIM had 566 million shares of common stock outstanding at the end of fiscal 2009 (Table 2.2). Investors actually provided the company with a total capital of US$2,208 million through capital stocks over the years (Table 2.1). RIM has accumulated total retained earnings of US$3,546 million since it was incorporated. RIM has also received a total of US$120 million in paid-in capital from its employees through stock purchase options. The combined net stockholders' equity was US$5,874 million, and these belong to RIM's common stockholders (Table 2.1).

- On average, stockholders have a total investment of US$10.38 per share (US$5,874 million/566 million shares) in the company. The US$10.38 figure is known as the stock's book value. In May 2009, RIM's stock traded in the general price range from US$70 to US$80 per share at the NASDAQ Stock Exchange (www.nasdaq.com: RIMM). Note that this market price is quite different from the stock's book value. Many factors affect the market price, the most important one being how investors expect the company to perform in the future. The company's excellent performance over the past few years has had a major influence on the market value of its stock.

The Income Statement. RIM's net revenue was US$11,065 million in 2009, compared with US$6,009 million in 2008, a gain of 84% (Table 2.2). Income from operations (operating profit) rose 57% to US$2,722 million, and net income was up 46% to US$1,893 million.

- RIM issued no preferred stock, so there is no required cash dividend. Therefore, the entire net income of US$1,893 million belongs to the common stockholders.

- Earnings per common share climbed by 45% to $3.35 (Table 2.2). RIM has been retaining all net income over the years for reinvestment in the firm. We can see that RIM has generated a net income of US$1,893 million in fiscal 2009 for common stockholders. As shown in Table 2.1, the balance of retained earnings at the end of fiscal 2008 was US$1,653 million. After the net income of US$1,893 million generated in fiscal 2009 was kept as retained earnings, the total retained earnings grew to US$3,546 million by the end of fiscal 2009.

2.2.3 The Cash Flow Statement

The income statement explained in the previous section indicates only whether the company was making or losing money during the reporting period. Therefore, the emphasis was on determining the net income (profits) generated by the firm's operating activities. However, the income statement ignores two other important business activities for the period: financing and investing activities. Therefore, we need another financial statement—the cash flow statement, which details how the company generated the cash it received and how the company used that cash during the reporting period.

Sources and Uses of Cash

The difference between the sources (inflows) and uses (outflows) of cash represents the net cash flow during the reporting period. This is a very important piece of information, because investors determine the value of an asset (or, indeed, of a whole firm) by the cash flows it generates. Certainly, a firm's net income is important, but cash flows are even more important, particularly because the company needs cash to pay dividends and to purchase the assets required to continue its operations. As mentioned in the previous section, the goal of the firm should be to maximize the price of its stock. Since the value of any asset depends on the cash flows produced by the asset, managers want to maximize the cash flows available to investors over the long run. Therefore, we should make investment decisions on the basis of cash flows rather than profits. For such investment decisions, it is necessary to convert profits (as determined in the income statement) to cash flows. Table 2.3 is RIM's statement of cash flows, released April 2, 2009.

Reporting Format

In preparing the cash flow statement such as that in Table 2.3, many companies identify the sources and uses of cash according to the types of business activities. There are three types of activities:

- **Operating activities.** We start with the net change in operating cash flows from the income statement. Here, operating cash flows represent those cash flows related to production and the sales of goods or services. All non-cash expenses are simply added back to net income (or after-tax profits). For example, an expense such as depreciation is only an accounting expense (a bookkeeping entry). Although we may charge depreciation against current income as an expense, it does not involve an actual cash outflow. The actual cash flow may have occurred when the asset was purchased. (Any adjustments in **working capital**[4] will also be listed here.) Detailed dicussions on depreciation will be given in Chapter 8.
- **Investing activities.** Once we determine the operating cash flows, we consider any cash flow transactions related to investment activities, which include purchasing new fixed assets (cash outflow), selling old equipment (cash inflow), and buying and selling financial assets.
- **Financing activities.** Finally, we detail cash transactions related to financing any capital used in business. For example, the company could borrow or sell more stock, resulting in cash inflows. Paying off existing debt will result in cash outflows.

By summarizing cash inflows and outflows from three activities for a given accounting period, we obtain the net change in the cash position of the company.

---

[4] The difference between the increase in current assets and the spontaneous increase in current liabilities is the **net change in net working capital**. If this change is positive, then additional financing, over and above the cost of the fixed assets, is needed to fund the increase in current assets. This will further reduce the cash flow from the operating activities.

TABLE 2.3 Consolidated Statements of Cash Flows for Research In Motion Limited, Incorporated under the Laws of Ontario (United States dollars, in thousands except per share data) (unaudited)

	For the three months ended February 28, 2009	For the year ended March 1, 2009
Cash flows from operating activities		
Net income	$ 518,259	$ 1,892,616
Items not requiring an outlay of cash:		
Amortization	116,125	327,896
Deferred income taxes	3,291	(36,623)
Income taxes payable	(456)	(6,897)
Stock-based compensation	8,900	38,100
Other	5,598	5,867
Net changes in working capital items	**(360,975)**	**(769,114)**
Net cash provided by operating activities	**290,742**	**1,451,845**
Cash flows from financing activities		
Issuance of common shares	2,183	27,024
Excess tax benefits from stock-based compensation	(52)	12,648
Repayment of debt	(14,061)	(14,305)
Net cash provided by financing activities	**(11,930)**	**25,367**
Cash flows from investing activities		
Acquisition of long-term investments	(67,326)	(507,082)
Proceeds on sale or maturity of long-term investments	127,048	431,713
Acquisition of capital assets	(251,932)	(833,521)
Acquisition of intangible assets	(221,964)	(687,913)
Business acquisitions	(48,425)	(48,425)
Acquisition of short-term investments	(250,018)	(917,316)
Proceeds on sale or maturity of short-term investments	170,178	739,021
Net cash used in investing activities	**(542,439)**	**(1,823,523)**
Effect of foreign exchange loss on cash and cash equivalents	(2,541)	(2,541)
Net decrease in cash and cash equivalents for the period	**(266,168)**	**(348,852)**
Cash and cash equivalents, beginning of period	**1,101,714**	**1,184,398**
Cash and cash equivalents, end of period	**$ 835,546**	**$ 835,546**
As at	**February 28, 2009**	**November 29, 2008**
Cash and cash equivalents	$ 835,546	$ 1,101,714
Short-term investments	682,666	574,279
Long-term investments	720,635	812,638
	$ 2,238,847	**$ 2,488,631**

Source: Press Release, Research In Motion Ltd., April 2, 2009. Reprinted with permission.

EXAMPLE 2.2 Understanding RIM's Cash Flow Statement

As shown in Table 2.3, RIM's net cash provided by **operating activities** during fiscal 2009 amounted to US\$1,452 million. Note that this is significantly lower than the net income of US\$1,893 million earned during the reporting period as shown in Table 2.2. Where did the extra income go?

- The main reason for the difference lies in the accrual-basis accounting principle used by RIM. In accrual-basis accounting, an accountant recognizes the impact of a business event as it occurs. When the business performs a service, makes a sale, or incurs an expense, the accountant enters the transaction into the books, regardless of whether cash has or has not been received or paid. For example, an increase in trade receivables of US\$937 million (US\$2,112 million in fiscal 2009 minus US\$1,175 million in fiscal 2008) represents the amount of total sales on credit (Table 2.1). Since the US\$937 million figure was included in the total sales (but this money has not been received yet) in determining the net income, we need to subtract it to determine the company's true cash position at the end of fiscal 2009. Similarly other receivables and accounts payables need to be considered. These and several other items need to be included in the calculation of net changes in working capital items.

- To find the net changes in working capital items during the year, we need to take into account the following changes from fiscal 2008 to fiscal 2009 (Table 2.1):

Trade receivables:	−US\$937 million
Other receivables:	−US\$83 million
Inventory:	−US\$286 million
Other current assets:	−US\$51 million
Accounts payable:	+US\$177 million
Accrued liabilities:	+US\$548 million
Income taxes payable:	−US\$114 million
Deferred revenue:	+US\$17 million
Others:	US\$40 million

Adding these changes together will give −US\$769 million in the net changes in working capital items. The negative sign denotes that this additional amount of cash is tied up in the receivables and the inventory the company is holding at the end of fiscal 2009.

- To find the net cash provided by operating activities, we also need to add back the non-cash items that were used in calculation of the net income. These non-cash items include depreciation or amortization of all fixed assets including capital assets and intangible assets, stock-based compensation, and so on. For fiscal 2009, RIM reported US\$328 million in total amortization (Table 2.3). The net cash provided by operating activities in the amount of US\$1,452 million (Table 2.3) is equal to the net income adjusted for non-cash items and the net changes in working capital items: US\$1,893 million + US\$328 million − US\$769 million = US\$1,452 million.

(Continued)

- **Financing activities** produced a net inflow of US$25 million. This net number is the result of capital inflow of US$27 million from issuance of common stocks, excess tax benefits of US$13 million from stock-based compensation, and cash outflow of US$14 million for repayment of debt.

- As to **investment activities**, RIM reported in Table 2.3 an investment outflow of US$507 million in purchase of long-term investments, an inflow of US$432 million from sale or maturity of long-term investments, an outflow of US$834 million in new capital assets, an outflow of US$688 million in purchase of intangible assets, US$48 million in business acquisitions, US$917 million in purchase of short-term investments, and an inflow of US$739 million in sale or maturity of short-term investments. The net cash flow provided from these investing activities amounted to US$1,824 million, which means an outflow of money.

- Finally, there was the effect of exchange rate changes on cash for foreign sales. This amounted to a net decrease of US$2.5 million (Table 2.3). RIM is exposed to foreign exchange risk as a result of transactions in currencies other than its functional currency, the U.S. dollar. The majority of RIM's revenues in fiscal 2009 was transacted in U.S. dollars. Portions of the revenues were denominated in British pounds, Canadian dollars, and euros. Purchases of raw materials were primarily transacted in U.S. dollars. Other expenses, consisting of the majority of salaries, certain operating costs and manufacturing overhead were incurred primarily in Canadian dollars. Fewer than 10% of the sales revenues were generated in Canada. More than 50% were from sales in the US.

- Together, the three types of activities generated a total cash outflow of US$349 million. With the initial cash balance of US$1,184 million, the ending cash balance thus decreased to US$836 million. This same amount denotes RIM's cash position, as shown in the cash and cash equivalents account in the balance sheet. Although this is already known from the balance sheet, one cannot explain this change in cash without the aid of a cash flow statement.

2.3 Using Ratios to Make Business Decisions

As we have seen in RIM's financial statements, the purpose of accounting is to provide information for decision making. Accounting information tells what happened at a particular point in time. In that sense, financial statements are essentially historical documents. However, most users of financial statements are concerned about what will happen in the future. For example,

- Stockholders are concerned with future earnings and dividends.
- Creditors are concerned with the company's ability to repay its debts.
- Managers are concerned with the company's ability to finance future expansion.
- Engineers are concerned with planning actions that will influence the future course of business events.

Although financial statements are historical documents, they can still provide valuable information bearing on all of these concerns. An important part of financial analysis is the calculation and interpretation of various financial ratios. In this section, we consider some of the ratios that analysts typically use in attempting to predict the future course of events in business organizations. We may group these ratios into five categories (debt management, liquidity, asset management, profitability, and market trend) as outlined in Figure 2.5. In all financial ratio calculations, to be illustrated below, we will use the 2009 financial statements for Research In Motion Ltd., as summarized in Table 2.4.

2.3.1 Debt Management Analysis

All businesses need assets to operate. To acquire assets, the firm must raise capital. When the firm finances its long-term needs externally, it may obtain funds from the capital markets. Capital comes in two forms: **debt** and **equity**. Debt capital is capital borrowed from financial institutions. Equity capital is capital obtained from the owners of the company.

The basic methods of financing a company's debt are through bank loans and the sale of bonds. For example, suppose a firm needs $10,000 to purchase a computer. In this situation, the firm would borrow money from a bank and repay the loan, together with the interest specified, in a few years. This kind of financing is known as *short-term debt financing*. Now suppose that the firm needs $100 million for a construction project. Normally, it would be very expensive (or require a substantial mortgage) to borrow the money directly from a bank. In this situation, the company would go public to borrow money on a long-term basis. When investors lend capital to a company and the company consents to repay the loan at an agreed-upon interest rate, the investor is the creditor of

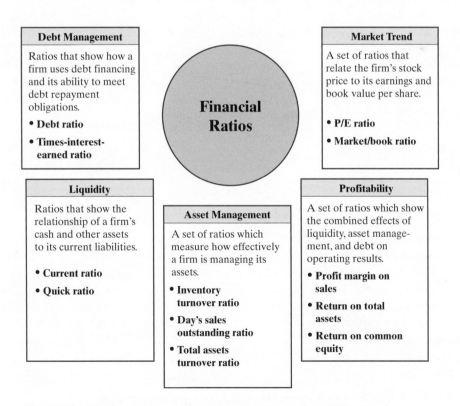

Figure 2.5 Types of ratios used in evaluating a firm's financial health.

TABLE 2.4 Summary of RIM's Financial Statements (US$, in thousands)

	February 28, 2009	March 1, 2008
Balance Sheet		
Cash and cash equivalents	835,546	1,184,398
Accounts receivables, net	2,269,845	1,249,381
Inventories	682,400	396,267
Total current assets	4,841,586	3,477,354
Total assets	8,101,372	5,511,187
Total current liabilities	2,115,351	1,474,431
Long-term debt	0	7,259
Total liabilities	2,227,244	1,577,621
Capital stock	2,208,235	2,169,856
Retained earnings	3,545,710	1,653,094
Total stockholders' equity	5,874,128	3,933,566
Income Statement		
Net revenue	11,065,186	6,009,395
Gross income (margin)	5,097,298	3,080,581
Operating income (margin)	2,722,096	1,731,159
Net income (margin)	1,892,616	1,293,867
Statements of Retained Earnings		
Beginning retained earnings	1,653,094	359,227
Net income	1,892,616	1,293,867
Ending retained earnings	3,545,710	1,653,094
Statement of Cash Flows		
Net cash from operating activities	1,451,845	1,576,759
Net cash used in financing activities	25,367	80,398
Net cash used in investing activities	(1,823,523)	(1,153,937)
Effect of exchange rate changes	(2,541)	4,034
Net change in cash and cash equivalents	(348,852)	507,254
Beginning cash position	1,184,398	677,144
Ending cash position	835,546	1,184,398

Source: Press Release, Research In Motion Ltd., April 2, 2009. Reprinted with permission.

the corporation. The document that records the nature of the arrangement between the issuing company and the investor is called a **bond**. Raising capital by issuing a bond is called *long-term debt financing*.

Similarly, there are different types of equity capital. For example, the equity of a proprietorship represents the money provided by the owner. For a corporation, equity capital comes in two forms: *preferred stock* and *common stock*. Investors provide capital to a corporation and the company agrees to endow the investor with fractional owner-ship in the corporation. Preferred stock pays a stated *dividend*, much like the interest payment on bonds. However, the dividend is not a legal liability until the company de-clares it. Preferred stockholders have preference over common stockholders as regards the receipt of dividends if the company has to liquidate its assets. We can examine the extent to which a company uses debt financing (or financial leverage) in the operation of its business if we

- Check the balance sheet to determine the extent to which borrowed funds have been used to finance assets, and
- Review the income statement to see the extent to which fixed charges (interests) are covered by operating profits.

Two essential indicators of a business's ability to pay its long-term liabilities are the *debt ratio* and the *times-interest-earned ratio*.

Debt Ratio

The relationship between total liabilities and total assets, generally called the **debt ratio**, tells us the proportion of the company's assets that it has financed with debt:

$$\text{Debt ratio} = \frac{\text{Total liabilites}}{\text{Total assets}}$$

$$= \frac{2{,}227{,}244}{8{,}101{,}372} = 27.49\%.$$

Debt ratio: A ratio that indi-cates what proportion of debt a *company* has relative to its total assets.

Total liabilities includes both current liabilities and long-term debt. If the debt ratio is unity, then the company has used debt to finance all of its assets. As of February 28, 2009, RIM's debt ratio was 27.49%; this means that its creditors have supplied close to 27% of the firm's total financing. Certainly, most creditors prefer low debt ratios, be-cause the lower the ratio, the greater is the cushion against creditors' losses in case of liquidation. If a company seeking financing already has large liabilities, then additional debt payments may be too much for the business to handle. For such a highly leveraged company, creditors generally charge higher interest rates on new borrowing to help protect themselves.

Times-Interest-Earned Ratio

The most common measure of the ability of a company's operations to provide protec-tion to the long-term creditor is the times-interest-earned ratio. We find this ratio by dividing earnings before interest and income taxes (EBIT) by the yearly interest charges that must be met. Interest charges are not usually listed as a separate item in the income statement, but they are usually detailed in the firm's annual reports. Since RIM's annual report for fiscal 2009 has not been released at the time of preparing this material, we will use its 2008 annual report to illustrate the calculation of this ratio. The total interest expenses[5] paid by RIM during fiscal 2008 were US$518,000. Its income before income taxes for fiscal 2008 was US$1,810,520,000 (Table 2.2). Then, we have

$$\text{Times-interest-earned ratio} = \frac{\text{EBIT}}{\text{Interest expense}}$$

$$= \frac{\$1,810,520,000 + \$518,000}{\$518,000} = 3496 \text{ times.}$$

The times-interest-earned ratio measures the extent to which operating income can decline before the firm is unable to meet its annual interest costs. Failure to meet this obligation can bring legal action by the firm's creditors, possibly resulting in the company's bankruptcy. Note that we use the earnings before interest and income taxes, rather than net income, in the numerator. Because RIM must pay interest with pre-tax dollars, RIM's ability to pay current interest is not affected by income taxes. Only those earnings remaining after all interest charges are subject to income taxes. For RIM, the times-interest-earned ratio for 2008 would be 3496 times. This ratio for RIM is extremely high in comparison with most companies. This is not surprising because RIM had a very low long-term debt (US$7 million) at the end of fiscal 2008. If this ratio is below 1 for a company, it means that the cash flow generated in a fiscal-year is not even enough to pay the interest charges on its debt, to say nothing of making debt principal repayment or generating wealth for its owners. When this ratio is below 2.5, it is often a warning sign that the company is in trouble. Many good companies maintain this ratio well above 10. Take Dell Corporation as an example. Dell is a manufacturer of a wide range of computer systems, including desktops, notebooks, and workstations. Its headquarters is located in Round Rock, Texas. In its fiscal-year 2009 (ended on January 30, 2009), Dell[6] reported US$3,324 million in income before income taxes and US$74 million in interest expenses paid. Dell's times-interest-earned ratio for fiscal 2009 was 46.

2.3.2 Liquidity Analysis

If you were one of the many suppliers to RIM, your primary concern would be whether RIM will be able to pay off its debts as they come due over the next year or so. Short-term creditors want to be repaid on time. Therefore, they focus on RIM's cash flows and on its working capital, as these are the company's primary sources of cash in the near future. The excess of current assets over current liabilities is known as **working capital**, a figure that indicates the extent to which current assets can be converted to cash to meet current obligations. Therefore, we view a firm's net working capital as a measure of its *liquidity* position. In general, the larger the working capital, the better able the business is to pay its debt.

The **current ratio** measures a company's ability to pay its short-term obligations.

Current Ratio

We calculate the **current ratio** by dividing current assets by current liabilities:

[5] Research In Motion, 2008 Annual Report (www.rim.com).
[6] Dell Inc., Form 10-K for Fiscal Year 2009 (www.dell.com, About Dell, Investors, Financial Reporting).

$$\text{Current ratio} = \frac{\text{Current assets}}{\text{Current liabilities}}$$

$$= \frac{\$4{,}841{,}586}{\$2{,}115{,}351} = 2.29 \text{ times.}$$

If a company is getting into financial difficulty, it begins paying its bills (accounts payable) more slowly, borrowing from its bank, and so on. If current liabilities are rising faster than current assets, the current ratio will fall, and that could spell trouble. What is an acceptable current ratio? The answer depends on the nature of the industry. The general rule of thumb calls for a current ratio of 2 to 1. This rule, of course, is subject to many exceptions, depending heavily on the composition of the assets involved. RIM had a very healthy current ratio in 2008.

Quick (Acid-Test) Ratio

The quick ratio tells us whether a company could pay all of its current liabilities if they came due immediately. We calculate the quick ratio by deducting inventories from current assets and then dividing the remainder by current liabilities:

$$\text{Quick ratio} = \frac{\text{Current assets} - \text{Inventories}}{\text{Current liabilities}}$$

$$= \frac{\$4{,}841{,}586 - \$682{,}400}{\$2{,}115{,}351} = 1.97 \text{ times.}$$

The quick ratio measures how well a company can meet its obligations without having to liquidate or depend too heavily on its inventory. Inventories are typically the least liquid of a firm's current assets; hence, they are the assets on which losses are most likely to occur in case of liquidation. We can see that RIM's quick ratio is also much higher than 1 (actually almost equal to 2). This indicates that RIM's current assets, including cash, cash equivalents, short-term investments, and accounts receivables and excluding inventory, are almost twice as much as its current liabilities. In case RIM has to pay all its current liabilities right away, it does not need to wait to sell its inventory, as its current assets excluding inventory are more than enough.

2.3.3 Asset Management Analysis

The ability to sell inventory and collect accounts receivables is fundamental to business success. Therefore, the third group of ratios measures how effectively the firm is managing its assets. We will review three ratios related to a firm's asset management: (1) the inventory turnover ratio, (2) the day's sales outstanding ratio, and (3) the total asset turnover ratio. The purpose of these ratios is to answer this question: Does the total amount of each type of asset, as reported on the balance sheet, seem reasonable in view of current and projected sales levels? The acquisition of any asset requires the use of funds. On the one hand, if a firm has too many assets, its cost of capital will be too high; hence, its profits will be depressed. On the other hand, if assets are too low, the firm is likely to lose profitable sales.

Inventory Turnover

The **inventory turnover** ratio measures how many times the company sold and replaced its inventory over a specific period—for example, during the year. We compute the ratio

Inventory turnover: A ratio that shows how many times the inventory of a firm is sold and replaced over a specific period.

by dividing sales by the average level of inventories on hand. We compute the average inventory figure by taking the average of the beginning and ending inventory figures. Since RIM has a beginning inventory figure of US$396 million and an ending inventory figure of US$682 million, its average inventory for the year would be US$539 million, or ($US396 + US$682)/2. Then we compute RIM's inventory turnover for 2009 as follows:

$$\text{Inventory turnover ratio} = \frac{\text{Sales}}{\text{Average inventory balance}}$$

$$= \frac{\$11,065,186}{\$539,000} = 20.53 \text{ times.}$$

As a rough approximation, RIM was able to sell and restock its inventory 20.53 times per year. For comparison, Nokia, the number-one phone maker globally, had an inventory turnover ratio of 18.75 during its fiscal-year 2008 (ended December 31, 2008).[7] This ratio varies from industry to industry. Within the same industry, the higher this ratio is, the better the company's operation is. This means that less working capital is tied up in inventory. Excessive inventory is unproductive and it represents investment with a low or zero rate of return.

Day's Sales Outstanding (Accounts Receivable Turnover)

Average collection period is often used to help determine if a company is trying to disguise weak sales.

The day's sales outstanding (DSO) is a rough measure of how many times a company's accounts receivable have been turned into cash during the year. We determine this ratio, also called the **average collection period**, by dividing accounts receivable by average sales per day. In other words, the DSO indicates the average length of time the firm must wait after making a sale before receiving cash. For RIM's fiscal 2009,

$$\text{DSO} = \frac{\text{Receivables}}{\text{Average sales per day}} = \frac{\text{Receivables}}{\text{Annual sales/365}}$$

$$= \frac{\$2,269,845}{\$11,065,186 \ / \ 365}$$

$$= 74.87 \text{ days.}$$

Thus, on average, it takes RIM 74.87 days to collect on a credit sale. On the other hand during its fiscal-year 2008, Nokia's average collection period was 67.98 days. Whether the average of 74.87 days taken to collect an account is good or bad depends on the credit terms RIM is offering its customers. If the credit terms are 60 days, we can say that RIM's customers, on the average, are not paying their bills on time. In order to improve their working-capital position, most customers tend to withhold payment for as long as the credit terms will allow and may even go over a few days. The long collection period may signal either that customers are in financial trouble or that the company manages its credit poorly.

---

[7] Nokia Corporation, Annual Accounts 2008 (www.nokia.com, Investors).

Total Assets Turnover

The total assets turnover ratio measures how effectively the firm uses its total assets in generating its revenues. It is the ratio of sales to all the firm's assets:

$$\text{Total assets turnover ratio} = \frac{\text{Sales}}{\text{Total assets}}$$

$$= \frac{\$11,065,186}{\$8,101,372} = 1.37 \text{ times.}$$

Asset turnover is a measure of how well assets are being used to produce revenue.

RIM's ratio of 1.37 times, compared with Nokia's 1.28, is about 7% faster, indicating that RIM is using its total assets about 7% more intensively than Nokia is. Actually, the total asset turnover ratios of RIM and Nokia are quite comparable. They are leading portable voice/data device manufacturers in the world. Within the same industry, if the turnover ratio for one company is much lower than those of its competitors, this indicates that the asset of that company is not being efficiently utilized to generate sales.

2.3.4 Profitability Analysis

One of the most important goals for any business is to earn a profit. The ratios examined thus far provide useful clues about the effectiveness of a firm's operations, but the profitability ratios show the combined effects of liquidity, asset management, and debt on operating results. Therefore, ratios that measure profitability play a large role in decision making.

Profit Margin on Sales

We calculate the profit margin on sales by dividing net income by sales. This ratio indicates the profit per dollar of sales:

$$\text{Profit margin on sales} = \frac{\text{Net income available to common stockholders}}{\text{Sales}}$$

$$= \frac{\$1,892,618}{\$11,065,186} = 17.10\%.$$

The **profit margin** measures how much out of every dollar of sales a company actually keeps in *earnings*.

Thus, RIM's profit margin is equivalent to 17.10 cents for each dollar of sales generated. RIM's profit margin is much higher than Nokia's profit margin of 7.86% (for its fiscal-year 2008). This indicates that although Nokia's revenue of approximately US$71.0 billion in its fiscal 2008 is much higher than RIM's sales of US$11.1 billion in its fiscal 2009, Nokia's operation in terms of wealth generation for its investors is less efficient than RIM's. One of the reasons is that a major portion of RIM's business is other enterprises while Nokia's cellphones directly target the consumers. Generally speaking, the level of debt may also affect the profit margin. If two firms have identical operations in the sense that their sales, operating costs, and earnings before income taxes are the same, but if one company uses more debt than the other, it will have higher interest charges. These interest charges will pull net income down, and since sales are constant, the result will be a relatively lower profit margin.

Return on Total Assets

The return on total assets—or simply, return on assets (ROA)—measures a company's success in using its assets to earn a profit. The ratio of net income to total assets measures the return on total assets after interest and taxes:

$$\text{Return on total assets} = \frac{\text{Net income} + \text{interest expense}(1 - \text{tax rate})}{\text{Average total assets}}$$

$$= \frac{\$1,294 + \$518(1 - 32.47\%)}{(\$3,089 + \$5,511)/2} = 38.23\%.$$

Adding interest expenses back to net income results in an adjusted earnings figure that shows what earnings would have been if the assets had been acquired solely by selling shares of stock. For fiscal 2008, RIM's net income was US$1,294 million, its total interest expenses were US$518 million, its total assets were US$5,511 million and US$3,089 million at the end and the beginning of the year, respectively, and its average tax rate was estimated to be 32.47% (dividing current income taxes by income before income taxes). Thus, RIM's return on total assets in its fiscal 2008 was 38.23%. In comparison, Nokia's return on total assets in its fiscal 2008 was 10.39%. This shows that RIM had much higher success than Nokia in using its total assets to earn profit.

Return on Common Equity

*The **return on equity** reveals how much profit a company generates with the money its shareholders have invested in it.*

Another popular measure of profitability is rate of return on common equity. This ratio shows the relationship between net income and common stockholders' investment in the company—that is, how much income is earned for every $1 invested by the common stockholders. To compute the return on common equity, we first subtract preferred dividends from net income, yielding the net income available to common stockholders. We then divide this net income available to common stockholders by the average common stockholders' equity during the year. We compute average common equity by using the beginning and ending balances. Note that RIM did not issue preferred stocks. At the beginning of fiscal 2009, RIM's common equity balance (total stockholders' equity) was US$3,934 million; at the end of fiscal 2009, the balance was US$5,874 million. The average balance was then simply US$4,904 million. With a net income of US$1,893 million, we have

$$\text{Return on common equity} = \frac{\text{Net income available to common stockholders}}{\text{Average common equity}}$$

$$= \frac{\$1,893}{\$4,904}$$

$$= 38.60\%$$

The rate of return on common equity for RIM was 38.60% during its fiscal 2009. In comparison, Nokia's return on common equity amounted to 26.84% during its own fiscal 2008. RIM's return on common equity was much higher.

To learn more about what management can do to increase the return on common equity, or ROE, we may rewrite the ROE in terms of the following three components:

$$ROE = \frac{\text{Net income}}{\text{Stockholders' equity}}$$

$$= \frac{\text{Net income}}{\text{Sales}} \times \frac{\text{Sales}}{\text{Assets}} \times \frac{\text{Assets}}{\text{Stockholders' equity}}.$$

The three principal components can be described as the profit margin, asset turnover, and **financial leverage**, respectively, so that

$$ROE = (\text{Profit margin}) \times (\text{Asset turnover}) \times (\text{Financial leverage})$$

$$= (17.10\%) \times (1.37) \times \left(\frac{8,101,372}{4,903,847}\right)$$

$$= 38.70\%.$$

Financial leverage: The degree to which an *investor* or business is utilizing borrowed money.

This expression tells us that management has only three key ratios for controlling a company's ROE: (1) the earnings from sales (the profit margin); (2) the revenue generated from each dollar of assets employed (asset turnover); and (3) the amount of equity used to finance the assets in the operation of the business (financial leverage).

2.3.5 Market Value Analysis

When you purchase a company's stock, what are your primary factors in valuing the stock? In general, investors purchase stock to earn a return on their investment. This return consists of two parts: (1) gains (or losses) from selling the stock at a price that differs from the investors' purchase price and (2) dividends—the periodic distributions of profits to stockholders. The market value ratios, such as the price-to-earnings ratio and the market-to-book ratio, relate the firm's stock price to its earnings and book value per share, respectively. These ratios give management an indication of what investors think of the company's past performance and future prospects. If the firm's asset and debt management is sound and its profit is rising, then its market value ratios will be high, and its stock price will probably be as high as can be expected.

Price-to-Earnings Ratio

The price-to-earnings (*P/E*) ratio shows how much investors are willing to pay per dollar of reported profits. RIM's stock sold for US$72.03 in mid-May 2009 on the NASDAQ Stock Exchange, so with an earnings per share ratio (EPS) of US$3.35 (Table 2.1), its *P/E* ratio was 21.50:

The *P/E* ratio shows how much investors are willing to pay per dollar of earnings.

$$P/E \text{ ratio} = \frac{\text{Price per share}}{\text{Earnings per share}}$$

$$= \frac{\$72.03}{\$3.35} = 21.50$$

That is, the stock was selling for about 21.50 times its current earnings per share. In general, *P/E* ratios are higher for firms with high growth prospects, other things held constant, but they are lower for firms with lower expected earnings. However, all stocks with high *P/E* ratios carry high risk whenever the expected growths fail to materialize. Any slight earnings disappointment tends to punish the market price significantly.

Book Value per Share

Another ratio frequently used in assessing the well-being of the common stockholders is the book value per share, which measures the amount that would be distributed to holders of each share of common stock if all assets were sold at their balance-sheet carrying amounts and if all creditors were paid off. We compute the book value per share for RIM's common stock as of the end of its fiscal 2008 as follows:

$$\text{Book value per share} = \frac{\text{Total stockholders' equity} - \text{preferred stock}}{\text{Shares outstanding}}$$

$$= \frac{\$3,933,566 - \$0}{562,652} = \text{US}\$6.99/\text{share}$$

If we compare this book value with the current market price of US$72.03, then we may say that the stock appears to be overpriced. Once again, though, market prices reflect expectations about future earnings and dividends, whereas book value largely reflects the results of events that occurred in the past. Therefore, the market value of a stock tends to exceed its book value. Table 2.5 summarizes the financial ratios for RIM in comparison to its direct competitor Nokia.

2.3.6 Limitations of Financial Ratios in Business Decisions

Business decisions are made in a world of uncertainty. As useful as ratios are, they have limitations. We can draw an analogy between their use in decision making and a physician's use of a thermometer. A reading of 40°C indicates that something is wrong with the patient,

TABLE 2.5 Comparisons of RIM's Key Financial Ratios with Those of Nokia

Category	Financial Ratios	RIM	Nokia
Debt Management			
	Debt ratio	27.49%	58.29%
	Times-interest-earned ratio	3496	33.06
Liquidity			
	Current ratio	2.29	1.20
	Quick ratio	1.97	1.08
Asset Management			
	Inventory turnover	20.53	18.75
	Day's sales outstanding	74.87	67.98
	Total assets turnover	1.37	1.28
Profitability			
	Profit margin	17.10%	7.86%
	Return on total asset	38.23%	10.39%
	Return on common equity	38.60%	26.84%
Market Trend			
	P/E ratio	21.50	10.37
	Book value per share	6.99	3.84

but the temperature alone does not indicate what the problem is or how to cure it. In other words, ratio analysis is useful, but analysts should be aware of ever-changing market conditions and make adjustments as necessary. It is also difficult to generalize about whether a particular ratio is "good" or "bad." For example, a high current ratio may indicate a strong liquidity position, which is good, but holding too much cash in a bank account (which will increase the current ratio) may not be the best utilization of funds. Ratio analysis based on any one year may not represent the true business condition. It is important to analyze trends in various financial ratios, as well as their absolute levels, for trends give clues as to whether the financial situation is likely to improve or deteriorate. To do a **trend analysis**, one simply plots a ratio over time. As a typical engineering student, your judgment in interpreting a set of financial ratios is understandably weak at this point, but it will improve as you encounter many facets of business decisions in the real world. Again, accounting is a language of business, and as you speak it more often, it can provide useful insights into a firm's operations.

> **Trend analysis** is based on the idea that what has happened in the past gives traders an idea of what will happen in the future.

SUMMARY

The primary purposes of this chapter were (1) to describe the basic financial statements, (2) to present some background information on cash flows and corporate profitability, and (3) to discuss techniques used by investors and managers to analyze financial statements. Following are some concepts we covered:

- Before making any major financial decisions, it is important to understand their impact on your net worth. Your net-worth statement is a snapshot of where you stand financially at a given point in time.

- The three basic financial statements contained in the annual report are the balance sheet, the income statement, and the statement of cash flows. Investors use the information provided in these statements to form expectations about future levels of earnings and dividends and about the firm's risk-taking behaviour.

- A firm's balance sheet shows a snapshot of a firm's financial position at a particular point in time through three categories: (1) assets the firm owns, (2) liabilities the firm owes, and (3) owners' equity, or assets less liabilities.

- A firm's income statement reports the results of operations over a period of time and shows earnings per share as its "bottom line." The main items are (1) revenues and gains, (2) expenses and losses, and (3) net income or net loss (revenue less expenses).

- A firm's statement of cash flows reports the impact of operating, investing, and financing activities on cash flows over an accounting period.

- The purpose of calculating a set of financial ratios is twofold: (1) to examine the relative strengths and weaknesses of a company compared with those of other companies in the same industry and (2) to learn whether the company's position has been improving or deteriorating over time.

- Liquidity ratios show the relationship of a firm's current assets to its current liabilities and thus its ability to meet maturing debts. Two commonly used liquidity ratios are the current ratio and the quick (acid-test) ratio.

- Asset management ratios measure how effectively a firm is managing its assets. Some of the major ratios are inventory turnover, fixed assets turnover, and total assets turnover.

- Debt management ratios reveal (1) the extent to which a firm is financed with debt and (2) the firm's likelihood of defaulting on its debt obligations. In this category are the debt ratio and the times-interest-earned ratio.

- Profitability ratios show the combined effects of liquidity, asset management, and debt management policies on operating results. Profitability ratios include the profit margin on sales, the return on total assets, and the return on common equity.

- Market value ratios relate the firm's stock price to its earnings and book value per share, and they give management an indication of what investors think of the company's past performance and future prospects. Market value ratios include the price-to-earnings ratio and the book value per share.

- Trend analysis, in which one plots a ratio over time, is important, because it reveals whether the firm's ratios are improving or deteriorating over time.

PROBLEMS

2.1   Consider the balance-sheet entries for War Eagle Corporation in Table P2.1.

(a) Compute the firm's
   Current assets: $_____
   Current liabilities: $_____
   Working capital: $_____
   Shareholders' equity: $_____

(b) If the firm had a net income of $500,000 after taxes, what is the earnings per share of common stock?

(c) When the firm issued its common stock, what was the market price of the stock per share?

2.2   A chemical processing firm is planning on adding a duplicate polyethylene plant at another location. The financial information for the first project year is shown in Table P2.2.

(a) Compute the new working-capital requirement during the project period.

(b) What is the taxable income during the project period?

(c) What is the net income during the project period?

(d) Compute the net cash flow from the project during the first year.

2.3   Table P2.3 shows financial statements for Nano Networks Inc. The closing stock price for Nano Networks was $56.67 (split adjusted) on December 31, 2009. On the basis of the financial data presented, compute the various financial ratios and make an informed analysis of Nano's financial health. Note that the balance sheet and the income statement entries in this problem are not complete. Only relevant entries are listed. Do not attempt to add individual entries to confirm either current assets or current liabilities.

(a) Debt ratio

(b) Times-interest-earned ratio

(c) Current ratio

TABLE P2.1

Balance Sheet Statement as of December 31, 2009		
Assets:		
Cash		$ 150,000
Marketable securities		200,000
Accounts receivables		150,000
Inventories		50,000
Prepaid taxes and insurance		30,000
Manufacturing plant at cost	$ 600,000	
Less accumulated depreciation	100,000	
Net fixed assets		500,000
Goodwill		20,000
Liabilities and shareholders' equity:		
Notes payable		50,000
Accounts payable		100,000
Income taxes payable		80,000
Long-term mortgage bonds		400,000
Preferred stock, 6%, $100 par value (1000 shares)		100,000
Common stock, $15 par value (10,000 shares)		150,000
Capital surplus		150,000
Retained earnings		70,000

(d) Quick (acid-test) ratio
(e) Inventory turnover ratio
(f) Day's sales outstanding
(g) Total assets turnover
(h) Profit margin on sales
(i) Return on total assets
(j) Return on common equity
(k) Price-to-earnings ratio
(l) Book value per share on common stock

TABLE P2.2 Financial Information for First Project Year

Sales		$1,500,000
Manufacturing costs		
Direct materials	$ 150,000	
Direct labour	200,000	
Overhead	100,000	
Depreciation	200,000	
Operating expenses		150,000
Equipment purchase		400,000
Borrowing to finance equipment		200,000
Increase in inventories		100,000
Decrease in accounts receivable		20,000
Increase in wages payable		30,000
Decrease in notes payable		40,000
Income taxes		272,000
Interest payment on financing		20,000

2.4 The balance sheet in Table P2.4 summarizes the financial conditions for Flex Inc., an electronic outsourcing contractor, for fiscal-year 2009. Unlike Nano Networks in Problem 2.3, Flex has reported a profit for several years running. Compute the various financial ratios and interpret the firm's financial health during fiscal-year 2009. Note that the balance sheet and the income statement entries in this problem are not complete. Only relevant entries are listed. Do not attempt to add individual entries to confirm either current assets or current liabilities.

(a) Debt ratio

(b) Times-interest-earned ratio

(c) Current ratio

(d) Quick (acid-test) ratio

(e) Inventory turnover ratio

(f) Day's sales outstanding

(g) Total assets turnover

(h) Profit margin on sales

(i) Return on total assets

(j) Return on common equity

(k) Price-to-earnings ratio. Assume a stock price of US$65 per share.

(l) Book value per share. Assume that 247,004,200 shares were outstanding.

TABLE P2.3 Balance Sheet for Nano Networks Inc.

	Dec. 2009 U.S. $ (000) (Year)	Dec. 2008 U.S. $ (000) (Year)
Balance Sheet Summary		
Cash	158,043	20,098
Securities	285,116	0
Receivables	24,582	8,056
Allowances	632	0
Inventory	0	0
Current assets	377,833	28,834
Property and equipment, net	20,588	10,569
Depreciation	8,172	2,867
Total assets	513,378	36,671
Current liabilities	55,663	14,402
Bonds	0	0
Preferred stock	0	0
Common stock	2	1
Total stockholders' equity	457,713	17,064
Total liabilities and equity	513,378	36,671
Income Statement Summary		
Total revenues	102,606	3,807
Cost of sales	45,272	4,416
Other expenses	71,954	31,661
Interest income	8,011	1,301
Income pre-tax	−6,609	−69
Income tax	2,425	2
Income continuing	−9,034	−30,971
Net income	**− 9,034**	**− 30,971**
EPS primary	**− $0.1**	**− $0.80**
EPS diluted	−$0.10	−$0.80
	−$0.05	−$0.40
	(split adjusted)	(split adjusted)

(*Continued*)

TABLE P2.4 Balance Sheet for Flex Inc.

	Aug. 2009 U.S. $ (000) (12 mos.)	Aug. 2008 U.S. $ (000) (Year)
Balance Sheet Summary		
Cash	1,325,637	225,228
Securities	362,769	83,576
Receivables	1,123,901	674,193
Allowances	−5,580	−3,999
Inventory	1,080,083	788,519
Current assets	3,994,084	1,887,558
Property and equipment, net	1,186,885	859,831
Depreciation	533,311	−411,792
Total assets	4,834,696	2,410,568
Current liabilities	1,113,186	840,834
Bonds	922,653	385,519
Preferred stock	0	0
Common stock	271	117
Other stockholders' equity	2,792,820	1,181,209
Total liabilities and equity	4,834,696	2,410,568
Income Statement Summary		
Total revenues	8,391,409	5,288,294
Cost of sales	7,614,589	4,749,988
Other expenses	335,808	237,063
Loss provision	2,143	2,254
Interest expense	36,479	24,759
Income pre-tax	432,342	298,983
Income tax	138,407	100,159
Income continuing	293,935	198,159
Net income	**293,935**	**198,159**
EPS primary	**$1.19**	**$1.72**
EPS diluted	$1.13	$1.65

2.5 J. C. Olson & Co. had earnings per share of $8 in year 2008, and it paid a $4 dividend. Book value per share at year's end was $80. During the same period, the total retained earnings increased by $24 million. Olson has no preferred stock, and no new common stock was issued during the year. If Olson's year-end debt (which equals its total liabilities) was $240 million, what was the company's year-end debt-to-asset ratio?

2.6 If Company A uses more debt than Company B and both companies have identical operations in terms of sales, operating costs, and so on, which of the following statements is *true*?

(a) Company B will definitely have a higher current ratio.

(b) Company B has a higher profit margin on sales than Company A.

(c) Both companies have identical profit margins on sales.

(d) Company B's return on total assets would be higher.

2.7 You are looking to buy stock in a high-growth company. Which of the following ratios best indicates the company's growth potential?

(a) Debt ratio

(b) Price-to-earnings ratio

(c) Profit margin

(d) Total asset turnover

2.8 Which of the following statements is *incorrect*?

(a) The quickest way to determine whether the firm has too much debt is to calculate the debt-to-equity ratio.

(b) The best rule of thumb for determining the firm's liquidity is to calculate the current ratio.

(c) From an investor's point of view, the rate of return on common equity is a good indicator of whether a firm is generating an acceptable return to the investor.

(d) The operating margin is determined by expressing net income as a percentage of total sales.

2.9 Consider the following financial data for Northgate Corporation:
- Cash and marketable securities, $100
- Total fixed assets, $280
- Annual sales, $1200
- Net income, $358
- Inventory, $180
- Current liabilities, $134
- Current ratio, 3.2
- Average correction period, 45 days
- Average common equity, $500

On the basis of these financial data, determine the firm's *return on (common) equity*.

(a) 141.60%

(b) 71.6%

(c) 76.0%

(d) 30%

Short Case Studies

ST2.1 Suncor Energy Inc. (TSE: SU) pioneered commercial development of Canada's Athabasca oil sands—one of the world's largest petroleum resource basins in the 1960s. Since then, Suncor has grown to become a major North American energy producer and marketer with a team of more than 6000 employees. Suncor's products include natural gas, refined oil products, wind power, and ethanol. It has businesses in Alberta, Ontario, and Colorado. Suncor's annual reports can be found at www.suncor.com (Investor Centre, Financial Reports).

Imperial Oil Ltd. (TSE: IMO; AMEX: IMO) is Canada's largest petroleum company. The company is engaged in the exploration, production, and sale of crude oil and natural gas. It is controlled by U.S.–based ExxonMobil, which owns 69.6% of its stock. Imperial owns 25% of Syncrude Canada Ltd., the world's largest oil sands operation. Imperial Oil operates service stations in Canada under the trade name Esso as well as other brand names. Imperial Oil's annual reports can be found at www.imperialoil.ca (Investors, Publications, Annual Report).

Examine the annual reports of Suncor and Imperial Oil and answer the following questions:

(a) On the basis of the most recent financial statements, comment on each company's financial performance in the following areas:
   • Asset management
   • Liquidity
   • Debt management
   • Profitability
   • Market trend

(b) Check the current stock prices for both companies. Based on your analysis in part (a), which company would you bet your money on and why?

ST2.2 Sirius XM Radio Inc. (NASDAQ: SIRI) is the holding company of two satellite radio services operating in the United States and Canada: Sirius Satellite Radio and XM Satellite Radio. CC Media Holdings Inc. (OTCBB: CCMO), the parent company of Clear Channel Communications Inc. and Clear Channel Outdoor Holdings Inc. (NYSE: CCO), is a global media and entertainment company specializing in mobile and on-demand entertainment and information services for local communities and premiere opportunities for advertisers. The company's businesses include radio and outdoor displays. Compare Sirius XM Radio Inc. (www.sirius.com, Investor Relations, Reports and Filings, Annual Reports) and CC Media Holdings Inc. (www.clearchannel.com, Investor Relations, 10-K Filings) using a thorough financial ratios analysis.

(a) For each company, compute all the ratios listed in Figure 2.5 (i.e., debt management, liquidity, asset management, market trend, and profitability) for fiscal-year 2008.

(b) Compare and contrast the two companies, using the ratios you calculated from part (a).

(c) Carefully read and summarize the "risk management" or "hedging" practices described in the financial statements of each company.

(d) If you were a mutual-fund manager and could invest in only one of these companies, which one would you select and why? Be sure to justify your answer by using your results from parts (a), (b), and (c).

On the Companion Website that accompanies this text, you will find Excel templates and exercises, as well as the following analysis tools: Cash Flow Analyzer, Depreciation Analysis, Loan Analysis, and Interest Tables.

THREE

Time Value of Money and Economic Equivalence

Millionaire Life is a promotional lottery across Canada that offers a top prize of $1 million a year for 25 years, four prizes of $1 million each, and 20 prizes of $100,000 each. This lottery game is administered by the Interprovincial Lottery Corporation (ILC), an alliance of five regional (some provincial) lottery corporations that covers all of Canada. The 2009 promotion ran from January 26, 2009, to February 28, 2009, and the top prize was won by Faye Lepage of Edmonton, Alberta. She opted for the lump sum of $17 million instead of the annuity of $1 million a year for 25 years.

According to the official Millionaire Life Game Conditions,[1] the grand prize winner has the option to receive a single payment of $17 million in lieu of the payment or annuity of $1 million per year for 25 years. The single cash payment option is the default option if the winner does not make a specific selection. In the event that the grand prize winner who has selected the annuity option dies prior to the

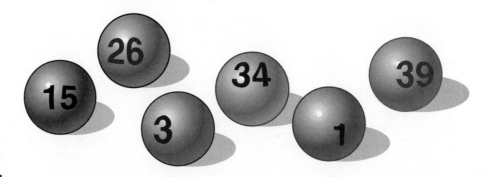

[1] British Columbia Lottery Corporation (www.bclc.com).

end of the 25-year period over which the annuity payments are to be made, ILC shall pay to the executor(s) or administrator(s) of such deceased winner the commuted present value, as determined by ILC, of the annuity payments remaining to be paid.

You may wonder why the value of one option, $17 million paid immediately, is so much lower than the $25 million paid over 25 years at $1 million per year. Isn't receiving the $25 million overall a lot better than receiving $17 million now? The answer to your question involves the principles we will discuss in this chapter, namely, the operation of interest and the time value of money.

Some past winners selected the single payment option stating that they wouldn't live another 25 years. However, the amount to be received can be left as gifts to the winners' surviving children and relatives. The two options can be better evaluated using the equivalence principles to be introduced in this chapter.

In engineering economic analysis, the principles discussed in this chapter are regarded as the underpinning for nearly all project investment analysis. This is because we always need to account for the effect of interest operating on sums of cash over time. Interest formulas allow us to place different cash flows received at different times in the same time frame and to compare them. As will become apparent, almost our entire study of engineering economic analysis is built on the principles introduced in this chapter.

CHAPTER LEARNING OBJECTIVES

After completing this chapter, you should understand the following concepts:

- The time value of money.

- The difference between simple interest and compound interest.

- The meaning of economic equivalence and why we need it in economic analysis.

- How to compare different money series by means of the concept of economic equivalence.

- The interest operation and the types of interest formulas used to facilitate the calculation of economic equivalence.

3.1 Interest: The Cost of Money

Most of us are familiar in a general way with the concept of interest. We know that money left in a savings account earns interest, so that the balance over time is greater than the sum of the deposits. We also know that borrowing to buy a car means repaying an amount over time, that that amount includes interest, and that it is therefore greater than the amount borrowed. What may be unfamiliar to us is the idea that, in the financial world, money itself is a commodity and, like other goods that are bought and sold, money costs money.

Market interest rate: Interest rate quoted by financial institutions.

The cost of money is established and measured by a **market interest rate**, a percentage that is periodically applied and added to an amount (or varying amounts) of money over a specified length of time. When money is borrowed, the interest paid is the charge to the borrower for the use of the lender's property; when money is lent or invested, the interest earned is the lender's gain from providing a good to another (Figure 3.1). **Interest**, then, may be defined as the cost of having money available for use. In this section, we examine how interest operates in a free-market economy and we establish a basis for understanding the more complex interest relationships that follow later on in the chapter.

Charge or Cost to Borrower

Interest Rate
8%

Profit or Earning to Lender

Figure 3.1 The meaning of *interest rate* to the lender (bank) and to the borrower.

	Cost of Refrigerator	Account Value
Case 1: Earning power exceeds inflation	$N = 0$ \$100 $N = 1$ \$108 (earning rate $= 8\%$)	$N = 0$ \$100 $N = 1$ \$106 (inflation rate $= 6\%$)
Case 2: Inflation exceeds earning power	$N = 0$ \$100 $N = 1$ \$104 (earning rate $= 4\%$)	$N = 0$ \$100 $N = 1$ \$106 (inflation rate $= 6\%$)

Figure 3.2 Gains achieved or losses incurred by delaying consumption.

3.1.1 The Time Value of Money

The "**time value of money**" seems like a sophisticated concept, yet it is a concept that you grapple with every day. Should you buy something today or save your money and buy it later? Here is a simple example of how your buying behaviour can have varying results: Pretend you have \$100, and you want to buy a \$100 refrigerator for your dorm room. If you buy it now, you are broke. Suppose that you can invest money at 6% interest, but the price of the refrigerator increases only at an annual rate of 4% due to inflation. In a year you can still buy the refrigerator, and you will have \$2 left over. Well, if the price of the refrigerator increases at an annual rate of 8% instead, you will not have enough money (you will be \$2 short) to buy the refrigerator a year from now. In that case, you probably are better off buying the refrigerator now. The situation is summarized in Figure 3.2.

Clearly, the rate at which you earn interest should be higher than the inflation rate to make any economic sense of the delayed purchase. In other words, in an inflationary economy, your purchasing power will continue to decrease as you further delay the purchase of the refrigerator. In order to make up this future loss in purchasing power, your earning interest rate should be sufficiently larger than the anticipated inflation rate. After all, time, like money, is a finite resource. There are only 24 hours in a day, so time has to be budgeted, too. What this example illustrates is that we must connect the "earning power" and the "purchasing power" to the concept of time.

When we deal with large amounts of money, long periods of time, or high interest rates, the change in the value of a sum of money over time becomes extremely significant. For example, at a current annual interest rate of 10%, \$1 million will earn \$100,000 in interest in a year; thus, to wait a year to receive \$1 million clearly involves a significant sacrifice. When deciding among alternative proposals, we must take into account the operation of interest and the time value of money in order to make valid comparisons of different amounts at various times.

The way interest operates reflects the fact that money has a time value. This is why amounts of interest depend on lengths of time; interest rates, for example, are typically given in terms of a percentage per year. We may define the principle of the time value of money as follows: The economic value of a sum depends on when it is received. Because money has both **earning** as well as **purchasing power** over time, as shown in Figure 3.3 (it can be put to work, earning more money for its owner), a dollar received today has a greater value than a dollar received at some future time.

The time value of money: The idea that a dollar today is worth more than a dollar in the future because the dollar received today can earn interest.

Purchasing power: The value of a currency expressed in terms of the amount of goods or services that one unit of money can buy.

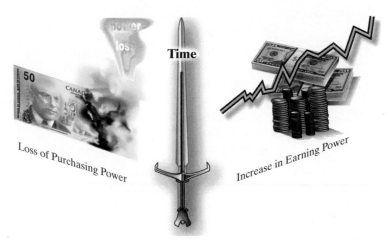

Figure 3.3 The time value of money. This is a two-edged sword whereby earning power grows, but purchasing power decreases, as time goes by.

When lending or borrowing interest rates are quoted by financial institutions on the marketplace, those interest rates reflect the desired amounts to be earned, as well as any protection from loss in the future purchasing power of money because of inflation. (If we want to know the true desired earnings in isolation from inflation, we can determine the real interest rate. We consider this issue in Chapter 14. The earning power of money and its loss of value because of inflation are calculated by different analytical techniques.) In the meantime, we will assume that, unless otherwise mentioned, *the interest rate used in this book reflects the market interest rate,* which takes into account the earning power, as well as the effect of inflation perceived in the marketplace. We will also assume that all cash flow transactions are given in terms of **actual dollars**, with the effect of inflation, if any, reflected in the amount.

Actual dollars: The cash flow measured in terms of the dollars at the time of the transaction.

3.1.2 Elements of Transactions Involving Interest

Many types of transactions (e.g., borrowing or investing money or purchasing machinery on credit) involve interest, but certain elements are common to all of these types of transactions:

- An initial amount of money in transactions involving debt or investments is called the **principal**.
- The **interest rate** measures the cost or price of money and is expressed as a percentage per period of time.
- A period of time, called the **interest period**, determines how frequently interest is calculated. (Note that even though the length of time of an interest period can vary, interest rates are frequently quoted in terms of an annual percentage rate. We will discuss this potentially confusing aspect of interest in Chapter 4.)
- A specified length of time marks the duration of the transaction and thereby establishes a certain **number of interest periods**.
- A **plan for receipts or disbursements** yields a particular cash flow pattern over a specified length of time. (For example, we might have a series of equal monthly payments that repay a loan.)
- A **future amount of money** results from the cumulative effects of the interest rate over a number of interest periods.

For the purposes of calculation, these elements are represented by the following variables:

A_n = A discrete payment or receipt occurring at the end of some interest period.

i = The interest rate per interest period.

N = The total number of interest periods.

P = A sum of money at a time chosen as time zero for purposes of analysis; sometimes referred to as the **present value** or **present worth**.

F = A future sum of money at the end of the analysis period. This sum may be specified as F_N.

A = An end-of-period payment or receipt in a uniform series that continues for N periods. This is a special situation where $A_1 = A_2 = \cdots = A_N$.

V_n = An equivalent sum of money at the end of a specified period n that considers the effect of the time value of money. Note that $V_0 = P$ and $V_N = F$.

Present value: The amount that a future sum of money is worth today, given a specified rate of return.

Because frequent use of these symbols will be made in this text, it is important that you become familiar with them. Note, for example, the distinction between A, A_n, and A_N. The symbol A_n refers to a specific payment or receipt, at the end of period n, in any series of payments. A_N is the final payment in such a series, because N refers to the total number of interest periods. A refers to any series of cash flows in which all payments or receipts are equal.

Example of an Interest Transaction

As an example of how the elements we have just defined are used in a particular situation, let us suppose that an electronics manufacturing company buys a machine for $25,000 and borrows $20,000 from a bank at a 9% annual interest rate. In addition, the company pays a $200 loan origination fee when the loan commences. The bank offers two repayment plans, one with equal payments made at the end of every year for the next five years, the other with a single payment made after the loan period of five years. These two payment plans are summarized in Table 3.1.

- In Plan 1, the principal amount P is $20,000, and the interest rate i is 9%. The interest period is one year, and the duration of the transaction is five years, which means there are five interest periods ($N = 5$). It bears repeating that whereas one year is a common interest period, interest is frequently calculated at other intervals: monthly, quarterly, or

TABLE 3.1 Repayment Plans for Example Given in Text (for $N = 5$ years and $i = 9\%$)

End of Year	Receipts	Payments Plan 1	Payments Plan 2
Year 0	$20,000.00	$ 200.00	$ 200.00
Year 1		5,141.85	0
Year 2		5,141.85	0
Year 3		5,141.85	0
Year 4		5,141.85	0
Year 5		5,141.85	30,772.48

$P = \$20,000,\ A = \$5,141.85,\ F = \$30,772.48$

Note: You actually borrow $19,800 with the origination fee of $200, but you pay back on the basis of $20,000.

semiannually, for instance. For this reason, we used the term **period** rather than **year** when we defined the preceding list of variables. The receipts and disbursements planned over the duration of this transaction yield a cash flow pattern of five equal payments A of $5141.85 each, paid at year's end during years 1 through 5. (You'll have to accept these amounts on faith for now—the next section presents the formula used to arrive at the amount of these equal payments, given the other elements of the problem.)

• Plan 2 has most of the elements of Plan 1, except that instead of five equal repayments, we have a grace period followed by a single future repayment F of $30,772.78.

Cash Flow Diagrams

Problems involving the time value of money can be conveniently represented in graphic form with a cash flow diagram (Figure 3.4). **Cash flow diagrams** represent time by a horizontal line marked off with the number of interest periods specified. The cash flows over time are represented by arrows at relevant periods: Upward arrows denote positive flows (receipts), downward arrows negative flows (disbursements). Note, too, that the arrows actually represent **net cash flows**: Two or more receipts or disbursements made at the same time are summed and shown as a single arrow. For example, $20,000 received during the same period as a $200 payment would be recorded as an upward arrow of $19,800. Also, the lengths of the arrows can suggest the relative values of particular cash flows.

Cash flow diagrams function in a manner similar to free-body diagrams or circuit diagrams, which most engineers frequently use: Cash flow diagrams give a convenient summary of all the important elements of a problem, as well as a reference point to determine whether the statement of the problem has been converted into its appropriate parameters. The text frequently uses this graphic tool, and you are strongly encouraged to develop the habit of using well-labelled cash flow diagrams as a means to identify and summarize pertinent information in a cash flow problem. Similarly, a table such as Table 3.1 can help you organize information in another summary format.

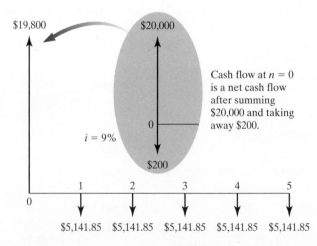

Figure 3.4 A cash flow diagram for Plan 1 of the loan repayment example summarized in Table 3.1.

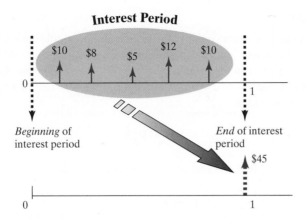

Figure 3.5 Any cash flows occurring during the interest period are summed to a single amount and placed at the end of the interest period.

End-of-Period Convention

In practice, cash flows can occur at the beginning or in the middle of an interest period—or indeed, at practically any point in time. One of the simplifying assumptions we make in engineering economic analysis is the **end-of-period convention**, which is the practice of placing all cash flow transactions at the end of an interest period. (See Figure 3.5.) This assumption relieves us of the responsibility of dealing with the effects of interest within an interest period, which would greatly complicate our calculations. We also need to note that the beginning of a period is the same point as the end of the previous period.

It is important to be aware of the fact that, like many of the simplifying assumptions and estimates we make in modelling engineering economic problems, the end-of-period convention inevitably leads to some discrepancies between our model and real-world results.

Suppose, for example, that $100,000 is deposited during the first month of the year in an account with an interest period of one year and an interest rate of 10% per year. In such a case, the difference of one month would cause an interest income loss of $10,000. This is because, under the end-of-period convention, the $100,000 deposit made during the interest period is viewed as if the deposit were made at the end of the year, as opposed to 11 months earlier. This example gives you a sense of why financial institutions choose interest periods that are less than one year, even though they usually quote their rate as an annual percentage.

Armed with an understanding of the basic elements involved in interest problems, we can now begin to look at the details of calculating interest.

> **End-of-period convention:** Unless otherwise mentioned, all cash flow transactions occur at the end of an interest period.

3.1.3 Methods of Calculating Interest

Money can be lent and repaid in many ways, and, equally, money can earn interest in many different ways. Usually, however, at the end of each interest period, the interest earned on the principal amount is calculated according to a specified interest rate. The two computational schemes for calculating this earned interest are said to yield either **simple interest** or **compound interest**. Engineering economic analysis uses the compound-interest scheme almost exclusively.

> **Simple interest:** The interest rate is applied only to the original principal amount in computing the amount of interest.

Simple Interest

Simple interest is interest earned on only the principal amount during each interest period. In other words, with simple interest, the interest earned during each interest period does not earn additional interest in the remaining periods, *even though you do not withdraw it.*

In general, for a deposit of P dollars at a simple interest rate of i for N periods, the total earned interest would be

$$I = (iP)N. \tag{3.1}$$

The total amount available at the end of N periods thus would be

$$F = P + I = P(1 + iN). \tag{3.2}$$

Simple interest is commonly used with add-on loans or bonds. (See Chapter 4.)

Compound Interest

Under a compound-interest scheme, the interest earned in each period is calculated on the basis of the total amount at the end of the previous period. This total amount includes the original principal plus the accumulated interest that has been left in the account. In this case, you are, in effect, increasing the deposit amount by the amount of interest earned. In general, if you deposited (invested) P dollars at interest rate i, you would have $P + iP = P(1 + i)$ dollars at the end of one period. If the entire amount (principal and interest) is reinvested at the same rate i for another period, at the end of the second period you would have

$$P(1 + i) + i[P(1 + i)] = P(1 + i)(1 + i)$$

$$= P(1 + i)^2.$$

Continuing, we see that the balance after the third period is

$$P(1 + i)^2 + i[P(1 + i)^2] = P(1 + i)^3.$$

This interest-earning process repeats, and after N periods the total accumulated value (balance) F will grow to

$$F = P(1 + i)^N. \tag{3.3}$$

EXAMPLE 3.1 Compound Interest

Suppose you deposit $1000 in a bank savings account that pays interest at a rate of 10% compounded annually. Assume that you don't withdraw the interest earned at the end of each period (one year), but let it accumulate. How much would you have at the end of year 3?

SOLUTION

Given: $P = \$1000$, $N = 3$ years, and $i = 10\%$ per year.
Find: F.

Applying Eq. (3.3) to our three-year, 10% case, we obtain

$$F = \$1000(1 + 0.10)^3 = \$1331.$$

The total interest earned is $331, which is $31 more than was accumulated under the simple-interest method (Figure 3.6). We can keep track of the interest accruing process more precisely as follows:

Period	Amount at Beginning of Interest Period	Interest Earned for Period	Amount at End of Interest Period
1	$1000	$1000(0.10)	$1100
2	1100	1100(0.10)	1210
3	1210	1210(0.10)	1331

COMMENTS: At the end of the first year, you would have $1000, plus $100 in interest, or a total of $1100. In effect, at the beginning of the second year, you would be depositing $1100, rather than $1000. Thus, at the end of the second year, the interest earned would be $0.10(\$1100) = \$110$, and the balance would be $\$1100 + \$110 = \$1210$. This is the amount you would be depositing at the beginning of the third year, and the interest earned for that period would be $0.10(\$1210) = \$121$. With a beginning principal amount of $1210 plus the $121 interest, the total balance would be $1331 at the end of year 3.

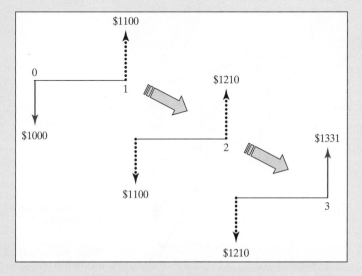

Figure 3.6 The process of computing the balance when $1000 at 10% is deposited for three years (Example 3.1).

3.1.4 Simple Interest versus Compound Interest

From Eq. (3.3), the total interest earned over N periods is

$$I = F - P = P[(1 + i)^N - 1]. \tag{3.4}$$

Compared with the simple-interest scheme, the additional interest earned with compound interest is

$$\Delta I = P[(1 + i)^N - 1] - (iP)N \tag{3.5}$$

$$= P[(1 + i)^N - (1 + iN)]. \tag{3.6}$$

As either i or N becomes large, the difference in interest earnings also becomes large, so the effect of compounding is further pronounced. Note that, when $N = 1$, compound interest is the same as simple interest.

Using Example 3.1, we can illustrate the difference between compound interest and simple interest. Under the simple-interest scheme, you earn interest only on the principal amount at the end of each interest period. Under the compounding scheme, you earn interest on the principal, as well as interest on interest.

Figure 3.7 illustrates the fact that compound interest is a sum of simple interests earned on the original principal, as well as periodic simple interests earned on a series of simple interests.

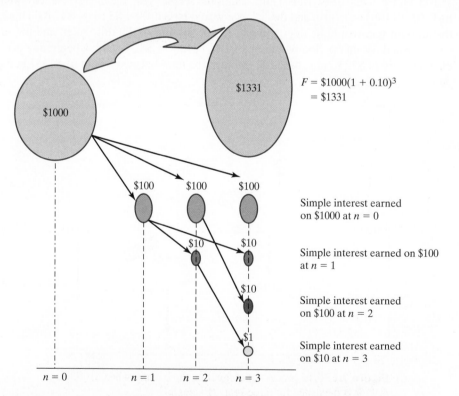

Figure 3.7 The relationship between simple interest and compound interest.

EXAMPLE 3.2 Comparing Simple with Compound Interest

In 1626, Peter Minuit of the Dutch West India Company paid $24 to purchase Manhattan Island in New York from the Native Americans. In retrospect, if Minuit had invested the $24 in a savings account that earned 8% interest, how much would it be worth in 2009?

SOLUTION

Given: $P = \$24$, $i = 8\%$ per year, and $N = 383$ years.

Find: F, based on (a) 8% simple interest and (b) 8% compound interest.

(a) With 8% simple interest,

$$F = \$24[1 + (0.08)(383)] = \$759.36.$$

(b) With 8% compound interest,

$$F = \$24(1 + 0.08)^{383} = \$151,883,149,141,875.$$

COMMENTS: The significance of compound interest is obvious in this example. Many of us can hardly comprehend the magnitude of $152 trillion. In 2009, the total population in the United States was estimated to be around 306 million. If the money were distributed equally among the population, each individual would receive $496,350. Certainly, there is no way of knowing exactly how much Manhattan Island is worth today, but most real-estate experts would agree that the value of the island is nowhere near $152 trillion. (Note that the U.S. national debt as of April 7, 2009, was estimated to be $11.15 trillion.)

3.2 Economic Equivalence

The observation that money has a time value leads us to an important question: If receiving $100 today is not the same thing as receiving $100 at any future point, how do we measure and compare various cash flows? How do we know, for example, whether we should prefer to have $20,000 today and $50,000 10 years from now, or $8000 each year for the next 10 years? In this section, we describe the basic analytical techniques for making these comparisons. Then, in Section 3.3, we will use these techniques to develop a series of formulas that can greatly simplify our calculations.

3.2.1 Definition and Simple Calculations

The central question in deciding among alternative cash flows involves comparing their economic worth. This would be a simple matter if, in the comparison, we did not need to consider the time value of money: We could simply add the individual payments within a cash flow, treating receipts as positive cash flows and payments (disbursements) as negative cash flows. The fact that money has a time value, however, makes our calculations more complicated. We need to know more than just the size of a payment in order to determine its economic effect completely. In fact, as we will see in this section, we need to know several things:

- The magnitude of the payment.
- The direction of the payment: Is it a receipt or a disbursement?
- The timing of the payment: When is it made?
- The interest rate in operation during the period under consideration.

It follows that, to assess the economic impact of a series of payments, we must consider the impact of each payment individually.

Economic equivalence: The process of comparing two different cash amounts at different points in time.

Calculations for determining the economic effects of one or more cash flows are based on the concept of economic equivalence. **Economic equivalence** exists between cash flows that have the same economic effect and could therefore be traded for one another in the financial marketplace, which we assume to exist.

Economic equivalence refers to the fact that a cash flow—whether a single payment or a series of payments—can be converted to an *equivalent* cash flow at any point in time. For example, we could find the equivalent future value F of a present amount P at interest rate i at period n; or we could determine the equivalent present value P of N equal payments A.

The preceding strict concept of equivalence, which limits us to converting a cash flow into another equivalent cash flow, may be extended to include the comparison of alternatives. For example, we could compare the value of two proposals by finding the equivalent value of each at any common point in time. If financial proposals that appear to be quite different turn out to have the same monetary value, then we can be *economically indifferent* to choosing between them: In terms of economic effect, one would be an even exchange for the other, so no reason exists to prefer one over the other in terms of their economic value.

A way to see the concepts of equivalence and economic indifference at work in the real world is to note the variety of payment plans offered by lending institutions for consumer loans. Table 3.2 extends the example we developed earlier to include three different repayment plans for a loan of $20,000, including a $200 loan orgination fee, for five years at 9% interest. You will notice, perhaps to your surprise, that the three plans require

TABLE 3.2 Typical Repayment Plans for a Bank Loan of $20,000 (for $N = 5$ years and $i = 9\%$)

	Repayments		
	Plan 1	**Plan 2**	**Plan 3**
Year 1	$ 5,141.85	0	$ 1,800.00
Year 2	5,141.85	0	1,800.00
Year 3	5,141.85	0	1,800.00
Year 4	5,141.85	0	1,800.00
Year 5	5,141.85	$30,772.48	21,800.00
Total payments	$25,709.25	$30,772.48	$29,000.00
Total interest paid	$ 5,709.25	$10,772.48	$ 9,000.00

Plan 1: Equal annual installments; Plan 2: End-of-loan-period repayment of principal and interest; Plan 3: Annual repayment of interest and end-of-loan repayment of principal

significantly different repayment patterns and different total amounts of repayment. However, because money has a time value, these plans are equivalent, and economically the bank is indifferent to a consumer's choice of plan. We will now discuss how such equivalence relationships are established.

Equivalence Calculations: A Simple Example

Equivalence calculations can be viewed as an application of the compound-interest relationships we developed in Section 3.1. Suppose, for example, that we invest $1000 at 12% annual interest for five years. The formula developed for calculating compound interest, $F = P(1 + i)^N$ (Eq. 3.3), expresses the equivalence between some present amount P and a future amount F, for a given interest rate i and a number of interest periods N. Therefore, at the end of the investment period, our sums grow to

$$\$1000(1 + 0.12)^5 = \$1762.34.$$

Thus, we can say that at 12% interest, $1000 received now is equivalent to $1762.34 received in five years and that we could trade $1000 now for the promise of receiving $1762.34 in five years. Example 3.3 further demonstrates the application of this basic technique.

EXAMPLE 3.3 Equivalence

Suppose you are offered the alternative of receiving either $3000 at the end of five years or P dollars today. There is no question that the $3000 will be paid in full (no risk). Because you have no current need for the money, you would deposit the P dollars in an account that pays 8% interest. What value of P would make you indifferent to your choice between P dollars today and the promise of $3000 at the end of five years?

STRATEGY: Our job is to determine the present amount that is economically equivalent to $3000 in five years, given the investment potential of 8% per year. Note that the statement of the problem assumes that you would exercise the option of using the earning power of your money by depositing it. The "indifference" ascribed to you refers to economic indifference; that is, in a marketplace where 8% is the applicable interest rate, you could trade one cash flow for the other.

SOLUTION

Given: $F = \$3000$, $N = 5$ years, and $i = 8\%$ per year.
Find: P.

Equation: Eq. (3.3), $F = P(1 + i)^N$.
Rearranging terms to solve for P gives

$$P = \frac{F}{(1 + i)^N}.$$

Substituting yields

$$P = \frac{\$3000}{(1 + 0.08)^5} = \$2042.$$

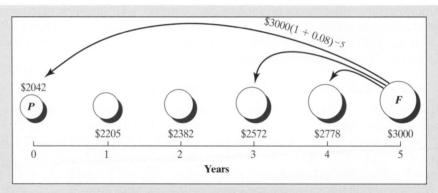

Figure 3.8 Various dollar amounts that will be economically equivalent to $3000 in five years, given an interest rate of 8% (Example 3.3).

We summarize the problem graphically in Figure 3.8.

COMMENTS: In this example, it is clear that if P is anything less than $2042, you would prefer the promise of $3000 in five years to P dollars today; if P is greater than $2042, you would prefer P. As you may have already guessed, at a lower interest rate, P must be higher to be equivalent to the future amount. For example, at $i = 4\%$, $P = \$2466$.

3.2.2 Equivalence Calculations: General Principles

In spite of their numerical simplicity, the examples we have developed reflect several important general principles, which we will now explore.

Principle 1: Equivalence Calculations Made to Compare Alternatives Require a Common Time Basis

Common base period: To establish an economic equivalence between two cash flow amounts, a common base period must be selected.

Just as we must convert fractions to common denominators to add them together, we must also convert cash flows to a common basis to compare their value. One aspect of this basis is the choice of a single point in time at which to make our calculations. In Example 3.3, if we had been given the magnitude of each cash flow and had been asked to determine whether they were equivalent, we could have chosen a reference point and used the compound interest formula to find the value of each cash flow at that point. As you can readily see, the choice of $n = 0$ or $n = 5$ would make our problem simpler because we need to make only one set of calculations: At 8% interest, either convert $2042 at time 0 to its equivalent value at time 5, or convert $3000 at time 5 to its equivalent value at time 0. (To see how to choose a different reference point, take a look at Example 3.4.)

When selecting a point in time at which to compare the value of alternative cash flows, we commonly use either the present time, which yields what is called the **present worth** of the cash flows, or some point in the future, which yields their **future worth**. The choice of the point in time often depends on the circumstances surrounding a particular decision, or it may be chosen for convenience. For instance, if the present worth is known for the first two of three alternatives, all three may be compared simply by calculating the present worth of the third.

EXAMPLE 3.4 Equivalent Cash Flows Are Equivalent at Any Common Point in Time

In Example 3.3, we determined that, given an interest rate of 8% per year, receiving $2042 today is equivalent to receiving $3000 in five years. Are these cash flows also equivalent at the end of year 3?

STRATEGY: This problem is summarized in Figure 3.9. The solution consists of solving two equivalence problems: (1) What is the future value of $2042 after three years at 8% interest (part (a) of the solution)? (2) Given the sum of $3000 after five years and an interest rate of 8%, what is the equivalent sum after 3 years (part (b) of the solution)?

SOLUTION

Given:

(a) $P = \$2042$; $i = 8\%$ per year; $N = 3$ years.
(b) $F = \$3000$; $i = 8\%$ per year; $N = 5 - 3 = 2$ years.

Find: (1) V_3 for part (a); (2) V_3 for part (b). (3) Are these two values equivalent?

Equation:

(a) $F = P(1 + i)^N$.
(b) $P = F(1 + i)^{-N}$.

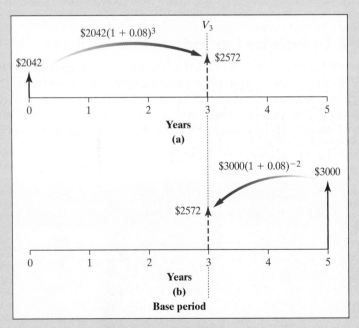

Figure 3.9   Selection of a base period for an equivalence calculation (Example 3.4).

Notation: The usual terminology of F and P is confusing in this example, since the cash flow at $n = 3$ is considered a future sum in part (a) of the solution and a past cash flow in part (b) of the solution. To simplify matters, we are free to arbitrarily designate a reference point $n = 3$ and understand that it needs not to be now or the present. Therefore, we assign the equivalent cash flow at $n = 3$ to a single variable, V_3.

1. The equivalent worth of $2042 after three years is

$$V_3 = 2042(1 + 0.08)^3$$
$$= \$2572.$$

2. The equivalent worth of the sum $3000 two years earlier is

$$V_3 = F(1 + i)^{-N}$$
$$= \$3000(1 + 0.08)^{-2}$$
$$= \$2572.$$

(Note that $N = 2$ because that is the number of periods during which discounting is calculated in order to arrive back at year 3.)

3. While our solution doesn't strictly prove that the two cash flows are equivalent at any time, they will be equivalent at any time as long as we use an interest rate of 8%.

Principle 2: Equivalence Depends on Interest Rate

The equivalence between two cash flows is a function of the magnitude and timing of individual cash flows and the interest rate or rates that operate on those flows. This principle is easy to grasp in relation to our simple example: $1000 received now is equivalent to $1762.34 received five years from now only at a 12% interest rate. Any change in the interest rate will destroy the equivalence between these two sums, as we will demonstrate in Example 3.5.

EXAMPLE 3.5 Changing the Interest Rate Destroys Equivalence

In Example 3.3, we determined that, given an interest rate of 8% per year, receiving $2042 today is equivalent to receiving $3000 in five years. Are these cash flows equivalent at an interest rate of 10%?

SOLUTION

Given: $P = \$2042$, $i = 10\%$ per year, and $N = 5$ years.
Find: F: Is it equal to $3000?

We first determine the base period under which an equivalence value is computed. Since we can select any period as the base period, let's select $N = 5$. Then we need to calculate the equivalent value of $2042 today five years from now.

$$F = \$2042(1 + 0.10)^5 = \$3289.$$

Since this amount is greater than $3000, the change in interest rate destroys the equivalence between the two cash flows.

Principle 3: Equivalence Calculations May Require the Conversion of Multiple Payment Cash Flows to a Single Cash Flow

In all the examples presented thus far, we have limited ourselves to the simplest case of converting a single payment at one time to an equivalent single payment at another time. Part of the task of comparing alternative cash flow series involves moving each individual cash flow in the series to the same single point in time and summing these values to yield a single equivalent cash flow. We perform such a calculation in Example 3.6.

EXAMPLE 3.6 Equivalence Calculations With Multiple Payments

Suppose that you borrow $1000 from a bank for three years at 10% annual interest. The bank offers two options: (1) repaying the interest charges for each year at the end of that year and repaying the principal at the end of year 3 or (2) repaying the loan all at once (including both interest and principal) at the end of year 3. The repayment schedules for the two options are as follows:

Options	Year 1	Year 2	Year 3
• Option 1: End-of-year repayment of interest, and principal repayment at end of loan	$100	$100	$1100
• Option 2: One end-of-loan repayment of both principal and interest	0	0	1331

Determine whether these options are equivalent, assuming that the appropriate interest rate for the comparison is 10%.

STRATEGY: Since we pay the principal after three years in either plan, the repayment of principal can be removed from our analysis. This is an important point: *We can ignore the common elements of alternatives being compared so that we can focus entirely on comparing the interest payments.* Notice that under Option 1, we will pay a total of $300 interest, whereas under Option 2, we will pay a total of $331. Before concluding that we prefer Option 2, remember that a comparison of the two cash flows is based on a *combination of payment amounts and the timing of those payments*. To make our comparison, we must compare the equivalent value of each option at a single point in time. Since Option 2 is already a single payment at $n = 3$ years, it is simplest to convert the cash flow pattern of Option 1 to a single value at $n = 3$. To do this, we must convert the three disbursements of Option 1 to their respective equivalent values at $n = 3$. At that point, since they share a time in common, we can simply sum them in order to compare them with the $331 sum in Option 2.

SOLUTION

Given: Interest payment series; $i = 10\%$ per year.

Find: A single future value F of the flows in Option 1.

Equation: $F = P(1 + i)^N$, applied to each disbursement in the cash flow diagram. N in Eq. (3.3) is the number of interest periods during which interest is in effect, and n is the period number (i.e., for year 1, $n = 1$). We determine the value of F by finding the interest period for each payment. Thus, for each payment in the series, N can be calculated by subtracting n from the total number of years of the loan (3). That is, $N = 3 - n$. Once the value of each payment has been found, we sum the payments:

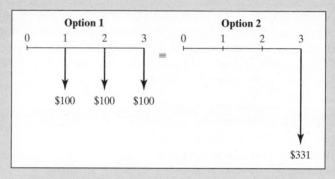

F_3 for $100 at $n = 1 : \$100(1 + .10)^{3-1} = \$121$;

F_3 for $100 at $n = 2 : \$100(1 + .10)^{3-2} = \$110$;

F_3 for $100 at $n = 3 : \$100(1 + .10)^{3-3} = \underline{\$100}$;

Total $= \$331$.

By converting the cash flow in Option 1 to a single future payment at year 3, we can compare Options 1 and 2. We see that the two interest payments are equivalent. Thus, the bank would be economically indifferent to a choice between the two plans. Note that the final interest payment in Option 1 does not accrue any compound interest.

Principle 4: Equivalence Is Maintained Regardless of Point of View

As long as we use the same interest rate in equivalence calculations, equivalence can be maintained regardless of point of view. In Example 3.6, the two options were equivalent at an interest rate of 10% from the banker's point of view. What about from a borrower's point of view? Suppose you borrow $1000 from a bank and deposit it in another bank that pays 10% interest annually. Then you make future loan repayments out of this savings account. Under Option 1, your savings account at the end of year 1 will show a balance of $1100 after the interest earned during the first period has been credited. Now you withdraw $100 from this savings account (the exact amount required to pay the loan interest during the first year), and you make the first-year interest payment to the bank. This leaves only $1000 in your savings account. At the end of year 2, your savings account will earn another interest payment in the amount of $1000(0.10) = $100, making an end-of-year balance of $1100. Now you withdraw another $100 to make the required loan interest payment. After this payment, your remaining balance will be $1000. This balance will grow again at 10%, so you will have $1100 at the end of year 3. After making the last loan payment ($1100), you will have no money left in either account. For Option 2, you can keep track of the yearly account balances in a similar fashion. You will find that you reach a zero balance after making the lump-sum payment of $1331. If the borrower had used the same interest rate as the bank, the two options would be equivalent.

3.2.3 Looking Ahead

The preceding examples should have given you some insight into the basic concepts and calculations involved in the concept of economic equivalence. Obviously, the variety of financial arrangements possible for borrowing and investing money is extensive, as is the variety of time-related factors (e.g., maintenance costs over time, increased productivity over time, etc.) in alternative proposals for various engineering projects. It is important to recognize that even the most complex relationships incorporate the basic principles we have introduced in this section.

In the remainder of the chapter, we will represent all cash flow diagrams either in the context of an initial deposit with a subsequent pattern of withdrawals or in an initial borrowed amount with a subsequent pattern of repayments. If we were limited to the methods developed in this section, a comparison between the two payment options would involve a large number of calculations. Fortunately, in the analysis of many transactions, certain cash flow patterns emerge that may be categorized. For many of these patterns, we can derive formulas that can be used to simplify our work. In Section 3.3, we develop these formulas.

3.3 Development of Interest Formulas

Now that we have established some working assumptions and notations and have a preliminary understanding of the concept of equivalence, we will develop a series of interest formulas for use in comparisons of more complex cash flows.

As we begin to compare series of cash flows instead of single payments, the required analysis becomes more complicated. However, when patterns in cash flow transactions can be identified, we can take advantage of these patterns by developing concise expressions for computing either the present or future worth of the series. We will classify five major categories of cash flow transactions, develop interest formulas for them, and present several working examples of each type. Before we give the details, however, we briefly describe the five types of cash flows in the next subsection.

3.3.1   The Five Types of Cash Flows

Whenever we identify patterns in cash flow transactions, we may use those patterns to develop concise expressions for computing either the present or future worth of the series. For this purpose, we will classify cash flow transactions into five categories: (1) a single cash flow, (2) a uniform series, (3) a linear gradient series, (4) a geometric gradient series, and (5) an irregular series. To simplify the description of various interest formulas, we will use the following notation:

1. **Single Cash Flow:** The simplest case involves the equivalence of a single present amount and its future worth. Thus, the single-cash-flow formulas deal with only two amounts: a single present amount P and its future worth F (Figure 3.10a). You have already seen the derivation of one formula for this situation in Section 3.1.3, which gave us Eq. (3.3):

$$F = P(1 + i)^N.$$

2. **Equal (Uniform) Series:** Probably the most familiar category includes transactions arranged as a series of equal cash flows at regular intervals, known as an **equal payment series** (or **uniform series**) (Figure 3.10b). For example, this category describes the cash flows of the common installment loan contract, which arranges the repayment of a loan in equal periodic installments. The equal-cash-flow formulas deal with the equivalence relations P, F, and A (the constant amount of the cash flows in the series).

3. **Linear Gradient Series:** While many transactions involve series of cash flows, the amounts are not always uniform; they may, however, vary in some regular way. One common pattern of variation occurs when each cash flow in a series increases (or decreases) by a fixed amount (Figure 3.10c). A five-year loan repayment plan might specify, for example, a series of annual payments that increase by $500 each year. We call this type of cash flow pattern a **linear gradient series** because its cash flow diagram produces an ascending (or descending) straight line, as you will see in Section 3.3.5. In addition to using P, F, and A, the formulas employed in such problems involve a *constant amount* G of the change in each cash flow.

4. **Geometric Gradient Series:** Another kind of gradient series is formed when the series in a cash flow is determined not by some fixed amount like $500, but by some fixed *rate*, expressed as a percentage. For example, in a five-year financial plan for a project, the cost of a particular raw material might be budgeted to increase at a rate of 4% per year. The curving gradient in the diagram of such a series suggests its name: a **geometric gradient series** (Figure 3.10d). In the formulas dealing with such series, the rate of change is represented by a lowercase g.

5. **Irregular (Mixed) Series:** Finally, a series of cash flows may be irregular, in that it does not exhibit a regular overall pattern (Figure 3.10e). Even in such a series, however, one or more of the patterns already identified may appear over segments of time in the total length of the series. The cash flows may be equal, for example, for five consecutive periods in a 10-period series. When such patterns appear, the formulas for dealing with them may be applied and their results included in calculating an equivalent value for the entire series.

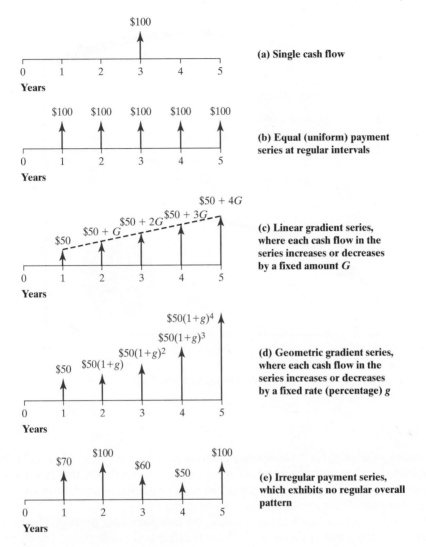

$100

(a) Single cash flow

$100 $100 $100 $100 $100

(b) Equal (uniform) payment series at regular intervals

$50 + 4G

$50 + 2G $50 + 3G

$50 + G

$50

(c) Linear gradient series, where each cash flow in the series increases or decreases by a fixed amount G

$50(1+g)^4

$50(1+g)^3

$50(1+g)^2

$50 $50(1+g)

(d) Geometric gradient series, where each cash flow in the series increases or decreases by a fixed rate (percentage) g

$70 $100 $60 $50 $100

(e) Irregular payment series, which exhibits no regular overall pattern

Figure 3.10   Five types of cash flows: (a) single cash flow, (b) equal (uniform) payment series, (c) linear gradient series, (d) geometric gradient series, and (e) irregular payment series.

3.3.2 Single-Cash-Flow Formulas

We begin our coverage of interest formulas by considering the simplest of cash flows: single cash flows.

Compound Amount Factor

Given a present sum P invested for N interest periods at interest rate i, what sum will have accumulated at the end of the N periods? You probably noticed right away that this description matches the case we first encountered in describing compound interest. To solve for F (the future sum), we use Eq. (3.3):

$$F = P(1 + i)^N.$$

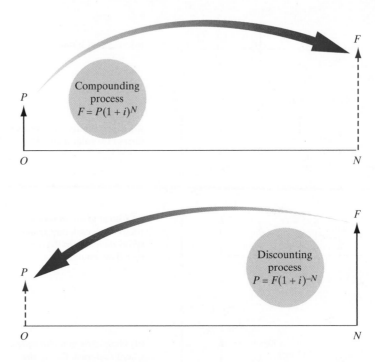

Figure 3.11 Equivalence relation between P and F.

Because of its origin in the compound-interest calculation, the factor $(1 + i)^N$ is known as the **compound-amount factor**. Like the concept of equivalence, this factor is one of the foundations of engineering economic analysis. Given the compound-amount factor, all the other important interest formulas can be derived.

Compounding process: the process of computing the future value of a current sum.

This process of finding F is often called the **compounding process**. The cash flow transaction is illustrated in Figure 3.11. (Note the time-scale convention: The first period begins at $n = 0$ and ends at $n = 1$.) If a calculator is handy, it is easy enough to calculate $(1 + i)^N$ directly.

Interest Tables

Interest formulas such as the one developed in Eq. (3.3), $F = P(1 + i)^N$ allow us to substitute known values from a particular situation into the equation and to solve for the unknown. Before the hand calculator was developed, solving these equations was very tedious. With a large value of N, for example, one might need to solve an equation such as $F = \$20,000(1 + 0.12)^{15}$. More complex formulas required even more involved calculations. To simplify the process, tables of compound-interest factors were developed, and these tables allow us to find the appropriate factor for a given interest rate and the number of interest periods. Even with hand calculators, it is still often convenient to use such tables, and they are included in this text in Appendix A. Take some time now to become familiar with their arrangement and, if you can, locate the compound-interest factor for the example just presented, in which we know P. Remember that, to find F, we need to know the factor by which to multiply $20,000 when the interest rate i is 12% and the number of periods is 15:

$$F = \$20,000\underbrace{(1 + 0.12)^{15}}_{5.4736} = \$109,472.$$

Factor Notation

As we continue to develop interest formulas in the rest of this chapter, we will express the resulting compound-interest factors in a conventional notation that can be substituted in a formula to indicate precisely which table factor to use in solving an equation. In the preceding example, for instance, the formula derived as Eq. (3.3) is $F = P(1 + i)^N$. In ordinary language, this tells us that, to determine what future amount F is equivalent to a present amount P, we need to multiply P by a factor expressed as 1 plus the interest rate, raised to the power given by the number of interest periods. To specify how the interest tables are to be used, we may also express that factor in functional notation as $(F/P, i, N)$, which is read as "Find F, Given P, i, and N." This is known as the **single-payment compound-amount factor**. When we incorporate the table factor into the formula, it is expressed as

$$F = P(1 + i)^N = P(F/P, i, N).$$

Thus, in the preceding example, where we had $F = \$20,000(1.12)^{15}$, we can write $F = \$20,000(F/P, 12\%, 15)$. The table factor tells us to use the 12% interest table and find the factor in the F/P column for $N = 15$. Because using the interest tables is often the easiest way to solve an equation, this factor notation is included for each of the formulas derived in the sections that follow.

EXAMPLE 3.7 Single Amounts: Find *F*, Given *i*, *N*, and *P*

If you had $2000 now and invested it at 10%, how much would it be worth in eight years (Figure 3.12)?

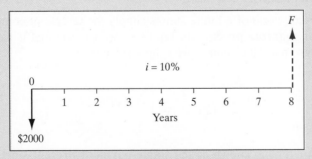

Figure 3.12 A cash flow diagram from the investor's point of view (Example 3.7).

SOLUTION

Given: $P = \$2000$, $i = 10\%$ per year, and $N = 8$ years.
Find: F.

We can solve this problem in any of three ways:

1. **Using a calculator.** You can simply use a calculator to evaluate the $(1 + i)^N$ term (financial calculators are preprogrammed to solve most future-value problems):

$$F = \$2000(1 + 0.10)^8$$
$$= \$4287.18.$$

2. **Using compound-interest tables.** The interest tables can be used to locate the compound-amount factor for $i = 10\%$ and $N = 8$. The number you get can be substituted into the equation. Compound-interest tables are included as Appendix A of this book. From the tables, we obtain

$$F = \$2000(F/P, 10\%, 8) = \$2000(2.1436) = \$4287.20.$$

This is essentially identical to the value obtained by the direct evaluation of the single-cash-flow compound-amount factor. This slight difference is due to rounding errors.

3. **Using Excel.** Many financial software programs for solving compound-interest problems are available for use with personal computers. Excel provides financial functions to evaluate various interest formulas, where the future-worth calculation looks like the following:

$$=FV(10\%,8,0,-2000)$$

Present-Worth Factor

Discounting process: A process of calculating the present value of a future amount.

Finding the present worth of a future sum is simply the reverse of compounding and is known as the **discounting process**. In Eq. (3.3), we can see that if we were to find a present sum P, given a future sum F, we simply solve for P:

$$P = F\left[\frac{1}{(1 + i)^N}\right] = F(P/F, i, N). \tag{3.7}$$

The factor $1/(1 + i)^N$ is known as the **single-payment present-worth factor** and is designated $(P/F, i, N)$. Tables have been constructed for P/F factors and for various values of i and N. The interest rate i and the P/F factor are also referred to as the **discount rate** and **discounting factor**, respectively.

EXAMPLE 3.8 Single Amounts: Find *P*, Given *F*, *i*, and *N*

Suppose that $1000 is to be received in five years. At an annual interest rate of 12%, what is the present worth of this amount?

SOLUTION

Given: $F = \$1000$, $i = 12\%$ per year, and $N = 5$ years.

Find: *P*.

$$P = \$1000(1 + 0.12)^{-5} = \$1000(0.5674) = \$567.40.$$

Using a calculator may be the best way to make this simple calculation. To have $1000 in your savings account at the end of five years, you must deposit $567.40 now.

We can also use the interest tables to find that

$$\overset{(0.5674)}{P = \$1000\overbrace{(P/F, 12\%, 5)}} = \$567.40.$$

Again, you could use a financial calculator or a computer to find the present worth. With Excel, the present-value calculation looks like the following:

$$=PV(12\%,5,0,-1000)$$

Note that, in Excel format, we enter the present value (*P*) as a negative number, indicating a cash outflow.

Solving for Time and Interest Rates

At this point, you should realize that the compounding and discounting processes are reciprocals of one another and that we have been dealing with one equation in two forms:

$$\text{Future-value form: } F = P(1 + i)^N;$$
$$\text{Present-value form: } P = F(1 + i)^{-N}.$$

There are four variables in these equations: *P*, *F*, *N*, and *i*. If you know the values of any three, you can find the value of the fourth. Thus far, we have always given you the interest rate *i* and the number of years *N*, plus either *P* or *F*. In many situations, though, you will need to solve for *i* or *N*, as we discuss next.

EXAMPLE 3.9 Solving for *i*

Suppose you buy a share for $10 and sell it for $20. Then your profit is $10. If that happens within a year, your rate of return is an impressive 100% ($10/$10 = 1). If it takes five years, what would be the average annual rate of return on your investment? (See Figure 3.13.)

SOLUTION

Given: $P = \$10$, $F = \$20$, and $N = 5$.

Find: *i*.

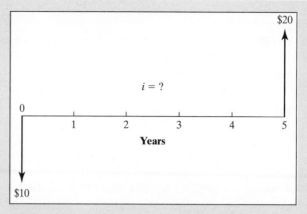

Figure 3.13 Cash flow diagram (Example 3.9).

Here, we know P, F, and N, but we do not know i, the interest rate you will earn on your investment. This type of rate of return is a lot easier to calculate because you make only a one-time lump-sum investment. Problems such as this are solved as follows:

$$F = P(1 + i)^N;$$
$$\$20 = \$10(1 + i)^5; \text{ solve for } i.$$

- **Method 1.** Go through a trial-and-error process in which you insert different values of i into the equation until you find a value that "works" in the sense that the right-hand side of the equation equals $20. The solution is $i = 14.87\%$. The trial-and-error procedure is extremely tedious and inefficient for most problems, so it is not widely practised in the real world.

- **Method 2.** You can solve the problem by using the interest tables in Appendix A. Now look across the $N = 5$ row, under the $(F/P, i, 5)$ column, until you can locate the value of 2:

$$\$20 = \$10(1 + i)^5;$$
$$2 = (1 + i)^5 = (F/P, i, 5).$$

This value is close to the 15% interest table with $(F/P, 15\%, 5) = 2.0114$, so the interest rate at which $10 grows to $20 over five years is very close to 15%. This procedure will be very tedious for fractional interest rates or when N is not a whole number, because you may have to approximate the solution by linear interpolation.

- **Method 3.** The most practical approach is to use either a financial calculator or an electronic spreadsheet such as Excel. A financial function such as RATE($N,0,P,F$) allows us to calculate an unknown interest rate. The precise command statement would be

$$= \textbf{RATE}(5,0,-10,20) = 14.87\%.$$

Note that, in Excel format, we enter the present value (P) as a negative number, indicating a cash outflow.

	A	B	C
1	P	−10	
2	F	20	
3	N	5	
4	i	14.87%	
5			

=RATE(5,0,−10,20)

EXAMPLE 3.10 Single Amounts: Find *N*, Given *P*, *F*, and *i*

You have just purchased 100 shares of General Electric stock at $60 per share. You will sell the stock when its market price has doubled. If you expect the stock price to increase 20% per year, how long do you anticipate waiting before selling the stock (Figure 3.14)?

SOLUTION

Given: $P = \$6000$, $F = \$12,000$, and $i = 20\%$ per year.
Find: N (years).

Using the single-payment compound-amount factor, we write

$$F = P(1 + i)^N = P(F/P, i, N);$$

$$\$12,000 = \$6000(1 + 0.20)^N = \$6000(F/P, 20\%, N);$$

$$2 = (1.20)^N = (F/P, 20\%, N).$$

Again, we could use a calculator or a computer spreadsheet program to find N.

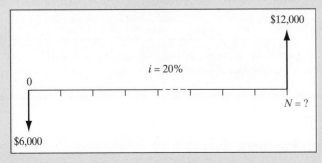

Figure 3.14   Cash flow diagram (Example 3.10).

1. **Using a calculator.** Solving for N gives

$$\log 2 = N \log 1.20,$$

or

$$N = \frac{\log 2}{\log 1.20}$$

$$= 3.80 \approx 4 \text{ years.}$$

2. **Using Excel.** Within Excel, the financial function NPER($i,0,P,F$) computes the number of compounding periods it will take an investment (P) to grow to a future value (F), earning a fixed interest rate (i) per compounding period. In our example, the Excel command would look like this:

$$=\text{NPER}(20\%,0,-6000,12000)$$

$$= 3.801784.$$

Rule of 72:
Rule giving the approximate number of years that it will take for your investment to double.

COMMENTS: A very handy rule of thumb, called the **Rule of 72**, estimates approximately how long it will take for a sum of money to double. The rule states that, to find the time it takes for a present sum of money to grow by a factor of two, we divide 72 by the interest rate expressed in percentage points. In our example, the interest rate is 20%. Therefore, the Rule of 72 indicates 72/20 = 3.60, or roughly 4 years, for a sum to double. This is, in fact, relatively close to our exact solution. Figure 3.15 illustrates the number of years required to double an investment at various interest rates.

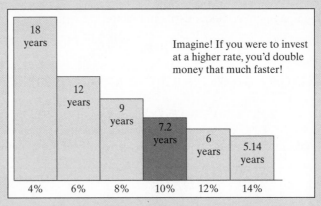

Figure 3.15 Number of years required to double an initial investment at various interest rates.

3.3.3 Uneven Payment Series

A common cash flow transaction involves a series of disbursements or receipts. Familiar examples of series payments are payment of installments on car loans and home mortgage payments. Payments on car loans and home mortgages typically involve identical sums to be paid at regular intervals. However, there is no clear pattern over the series; we call the transaction an uneven cash flow series.

We can find the present worth of any uneven stream of payments by calculating the present value of each individual payment and summing the results. Once the present worth is found, we can make other equivalence calculations (e.g., future worth can be calculated by using the interest factors developed in the previous section).

EXAMPLE 3.11   Present Values of an Uneven Series by Decomposition Into Single Payments

Wilson Technology, a growing machine shop, wishes to set aside money now to invest over the next four years in automating its customer service department. The company can earn 10% on a lump sum deposited now, and it wishes to withdraw the money in the following increments:

- **Year 1:** $25,000, to purchase a computer and database software designed for customer service use;
- **Year 2:** $3000, to purchase additional hardware to accommodate anticipated growth in use of the system;
- **Year 3:** No expenses; and
- **Year 4:** $5000, to purchase software upgrades.

How much money must be deposited now to cover the anticipated payments over the next four years?

STRATEGY: This problem is equivalent to asking what value of P would make you indifferent in your choice between P dollars today and the future expense stream of ($25,000, $3000, $0, $5000). One way to deal with an uneven series of cash flows is to calculate the equivalent present value of each single cash flow and to sum the present values to find P. In other words, the cash flow is broken into three parts as shown in Figure 3.16.

SOLUTION

Given: Uneven cash flow in Figure 3.16, with $i = 10\%$ per year.
Find: P.

$$P = \$25{,}000(P/F, 10\%, 1) + \$3000(P/F, 10\%, 2)$$
$$+ \ \$5000(P/F, 10\%, 4)$$
$$= \$28{,}622.$$

COMMENTS: To see if $28,622 is indeed sufficient, let's calculate the balance at the end of each year. If you deposit $28,622 now, it will grow to (1.10)($28,622), or $31,484, at the end of year 1. From this balance, you pay out $25,000. The remaining

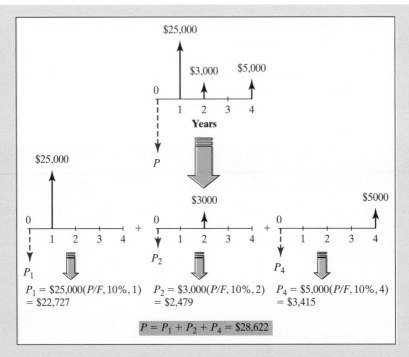

Figure 3.16 Decomposition of uneven cash flow series (Example 3.11).

balance, $6484, will again grow to (1.10)($6484), or $7132, at the end of year 2. Now you make the second payment ($3000) out of this balance, which will leave you with only $4132 at the end of year 2. Since no payment occurs in year 3, the balance will grow to $(1.10)^2($4132)$, or $5000, at the end of year 4. The final withdrawal in the amount of $5000 will deplete the balance completely.

EXAMPLE 3.12 Calculating the Actual Worth of a Long-Term Contract of Henrik Zetterberg with the Detroit Red Wings

On January 28, 2009, the Detroit Red Wings of the National Hockey League signed a 12-year contract extension worth US$73 million with centre Henrik Zetterberg. This extension marked the longest deal given in the history of the Red Wings.

According to LetsGoWings, an independent, fan-operated website, the annual payment schedule of this contract is as follows:

Season	Salary payment
2009/10	7,400,000
2010/11	7,750,000
2011/12	7,750,000
2012/13	7,750,000
2013/14	7,500,000
2014/15	7,500,000
2015/16	7,500,000
2016/17	7,500,000
2017/18	7,500,000
2018/19	3,350,000
2019/20	1,000,000
2020/21	1,000,000[2]

How much was Zetterberg's contract worth at the time of signing? Assume that Zetterberg's interest rate is 6% per year.

SOLUTION

Given: Payment series given in Figure 3.17, with $i = 6\%$ per year.
Find: P.

The actual worth of the contract at the time of signing is the present equivalent of all future payments:

$$
\begin{aligned}
P_{contract} &= \$7,400,000(P/F, 6\%, 1) \ + \ \$7,750,000 \ (P/F, 6\%, 2) \\
&\quad + \ \$7,750,000 \ (P/F, 6\%, 3) \ + \ \cdots\cdots \\
&\quad + \ \$1,000,000 \ (P/F, 6\%, 12) \\
&= \$54,443.159
\end{aligned}
$$

COMMENTS: Note that the actual contract is worth less than the published figure of $73 million. The "brute force" approach of considering one cash flow at a time using the P/F factor will always work for present equivalent calculations, but it is slow and subject to error in manual calculations because of the many factors that must be included in the calculations. In the following sections, we will develop more efficient methods for cash flows with certain patterns.

[2] Adapted from Red Wings Salary Cap Chart at www.letsgowings.com.

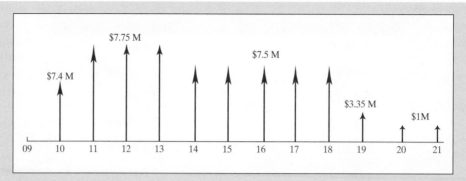

Figure 3.17 Cash flow diagram for Henrik Zetterberg's contract with the Detroit Red Wings.

3.3.4 Equal Payment Series

As we learned in Example 3.12, the present worth of a stream of future cash flows can always be found by summing the present worth of each of the individual cash flows. However, if cash flow regularities are present within the stream (such as we just saw in the prorated bonus payment series in Example 3.12) then the use of shortcuts, such as finding the present worth of a uniform series, may be possible. We often encounter transactions in which a uniform series of payments exists. Rental payments, bond interest payments, and commercial installment plans are based on uniform payment series.

Compound-Amount Factor: Find *F*, Given *A*, *i*, and *N*

Suppose we are interested in the future amount *F* of a fund to which we contribute *A* dollars each period and on which we earn interest at a rate of *i* per period. The contributions are made at the end of each of *N* equal periods. These transactions are graphically illustrated in Figure 3.18. Looking at this diagram, we see that if an amount *A* is invested at the end of each period, for *N* periods, the total amount *F* that can be withdrawn at the end of the *N* periods will be the sum of the compound amounts of the individual deposits.

As shown in Figure 3.18, the *A* dollars we put into the fund at the end of the first period will be worth $A(1 + i)^{N-1}$ at the end of *N* periods. The *A* dollars we put into the fund at the end of the second period will be worth $A(1 + i)^{N-2}$, and so forth. Finally, the last *A* dollars that we contribute at the end of the *N*th period will be worth exactly *A* dollars at that time. This means that there exists a series of the form

$$F = A(1 + i)^{N-1} + A(1 + i)^{N-2} + \cdots + A(1 + i) + A,$$

or, expressed alternatively,

$$F = A + A(1 + i) + A(1 + i)^2 + \cdots + A(1 + i)^{N-1}. \tag{3.8}$$

Multiplying Eq. (3.8) by $(1 + i)$ results in

$$(1 + i)F = A(1 + i) + A(1 + i)^2 + \cdots + A(1 + i)^N. \tag{3.9}$$

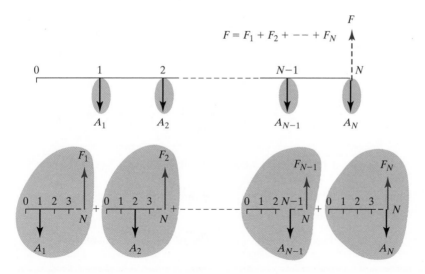

Figure 3.18   The future worth of a cash flow series obtained by summing the future-worth figures of each of the individual flows.

Subtracting Eq. (3.8) from Eq. (3.9) to eliminate common terms gives us

$$F(1 + i) - F = -A + A(1 + i)^N.$$

Solving for F yields

$$F = A\left[\frac{(1 + i)^N - 1}{i}\right] = A(F/A, i, N). \tag{3.10}$$

The bracketed term in Eq. (3.10) is called the **equal payment series compound-amount factor**, or the **uniform series compound-amount factor**; its factor notation is (F/A, i, N). This interest factor has been calculated for various combinations of i and N in the interest tables.

EXAMPLE 3.13 Uniform Series: Find F, Given i, A, and N

Suppose you make an annual contribution of $3000 to your savings account at the end of each year for 10 years. If the account earns 7% interest annually, how much can be withdrawn at the end of 10 years (Figure 3.19)?

SOLUTION

Given: $A = \$3000$, $N = 10$ years, and $i = 7\%$ per year.
Find: F.

$$F = \$3000(F/A, 7\%, 10)$$
$$= \$3000(13.8164)$$
$$= \$41,449.20.$$

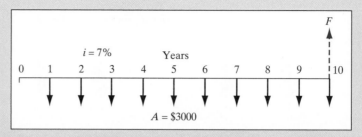

Figure 3.19 Cash flow diagram (Example 3.13).

To obtain the future value of the annuity with the use of Excel, we may use the following financial command:

$$=\text{FV}(7\%,10,-3000,0,0)$$

EXAMPLE 3.14 Handling Time Shifts in a Uniform Series

In Example 3.13, the first deposit of the 10-deposit series was made at the end of period 1 and the remaining nine deposits were made at the end of each following period. Suppose that all deposits were made at the beginning of each period instead. How would you compute the balance at the end of period 10?

SOLUTION

Given: Cash flow as shown in Figure 3.20, and $i = 7\%$ per year.
Find: F_{10}.

Compare Figure 3.20 with Figure 3.19: Each payment has been shifted to one year earlier; thus, each payment would be compounded for one extra year. Note that with the end-of-year deposit, the ending balance (F) was $41,449.20. With the beginning-of-year deposit, the same balance accumulates by the end of period 9. This balance can earn interest for one additional year. Therefore, we can easily calculate the resulting balance as

$$F_{10} = \$41{,}449.20(1.07) = \$44{,}350.64.$$

The annuity due can be easily evaluated with the following financial command available on Excel:

$$=\textbf{FV}(7\%,10,-3000,0,1)$$

COMMENTS: Another way to determine the ending balance is to compare the two cash flow patterns. By adding the $3000 deposit at period 0 to the original cash flow

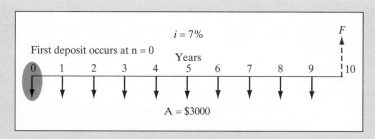

Figure 3.20   Cash flow diagram (Example 3.14).

and subtracting the $3000 deposit at the end of period 10, we obtain the second cash flow. Therefore, the ending balance can be found by making the following adjustment to the $41,449.20:

$$F_{10} = \$41,449.20 + \$3000(F/P, 7\%, 10) - \$3000 = \$44,350.64.$$

Sinking-Fund Factor: Find *A*, Given *F*, *i*, and *N*

If we solve Eq. (3.10) for *A*, we obtain

$$A = F\left[\frac{i}{(1+i)^N - 1}\right] = F(A/F, i, N). \tag{3.11}$$

The term within the brackets is called the **equal payment series sinking-fund factor**, or **sinking fund** factor, and is referred to by the notation (*A/F, i, N*). A sinking fund is an interest-bearing account into which a fixed sum is deposited each interest period; it is commonly established for the purpose of replacing fixed assets or retiring corporate bonds.

Sinking fund: A means of repaying funds advanced through a bond issue. This means that *every period, a company will pay back a portion of its bonds.*

EXAMPLE 3.15 Combination of a Uniform Series and a Single Present and Future Amount

To help you reach a $5000 goal five years from now, your father offers to give you $500 now. You plan to get a part-time job and make five additional deposits, one at the end of each year. (The first deposit is made at the end of the first year.) If all your money is deposited in a bank that pays 7% interest, how large must your annual deposit be?

STRATEGY: If your father reneges on his offer, the calculation of the required annual deposit is easy because your five deposits fit the standard end-of-period pattern for a uniform series. All you need to evaluate is

$$A = \$5000(A/F, 7\%, 5) = \$5000(0.1739) = \$869.50$$

If you do receive the $500 contribution from your father at $n = 0$, you may divide the deposit series into two parts: one contributed by your father at $n = 0$ and five equal annual deposit series contributed by yourself. Then you can use the F/P factor to find how much your father's contribution will be worth at the end of year 5 at a 7% interest rate. Let's call this amount F_c. The future value of your five annual deposits must then make up the difference, $\$5000 - F_c$.

SOLUTION

Given: Cash flow as shown in Figure 3.21, with $i = 7\%$ per year, and $N = 5$ years. Find: A.

$$
\begin{aligned}
A &= (\$5000 - F_C)(A/F, 7\%, 5) \\
&= [\$5000 - \$500(F/P, 7\%, 5)](A/F, 7\%, 5) \\
&= [\$5000 - \$500(1.4026)](0.1739) \\
&= \$747.55
\end{aligned}
$$

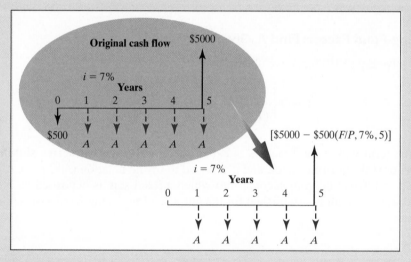

Figure 3.21 Cash flow diagram (Example 3.15).

EXAMPLE 3.16 Comparison of Three Different Investment Plans

Consider three investment plans at an annual interest rate of 9.38% (Figure 3.22):

- **Investor A.** Invest $2000 per year for the first 10 years of your career. At the end of 10 years, make no further investments, but reinvest the amount accumulated at the end of 10 years for the next 31 years.
- **Investor B.** Do nothing for the first 10 years. Then start investing $2000 per year for the next 31 years.
- **Investor C.** Invest $2000 per year for the entire 41 years.

Note that all investments are made at the *beginning* of each year; the first deposit will be made at the beginning of age 25 ($n = 0$), and you want to calculate the balance at the age of 65 ($n = 41$).

STRATEGY: Since the investments are made at the beginning of each year, we need to use the procedure outlined in Example 3.14. In other words, each deposit has one extra interest-earning period.

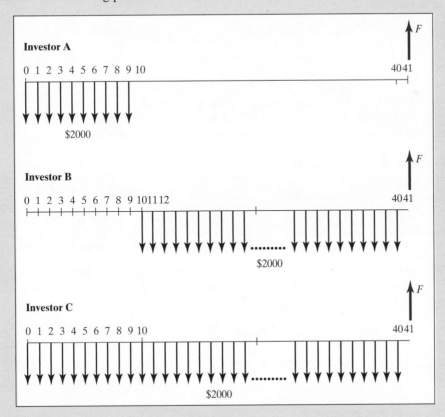

Figure 3.22 Cash flow diagrams for three investment options (Example 3.16).

SOLUTION

Given: Three different deposit scenarios with $i = 9.38\%$ and $N = 41$ years.
Find: Balance at the end of 41 years (or at the age of 65).

- Investor A:

$$F_{65} = \overbrace{\$2000(F/A, 9.38\%, 10)(1.0938)}^{\text{Balance at the end of 10 years}}(F/P, 9.38\%, 31)$$

$$\underbrace{\qquad\qquad\qquad}_{\$33,845}$$

$$= \$545,216.$$

- Investor B:

$$F_{65} = \underbrace{\$2000(F/A, 9.38\%, 31)}_{\$322,159}(1.0938)$$

$$= \$352,377.$$

- Investor C:

$$F_{65} = \underbrace{\$2000(F/A, 9.38\%, 41)}_{\$820,620}(1.0938)$$

$$= \$897,594.$$

If you know how your balance changes at the end of each year, you may want to construct a tableau such as the one shown in Table 3.3. Note that, due to rounding errors, the final balance figures are slightly off from those calculated by interest formulas.

TABLE 3.3 How Time Affects the Value of Money

		Investor A		Investor B		Investor C	
Age	Years	Contribution	Year-End Value	Contribution	Year-End Value	Contribution	Year-End Value
25	1	$2,000	$ 2,188	$0	$0	$2,000	$ 2,188
26	2	$2,000	$ 4,580	$0	$0	$2,000	$ 4,580
27	3	$2,000	$ 7,198	$0	$0	$2,000	$ 7,198
28	4	$2,000	$10,061	$0	$0	$2,000	$10,061
29	5	$2,000	$13,192	$0	$0	$2,000	$13,192
30	6	$2,000	$16,617	$0	$0	$2,000	$16,617
31	7	$2,000	$20,363	$0	$0	$2,000	$20,363
32	8	$2,000	$24,461	$0	$0	$2,000	$24,461
33	9	$2,000	$28,944	$0	$0	$2,000	$28,944
34	10	$2,000	$33,846	$0	$0	$2,000	$33,846

(Continued)

TABLE 3.3 (Continued)

		Investor A		Investor B		Investor C	
Age	Years	Contribution	Year-End Value	Contribution	Year-End Value	Contribution	Year-End Value
35	11	$0	$ 37,021	$2,000	$ 2,188	$ 2,000	$ 39,209
36	12	$0	$ 40,494	$2,000	$ 4,580	$ 2,000	$ 45,075
37	13	$0	$ 44,293	$2,000	$ 7,198	$ 2,000	$ 51,490
38	14	$0	$ 48,448	$2,000	$ 10,061	$ 2,000	$ 58,508
39	15	$0	$ 52,992	$2,000	$ 13,192	$ 2,000	$ 66,184
40	16	$0	$ 57,963	$2,000	$ 16,617	$ 2,000	$ 74,580
41	17	$0	$ 63,401	$2,000	$ 20,363	$ 2,000	$ 83,764
42	18	$0	$ 69,348	$2,000	$ 24,461	$ 2,000	$ 93,809
43	19	$0	$ 75,854	$2,000	$ 28,944	$ 2,000	$104,797
44	20	$0	$ 82,969	$2,000	$ 33,846	$ 2,000	$116,815
45	21	$0	$ 90,752	$2,000	$ 39,209	$ 2,000	$129,961
46	22	$0	$ 99,265	$2,000	$ 45,075	$ 2,000	$144,340
47	23	$0	$108,577	$2,000	$ 51,490	$ 2,000	$160,068
48	24	$0	$118,763	$2,000	$ 58,508	$ 2,000	$177,271
49	25	$0	$129,903	$2,000	$ 66,184	$ 2,000	$196,088
50	26	$0	$142,089	$2,000	$ 74,580	$ 2,000	$216,670
51	27	$0	$155,418	$2,000	$ 83,764	$ 2,000	$239,182
52	28	$0	$169,997	$2,000	$ 93,809	$ 2,000	$263,807
53	29	$0	$185,944	$2,000	$104,797	$ 2,000	$290,741
54	30	$0	$203,387	$2,000	$116,815	$ 2,000	$320,202
55	31	$0	$222,466	$2,000	$129,961	$ 2,000	$352,427
56	32	$0	$243,335	$2,000	$144,340	$ 2,000	$387,675
57	33	$0	$266,162	$2,000	$160,068	$ 2,000	$426,229
58	34	$0	$291,129	$2,000	$177,271	$ 2,000	$468,400
59	35	$0	$318,439	$2,000	$196,088	$ 2,000	$514,527
60	36	$0	$348,311	$2,000	$216,670	$ 2,000	$564,981
61	37	$0	$380,985	$2,000	$239,182	$ 2,000	$620,167
62	38	$0	$416,724	$2,000	$263,807	$ 2,000	$680,531
63	39	$0	$455,816	$2,000	$290,741	$ 2,000	$746,557
64	40	$0	$498,574	$2,000	$320,202	$ 2,000	$818,777
65	41	$0	$545,344	$2,000	$352,427	$ 2,000	$897,771
		$20,000		$62,000		$82,000	
Value at 65			**$545,344**		**$352,427**		**$897,771**
Less Total Contributions			**$20,000**		**$ 62,000**		**$82,000**
Net Earnings			**$525,344**		**$290,427**		**$815,771**

Source: Adapted from *Making Money Work for You*, UNH Cooperative Extension.

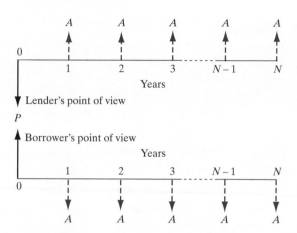

Figure 3.23 A loan cash flow diagram.

Capital Recovery Factor (Annuity Factor): Find A, Given P, i, and N

We can determine the amount of a periodic payment A if we know P, i, and N. Figure 3.23 illustrates this situation. To relate P to A, recall the relationship between P and F in Eq. (3.3), $F = P(1 + i)^N$. Replacing F in Eq. (3.11) by $P(1 + i)^N$, we get

$$A = P(1 + i)^N \left[\frac{i}{(1 + i)^N - 1} \right],$$

or

$$A = P \left[\frac{i(1 + i)^N}{(1 + i)^N - 1} \right] = P(A/P, i, N). \qquad (3.12)$$

Capital recovery factor:
Commonly used to determine the revenue requirements needed to address the up-front capital costs for projects.

Now we have an equation for determining the value of the series of end-of-period payments A when the present sum P is known. The portion within the brackets is called the **equal payment series capital recovery factor**, or simply **capital recovery factor**, which is designated $(A/P, i, N)$. In finance, this A/P factor is referred to as the **annuity factor** and indicates a series of payments of a fixed, or constant, amount for a specified number of periods.

Annuity:
An annuity is essentially a level stream of *cash flows* for a fixed period of time.

EXAMPLE 3.17 Uniform Series: Find A, Given P, i, and N

BioGen Company, a small biotechnology firm, has borrowed $250,000 to purchase laboratory equipment for gene splicing. The loan carries an interest rate of 8% per year and is to be repaid in equal installments over the next six years. Compute the amount of the annual installment (Figure 3.24).

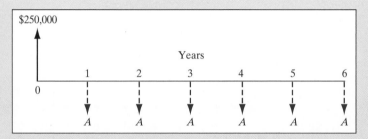

Figure 3.24 A loan cash flow diagram from BioGen's point of view.

SOLUTION

Given: $P = \$250,000$, $i = 8\%$ per year, and $N = 6$ years.
Find: A.

$$A = \$250,000(A/P, 8\%, 6)$$
$$= \$250,000(0.2163)$$
$$= \$54,075.$$

Here is an Excel solution using annuity function commands:

$$=\textbf{PMT}(i, N, P)$$
$$=\textbf{PMT}(8\%, 6, -250000)$$
$$=\$54,075.$$

EXAMPLE 3.18 Deferred Loan Repayment

In Example 3.17, suppose that BioGen wants to negotiate with the bank to defer the first loan repayment until the end of year 2 (but still desires to make six equal install-ments at 8% interest). If the bank wishes to earn the same profit, what should be the annual installment, also known as **deferred annuity** (Figure 3.25)?

SOLUTION

Given: $P = \$250,000$, $i = 8\%$ per year, and $N = 6$ years, but the first payment occurs at the end of year 2.
Find: A.

By deferring the loan for a year, the bank will add the interest accrued during the first year to the principal. In other words, we need to find the equivalent worth P' of $250,000 at the end of year 1:

$$P' = \$250,000(F/P, 8\%, 1)$$
$$= \$270,000.$$

Deferred annuity: A type of annuity contract that delays payments of income, installments, or a lump sum until the investor elects to receive them.

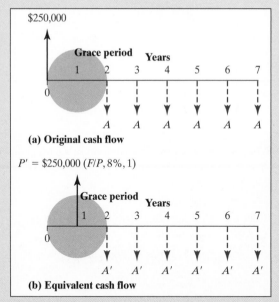

Grace period:
The additional period of time a lender provides for a borrower to make payment on a debt without penalty.

(a) Original cash flow

$P' = \$250,000\,(F/P, 8\%, 1)$

(b) Equivalent cash flow

Figure 3.25 A deferred loan cash flow diagram from BioGen's point of view (Example 3.17).

In fact, BioGen is borrowing $270,000 for six years. To retire the loan with six equal installments, the deferred equal annual payment on P' will be

$$A' = \$270,000(A/P, 8\%, 6)$$
$$= \$58,401.$$

By deferring the first payment for one year, BioGen needs to make additional payments of $4326 in each year.

Present-Worth Factor: Find *P*, Given *A*, *i*, and *N*

What would you have to invest now in order to withdraw *A* dollars at the end of each of the next *N* periods? In answering this question, we face just the opposite of the equal payment capital recovery factor situation: *A* is known, but *P* has to be determined. With the capital recovery factor given in Eq. (3.12), solving for *P* gives us

$$P = A\left[\frac{(1+i)^N - 1}{i(1+i)^N}\right] = A(P/A, i, N). \tag{3.13}$$

The bracketed term is referred to as the **equal payment series present-worth factor** and is designated $(P/A, i, N)$.

EXAMPLE 3.19 Uniform Series: Find *P*, Given *A*, *i*, and *N*

Let us revisit the Millionaire Life example, introduced in the chapter opening. Suppose that you had selected the annual payment option. Let's see how your decision stands up against the $17 million cash prize option. If you could invest your money

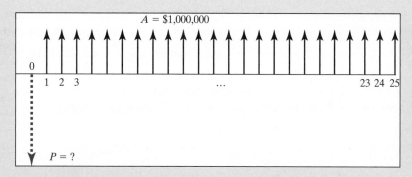

Figure 3.26   A cash flow diagram for the annual payments of your lottery winnings (Example 3.19).

at 8% interest, what is the present worth of the 25 future payments at $1,000,000 per year for 25 years?

SOLUTION

Given: $i = 8\%$ per year, $A = \$1,000,000$, and $N = 19$ years.
Find: P.

- Using interest factor:

$$P = \$1,000,000(P/A, 8\%, 25) = \$1,000,000 \times 10.6748$$
$$= \$10,674,800.$$

- Using Excel:

$$=PV(8\%,25,-1000000) = \$10,674,776.19$$

COMMENTS: Based on these calculations, we can see that the present worth of the $25 million spread over 25 years is only $10,674,800, much lower than the cash prize of $17 million. But this is based on the assumption that you are able to earn 8% return on your investment. At this point, we may be interested in knowing at just what rate of return your decision of receiving $25 million in 25 years would in fact make sense. Since we know that $P = \$17,000,000$, $N = 25$, and $A = \$1,000,000$, we solve for i.

If you know the cash flows and the PV (or FV) of a cash flow stream, you can determine the interest rate. In this case, you are looking for the interest rate that causes the P/A factor to equal $(P/A, i, 25) = (\$17,000,000/\$1,000,000) = 17$. Since we are dealing with an annuity, we could proceed as follows:

- With a financial calculator, enter $N = 25$, $PV = \$17,000,000$, $PMT = -1,000,000$, and then press the i key to find $i = 3.22\%$.

- To use the interest tables, first recognize that $\$17,000,000 = \$1,000,000 \times (P/A, i, 25)$ or $(P/A, i, 25) = 17$. Look up 17 or a close value in Appendix A.

In the *P/A* column with $N = 25$ in the 4% interest table, you will find that $(P/A, 4\%, 25) = 15.6221$. If you look up the 3% interest table, you find that $(P/A, 3\%, 25) = 17.4131$, indicating that the interest rate should be between 3% and 4%.

- To use Excel's financial command, you simply evaluate the following command to solve the unknown interest rate problem for an annuity:

$$= \textbf{RATE}(N,A,P,F,type,guess)$$
$$= \textbf{RATE}(25, 1000000, -17000000, 0, 0, 10\%)$$
$$= 3.22\%$$

This shows that you would be indifferent to the two choices if your interest rate is 3.22%.

3.3.5 Linear Gradient Series

Engineers frequently encounter situations involving periodic payments that increase or decrease by a constant amount (G) from period to period. These situations occur often enough to warrant the use of special equivalence factors that relate the arithmetic gradient to other cash flows. Figure 3.27 illustrates a **strict gradient series**, $A_n = (n - 1)G$. Note that the origin of the series is at the end of the first period with a zero value. The gradient G can be either positive or negative. If $G > 0$, the series is referred to as an *increasing* gradient series. If $G < 0$, it is a *decreasing* gradient series.

Unfortunately, the strict form of the increasing or decreasing gradient series does not correspond with the form that most engineering economic problems take. A typical problem involving a linear gradient includes an initial payment during period 1 that increases by G during some number of interest periods, a situation illustrated in Figure 3.28. This contrasts with the strict form illustrated in Figure 3.27, in which no payment is made during period 1 and the gradient is added to the previous payment beginning in period 2.

Gradient Series as Composite Series

In order to utilize the strict gradient series to solve typical problems, we must view cash flows as shown in Figure 3.28 as a **composite series**, or a set of two cash flows, each corresponding to a form that we can recognize and easily solve: a uniform series of N

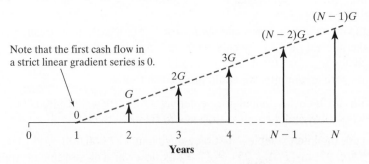

Figure 3.27 A cash flow diagram for a strict gradient series.

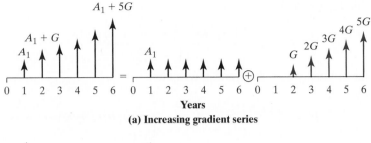

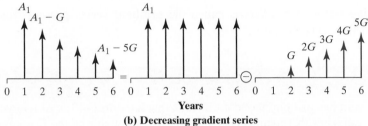

Figure 3.28 Two types of linear gradient series as composites of a uniform series of N payments of A_1 and the gradient series of increments of constant amount G.

payments of amount A_1 and a gradient series of increments of constant amount G. The need to view cash flows that involve linear gradient series as composites of two series is very important in solving problems, as we shall now see.

Present-Worth Factor: Linear Gradient: Find P, Given G, N, and i

How much would you have to deposit now to withdraw the gradient amounts specified in Figure 3.27? To find an expression for the present amount P, we apply the single-payment present-worth factor to each term of the series and obtain

$$P = 0 + \frac{G}{(1 + i)^2} + \frac{2G}{(1 + i)^3} + \cdots + \frac{(N - 1)G}{(1 + i)^N},$$

or

$$P = \sum_{n=1}^{N}(n - 1)G(1 + i)^{-n}. \tag{3.14}$$

Letting $G = a$ and $1/(1 + i) = x$ yields

$$P = 0 + ax^2 + 2ax^3 + \cdots + (N - 1)ax^N$$

$$= ax[0 + x + 2x^2 + \cdots + (N - 1)x^{N-1}]. \tag{3.15}$$

Since an arithmetic–geometric series $\{0, x, 2x^2, \ldots, (N - 1)x^{N-1}\}$ has the finite sum

$$0 + x + 2x^2 + \cdots + (N - 1)x^{N-1} = x\left[\frac{1 - Nx^{N-1} + (N - 1)x^N}{(1 - x)^2}\right],$$

we can rewrite Eq. (3.15) as

$$P = ax^2 \left[\frac{1 - Nx^{N-1} + (N-1)x^N}{(1-x)^2} \right]. \tag{3.16}$$

Replacing the original values for A and x, we obtain

$$P = G \left[\frac{(1+i)^N - iN - 1}{i^2(1+i)^N} \right] = G(P/G, i, N). \tag{3.17}$$

The resulting factor in brackets is called the **gradient series present-worth factor**, which we denote as $(P/G, i, N)$.

EXAMPLE 3.20 Linear Gradient: Find *P*, Given *A₁*, *G*, *i*, and *N*

A textile mill has just purchased a lift truck that has a useful life of five years. The engineer estimates that maintenance costs for the truck during the first year will be $1000. As the truck ages, maintenance costs are expected to increase at a rate of $250 per year over the remaining life. Assume that the maintenance costs occur at the end of each year. The firm wants to set up a maintenance account that earns 12% annual interest. All future maintenance expenses will be paid out of this account. How much does the firm have to deposit in the account now?

SOLUTION

Given: $A_1 = \$1000$, $G = \$250$, $i = 12\%$ per year, and $N = 5$ years.
Find: P.

Asking how much the firm has to deposit now is equivalent to asking what the equivalent present worth for this maintenance expenditure is if 12% interest is used. The cash flow may be broken into two components as shown in Figure 3.29.

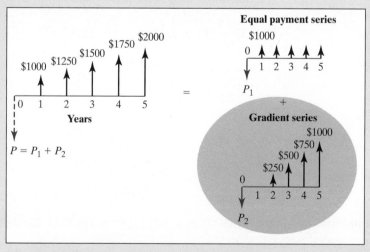

Figure 3.29 Cash flow diagram (Example 3.20).

The first component is an equal payment series (A_1), and the second is a linear gradient series (G). We have

$$P = P_1 + P_2$$

$$P = A_1(P/A, 12\%, 5) + G(P/G, 12\%, 5)$$

$$= \$1000(3.6048) + \$250(6.397)$$

$$= \$5204.$$

Note that the value of N in the gradient factor is 5, not 4. This is because, by definition of the series, the first gradient value begins at period 2.

COMMENTS: As a check, we can compute the present worth of the cash flow by using the $(P/F, 12\%, n)$ factors:

Period (n)	Cash Flow	(P/F, 12%, n)	Present Worth
1	$1000	0.8929	$ 892.90
2	1250	0.7972	996.50
3	1500	0.7118	1067.70
4	1750	0.6355	1112.13
5	2000	0.5674	1134.80
		Total	$5204.03

The slight difference is caused by a rounding error.

Gradient-to-Equal-Payment Series Conversion Factor: Find A, Given G, i, and N

We can obtain an equal payment series equivalent to the gradient series, as depicted in Figure 3.30, by substituting Eq. (3.17) for P into Eq. (3.12) to obtain

$$A = G\left[\frac{(1 + i)^N - iN - 1}{i[(1 + i)^N - 1]}\right] = G(A/G, i, N), \qquad (3.18)$$

where the resulting factor in brackets is referred to as the **gradient-to-equal-payment series conversion factor** and is designated $(A/G, i, N)$.

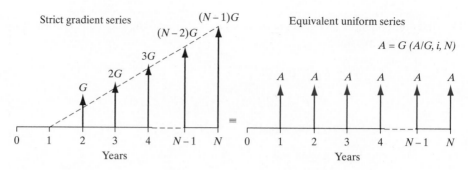

Figure 3.30

EXAMPLE 3.21 Linear Gradient: Find A, Given A₁, G, i, and N

John and Barbara have just opened two savings accounts at their credit union. The accounts earn 10% annual interest. John wants to deposit $1000 in his account at the end of the first year and increase this amount by $300 for each of the next five years. Barbara wants to deposit an equal amount each year for the next six years. What should be the size of Barbara's annual deposit so that the two accounts will have equal balances at the end of six years (Figure 3.31)?

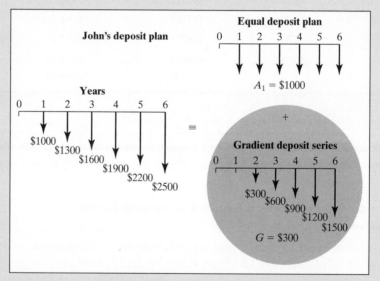

Figure 3.31 John's deposit series viewed as a combination of uniform and gradient series (Example 3.21).

SOLUTION

Given: $A_1 = \$1000$, $G = \$300$, $i = 10\%$, and $N = 6$.
Find: A.

Since we use the end-of-period convention unless otherwise stated, this series begins at the end of the first year and the last contribution occurs at the end of the sixth year. We can separate the constant portion of $1000 from the series, leaving the gradient series of $0, 300, 600, \ldots, 1500$.

To find the equal payment series beginning at the end of year 1 and ending at year 6 that would have the same present worth as that of the gradient series, we may proceed as follows:

$$A = \$1000 + \$300(A/G, 10\%, 6)$$
$$= \$1000 + \$300(2.2236)$$
$$= \$1667.08.$$

Barbara's annual contribution should be $1667.08.

COMMENTS: Alternatively, we can compute Barbara's annual deposit by first computing the equivalent present worth of John's deposits and then finding the equivalent uniform annual amount. The present worth of this combined series is

$$P = \$1000(P/A, 10\%, 6) + \$300(P/G, 10\%, 6)$$
$$= \$1000(4.3553) + \$300(9.6842)$$
$$= \$7260.56.$$

The equivalent uniform deposit is

$$A = \$7260.56(A/P, 10\%, 6) = \$1667.02.$$

(The slight difference in cents is caused by a rounding error.)

EXAMPLE 3.22 Declining Linear Gradient: Find *F*, Given *A₁*, *G*, *i*, and *N*

Suppose that you make a series of annual deposits into a bank account that pays 10% interest. The initial deposit at the end of the first year is $1200. The deposit amounts decline by $200 in each of the next four years. How much would you have immediately after the fifth deposit?

SOLUTION

Given: Cash flow shown in Figure 3.32, $i = 10\%$ per year, and $N = 5$ years.

Find: *F*.

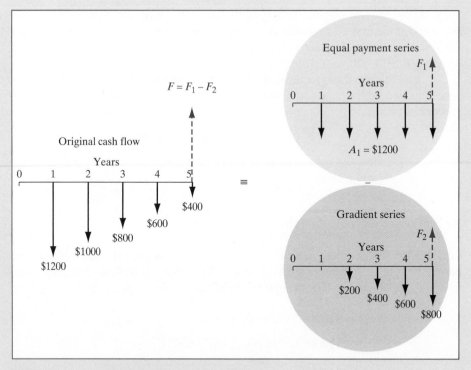

Figure 3.32 A series of decreasing gradient deposits viewed as a combination of uniform and a strict gradient series (Example 3.22).

The cash flow includes a decreasing gradient series. Recall that we derived the linear gradient factors for an increasing gradient series. For a decreasing gradient series, the solution is most easily obtained by separating the flow into two components: a uniform series and an increasing gradient that is *subtracted* from the uniform series (Figure 3.32). The future value is

$$
\begin{aligned}
F &= F_1 - F_2 \\
&= A_1(F/A, 10\%, 5) - \$200(P/G, 10\%, 5)(F/P, 10\%, 5) \\
&= \$1200(6.105) - \$200(6.862)(1.611) \\
&= \$5115.
\end{aligned}
$$

Compound growth: The year-over-year growth rate of an investment over a specified period of time. Compound growth is an imaginary number that describes the rate at which an investment grew as though it had grown at a steady rate.

3.3.6 Geometric Gradient Series

Many engineering economic problems—particularly those relating to construction costs—involve cash flows that increase or decrease over time, not by a constant *amount* (as with a linear gradient), but rather by a constant percentage (a **geometric gradient**). This kind of cash flow is called **compound growth**. Price changes caused by inflation are a good example of a geometric gradient series. If we use g to designate the percentage change in a payment from one period to the next, the magnitude of the nth payment, A_n, is related to the first payment A_1 by the formula

$$A_n = A_1(1 + g)^{n-1}, n = 1, 2, \ldots, N. \tag{3.19}$$

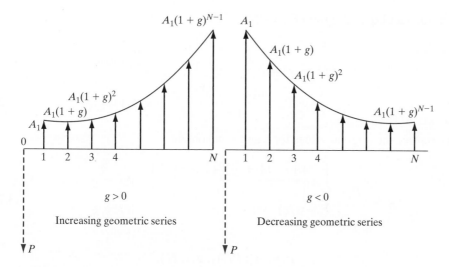

Figure 3.33 A geometrically increasing or decreasing gradient series at a constant rate g.

The variable g can take either a positive or a negative sign, depending on the type of cash flow. If $g > 0$, the series will increase, and if $g < 0$, the series will decrease. Figure 3.33 illustrates the cash flow diagram for this situation.

Present-Worth Factor: Find P, Given A_1, g, i, and N

Notice that the present worth of any cash flow A_n at interest rate i is

$$P_n = A_n(1 + i)^{-n} = A_1(1 + g)^{n-1}(1 + i)^{-n}.$$

To find an expression for the present amount P for the entire series, we apply the **single-payment present-worth factor** to each term of the series:

$$P = \sum_{n=1}^{N} A_1(1 + g)^{n-1}(1 + i)^{-n}. \tag{3.20}$$

Bringing the constant term $A_1(1 + g)^{-1}$ outside the summation yields

$$P = \frac{A_1}{(1 + g)} \sum_{n=1}^{N} \left[\frac{1 + g}{1 + i}\right]^n. \tag{3.21}$$

Let $a = \dfrac{A_1}{1 + g}$ and $x = \dfrac{1 + g}{1 + i}$. Then, rewrite Eq. (3.21) as

$$P = a(x + x^2 + x^3 + \cdots + x^N). \tag{3.22}$$

Since the summation in Eq. (3.22) represents the first N terms of a geometric series, we may obtain the closed-form expression as follows: First, multiply Eq. (3.22) by x to get

$$xP = a(x^2 + x^3 + x^4 + \cdots + x^{N+1}). \tag{3.23}$$

Then, subtract Eq. (3.23) from Eq. (3.22):

$$P - xP = a(x - x^{N+1})$$

$$P(1 - x) = a(x - x^{N+1})$$

$$P = \frac{a(x - x^{N+1})}{1 - x} \qquad (x \neq 1). \qquad (3.24)$$

If we replace the original values for a and x, we obtain

$$P = \begin{cases} A_1\left[\dfrac{1 - (1 + g)^N(1 + i)^{-N}}{i - g}\right] & \text{if } i \neq g \\ NA_1/(1 + i) & \text{if } i = g \end{cases} \qquad (3.25)$$

or

$$P = A_1(P/A_1, g, i, N).$$

The factor within brackets is called the **geometric-gradient-series present-worth factor** and is designated $(P/A_1, g, i, N)$. In the special case where $i = g$, Eq. (3.21) becomes $P = [A_1/(1 + i)]N$.

EXAMPLE 3.23 Geometric Gradient: Find P, Given A_1, g, i, and N

Ansell Inc., a medical device manufacturer, uses compressed air in solenoids and pressure switches in its machines to control various mechanical movements. Over the years, the manufacturing floor has changed layouts numerous times. With each new layout, more piping was added to the compressed-air delivery system to accommodate new locations of manufacturing machines. None of the extra, unused old pipe was capped or removed; thus, the current compressed-air delivery system is inefficient and fraught with leaks. Because of the leaks, the compressor is expected to run 70% of the time that the plant will be in operation during the upcoming year. This will require 260 kWh of electricity at a rate of $0.05/kWh. (The plant runs 250 days a year, 24 hours per day.) If Ansell continues to operate the current air delivery system, the compressor run time will increase by 7% per year for the next five years because of ever-worsening leaks. (After five years, the current system will not be able to meet the plant's compressed-air requirement, so it will have to be replaced.) If Ansell decides to replace all of the old piping now, it will cost $28,570.

The compressor will still run the same number of days; however, it will run 23% less (or will have $70\%(1 - 0.23) = 53.9\%$ usage during the day) because of the reduced air pressure loss. If Ansell's interest rate is 12%, is the machine worth fixing now?

SOLUTION

Given: Current power consumption, $g = 7\%$, $i = 12\%$, and $N = 5$ years.
Find: A_1 and P.

Step 1: We need to calculate the cost of power consumption of the current piping system during the first year:

$$\text{Power cost} = \% \text{ of day operating}$$
$$\times \text{ days operating per year}$$
$$\times \text{ hours per day}$$
$$\times \text{ kWh} \times \text{\$/kWh}$$
$$= (70\%) \times (250 \text{ days/year}) \times (24 \text{ hours/day})$$
$$\times (260 \text{ kWh}) \times (\$0.05/\text{kWh})$$
$$= \$54{,}440.$$

Step 2: Each year, the annual power cost will increase at the rate of 7% over the previous year's cost. The anticipated power cost over the five-year period is summarized in Figure 3.34. The equivalent present lump-sum cost at 12% for this geometric gradient series is

$$P_{\text{Old}} = \$54{,}440(P/A_1, 7\%, 12\%, 5)$$
$$= \$54{,}440\left[\frac{1 - (1 + 0.07)^5(1 + 0.12)^{-5}}{0.12 - 0.07}\right]$$
$$= \$222{,}283.$$

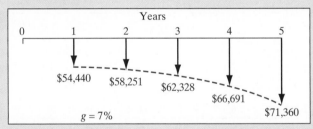

Figure 3.34   Annual power cost series if repair is not performed.

Step 3: If Ansell replaces the current compressed-air system with the new one, the annual power cost will be 23% less during the first year and will remain at that level over the next five years. The equivalent present lump-sum cost at 12% is then

$$P_{\text{New}} = \$54{,}440(1 - 0.23)(P/A, 12\%, 5)$$
$$= \$41{,}918.80(3.6048)$$
$$= \$151{,}109.$$

Step 4: The net cost for not replacing the old system now is $71,174 (= $222,283 − $151,109). Since the new system costs only $28,570, the replacement should be made now.

COMMENTS: In this example, we assumed that the cost of removing the old system was included in the cost of installing the new system. If the removed system has some salvage value, replacing it will result in even greater savings. We will consider many types of replacement issues in Chapter 11.

EXAMPLE 3.24 Geometric Gradient: Find A_1, Given F, g, i, and N

Jimmy Carpenter, a self-employed individual, is opening a retirement account at a bank. His goal is to accumulate $1,000,000 in the account by the time he retires from work in 20 years' time. A local bank is willing to open a retirement account that pays 8% interest compounded annually throughout the 20 years. Jimmy expects that his annual income will increase 6% yearly during his working career. He wishes to start with a deposit at the end of year 1 (A_1) and increase the deposit at a rate of 6% each year thereafter. What should be the size of his first deposit (A_1)? The first deposit will occur at the end of year 1, and subsequent deposits will be made at the end of each year. The last deposit will be made at the end of year 20.

SOLUTION

Given: $F = \$1,000,000$, $g = 6\%$ per year, $i = 8\%$ per year, and $N = 20$ years.
Find: A_1 as in Figure 3.35.

We have

$$F = A_1(P/A_1, 6\%, 8\%, 20)\,(F/P, 8\%, 20)$$

$$= A_1(72.6911).$$

Solving for A_1 yields

$$A_1 = \$1,000,000/72.6911 = \$13,757.$$

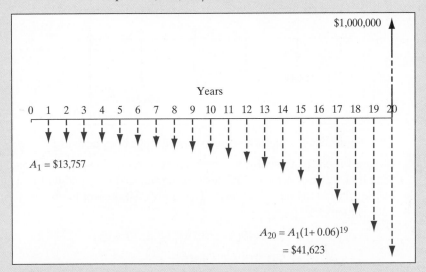

Figure 3.35 Jimmy Carpenter's retirement plan (Example 3.24).

Table 3.4 summarizes the interest formulas developed in this section and the cash flow situations in which they should be used. Recall that these formulas are applicable only to situations where the interest (compounding) period is the same as the payment period (e.g., annual compounding with annual payment). Also, we present some useful Excel financial commands in the table.

TABLE 3.4 Summary of Discrete Compounding Formulas With Discrete Payments

Flow Type	Factor Notation	Formula	Excel Command	Cash Flow Diagram
S I N G L E	Compound amount $(F/P, i, N)$	$F = P(1 + i)^N$	=FV$(i, N, 0, P)$	
	Present worth $(P/F, i, N)$	$P = F(1 + i)^{-N}$	=PV$(i, N, 0, F)$	
E Q U A L P A Y M E N T	Compound amount $(F/A, i, N)$	$F = A\left[\dfrac{(1 + i)^N - 1}{i}\right]$	= FV(i, N, A)	
	Sinking fund $(A/F, i, N)$	$A = F\left[\dfrac{i}{(1 + i)^N - 1}\right]$	=PMT$(i, N, 0, F)$	
S E R I E S	Present worth $(P/A, i, N)$	$P = A\left[\dfrac{(1 + i)^N - 1}{i(1 + i)^N}\right]$	=PV(i, N, A)	
	Capital recovery $(A/P, i, N)$	$A = P\left[\dfrac{i(1 + i)^N}{(1 + i)^N - 1}\right]$	=PMT(i, N, P)	
G R A D I E N T	Linear gradient Present worth $(P/G, i, N)$	$P = G\left[\dfrac{(1 + i)^N - iN - 1}{i^2(1 + i)^N}\right]$		
	Annual worth $(A/G, i, N)$	$A = G\left[\dfrac{(1 + i)^N - iN - 1}{i\left[(1 + i)^N - 1\right]}\right]$		
S E R I E S	Geometric gradient Present worth $(P/A_1, g, i, N)$	$P = \begin{cases} A_1\left[\dfrac{1 - (1 + g)^N(1 + i)^{-N}}{i - g}\right] \\ A_1\left(\dfrac{N}{1 + i}\right), (\text{if } i = g) \end{cases}$		

TABLE 3.5 Limiting Forms of Discrete Compounding Formulas With Discrete Payments

	Limit as $N \to \infty$ (*i* is specified)		Limit as $i \to 0$ (*N* is specified)	
(F/P, i, N)	∞		1	
(P/F, i, N)	0		1	
(F/A, i, N)	∞		N	
(A/F, i, N)	0		$\dfrac{1}{N}$	
(P/A, i, N)	$\dfrac{1}{i}$		N	
(A/P, i, N)	i		$\dfrac{1}{N}$	
(P/G, i, N)	$\dfrac{1}{i^2}$		∞	
(P/A$_1$, g, i, N)	∞	$g > i$	$\dfrac{(1 + g) - 1}{g}$	$i \neq g$
	$\dfrac{1}{i - g}$	$g < i$		
	∞	$i = g$	N	$i = g$

3.3.7 Limiting Forms of Interest Formulas

When the interest rate, i, is very small or the number of interest periods, N, is very large, the interest formulas in Table 3.4 reduce to the limiting forms summarized in Table 3.5. These limits are readily established by taking the limit as $i \to 0$ or as $N \to \infty$ of the formulas in Table 3.4. When indeterminate forms arise in the analysis, the usual calculus methods are used to establish the correct limit.

As we will see in later chapters, there are situations that satisfy the interest rate condition or the number of periods condition. If the interest factors required in such an analysis have finite limits, use of the limiting forms in Table 3.5 can save considerable effort.

3.4 Unconventional Equivalence Calculations

In the preceding section, we occasionally presented two or more methods of attacking problems even though we had standard interest factor equations by which to solve them. It is important that you become adept at examining problems from unusual angles and that you seek out unconventional solution methods because not all cash flow problems conform to the neat patterns for which we have discovered and developed equations. Two categories of problems that demand unconventional treatment are composite (mixed) cash flows and problems in which we must determine the interest rate implicit in a financial contract. We will begin this section by examining instances of composite cash flows.

3.4.1 Composite Cash Flows

Although many financial decisions do involve constant or systematic changes in cash flows, others contain several components of cash flows that do not exhibit an overall pattern. Consequently, it is necessary to expand our analysis to deal with these mixed types of cash flows.

To illustrate, consider the cash flow stream shown in Figure 3.36. We want to compute the equivalent present worth for this mixed payment series at an interest rate of 15%. Three different methods are presented.

Method 1. A "brute force" approach is to multiply each payment by the appropriate $(P/F, 10\%, n)$ factors and then to sum these products to obtain the present worth of the cash flows, $543.72. Recall that this is exactly the same procedure we used to solve the category of problems called the uneven payment series, described in Section 3.3.3. Figure 3.36 illustrates this computational method.

Method 2. We may group the cash flow components according to the type of cash flow pattern that they fit, such as the single payment, equal payment series, and so forth, as shown in Figure 3.37. Then the solution procedure involves the following steps:

- **Group 1:** Find the present worth of $50 due in year 1:

$$\$50(P/F, 15\%, 1) = \$43.48.$$

- **Group 2:** Find the equivalent worth of a $100 equal payment series at year 1 (V_1), and then bring this equivalent worth at year 0 again:

$$\underbrace{\$100(P/A, 15\%, 3)}_{V_1}(P/F, 15\%, 1) = \$198.54.$$

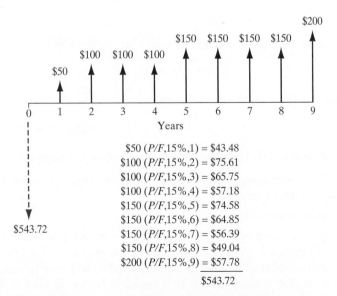

$50 (P/F,15%,1) = $43.48
$100 (P/F,15%,2) = $75.61
$100 (P/F,15%,3) = $65.75
$100 (P/F,15%,4) = $57.18
$150 (P/F,15%,5) = $74.58
$150 (P/F,15%,6) = $64.85
$150 (P/F,15%,7) = $56.39
$150 (P/F,15%,8) = $49.04
$200 (P/F,15%,9) = $57.78
$543.72

Figure 3.36 Equivalent present worth calculation using only P/F factors (Method 1 "Brute Force" Approach).

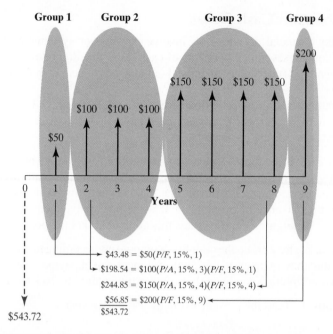

Figure 3.37 Equivalent present-worth calculation for an uneven payment series, using *P/F* and *P/A* factors (Method 2: grouping approach).

- **Group 3:** Find the equivalent worth of a $150 equal payment series at year 4 (V_4), and then bring this equivalent worth at year 0.

$$\underbrace{\$150(P/A, 15\%, 4)}_{V_4}(P/F, 15\%, 4) = \$244.85.$$

- **Group 4:** Find the equivalent present worth of the $200 due in year 9:

$$\$200(P/F, 15\%, 9) = \$56.85.$$

- **Group total—sum the components:**

$$P = \$43.48 + \$198.54 + \$244.85 + \$56.85 = \$543.72$$

A pictorial view of this computational process is given in Figure 3.37.

Method 3. In computing the present worth of the equal payment series components, we may use an alternative method.

- **Group 1:** Same as in Method 2.
- **Group 2:** Recognize that a $100 equal payment series will be received during years 2 through 4. Thus, we could determine the value of a four-year annuity, subtract the value of a one-year annuity from it, and have remaining the value of a four-year annuity whose first payment is due in year 2. This result is achieved by subtracting the (*P/A*, 15%, 1) for a one-year, 15% annuity from that for a four-year annuity and then multiplying the difference by $100:

$$\$100[(P/A, 15\%, 4) - (P/A, 15\%, 1)] = \$100(2.8550 - 0.8696)$$
$$= \$198.54.$$

Thus, the equivalent present worth of the annuity component of the uneven stream is $198.54.

- **Group 3:** We have another equal payment series that starts in year 5 and ends in year 8.

$$\$150[(P/A, 15\%, 8) - (P/A, 15\%, 4)] = \$150(4.4873 - 2.8550)$$
$$= \$244.85.$$

- **Group 4:** Same as Method 2.
- **Group total—sum the components:**

$$P = \$43.48 + \$198.54 + \$244.85 + \$56.85 = \$543.72.$$

Either the "brute force" approach of Figure 3.35 or the method utilizing both $(P/A, i, n)$ and $(P/F, i, n)$ factors can be used to solve problems of this type. However, Method 2 or Method 3 is much easier if the annuity component runs for many years. For example, the alternative solution would be clearly superior for finding the equivalent present worth of a stream consisting of $50 in year 1, $200 in years 2 through 19, and $500 in year 20.

Also, note that in some instances we may want to find the equivalent value of a stream of payments at some point other than the present (year 0). In this situation, we proceed as before, but compound and discount to some other point in time—say, year 2, rather than year 0. Example 3.25 illustrates the situation.

EXAMPLE 3.25 Cash Flows With Subpatterns

The two cash flows in Figure 3.38 are equivalent at an interest rate of 12% compounded annually. Determine the unknown value C.

SOLUTION

Given: Cash flows as in Figure 3.38; $i = 12\%$ per year.
Find: C.

- **Method 1.** Compute the present worth of each cash flow at time 0:

$$P_1 = \$100(P/A, 12\%, 2) + \$300(P/A, 12\%, 3)(P/F, 12\%, 2)$$
$$= \$743.42;$$

$$P_2 = C(P/A, 12\%, 5) - C(P/F, 12\%, 3)$$
$$= 2.8930C.$$

Since the two flows are equivalent, $P_1 = P_2$, and we have

$$743.42 = 2.8930C.$$

Solving for C, we obtain $C = \$256.97$.

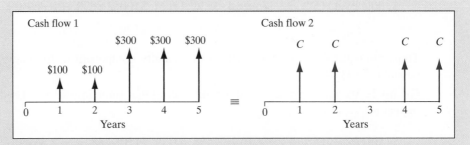

Figure 3.38 Equivalence calculation (Example 3.25).

- **Method 2.** We may select a time point other than 0 for comparison. The best choice of a base period is determined largely by the cash flow patterns. Obviously, we want to select a base period that requires the minimum number of interest factors for the equivalence calculation. Cash flow 1 represents a combined series of two equal payment cash flows, whereas cash flow 2 can be viewed as an equal payment series with the third payment missing. For cash flow 1, computing the equivalent worth at period 5 will require only two interest factors:

$$V_{5,1} = \$100(F/A, 12\%, 5) + \$200(F/A, 12\%, 3)$$
$$= \$1310.16.$$

For cash flow 2, computing the equivalent worth of the equal payment series at period 5 will also require two interest factors:

$$V_{5,2} = C(F/A, 12\%, 5) - C(F/P, 12\%, 2)$$
$$= 5.0984C.$$

Therefore, the equivalence would be obtained by letting $V_{5,1} = V_{5,2}$:

$$\$1310.16 = 5.0984C.$$

Solving for C yields $C = \$256.97$, which is the same result obtained from Method 1. The alternative solution of shifting the time point of comparison will require only four interest factors, whereas Method 1 requires five interest factors.

EXAMPLE 3.26 Establishing a University Fund

A couple with a newborn daughter wants to save for their child's university expenses in advance. The couple can establish a university fund that pays 7% annual interest. Assuming that the child enters university at age 18, the parents estimate that an amount of $40,000 per year (actual dollars) will be required to support the child's university expenses for four years. Determine the equal annual amounts the couple must save until they send their child to university. (Assume that the first deposit will be made on the child's first birthday and the last deposit on the child's 18th birthday. The first withdrawal will be made at the beginning of first year, which also is the child's 18th birthday.)

SOLUTION

Given: Deposit and withdrawal series shown in Figure 3.39; $i = 7\%$ per year.
Find: Unknown annual deposit amount (X).

- **Method 1.**   Establish economic equivalence at period 0:

    Step 1: Find the equivalent single lump-sum deposit now:

    $$P_{\text{Deposit}} = X(P/A, 7\%, 18)$$

    $$= 10.0591X.$$

    Step 2: Find the equivalent single lump-sum withdrawal now:

    $$P_{\text{Withdrawal}} = \$40,000(P/A, 7\%, 4)(P/F, 7\%, 17)$$

    $$= \$42,892.$$

    Step 3: Since the two amounts are equivalent, by equating $P_{\text{Deposit}} = P_{\text{Withdrawal}}$, we obtain X:

    $$10.0591X = \$42,892$$

    $$X = \$4264.$$

- **Method 2.**   Establish the economic equivalence at the child's 18th birthday:

    Step 1: Find the accumulated deposit balance on the child's 18th birthday:

    $$V_{18} = X(F/A, 7\%, 18)$$

    $$= 33.9990X.$$

    Step 2: Find the equivalent lump-sum withdrawal on the child's 18th birthday:

    $$V_{18} = \$40,000 + \$40,000(P/A, 7\%, 3)$$

    $$= \$144,972.$$

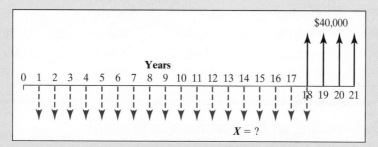

Figure 3.39   Establishing a university fund (Example 3.26). Note that the $40,000 figure represents the actual anticipated expenditures considering the future inflation.

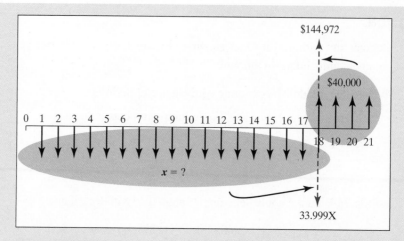

Figure 3.40   An alternative equivalence calculation (Example 3.26).

Step 3: Since the two amounts must be the same, we obtain

$$33.9990X = \$144,972$$

$$X = \$4264.$$

The computational steps are summarized in Figure 3.40. In general, the second method is the more efficient way to obtain an equivalence solution to this type of decision problem.

COMMENTS: To verify whether the annual deposits of $4264 over 18 years would be sufficient to meet the child's university expenses, we can calculate the actual year-by-year balances: With the 18 annual deposits of $4264, the balance on the child's 18th birthday is

$$\$4264(F/A, 7\%, 18) = \$144,972.$$

From this balance, the couple will make four annual tuition payments:

Year N	Beginning Balance	Interest Earned	Tuition Payment	Ending Balance
First	$144,972	$ 0	$40,000	$104,972
Second	104,972	7,348	40,000	72,320
Third	72,320	5,062	40,000	37,382
Fourth	37,382	2,618	40,000	0

3.4.2 Determining an Interest Rate to Establish Economic Equivalence

Thus far, we have assumed that, in equivalence calculations, a typical interest rate is given. Now we can use the same interest formulas that we developed earlier to determine interest rates that are explicit in equivalence problems. For most commercial loans, interest rates are already specified in the contract. However, when making some investments in financial assets, such as stocks, you may want to know the rate of growth (or rate of return) at which your asset is appreciating over the years. (This kind of calculation is the basis of rate-of-return analysis, which is covered in Chapter 5.) Although we can use interest tables to find the rate that is implicit in single payments and annuities, it is more difficult to find the rate that is implicit in an uneven series of payments. In such cases, a trial-and-error procedure or computer software may be used. To illustrate, consider Example 3.27.

EXAMPLE 3.27 Calculating an Unknown Interest Rate With Multiple Factors

You may have already won $2 million! Just peel the game piece off the Instant Winner Sweepstakes ticket and mail it to us along with your order for subscriptions to your two favourite magazines. As a grand prize winner, you may choose between a $1 million cash prize paid immediately or $100,000 per year for 20 years—that's $2 million!

Suppose that, instead of receiving one lump sum of $1 million, you decide to accept the 20 annual installments of $100,000. If you are like most jackpot winners, you will be tempted to spend your winnings to improve your lifestyle during the first several years. Only after you get this type of spending "out of your system" will you save later sums for investment purposes. Suppose that you are considering the following two options:

Option 1: You save your winnings for the first seven years and then spend every cent of the winnings in the remaining 13 years.

Option 2: You do the reverse, spending for seven years and then saving for 13 years.

If you can save winnings at 7% interest, how much would you have at the end of 20 years, and what interest rate on your savings will make these two options equivalent? (Cash flows into savings for the two options are shown in Figure 3.41.)

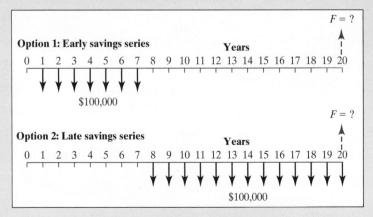

Figure 3.41 Equivalence calculation (Example 3.27).

SOLUTION

Given: Cash flows in Figure 3.41.

Find: (a) F and (b) i at which the two flows are equivalent.

(a) In Option 1, the net balance at the end of year 20 can be calculated in two steps: Find the accumulated balance at the end of year 7 (V_7) first; then find the equivalent worth of V_7 at the end of year 20. For Option 2, find the equivalent worth of the 13 equal annual deposits at the end of year 20. We thus have

$$F_{\text{Option 1}} = \$100,000(F/A, 7\%, 7)(F/P, 7\%, 13)$$
$$= \$2,085,485;$$
$$F_{\text{Option 2}} = \$100,000(F/A, 7\%, 13)$$
$$= \$2,014,064.$$

Option 1 accumulates \$71,421 more than Option 2.

(b) To compare the alternatives, we may compute the present worth for each option at period 0. By selecting period 7, however, we can establish the same economic equivalence with fewer interest factors. As shown in Figure 3.42, we calculate the equivalent value V_7 for each option at the end of period 7, remembering that the end of period 7 is also the beginning of period 8. (Recall from Example 3.4 that the choice of the point in time at which to compare two cash flows for equivalence is arbitrary.)

- For Option 1,

$$V_7 = \$100,000(F/A, i, 7).$$

- For Option 2,

$$V_7 = \$100,000(P/A, i, 13).$$

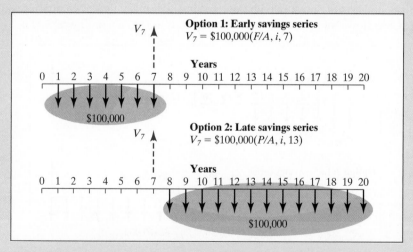

Figure 3.42 Establishing an economic equivalence at period 7 (Example 3.27).

We equate the two values:

$$\$100{,}000(F/A, i, 7) = \$100{,}000(P/A, i, 13);$$

$$\frac{(F/A, i, 7)}{(P/A, i, 13)} = 1.$$

Here, we are looking for an interest rate that gives a ratio of unity. When using the interest tables, we need to resort to a trial-and-error method. Suppose that we guess the interest rate to be 6%. Then

$$\frac{(F/A, 6\%, 7)}{(P/A, 6\%, 13)} = \frac{8.3938}{8.8527} = 0.9482.$$

This is less than unity. To increase the ratio, we need to use a value of i such that it increases the $(F/A, i, 7)$ factor value, but decreases the $(P/A, i, 13)$ value. This will happen if we use a larger interest rate. Let's try $i = 7\%$:

$$\frac{(F/A, 7\%, 7)}{(P/A, 7\%, 13)} = \frac{8.6540}{8.3577} = 1.0355.$$

Now the ratio is greater than unity.

Interest Rate	(F/A, i, 7)/(P/A, i, 13)
6%	0.9482
?	1.0000
7%	1.0355

As a result, we find that the interest rate is between 6% and 7% and may be approximated by **linear interpolation** as shown in Figure 3.43:

$$i = 6\% + (7\% - 6\%)\left[\frac{1 - 0.9482}{1.0355 - 0.9482}\right]$$

$$= 6\% + 1\%\left[\frac{0.0518}{0.0873}\right]$$

$$= 6.5934\%$$

Interpolation is a method of constructing new data points from a discrete set of known data points.

At 6.5934% interest, the two options are equivalent, and you may decide to indulge your desire to spend like crazy for the first 7 years. However, if you could obtain a higher interest rate, you would be wiser to save for 7 years and spend for the next 13.

COMMENTS: This example demonstrates that finding an interest rate is an iterative process that is more complicated and generally less precise than finding an equivalent worth at a known interest rate. Since computers and financial calculators can speed the process of finding unknown interest rates, such tools are highly recommended for these types of problem solving. With Excel, a more precise break-even value of 6.60219% is found by using the Goal Seek function.[3]

---

[3] Goal Seek can be used when you know the result of a formula, but not the input value required by the formula to decide the result. You can change the value of a specified cell until the formula that is dependent on the changed cell returns the result you want. Goal Seek is found under the Tools menu.

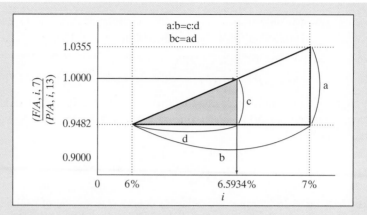

Figure 3.43 Linear interpolation to find an unknown interest rate (Example 3.27).

In Figure 3.44, the cell that contains the formula that you want to settle is called the **Set cell** (**F11 = F7-F9**). The value you want the formula to change to is called **To value (0)** and the part of the formula that you wish to change is called **By changing cell (F5, interest rate)**. The **Set cell** MUST always contain a formula or a function, whereas the **Changing cell** must contain a value only, not a formula or function.

	A	B	C	D	E	F	G	H	I	J
1	Example 3.27									
2	Year	Option 1	Option 2							
3	0									
4	1	$ (100,000)								
5	2	$ (100,000)			Interest rate	6.60206%				
6	3	$ (100,000)								
7	4	$ (100,000)			FV of Option 1	$ 1,962,870.01				
8	5	$ (100,000)								
9	6	$ (100,000)			FV of Option 2	$1,962,870.01		=FV(F5,7,-100000)*(1+F5)^(13)		
10	7	$ (100,000)								
11	8		$ (100,000)		Target cell	$ (0.00)		=FV(F5,13,-100000)		
12	9		$ (100,000)							
13	10		$ (100,000)							
14	11		$ (100,000)							
15	12		$ (100,000)							
16	13		$ (100,000)			Goal Seek				
17	14		$ (100,000)							
18	15		$ (100,000)			Set cell: F11				
19	16		$ (100,000)			To value: 0				
20	17		$ (100,000)			By changing cell: F5				
21	18		$ (100,000)							
22	19		$ (100,000)			OK Cancel				
23	20		$ (100,000)							
24										

Figure 3.44 Using the Goal Seek function in Excel to find the break-even interest rate (Example 3.27). As soon as you select OK you will see that Goal Seek recalculates your formula. You then have two options, OK or Cancel. If you select OK the new term will be inserted into your worksheet. If you select Cancel, the Goal Seek box will disappear, and your worksheet will be in its original state.

SUMMARY

- Money has time value because it can earn more money over time. A number of terms involving the time value of money were introduced in this chapter:

  Interest is the cost of money. More specifically, it is a cost to the borrower and an earning to the lender, above and beyond the initial sum borrowed or loaned.

  Interest rate is a percentage periodically applied to a sum of money to determine the amount of interest to be added to that sum.

  Simple interest is the practice of charging an interest rate only to the initial sum.

  Compound interest is the practice of charging an interest rate to the initial sum *and* to any previously accumulated interest that has not been withdrawn from the initial sum. Compound interest is by far the most commonly used system in the real world.

  Economic equivalence exists between individual cash flows and/or patterns of cash flows that have the same worth. Even though the amounts and timing of the cash flows may differ, the appropriate interest rate may make them equivalent.

- The compound-interest formula, perhaps the single most important equation in this text, is

$$F = P(1 + i)^N,$$

  where P is the present sum, i is the interest rate, N is the number of periods for which interest is compounded, and F is the resulting future sum. All other important interest formulas are derived from this one.

- **Cash flow diagrams** are visual representations of cash inflows and outflows along a timeline. They are particularly useful for helping us detect which of the following five patterns of cash flow is represented by a particular problem:

  1. **Single payment.** A single present or future cash flow.

  2. **Uniform series.** A series of cash flows of equal amounts at regular intervals.

  3. **Linear gradient series.** A series of cash flows increasing or decreasing by a fixed amount at regular intervals.

  4. **Geometric gradient series.** A series of cash flows increasing or decreasing by a fixed percentage at regular intervals.

  5. **Irregular series.** A series of cash flows exhibiting no overall pattern. However, patterns might be detected for portions of the series.

- **Cash flow patterns** are significant because they allow us to develop **interest formulas**, which streamline the equivalence calculations. Table 3.4 summarizes the important interest formulas that form the foundation for all other analyses you will conduct in engineering economic analysis.

PROBLEMS

3.1   You deposit $5000 in a savings account that earns 8% simple interest per year. What is the minimum number of years you must wait to double your balance? Suppose instead that you deposit the $5000 in another savings account that earns

7% interest compounded yearly. How many years will it take now to double your balance?

3.2 Compare the interest earned by $1000 for five years at 8% simple interest with that earned by the same amount for five years at 8% compounded annually.

3.3 You are considering investing $3000 at an interest rate of 8% compounded annually for five years or investing the $3000 at 9% per year simple interest for five years. Which option is better?

3.4 You are about to borrow $10,000 from a bank at an interest rate of 9% compounded annually. You are required to make five equal annual repayments in the amount of $2571 per year, with the first repayment occurring at the end of year 1. Show the interest payment and principal payment in each year.

3.5 Suppose you have the alternative of receiving either $12,000 at the end of five years or P dollars today. Currently you have no need for money, so you would deposit the P dollars in a bank that pays 5% interest. What value of P would make you indifferent in your choice between P dollars today and the promise of $12,000 at the end of five years?

3.6 Suppose that you are obtaining a personal loan from your uncle in the amount of $20,000 (now) to be repaid in two years to cover some of your college expenses. If your uncle usually earns 8% interest (annually) on his money, which is invested in various sources, what minimum lump-sum payment two years from now would make your uncle happy?

3.7 What will be the amount accumulated by each of these present investments?
(a) $5000 in 8 years at 5% compounded annually
(b) $2250 in 12 years at 3% compounded annually
(c) $8000 in 31 years at 7% compounded annually
(d) $25,000 in 7 years at 9% compounded annually

3.8 What is the present worth of these future payments?
(a) $5500 6 years from now at 10% compounded annually
(b) $8000 15 years from now at 6% compounded annually
(c) $30,000 5 years from now at 8% compounded annually
(d) $15,000 8 years from now at 12% compounded annually

3.9 For an interest rate of 13% compounded annually, find
(a) How much can be lent now if $10,000 will be repaid at the end of five years?
(b) How much will be required in four years to repay a $25,000 loan received now?

3.10 How many years will it take an investment to triple itself if the interest rate is 12% compounded annually?

3.11 You bought 300 shares of Microsoft (MSFT) stock at $2,600 on December 31, 2005. Your intention is to keep the stock until it doubles in value. If you expect 15% annual growth for MSFT stock, how many years do you anticipate holding onto the stock? Compare your answer with the solution obtained by the Rule of 72 (discussed in Example 3.10).

3.12 From the interest tables in the text, determine the values of the following factors by interpolation, and compare your answers with those obtained by evaluating the F/P factor or the P/F factor:

(a) The single-payment compound-amount factor for 38 periods at 9.5% interest

(b) The single-payment present-worth factor for 47 periods at 8% interest

3.13 If you desire to withdraw the following amounts over the next five years from a savings account that earns 8% interest compounded annually, how much do you need to deposit now?

N	Amount
2	$32,000
3	43,000
4	46,000
5	28,000

3.14 If $1500 is invested now, $1800 two years from now, and $2000 four years from now at an interest rate of 6% compounded annually, what will be the total amount in 15 years?

3.15 A local newspaper headline blared, "Shawn Smith Signed for $30 Million." A reading of the article revealed that on April 1, 2009, Shawn Smith, the former record-breaking centre from Hockey University, signed a $30 million package with the Edmonton Ice Knights. The terms of the contract were $3 million immediately, $2.4 million per year for the first five years (with the first payment after one year), and $3 million per year for the next five years (with the first payment at year 6). If Shawn's interest rate is 8% per year, what would his contract be worth at the time he signs it?

3.16 How much invested now at 6% would be just sufficient to provide three payments, with the first payment in the amount of $7000 occurring two years hence, then $6000 five years hence, and finally $5000 seven years hence?

3.17 What is the future worth of a series of equal deposits of $1000 for 10 years in a savings account that earns 7% annual interest if

(a) All deposits are made at the *end* of each year?

(b) All deposits are made at the *beginning* of each year?

3.18 What is the future worth of the following series of payments?

(a) $3000 at the end of each year for 5 years at 7% compounded annually

(b) $4000 at the end of each year for 12 years at 8.25% compounded annually

(c) $5000 at the end of each year for 20 years at 9.4% compounded annually

(d) $6000 at the end of each year for 12 years at 10.75% compounded annually

3.19 What equal annual series of payments must be paid into a sinking fund to accumulate the following amounts?

(a) $22,000 in 13 years at 6% compounded annually

(b) $45,000 in 8 years at 7% compounded annually

(c) $35,000 in 25 years at 8% compounded annually

(d) $18,000 in 8 years at 14% compounded annually

3.20 Part of the income that a machine generates is put into a sinking fund to replace the machine when it wears out. If $1500 is deposited annually at 7% interest, how many years must the machine be kept before a new machine costing $30,000 can be purchased?

3.21 A no-load (commission-free) mutual fund has grown at a rate of 11% compounded annually since its beginning. If it is anticipated that it will continue to grow at that rate, how much must be invested every year so that $15,000 will be accumulated at the end of five years?

3.22 What equal annual payment series is required to repay the following present amounts?

(a) $10,000 in 5 years at 5% interest compounded annually

(b) $5500 in 4 years at 9.7% interest compounded annually

(c) $8500 in 3 years at 2.5% interest compounded annually

(d) $30,000 in 20 years at 8.5% interest compounded annually

3.23 You have borrowed $25,000 at an interest rate of 16%. Equal payments will be made over a three-year period. (The first payment will be made at the end of the first year.) What will the annual payment be, and what will the interest payment be for the second year?

3.24 What is the present worth of the following series of payments?

(a) $800 at the end of each year for 12 years at 5.8% compounded annually

(b) $2500 at the end of each year for 10 years at 8.5% compounded annually

(c) $900 at the end of each year for 5 years at 7.25% compounded annually

(d) $5500 at the end of each year for 8 years at 8.75% compounded annually

3.25 From the interest tables in Appendix A, determine the values of the following factors by interpolation and compare your results with those obtained from evaluating the *A/P* and *P/A* interest formulas:

(a) The capital recovery factor for 38 periods at 6.25% interest

(b) The equal payment series present-worth factor for 85 periods at 9.25% interest

3.26 An individual deposits an annual bonus into a savings account that pays 8% interest compounded annually. The size of the bonus increases by $2000 each year, and the initial bonus amount was $5000. Determine how much will be in the account immediately after the fifth deposit.

3.27 Five annual deposits in the amounts of $3000, $2500, $2000, $1500, and $1000, in that order, are made into a fund that pays interest at a rate of 7% compounded annually. Determine the amount in the fund immediately after the fifth deposit.

3.28 Compute the value of *P* in the accompanying cash flow diagram, assuming that $i = 9\%$.

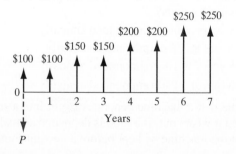

3.29 What is the equal payment series for 12 years that is equivalent to a payment series of $15,000 at the end of the first year, decreasing by $1000 each year over 12 years? Interest is 8% compounded annually.

3.30 Suppose that an oil well is expected to produce 100,000 barrels of oil during its first year in production. However, its subsequent production (yield) is expected to decrease by 10% over the previous year's production. The oil well has a proven reserve of 1,000,000 barrels.

(a) Suppose that the price of oil is expected to be $60 per barrel for the next several years. What would be the present worth of the anticipated revenue stream at an interest rate of 12% compounded annually over the next seven years?

(b) Suppose that the price of oil is expected to start at $60 per barrel during the first year, but to increase at the rate of 5% over the previous year's price. What would be the present worth of the anticipated revenue stream at an interest rate of 12% compounded annually over the next seven years?

(c) Consider part (b) again. After three years' production, you decide to sell the oil well. Assume that the well's remaining useful life is four more years. What would be a fair price?

3.31 A city engineer has estimated the annual toll revenues from a newly proposed highway construction over 20 years as follows:

$$A_n = (\$2{,}000{,}000)(n)(1.06)^{n-1},$$

$$n = 1, 2, \ldots, 20.$$

To validate the bond, the engineer was asked to present the estimated total present value of toll revenue at an interest rate of 6%. Assuming annual compounding, find the present value of the estimated toll revenue.

3.32 What is the amount of 10 equal annual deposits that can provide five annual withdrawals when a first withdrawal of $5000 is made at the end of year 11 and subsequent withdrawals increase at the rate of 8% per year over the previous year's withdrawal if

(a) The interest rate is 9% compounded annually?

(b) The interest rate is 6% compounded annually?

3.33 By using only those factors given in interest tables, find the values of the factors that follow, which are not given in your tables. Show the relationship between the factors by using factor notation, and calculate the value of the factor. Then compare the solution you obtained by using the factor formulas with a direct calculation of the factor values.

Example: $(F/P, 8\%, 38) = (F/P, 8\%, 30)(F/P, 8\%, 8) = 18.6253$

(a) $(P/F, 8\%, 67)$

(b) $(A/P, 8\%, 42)$

(c) $(P/A, 8\%, 135)$

3.34 Prove the following relationships among interest factors:

(a) $(F/P, i, N) = i(F/A, i, N) + 1$

(b) $(P/F, i, N) = 1 - (P/A, i, N)i$

(c) $(A/F, i, N) = (A/P, i, N) - i$

(d) $(A/P, i, N) = i/[1 - (P/F, i, N)]$

(e) $(P/A, i, N \rightarrow \infty) = 1/i$

(f) $(A/P, i, N \rightarrow \infty) = i$

(g) $(P/G, i, N \rightarrow \infty) = 1/i^2$

3.35 Find the present worth of the cash receipts where $i = 12\%$ compounded annually with only four interest factors.

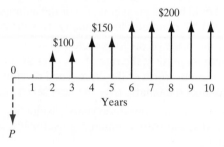

3.36 Find the equivalent present worth of the cash receipts where $i = 8\%$. In other words, how much do you have to deposit now (with the second deposit in the amount of $200 at the end of the first year) so that you will be able to withdraw $200 at the end of second year, $120 at the end of third year, and so forth if the bank pays you a 8% annual interest on your balance?

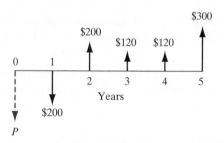

3.37 What value of A makes two annual cash flows equivalent at 13% interest compounded annually?

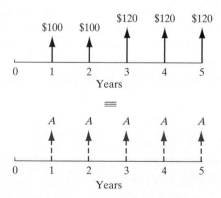

3.38 The two cash flow transactions shown in the accompanying cash flow diagram are said to be equivalent at 6% interest compounded annually. Find the unknown value of X that satisfies the equivalence.

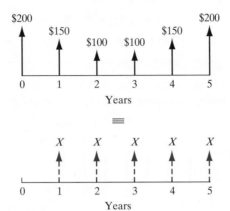

3.39 Solve for the present worth of this cash flow using at most three interest factors at 10% interest compounded annually.

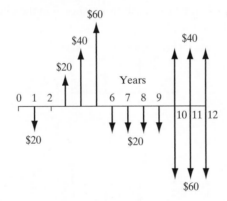

3.40 From the accompanying cash flow diagram, find the value of C that will establish the economic equivalence between the deposit series and the withdrawal series at an interest rate of 8% compounded annually.

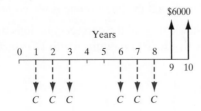

(a) $1335
(b) $862
(c) $1283
(d) $828

3.41 The following equation describes the conversion of a cash flow into an equivalent equal payment series with $N = 10$:

$$A = [800 + 20(A/G, 6\%, 7)]$$
$$\times \ (P/A, 6\%, 7)(A/P, 6\%, 10)$$
$$+ \ [300(F/A, 6\%, 3) - 500](A/F, 6\%, 10)$$

Reconstruct the original cash flow diagram.

3.42 Consider the cash flow shown in the accompanying diagram. What value of C makes the inflow series equivalent to the outflow series at an interest rate of 10%?

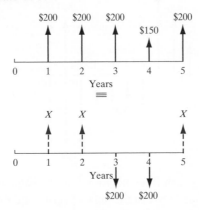

3.43 Find the value of X so that the two cash flows shown in the diagram are equivalent for an interest rate of 8%.

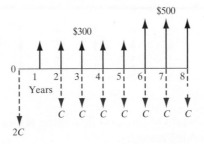

3.44 What single amount at the end of the fifth year is equivalent to a uniform annual series of $3000 per year for 10 years if the interest rate is 9% compounded annually?

3.45 From the following list, identify all the correct equations used in computing either the equivalent present worth (P) or future worth (F) for the cash flow shown at $i = 10\%$.

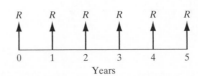

(1) $P = R(P/A, 10\%, 6)$
(2) $P = R + R(P/A, 10\%, 5)$
(3) $P = R(P/F, 10\%, 5) + R(P/A, 10\%, 5)$
(4) $F = R(F/A, 10\%, 5) + R(F/P, 10\%, 5)$
(5) $F = R + R(F/A, 10\%, 5)$
(6) $F = R(F/A, 10\%, 6)$
(7) $F = R(F/A, 10\%, 6) - R$

3.46 On the day his baby was born, a father decided to establish a savings account for the child's university education. Any money that is put into the account will earn an interest rate of 8% compounded annually. The father will make a series of annual deposits in equal amounts on each of his child's birthdays from the 1st through the 18th, so that the child can make four annual withdrawals from the account in the amount of $30,000 on each birthday. Assuming that the first withdrawal will be made on the child's 18th birthday, which of the following equations are correctly used to calculate the required annual deposit?

(1) $A = (\$30{,}000 \times 4)/18$
(2) $A = \$30{,}000(F/A, 8\%, 4) \times (P/F, 8\%, 21)(A/P, 8\%, 18)$
(3) $A = \$30{,}000(P/A, 8\%, 18) \times (F/P, 8\%, 21)(A/F, 8\%, 4)$
(4) $A = [\$30{,}000(P/A, 8\%, 3) + \$30{,}000](A/F, 8\%, 18)$
(5) $A = \$30{,}000[(P/F, 8\%, 18) + (P/F, 8\%, 19) + (P/F, 8\%, 20) + (P/F, 8\%, 21)](A/P, 8\%, 18)$

3.47 Find the equivalent equal payment series (A) using an A/G factor such that the two cash flows are equivalent at 10% compounded annually.

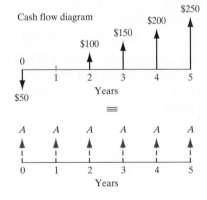

Cash flow diagram

3.48 Consider the following cash flow:

Year End	Payment
0	$500
1–5	$1000

In computing F at the end of year 5 at an interest rate of 12%, which of the following equations is *in*correct?

(a) $F = \$1000(F/A, 12\%, 5) - \$500(F/P, 12\%, 5)$

(b) $F = \$500(F/A, 12\%, 6) + \$500(F/A, 12\%, 5)$

(c) $F = [\$500 + \$1000(P/A, 12\%, 5)] \times (F/P, 12\%, 5)$

(d) $F = [\$500(A/P, 12\%, 5) + \$1000] \times (F/A, 12\%, 5)$

3.49 Consider the cash flow series given. In computing the equivalent worth at $n = 4$, which of the following equations is *in*correct?

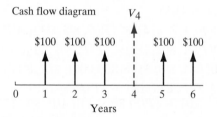

(a) $V_4 = [\$100(P/A, i, 6) - \$100(P/F, i, 4)](F/P, i, 4)$

(b) $V_4 = \$100(F/A, i, 3) + \$100(P/A, i, 2)$

(c) $V_4 = \$100(F/A, i, 4) - \$100 + \$100(P/A, i, 2)$

(d) $V_4 = [\$100(F/A, i, 6) - \$100(F/P, i, 2)](P/F, i, 2)$

3.50 Henry Cisco is planning to make two deposits: $25,000 now and $30,000 at the end of year 6. He wants to withdraw C each year for the first six years and $(C + \$1000)$ each year for the next six years. Determine the value of C if the deposits earn 10% interest compounded annually.

(a) $7711

(b) $5794

(c) $6934

(d) $6522

3.51 At what rate of interest compounded annually will an investment double itself in five years?

3.52 Determine the interest rate (i) that makes the pairs of cash flows shown economically equivalent.

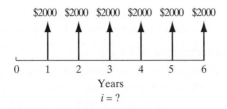

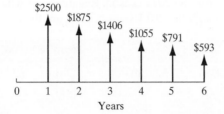

3.53 You have $10,000 available for investment in stock. You are looking for a growth stock whose value can grow to $35,000 over five years. What kind of growth rate are you looking for?

3.54 How long will it take to save $1 million if you invest $2000 each year at 6%?

Short Case Studies

ST3.1 Read the following letter from a magazine publisher:

Dear Parent:

Currently your *Better Parenting* subscription will expire with your 24-month issue. To renew on an annual basis until your child reaches 72 months old would cost you a total of $71.80 ($17.95 per year). We feel it is so important for you to continue receiving this material until the 72nd month that we offer you an opportunity to renew now for $57.44. Not only is this a savings of 20% over the regular rate, but it is an excellent inflation hedge for you against increasing rates in the future. Please act now by sending $57.44.

(a) If your money is worth 6% per year, determine whether this offer can be of any value (assuming that the subscription cost does not increase over the next three years).

(b) What rate of interest would make you indifferent between the two renewal options?

ST3.2 Millionaire Life is a promotional lottery game offered across Canada (see British Columbia Lottery Corporation at www.bclc.com). The prizes included a top prize of $1 million a year for 25 years, four prizes of $1 million, 36 prizes of $100,000, 16 prizes of $1000, and 180,507 prizes of $20. The winner of the top prize had the option of receiving $1 million a year for 25 years or a single lump sum of $17 million. All other prizes were cash prizes. Suppose that the remaining lottery proceeds can be invested to generate an interest rate of 6% per year.

(a) If the winner of the top prize selected to receive $1 million a year for 25 years, the first payment occurred right away, and the last payment occurs 24 years later, what is the balance of lottery proceeds right after the last payout?

(b) If the winner of the top prize selected to receive a single lump sum of $17 million, what is the balance of lottery proceeds 24 years after this payout?

ST3.3 A local newspaper carried the following story:

Toronto Wolves defenceman John Young will earn either $11,406,000 over 12 years or $8,600,000 over 6 years. Young must declare which plan he prefers. The $11 million package is deferred through the year 2017, while the nondeferred arrangement ends after the 2011 season. Regardless of which plan is chosen, Young will be playing through the 2011 season. Here are the details of the two plans:

Deferred Plan		Nondeferred Plan	
2006	$2,000,000	2006	$2,000,000
2007	566,000	2007	900,000
2008	920,000	2008	1,000,000
2009	930,000	2009	1,225,000
2010	740,000	2010	1,500,000
2011	740,000	2011	1,975,000
2012	740,000		
2013	790,000		
2014	540,000		
2015	1,040,000		
2016	1,140,000		
2017	1,260,000		
Total	$11,406,000	Total	$8,600,000

(a) As it happened, Young ended up with the nondeferred plan. In retrospect, if Young's interest rate was 6%, did he make a wise decision in 2006?

(b) At what interest rate would the two plans be economically equivalent?

ST3.4 Fairmont Textile has a plant in which employees have been having trouble with carpal tunnel syndrome (CTS, an inflammation of the nerves that pass through the carpal tunnel, a tight space at the base of the palm), resulting from long-term repetitive activities, such as years of sewing. It seems as if 15 of the employees working in this facility developed signs of CTS over the last five years. Healthco, the company's insurance firm, has been increasing Fairmont's liability insurance steadily because of this problem. Healthco is willing to lower the insurance premiums to $16,000 a year (from the current $30,000 a year) for the next five years if Fairmont implements an acceptable CTS-prevention program that includes making the employees aware of CTS and how to reduce the chances of it developing. What would be the maximum amount that Fairmont should invest in the program to make it worthwhile? The firm's interest rate is 12% compounded annually.

ST3.5 Kersey Manufacturing Co., a small fabricator of plastics, needs to purchase an extrusion moulding machine for $120,000. Kersey will borrow money from a bank at an interest rate of 9% over five years. Kersey expects its product sales to be slow during the first year, but to increase subsequently at an annual rate of 10%. Kersey therefore arranges with the bank to pay off the loan on a "balloon scale," which results in the lowest payment at the end of the first year and each subsequent payment being just 10% over the previous one. Determine the five annual payments.

ST3.6 In March 2005, Adidas put on sale what it billed as the world's first computerized "smart shoe." But consumers would decide whether to accept the bionic running shoe's $250 price tag—four times the average shoe price at stores such as Foot Locker. Adidas used a sensor, a microprocessor, and a motorized cable system to automatically adjust the shoe's cushioning. The sensor under the heel measures compression and decides whether the shoe is too soft or firm. That information is sent to the microprocessor and, while the shoe is in the air, the cable adjusts the heel cushion. The whole system weighs fewer than 40 grams. Adidas's computer-driven shoe—three years in the making—was the latest innovation in the $16.4 billion U.S. sneaker industry. The top-end running shoe from New Balance listed for $199.99. With runners typically replacing shoes by 800 kilometres, the $250 Adidas could push costs to 31 cents per kilometre. Adidas spent an estimated $20 million on the rollout.[4]

The investment required to develop a full-scale commercial rollout cost Adidas $70 million (including the $20 million ad campaign), which would be financed at an interest rate of 10%. With a price tag of $250 per pair of shoes, Adidas would have about $100 net cash profit from each sale. The product would have a five-year market life. Assuming that the annual demand for the product would remain constant over the market life, how many units does Adidas have to sell each year to pay off the initial investment and interest?

ST3.7 Recently a National Hockey League player agreed to an eight-year, $50 million contract that at the time made him one of the highest paid young players in professional hockey history. The contract included a signing bonus of $11 million. The agreement called for annual salaries of $2.5 million in 2008, $1.75 million in 2009, $4.15 million in 2010, $4.90 million in 2011, $5.25 million in 2012, $6.2 million in 2013, $6.75 million in 2014, and $7.5 million in 2015. The $11 million signing bonus was prorated over the course of the contract, so that an additional $1.375 million was paid each year over the eight-year contract period. Table ST3.7 shows the net annual payment schedule, with the salary paid at the beginning of each season.

(a) How much was the player's contract actually worth at the time of signing?

(b) For the signing bonus portion, suppose that the player was allowed to take either the prorated payment option as just described or a lump-sum payment option in the amount of $8 million at the time he signed the contract. Should he have taken the lump-sum option instead of the prorated one? Assume that his interest rate is 6%.

---

[4] Michael McCarthy, "Adidas puts computer on new footing," *USA Today*, March 3, 2005, section 5B.

TABLE ST3.7 Net Annual Payment Schedule

Beginning of Season	Prorated Contract Salary	Actual Signing Bonus	Annual Payment
2008	$2,500,000	$1,375,000	$3,875,000
2009	1,750,000	1,375,000	3,125,000
2010	4,150,000	1,375,000	5,525,000
2011	4,900,000	1,375,000	6,275,000
2012	5,250,000	1,375,000	6,625,000
2013	6,200,000	1,375,000	7,575,000
2014	6,750,000	1,375,000	8,125,000
2015	7,500,000	1,375,000	8,875,000

ST3.8 Payday is a lottery game offered in many provinces and territories in Canada (for example, see Western Canada Lottery Corporation at www.wclc.com). The grand prize of the game is $1000 a week for life or a single cash payment prize of $675,000. There may be multiple winners of the grand prize in each draw. However, the game condition limits the number of $1000 A Week for Life (or the $675,000 cash payout) prizes per draw to a total of four prizes. For each draw, should the aggregate number of grand prize winners exceed four, each winner in the grand prize category will be paid a single cash amount determined by dividing $2,700,000 by the number of winning sets of numbers in this grand prize category for that draw. Analyze the trade-offs between the two options of the grand prize, namely, a single cash prize of $675,000 or $1000 a week for life (you can assume that the annuity payment is $52,000 per year for life). How does a player choose between these two options depending on his or her age? How does interest rate affect the decision?

On the Companion Website that accompanies this text, you will find Excel templates and exercises, as well as the following analysis tools: Cash Flow Analyzer, Depreciation Analysis, Loan Analysis, and Interest Tables.

FOUR

Understanding Money and Its Management

Don't Handcuff Your Mortgage. [1] Would you like to pay an extra $300 a month on your mortgage? Unlikely. But that has not stopped a number of Canadians with the deal of a lifetime on a variable-rate mortgage from switching over to a more expensive fixed-rate mortgage. A fear of rising mortgage interest rates is driving this rash decision.

The interest rate charged on a variable-rate mortgage is tied to the prime rate. The prime rate is the one used by banks for lending to the most creditworthy borrowers. The prime rate is adjusted from time to time. The prime rate at TD Canada Trust was 2.25% effective April 22, 2009. Before summer 2008, many variable-rate mortgages were offered at prime minus a set percentage. At one time, some banks offered variable-rate mortgages at prime minus 0.90%. With a prime rate of 2.25%, the interest rate charged on such a variable-rate mortgage was only 1.35%. In summer 2008, the variable-rate mortgage changed from a Prime Minus offering to a Prime Plus offering due to credit crunch and the U.S. credit crisis. The best variable-rate mortgage one can get these days is prime plus 0.60%, or 2.85% with the prime rate at 2.25%.

Changing Rates		
	Five-Year Fixed Mortgage Rate	Prime Rate
February 2008	5.89%	5.75%
August 2008	5.65%	4.75%
February 2009	4.29%	3.00%
June 2009	4.00%	2.25%

[1] Garry Marr, "Don't Handcuff Your Mortgage," *Financial Post*, June 13. 2009.

In contrast to variable-rate mortgages, a fixed-rate mortgage charges a fixed interest rate over a specified period of time. For example, the current five-year fixed-rate mortgage carries an interest rate of 4%. A fixed-rate mortgage provides the protection that this rate is guaranteed even though the prime rate may continue to rise. However, note that there is still a huge difference between the variable rate of 1.35% and the fixed rate of 4%. Even the prime rate has increased recently; if you have the deal of prime minus 0.90% in your pocket, what is the rush of converting your variable-rate mortgage to a fixed-rate mortgage?

The average selling price of a home in May 2009 in Canada was $306,366. Based on a 20% down payment and a 25-year amortization, your monthly payment would be $962.61 at the 1.35% interest rate. If you convert it to a five-year fixed rate mortgage with an interest rate of 4%, your monthly payment will jump to $1289.24, which is an increase of more than $300 in your monthly mortgage payment!

In this chapter, we will consider several concepts crucial to managing money. In Chapter 3, we examined how time affects the value of money, and we developed various interest formulas for that purpose. Using these basic formulas, we will now extend the concept of equivalence to determine interest rates that are implicit in many financial contracts. To this end, we will introduce several examples in the area of loan transactions. For example, many commercial loans require that interest compound more frequently than once a year—for instance, monthly or quarterly. To consider the effect of more frequent compounding, we must begin with an understanding of the concepts of nominal and effective interest.

CHAPTER LEARNING OBJECTIVES

After completing this chapter, you should understand the following concepts:

- The difference between the nominal interest rate and the effective interest rate.
- The procedure for computing the effective interest rate, based on a payment period.
- How commercial loans and mortgages are structured in terms of interest and principal payments.
- The basics of investing in bonds.

4.1 Nominal and Effective Interest Rates

In all the examples in Chapter 3, the implicit assumption was that payments are received *once a year*, or *annually*. However, some of the most familiar financial transactions, both personal and in engineering economic analysis, involve payments that are not based on one annual payment—for example, monthly mortgage payments and quarterly earnings on savings accounts. Thus, if we are to compare different cash flows with different compounding periods, we need to evaluate them on a common basis. This need has led to the development of the concepts of the **nominal interest rate** and the **effective interest rate**.

4.1.1 Nominal Interest Rates

Take a closer look at the billing statement from any of your credit cards. Or if you financed a new car recently, examine the loan contract. You will typically find the interest that the bank charges on your unpaid balance. Even if a financial institution uses a unit of time other than a year—say, a month or a quarter (e.g., when calculating interest payments)—the institution usually quotes the interest rate on an *annual basis*. Many banks, for example, state the interest arrangement for credit cards in this way:

18% compounded monthly.

This statement simply means that each month the bank will charge 1.5% interest on an unpaid balance. We say that 18% is the **nominal interest rate** or **annual percentage rate (APR)**, and the compounding frequency is monthly (or 12 times a year). As shown in Figure 4.1, to obtain the interest rate per compounding period, we divide, for example, 18% by 12, to get 1.5% per month.

> **Annual percentage rate (APR)** is the yearly cost of a loan, including interest, insurance, and the origination fee, expressed as a percentage.

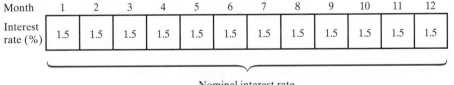

Figure 4.1 The nominal interest rate is determined by summing the individual interest rates per period.

Although the annual percentage rate, or APR, is commonly used by financial institutions and is familiar to many customers, the APR does not explain precisely the amount of interest that will accumulate in a year. To explain the true effect of more frequent compounding on annual interest amounts, we will introduce the term *effective interest rate*, commonly known as *annual effective yield*, or **annual percentage yield (APY)**.

Annual percentage yield (APY) is the rate actually earned or paid in one year, taking into account the affect of compounding.

4.1.2 Effective Annual Interest Rates

The **effective annual interest rate** is the rate that truly represents the interest earned in a year. For instance, in our credit card example, the bank will charge 1.5% interest on any unpaid balance at the end of each month. Therefore, the 1.5% rate represents the effective interest rate per month. On a yearly basis, you are looking for a cumulative rate—1.5% each month for 12 months. This cumulative rate predicts the actual interest payment on your outstanding credit card balance.

Suppose you deposit $10,000 in a savings account that pays you at an interest rate of 9% compounded quarterly. Here, 9% represents the nominal interest rate, and the interest rate per quarter is 2.25% (9%/4). The following is an example of how interest is compounded when it is paid quarterly:

End of Period	Base Amount	Interest Earned 2.25% × (Base Amount)	New Base
First quarter	$10,000.00	2.25% × $10,000.00 = $225.00	$10,225.00
Second quarter	$10,225.00	2.25% × $10,225.00 = $230.06	$10,455.06
Third quarter	$10,455.06	2.25% × $10,455.06 = $225.24	$10,690.30
Fourth quarter	$10,690.30	2.25% × $10,690.30 = $240.53	$10,930.83

Clearly, you are earning more than 9% per year on your original deposit. In fact, you are earning 9.3083% per year ($930.83/$10,000). We could calculate the total annual interest payment for a principal amount of $10,000 with the formula given in Eq. (3.3). If $P = \$10,000$, $i = 2.25\%$, and $N = 4$, we obtain

$$F = P(1 + i)^N$$

$$= \$10,000(1 + 0.0225)^4$$

$$= \$10,930.83.$$

The implication is that, for each dollar deposited, you are earning an equivalent annual interest of 9.38 cents. In terms of an effective annual interest rate (i_a), the interest payment can be rewritten as a percentage of the principal amount:

$$i_a = \$930.83/\$10,000 = 0.093083, \text{ or } 9.3083\%.$$

In other words, earning 2.25% interest per quarter for four quarters is equivalent to earning 9.3083% interest just one time each year.

Table 4.1 shows effective interest rates at various compounding intervals for 4%–12% APR. As you can see, depending on the frequency of compounding, the effective interest earned or paid by the borrower can differ significantly from the APR. Therefore, truth-in-lending laws require that financial institutions quote both the nominal interest rate and the compounding frequency (i.e., the effective interest) when you deposit or borrow money.

Certainly, more frequent compounding increases the amount of interest paid over a year at the same nominal interest rate. Assuming that the nominal interest rate is r, and M compounding periods occur during the year, we can calculate the effective annual interest rate

$$i_a = \left(1 + \frac{r}{M}\right)^M - 1. \tag{4.1}$$

When $M = 1$, we have the special case of annual compounding. Substituting $M = 1$ in Eq. (4.1) reduces it to $i_a = r$. That is, when compounding takes place once annually, the effective interest is equal to the nominal interest. Thus, in the examples in Chapter 3, in which only annual interest was considered, we were, by definition, using effective interest rates.

TABLE 4.1 Nominal and Effective Interest Rates With Different Compounding Periods

	Effective Rates				
Nominal Rate	Compounding Annually	Compounding Semiannually	Compounding Quarterly	Compounding Monthly	Compounding Daily
4%	4.00%	4.04%	4.06%	4.07%	4.08%
5	5.00	5.06	5.09	5.12	5.13
6	6.00	6.09	6.14	6.17	6.18
7	7.00	7.12	7.19	7.23	7.25
8	8.00	8.16	8.24	8.30	8.33
9	9.00	9.20	9.31	9.38	9.42
10	10.00	10.25	10.38	10.47	10.52
11	11.00	11.30	11.46	11.57	11.62
12	12.00	12.36	12.55	12.68	12.74

EXAMPLE 4.1 Determining the Compounding Frequency

The following table summarizes interest rates on several term deposits (TDs) and guaranteed investment certificates (GICs) offered by TD Canada Trust during December 2008:

Product	Minimum	Rate	APY*
3-Month TD	$5000	0.95%	0.95%
1-Year TD	$1000	0.95%	0.95%
1-Year Money Market GIC	$1000	1.50%	1.51%
1+1 GIC	$1000	1st year: 1.80% 2nd year: 5.00%	3.388%
2-Year Premium Rate Redeemable GIC	$1000	3.50%	3.53%

*Annual percentage yield = effective annual interest rate (i_a).

In the table, no mention is made of specific compounding frequencies. Take the 2-Year Premium Rate Redeemable GIC as an example: (a) find the compounding frequency assumed; (b) find the total balance two years later for a deposit amount of $100,000.

SOLUTION

Given: $r = 3.50\%$ per year, i_a (APY) = 3.53%, $P = \$100,000$, and $N = 2$ years.
Find: M and the balance at the end of two years.

(a) The nominal interest rate is 3.50% per year, and the effective annual interest rate (yield) is 3.53%. Using Eq. (4.1), we obtain the expression

$$0.0353 = \left(1 + \frac{0.0350}{M}\right)^M - 1,$$

or

$$1.0353 = \left(1 + \frac{0.0350}{M}\right)^M.$$

By trial and error, we find that $M = 2$, which indicates semiannual compounding. Thus, the 2-Year Premium Rate Redeemable GIC earns 3.50% interest compounded semiannually.

Normally, if the GIC is not cashed at maturity, it will be renewed automatically at a new going interest rate. Similarly, we can find the compounding frequencies for the other TDs and GICs.

(b) If you purchase the 2-Year Premium Rate Redeemable GIC, it will earn 3.5% interest compounded semi-annually. This means that your GIC earns an effective annual interest of 3.53%:

$$F = P(1 + i_a)^N$$
$$= \$100,000(1 + 0.0353)^2$$
$$= \$100,000\left(1 + \frac{0.035}{2}\right)^{2\times2}$$
$$= \$107,185.$$

4.1.3 Effective Interest Rates per Payment Period

We can generalize the result of Eq. (4.1) to compute the effective interest rate for periods of *any duration*. As you will see later, the effective interest rate is usually computed on the basis of the payment (transaction) period. For example, if cash flow transactions occur quarterly, but interest is compounded monthly, we may wish to calculate the effective interest rate on a quarterly basis. To do this, we may redefine Eq. (4.1) as

$$i = \left(1 + \frac{r}{M}\right)^C - 1$$

$$= \left(1 + \frac{r}{CK}\right)^C - 1, \tag{4.2}$$

where

M = the number of compounding periods per year,

C = the number of compounding periods per payment period, and

K = the number of payment periods per year.

Note that $M = CK$ in Eq. (4.2). For the special case of annual payments with annual compounding, we obtain $i = i_a$ with $C = M$ and $K = 1$.

EXAMPLE 4.2 Effective Rate per Payment Period

Suppose that you make quarterly deposits in a savings account that earns 9% interest compounded monthly. Compute the effective interest rate per quarter.

SOLUTION

Given: $r = 9\%$, C = three compounding periods per quarter, K = four quarterly payments per year, and $M = 12$ compounding periods per year.
Find: i.

Using Eq. (4.2), we compute the effective interest rate per quarter as

$$i = \left(1 + \frac{0.09}{12}\right)^3 - 1$$

$$= 2.27\%.$$

COMMENTS: Figure 4.2 illustrates the relationship between the nominal and effective interest rates per payment period.

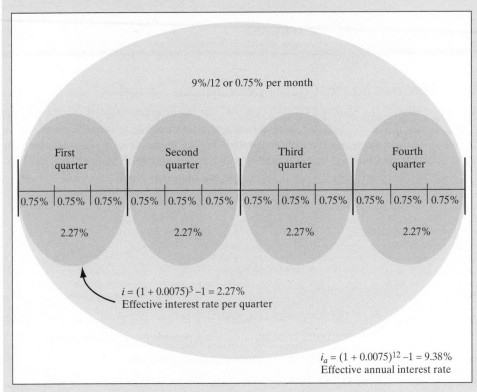

Figure 4.2　Functional relationships among r, i, and i_a, where interest is calculated based on 9% compounded monthly and payments occur quarterly (Example 4.2).

Continuous compounding: The process of calculating interest and adding it to existing principal and interest at infinitely short time intervals.

4.1.4 Continuous Compounding

To be competitive on the financial market or to entice potential depositors, some financial institutions offer frequent compounding. As the number of compounding periods (M) becomes very large, the interest rate per compounding period (r/M) becomes very small. As M approaches infinity and r/M approaches zero, we approximate the situation of **continuous compounding**.

By taking limits on both sides of Eq. (4.2), we obtain the effective interest rate per payment period as

$$i = \lim_{CK \to \infty} \left[\left(1 + \frac{r}{CK} \right)^C - 1 \right]$$

$$= \lim_{CK \to \infty} \left(1 + \frac{r}{CK} \right)^C - 1$$

$$= (e^r)^{1/K} - 1.$$

Therefore, the effective interest rate per payment period is

$$i = e^{r/K} - 1. \tag{4.3}$$

To calculate the effective annual interest rate for continuous compounding, we set K equal to unity, resulting in

$$i_a = e^r - 1. \tag{4.4}$$

As an example, the effective annual interest rate for a nominal interest rate of 12% compounded continuously is $i_a = e^{0.12} - 1 = 12.7497\%$.

EXAMPLE 4.3 Calculating an Effective Interest Rate With Quarterly Payment

SOLUTION

Find the effective interest rate per *quarter* at a nominal rate of 8% compounded (a) quarterly, (b) monthly, (c) weekly, (d) daily, and (e) continuously.

Given: $r = 8\%, M, C,$ and $K = 4$ quarterly payments per year.
Find: i.

(a) Quarterly compounding:

   $r = 8\%, M = 4, C = 1$ compounding period per quarter, and $K = 4$ payments per year. Then

   $$i = \left(1 + \frac{0.08}{4} \right)^1 - 1 = 2.00\%.$$

1 compounding period

(b) Monthly compounding:

   $r = 8\%, M = 12, C = 3$ compounding periods per quarter, and $K = 4$ payments per year. Then

   $$i = \left(1 + \frac{0.08}{12} \right)^3 - 1 = 2.013\%.$$

3 compounding periods

(c) Weekly compounding:

$r = 8\%$, $M = 52$, $C = 13$ compounding periods per quarter, and $K = 4$ payments per year. Then

$$i = \left(1 + \frac{0.08}{52}\right)^{13} - 1 = 2.0186\%.$$

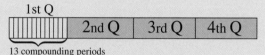

13 compounding periods

(d) Daily compounding:

$r = 8\%$, $M = 365$, $C = 91.25$ days per quarter, and $K = 4$. Then

$$i = \left(1 + \frac{0.08}{365}\right)^{91.25} - 1 = 2.0199\%.$$

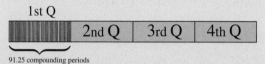

91.25 compounding periods

(e) Continuous compounding:

$r = 8\%$, $M \to \infty$, $C \to \infty$, and $K = 4$. Then, from Eq. (4.3),

$$i = e^{0.08/4} - 1 = 2.0201\%.$$

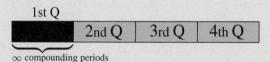

∞ compounding periods

COMMENTS: Note that the difference between daily compounding and continuous compounding is often negligible. Many banks offer continuous compounding to entice deposit customers, but the extra benefits are small.

4.2 Equivalence Calculations With Effective Interest Rates

All the examples in Chapter 3 assumed annual payments and annual compounding. However, a number of situations involve cash flows that occur at intervals that are not the same as the compounding intervals often used in practice. Whenever payment and compounding periods differ from each other, *one or the other must be transformed so that both conform to the same unit of time.* For example, if payments occur quarterly and compounding occurs monthly, the most logical procedure is to calculate the effective interest rate per quarter. By contrast, if payments occur monthly and compounding occurs quarterly, we may be able to find the equivalent monthly interest rate. The bottom line is that, to proceed with equivalency analysis, the compounding and payment periods must be in the same order.

4.2.1 When Payment Period Is Equal to Compounding Period

Whenever the compounding and payment periods are equal ($M = K$), whether the interest is compounded annually or at some other interval, the following solution method can be used:

1. Identify the number of compounding periods (M) per year.
2. Compute the effective interest rate per payment period—that is, using Eq. (4.2). Then, with $C = 1$ and $K = M$, we have

$$i = \frac{r}{M}.$$

3. Determine the number of compounding periods:

$$N = M \times (\text{number of years}).$$

EXAMPLE 4.4 Calculating Auto Loan Payments

Suppose you want to buy a car. You have surveyed the dealers' newspaper advertisements, and the following one has caught your attention:

University Graduate Special: New 2010 Nissan Altima 2.5S with 175 HP, CVT transmission package, A/C, and cruise control

MSRP:	$23,798
Dealer's discount:	$1,143
Manufacturer rebate	$800
university graduate cash:	$500
Sale price:	$21,355

MSRP: Manufacturer's Suggested Retail Price.

Price and payment is plus tax, title, customer service fee, with approved credit for 72 months at 6.25% APR.

You can afford to make a down payment of $6355 (and taxes and insurance as well), so the net amount to be financed is $15,000. What would the monthly payment be (Figure 4.3)?

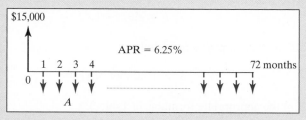

Figure 4.3   A car loan cash transaction (Example 4.4).

DISCUSSION: The advertisement does not specify a compounding period, but in automobile financing, the interest and the payment periods are almost always monthly. Thus, the 6.25% APR means 6.25% compounded monthly.

SOLUTION

Given: $P = \$15,000$, $r = 6.25\%$ per year, K = 12 payments per year, $N = 72$ months, and $M = 12$ compounding periods per year.

Find: A.

In this situation, we can easily compute the monthly payment with Eq. (3.12):

$$i = 6.25\%/12 = 0.5208\% \text{ per month},$$

$$N = (12)(6) = 72 \text{ months},$$

$$A = \$15,000(A/P, 0.5208\%, 72) = \$250.37.$$

4.2.2 Compounding Occurs at a Different Rate Than That at Which Payments Are Made

We will consider two situations: (1) compounding is more frequent than payments and (2) compounding is less frequent than payments.

Compounding Is More Frequent Than Payments

The computational procedure for compounding periods and payment periods that cannot be compared is as follows:

1. Identify the number of compounding periods per year (M), the number of payment periods per year (K), and the number of compounding periods per payment period (C).

2. Compute the effective interest rate per payment period:
   - For discrete compounding, compute

$$i = \left(1 + \frac{r}{M}\right)^c - 1.$$

   - For continuous compounding, compute

$$i = e^{r/K} - 1.$$

3. Find the total number of payment periods:

$$N = K \times (\text{number of years}).$$

4. Use i and N in the appropriate formulas in Table 3.4.

EXAMPLE 4.5 Compounding Occurs More Frequently Than Payments Are Made (Discrete-Compounding Case)

Suppose you make equal quarterly deposits of $1500 into a fund that pays interest at a rate of 6% compounded monthly, as shown in Figure 4.4. Find the balance at the end of year 2.

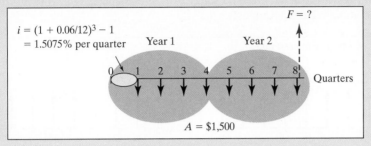

Figure 4.4   Quarterly deposits with monthly compounding (Example 4.5).

SOLUTION

Given: $A = \$1500$ per quarter, $r = 6\%$ per year, $M = 12$ compounding periods per year, and $N = 8$ quarters.

Find: F.

We follow the aforementioned procedure for noncomparable compounding and payment periods:

1. Identify the parameter values for M, K, and C, where

$$M = 12 \text{ compounding periods per year,}$$
$$K = 4 \text{ payment periods per year,}$$
$$C = 3 \text{ compounding periods per payment period.}$$

2. Use Eq. (4.2) to compute the effective interest:

$$i = \left(1 + \frac{0.06}{12}\right)^3 - 1$$

$$= 1.5075\% \text{ per quarter.}$$

3. Find the total number of payment periods, N:

$$N = K(\text{number of years}) = 4(2) = 8 \text{ quarters.}$$

4. Use i and N in the appropriate equivalence formulas:

$$F = \$1500(F/A, 1.5075\%, 8) = \$12{,}652.60.$$

COMMENT: No 1.5075% interest table appears in Appendix A, but the interest factor can still be evaluated by $F = \$1500(A/F, 0.5\%, 3)(F/A, 0.5\%, 24)$, where the first interest factor finds its equivalent monthly payment and the second interest factor converts the monthly payment series to an equivalent lump-sum future payment.

EXAMPLE 4.6 Compounding Occurs More Frequently Than Payments Are Made (Continuous-Compounding Case)

A series of equal quarterly receipts of $500 extends over a period of five years as shown in Figure 4.5. What is the present worth of this quarterly payment series at 8% interest compounded continuously?

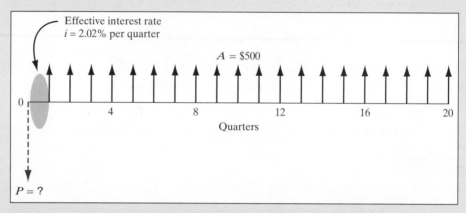

Figure 4.5 A present-worth calculation for an equal payment series with an interest rate of 8% compounded continuously (Example 4.6).

Discussion: A question that is equivalent to the preceding one is "How much do you need to deposit now in a savings account that earns 8% interest compounded continuously so that you can withdraw $500 at the end of each quarter for five years?" Since the payments are quarterly, we need to compute i per quarter for the equivalence calculations:

$$i = e^{r/K} - 1 = e^{0.08/4} - 1$$
$$= 2.02\% \text{ per quarter}$$

$$N = (4 \text{ payment periods per year})(5 \text{ years})$$
$$= 20 \text{ quarterly periods.}$$

SOLUTION

Given: $i = 2.02\%$ per quarter, $N = 20$ quarters, and $A = \$500$ per quarter.
Find: P.

Using the $(P/A, i, N)$ factor with $i = 2.02\%$ and $N = 20$, we find that

$$P = \$500(P/A, 2.02\%, 20)$$
$$= \$500(16.3199)$$
$$= \$8,159.96.$$

Compounding Is Less Frequent Than Payments

The next two examples contain identical parameters for savings situations in which compounding occurs less frequently than payments. However, two different underlying assumptions govern how interest is calculated. In Example 4.7, the assumption is that, whenever a deposit is made, it starts to earn interest. In Example 4.8, the assumption is that the deposits made within a quarter do not earn interest until the end of that quarter. As a result, in Example 4.7 we transform the compounding period to conform to the payment period, and in Example 4.8 we lump several payments together to match the compounding period. In the real world, which assumption is applicable depends on the transactions and the financial institutions involved. The accounting methods used by many firms record cash transactions that occur within a compounding period as if they had occurred at the end of that period. For example, when cash flows occur daily, but the compounding period is monthly, the cash flows within each month are summed (ignoring interest) and treated as a single payment on which interest is calculated.

   Note: *In this textbook, we assume that whenever the time point of a cash flow is specified, one cannot move it to another time point without considering the time value of money (i.e., the practice demonstrated in Example 4.7 should be followed).*

EXAMPLE 4.7 Compounding Is Less Frequent Than Payments: Effective Interest Rate per Payment Period

Suppose you make $500 monthly deposits to a registered retirement savings plan (RRSP) that pays interest at a rate of 10% compounded quarterly. Compute the balance at the end of 10 years.

SOLUTION

Given: $r = 10\%$ per year, $M = 4$ quarterly compounding periods per year, $K = 12$ payment periods per year, $A = \$500$ per month, $N = 120$ months, and interest is accrued on deposits made during the compounding period.

Find: i, F.

As in the case of Example 4.5, the procedure for noncomparable compounding and payment periods is followed:

1. The parameter values for $M, K,$ and C are

$$M = 4 \text{ compounding periods per year,}$$

$$K = 12 \text{ payment periods per year,}$$

$$C = \tfrac{1}{3} \text{ compounding period per payment period.}$$

2. As shown in Figure 4.6, the effective interest rate per payment period is calculated with Eq. (4.2):

$$i = 0.826\% \text{ per month.}$$

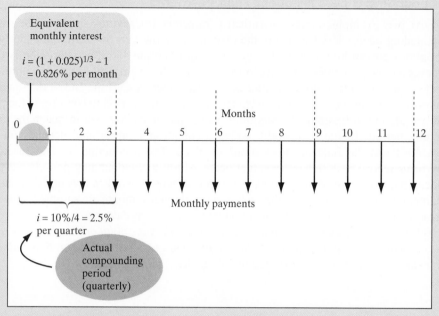

Figure 4.6 Calculation of equivalent monthly interest rate when the quarterly interest rate is specified (Example 4.7).

3. Find N:

$$N = (12)(10) = 120 \text{ payment periods.}$$

4. Use i and N in the appropriate equivalence formulas (Figure 4.7):

$$F = \$500(F/A, 0.826\%, 120)$$
$$= \$101,907.89.$$

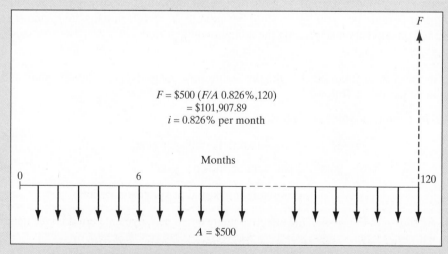

Figure 4.7 Cash flow diagram (Example 4.7).

EXAMPLE 4.8 **Compounding Is Less Frequent Than Payment: Summing Cash Flows to the End of the Compounding Period**

Some financial institutions will not pay interest on funds deposited after the start of the compounding period. To illustrate, consider Example 4.7 again. Suppose that money deposited during a quarter (the compounding period) will not earn any interest (Figure 4.8). Compute what F will be at the end of 10 years.

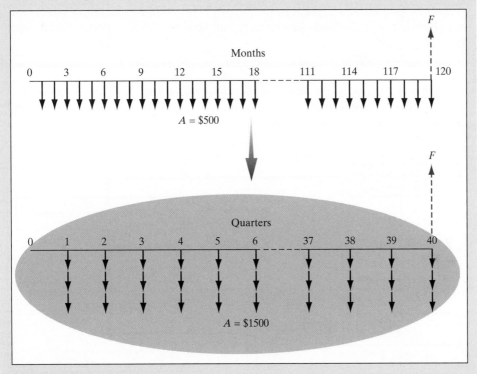

Figure 4.8 Transformed cash flow diagram created by summing monthly cash flows to the end of the quarterly compounding period (Example 4.8).

SOLUTION

Given: Same as for Example 4.7; however, no interest on flow during the compounding period.

Find: F.

In this case, the three monthly deposits during each quarterly period will be placed at the end of each quarter. Then the payment period coincides with the interest period, and we have

$$i = \frac{10\%}{4} = 2.5\% \text{ per quarter,}$$

$$A = 3(\$500) = \$1500 \text{ per quarter,}$$

$$N = 4(10) = 40 \text{ payment periods,}$$
$$F = \$1500(F/A, 2.5\%, 40) = \$101{,}103.83.$$

COMMENTS: In Example 4.8, the balance will be $804.06 less than in Example 4.7, a fact that is consistent with our understanding that increasing the frequency of compounding increases the future value of money. Some financial institutions follow the practice illustrated in Example 4.7. As an investor, you should reasonably ask yourself whether it makes sense to make deposits in an interest-bearing account more frequently than interest is paid. In the interim between interests compounding, you may be tying up your funds prematurely and forgoing other opportunities to earn interest.

Figure 4.9 is a decision chart that allows you to sum up how you can proceed to find the effective interest rate per payment period, given the various possible compounding and interest arrangements.

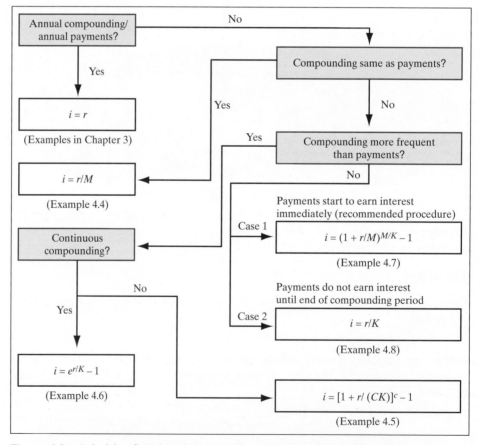

Figure 4.9 A decision flowchart demonstrating how to compute the effective interest rate i per payment period.

4.3 Equivalence Calculations With Continuous Payments

As we have seen so far, interest can be compounded annually, semiannually, monthly, or even continuously. Discrete compounding is appropriate for many financial transactions such as mortgages, bonds, and installment loans, which require payments or receipts at discrete times. In most businesses, however, transactions occur continuously throughout the year. In these circumstances, we may describe the financial transactions as having a continuous flow of money, for which continuous compounding and discounting are more realistic. This section illustrates how one establishes economic equivalence between cash flows under continuous compounding.

Continuous cash flows represent situations in which money flows continuously and at a known rate throughout a given period. In business, many daily cash flow transactions can be viewed as continuous. An advantage of the continuous-flow approach is that it more closely models the realities of business transactions. Costs for labour, for carrying inventory, and for operating and maintaining equipment are typical examples. Others include capital improvement projects that conserve energy or water or that process steam. Savings on these projects can occur continuously.

4.3.1 Single-Payment Transactions

First we will illustrate how single-payment formulas for continuous compounding and discounting are derived. Suppose that you invested P dollars at a nominal rate of $r \%$ interest for N years. If interest is compounded continuously, the effective annual interest is $i = e^r - 1$. The future value of the investment at the end of N years is obtained with the F/P factor by substituting $e^r - 1$ for i:

$$
\begin{aligned}
F &= P(1 + i)^N \\
&= P(1 + e^r - 1)^N \\
&= Pe^{rN}.
\end{aligned}
$$

This implies that \$1 invested now at an interest rate of $r\%$ compounded continuously accumulates to e^{rN} dollars at the end of N years. Correspondingly, the present value of F due N years from now and discounted continuously at an interest rate of $r\%$ is equal to

$$
P = Fe^{-rN}.
$$

We can say that the present value of \$1 due N years from now and discounted continuously at an annual interest rate of $r\%$ is equal to e^{-rN} dollars.

4.3.2 Continuous-Funds Flow

Suppose that an investment's future cash flow per unit of time (e.g., per year) can be expressed by a continuous function ($f(t)$) that can take any shape. Suppose also that the investment promises to generate cash of $f(t)\Delta t$ dollars between t and $t + \Delta t$, where t is a point in the time interval $0 \leq t \leq N$ (Figure 4.10). If the nominal interest rate is a

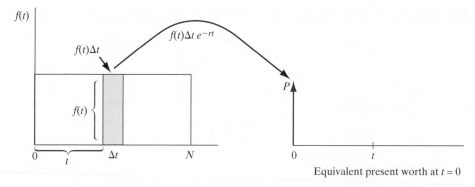

Figure 4.10 Finding an equivalent present worth of a continuous-flow payment function $f(t)$ at a nominal rate of $r\%$.

constant r during this interval, the present value of the cash stream is given approximately by the expression

$$\Sigma(f(t)\Delta t)e^{-rt},$$

where e^{-rt} is the discounting factor that converts future dollars into present dollars. With the project's life extending from 0 to N, we take the summation over all subperiods (compounding periods) in the interval from 0 to N. As the interval is divided into smaller and smaller segments (i.e., as Δt approaches zero), we obtain the expression for the present value by the integral

$$P = \int_0^N f(t)e^{-rt}\, dt. \tag{4.5}$$

Similarly, the expression for the future value of the cash flow stream is given by the equation

$$F = Pe^{rN} = \int_0^N f(t)e^{r(N-t)}\, dt, \tag{4.6}$$

where $e^{r(N-t)}$ is the compounding factor that converts present dollars into future dollars. It is important to observe that the time unit is the *year*, because the effective interest rate is expressed in terms of a year. Therefore, all time units in equivalence calculations must be converted into years. Table 4.2 summarizes some typical continuous cash functions that can facilitate equivalence calculations.[2]

---

[2] Chan S. Park and Gunter P. Sharp-Bette, *Advanced Engineering Economics*. New York: John Wiley & Sons, 1990. Reprinted by permission of John Wiley & Sons, Inc.

TABLE 4.2 Summary of Interest Factors for Typical Continuous Cash Flows With Continuous Compounding

Type of Cash Flow	Cash Flow Function	Find	Given	Algebraic Notation	Factor Notation
Uniform (step) $f(t) = \bar{A}$		P	$\bar{A}$	$\bar{A}\left[\dfrac{e^{rN} - 1}{re^{rN}}\right]$	$(P/\bar{A}, r, N)$
		$\bar{A}$	P	$P\left[\dfrac{re^{rN}}{e^{rN} - 1}\right]$	$(\bar{A}/P, r, N)$
		F	$\bar{A}$	$\bar{A}\left[\dfrac{e^{rN} - 1}{r}\right]$	$(F/\bar{A}, r, N)$
		$\bar{A}$	F	$F\left[\dfrac{r}{e^{rN} - 1}\right]$	$(\bar{A}/P, r, N)$
Gradient (ramp) $f(t) = Gt$		P	G	$\dfrac{G}{r^2}(1 - e^{-rN}) - \dfrac{G}{r}(Ne^{-rN})$	
Decay $f(t) = ce^{-jt}$ $j^t = $ decay rate with time		P	c, j	$\dfrac{c}{r + j}(1 - e^{-(r+j)N})$	

EXAMPLE 4.9 Comparison of Daily Flows and Daily Compounding With Continuous Flows and Continuous Compounding

Consider a situation in which money flows daily. Suppose you own a retail shop and generate $200 cash each day. You establish a special business account and deposit your daily cash flows in an account for 15 months. The account earns an interest rate of 6%. Compare the accumulated cash values at the end of 15 months, assuming

(a) Daily compounding and
(b) Continuous compounding.

SOLUTION

(a) With daily compounding:

Given: $A = \$200$ per day $r = 6\%$ per year, $M = 365$ compounding periods per year, and $N = 455$ days.

Find: F.

Assuming that there are 455 days in the 15-month period, we find that

$$i = 6\%/365$$

$$= 0.01644\% \text{ per day},$$

$$N = 455 \text{ days}.$$

The balance at the end of 15 months will be

$$F = \$200(F/A, 0.01644\%, 455)$$

$$= \$200(472.4095)$$

$$= \$94,482.$$

(b) With continuous compounding:

Now we approximate this discrete cash flow series by a uniform continuous cash flow function as shown in Figure 4.11. In this situation, an amount flows at the rate of $\overline{A}$ per year for N years.

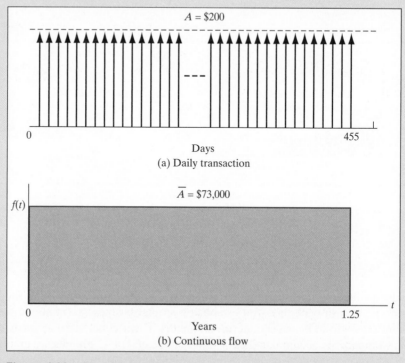

Figure 4.11 Comparison between daily transaction and continuous-funds flow transaction (Example 4.9).

Note: *Our time unit is a year.* Thus, a 15-month period is 1.25 years. Then the cash flow function is expressed as

$$f(t) = \overline{A}, 0 \le t \le 1.25$$

$$= \$200(365)$$

$$= \$73,000 \text{ per year.}$$

Given: $\overline{A} = \$73,000$ per year, $r = 6\%$ per year, compounded continuously, and $N = 1.25$ years.

Find: F.

Substituting these values back into Eq. (4.6) yields

$$F = \int_0^{1.25} 73,000 e^{0.06(1.25-t)} \, dt$$

$$= \$73,000 \left[\frac{e^{0.075} - 1}{0.06} \right]$$

$$= \$94,759.$$

The factor in the bracket is known as the **funds flow compound amount factor** and is designated $(F/\overline{A}, r, N)$ as shown in Table 4.2. Notice that the difference between the two methods is only \$277 (less than 0.3%).

COMMENTS: As shown in this example, the differences between discrete daily compounding and continuous compounding have no practical significance in most cases. Consequently, as a mathematical convenience, instead of assuming that money flows in discrete increments at the end of each day, we could assume that money flows continuously at a uniform rate during the period in question. This type of cash flow assumption is common practice in the chemical industry.

4.4 Changing Interest Rates

Up to this point, we have assumed a constant interest rate in our equivalence calculations. When an equivalence calculation extends over several years, more than one interest rate may be applicable to properly account for the time value of money. That is to say, over time, interest rates available in the financial marketplace fluctuate, and a financial institution that is committed to a long-term loan may find itself in the position of losing the opportunity to earn higher interest because some of its holdings are tied up in a lower interest loan. The financial institution may attempt to protect itself from such lost earning opportunities by building gradually increasing interest rates into a long-term loan at the outset. Adjustable-rate mortgage (ARM) loans are perhaps the most common examples of variable interest rates. In this section, we will consider variable interest rates in both single payments and a series of cash flows.

4.4.1 Single Sums of Money

To illustrate the mathematical operations involved in computing equivalence under changing interest rates, first consider the investment of a single sum of money, P, in a savings account for N interest periods. If i_n denotes the interest rate appropriate during period n, then the future worth equivalent for a single sum of money can be expressed as

$$F = P(1 + i_1)(1 + i_2) \cdots (1 + i_{N-1})(1 + i_N),\qquad(4.7)$$

and solving for P yields the inverse relation

$$P = F[(1 + i_1)(1 + i_2) \cdots (1 + i_{N-1})(1 + i_N)]^{-1}.\qquad(4.8)$$

EXAMPLE 4.10 Changing Interest Rates With a Lump-Sum Amount

Suppose you deposit $2000 in a registered retirement savings plan (RRSP) that pays interest at 6% compounded monthly for the first two years and 9% compounded monthly for the next three years. Determine the balance at the end of five years (Figure 4.12).

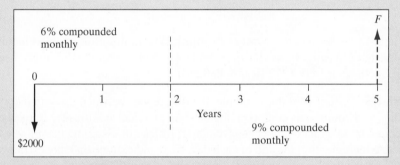

Figure 4.12   Changing interest rates (Example 4.10).

SOLUTION

Given: $P = \$2000$, $r = 6\%$ per year for first two years, 9% per year for last three years, $M = 4$ compounding periods per year, $N = 20$ quarters.
Find: F.

We will compute the value of F in two steps. First we will compute the balance B_2 in the account at the end of two years. With 6% compounded quarterly, we have

$$i = 6\%/12 = 0.5\%$$

$$N = 12(2) = 24 \text{ months}$$

$$B_2 = \$2000(F/P, 0.5\%, 24)$$

$$= \$2000(1.12716)$$

$$= \$2254.$$

Since the fund is not withdrawn, but reinvested at 9% compounded monthly, as a second step we compute the final balance as follows:

$$i = 9\%/12 = 0.75\%$$

$$N = 12(3) = 36 \text{ months}$$

$$F = B_2(F/P, 0.75\%, 36)$$

$$= \$2254(1.3086)$$

$$= \$2950.$$

4.4.2 Series of Cash Flows

The phenomenon of changing interest rates can easily be extended to a series of cash flows. In this case, the present worth of a series of cash flows can be represented as

$$P = A_1(1 + i_1)^{-1} + A_2[(1 + i_1)^{-1}(1 + i_2)^{-1}] + \ldots$$
$$+ A_N[(1 + i_1)^{-1}(1 + i_2)^{-1} \ldots (1 + i_N)^{-1}]. \quad (4.9)$$

The future worth of a series of cash flows is given by the inverse of Eq. (4.9):

$$F = A_1[(1 + i_2)(1 + i_3) \ldots (1 + i_N)]$$
$$+ A_2[(1 + i_3)(1 + i_4) \ldots (1 + i_N)] + \ldots + A_N. \quad (4.10)$$

The uniform series equivalent is obtained in two steps. First, the present-worth equivalent of the series is found from Eq. (4.9). Then A is obtained after establishing the following equivalence equation:

$$P = A(1 + i_1)^{-1} + A[(1 + i_1)^{-1}(1 + i_2)^{-1}] + \ldots$$
$$+ A[(1 + i_1)^{-1}(1 + i_2)^{-1} \ldots (1 + i_N)^{-1}]. \quad (4.11)$$

EXAMPLE 4.11 Changing Interest Rates With Uneven Cash Flow Series

Consider the cash flow in Figure 4.13 with the interest rates indicated, and determine the uniform series equivalent of the cash flow series.

DISCUSSION: In this problem and many others, the easiest approach involves collapsing the original flow into a single equivalent amount, for example, at time zero, and then converting the single amount into the final desired form.

SOLUTION

Given: Cash flows and interest rates as shown in Figure 4.13; $N = 3$.
Find: A.

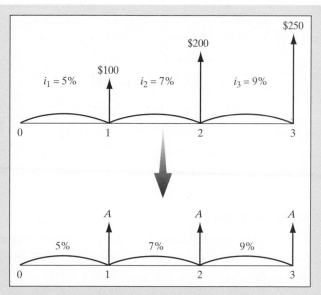

Figure 4.13 Equivalence calculation with changing interest rates (Example 4.11).

Using Eq. (4.9), we find the present worth:

$$P = \$100(P/F, 5\%, 1) + \$200(P/F, 5\%, 1)(P/F, 7\%, 1)$$
$$+ \$250(P/F, 5\%, 1)(P/F, 7\%, 1)(P/F, 9\%, 1)$$
$$= \$477.41.$$

Then we obtain the uniform series equivalent as follows:

$$\$477.41 = A(P/F, 5\%, 1) + A(P/F, 5\%, 1)(P/F, 7\%, 1)$$
$$+ A(P/F, 5\%, 1)(P/F, 7\%, 1)(P/F, 9\%, 1)$$
$$= 2.6591A$$
$$A = \$179.54.$$

4.5 Debt Management

Credit card debt and commercial loans are among the most significant financial transactions involving interest. Many types of loans are available, but here we will focus on those most frequently used by individuals and in business.

4.5.1 Commercial Loans

One of the most important applications of compound interest involves loans that are paid off in **installments** over time. If the loan is to be repaid in equal periodic amounts (weekly, monthly, quarterly, or annually), it is said to be an **amortized loan**.

Examples of installment loans include automobile loans, loans for appliances, home mortgage loans, and the majority of business debts other than very short-term loans. Most commercial loans have interest that is compounded monthly. With an auto loan, a local bank or a dealer advances you the money to pay for the car, and you repay the principal plus interest in monthly installments, usually over a period of three to five years. The car is your collateral. If you don't keep up with your payments, the lender can repossess the car and keep all the payments you have made.

Two things determine what borrowing will cost you: the finance charge and the length of the loan. The cheapest loan is not the one with the lowest payments or even the one with the lowest interest rate. Instead, you have to look at (1) the total cost of borrowing, which depends on the interest rate plus fees, and (2) the term, or length of time it takes you to repay the loan. While you probably cannot influence the rate or the fees, you may be able to arrange for a shorter term.

- **The annual percentage rate** (APR) is set by lenders, who are required to tell you what a loan will actually cost per year, expressed as an APR. Some lenders charge lower interest, but add high fees; others do the reverse. Combining the fees with a year of interest charges to give you the true annual interest rate, the APR allows you to compare these two kinds of loans on equal terms.

- **Fees** are the expenses the lender will charge to lend the money. The application fee covers processing expenses. Attorney fees pay the lender's attorney. Credit search fees cover researching your credit history. **Origination fees** cover administrative costs and sometimes appraisal fees. All these fees add up very quickly and can substantially increase the cost of your loan.

  Origination fees: A fee charged by a lender for processing a loan application, expressed as a percentage of the mortgage amount.

- **Finance charges** are the cost of borrowing. For most loans, they include all the interest, fees, service charges, points, credit-related insurance premiums, and any other charges.

- **The periodic interest rate** is the interest the lender will charge on the amount you borrow. If lender also charges fees, the periodic interest rate will not be the true interest rate.

- **The term of your loan** is crucial in determining its cost. Shorter terms mean squeezing larger amounts into fewer payments. However, they also mean paying interest for fewer years, saving a lot of money.

Amortized Installment Loans

So far, we have considered many instances of amortized loans in which we calculated present or future values of the loans or the amounts of the installment payments. An additional aspect of amortized loans that will be of great interest to us is calculating the amount of interest versus the portion of the principal that is paid off in each installment. As we shall explore more fully in Chapter 10, the interest paid on a loan is an important element in calculating taxable income and has repercussions for both personal and business loan transactions. For now, we will focus on several methods of calculating interest and principal paid at any point in the life of the loan.

In calculating the size of a monthly installment, lending institutions may use two types of schemes. The first is the conventional amortized loan, based on the compound-interest method, and the other is the add-on loan, based on the simple-interest concept. We explain each method in what follows, but it should be understood that the amortized loan is the most common in various commercial lending.

In a typical amortized loan, the amount of interest owed for a specified period is calculated on the basis of the remaining balance on the loan at the beginning of the period.

A set of formulas has been developed to compute the remaining loan balance, interest payment, and principal payment for a specified period. Suppose we borrow an amount P at an interest rate i and agree to repay this principal sum P, including interest, in equal payments A over N periods. Then the size of the payment is $A = P(A/P, i, N)$, and each payment is divided into an amount that is interest and a remaining amount that goes toward paying the principal.

Let

$$B_n = \text{Remaining balance at the end of period } n, \text{ with } B_0 = P,$$

$$I_n = \text{Interest payment in period } n, \text{ where } I_n = B_{n-1}i,$$

$$PP_n = \text{Principal payment in period } n.$$

Then each payment can be defined as

$$A = PP_n + I_n. \tag{4.12}$$

The interest and principal payments for an amortized loan can be determined in several ways; two are presented here. No clear-cut reason is available to prefer one method over the other. Method 1, however, may be easier to adopt when the computational process is automated through a spreadsheet application, whereas Method 2 may be more suitable for obtaining a quick solution when a period is specified. You should become comfortable with at least one of these methods; pick the one that comes most naturally to you.

Method 1: Tabular Method. The first method is tabular. The interest charge for a given period is computed progressively on the basis of the remaining balance at the beginning of that period. Example 4.12 illustrates the process of creating a loan repayment schedule based on an iterative approach.

EXAMPLE 4.12 Loan Balance, Principal, and Interest: Tabular Method

Suppose you secure a home improvement loan in the amount of $5000 from a local bank. The loan officer computes your monthly payment as follows:

$$\text{Contract amount} = \$5000,$$
$$\text{Contract period} = 24 \text{ months},$$
$$\text{Annual percentage rate} = 12\%,$$
$$\text{Monthly installments} = \$235.37.$$

Figure 4.14 is the cash flow diagram. Construct the loan payment schedule by showing the remaining balance, interest payment, and principal payment at the end of each period over the life of the loan.

SOLUTION

Given: $P = \$5000$, $A = \$235.37$ per month, $r = 12\%$ per year, $M = 12$ compounding periods per year, and $N = 24$ months.
Find: B_n and I_n for $n = 1$ to 24.

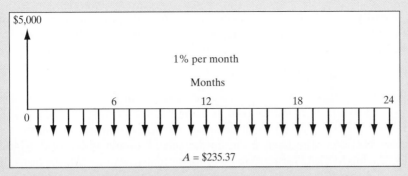

Figure 4.14 Cash flow diagram of the home improvement loan with an APR of 12% (Example 4.12).

TABLE 4.3 Creating a Loan Repayment Schedule With Excel (Example 4.12)

Payment No.	Size of Payment	Principal Payment	Interest Payment	Loan Balance
1	$235.37	$185.37	$50.00	$4814.63
2	235.37	187.22	48.15	4627.41
3	235.37	189.09	46.27	4438.32
4	235.37	190.98	44.38	4247.33
5	235.37	192.89	42.47	4054.44
6	235.37	194.83	40.54	3859.62
7	235.37	196.77	38.60	3662.85
8	235.37	198.74	36.63	3464.11
9	235.37	200.73	34.64	3263.38
10	235.37	202.73	32.63	3060.65
11	235.37	204.76	30.61	2855.89
12	235.37	206.81	28.56	2649.08
13	235.37	208.88	26.49	2440.20
14	235.37	210.97	24.40	2229.24
15	235.37	213.08	22.29	2016.16
16	235.37	215.21	20.16	1800.96
17	235.37	217.36	18.01	1583.60
18	235.37	219.53	15.84	1364.07
19	235.37	221.73	13.64	1142.34
20	235.37	223.94	11.42	918.40
21	235.37	226.18	9.18	692.21
22	235.37	228.45	6.92	463.77
23	235.37	230.73	4.64	233.04
24	235.37	233.04	2.33	0.00

We can easily see how the bank calculated the monthly payment of $235.37. Since the effective interest rate per payment period on this loan is 1% per month, we establish the following equivalence relationship:

$$235.37(P/A, 1\%, 24) = 235.37(21.2431) = 5000.$$

The loan payment schedule can be constructed as in Table 4.3. The interest due at $n = 1$ is $50.00, 1% of the $5000 outstanding during the first month. The $185.37 left over is applied to the principal, reducing the amount outstanding in the second month to $4814.63. The interest due in the second month is 1% of $4814.63, or $48.15, leaving $187.22 for repayment of the principal. At $n = 24$, the last $235.37 payment is just sufficient to pay the interest on the unpaid principal of the loan and to repay the remaining principal. Figure 4.15 illustrates the ratios between the interest and principal payments over the life of the loan.

COMMENTS: Certainly, constructing a loan repayment schedule such as that in Table 4.3 can be tedious and time consuming, unless a computer is used. As you can see in the Companion Website for this book, you can download an Excel file that creates the loan repayment schedule, on which you can make any adjustment to solve a typical loan problem of your choice.

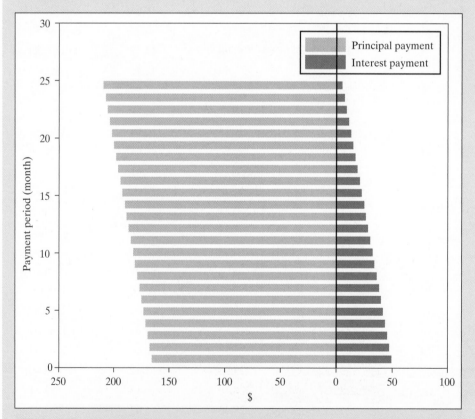

Figure 4.15 The proportions of principal and interest payments over the life of the loan (monthly payment = $235.37) (Example 4.12).

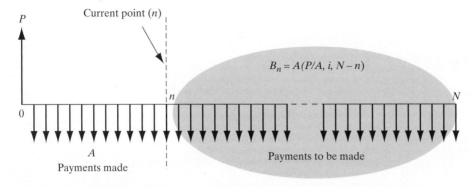

Figure 4.16   Calculating the remaining loan balance on the basis of Method 2.

Method 2: Remaining-Balance Method. Alternatively, we can derive B_n by computing the equivalent payments remaining after the nth payment. Thus, the balance with $N - n$ payments remaining is

$$B_n = A(P/A, i, N - n),\tag{4.13}$$

and the interest payment during period n is

$$I_n = (B_{n-1})i = A(P/A, i, N - n + 1)i,\tag{4.14}$$

where $A(P/A, i, N - n + 1)$ is the balance remaining at the end of period $n - 1$ and

$$PP_n = A - I_n = A - A(P/A, i, N - n + 1)i$$
$$= A[1 - (P/A, i, N - n + 1)i].$$

Knowing the interest factor relationship $(P/F, i, n) = 1 - (P/A, i, n)i$ from Table 3.4, we obtain

$$PP_n = A(P/F, i, N - n + 1).\tag{4.15}$$

As we can see in Figure 4.16, this method provides more concise expressions for computing the balance of the loan.

EXAMPLE 4.13 Loan Balances, Principal, and Interest: Remaining-Balance Method

Consider the home improvement loan in Example 4.12, and

(a) For the sixth payment, compute both the interest and principal payments.
(b) Immediately after making the sixth monthly payment, you would like to pay off the remainder of the loan in a lump sum. What is the required amount?

SOLUTION

(a) Interest and principal payments for the sixth payment.
    Given: (as for Example 4.12)

Find: I_6 and PP_6.

Using Eqs. (4.14) and (4.15), we compute

$$I_6 = \$235.37(P/A, 1\%, 19)(0.01)$$
$$= (\$4,054.44)(0.01)$$
$$= \$40.54.$$
$$PP_6 = \$235.37(P/F, 1\%, 19) = \$194.83,$$

or we simply subtract the interest payment from the monthly payment:

$$PP_6 = \$235.37 - \$40.54 = \$194.83.$$

(b) Remaining balance after the sixth payment.

The lower half of Figure 4.17 shows the cash flow diagram that applies to this part of the problem. We can compute the amount you owe after you make the sixth payment by calculating the equivalent worth of the remaining 18 payments at the end of the sixth month, with the time scale shifted by 6:

Given: $A = \$235.37$, $i = 1\%$ per month, and $N = 18$ months.

Find: Balance remaining after six months (B_6).

$$B_6 = \$235.37(P/A, 1\%, 18) = \$3859.62.$$

If you desire to pay off the remainder of the loan at the end of the sixth payment, you must come up with $3859.62. To verify our results, compare this answer with the value given in Table 4.3.

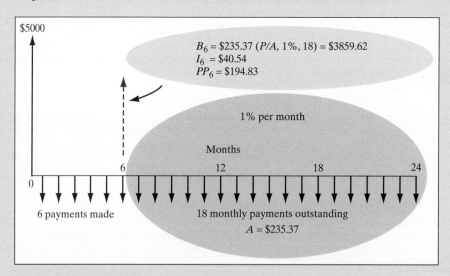

Figure 4.17 Computing the outstanding loan balance after making the sixth payment on the home improvement loan (Example 4.13).

Add-On Interest Loans

The add-on loan is totally different from the popular amortized loan. In the add-on loan, the total interest to be paid is precalculated and added to the principal. The principal and this precalculated interest amount are then paid together in equal installments. In such a case, the interest rate quoted is not the effective interest rate, but what is known as **add-on interest**. If you borrow P for N years at an add-on rate of i, with equal payments due at the end of each month, a typical financial institution might compute the monthly installment payments as follows:

$$\text{Total add-on interest} = P(i)(N),$$

$$\text{Principal plus add-on interest} = P + P(i)(N) = P(1 + iN)$$

$$\text{Monthly installments} = \frac{P(1 + iN)}{(12 \times N)}. \tag{4.16}$$

Add-on interest: A method of computing interest whereby interest charges are made for the entire principal amount for the entire term, regardless of any repayments of principal made.

Notice that the add-on interest is *simple interest*. Once the monthly payment is determined, the financial institution computes the APR on the basis of this payment, and you will be told what the value will be. Even though the add-on interest is specified along with the APR, many ill-informed borrowers think that they are actually paying the add-on rate quoted for this installment loan. To see how much interest you actually pay under a typical add-on loan arrangement, consider Example 4.14.

EXAMPLE 4.14 Effective Interest Rate for an Add-On Interest Loan

Consider again the home improvement loan problem in Example 4.12. Suppose that you borrow $5000 with an add-on rate of 12% for two years. You will make 24 equal monthly payments.

(a) Determine the amount of the monthly installment.

(b) Compute the nominal and the effective annual interest rate on the loan.

SOLUTION

Given: Add-on rate = 12% per year, loan amount (P) = $5000, and N = 2 years. Find: (a) A and (b) i_a and i.

(a) First we determine the amount of add-on interest:

$$iPN = (0.12)(\$5000)(2) = \$1200.$$

Then we add this simple-interest amount to the principal and divide the total amount by 24 months to obtain A:

$$A = \frac{(\$5000 + \$1200)}{24} = \$258.33.$$

(b) Putting yourself in the lender's position, compute the APR value of the loan just described. Since you are making monthly payments with monthly compounding, you need to find the effective interest rate that makes the present $5000 sum equivalent to 24 future monthly payments of $258.33. In this situation, we are solving for i in the equation

$$\$258.33 = \$5000(A/P, i, 24),$$

or

$$(A/P, i, 24) = 0.0517.$$

You know the value of the *A/P* factor, but you do not know the interest rate i. As a result, you need to look through several interest tables and determine i by interpolation. A more effective approach is to use Excel's RATE function with the following parameters:

$$=\text{RATE(N,A,P,F,type,guess)}$$

$$=\text{RATE}(24, 258.33, -5000, 0, 0, 1\%) \rightarrow 1.7975\%$$

The nominal interest rate for this add-on loan is $1.7975 \times 12 = 21.57\%$, and the effective annual interest rate is $(1 + 0.01975)^{12} - 1 = 26.45\%$, rather than the 12% quoted add-on interest. When you take out a loan, you should not confuse the add-on interest rate stated by the lender with the actual interest cost of the loan.

COMMENTS: In the real world, truth-in-lending laws require that APR information always be provided in mortgage and other loan situations, so you would not have to calculate nominal interest as a prospective borrower (although you might be interested in calculating the actual or effective interest). However, in later engineering economic analyses, you will discover that solving for implicit interest rates, or rates of return on investment, is performed regularly. Our purpose in this text is to periodically give you some practice with this type of problem, even though the scenario described does not exactly model the real-world information you would be given.

4.5.2 Loan versus Lease Financing

When, for example, you choose a car, you also choose how to pay for it. If you do not have the cash on hand to buy a new car outright—and most of us don't—you can consider taking out a loan or leasing the car to spread the payments over time. Deciding whether to pay cash, take a loan, or sign a lease depends on a number of personal as well as economic factors. Leasing is an option that lets you pay for the portion of a vehicle you expect to use over a specified term, plus a charge for rent, taxes, and fees. For instance, you might want a $20,000 vehicle. Suppose that vehicle might be worth about $9000 at the end of your three-year lease. (This is called the residual value.)

- If you have enough money to buy the car, you could purchase it in cash. If you pay cash, however, you will lose the opportunity to earn interest on the amount you spend. That could be substantial if you know of an investment that is paying a good return.

- If you purchase the vehicle using debt financing, your monthly payments will be based on the entire $20,000 value of the vehicle. You will continue to own the vehicle at the end of your financing term, but the interest you will pay on the loan will drive up the real cost of the car.
- If you lease the same vehicle, your monthly payments will be based on the amount of the vehicle you expect to "use up" over the term of the lease. This value ($11,000 in our example) is the difference between the original cost ($20,000) and the estimated value at the end of the lease ($9000). With leasing, the length of your lease agreement, the monthly payments, and the yearly mileage allowance can be tailored to your driving needs. The greatest financial appeal for leasing is low initial outlay costs: Usually you pay a leasing administrative fee, one month's lease payment, and a refundable security deposit. The terms of your lease will include a specific mileage allowance; if you put additional miles on your car, you will have to pay more for each extra mile.

Discount Rate to Use in Comparing Different Financing Options

In calculating the net cost of financing a car, we need to decide which interest rate to use in discounting the loan repayment series. The dealer's (bank's) interest rate is supposed to reflect the time value of money of the dealer (or the bank) and is factored into the required payments. However, the correct interest rate to use in comparing financing options is the interest rate that reflects *your* time value of money. For most individuals, this interest rate might be equivalent to the savings rate from their deposits. To illustrate, consider Example 4.15, in which we compare three auto financing options.

EXAMPLE 4.15 Financing Your Vehicle: Paying Cash, Taking a Loan, or Leasing

Suppose you intend to own or lease a vehicle for 42 months. Consider the following three ways of financing the vehicle—say, a 2009 BMW 323i Sedan:

- **Option A:** Purchase the vehicle at the normal price of $32,508, and pay for the vehicle over 42 months with equal monthly payments at 5.65% APR financing.
- **Option B:** Purchase the vehicle at a discount price of $31,020 to be paid immediately.
- **Option C:** Lease the vehicle for 42 months.

The accompanying chart lists the items of interest under each option. For each option, license, title, and registration fees, as well as taxes and insurance, are extra.

For the lease option, the lessee must come up with $1507.76 at signing. This cash due at signing includes the first month's lease payment of $513.76 and a $994 administrative fee. The lease rate is based on 60,000 kilometres over the life of the contract. There will be a surcharge at the rate of 18 cents per kilometre for any additional kilometres driven over 60,000. No security deposit is required; however, a $395 disposition fee is due at the end of the lease, at which time the lessee has the option to purchase the car for $17,817. The lessee is also responsible for excessive

Item	Option A Debt Financing	Option B Paying Cash	Option C Lease Financing
Down payment	$4,500	$0	$0
APR(%)	5.65%		
Monthly payment	$736.53		$513.76
Length	42 months		42 months
Fees			$994
Cash due at end of lease			$395
Purchase option at end of lease			$17,817
Cash due at signing	$4500	$31,020	$1,507.76

wear and use. If the funds that would be used to purchase the vehicle are presently earning 4.5% annual interest compounded monthly, which financing option is a better choice?

DISCUSSION: With a lease payment, you pay for the portion of the vehicle you expect to use. At the end of the lease, you simply return the vehicle to the dealer and pay the agreed-upon disposal fee. With traditional financing, your monthly payment is based on the entire $32,508 value of the vehicle, and you will own the vehicle at the end of your financing terms. Since you are comparing the options over 42 months, you must explicitly consider the unused portion (resale value) of the vehicle at the end of the term. In other words, you must consider the resale value of the vehicle in order to figure out the net cost of owning it. *As the resale value, you could use the $17,817 quoted by the dealer in the lease option.* Then you have to ask yourself if you can get that kind of resale value after 42 months' ownership.

Note that the 5.65% APR represents the dealer's interest rate used in calculating the loan payments. With 5.65% interest, your monthly payments will be $A = (\$32,508 - \$4500)(A/P, 5.65\%/12, 42) = \$736.53$. Note also, however, that the 4.5% APR represents your earning opportunity rate. In other words, if you do not buy the car, your money continues to earn 4.5% APR. Therefore, 4.5% represents your opportunity cost of purchasing the car. So which interest rate do you use in your analysis? Clearly, the 4.5% rate is the appropriate one to use.

SOLUTION

Given: Financial facts shown in Figure 4.18, $r = 4.5\%$, payment period = monthly, and compounding period = monthly.

Find: The most economical financing option, under the assumption that you will be able to sell the vehicle for $17,817 at the end of 42 months.

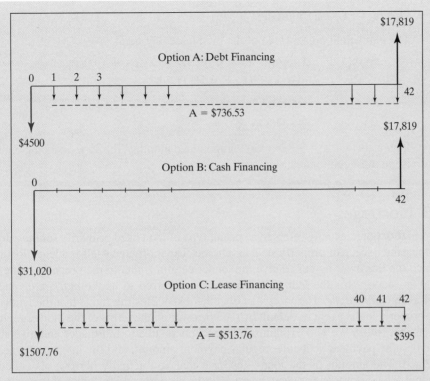

Figure 4.18 Comparing different financing options.

For each option, we will calculate the net equivalent total cost at $n = 0$. Since the loan payments occur monthly, we need to determine the effective interest rate per month, which is 4.5%/12.

- **Option A: Conventional debt financing**

  The equivalent present cost of the total loan payments is

  $$P_1 = \$4500 + \$736.53(P/A, 4.5\%/12, 42)$$
  $$= \$33,071.77.$$

  The equivalent present worth of the resale value of the car is

  $$P_2 = \$17,817(P/F, 4.5\%/12, 42) = \$15,225.13.$$

  The equivalent net financing cost is

  $$P_{\text{Option A}} = \$33,071.77 - \$15,225.13$$
  $$= \$17,846.64.$$

- **Option B: Cash financing**

  $$P_{\text{Option B}} = \$31,020 - \$17,817(P/F, 4.5\%/12, 42)$$
  $$= \$31,020 - \$15,225.13$$
  $$= \$15,844.87.$$

- **Option C: Lease financing**

  The equivalent present cost of the total lease payments is

  $$P_{\text{Option C}} = \$1507.76 + \$513.76(P/A, 4.5\%/12, 41)$$
  $$+ \$395(P/F, 4.5\%/12, 42)$$
  $$= \$1507.76 + \$19,490.96 + \$337.54$$
  $$= \$21,336.26$$

  It appears that, at 4.5% interest compounded monthly, the cash financing option is the most economical one.

4.5.3 Mortgages

Mortgage: A loan to finance the purchase of real estate, usually with specified payment periods and interest rates.

The term **mortgage** usually refers to a special type of loan used primarily for the purpose of purchasing a piece of property such as a home. The mortgage itself is a legal document in which the borrower agrees to give the lender certain rights to the property being purchased as security for the loan. The borrower is referred to as the *mortgagor* and the lender as the *mortgagee*. The mortgage document specifies the rights that the lender has to the property in the event of default by the borrower on the terms of the mortgage. In the sections that follow we will explain some concepts related to mortgages and give an example on how payment schedule, regular payment amount, and the interest charges are calculated. The amount of the loan—the cash that you actually borrow—is called the **principal**. The difference between the price of the property and the amount that one owes on the mortgage is called the purchaser's **equity**.

Types of Mortgages

Canadian banks, credit unions, trust companies, mortgage companies, private lenders, and others offer four main types of mortgages. These four types of mortgages are briefly described below.

1. **Conventional mortgage:** The conventional mortgage is the most common type of financing for principal residence or residential investment property. In this type of mortgage, the loan amount generally does not exceed 80% of the appraised value or the purchase price of the property, whichever is lower. The purchaser is responsible for raising the other 20% as a down payment.

2. **High-ratio mortgage:** If a potential buyer is unable to raise the necessary 20% funding to complete the purchase of the property, then he or she may obtain a high-ratio mortgage. Essentially, these are conventional mortgages that exceed the 80% referred to above. High-ratio mortgages are available for up to 95% of the purchase price or of the appraisal, whichever is lower. These mortgages must, by law, be insured. Lenders are insured against loss by an insurer, the Canada Mortgage and Housing Corporation (CMHC). Borrowers must pay an application fee, which usually includes the property appraisal fee to CMHC and an insurance fee. The insurance fee is usually added to the principal amount of the mortgage, although it may be paid in cash.

3. **Collateral mortgage:** A collateral mortgage provides backup protection of a loan that is filed against a property. It is secondary to a main form of security taken by

the lender for the loan, for example, a promissory note. When the promissory note is paid off, the collateral mortgage is automatically discharged. The money borrowed may be used for the purchase of the property itself or for other purposes, such as home improvements and other investments.

Terms and Conditions of Mortgages

To make the best mortgage decision, one has to consider many factors. The key factors are amortization, term of the mortgage, whether the mortgage is open or closed, interest rate, payment schedule, prepayment privilege, and portability. A brief explanation of each of these concepts is provided below:

1. **Amortization:** Amortization refers to the number of years it would take to repay a mortgage loan in full for a given interest rate and payment schedule. The usual amortization period is 25 years, although there is a wide range of choices. The longer the amortization period, the smaller the regular payment (usually monthly) and the larger the total interest payments.

2. **Term of the mortgage:** Term refers to the number of months or years that the mortgage—the legal document—covers. Terms may vary from 6 months to 10 years. At the end of a term, the unpaid principal is due and payable. One has the option to renew the mortgage with the same bank or refinance it through a different lending institution.

3. **Interest rate:** By law, mortgages must contain a statement showing, among other things, the rate of interest calculated annually or semiannually. Mortgage interest has traditionally been quoted as a nominal annual rate based on semiannual compounding. A *fixed-rate mortgage* is one where the rate of interest is fixed for the whole term. A **variable-rate mortgage (VRM)** is one where the rate of interest varies according to the prime rate set by the lender. This prime rate may change from time to time, though not frequently. For both a fixed-rate mortgage and a variable-rate mortgage, the required regular payment amount does not change within the term. However, the interest portion of the regular payment amount for a payment period is dependent on the outstanding balance at the beginning of the period and the interest rate applicable for the period.

   > **VRM:** A mortgage with an interest rate that may change, usually in response to changes in the prime rate.

4. **Open or closed mortgage:** An *open mortgage* provides the borrower with the flexibility to repay the loan more quickly. One can pay off the mortgage in full at any time before the term is over without any penalty or extra charges. Because of this flexibility, the interest rate for an open mortgage is higher than a closed mortgage, when other conditions remain the same. *A closed mortgage* does not allow the borrower to repay the loan more quickly than agreed upon. Payments must be made as specified in the agreement. If the borrower wants to pay off the mortgage before the term is over, a penalty charge is applied. The penalty charge is often equal to the greater of (1) three months' interest on the amount of the prepayment; and (2) the interest rate differential. The "interest rate differential" refers to the amount, if any, by which the existing interest rate exceeds the interest rate at which the lender would lend to the same borrower for a term commencing on the prepayment date and expiring at the existing term date. However, even for a closed mortgage, there may be some prepayment privileges such that no penalty will be levied.

5. **Payment schedule:** Most mortgage loans are amortized. Constant regular payments are made to pay off the principal. Monthly payments are the most common,

although some mortgages may be paid weekly, bi-weekly, semimonthly, quarterly, semiannually, or annually.

6. **Prepayment privileges:** Many financial institutions offer closed mortgages with some prepayment privileges. For example, the borrower taking a closed mortgage may be allowed to make a prepayment of up to 10% of the original principal every calendar year or every anniversary. Another privilege allows the borrower to increase the regular payment by up to 100% once per year. If you are able to take advantage of these prepayment privileges, you may dramatically decrease the actual amortization period.

7. **Portability:** Some lenders offer mortgages with a feature called portability. This feature means that a borrower can sell one home and buy another during the term of the mortgage. The mortgage can be transferred from one property to the other without penalty.

Throughout this book, we will assume that the quoted mortgage interest rates are based on semiannual compounding. In the following, we use Example 4.16 to illustrate various calculations relevant to mortgage financing.

EXAMPLE 4.16 Closed Mortgage With Prepayment Privileges

John Montgomery is considering buying a $125,000 home with a $25,000 down payment. He can get a conventional mortgage in the amount of $100,000 with a three-year term at 8% per annum from TD Canada Trust. He has selected an amortization period of 25 years. The mortgage is a closed mortgage with the following prepayment privileges:

• Once each calendar year, on any regular payment date, John can prepay on account of principal a sum not more than 10% of the original borrowed amount, without notice or charge. If this privilege is not exercised in a certain year, it cannot be carried forward to the following years.

• Once each calendar year, on any regular payment date, on written notice, John, without charge, can increase the amount of the regular installment of principal and interest. The total of such increases cannot exceed 100% of the installment of principal and interest set out in the mortgage document. If the regular installment has been increased, the mortgagor may decrease the installments to an amount not less than the installment of principal and interest set out in the mortgage document, on written notice, without charge.

• On any regular payment date, John can prepay the whole or any part of the principal amount then outstanding on payment of an amount equal to the greater of

  (1) 3 months' interest, at the rate specified in the mortgage document, on the amount prepaid or

  (2) the amount, if any, by which interest at the rate specified in the mortgage document exceeds interest at the prevailing rate, calculated on the amount of principal prepayment, for a term commencing on the date of prepayment and expiring on the maturity date of the mortgage.

Answer the following questions regarding the mortgage in consideration:

(a) What is the amount of his regular payment if he chooses to pay weekly, semi-monthly, or monthly?

(b) What would the balance be at the end of the term for each of the three payment frequencies if the calculated regular payment amount is followed exactly?

(c) Assume that John has selected the option of monthly payment because he receives only one salary payment per month. In year 2, he increases his monthly payment by 50%. In year 3, he doubles his calculated monthly payment in (a). What is the balance of the mortgage at the time of renewal?

(d) In addition to (c), if he makes lump sum payments of $8000 and $10,000 at the first and the second anniversaries, respectively, what is the balance of the mortgage at the time of renewal?

(e) After John has made only the calculated monthly payments for one year, the interest rate for a two-year term has dropped to 6%. What would be the total penalty charge if he chooses to pay off his mortgage completely? What if the prevailing rate for a two-year mortgage has increased to 9%?

SOLUTION

Given: $P = \$100,000$, $r = 8\%$ per year, $M = 2$ compounding periods per year, amortization $= 25$ years, term $= 3$ years

Find: (a) The regular payment amounts on the following payment schedules: weekly, semimonthly, and monthly.

Figure 4.19(a) shows the cash flow diagram of the mortgage for the complete amortization period. In this figure, N is the number of payment periods in the amortization period.

- For weekly payment: $N = (52)(25) = 1300$ weeks

$$i_{wk} = (1 + r/M)^C - 1 = (1 + 0.08/2)^{1/26} - 1 = 0.1510\%$$

$$A_{wk} = \$100,000\ (A/P, 0.1510\%, 1300) = \$175.68$$

- For semimonthly payment: $N = (25)(24) = 600$ half-months

$$i_{1/2\ mon} = (1 + 0.08/2)^{1/12} - 1 = 0.3274\%$$

$$A_{1/2\ mon} = \$100,000\ (A/P, 0.3274\%, 600) = \$380.98$$

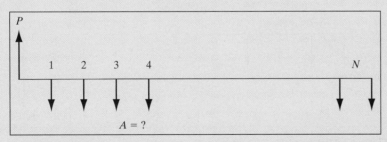

Figure 4.19(a) Cash flow diagram for the complete amortization (Example 4.16(a)).

- For monthly payment: $N = (25)(12) = 300$ months

  $i_{mon} = (1 + 0.08/2)^{1/6} - 1 = 0.6558\%$

  $A_{mon} = \$100,000\ (A/P, 0.6558\%, 300) = \$763.20$

COMMENTS: The more frequent the payments, the smaller the total amount paid per month (i.e., $A_{mon} > 2 \times A_{1/2\ mon} > 4.33 \times A_{wk}$).

(b) What are the end-of-term balances for weekly, semimonthly, and monthly payments?

Figure 4.19(b) shows the cash flow diagram of the mortgage for the first term of three years. In this figure, n is the number of payment periods in a term, A is the calculated regular payment amount, and B is the end-of-term balance to be computed.

- For weekly payment: $n = (3)(52) = 156$ weeks

  $i_{wk} = 0.1510\%$

  $\begin{aligned} B_{wk} &= P(F/P, i_{wk}, n) - A_{wk}(F/A, i_{wk}, n) \\ &= \$100,000(F/P, 0.1510\%, 156) - \$175.68(F/A, 0.1510\%, 156) \\ &= \$95,655.93 \end{aligned}$

- For semimonthly payment: $n = (3)(24) = 72$ half-months

  $i_{1/2\ mon} = 0.3274\%$

  $\begin{aligned} B_{1/2\ mon} &= \$100,000(F/P, 0.3274\%, 72) - \$380.98(F/A, 0.3274\%, 72) \\ &= \$95,655.54 \end{aligned}$

- For monthly payment: $n = (3)(12) = 36$ months

  $i_{mon} = 0.6558\%$

  $\begin{aligned} B_{mon} &= \$100,000(F/P, 0.6558\%, 36) - \$763.20(F/A, 0.6558\%, 36) \\ &= \$95,655.54 \end{aligned}$

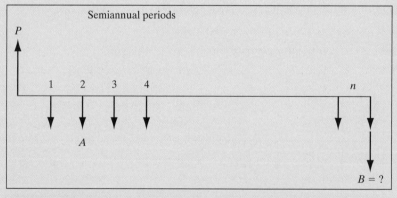

Figure 4.19(b) Cash flow diagram for the first term of the mortgage (Example 4.16(b)).

COMMENTS: Note that the end-of-term balances are the same (ignoring the round-off errors) for all three payment options. This is because all three options will pay off the mortgage loan in exactly 25 years, if the calculated regular payment amounts are made for 25 years. The difference between the three payment schedules then is the total amount of interest paid. As shown in part (a), more money is paid out over the whole term when the frequency of payments decreases. Since the same amount of principal is paid off over the three-year term, the excess amount corresponds to additional interest paid.

(c) What is the end-of-term balance, when monthly payments and some prepayment privileges are used?

Figure 4.19(c) shows the cash flow diagram of the mortgage for the first term of three years when the amount of regular payment increases from year to year. In this figure, B is the end of term balance to be calculated.

$$\begin{aligned}
B_{(c)} &= \$100,000(F/P, 0.6558\%, 36) \\
&\quad - \$763.20(F/A, 0.6558\%, 36) \\
&\quad - \$381.60(F/A, 0.6558\%, 24) \\
&\quad - \$381.60(F/A, 0.6558\%, 12) \\
&= \$\ 81,023.51
\end{aligned}$$

(d) What is the end-of-term balance with some additional lump sum payments?

Notice that the lump sum payments are within the prepayment privilege limits. We can utilize the result calculated in (c), as shown in Figure 4.19(d).

$$\begin{aligned}
B_{(d)} &= \$81,023.51 - \$8000(F/P, 0.6558\%, 24) \\
&\quad - \$10,000(F/P, 0.6558\%, 12) \\
&= \$60,848.71
\end{aligned}$$

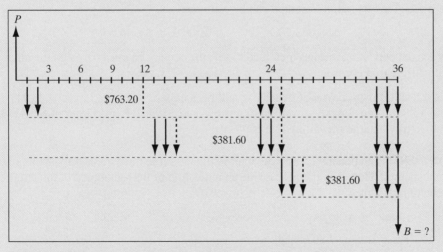

Figure 4.19(c) Equivalent cash flow diagram (Example 4.16(c)).

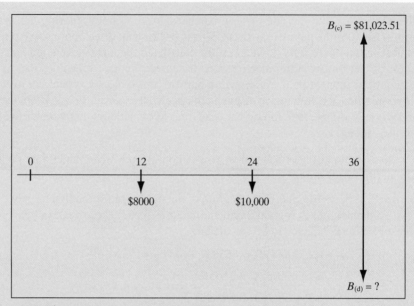

$B_{(c)} = \$81,023.51$

0   12   24   36

$8000 $10,000

$B_{(d)} = ?$

Figure 4.19(d)   Cash flow diagram for Example 4.16(d).

(e)   What are the prepayment penalties?

To find the prepayment penalty, we need to calculate the total prepayment amount first (i.e., the balance of the loan after monthly payments have been made for a year).

$$B = \$100,000(F/P, 0.6558\%, 12) - \$763.20(F/A, 0.6558\%, 12)$$
$$= \$98,663.79$$

Three months' simple interest = $\$98,663.79 \times 0.6558\% \times 3 = \$1941.11$.

When the prevailing rate has dropped to 6% and there are two more years left in the term, the interest rate differential can be estimated as

$$\$98,663.79 \times (8\% - 6\%) \times 2 = \$3946.55.$$

When the prevailing rate has increased to 9%, the penalty charged based on the interest rate differential is zero.

As a result, the prepayment penalty would be $3946.55 (i.e., the larger of $1941.11 and $3946.55) when the prevailing rate is 6% and $1941.11 (the larger of $1941. 11 and zero) when the prevailing rate is 9%.

COMMENT: The penalty charge calculation illustrated in this case is only an approximate method. However, it does provide an indication of the magnitude of the required penalty payment.

4.6 Investment in Bonds

Bonds are a specialized form of a loan in which the creditor—usually a business or the federal, provincial, or local government—promises to pay a stated amount of interest at specified intervals for a defined period and then to repay the principal at a specific date, known as the maturity date of the bond. Bonds are an important financial instrument by which the business world may raise funds to finance projects.

In the case of bonds, the lenders are investors (known as bondholders) who may be individuals or other businesses. Bonds can be a significant investment opportunity. In addition to the interest they earn, once purchased by the initial bondholder, they may be sold again for amounts other than their stated face value and thus enhance the bondholder's opportunity to increase the return on his or her initial investment. Given these complications, the concept of economic equivalence can be important in determining the worth of bonds. We will illustrate some typical bond investment problems in the context of economic equivalence.

4.6.1 Bond Terminology

Before explaining how bond values are determined, some of the terms associated with a typical bond will be defined.

Par value (or face value): Individual bonds are normally issued in even denominations, such as $1000 or multiples of $1000. The stated face value on the individual bond is termed the *par value*. For this type of bond, it is usually set at $1000.

Maturity date: Bonds generally have a specified *maturity date* on which the par value is to be repaid. Depending on the time remaining to maturity, bonds can be classified into the following categories: short-term bonds (maturing within three years), medium-term bonds (maturing from three to 10 years), and long-term bonds (maturing in more than 10 years). The term status of a bond depends on the time remaining to maturity, and the bond will change its term status as the maturity date approaches.

Coupon rate: The interest paid on the par value of a bond is called the annual *coupon rate*. For example, if you hold a 20-year $1000 bond paying 8% semiannually to maturity, you will receive an interest payment of $1000 x 8%/2 = $40 every six months for 20 years and then $1000 at its maturity.

Discount or premium bond: A bond that sells below its par value is called a *discount bond*. When a bond sells above its par value, it is called a *premium bond*. A bond that is purchased for its par value is said to have been bought *at par*. The price one has to pay to purchase a bond is called the bond's *market value*.

A call or redemption feature: Issuers of bonds sometimes reserve the right to pay off the bonds before the maturity date. This privilege is known as the *call feature* or *redemption feature* and a bond bearing this clause is known as a *callable bond* or a *redeemable bond*. As a rule, the issuer needs to give prior notice to the bondholders to call the bond.

Bearer bonds and registered bonds: A bond in bearer form does not have the name of the owner printed on the bond. It has interest coupons attached. The holder of the bond is presumed to be the owner of the bond. A fully registered bond shows the name of the current owner and interest is paid by cheque or direct deposit in the owner's bank account.

4.6.2 Types of Bonds

We can divide bonds into government bonds and corporate bonds. In the following, we provide a brief description of bonds in both categories.

There are three main issuers of government bonds: federal government, provinces, and municipalities. The federal government has very broad taxation power and thus has guaranteed income. Securities issued by the federal government, therefore, have the highest credit rating of any bonds issued in Canada. The federal government is also the largest issuer in the Canadian bond market.

Government of Canada Treasury bills: Treasury bills are short-term debts. Depending on the amount of money invested, they may have terms to maturity of 30 days, 60 days, 90 days, 180 days, and one year. Treasury bills are sold at a discount (below the par value) and mature at the par value. The difference between the purchase price and par at maturity gives the investor the yield to maturity. There are no interest payments. Large financial institutions buy treasury bills at the bi-weekly auctions by the government. They then offer these treasury bills to individual investors. Individual investors cannot buy treasury bills directly from the government. The prices of treasury bills vary from day to day. Treasury bills may be bought or sold at market prices. Their denominations begin as low as $1000.

Marketable bonds: Compared to treasury bills, marketable bonds have a longer term to maturity that can range from one year to 30 years. They have a fixed semiannual interest payment. Marketable bonds are issued and offered to large financial institutions at auctions on a scheduled basis by the government. These financial institutions will then offer these bonds to individual investors. Marketable bonds can be bought or sold at market prices. Their market prices may change from day to day. Their denominations also begin as low as $1000.

Canada premium bonds: Canada premium bonds have a fixed interest rate. The government guarantees both the principal and the interest rate. You may buy them at any bank, credit union, or most investment dealers. Their prices are fixed. You can only redeem your bonds each year on the anniversary date and during the 30 days thereafter. Canada premium bonds have two forms. You may choose the regular interest bonds (R-bonds), wherein interest can be deposited directly into a banking account or paid by cheque. You can also purchase the compound interest bonds (C-bonds). In this case, your interest will be automatically reinvested to purchase more C-bonds until your bonds reach maturity. Compound interest bonds are available in denominations as small as $100 while regular interest bonds are available starting at $300.

Canada savings bonds: Canada savings bonds offer minimum guaranteed interest rates that may be increased depending on the market conditions. The government guarantees both the principal and the minimum interest rate. Canada savings bonds can be redeemed at any time. The prices of the bonds are fixed. If you cash your bonds within three months of the issue date, you will receive the full face value of the bond only with no interest. If you cash them after that date, you'll receive the full face value plus all the interest earned for each full month elapsed since the issue date. Like the Canada premium bonds, there are also two forms of Canada savings bonds. They are available in similar denominations.

Provincial bonds: Similar to the federal government, the provincial governments also issue debt directly and may guarantee them through agencies under their jurisdictions.

Provincial bonds are issued to raise funds for major public capital projects. All provinces have statutes covering the use of funds raised through the issue of bonds. The bonds may be available in a wide range of denominations from a few hundred dollars to thousands of dollars. The terms, interest rates, and other conditions depend on the debt issues.

Municipal bonds: Municipalities are responsible for construction and maintenance of streets, sewers, waterworks, schools, and other services for individual communities. To provide these essential services and facilities, a municipality needs to take on capital projects from time to time. One method to raise the required capital for such projects is to issue bonds. Investors need to check with the issuers for further details such as denominations, interest rates, maturity dates, and guarantees.

Like governments, corporations may also issue bonds to borrow money. These securities have various terms, interest rates, and payment arrangements at the discretion of the issuer. Bonds for larger corporations are publicly traded.

Mortgage bonds: A mortgage is a legal document specifying that the borrower's fixed assets such as land, buildings, and equipment are pledged as security for a loan and the lender is entitled to ownership or partial ownership of the assets if the borrower fails to pay interest or repay principal when due. The lender holds the mortgage document until the loan is paid off. The mortgage bond is very similar to a home mortgage. Both are issued to protect the lender if the borrower fails to repay the loan. Mortgage bonds are issued so that the company can borrow money from many individuals who purchase the bonds. Instead of creating a mortgage document between each individual buyer and the company issuing the bonds, the mortgage document is deposited with a trust company on behalf of all individual buyers. This mechanism is a mortgage bond.

First mortgage bonds, second mortgage bonds, and general mortgage bonds: A first mortgage bond has the first priority on charging the company's assets and earnings. It may be closed or open. If it is closed, no additional first mortgage bonds may be issued. If it is open, additional first mortgage bonds may be issued under certain conditions, usually when new assets are acquired. A second mortgage bond has the right to the assets and earnings of the company after the claims of the first mortgage bonds have been satisfied. A general mortgage bond ranks after the first mortgage bonds and other prior mortgage bonds.

Corporate debentures and notes: These are debt securities that are not backed by specific fixed assets or other property. They rank after all mortgage bonds. Their only security is the owners' investment and the retained earnings of the company.

Convertible bonds and debentures: Convertible bonds or debentures may be exchanged for the company's stocks at the holder's option. These bonds or debentures have a fixed interest rate and maturity date. Their values may also increase because of the holder's right to convert them into common stocks at stated prices over stated periods.

4.6.3 Bond Valuation

You can trade bonds on the market just as you do stocks. Once a bond is purchased, you may keep it until it matures or for a variable number of interest periods before selling it. You can purchase or sell bonds at prices other than face value, depending on the economic environment. Furthermore, bond prices change over time because of the risk of

Yield to maturity (YTM): Yield that would be realized on a bond or other fixed income security if the bond was held until the maturity date.

Current yield: Annual income (interest) divided by the current market price of the bond.

nonpayment of interest or par value, supply and demand, and the economic outlook. These factors affect the **yield to maturity** (or **return on investment**) of the bond:

- The **yield to maturity** represents the actual interest earned from a bond over the holding period. In other words, the yield to maturity on a bond is the interest rate that establishes the equivalence between all future interest and face-value receipts and the market price of the bond.
- The **current yield** of a bond is the annual interest earned, as a percentage of the current market price. The current yield provides an indication of the annual return realized from investment in the bond. To illustrate, we will explain these values with numerical examples shortly.

Example 4.17 illustrates how you calculate the yield to maturity and the current yield, considering both the purchase price and the capital gain for a new issue.

EXAMPLE 4.17 Yield to Maturity and Current Yield

Consider buying a $1000 corporate bond at the market price of $996.25. The interest will be paid semiannually, the interest rate per payment period will be simply 4.8125%, and 20 interest payments over 10 years are required. We show the resulting cash flow to the investor in Figure 4.20. Find (a) the yield to maturity and (b) the current yield.

DISCUSSION

- **Mortgage bond:** The company wants to issue bonds totalling $100 million, and the company will back the bonds with specific pieces of property, such as buildings.
- **Par value:** The bond has a par value of $1000.
- **Maturity date:** The bonds have a 10-year maturity at the time of their issue.
- **Coupon rate:** The coupon rate is 9.625%, and interest is payable semiannually. For example, the bonds have a $1000 par value, and they pay $96.25 in simple interest $\left(9\frac{5}{8}\%\right)$ each year ($48.13 every six months).
- **Discount bond:** At 99.625%, or a 0.375% discount, the bonds are offered at less than their par value.

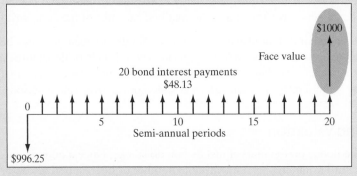

Figure 4.20 A typical cash flow transaction associated with an investment in the corporate bond.

SOLUTION

Given: Initial purchase price = $996.25, coupon rate = 9.625% per year paid semi-annually, and 10-year maturity with a par value of $1000.

Find: (a) Yield to maturity and (b) current yield.

(a) *Yield to maturity.* We find the yield to maturity by determining the interest rate that makes the present worth of the receipts equal to the market price of the bond:

$$\$996.25 = \$48.13(P/A, i, 20) + \$1000(P/F, i, 20).$$

The value of i that makes the present worth of the receipts equal to $996.25 lies between 4.5% and 5%. We could use a linear interpolation to find the yield to maturity, but it is much easier to obtain with Excel's **Goal Seek** function. As shown in Figure 4.21, solving for i yields $i = 4.8422\%$.

Note that this 4.8422% is the yield to maturity *per semiannual period*. The nominal (annual) yield is $2(4.8422) = 9.6844\%$ compounded semiannually. Compared with the coupon rate of $\left(9\frac{5}{8}\%\right)$ (or 9.625%), purchasing the bond with the price discounted at 0.375% brings about an additional 0.0594% yield. The effective annual interest rate is then

$$i_a = (1 + 0.048422)^2 - 1 = 9.92\%.$$

The 9.92% represents the **effective annual yield** to maturity on the bond. Notice that when you purchase a bond at par value and sell at par value, the yield to maturity will be the *same* as the coupon rate of the bond.

Until now, we have observed differences in nominal and effective interest rates because of the frequency of compounding. In the case of bonds, the reason

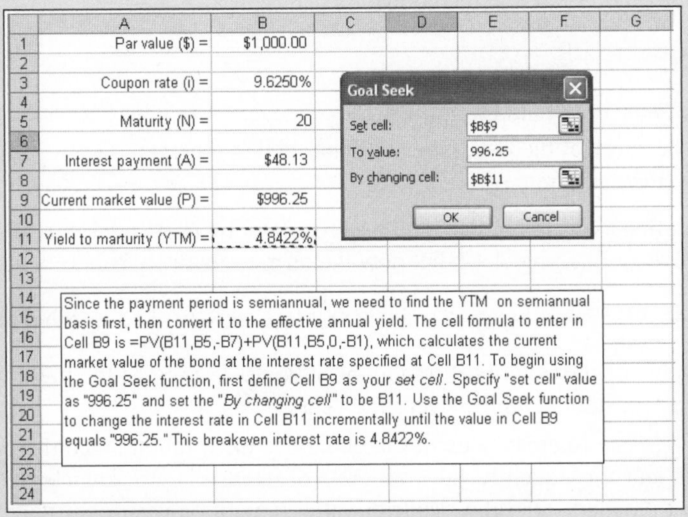

Figure 4.21 Finding the yield to maturity of the bond with Excel's Goal Seek function.

is different: The stated (par) value of the bond and the actual price for which it is sold are not the same. We normally state the nominal interest as a percentage of par value. However, when the bond is sold at a discount, the same nominal interest on a smaller initial investment is earned; hence, your effective interest earnings are greater than the stated nominal rate.

(b) *Current yield.* For our example, we compute the current yield as follows:

$$\frac{\$48.13}{996.25} = 4.83\% \text{ per semiannual period,}$$

$$4.83\% \times 2 = 9.66\% \text{ per year (nominal current yield),}$$

$$i_a = (1 + 0.0483)^2 - 1 = 9.90\%.$$

This effective current yield is 0.02% lower than the 9.92% yield to maturity we just computed. If the bond is selling at a discount, the current yield is smaller than the yield to maturity. If the bond is selling at a premium, the current yield is larger than the yield to maturity. A significant difference between the yield to maturity and the current yield of a bond can exist because the market price of a bond may be more or less than its face value. Moreover, both the current yield and the yield to maturity may differ considerably from the stated coupon value of the bond.

EXAMPLE 4.18 Bond Value Over Time

Consider again the bond investment introduced in Example 4.17. If the yield to maturity remains constant at 9.68%, (a) what will be the value of the bond one year after it was purchased? (b) If the market interest rate drops to 9% a year later, what would be the market price of the bond?

SOLUTION

Given: The same data as in Example 4.17.
Find: (a) The value of the bond one year later and (b) the market price of the bond a year later at the going rate of 9% interest.

(a) We can find the value of the bond one year later by using the same valuation procedure as in Example 4.17, but now the term to maturity is only nine years:

$$\$48.13(P/A, 4.84\%, 18) + \$1000(P/F, 4.84\%, 18) = \$996.80.$$

The value of the bond will remain at $996.80 as long as the yield to maturity remains constant at 9.68% over nine years.

(b) Now suppose interest rates in the economy have fallen since the bonds were issued, and consequently, the going rate of interest is 9%. Then both the coupon interest payments and the maturity value remain constant, but now 9% values have to be used to calculate the value of the bond. The value of the bond at the end of the first year would be

$$\$48.13(P/A, 4.5\%, 18) + \$1000(P/F, 4.5\%, 18) = \$1038.06.$$

Thus, the bond would sell at a premium over its par value.

COMMENTS: The arithmetic of the bond price increase should be clear, but what is the logic behind it? We can explain the reason for the increase as follows: Because the going market interest rate for the bond has fallen to 9%, if we had $1000 to invest, we could buy new bonds with a coupon rate of 9%. These would pay $90 interest each year, rather than $96.80. We would prefer $96.80 to $90; therefore, we would be willing to pay more than $1000 for this company's bonds to obtain higher coupons. All investors would recognize these facts, and hence the bonds would be bid up in price to $1038.06. At that point, they would provide the same yield to maturity (rate of return) to a potential investor as would the new bonds, namely, 9%.

SUMMARY

- Interest is most frequently quoted by financial institutions as an **annual percentage rate**, or **APR**. However, compounding frequently occurs more often than once annually, and the APR does not account for the effect of this more frequent compounding. The situation leads to the distinction between nominal and effective interest:

- **Nominal interest** is a stated rate of interest for a given period (usually a year).

- **Effective interest** is the actual rate of interest, which accounts for the interest amount accumulated over a given period. The **effective rate** is related to the APR by the equation

$$i = \left(1 + \frac{r}{M}\right)^M - 1,$$

where r is the APR, M is the number of compounding periods, and i is the effective interest rate.

  In any equivalence problem, the interest rate to use is the effective interest rate per payment period, or

$$i = \left[1 + \frac{r}{CK}\right]^C - 1,$$

where C is the number of compounding periods per payment period, K is the number of payment periods per year, and r/K is the nominal interest rate per payment period. Figure 4.9 outlines the possible relationships between compounding and payment periods and indicates which version of the effective-interest formula to use.

- The equation for determining the effective interest of continuous compounding is

$$i = e^{r/K} - 1.$$

The difference in accumulated interest between continuous compounding and very frequent compounding ($M > 50$) is minimal.

- Cash flows, as well as compounding, can be continuous. Table 4.2 shows the interest factors to use for continuous cash flows with continuous compounding.

- Nominal (and hence effective) interest rates may fluctuate over the life of a cash flow series. Some forms of home mortgages and bond yields are typical examples.

- **Amortized loans** are paid off in equal installments over time, and most of these loans have interest that is compounded monthly.

- Under a typical **add-on loan**, the lender precalculates the total simple interest amount and adds it to the principal. The principal and this precalculated interest amount are then paid together in equal installments.

- The term **mortgage** refers to a special type of loan for buying a piece of property, such as a house or a commercial building. The cost of the mortgage will depend on many factors, including the amount and term of the loan and the frequency of payments, as well as other fees.

- Two types of mortgages are common: fixed-rate mortgages and variable-rate mortgages. Fixed-rate mortgages offer loans whose interest rates are fixed over the period of the contract, whereas variable-rate mortgages offer interest rates that fluctuate with market conditions. In general, the initial interest rate is lower for variable-rate mortgages, as the lenders have the flexibility to adjust the cost of the loans over the period of the contract.

- **Asset-backed bonds:** If a company backs its bonds with specific pieces of property, such as buildings, we call these types of bonds **mortgage bonds**, which indicate the terms of repayment and the particular assets pledged to the bondholders in case of default. It is much more common, however, for a corporation simply to pledge its overall assets. A **debenture bond** represents such a promise.

- **Par value:** Individual bonds are normally issued in even denominations of $1000 or multiples of $1000. The stated face value of an individual bond is termed the **par value**.

- **Maturity date:** Bonds generally have a specified **maturity** date on which the par value is to be repaid.

- **Coupon rate:** We call the interest paid on the par value of a bond the **annual coupon rate**. The time interval between interest payments could be of any duration, but a semiannual period is the most common.

- **Discount or premium bond:** A bond that sells below its par value is called a **discount bond**. When a bond sells above its par value, it is called a **premium bond**.

PROBLEMS

4.1 If your credit card calculates interest based on 12.5% APR compounded daily,
   (a) What are your monthly interest rate and annual effective interest rate?
   (b) If you current outstanding balance is $2000 and you skip payments for two months, what would be the total balance two months from now?

4.2 A department store has offered you a credit card that charges interest at 1.05% per month, compounded monthly. What is the nominal interest (annual percentage) rate for this credit card? What is the effective annual interest rate?

4.3 A local bank advertised the following information: interest 6.89%—effective annual yield 7.128%. No mention was made of the compounding period in the advertisement. Can you figure out the compounding scheme used by the bank?

4.4 College Financial Sources, which makes small loans to college students, offers to lend $500. The borrower is required to pay $40 at the end of each week for 16 weeks. Find the interest rate per week. What is the nominal interest rate per year? What is the effective interest rate per year?

4.5 A financial institution is willing to lend you $400. However, $450 is repaid at the end of one week.

(a) What is the nominal interest rate?

(b) What is the effective annual interest rate?

4.6 The Cadillac Motor Car Company is advertising a 24-month lease of a Cadillac DeVille for $520, payable at the beginning of each month. The lease requires a $2500 down payment, plus a $500 refundable security deposit. As an alternative, the company offers a 24-month lease with a single up-front payment of $12,780, plus a $500 refundable security deposit. The security deposit will be refunded at the end of the 24-month lease. Assuming an interest rate of 6%, compounded monthly, which lease is the preferred one?

4.7 As a typical middle-class consumer, you are making monthly payments on your home mortgage (9% annual interest rate), car loan (12%), home improvement loan (14%), and past-due charge accounts (18%). Immediately after getting a $100 monthly raise, your friendly mutual fund broker tries to sell you some investment funds with a guaranteed return of 10% per year. Assuming that your only other investment alternative is a savings account, should you buy?

4.8 A loan company offers money at 1.8% per month, compounded monthly.

(a) What is the nominal interest rate?

(b) What is the effective annual interest rate?

(c) How many years will it take an investment to triple if interest is compounded monthly?

(d) How many years will it take an investment to triple if the nominal rate is compounded continuously?

4.9 Suppose your savings account pays 9% interest compounded quarterly. If you deposit $10,000 for one year, how much would you have?

4.10 What will be the amount accumulated by each of these present investments?

(a) $5635 in 10 years at 5% compounded semiannually.

(b) $7500 in 15 years at 6% compounded quarterly.

(c) $38,300 in 7 years at 9% compounded monthly.

4.11 How many years will it take an investment to triple if the interest rate is 9% compounded

(a) Quarterly?   (b) Monthly?   (c) Continuously?

4.12 A series of equal quarterly payments of $5000 for 12 years is equivalent to what present amount at an interest rate of 9% compounded

(a) Quarterly?   (b) Monthly?   (c) Continuously?

4.13 What is the future worth of an equal payment series of $3000 each quarter for five years if the interest rate is 8% compounded continuously?

4.14 Suppose that $2000 is placed in a bank account at the end of each quarter over the next 15 years. What is the future worth at the end of 15 years when the interest rate is 6% compounded

(a) Quarterly? (b) Monthly? (c) Continuously?

4.15 A series of equal quarterly deposits of $1000 extends over a period of three years. It is desired to compute the future worth of this quarterly deposit series at 12% compounded monthly. Which of the following equations is correct?

(a) $F = 4(\$1000)(F/A, 12\%, 3)$. (b) $F = \$1000(F/A, 3\%, 12)$.

(c) $F = \$1000(F/A, 1\%, 12)$. (d) $F = \$1000(F/A, 3.03\%, 12)$.

4.16 If the interest rate is 8.5% compounded continuously, what is the required quarterly payment to repay a loan of $12,000 in five years?

4.17 What is the future worth of a series of equal monthly payments of $2500 if the series extends over a period of eight years at 12% interest compounded

(a) Quarterly? (b) Monthly? (c) Continuously?

4.18 Suppose you deposit $500 at the end of each quarter for five years at an interest rate of 8% compounded monthly. What equal end-of-year deposit over the five years would accumulate the same amount at the end of the five years under the same interest compounding? To answer the question, which of the following is correct?

(a) $A = [\$500(F/A, 2\%, 20)] \times (A/F, 8\%, 5)$.

(b) $A = \$500(F/A, 2.013\%, 4)$.

(c) $A = \$500\left(F/A, \dfrac{8\%}{12}, 20 \right) \times (A/F, 8\%, 5)$.

(d) None of the above.

4.19 A series of equal quarterly payments of $2000 for 15 years is equivalent to what future lump-sum amount at the end of 10 years at an interest rate of 8% compounded continuously?

4.20 What will be the required quarterly payment to repay a loan of $32,000 in five years, if the interest rate is 7.8% compounded continuously?

4.21 A series of equal quarterly payments of $4000 extends over a period of three years. What is the present worth of this quarterly payment series at 8.75% interest compounded continuously?

4.22 What is the future worth of the following series of payments?

(a) $6000 at the end of each six-month period for 6 years at 6% compounded semiannually.

(b) $42,000 at the end of each quarter for 12 years at 8% compounded quarterly.

(c) $75,000 at the end of each month for 8 years at 9% compounded monthly.

4.23 What equal series of payments must be paid into a sinking fund to accumulate the following amount?

(a) $21,000 in 10 years at 6.45% compounded semiannually when payments are semiannual.

(b) $9000 in 15 years at 9.35% compounded quarterly when payments are quarterly.

(c) $24,000 in 5 years at 6.55% compounded monthly when payments are monthly.

4.24 You have a habit of drinking a cup of Starbucks coffee ($2.50 a cup) on the way to work every morning. If, instead, you put the money in the bank for 30 years, how much would you have at the end of that time, assuming that your account earns 5% interest compounded *daily*? Assume also that you drink a cup of coffee every day, including weekends.

4.25 John Jay is purchasing a $24,000 automobile, which is to be paid for in 48 monthly installments of $543.35. What effective annual interest is he paying for this financing arrangement?

4.26 A loan of $12,000 is to be financed to assist in buying an automobile. On the basis of monthly compounding for 42 months, the end-of-the-month equal payment is quoted as $445. What nominal interest rate in percentage is being charged?

4.27 Suppose a young newlywed couple is planning to buy a home two years from now. To save the down payment required at the time of purchasing a home worth $220,000 (let's assume that the down payment is 10% of the sales price, or $22,000), the couple decides to set aside some money from each of their salaries at the end of every month. If each of them can earn 6% interest (compounded monthly) on his or her savings, determine the equal amount this couple must deposit each month until the point is reached where the couple can buy the home.

4.28 What is the present worth of the following series of payments?

(a) $1500 at the end of each six-month period for 12 years at 8% compounded semiannually.

(b) $2500 at the end of each quarter for 8 years at 8% compounded quarterly.

(c) $3800 at the end of each month for 5 years at 9% compounded monthly.

4.29 What is the amount of the quarterly deposits A such that you will be able to withdraw the amounts shown in the cash flow diagram if the interest rate is 8% compounded quarterly?

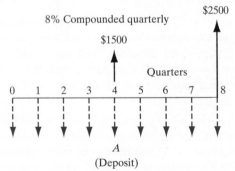

4.30 Georgi Rostov deposits $15,000 in a savings account that pays 6% interest compounded monthly. Three years later, he deposits $14,000. Two years after the $14,000 deposit, he makes another deposit in the amount of $12,500. Four years after the $12,500 deposit, half of the accumulated funds is transferred to a fund that pays 8% interest compounded quarterly. How much money will be in each account six years after the transfer?

4.31 A man is planning to retire in 25 years. He wishes to deposit a regular amount every three months until he retires, so that, beginning one year following his retirement, he will receive annual payments of $60,000 for the next 10 years. How much must he deposit if the interest rate is 6% compounded quarterly?

4.32 You borrowed $15,000 for buying a new car from a bank at an interest rate of 12% compounded monthly. This loan will be repaid in 48 equal monthly installments over four years. Immediately after the 20th payment, you desire to pay the remainder of the loan in a single payment. Compute this lump-sum amount of that time.

4.33 A building is priced at $125,000. If a down payment of $25,000 is made and a payment of $1000 every month thereafter is required, how many months will it take to pay for the building? Interest is charged at a rate of 9% compounded semi-annually.

4.34 You obtained a loan of $20,000 to finance an automobile. Based on monthly compounding over 24 months, the end-of-the-month equal payment was figured to be $922.90. What APR was used for this loan?

4.35 *The Engineering Economist* (a professional journal) offers three types of subscriptions, payable in advance: one year at $66, two years at $120, and three years at $160. If money can earn 6% interest compounded monthly, which subscription should you take? (Assume that you plan to subscribe to the journal over the next three years.)

4.36 A couple is planning to finance its three-year-old son's university education. Money can be deposited at 6% compounded quarterly. What quarterly deposit must be made from the son's 3rd birthday to his 18th birthday to provide $50,000 on each birthday from the 18th to the 21st? (Note that the last deposit is made on the date of the first withdrawal.)

4.37 Sam Salvetti is planning to retire in 15 years. Money can be deposited at 8% compounded quarterly. What quarterly deposit must be made at the end of each quarter until Sam retires so that he can make a withdrawal of $25,000 semiannually over the first five years of his retirement? Assume that his first withdrawal occurs at the end of six months after his retirement.

4.38 Michelle Hunter received $250,000 from an insurance company after her husband's death. Michelle wants to deposit this amount in a savings account that earns interest at a rate of 6% compounded monthly. Then she would like to make 120 equal monthly withdrawals over the 10-year period such that, when she makes the last withdrawal, the savings account will have a balance of zero. How much can she withdraw each month?

4.39 Anita Tahani, who owns a travel agency, bought an old house to use as her business office. She found that the ceiling was poorly insulated and that the heat loss could be cut significantly if 15 centimetres of foam insulation were installed. She estimated that with the insulation, she could cut the heating bill by $40 per month and the air-conditioning cost by $25 per month. Assuming that the summer season is three months (June, July, and August) of the year and that the winter season is another three months (December, January, and February) of the year, how much can Anita spend on insulation if she expects to keep the property for five years? Assume that neither heating nor air-conditioning would be required during the fall and spring seasons. If she decides to install the insulation, it will be done at the beginning of May. Anita's interest rate is 9% compounded monthly.

4.40 A new chemical production facility that is under construction is expected to be in full commercial operation one year from now. Once in full operation, the facility will generate \$63,000 cash profit daily over the plant's service life of 12 years. Determine the equivalent present worth of the future cash flows generated by the facility at the beginning of commercial operation, assuming

(a) 12% interest compounded daily, with the daily flows.

(b) 12% interest compounded continuously, with the daily flow series approximated by a uniform continuous cash flow function.

Also, compare the difference between (a) discrete (daily) and (b) continuous compounding.

4.41 Income from a project is expected to decline at a constant rate from an initial value of \$500,000 at time 0 to a final value of \$40,000 at the end of year 3. If interest is compounded continuously at a nominal annual rate of 11%, determine the present value of this continuous cash flow.

4.42 A sum of \$80,000 will be received uniformly over a five-year period beginning two years from today. What is the present value of this deferred-funds flow if interest is compounded continuously at a nominal rate of 9%?

4.43 A small chemical company that produces an epoxy resin expects its production volume to decay exponentially according to the relationship

$$y_t = 5e^{-0.25t},$$

where y_t is the production rate at time t. Simultaneously, the unit price is expected to increase linearly over time at the rate

$$u_t = \$55(1 + 0.09t).$$

What is the expression for the present worth of sales revenues from $t = 0$ to $t = 20$ at 12% interest compounded continuously?

4.44 Consider the accompanying cash flow diagram, which represents three different interest rates applicable over the five-year time span shown.

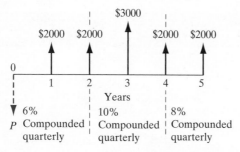

(a) Calculate the equivalent amount P at the present time.

(b) Calculate the single-payment equivalent to F at $n = 5$.

(c) Calculate the equal-payment-series cash flow A that runs from $n = 1$ to $n = 5$.

4.45 Consider the cash flow transactions depicted in the accompanying cash flow diagram, with the changing interest rates specified.

(a) What is the equivalent present worth? (In other words, how much do you have to deposit now so that you can withdraw \$300 at the end of year 1, \$300 at the end of year 2, \$500 at the end of year 3, and \$500 at the end of year 4?)

(b) What is the single effective annual interest rate over four years?

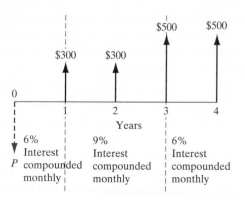

4.46 Compute the future worth of the cash flows with the different interest rates specified. The cash flows occur at the end of each year over four years.

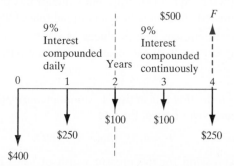

4.47 An automobile loan of $20,000 at a nominal rate of 9% compounded monthly for 48 months requires equal end-of-month payments of $497.70. Complete the following table for the first six payments, as you would expect a bank to calculate the values:

End of Month (n)	Interest Payment	Repayment of Principal	Remaining Loan Balance
1			$19,652.30
2			
3		$352.94	
4	$142.12		
5	$139.45		
6			$17,874.28

4.48 Mr. Smith wants to buy a new car that will cost $18,000. He will make a down payment in the amount of $8000. He would like to borrow the remainder from a bank at an interest rate of 9% compounded monthly. He agrees to pay off the loan monthly for a period of two years. Select the correct answer for the following questions:

(a) What is the amount of the monthly payment A?

    (i) $A = \$10,000(A/P, 0.75\%, 24)$.

    (ii) $A = \$10,000(A/P, 9\%, 2)/12$.

(iii) $A = \$10,000(A/F, 0.75\%, 24)$.

(iv) $A = \$12,500(A/F, 9\%, 2)/12$.

(b) Mr. Smith has made 12 payments and wants to figure out the balance remaining immediately after the 12th payment. What is that balance?

(i) $B_{12} = 12A$.

(ii) $B_{12} = A(P/A, 9\%, 1)/12$.

(iii) $B_{12} = A(P/A, 0.75\%, 12)$.

(iv) $B_{12} = 10,000 - 12A$.

4.49 Tony Wu is considering purchasing a used automobile. The price, including the title and taxes, is \$12,345. Tony is able to make a \$2345 down payment. The balance, \$10,000, will be borrowed from his credit union at an interest rate of 8.48% compounded daily. The loan should be paid in 36 equal monthly payments. Compute the monthly payment. What is the total amount of interest Tony has to pay over the life of the loan?

4.50 Suppose you are in the market for a new car worth \$18,000. You are offered a deal to make a \$1800 down payment now and to pay the balance in equal end-of-month payments of \$421.85 over a 48-month period. Consider the following situations:

(a) Instead of going through the dealer's financing, you want to make a down payment of \$1800 and take out an auto loan from a bank at 11.75% compounded monthly. What would be your monthly payment to pay off the loan in four years?

(b) If you were to accept the dealer's offer, what would be the effective rate of interest per month the dealer charges on your financing?

4.51 Bob Pearson borrowed \$25,000 from a bank at an interest rate of 10% compounded monthly. The loan will be repaid in 36 equal monthly installments over three years. Immediately after his 20th payment, Bob desires to pay the remainder of the loan in a single payment. Compute the total amount he must pay. Use semiannual compounding.

4.52 You plan to buy a \$200,000 home with a 10% down payment. The bank you want to finance the loan suggests two options: a 20-year mortgage at 9% APR and a 30-year mortgage at 10% APR. What is the difference in monthly payments (for the first 20 years) between these two options? Use semiannual compounding.

4.53 David Kapamagian borrowed money from a bank to finance a small fishing boat. The bank's terms allowed him to defer payments (including interest) on the loan for six months and to make 36 equal end-of-month payments thereafter. The original bank loan was for \$4800, with an interest rate of 12% compounded monthly. After 16 monthly payments, David found himself in a financial bind and went to a loan company for assistance in lowering his monthly payments. Fortunately, the loan company offered to pay his debts in one lump sum if he would pay the company \$104 per month for the next 36 months. What monthly rate of interest is the loan company charging on this transaction?

4.54 You are buying a home for \$250,000.

(a) If you make a down payment of \$50,000 and take out a mortgage on the rest of the money at 8.5% compounded semiannually, what will be your monthly payment to retire the mortgage in 15 years?

(b) Consider the seventh payment. How much will the interest and principal payments be?

4.55 With a $350,000 home mortgage loan with a 20-year term at 9% APR compounded semiannually, compute the total payments on principal and interest over the first five years of ownership, given monthly payments.

4.56 A lender requires that monthly mortgage payments be no more than 25% of gross monthly income with a maximum amortization of 30 years. If you can make only a 15% down payment, what is the minimum monthly income needed to purchase a $400,000 house when the interest rate is 9% compounded semiannually?

4.57 To buy a $150,000 house, you take out a 9% (APR) mortgage for $120,000. Five years later, you sell the house for $185,000 (after all other selling expenses). What equity (the amount that you can keep before tax) would you realize with a 30-year amortization?

4.58 Just after their 14th monthly payment,
- Family *A* had a balance of $80,000 on a 9%, 30-year mortgage;
- Family *B* had a balance of $80,000 on a 9%, 15-year mortgage; and
- Family *C* had a balance of $80,000 on a 9%, 20-year mortgage.

How much interest did each family pay on the 15th payment?

4.59 Home mortgage lenders in the United States usually charge points on a loan to avoid exceeding a legal limit on interest rates or to be competitive with other lenders. As an example, for a two-point loan, the lender would lend only $98 for each $100 borrowed. The borrower would receive only $98, but would have to make payments just as if he or she had received $100. Suppose that you receive a loan of $130,000, payable at the end of each month for 30 years with an interest rate of 9% compounded monthly, but you have been charged three points. What is the effective interest rate on this home mortgage loan?

4.60 A restaurant is considering purchasing a lot adjacent to its business to provide adequate parking space for its customers. The restaurant needs to borrow $35,000 to secure the lot. A deal has been made between a local bank and the restaurant so that the restaurant would pay the loan back over a five-year period with the following payment terms: 15%, 20%, 25%, 30%, and 35% of the initial loan at the end of first, second, third, fourth, and fifth years, respectively.

(a) What rate of interest is the bank earning from this loan?

(b) What would be the total interest paid by the restaurant over the five-year period?

4.61 Don Harrison's current salary is $60,000 per year, and he is planning to retire 25 years from now. He anticipates that his annual salary will increase by $3,000 each year (to $60,000 the first year, $63,000 the second year, $66,000 the third year, and so forth), and he plans to deposit 5% of his yearly salary into a retirement fund that earns 7% interest compounded daily. What will be the amount accumulated at the time of Don's retirement?

4.62 Consider the following two options for financing a car:
- **Option A.** Purchase the vehicle at the normal price of $26,200 and pay for the vehicle over three years with equal monthly payments at 1.9% APR financing.
- **Option B.** Purchase the vehicle for a discount price of $24,048, to be paid immediately. The funds that would be used to purchase the vehicle are presently earning 5% annual interest compounded monthly.

(a) What is the meaning of the APR of 1.9% quoted by the dealer?

(b) Under what circumstances would you prefer to go with the dealer's financing?

(c) Which interest rate (the dealer's interest rate or the savings rate) would you use in comparing the two options?

4.63 Katerina Unger wants to purchase a set of furniture worth $3000. She plans to finance the furniture for two years. The furniture store tells Katerina that the interest rate is only 1% per month, and her monthly payment is computed as follows:

- Installment period = 24 months.
- Interest = $24(0.01)(\$3000) = \$720$.
- Loan processing fee = $25.
- Total amount owed = $\$3000 + \$720 + \$25 = \$3745$.
- Monthly payment = $\$3745/24 = \$156.04$ per month.

(a) What is the annual effective interest rate that Katerina is paying for her loan transaction? What is the nominal interest (annual percentage rate) for the loan?

(b) Katerina bought the furniture and made 12 monthly payments. Now she wants to pay off the remaining installments in one lump sum (at the end of 12 months). How much does she owe the furniture store?

4.64 You purchase a piece of furniture worth $5000 on credit through a local furniture store. You are told that your monthly payment will be $146.35, including an acquisition fee of $25, at a 10% add-on interest rate over 48 months. After making 15 payments, you decide to pay off the balance. Compute the remaining balance, based on the conventional amortized loan.

4.65 Kathy Stonewall bought a new car for $15,458. A dealer's financing was available at an interest rate of 11.5% compounded monthly. Dealer financing required a 10% down payment and 60 equal monthly payments. Because the interest rate was rather high, Kathy checked her credit union for possible financing. The loan officer at the credit union quoted a 9.8% interest rate for a new-car loan and 10.5% for a used car. But to be eligible for the loan, Kathy has to be a member of the union for at least six months. Since she joined the union two months ago, she has to wait four more months to apply for the loan. Consequently, she decided to go ahead with the dealer's financing, and four months later she refinanced the balance through the credit union at an interest rate of 10.5%.

(a) Compute the monthly payment to the dealer.

(b) Compute the monthly payment to the union.

(c) What is the total interest payment on each loan?

4.66 A house can be purchased for $155,000, and you have $25,000 cash for a down payment. You are considering the following two financing options:

- **Option 1.** Getting a new standard mortgage with a 7.5% (APR) interest and a 30-year amortization.
- **Option 2.** Assuming the seller's old mortgage, which has an interest rate of 5.5% (APR), a remaining amortization of 25 years (the original amortization was 30 years), a remaining balance of $97,218, and payments of $593 per month. You can obtain a second mortgage for the remaining balance ($32,782) from your credit union at 9% (APR) with a 10-year repayment period.

(a) What is the effective interest rate of the combined mortgage?

(b) Compute the monthly payments for each option over the life of the mortgage.

(c) Compute the total interest payment for each option.

(d) What homeowner's interest rate makes the two financing options equivalent?

4.67 A loan of $10,000 is to be financed over a period of 24 months. The agency quotes a nominal rate of 8% for the first 12 months and a nominal rate of 9% for any remaining unpaid balance after 12 months, compounded monthly. Based on these rates, what equal end-of-the-month payment for 24 months would be required to repay the loan with interest?

4.68 Emily Wang financed her office furniture from a furniture dealer. The dealer's terms allowed her to defer payments (including interest) for six months and to make 36 equal end-of-month payments thereafter. The original note was for $15,000, with interest at 9% compounded monthly. After 26 monthly payments, Emily found herself in a financial bind and went to a loan company for assistance. The loan company offered to pay her debts in one lump sum if she would pay the company $186 per month for the next 30 months.

(a) Determine the original monthly payment made to the furniture store.

(b) Determine the lump-sum payoff amount the loan company will make.

(c) What monthly rate of interest is the loan company charging on this loan?

4.69 If you borrow $120,000 with a 30-year term at a 9% (APR) variable rate compounded monthly and the interest rate can be changed every five years,

(a) What is the initial monthly payment?

(b) If the lender's interest rate is 9.75% (APR) at the end of five years, what will the new monthly payments be?

4.70 The Jimmy Corporation issued a new series of bonds on January 1, 1996. The bonds were sold at par ($1000), have a 12% coupon rate, and mature in 30 years, on December 31, 2025. Coupon interest payments are made semiannually (on June 30 and December 31).

(a) What was the yield to maturity (YTM) of the bond on January 1, 1996?

(b) Assuming that the level of interest rates had fallen to 9%, what was the price of the bond on January 1, 2001, five years later?

(c) On July 1, 2001, the bonds sold for $922.38. What was the YTM at that date? What was the current yield at that date?

4.71 A $1000, 9.50% semiannual bond is purchased for $1010. If the bond is sold at the end of three years and six interest payments, what should the selling price be to yield a 10% return on the investment?

4.72 Mr. Gonzalez wishes to sell a bond that has a face value of $1000. The bond bears an interest rate of 8%, with bond interests payable semiannually. Four years ago, $920 was paid for the bond. At least a 9% return (yield) on the investment is desired. What must be the minimum selling price?

4.73 Suppose you have the choice of investing in (1) a zero-coupon bond, which costs $513.60 today, pays nothing during its life, and then pays $1000 after five years, or (2) a bond that costs $1000 today, pays $113 in interest semiannually, and matures at the end of five years. Which bond would provide the higher yield?

4.74 Suppose you were offered a 12-year, 15% coupon, $1000 par value bond at a price of $1298.68. What rate of interest (yield to maturity) would you earn if you bought the bond and held it to maturity (at semiannual interest)?

4.75 The Diversified Products Company has two bond issues outstanding. Both bonds pay $100 semiannual interest, plus $1000 at maturity. Bond A has a remaining maturity of 15 years, bond B a maturity of one year. What is the value of each of these bonds now, when the going rate of interest is 9%?

4.76 The AirJet Service Company's bonds have four years remaining to maturity. Interest is paid annually, the bonds have a $1000 par value, and the coupon interest rate is 8.75%.

   (a) What is the yield to maturity at a current market price of $1108?

   (b) Would you pay $935 for one of these bonds if you thought that the market rate of interest was 9.5%?

4.77 Suppose Ford sold an issue of bonds with a 15-year maturity, a $1000 par value, a 12% coupon rate, and semiannual interest payments.

   (a) Two years after the bonds were issued, the going rate of interest on bonds such as these fell to 9%. At what price would the bonds sell?

   (b) Suppose that, two years after the bonds' issue, the going interest rate had risen to 13%. At what price would the bonds sell?

   (c) Today, the closing price of the bond is $783.58. What is the current yield?

4.78 Suppose you purchased a corporate bond with a 10-year maturity, a $1000 par value, a 10% coupon rate, and semiannual interest payments. All this means that you receive a $50 interest payment at the end of each six-month period for 10 years (20 times). Then, when the bond matures, you will receive the principal amount (the face value) in a lump sum. Three years after the bonds were purchased, the going rate of interest on new bonds fell to 6% (or 6% compounded semiannually). What is the current market value (P) of the bond (three years after its purchase)?

Short Case Studies

ST4.1 Jim Norton, a third-year engineering student, was mailed two credit card applications from two different banks. Each bank offered a different annual fee and finance charge.

   Jim expects his average monthly balance after payment to the bank to be $300 and plans to keep the credit card he chooses for only 24 months. (After graduation, he will apply for a new card.) Jim's interest rate on his savings account is 6% compounded daily. The following table lists the terms of each bank:

Terms	Bank A	Bank B
Annual fee	$20	$30
Finance charge	1.55% monthly interest rate	16.5% annual percentage rate compounded monthly

   (a) Compute the effective annual interest rate for each card.

   (b) Which bank's credit card should Jim choose?

(c) Suppose Jim decided to go with Bank B and used the card for one year. The balance after one year is $1500. If he makes just a minimum payment each month (say, 5% of the unpaid balance), how long will it take to pay off the card debt? Assume that he will not make any new purchases on the card until he pays off the debt.

ST4.2 The following is a promotional pamphlet prepared by Trust Company Bank in Toronto, Ontario:

"Lower your monthly car payments as much as 48%." Now you can buy the car you want and keep the monthly payments as much as 48% lower than they would be if you financed with a conventional auto loan. Trust Company's *Alternative Auto Loan* (AAL) makes the difference. It combines the lower monthly payment advantages of leasing with tax and ownership of a conventional loan. And if you have your monthly payment deducted automatically from your Trust Company chequing account, you will save $\frac{1}{2}$% on your loan interest rate. Your monthly payments can be spread over 24, 36, or 48 months.

Amount Financed	Financing Period (months)	Monthly Alternative Auto Loan	Payment Conventional Auto Loan
	24	$249	$477
$10,000	36	211	339
	48	191	270
	24	498	955
$20,000	36	422	678
	48	382	541

The amount of the final payment will be based on the residual value of the car at the end of the loan. Your monthly payments are kept low because you make principal payments on only a portion of the loan and not on the residual value of the car. Interest is computed on the full amount of the loan. At the end of the loan period you may:

1. Make the final payment and keep the car.
2. Sell the car yourself, repay the note (remaining balance), and keep any profit you make.
3. Refinance the car.
4. Return the car to Trust Company in good working condition and pay only a return fee.

So, if you've been wanting a special car, but not the high monthly payments that could go with it, consider the *Alternative Auto Loan*. For details, ask at any Trust Company branch.

Note 1: The chart above is based on the following assumptions. Conventional auto loan 13.4% annual percentage rate. *Alternative Auto Loan* 13.4% annual percentage rate.

Note 2: The residual value is assumed to be 50% of sticker price for 24 months; 45% for 36 months. The amount financed is 80% of sticker price.

Note 3: Monthly payments are based on principal payments equal to the depreciation amount on the car and interest in the amount of the loan.

Note 4: The residual value of the automobile is determined by a published residual value guide in effect at the time your Trust Company's *Alternative Auto Loan* is originated.
Note 5: The minimum loan amount is $10,000 (Trust Company will lend up to 80% of the sticker price). Annual household income requirement is $50,000.
Note 6: Trust Company reserves the right of final approval based on customer's credit history.

(a) Show how the monthly payments were computed for the *Alternative Auto Loan* by the bank.

(b) Suppose that you have decided to finance a new car for 36 months from Trust Company. Suppose also that you are interested in owning the car (not leasing it). If you decided to go with the *Alternative Auto Loan*, you would make the final payment and keep the car at the end of 36 months. Assume that your opportunity cost rate (personal interest rate) is an interest rate of 8% compounded monthly. (You may view this opportunity cost rate as an interest rate at which you can invest your money in some financial instrument, such as a savings account.) Compare Trust Company's alternative option with the conventional option and make a choice between them.

ST4.3 A Registered Education Savings Plan (RESP) is a special savings account that can help you save for your child's post-secondary education. RESPs are registered by the Government of Canada so that savings for post-secondary education can grow tax-free until the beneficiary of the RESP enrols in studies beyond high school. If you open an RESP account, the federal government adds money to your RESP through the Canada Education Savings Grant (CESG). If you are an Alberta resident, you may be able to get additional money under the Alberta Centennial Education Savings (ACES) Plan.

You can put in up to a total of $50,000 for each child of yours named in an RESP. As soon as your child is born, you can open an RESP account for him or her and start to put money into this account every year until your child reaches 18 and attends a post-secondary institution. There is no annual contribution limit. The money in the RESP account can be invested into GICs, mutual funds, bonds, or stocks so that the income generated is tax-free. The CESG adds 20% of your own annual contribution up to a maximum of $400 per year to your RESP account. The maximum lifetime CESG each child can receive is $7200. The Government of Alberta will contribute $500 into the RESP of every child born to Alberta residents in 2005 and later. As well, grants of $100 are available to children of Alberta residents when they turn 8, 11, and 14 in 2005 and later.

Evaluate the benefits of RESP assuming different scenarios. For example, you may consider the scenario of contributing $2000 per year for a newborn baby for 18 years. Assume a moderate interest rate of 4% per year. How much will the RESP account accumulate when the child turns 18? Include all possible government grants. Ignore tax effects for now.

ST4.4 Suppose you are going to buy a home worth $210,000 and you make a down payment in the amount of $50,000. The balance will be borrowed from the Capital Savings and Loan Bank. The loan officer suggests the following two financing plans for the property:

- **Option 1:** A conventional fixed rate mortgage with an interest rate of 5.85%, a term of 5 years, and an amortization of 25 years.

- **Option 2:** A variable rate mortgage with an interest rate of prime plus 1%, that is, 3.25% as the prime rate is currently 2.25%, a term of 5 years, and an amortization of 25 years.

(a) Compute the monthly payment under each option.

(b) If you take the variable rate mortgage, which of the two monthly payments calculated under part (a) would you follow? Why?

(c) If the prime rate remains unchanged over the next five years, what would be the ending balance if you take the variable rate mortgage and follow each of the two monthly payments calculated in part (a)?

(d) Suppose that the prime rate will be adjusted up by a constant increment every six months. What will the prime rate have to be near the end of the term for the two options to be equivalent?

ST4.5 Common practices on mortgage loans are different from country to country. For example, in the United States, the compounding frequency is usually monthly. The term and the amortization of the mortgage are usually the same. This means that the interest rate for a fixed rate mortgage has to be specified for the entire amortization in the mortgage document. Consider the following advertisement seeking to sell a beachfront condominium at SunDestin, Florida:

$$95\% \text{ Financing } 8\tfrac{1}{8}\% \text{ interest!!}$$

Down Payment of 5%. Own a Gulf-Front Condominium for only $100,000 with a 30-year variable-rate mortgage. We're providing incredible terms: $95,000 mortgage (30 years), year 1 at 8.125%, year 2 at 10.125%, year 3 at 12.125%, and years 4 through 30 at 13.125%.

Even though the phrase "variable-rate mortgage" is used in this advertisement, the interest rate over the amortization of the mortgage does not depend on the prime rate at all. Actually the interest rates are specified for the entire amortization period.

(a) Compute the monthly payments for each year.

(b) Calculate the total interest paid over the life of the loan.

(c) Determine the equivalent single-effect annual interest rate for the loan.

On the Companion Website that accompanies this text, you will find Excel templates and exercises, as well as the following analysis tools: Cash Flow Analyzer, Depreciation Analysis, Loan Analysis, and Interest Tables.

Evaluation of Business and Engineering Assets

FIVE

Analysis of Independent Projects

Parking Meters Get Smarter[1]—Wireless Technology Turns Old-Fashioned Coin-Operated Device into a Sophisticated Tool for Catching Scofflaws and Raising Cash. Technology is taking much of the fun out of finding a place to park the car:

- In Pacific Grove, California, parking meters "know" when a car pulls out of the spot and quickly reset to zero—eliminating drivers' little joy of parking for free on someone else's quarters.
- In Montreal, when cars stay past their time limit, meters send real-time alerts to an enforcement officer's handheld device, reducing the number of people needed to monitor parking spaces—not to mention drivers' chances of getting away with violations.
- In Aspen, Colorado, wireless "in-car" meters may eliminate the need for curbside parking meters altogether: They dangle from the rearview mirror inside the car, ticking off a prepaid time.

Now, in cities from New York to Vancouver, the door is open to a host of wireless technologies seeking to improve the parking meter even further. Chicago and Sacramento, California, among others, are equipping enforcement vehicles with infrared cameras capable of scanning licence plates even at 50 kilometres an hour. Using a global positioning system, the cameras can tell which individual cars have parked too long in a two-hour parking zone. At a cost of $75,000 a camera, the system is an expensive upgrade of the old method of chalking tires and then coming back two hours later to see if the car has moved.

   The camera system, supplied by Canada's Autovu Technologies (now part of Montreal-based Genetec Inc.), also helps identify scofflaws and stolen vehicles by linking to a database of unpaid tickets and auto

[1] Christopher Conkey, "Parking Meters Get Smarter," *The Wall Street Journal*, June 30, 2005, p. B1.

thefts. Sacramento bought three cameras in August, and since then its practice of "booting," or immobilizing, cars with a lot of unpaid tickets has increased sharply. Revenue is soaring too. According to Howard Chan, Sacramento's parking director, Sacramento booted 189 cars and took in parking revenue of US$169,000 for the fiscal-year ended in June 2004; for fiscal 2005, the city expects to boot 805 cars and take in more than US$475,000.

In downtown Montreal, more than 400 "pay-by-space" meters, each covering 10 to 15 spaces, are a twist on regular multispace meters. Motorists park, then go to the meter to type in the parking-space number, and pay by card or coin. These meters, which cost about $9000 each, identify violators in real time for enforcement officers carrying handheld devices: A likeness of the block emerges on the screen, and cars parked illegally show up in red.

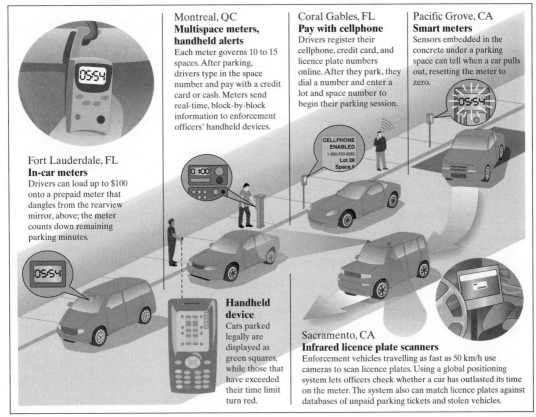

Montreal, QC
Multispace meters, handheld alerts
Each meter governs 10 to 15 spaces. After parking, drivers type in the space number and pay with a credit card or cash. Meters send real-time, block-by-block information to enforcement officers' handheld devices.

Coral Gables, FL
Pay with cellphone
Drivers register their cellphone, credit card, and licence plate numbers online. After they park, they dial a number and enter a lot and space number to begin their parking session.

CELLPHONE
ENABLED
1-800-555-5555
Lot 28
Space 6

Pacific Grove, CA
Smart meters
Sensors embedded in the concrete under a parking space can tell when a car pulls out, resetting the meter to zero.

Fort Lauderdale, FL
In-car meters
Drivers can load up to $100 onto a prepaid meter that dangles from the rearview mirror, above; the meter counts down remaining parking minutes.

Handheld device
Cars parked legally are displayed as green squares, while those that have exceeded their time limit turn red.

Sacramento, CA
Infrared licence plate scanners
Enforcement vehicles travelling as fast as 50 km/h use cameras to scan licence plates. Using a global positioning system lets officers check whether a car has outlasted its time on the meter. The system also can match licence plates against databases of unpaid parking tickets and stolen vehicles.

Parking czars in municipalities across the country are starting to realize parking meters' original goals: generating revenue and creating a continuous turnover of parking spaces on city streets. Clearly, their main question is "Would there be enough new revenues from installing the expensive parking monitoring devices?" or "How many devices could be installed to maximize the revenue streams?" From the device manufacturer's point of view, the question is "Would there be enough demand for these products to justify the investment required in new facilities and marketing?" If the manufacturer decides to go ahead and market the products, but the actual demand is far less than its forecast or the adoption of the technology is too slow, what would be the potential financial risk?

In Chapters 3 and 4, we presented the concept of the time value of money and developed techniques for establishing cash flow equivalence with compound-interest factors. That background provides a foundation for accepting or rejecting a capital investment: the economic evaluation of a project's desirability.

There are four common measures of investment attractiveness or profitability. Three of these are based on cash flow equivalence: (1) equivalent present worth (PW), (2) equivalent future worth (FW), and (3) equivalent annual worth (AE). Present worth represents a measure of future cash flow relative to the time point "now," with provisions that account for earning opportunities. Future worth is a measure of cash flow at some future planning horizon and offers a consideration of the earning opportunities of intermediate cash flows. Annual worth is a measure of cash flow in terms of equivalent equal payments made on an annual basis.

The fourth measure, rate of return, characterizes profitability or attractiveness of a project as a calculated yield, interest rate, or rate of return generated by the investment. The apparent simplicity of this concept makes it the preferred choice in many organizations. However, the calculation can involve certain complications that cannot be ignored. If applied properly all four methods must and do provide identical conclusions with respect to the attractiveness of any given investment.

Our treatment of measures of investment attractiveness extends over two chapters. In Chapter 5, we define independent investments and distinguish these from mutually exclusive investments. Payback period is introduced as a useful project screening tool, though it provides no true measure of profitability. The remainder of this chapter is devoted to the development of the four rigorous measures of investment attractiveness and the application of these to independent investments. Chapter 6 applies these same four measures in the selection of the best project from a group of mutually exclusive projects.

We must also recognize that one of the most important parts of the capital budgeting process is the estimation of relevant cash flows. For all examples in this chapter, and those in Chapter 6, however, net cash flows can be viewed as before-tax values or after-tax values for which tax effects have been incorporated. Since some organizations (e.g., governments and nonprofit organizations) are not subject to tax, the before-tax situation provides a valid base for this type of economic evaluation. Taking this after-tax view will allow us to focus on our main area of concern: the economic evaluation of investment projects. The procedures for determining after-tax net cash flows in taxable situations are developed in Chapters 9 and 10.

CHAPTER LEARNING OBJECTIVES

After completing this chapter, you should understand the following concepts:

- How firms screen potential investment opportunities.
- How firms evaluate the profitability of an investment project by considering the time value of money.
- How to determine the net present worth (cost), net annual worth (cost), net future worth (cost), and the internal rate of return of a project.
- How to determine the capital recovery cost when you purchase an asset.
- How to determine the unit cost or unit profit.
- The meaning of the rate of return.
- The various methods to compute the rate of return.
- How to resolve the multiple rates of return.
- How to make an accept or reject decision with each of the PW, FW, AE, and IRR criteria.

5.1 Describing Project Cash Flows

In Chapter 1, we described many engineering economic decision problems, but we did not provide suggestions on how to solve them. What do all engineering economic decision problems have in common? The answer is that they all involve two dissimilar types of amounts. First, there is the investment, which is usually made in a lump sum at the beginning of the project. Although not literally made "today," the investment is made at a specific point in time that, for analytical purposes, is called today, or time 0. Second, there is a stream of cash benefits that are expected to result from the investment over some years in the future.

5.1.1 Loan versus Project Cash Flows

An investment made in a fixed asset is similar to an investment made by a bank when it lends money. The essential characteristic of both transactions is that funds are committed today in the expectation of their earning a return in the future. In the case of the bank loan, the future return takes the form of interest plus repayment of the principal and is known as the **loan cash flow**. In the case of the fixed asset, the future return takes the form of cash generated by productive use of the asset. As shown in Figure 5.1, the

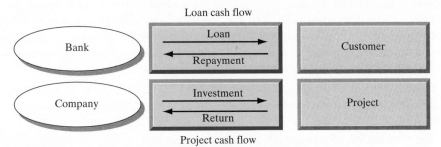

Figure 5.1   A bank loan versus an investment project.

representation of these future earnings, along with the capital expenditures and annual expenses (such as wages, raw materials, operating costs, maintenance costs, and income taxes), is the **project cash flow**. This similarity between the loan cash flow and the project cash flow brings us to an important conclusion: We can use the same equivalence techniques developed in Chapter 3 to measure economic worth. Example 5.1 illustrates a typical procedure for obtaining a project's cash flows.

EXAMPLE 5.1 Identifying Project Cash Flows

XL Chemicals is thinking of installing a computer process control system in one of its process plants. The plant is used about 40% of the time, or 3500 operating hours per year, to produce a proprietary demulsification chemical. During the remaining 60% of the time, it is used to produce other specialty chemicals. Annual production of the demulsification chemical amounts to 30,000 kilograms, and it sells for $15 per kilogram. The proposed computer process control system will cost $650,000 and is expected to provide the following specific benefits in the production of the demulsification chemical:

- First, the selling price of the product could be increased by $2 per kilogram because the product will be of higher purity, which translates into better demulsification.
- Second, production volumes will increase by 4000 kilograms per year as a result of higher reaction yields, without any increase in the quantities of raw material or in production time.
- Finally, the number of process operators can be reduced by one per shift, which represents a savings of $25 per hour. The new control system would result in additional maintenance costs of $53,000 per year and has an expected useful life of eight years.

Although the system is likely to provide similar benefits in the production of the other specialty chemicals manufactured in the process plant, these benefits have not yet been quantified.

SOLUTION

Given: The preceding cost and benefit information.
Find: Net cash flow in each year over the life of the new system.

Although we could assume that similar benefits are derivable from the production of the other specialty chemicals, let's restrict our consideration to the demulsification chemical and allocate the full initial cost of the control system and the annual maintenance costs to this chemical. (Note that you could logically argue that only 40% of these costs belong to this production activity.) The gross benefits are the additional revenues realized from the increased selling price and the extra production, as well as the cost savings resulting from having one fewer operator:

- Revenues from the price increases are 30,000 kilograms per year × $2/kilogram, or $60,000 per year. The added production volume at the new pricing adds revenues of 4000 kilograms per year × $17 per kilogram, or $68,000 per year.
- The elimination of one operator results in an annual savings of 3500 operating hours per year × $25 per hour, or $87,500 per year.

- The net benefits in each of the eight years that make up the useful lifetime of the new system are the gross benefits less the maintenance costs: ($60,000 + $68,000 + $87,500 − $53,000) = $162,500 per year.

Now we are ready to summarize a cash flow table as follows:

Year (n)	Cash Inflows (Benefits)	Cash Outflows (Costs)	Net Cash Flows
0	0	$650,000	−$650,000
1	215,500	53,000	162,500
2	215,500	53,000	162,500
⋮	⋮	⋮	⋮
8	215,500	53,000	162,500

COMMENTS: If the company purchases the computer process control system for $650,000 now, it can expect an annual savings of $162,500 for eight years. (Note that these savings occur in discrete lumps at the ends of years.) We also considered only the benefits associated with the production of the demulsification chemical. We could also have quantified some benefits attributable to the production of the other chemicals from this plant. Suppose that the demulsification chemical benefits alone justify the acquisition of the new system. Then it is obvious that, had we considered the benefits deriving from the other chemicals as well, the acquisition of the system would have been even more clearly justified.

We draw a cash flow diagram of this situation in Figure 5.2. Assuming that these cost savings and cash flow estimates are correct, should management give the go-ahead for installation of the system? If management decides not to purchase the computer control system, what should it do with the $650,000 (assuming that it has this amount in the first place)? The company could buy $650,000 of Treasury bills, or it could invest the amount in other cost-saving projects. How would the company compare cash flows that differ both in timing and amount for the alternatives it is considering? This is an extremely important question, because virtually every engineering investment decision involves a comparison of alternatives. Indeed, these are the types of questions this chapter is designed to help you answer.

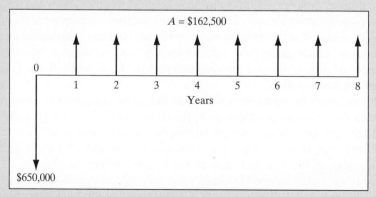

Figure 5.2 Cash flow diagram for the computer process control project described in Example 5.1.

5.1.2 Independent versus Mutually Exclusive Investment Projects

Most firms have a number of unrelated investment opportunities available. For example, in the case of XL Chemicals, other projects being considered in addition to the computer process control project in Example 5.1 are a new waste heat recovery boiler, a CAD system for the engineering department, and a new warehouse. The economic attractiveness of each of these projects can be measured, and a decision to accept or reject each project can be made without reference to any of the other projects. In other words, the decision regarding any one project has no effect on the decision to accept or reject another project. Such projects are said to be **independent**. In this chapter, we introduce methods for evaluating independent projects.

In Chapter 6, we will see that in many engineering situations we are faced with selecting the most economically attractive project from a number of alternative projects, all of which solve the same problem or meet the same need. It is unnecessary to choose more than one project in this situation, and the acceptance of one automatically entails the rejection of all of the others. Such projects are said to be **mutually exclusive**. We dedicate Chapter 6 to the comparison of mutually exclusive projects

As long as the total cost of all the independent projects found to be economically attractive is less than the investment funds available to a firm, all of these projects could proceed. However, this is rarely the case. The selection of those projects that should proceed when investment funds are limited is the subject of capital budgeting. Apart from Chapter 12, which deals with capital budgeting, the availability of funds will not be a consideration in accepting or rejecting projects dealt with in this book.

5.2 Initial Project Screening Method

Let's suppose that you are in the market for a new punch press for your company's machine shop, and you visit an equipment dealer. As you take a serious look at one of the punch press models in the display room, an observant equipment salesperson approaches you and says, "That press you are looking at is the state of the art in its category. If you buy that top-of-the-line model, it will cost a little bit more, but it will pay for itself in fewer than two years." Before studying the four measures of investment attractiveness, we will review a simple, but nonrigorous, method commonly used to screen capital investments. One of the primary concerns of most businesspeople is whether and when the money invested in a project can be recovered. The **payback method** screens projects on the basis of how long it takes for net receipts to equal investment outlays. This calculation can take one of two forms: either ignore time-value-of-money considerations or include them. The former case is usually designated the **conventional payback method**, the latter case the **discounted payback method**.

A common standard used to determine whether to pursue a project is that the project does not merit consideration unless its payback period is shorter than some specified period. (This time limit is determined largely by management policy. For example, a high-tech firm, such as a computer chip manufacturer, would set a short time limit for any new investment, because high-tech products rapidly become obsolete.) If the payback period is within the acceptable range, a formal project evaluation (such as a present-worth analysis) may begin. It is important to remember that **payback screening** is not an *end* in itself, but rather a method of screening out certain obviously unacceptable investment alternatives before progressing to an analysis of potentially acceptable ones.

5.2.1 Payback Period: The Time It Takes to Pay Back

Determining the relative worth of new production machinery by calculating the time it will take to pay back what it cost is the single most popular method of project screening. If a company makes investment decisions solely on the basis of the payback period, it considers only those projects with a payback period *shorter* than the maximum acceptable payback period. (However, because of shortcomings of the payback screening method, which we will discuss later, it is rarely used as the only decision criterion.)

Payback period: The length of time required to recover the cost of an investment.

What does the payback period tell us? One consequence of insisting that each proposed investment have a short payback period is that investors can assure themselves of being restored to their initial position within a short span of time. By restoring their initial position, investors can take advantage of additional, perhaps better, investment possibilities that may come along.

EXAMPLE 5.2 Conventional Payback Period for the Computer Process Control System Project

Consider the cash flows given in Example 5.1. Determine the payback period for this computer process control system project.

SOLUTION

Given: Initial cost = $650,000 and annual net benefits = $162,500.
Find: Conventional payback period.

Given a uniform stream of receipts, we can easily calculate the payback period by dividing the initial cash outlay by the annual receipts:

$$\text{Payback period} = \frac{\text{Initial cost}}{\text{Uniform annual benefit}} = \frac{\$650,000}{\$162,500}$$

$$= 4 \text{ years.}$$

If the company's policy is to consider only projects with a payback period of five years or less, this computer process control system project passes the initial screening.

In Example 5.2, dividing the initial payment by annual receipts to determine the payback period is a simplification we can make because the annual receipts are uniform. Whenever the expected cash flows vary from year to year, however, the payback period must be determined by adding the expected cash flows for each year until the sum is equal to or greater than zero. The significance of this procedure is easily explained. The cumulative cash flow equals zero at the point where cash inflows exactly match, or pay back, the cash outflows; thus, the project has reached the payback point. Similarly, if the cumulative cash flows are greater than zero, then the cash inflows exceed the cash outflows, and the project has begun to generate a profit, thus surpassing its payback point. To illustrate, consider Example 5.3.

EXAMPLE 5.3 Conventional Payback Period With Salvage Value

Autonumerics Company has just bought a new spindle machine at a cost of $105,000 to replace one that had a salvage value of $20,000. The projected annual after-tax savings via improved efficiency, which will exceed the investment cost, are as follows:

Period	Cash Flow	Cumulative Cash Flow
0	−$105,000 + $20,000	−$85,000
1	15,000	−70,000
2	25,000	−45,000
3	35,000	−10,000
4	45,000	35,000
5	45,000	80,000
6	35,000	115,000

SOLUTION

Given: Cash flow series as shown in Figure 5.3(a).
Find: Conventional payback period.

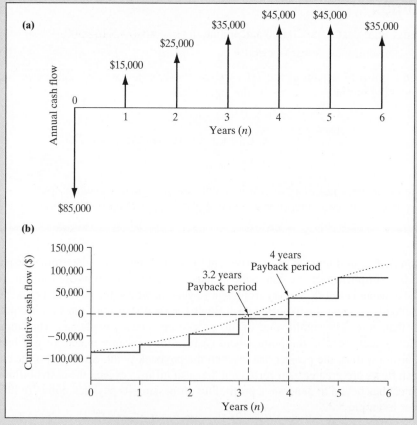

Figure 5.3 Illustration of conventional payback period (Example 5.3).

The salvage value of retired equipment becomes a major consideration in most justi-fication analysis. (In this example, the salvage value of the old machine should be taken into account, as the company already had decided to replace the old machine.) When used, the salvage value of the retired equipment is subtracted from the pur-chase price of new equipment, revealing a closer true cost of the investment. As we see from the cumulative cash flow in Figure 5.3(b), the total investment is recovered during year 4. If the firm's stated maximum payback period is three years, the proj-ect will not pass the initial screening stage.

COMMENTS: In Example 5.3, we assumed that cash flows occur only in discrete lumps at the ends of years. If instead cash flows occur continuously throughout the year, the payback period calculation needs adjustment. A negative balance of $10,000 remains at the start of year 4. If $45,000 is expected to be received as a more or less continuous flow during year 4, the total investment will be recovered two-tenths ($10,000/$45,000) of the way through the fourth year. Thus, in this situation, the payback period is 3.2 years. However, if we use the end-of-period cash flow con-vention, that is, cash flows are discrete, then the playback period is 4 years.

5.2.2 Benefits and Flaws of Payback Screening

The simplicity of the payback method is one of its most appealing qualities. Initial project screening by the method reduces the information search by focusing on that time at which the firm expects to recover the initial investment. The method may also eliminate some alternatives, thus reducing the firm's time spent analyzing. But the much-used payback method of equipment screening has a number of serious draw-backs. The principal objection to the method is that it fails to measure profitability (i.e., no "profit" is made during the payback period). Simply measuring how long it will take to recover the initial investment outlay contributes little to gauging the earning power of a project. (In other words, you already know that the money you borrowed for the drill press is costing you 12% per year; the payback method can't tell you how much your invested money is contributing toward the interest expense.) Also, because pay-back period analysis ignores differences in the timing of cash flows, it fails to recog-nize the difference between the present and future value of money. For example, although the payback on two investments can be the same in terms of numbers of years, a front-loaded investment is better because money available today is worth more than that to be gained later. Finally, because payback screening ignores all proceeds after the payback period, it does not allow for the possible advantages of a project with a longer economic life.

By way of illustration, consider the two investment projects listed in Table 5.1. Each requires an initial investment outlay of $90,000. Project 1, with expected annual cash pro-ceeds of $30,000 for the first three years, has a payback period of three years. Project 2 is expected to generate annual cash proceeds of $25,000 for six years; hence, its payback period is four years. If the company's maximum payback period is set to three years, then project 1 would pass the initial project screening, whereas project 2 would fail even though it is clearly the more profitable investment. With the methods to be introduced later in this chapter, we can confirm that Project 2 is much more profitable than Project 1.

TABLE 5.1 Investment Cash Flows for Two Competing Projects

n	Project 1	Project 2
0	−$90,000	−$90,000
1	30,000	25,000
2	30,000	25,000
3	30,000	25,000
4	1,000	25,000
5	1,000	25,000
6	1,000	25,000
	$ 3,000	$60,000

5.2.3 Discounted Payback Period

To remedy one of the shortcomings of the conventional payback period, we may modify the procedure so that it takes into account the time value of money—that is, the cost of funds (interest) used to support a project. This modified payback period is often referred to as the **discounted payback period**. In other words, we may define the discounted payback period as the number of years required to recover the investment from *discounted* cash flows.

Discounted payback period: The length of time required to recover the cost of an investment based on discounted cash flows.

For the project in Example 5.3, suppose the company requires a rate of return of 15%. To determine the period necessary to recover both the capital investment and the cost of funds required to support the investment, we may construct Table 5.2, showing cash flows and costs of funds to be recovered over the life of the project.

To illustrate, let's consider the cost of funds during the first year: With $85,000 committed at the beginning of the year, the interest in year 1 would be $12,750 ($85,000 × 0.15). Therefore, the total commitment grows to $97,750, but the $15,000 cash flow in year 1 leaves a net commitment of $82,750. The cost of funds during the second year would be $12,413 ($82,750 × 0.15), but with the $25,000 receipt from the project, the net commitment drops to $70,163. When this process repeats for the remaining years of the

TABLE 5.2 Payback Period Calculation Taking Into Account the Cost of Funds (Example 5.3)

Period	Cash Flow	Cost of Funds (15%)*		Cumulative Cash Flow
0	−$85,000	0		−$85,000
1	15,000	−$85,000(0.15) =	−$12,750	−82,750
2	25,000	−$82,750(0.15) =	−12,413	−70,163
3	35,000	−$70,163(0.15) =	−10,524	−45,687
4	45,000	−$45,687(0.15) =	−6,853	−7,540
5	45,000	−$7,540(0.15) =	−1,131	36,329
6	35,000	$36,329(0.15) =	5,449	76,778

*Cost of funds = Unrecovered beginning balance × interest rate.

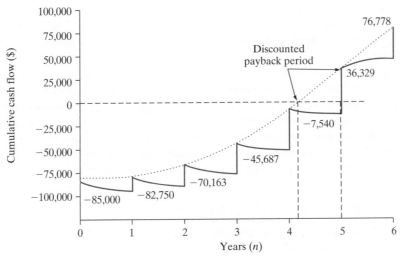

Figure 5.4 Illustration of discounted payback period.

project's life, we find that the net commitment to the project ends during year 5. Depending on which cash flow assumption we adopt, the project must remain in use about 4.2 years (assuming continuous cash flows) or 5 years (assuming year-end cash flows) in order for the company to cover its cost of capital and also recover the funds it has invested. Figure 5.4 illustrates this relationship.

The inclusion of effects stemming from time value of money has increased the payback period calculated for this example by a year. Certainly, this modified measure is an improved one, but it does not show the complete picture of the project's profitability either.

5.2.4 Where Do We Go From Here?

Should we abandon the payback method? Certainly not, but if you use payback screening exclusively to analyze capital investments, look again. You may be missing something that another method can help you spot. Therefore, it is illogical to claim that payback is either a good or bad method of justification. Clearly, it is not a measure of profitability. But when it is used to supplement other methods of analysis, it can provide useful information. For example, payback can be useful when a company needs a measure of the speed of cash recovery, when the company has a cash flow problem, when a product is built to last only for a short time, and when the machine the company is contemplating buying itself is known to have a short market life.

5.3 Present-Worth Analysis

Until the 1950s, the payback method was widely used as a means of making investment decisions. As flaws in this method were recognized, however, businesspeople began to search for methods to improve project evaluations. The result was the development of **discounted cash flow techniques (DCFs)**, which take into account the time value of money. One of the DCFs is the net-present-worth (NPW) method, also called the net-present-value (NPV) method, or present equivalent (PE) method. A capital investment

Discounted cash flow technique (DCF): A method of evaluating an investment by estimating future cash flows and taking into consideration the time value of money.

problem is essentially a problem of determining whether the anticipated cash inflows from a proposed project are sufficient to attract investors to invest funds in the project. In developing the NPW criterion, we will use the concept of cash flow equivalence discussed in Chapter 3.

As we observed, the most convenient point at which to calculate the equivalent values is often at time 0. Under the NPW criterion, the present worth of all cash inflows is compared against the present worth of all cash outflows associated with an investment project. The difference between the present worth of these cash flows, referred to as the **net present worth (NPW)**, **net present value** (NPV) determines whether the project is an acceptable investment. When two or more projects are under consideration, NPW analysis further allows us to select the best project by comparing their NPW figures.

Net present worth (NPW): The difference between the present value of cash inflows and the present value of cash outflows.

5.3.1 Net-Present-Worth Criterion

We will first summarize the basic procedure for applying the net-present-worth criterion to a typical investment project:

- Determine the interest rate that the firm wishes to earn on its investments. The interest rate you determine represents the rate at which the firm can always invest the money in its **investment pool**. This interest rate is often referred to as either a **required rate of return** or a **minimum attractive rate of return** (MARR). Usually, selection of the MARR is a policy decision made by top management. It is possible for the MARR to change over the life of a project, as we saw in Section 4.4, but for now we will use a single rate of interest in calculating the NPW.
- Estimate the service life of the project.
- Estimate the cash inflow for each period over the service life.
- Estimate the cash outflow over each service period.
- Determine the net cash flows (net cash flow = cash inflow − cash outflow).
- Find the present worth of each net cash flow at the MARR. Add up all the present-worth figures; their sum is defined as the project's NPW, given by

Investment pool operates like a mutual fund to earn a targeted return by investing the firm's money in various investment assets.

$$PW(i) = \text{NPW calculated at } i = \frac{A_0}{(1+i)^0} + \frac{A_1}{(1+i)^1} + \frac{A_2}{(1+i)^2} + \cdots$$

$$+ \frac{A_N}{(1+i)^N}$$

$$= \sum_{n=0}^{N} \frac{A_n}{(1+i)^n}$$

$$= \sum_{n=0}^{N} A_n(P/F, i, n), \tag{5.1}$$

where

$$A_n = \text{Net cash flow at end of period } n,$$
$$i = \text{MARR (or cost of capital)},$$
$$N = \text{Service life of the project}.$$

A_n will be positive if the corresponding period has a net cash inflow and negative if there is a net cash outflow.

- In this context, a positive NPW means that the equivalent worth of the inflows is greater than the equivalent worth of outflows, so the project makes a profit. Therefore, if the $PW(i)$ is positive for a single project, the project should be accepted; if the $PW(i)$ is negative, the project should be rejected. The decision rule is

If $PW(i) > 0$, accept the investment.

If $PW(i) = 0$, remain indifferent.

If $PW(i) < 0$, reject the investment.

If MARR is the minimum attractive rate of return, why are we indifferent to a project for which $PW(MARR) = 0$? There are a number of things to consider in this regard. First, most organizations have more investment opportunities than available funds so the projects that are approved have large positive PW values. Secondly, cost and benefit estimates are always subject to uncertainty. When the PW value is zero, there is no room to offset the effect if actual costs are slightly higher or actual benefits are slightly lower than estimated. In general, an organization would be satisfied if on average, across the totality of all projects implemented, it earned MARR. The reality is that some projects will ultimately provide a return equal to or greater than that originally promised, whereas others will do considerably more poorly than promised and may have an actual return well below MARR. Therefore, in the overall picture, a project that is at the margin $PW(MARR) = 0$, may not be particularly appealing.

Note that the decision rule is for the evaluation of independent projects where you can estimate the revenues as well as the costs associated with a project. There may be projects that cannot be avoided, e.g., the installation of pollution control equipment. In a case such as this, the project would be accepted even though its $PW < 0$. However, in this situation, we are interested in finding the best pollution control equipment. In Chapter 6, we will be addressing such issues via comparison of mutually exclusive investment projects.

EXAMPLE 5.4 Net Present Worth: Uniform Flows

Consider the investment cash flows associated with the computer process control project discussed in Example 5.1. If the firm's MARR is 15%, compute the NPW of this project. Is the project acceptable?

SOLUTION

Given: Cash flows in Figure 5.2 and MARR $= 15\%$ per year.

Find: NPW.

Since the computer process control project requires an initial investment of $650,000 at $n = 0$, followed by the eight equal annual savings of $162,000, we can easily determine the NPW as follows:

$$PW(15\%)_{\text{Outflow}} = \$650,000;$$
$$PW(15\%)_{\text{Inflow}} = \$162,500(P/A, 15\%, 8)$$
$$= \$729,190.$$

Then the NPW of the project is

$$PW(15\%) = PW(15\%)_{\text{Inflow}} - PW(15\%)_{\text{Outflow}}$$
$$= \$729{,}190 - \$650{,}000$$
$$= \$79{,}190,$$

or, from Eq. (5.1),

$$PW(15\%) = -\$650{,}000 + \$162{,}500(P/A, 15\%, 8)$$
$$= \$79{,}190.$$

Since $PW(15\%) > 0$, the project is acceptable.

Now let's consider an example in which the investment cash flows are not uniform over the service life of the project.

EXAMPLE 5.5 Net Present Worth: Uneven Flows

Tiger Machine Tool Company is considering acquiring a new metal-cutting machine. The required initial investment of $75,000 and the projected cash benefits[2] over the project's three-year life are as follows:

End of Year	Net Cash Flow
0	−$75,000
1	24,400
2	27,340
3	55,760

You have been asked by the president of the company to evaluate the economic merit of the acquisition. The firm's MARR is known to be 15%.

SOLUTION

Given: Cash flows as tabulated and MARR = 15% per year.
Find: NPW.

If we bring each flow to its equivalent at time zero, we find that

$$PW(15\%) = -\$75{,}000 + \$24{,}000(P/F, 15\%, 1) + \$27{,}340(P/F, 15\%, 2)$$
$$+ \$55{,}760(P/F, 15\%, 3)$$
$$= \$3553.$$

Since the project results in a surplus of $3553, the project is acceptable.

---

[2] As we stated at the beginning of this chapter, we treat net cash flows as before-tax values or as having their tax effects precalculated. Explaining the process of obtaining cash flows requires an understanding of income taxes and the role of depreciation, which are discussed in Chapter 8 and 9.

TABLE 5.3 Present-Worth Amounts at Varying Interest Rates (Example 5.5)

i(%)	PW(i)	i(%)	PW(i)
0	$32,500	20	−$3,412
2	27,743	22	−5,924
4	23,309	24	−8,296
6	19,169	26	−10,539
8	15,296	28	−12,662
10	11,670	30	−14,673
12	8,270	32	−16,580
14	5,077	34	−18,360
16	2,076	36	−20,110
17.45*	0	38	−21,745
18	−750	40	−23,302

*Break-even interest rate (also known as the rate of return).

In Example 5.5, we computed the NPW of a project at the interest rate of 15%. If we compute the NPW at other interest rate values, we obtain the results in Table 5.3. Plotting the NPW as a function of interest rate gives Figure 5.5, the present-worth profile. (You may use a spreadsheet program such as Excel to generate Table 5.3 or Figure 5.5.)

Figure 5.5 indicates that the investment project has a positive NPW if the interest rate is below 17.45% and a negative NPW if the interest rate is above 17.45%. As we

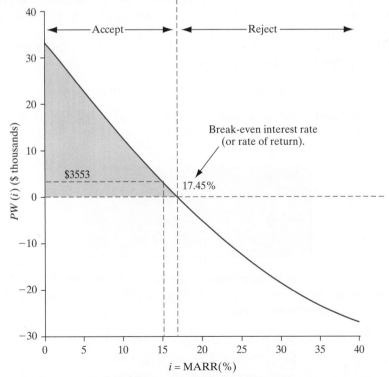

Figure 5.5 Present-worth profile described in Example 5.5.

will see later in this chapter, this **break-even interest rate** is known as the **internal rate of return**. If the firm's MARR is 15%, then the project has an NPW of $3553 and so may be accepted. The $3553 figure measures the equivalent immediate gain in present worth to the firm following acceptance of the project. By contrast, at $i = 20\%, \mathrm{PW}(20\%) = -\$3412$, and the firm should reject the project. (Note that either accepting or rejecting an investment is influenced by the choice of a MARR, so it is crucial to estimate the MARR correctly. We will defer this important issue until Section 5.3.3. For now, we will assume that the firm has an accurate MARR estimate available for use in investment analysis.)

5.3.2 Meaning of Net Present Worth

In present-worth analysis, we assume that all the funds in a firm's treasury can be placed in investments that yield a return equal to the MARR. We may view these funds as an **investment pool**. Alternatively, if no funds are available for investment, we assume that the firm can borrow them at the MARR (or cost of capital) from the capital market. In this section, we will examine these two views as we explain the meaning of the MARR in NPW calculations.

Investment Pool Concept

An investment pool is equivalent to a firm's treasury. All fund transactions are administered and managed by the firm's comptroller. The firm may withdraw funds from this investment pool for other investment purposes, but if left in the pool, these funds will earn the MARR. Thus, in investment analysis, net cash flows will be net cash flows relative to an investment pool. To illustrate the investment pool concept, we consider again the project in Example 5.5 that required an investment of $75,000.

If the firm did not invest in the project and left $75,000 in the investment pool for three years, these funds would grow as follows:

$$\$75,000(F/P, 15\%, 3) = \$114,066.$$

Suppose the company decided instead to invest $75,000 in the project described in Example 5.5. Then the firm would receive a stream of cash inflows during the project's life of three years in the following amounts:

Period (n)	Net Cash Flow (A_n)
1	$24,400
2	27,340
3	55,760

Since the funds that return to the investment pool earn interest at a rate of 15%, it is worthwhile to see how much the firm would benefit from its $75,000 investment. For this alternative, the returns after reinvestment are

$$\$24,400(F/P, 15\%, 2) = \$32,269,$$
$$\$27,340(F/P, 15\%, 1) = \$31,441,$$
$$\$55,760(F/P, 15\%, 0) = \$55,760.$$

These returns total $119,470. At the end of three years, the additional cash accumulation from investing in the project is

$$\$119,470 - \$114,066 = \$5404.$$

If we compute the equivalent present worth at time 0 of this net cash surplus, we obtain

$$\$5404(P/F, 15\%, 3) = \$3553,$$

which is exactly what we get when we compute the NPW of the project with Eq. (5.1). Clearly, on the basis of its positive NPW, the alternative of purchasing a new machine should be preferred to that of simply leaving the funds in the investment pool at the MARR. Thus, in PW analysis, any investment is assumed to be returned at the MARR. If a surplus exists at the end of the project, then PW(MARR) > 0. Figure 5.6 illustrates the reinvestment concept as it relates to the firm's investment pool.

Borrowed-Funds Concept

Suppose that the firm does not have $75,000 at the outset. In fact, the firm doesn't have to have an investment pool at all. Suppose further that the firm borrows all of its capital from a bank at an interest rate of 15%, invests in the project, and uses the proceeds from the investment to pay off the principal and interest on the bank loan. How much is left for the firm at the end of the project's life?

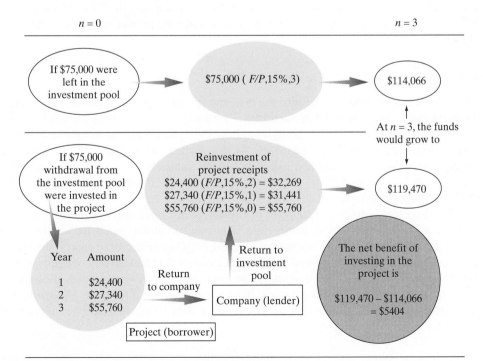

Figure 5.6 The concept of an investment pool with the company as a lender and the project as a borrower.

At the end of the first year, the interest on the project's use of the bank loan would be $75,000(0.15) = $11,250. Therefore, the total loan balance grows to $75,000(1.15) = $86,250. Then, the firm receives $24,400 from the project and applies the entire amount to repay the loan portion, leaving a balance due of

$$75,000(1 + 0.15) - $24,400 = $61,850.$$

This amount becomes the net amount the project is borrowing at the beginning of year 2, which is also known as the **project balance**. At the end of year 2, the bank debt grows to $61,850(1.15) = $71,128, but with the receipt of $27,340, the project balance is reduced to

$$71,128 - $27,340 = $43,788.$$

Similarly, at the end of year 3, the project balance becomes

$$43,788(1.15) = $50,356.$$

But with the receipt of $55,760 from the project, the firm should be able to pay off the remaining balance and come out with a surplus in the amount of $5404. This terminal project balance is also known as the **net future worth** of the project. In other words, the firm repays its initial bank loan and interest at the end of year 3, with a resulting profit of $5404. If we compute the equivalent present worth of this net profit at time 0, we obtain

$$PW(15\%) = $5404(P/F, 15\%, 3) = $3553.$$

The result is identical to the case in which we directly computed the NPW of the project at $i = 15\%$ shown in Example 5.5. Figure 5.7 illustrates the project balance as a function of time.[3]

5.3.3 Basis for Selecting the MARR

The basic principle used to determine the discount rate in project evaluations is similar to the bond evaluation discussed in Section 4.6.3. However, one needs to learn a lot more about sources of capital to fully understand how the MARR value is determined, which will be discussed in detail in Chapter 12. In this section, we will briefly introduce the concepts of sources of capital and risk. The MARR values for project evaluation in this chapter will be provided, thus, you don't need to worry about how to determine the MARR value until Chapter 12.

Cost of capital: The required return necessary to make an investment project worthwhile.

The first element to cover is the **cost of capital**, which is the required return necessary to make an investment project worthwhile. The cost of capital would include both the *cost of debt* (the interest rate associated with borrowing) and the *cost of equity* (the return that stockholders require for a company). Both the cost of debt and the cost of equity reflect the presence of inflation in the economy. *The cost of capital determines how a company can raise money (through issuing a stock, borrowing, or a mix of the two). Therefore, this is normally considered as the rate of return that a firm would receive if it invested its money someplace else with a similar risk.*

---

[3] Note that the sign of the project balance changes from negative to positive during year 3. The time at which the project balance becomes zero for the first time is known as the **discounted payback period**.

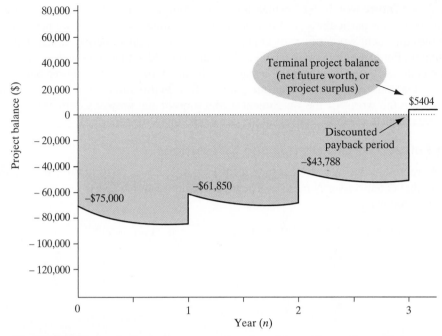

Figure 5.7 Project balance diagram as a function of time. (A negative project balance indicates the amount of the loan remaining to be paid off or the amount of investment to be recovered.)

The second element is a consideration of any additional risk associated with the project. If the project belongs to the normal risk category, the cost of capital may already reflect the risk premium. However, if you are dealing with a project with higher risk, the additional risk premium may be added onto the cost of capital.

In sum, the discount rate (**MARR**) to use for project evaluation would be equivalent to the firm's cost of capital for a project of normal risk, but could be much higher if you are dealing with a risky project. Chapter 12 will detail the analytical process of determining this discount rate. For now, we assume that such a rate is already known to us, and we will focus on the evaluation of the investment project.

MARR: This is based on the firm's cost of capital plus or minus a risk premium to reflect the project's specific risk characteristics.

5.4 Variations of Present-Worth Analysis

As variations of present-worth analysis, we will consider two additional measures of investment worth: **future-worth analysis** and **capitalized equivalent-worth analysis**. (The equivalent annual worth measure is another variation of the present-worth measure, but we will present it in Section 5.5.) Future-worth analysis calculates the future worth of an investment undertaken. Capitalized equivalent-worth analysis calculates the present worth of a project with a perpetual life span.

5.4.1 Future-Worth Analysis

The second evaluation method that derives from discounted cash flow techniques is the future worth analysis. Net present worth measures the surplus in an investment project at

Net future worth (NFW): The value of an asset or cash at a specified date in the future that is equivalent in value to a specified sum today.

time 0. **Net future worth (NFW)** measures this surplus at a time other than 0. Net-future-worth analysis is particularly useful in an investment situation in which we need to compute the equivalent worth of a project at the end of its investment period, rather than at its beginning. For example, it may take 7 to 10 years to build a nuclear power plant because of the complexities of engineering design and the many time-consuming regulatory procedures that must be followed to ensure public safety. In this situation, it is more common to measure the worth of the investment at the time of the project's commercialization (i.e., we conduct an NFW analysis at the end of the investment period).

Net-Future-Worth Criterion and Calculations

Let A_n represent the cash flow at time n for $n = 0, 1, 2, \ldots, N$ for a typical investment project that extends over N periods. Then the net-future-worth (NFW) expression at the end of period N is

$$\text{FW}(i) = A_0(1 + i)^N + A_1(1 + i)^{N-1} + A_2(1 + i)^{N-2} + \cdots + A_N$$

$$= \sum_{n=0}^{N} A_n(1 + i)^{N-n}$$

$$= \sum_{n=0}^{N} A_n(F/P, i, N - n). \tag{5.2}$$

As you might expect, the decision rule for the NFW criterion is the same as that for the NPW criterion: For a single project evaluation,

If $\text{FW}(i) > 0$, accept the investment.

If $\text{FW}(i) = 0$, remain indifferent to the investment.

If $\text{FW}(i) < 0$, reject the investment.

EXAMPLE 5.6 Net Future Worth: At the End of the Project

Consider the project cash flows in Example 5.5. Compute the NFW at the end of year 3 at $i = 15\%$.

SOLUTION

Given: Cash flows in Example 5.5 and MARR $= 15\%$ per year.
Find: NFW.

As seen in Figure 5.8, the NFW of this project at an interest rate of 15% would be

$$\text{FW}(15\%) = -\$75,000(F/P, 15\%, 3) + \$24,400(F/P, 15\%, 2)$$
$$+ \$27,340(F/P, 15\%, 1) + \$55,760$$
$$= \$5404.$$

Note that the net future worth of the project is equivalent to the terminal project balance as calculated in Section 5.3.2. Since FW(15%) > 0, the project is acceptable. We reach the same conclusion as with the present-worth analysis.

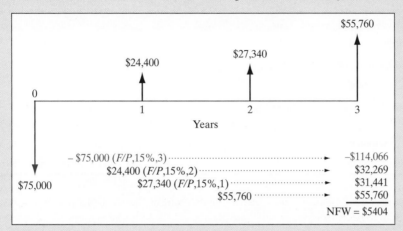

Figure 5.8 Future-worth calculation at the end of year 3 (Example 5.6).

EXAMPLE 5.7 Future Equivalent: At an Intermediate Time

Higgins Corporation (HC), a Montreal-based robot-manufacturing company, has developed a new advanced-technology robot called Helpmate, which incorporates advanced technology such as vision systems, tactile sensing, and voice recognition. These features allow the robot to roam the corridors of a hospital or office building without following a predetermined track or bumping into objects. HC's marketing department plans to target sales of the robot toward major hospitals. The robots will ease nurses' workloads by performing low-level duties such as delivering medicines and meals to patients.

- The firm would need a new plant to manufacture the Helpmates; this plant could be built and made ready for production in two years. It would require a 12-hectare site, which can be purchased for $1.5 million in year 0. Building construction would begin early in year 1 and continue throughout year 2. The building would cost an estimated $10 million, with a $4 million payment due to the contractor at the end of year 1, and with another $6 million payable at the end of year 2.

- The necessary manufacturing equipment would be installed late in year 2 and would be paid for at the end of year 2. The equipment would cost $13 million, including transportation and installation. When the project terminates, the land is expected to have an after-tax market value of $2 million, the building an after-tax value of $3 million, and the equipment an after-tax value of $3 million.

  For capital budgeting purposes, assume that the cash flows occur at the end of each year. Because the plant would begin operations at the beginning of year 3, the first operating cash flows would occur at the end of year 3. The Helpmate plant's estimated economic life is six years after completion, with the following expected after-tax operating cash flows in millions:

Calendar Year	'10	'11	'12	'13	'14	'15	'16	'17	'18
End of Year	0	1	2	3	4	5	6	7	8
After-tax cash flows									
A. Operating revenue				$6	$8	$13	$18	$14	$8
B. Investment									
Land	−1.5								+2
Building		−4	−6						+3
Equipment			−13						+3
Net cash flow	−$1.5	−$4	−$19	$6	$8	$13	$18	$14	$16

Compute the equivalent worth of this investment at the start of operations. Assume that HC's MARR is 15%.

SOLUTION

Given: Preceding cash flows and MARR = 15% per year.

Find: Equivalent worth of project at the end of calendar year 2.

One easily understood method involves calculating the present worth and then transforming it to the equivalent worth at the end of year 2. First, we can compute PW(15%) at time 0 of the project:

$$PW(15\%) = -\$1.5 - \$4(P/F, 15\%, 1) - \$19(P/F, 15\%, 2)$$
$$+ \$6(P/F, 15\%, 3) + \$8(P/F, 15\%, 4) + \$13(P/F, 15\%, 5)$$
$$+ \$18(P/F, 15\%, 6) + \$14(P/F, 15\%, 7) + \$16(P/F, 15\%, 8)$$
$$= \$13.91 \text{ million.}$$

Then, the equivalent project worth at the start of operation is

$$FW(15\%) = PW(15\%)(F/P, 15\%, 2)$$
$$= \$18.40 \text{ million.}$$

A second method brings all flows prior to year 2 up to that point and discounts future flows back to year 2. The equivalent worth of the earlier investment, when the plant begins full operation, is

$$-\$1.5(F/P, 15\%, 2) - \$4(F/P, 15\%, 1) - \$19 = -\$25.58 \text{ million,}$$

which produces an equivalent flow as shown in Figure 5.9. If we discount the future flows to the start of operation, we obtain

$$FW(15\%) = -\$25.58 + \$6(P/F, 15\%, 1) + \$8(P/F, 15\%, 2) + \cdots$$
$$+ \$16(F/P, 15\%, 6)$$
$$= \$18.40 \text{ million.}$$

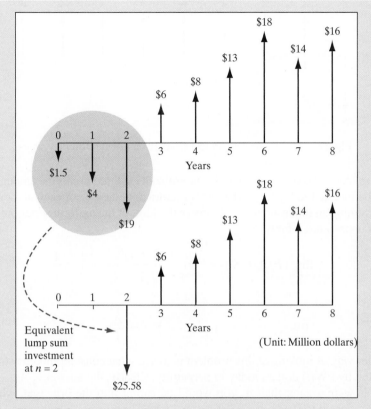

Figure 5.9 Cash flow diagram for the Helpmate project (Example 5.7).

COMMENTS: If another company is willing to purchase the plant and the right to manufacture the robots immediately after completion of the plant (year 2), HC would set the price of the plant at $43.98 million ($18.40 + $25.58) at a minimum.

5.4.2 Capitalized Equivalent Method

Another special case of the PW criterion is useful when the life of a proposed project is **perpetual** or the planning horizon is extremely long (say, 40 years or more). Many public projects, such as bridges, waterway structures, irrigation systems, and hydroelectric dams, are expected to generate benefits over an extended period (or forever). In this section, we will examine the **capitalized equivalent** (CE(i)) method for evaluating such projects.

Perpetual Service Life

Consider the cash flow series shown in Figure 5.10. How do we determine the PW for an infinite (or almost infinite) uniform series of cash flows or a repeated cycle of cash flows? The process of computing the PW cost for this infinite series is referred to as the **capitalization**

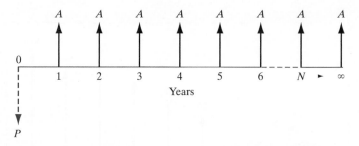

Figure 5.10   Equivalent present worth of an infinite cash flow series.

Capitalized cost related to car leasing, means the amount that is being financed. of the project cost. The cost, known as the **capitalized cost**, represents the amount of money that must be invested today to yield a certain return A at the end of each and every period forever, assuming an interest rate of i. Observe the limit of the uniform series present-worth factor as N approaches infinity:

$$\lim_{N\to\infty}(P/A, i, N) = \lim_{N\to\infty}\left[\frac{(1+i)^N - 1}{i(1+i)^N}\right] = \frac{1}{i}.$$

Thus,

$$\text{PW}(i) = A(P/A, i, N\to\infty) = \frac{A}{i}. \tag{5.3}$$

Another way of looking at this problem is to ask what constant income stream could be generated by $\text{PW}(i)$ dollars today in perpetuity. Clearly, the answer is $A = i\text{PW}(i)$. If withdrawals were greater than A, you would be eating into the principal, which would eventually reduce it to 0.

EXAMPLE 5.8 Capitalized Equivalent Cost

A major university has just completed a new engineering complex worth $50 million. A campaign targeting alumni is planned to raise funds for future maintenance costs, which are estimated at $2 million per year. Any unforeseen costs above $2 million per year would be obtained by raising tuition. Assuming that the university can create a trust fund that earns 8% interest annually, how much has to be raised now to cover the perpetual string of $2 million in annual costs?

SOLUTION

Given: $A = \$2$ million, $i = 8\%$ per year, and $N = \infty$.
Find: CE(8%).

The capitalized cost equation is

$$\text{CE}(i) = \frac{A}{i},$$

so

$$CE(8\%) = \$2,000,000/0.08$$
$$= \$25,000,000.$$

COMMENTS: It is easy to see that this lump-sum amount should be sufficient to pay maintenance expenses for the university forever. Suppose the university deposited $25 million in a bank that paid 8% interest annually. Then at the end of the first year, the $25 million would earn 8%($25 million) = $2 million interest. If this interest were withdrawn, the $25 million would remain in the account. At the end of the second year, the $25 million balance would again earn 8%($25 million) = $2 million. This annual withdrawal could be continued forever, and the endowment (gift funds) would always remain at $25 million.

Project's Service Life Is Extremely Long

The benefits of typical civil engineering projects, such as bridge and highway construction, although not perpetual, can last for many years. In this section, we will examine the use of the CE(i) criterion to approximate the NPW of engineering projects with long lives.

EXAMPLE 5.9 Comparison of Present Worth for Long Life and Infinite Life

Mr. Gaynor L. Bracewell amassed a small fortune developing real estate in British Columbia over the past 30 years. He sold more than 300 hectares of timber and farmland to raise $800,000, with which he built a small hydroelectric plant, known as Edgemont Hydro. The plant was a decade in the making. The design for Mr. Bracewell's plant, which he developed using his military training as a civil engineer, is relatively simple. A 7-metre-deep canal, blasted out of solid rock just above the higher of two dams on his property, carries water 350 metres along the river to a "trash rack," where leaves and other debris are caught. A 2-metre-wide pipeline capable of holding 1.5 million kilograms of liquid then funnels the water into the powerhouse at 3.5 metres per second, thereby creating the necessary thrust against the turbines.

Government regulations encourage private power development, and any electricity generated must be purchased by the power company that owns the provincial power grid. The plant can generate 6 million kilowatt hours per year. Suppose that, after paying income taxes and operating expenses, Mr. Bracewell's annual income from the hydroelectric plant will be $120,000. With normal maintenance, the plant is expected to provide service for at least 50 years. Figure 5.11 illustrates when and in what quantities Mr. Bracewell spent his $800,000 (not taking into account the time value of money) during the last 10 years. Was Mr. Bracewell's $800,000 investment a wise one? How long will he have to wait to recover his initial investment, and will he ever make a profit? Examine the situation by computing the project worth at varying interest rates.

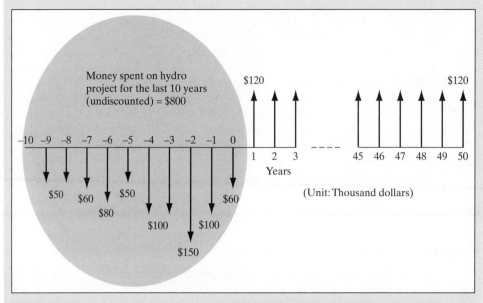

Figure 5.11 Net cash flow diagram for Mr. Bracewell's hydroelectric project (Example 5.9).

(a) If Mr. Bracewell's interest rate is 8%, compute the NPW (at time 0 in Figure 5.11) of this project with a 50-year service life and infinite service, respectively.

(b) Repeat part (a), assuming an interest rate of 12%.

SOLUTION

Given: Cash flow in Figure 5.11 (to 50 years or ∞) and $i = 8\%$ or 12%.
Find: NPW at time 0.

One of the main questions is whether Mr. Bracewell's plant will be profitable. Now we will compute the equivalent total investment and the equivalent worth of receiving future revenues at the start of power generation (i.e., at time 0).

(a) Let $i = 8\%$. Then
- with a plant service life of 50 years, we can make use of single-payment compound-amount factors in the invested cash flow to help us find the equivalent total investment at the start of power generation. Using K to indicate thousands, we obtain

$$V_1 = -\$50K(F/P, 8\%, 9) - \$50K(F/P, 8\%, 8)$$
$$- \$60K(F/P, 8\%, 7) \cdots - \$100K(F/P, 8\%, 1) - \$60K$$
$$= -\$1,101K.$$

The equivalent total benefit at the start of generation is

$$V_2 = \$120K(P/A, 8\%, 50) = \$1,468K.$$

Summing, we find the net equivalent worth at the start of power generation:

$$V_1 + V_2 = -\$1,101K + \$1,468K$$
$$= \$367K.$$

- With an infinite service life, the net equivalent worth is called the capitalized equivalent worth. The investment portion prior to time 0 is identical, so the capitalized equivalent worth is

$$CE(8\%) = -\$1,101K + \$120K/(0.08)$$
$$= \$399K.$$

Note that the difference between the infinite situation and the planning horizon of 50 years is only $32,000.

(b) Let $i = 12\%$. Then

- With a service life of 50 years, proceeding as we did in part (a), we find that the equivalent total investment at the start of power generation is

$$V_1 = -\$50K(F/P, 12\%, 9) - \$50K(F/P, 12\%, 8)$$
$$- \$60K(F/P, 12\%, 7) \cdots - \$100K(F/P, 12\%, 1) - 60K$$
$$= -\$1,299K.$$

Equivalent total benefits at the start of power generation are

$$V_2 = \$120K(P/A, 12\%, 50) = \$997K.$$

The net equivalent worth at the start of power generation is

$$V_1 + V_2 = -\$1,299K + \$997K$$
$$= -\$302K.$$

- With infinite cash flows, the capitalized equivalent worth at the current time is

$$CE(12\%) = -\$1,299K + \$120K/(0.12)$$
$$= -\$299K.$$

Note that the difference between the infinite situation and a planning horizon of 50 years is merely $3000, which demonstrates that we may approximate the present worth of long cash flows (i.e., 50 years or more) by using the capitalized equivalent value. The accuracy of the approximation improves as the interest rate increases (or the number of years is greater).

COMMENTS: At $i = 12\%$, Mr. Bracewell's investment is not a profitable one, but at 8% it is. This outcome indicates the importance of using the appropriate i in investment analysis. The issue of selecting an appropriate i will be presented again in Chapter 12.

5.5 Annual Equivalent-Worth Criterion

Annual equivalent worth analysis is the third equivalence method for evaluating the attractiveness or profitability of independent investment projects. While present worth analysis and future worth analysis measure project profitability referenced to time zero or some future point in time, annual equivalent worth (AE) represents it as an equivalent annual value.

Referencing the equivalence measurement to an annual time frame is convenient in situations in which the desired or preferred answer is an annual or unit value. For example, you are considering buying a new car and expect to drive 24,000 kilometres per year on business. What rate of reimbursement on a per kilometre basis would your employer need to provide to cover your annual equivalent costs of owning the car? Or consider a real estate developer who is planning to build a shopping centre of 50,000 square metres. What would be the minimum annual rental fee per square metre required to justify this investment? It should also be noted that since corporations issue annual reports and develop yearly budgets, annual measurements of costs and benefits may be more useful than overall values. Hence, management may insist on a consistent and annual reporting basis for investment projects.

In this section, we set forth the fundamental decision rule based on annual equivalent worth by considering both revenue and cost streams of a project. If revenue streams are irrelevant, then we make a decision solely on the basis of cost.

5.5.1 Fundamental Decision Rule

The **annual equivalent worth (AE)** criterion provides a basis for measuring the worth of an investment by determining equal payments on an annual basis. Knowing that any lump-sum cash amount can be converted into a series of equal annual payments, we may first find the net present worth (NPW) of the original series and then multiply this amount by the capital recovery factor:

$$AE(i) = PW(i)(A/P, i, N). \qquad (5.4)$$

The accept–reject selection rule for a single *revenue* project is as follows:

> If $AE(i) > 0$, accept the investment.
>
> If $AE(i) = 0$, remain indifferent to the investment.
>
> If $AE(i) < 0$, reject the investment.

Notice that the factor $(A/P, i, N)$ in Eq. (5.4) is positive for $-1 < i < \infty$, which indicates that the value of $AE(i)$ will be positive if, and only if, $PW(i)$ is positive. In other words, accepting a project that has a positive $AE(i)$ is equivalent to accepting a project that has a positive $PW(i)$. Therefore, the AE criterion for evaluating a project is consistent with the NPW criterion.

Example 5.10 illustrates how to find the equivalent annual worth for a proposed energy-savings project. As you will see, you first calculate the net present worth of the project and then convert this present worth into an equivalent annual basis.

EXAMPLE 5.10 Annual Equivalent Worth: A Single-Project Evaluation

A utility company is considering adding a second feedwater heater to its existing system unit to increase the efficiency of the system and thereby reduce fuel costs. The 150-MW unit will cost $1,650,000 and has a service life of 25 years. The expected salvage value of the unit is considered negligible. With the second unit installed, the efficiency of the system will improve from 55% to 56%. The fuel cost to run the feedwater is estimated at $0.05 kWh. The system unit will have a load factor of 85%, meaning that the system will run 85% of the year.

(a) Determine the equivalent annual worth of adding the second unit with an interest rate of 12%.

(b) If the fuel cost increases at the annual rate of 4% after the first year, what is the equivalent annual worth of having the second feedwater unit at $i = 12\%$?

DISCUSSION: Whenever we compare machines with different efficiency ratings, we need to determine the input powers required to operate the machines. Since the percent efficiency is equal to the ratio of the output power to the input power, we can determine the input power by dividing the output power by the motor's percent efficiency:

$$\text{Input power} = \frac{\text{output power}}{\text{percent efficiency}}.$$

For example, a 30-HP motor with 90% efficiency will require an input power of

$$\text{Input power} = \frac{(30 \text{ HP} \times 0.746 \text{ kW/HP})}{0.90}$$
$$= 24.87 \text{ kW}.$$

Therefore, energy consumption with and without the second unit can be calculated as follows:

- Before adding the second unit, $\dfrac{150,000 \text{ kW}}{0.55} = 272,727 \text{ kW}$

- After adding the second unit, $\dfrac{150,000 \text{ kW}}{0.56} = 267,857 \text{ kW}$

So the reduction in energy consumption is 4871 kW.

Since the system unit will operate only 85% of the year, the total annual operating hours are calculated as follows:

$$\text{Annual operating hours} = (365)(24)(0.85) = 7446 \text{ hours/year}.$$

SOLUTION

Given: $P = \$1,650,000$, $N = 25$ years, $S = 0$, annual fuel savings, and $i = 12\%$.
Find: AE of fuel savings due to improved efficiency.

(a) With the assumption of constant fuel cost over the service life of the second heater,

$$A_{\text{fuel savings}} = (\text{reduction in fuel requirement}) \times (\text{fuel cost})$$
$$\times (\text{operating hours per year})$$
$$= \left(\frac{150,000 \text{ kW}}{0.55} - \frac{150,000 \text{ kW}}{0.56} \right) \times (\$0.05/\text{kWh})$$
$$\times ((8760)(0.85) \text{ hours/year})$$
$$= (4871 \text{ kW}) \times (\$0.05/\text{kWh}) \times (7446 \text{ hours/year})$$
$$= \$1,813,473/\text{year};$$

$$PW(12\%) = -\$1,650,000 + \$1,813,473(P/A, 12\%, 25)$$
$$= \$12,573,321;$$

$$AE(12\%) = \$12,573,321(A/P, 12\%, 25)$$
$$= \$1,603,098.$$

(b) With the assumption of escalating energy costs at the annual rate of 4%, since the first year's fuel savings is already calculated in (a), we use it as A_1 in the geometric-gradient-series present-worth factor $(P/A_1, g, i, N)$:

$$A_1 = \$1,813,473$$
$$PW(12\%) = -\$1,650,000 + \$1,813,473(P/A_1, 4\%, 12\%, 25)$$
$$= \$17,463,697$$
$$AE(12\%) = \$17,463,697(A/P, 12\%, 25)$$
$$= \$2,226,621.$$

Clearly, either situation generates enough fuel savings to justify adding the second unit of the feedwater. Figure 5.12 illustrates the cash flow series associated with the required investment and fuel savings over the heater's service life of 25 years.

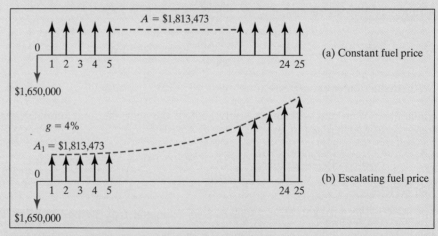

Figure 5.12   Cash flow diagram (Example 5.10).

5.5.2 Annual-Worth Calculation With Repeating Cash Flow Cycles

In some situations, a **cyclic cash flow pattern** may be observed over the life of the project. Unlike the situation in Example 5.10, where we first computed the NPW of the entire cash flow and then calculated the AE, we can compute the AE by examining the first cash flow cycle. Then we can calculate the NPW for the first cash flow cycle and derive the AE over that cycle. This shortcut method gives the same solution when the NPW of the entire project is calculated, and then the AE can be computed from this NPW.

EXAMPLE 5.11 Annual Equivalent Worth: Repeating Cash Flow Cycles

SOLEX Company is producing electricity directly from a solar source by using a large array of solar cells and selling the power to the local utility company. SOLEX decided to use amorphous silicon cells because of their low initial cost, but these cells degrade over time, thereby resulting in lower conversion efficiency and power output. The cells must be replaced every four years, which results in a particular cash flow pattern that repeats itself as shown in Figure 5.13. Determine the annual equivalent cash flows at $i = 12\%$.

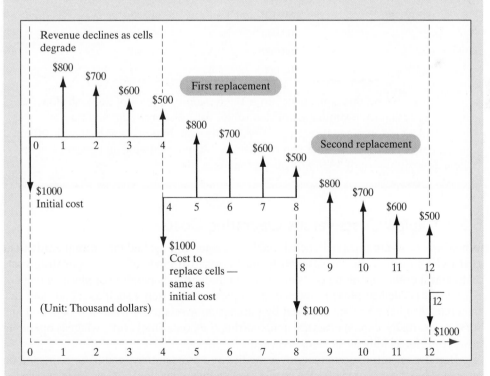

Figure 5.13 Conversion of repeating cash flow cycles into an equivalent annual payment (Example 5.11).

SOLUTION

Given: Cash flows in Figure 5.13 and $i = 12\%$.

Find: Annual equivalent benefit.

To calculate the AE, we need only consider one cycle over the four-year replacement period of the cells. For $i = 12\%$, we first obtain the NPW for the first cycle as follows:

$$PW(12\%) = -\$1,000,000$$
$$+ [(\$800,000 - \$100,000(A/G, 12\%, 4)](P/A, 12\%, 4)$$
$$= -\$1,000,000 + \$2,017,150.$$
$$= \$1,017,150.$$

Then we calculate the AE over the four-year life cycle:

$$AE(12\%) = \$1,017,150(A/P, 12\%, 4)$$
$$= \$334,880.$$

We can now say that the two cash flow series are equivalent:

Original Cash Flows		Annual Equivalent Flows	
n	A_n	n	A_n
0	−$1,000,000	0	0
1	800,000	1	$334,880
2	700,000	2	334,880
3	600,000	3	334,880
4	500,000	4	334,880

We can extend this equivalency over the remaining cycles of the cash flow. The reasoning is that each similar set of five values (one disbursement and four receipts) is equivalent to four annual receipts of $334,880 each. In other words, the $1 million investment in the solar project will recover the entire investment and generate equivalent annual savings of $334,880 over a four-year life cycle.

5.5.3 Capital Costs versus Operating Costs

When only costs are involved, the AE method is sometimes called the **annual equivalent cost (AEC)** method. In this case, costs must cover two kinds of costs: **operating costs** and **capital costs**. Operating costs are incurred through the operation of physical plant or equipment needed to provide service; examples include items such as labour and raw materials. Capital costs are incurred by purchasing assets to be used in production and service. Normally, capital costs are nonrecurring (i.e., one-time) costs, whereas operating costs recur for as long as an asset is owned.

> **Capital cost:** the amount of net investment.

Because operating costs recur over the life of a project, they tend to be estimated on an annual basis anyway, so, for the purposes of annual equivalent cost analysis, no special calculation is required. However, because capital costs tend to be one-time costs, in conducting an annual equivalent cost analysis we must translate this one-time cost into its annual equivalent over the life of the project. The annual equivalent of a capital cost is given a special name: **capital recovery cost**, designated CR(i).

Two general monetary transactions are associated with the purchase and eventual retirement of a capital asset: its initial cost (P) and its salvage value (S). Taking into account these sums, we calculate the capital recovery cost as follows:

$$\text{CR}(i) = P(A/P, i, N) - S(A/F, i, N). \tag{5.5}$$

Now, recall algebraic relationships between factors in Table 3.4, and notice that the factor (A/F, i, N) can be expressed as

$$(A/F, i, N) = (A/P, i, N) - i.$$

Then we may rewrite CR(i) as

$$\begin{aligned} \text{CR}(i) &= P(A/P, i, N) - S[(A/P, i, N) - i] \\ &= (P - S)(A/P, i, N) + iS. \end{aligned} \tag{5.6}$$

Since we are calculating the equivalent annual costs, we treat cost items with a positive sign. Then the salvage value is treated as having a negative sign in Eq. (5.6). We may interpret this situation thus: To obtain the machine, one borrows a total of P dollars, S dollars of which are returned at the end of the Nth year. The first term, $(P - S)(A/P, i, N)$, implies that the balance $(P - S)$ will be paid back in equal installments over the N-year period at a rate of i. The second term, iS, implies that simple interest in the amount iS is paid on S until it is repaid (Figure 5.14). Thus, the amount to be financed is $P - S$ (P/F, i, N), and the installments of this loan over the N-year period are

$$\begin{aligned} \text{AE}(i) &= [P - S(P/F, i, N)](A/P, i, N) \\ &= P(A/P, i, N) - S(P/F, i, N)(A/P, i, N) \\ &= [P(A/P, i, N) - S(A/F, i, N)] \\ &= \text{CR}(i). \end{aligned} \tag{5.7}$$

Capital recovery cost: The annual payment that will repay the cost of a fixed asset over the useful life of the asset and will provide an economic rate of return on the investment.

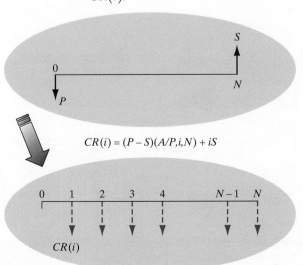

$$CR(i) = (P - S)(A/P, i, N) + iS$$

Figure 5.14 Capital recovery (ownership) cost calculation.

Therefore, CR(i) tells us what the bank would charge each year. Many auto leases are based on this arrangement, in that most require a guarantee of S dollars in salvage. From an industry viewpoint, CR(i) is the annual cost to the firm of owning the asset.

With this information, the amount of annual savings required to recover the capital and operating costs associated with a project can be determined, as Example 5.12 illustrates.

EXAMPLE 5.12 Capital Recovery Cost

Consider a machine that costs $20,000 and has a five-year useful life. At the end of the five years, it can be sold for $4000 after tax adjustment. The annual operating and maintenance (O&M) costs are about $500. If the firm could earn an after-tax revenue of $5000 per year with this machine, should it be purchased at an interest rate of 10%? (All benefits and costs associated with the machine are accounted for in these figures.)

SOLUTION

Given: $P = \$20,000$, $S = \$4000$, O&M $= \$500$, $A = \$5000$, $N = 5$ years, and $i = 10\%$ per year.

Find: AE, and determine whether to purchase the machine.

The first task is to separate cash flows associated with acquisition and disposal of the asset from the normal operating cash flows. Since the operating cash flows—the $5000 yearly revenue—are already given in equivalent annual flows, we need to convert only the cash flows associated with acquisition and disposal of the asset into equivalent annual flows (Figure 5.15). Using Eq. (5.6), we obtain

$$CR(i) = (P - S)(A/P, i, N) + iS$$
$$= (\$20,000 - \$4000)(A/P, 10\%, 5) + (0.10)\$4000$$
$$= \$4620.76,$$

$$O\&M(i) = \$500,$$

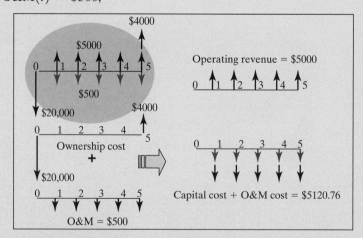

Figure 5.15 Separating ownership cost (capital cost) and operating cost from operating revenue, which must exceed the annual equivalent cost to make the project acceptable.

$$AEC(10\%) = CR(10\%) + O\&M(10\%)$$
$$= \$4620.76 + \$500$$
$$= \$5120.76,$$

$$AE(10\%) = \$5000 - \$5120.76$$
$$= -\$120.76.$$

This negative AE value indicates that the machine does not generate sufficient revenue to recover the original investment, so we must reject the project. In fact, there will be an equivalent loss of $120.76 per year over the life of the machine.

COMMENTS: We may interpret the value found for the annual equivalent cost as asserting that the annual operating revenues must be at least $5120.76 in order to recover the cost of owning and operating the asset. However, the annual operating revenues actually amount to only $5000, resulting in a loss of $120.76 per year. Therefore, the project is not worth undertaking.

5.5.4 Benefits of AE Analysis

In general, most engineering economic analysis problems can be solved by the present-worth methods that were introduced earlier in this chapter. However, some economic analysis problems can be solved more efficiently by annual-worth analysis.

Example 5.10 should look familiar to you: It is exactly the situation we encountered in Chapter 4 when we converted a mixed cash flow into a single present value and then into a series of equivalent cash flows. In the case of Example 5.10, you may wonder why we bother to convert NPW to AE at all, since we already know that the project is acceptable from NPW analysis. In fact, the example was mainly an exercise to familiarize you with the AE calculation.

However, in the real world, a number of situations can occur in which AE analysis is preferred, or even demanded, over NPW analysis. For example, corporations issue annual reports and develop yearly budgets. For these purposes, a company may find it more useful to present the annual cost or benefit of an ongoing project, rather than its overall cost or benefit. Following are some additional situations in which AE analysis is preferred:

1. **Consistency of report formats.** Financial managers commonly work with annual rather than overall costs in any number of internal and external reports. Engineering managers may be required to submit project analyses on an annual basis for consistency and ease of use by other members of the corporation and stockholders.

2. **Need for unit costs or profits.** In many situations, projects must be broken into unit costs (or profits) for ease of comparison with alternatives. Pricing the use of an asset and reimbursement analyses are key examples, and these will be discussed in the chapter.

3. **Life-cycle cost analysis.** When there is no need for estimating the revenue stream for a proposed project, we can consider only the cost streams of the project. In that case, it is common to convert this life-cycle cost (LCC) into an equivalent annual

cost for purposes of comparison. Industry has used the LCC to help determine which project will cost less over the life of a product. This topic will be further discussed in Chapter 6.

5.5.5 Unit Profit or Cost Calculation

In many situations, we need to know the unit profit (or cost) of operating an asset. To obtain this quantity, we may proceed as follows:

- Determine the number of units to be produced (or serviced) each year over the life of the asset.
- Identify the cash flow series associated with production or service over the life of the asset.
- Calculate the net present worth of the project cash flow series at a given interest rate, and then determine the equivalent annual worth.
- Divide the equivalent annual worth by the number of units to be produced or serviced during each year. When the number of units varies each year, you may need to convert them into equivalent annual units.

To illustrate the procedure, Example 5.13 uses the annual equivalent concept in estimating the savings per machine hour for the proposed acquisition of a machine.

EXAMPLE 5.13 Unit Profit per Machine Hour When Annual Operating Hours Remain Constant

Consider the investment in the metal-cutting machine of Example 5.5. Recall that this three-year investment was expected to generate an NPW of $3553. Suppose that the machine will be operated for 2000 hours per year. Compute the equivalent savings per machine hour at $i = 15\%$.

SOLUTION

Given: NPW = $3553, $N = 3$ years, $i = 15\%$ per year, and 2000 machine hours per year.
Find: Equivalent savings per machine hour.

We first compute the annual equivalent savings from the use of the machine. Since we already know the NPW of the project, we obtain the AE by the formula

$$AE(15\%) = \$3553(A/P, 15\%, 3) = \$1556.$$

With an annual usage of 2000 hours, the equivalent savings per machine hour would be

$$\text{Savings per machine hour} = \$1556/2000 \text{ hours} = \$0.78/\text{hour}.$$

COMMENTS: Note that we cannot simply divide the NPW ($3553) by the total number of machine hours over the three-year period (6000 hours) to obtain $0.59/hour. This $0.59 figure represents the instant savings in present worth for each hourly use of the equipment, but does not consider the time over which the savings occur. Once we have the annual equivalent worth, we can divide by the desired time unit if the compounding period is one year. If the compounding period is shorter, then the equivalent worth should be calculated for the compounding period.

EXAMPLE 5.14 Unit Profit per Machine Hour When Annual Operating Hours Fluctuate

Consider again Example 5.13, and suppose that the metal-cutting machine will be operated according to varying hours: 1500 hours the first year, 2500 hours the second year, and 2000 hours the third year. The total operating hours still remain at 6000 over three years. Compute the equivalent savings per machine hour at $i = 15\%$.

SOLUTION

Given: NPW = \$3553, $N = 3$ years, $i = 15\%$ per year, and operating hours of 1500 the first year, 2500 the second year, and 2000 the third year.

Find: Equivalent savings per machine hour.

As calculated in Example 5.13, the annual equivalent savings is \$1556. Let C denote the equivalent annual savings per machine hour, which we need to determine. Now, with varying annual usages of the machine, we can set up the equivalent annual savings as a function of C:

$$
\begin{aligned}
\text{Equivalent annual savings} = {} & [C(1500)(P/F, 15\%, 1) \\
& + C(2500)(P/F, 15\%, 2) \\
& + C(2000)(P/F, 15\%, 3)](A/P, 15\%, 3) \\
= {} & 1975.16C.
\end{aligned}
$$

We can equate this amount to the \$1556 we calculated in Example 5.13 and solve for C. This gives us

$$C = \$1556/1975.16 = \$0.79/\text{hour},$$

which is about a penny more than the \$0.78 we found in Example 5.13.

5.5.6 Pricing the Use of an Asset

Companies often need to calculate the cost of equipment that corresponds to a **unit of use** of that equipment. For example, if you own an asset such as a building, you would be interested in knowing the cost per square metre of owning and operating the asset. This information will be the basis for determining the rental fee for the asset. A familiar example is an employer's reimbursement of costs for the use of an employee's personal car for business purposes. If an employee's job depends on obtaining and using a personal vehicle on the employer's behalf, reimbursement on the basis of the employee's overall costs per kilometre seems fair.

EXAMPLE 5.15 Pricing an Apartment Rental Fee

Sunbelt Corporation, an investment company, is considering building a 50-unit apartment complex in a growing area in Fort McMurray, Alberta. Since the long-term growth potential of the town is excellent, it is believed that the company could average 85% full occupancy for the complex each year. If the following financial data are reasonably accurate estimates, determine the minimum monthly rent that should be charged if a 15% rate of return is desired:

- Land investment cost = $1,000,000
- Building investment cost = $2,500,000
- Annual upkeep cost = $150,000
- Property taxes and insurance = 5% of total initial investment
- Study period = 25 years
- Salvage value = Only land cost can be recovered in full.

SOLUTION

Given: Preceding financial data, study period = 25 years, and i = 15%.
Find: Minimum monthly rental charge.

First we need to determine the capital cost associated with ownership of the property:

$$\text{Total investment required} = \text{land cost} + \text{building cost} = \$3,500,000,$$

$$\text{Salvage value} = \$1,000,000 \text{ at the end of 25 years,}$$

$$CR(15\%) = (\$3,500,000 - \$1,000,000)(A/P, 15\%, 25)$$
$$+ (\$1,000,000)(0.15)$$
$$= \$536,749.$$

Second, the annual operating cost has two elements: (1) property taxes and insurance and (2) annual upkeep cost. Thus,

$$\text{O\&M cost} = (0.05)(\$3,500,000) + \$150,000$$
$$= \$325,000.$$

So the total annual equivalent cost is

$$AEC(15\%) = \$536,749 + \$325,000$$
$$= \$861,749,$$

which is the minimum annual rental required to achieve a 15% rate of return. Therefore, with annual compounding, the monthly rental amount per unit is

$$\text{Required monthly charge} = \frac{\$861,749}{(12 \times 50)(0.85)}$$
$$= \$1690.$$

COMMENTS: The rental charge that is exactly equal to the cost of owning and operating the building is known as the **break-even point**.

5.6 Rate of Return Analysis

Along with the NPW, FW, and the AE criteria, the fourth measure of investment worth is **rate of return**. As shown in earlier sections of this chapter, the NPW measure is easy to calculate and apply. Nevertheless, many engineers and financial managers prefer rate-of-return analysis to the NPW method because they find it intuitively more appealing to analyze investments in terms of percentage rates of return rather than dollars of NPW. Consider the following statements regarding an investment's profitability:

- This project will bring in a 15% rate of return on the investment.
- This project will result in a net surplus of $10,000 in the NPW.

Neither statement describes the nature of the investment project in any complete sense. However, the rate of return is somewhat easier to understand because many of us are so familiar with savings-and-loan interest rates, which are in fact rates of return.

Many different terms are used to refer to **rate of return**, including **yield** (i.e., the yield to maturity, commonly used in bond valuation), **internal rate of return**, and **marginal efficiency of capital**. First we will review three common definitions of rate of return. Then we will use the definition of internal rate of return as a measure of profitability for a single investment project throughout the text.

Yield: The annual rate of return on an investment, expressed as a percentage.

5.6.1 Return on Investment

There are several ways of defining the concept of a rate of return on investment. The first is based on a typical loan transaction, the second on the mathematical expression of the present-worth function, and the third on the project cash flow series.

Definition 1

The rate of return is the interest rate earned on the unpaid balance of an amortized loan.

Suppose that a bank lends $10,000 and $4021 is repaid at the end of each year for three years. How would you determine the interest rate that the bank charges on this transaction? As we learned in Chapter 3, you would set up the equivalence equation

$$\$10,000 = \$4021(P/A, i, 3)$$

and solve for i. It turns out that $i = 10\%$. In this situation, the bank will earn a return of 10% on its investment of $10,000. The bank calculates the balances over the life of the loan as follows:

Year	Unpaid Balance at Beginning of Year	Return on Unpaid Balance (10%)	Payment Received	Unpaid Balance at End of Year
0	−$10,000	$0	$0	−$10,000
1	−10,000	−1,000	+4,021	−6,979
2	−6,979	−698	+4,021	−3,656
3	−3,656	−366	+4,021	0

A negative balance indicates an unpaid balance. In other words, the customer still owes money to the bank.

Observe that, for the repayment schedule shown, the 10% interest is calculated only on each year's outstanding balance. In this situation, only part of the $4021 annual payment represents interest; the remainder goes toward repaying the principal. Thus, the three annual payments repay the loan itself and additionally provide a return of 10% on the *amount still outstanding each year*.

Note that when the last payment is made, the outstanding principal is eventually reduced to zero.[4] If we calculate the NPW of the loan transaction at its rate of return (10%), we see that

$$PW(10\%) = -\$10,000 + \$4021(P/A, 10\%, 3) = 0,$$

which indicates that the bank can break even at a 10% rate of interest. In other words, the rate of return becomes the rate of interest that equates the present value of future cash repayments to the amount of the loan. This observation prompts the second definition of rate of return.

Definition 2

The rate of return is the break-even interest rate i^* that equates the present worth of a project's cash outflows to the present worth of its cash inflows, or

$$PW(i^*) = PW_{\text{Cash inflows}} - PW_{\text{Cash outflows}}$$
$$= 0.$$

Note that the expression for the NPW is equivalent to

$$PW(i^*) = \frac{A_0}{(1 + i^*)^0} + \frac{A_1}{(1 + i^*)^1} + \dots + \frac{A_N}{(1 + i^*)^N} = 0. \qquad (5.8)$$

Here we know the value of A_n for each period, but not the value of i^*. Since it is the only unknown, however, we can solve for i^*. (Inevitably, there will be real and imaginery N values of i^* that satisfy this equation. In most project cash flows, you would be able to find a unique positive i^* that satisfies Eq. (5.8). However, you may encounter some cash flows that cannot be solved for a single rate of return greater than -100%. By the nature of the NPW function in Eq. (5.8), it is possible to have more than one rate of return for certain types of cash flows. For some cash flows, we may not find a specific rate of return at all.)[5]

Note that the formula in Eq. (5.8) is simply the NPW formula solved for the particular interest rate (i^*) at which PW(i) is equal to zero. By multiplying both sides of Eq. (5.8) by $(1 + i^*)^N$, we obtain

$$PW(i^*)(1 + i^*)^N = FW(i^*) = 0.$$

If we multiply both sides of Eq. (5.8) by the capital recovery factor $(A/P, i^*, N)$, we obtain the relationship AE(i^*) = 0. Therefore, the i^* of a project may be defined as the rate of interest that equates the present worth, future worth, and annual equivalent worth of the entire series of cash flows to zero.

---

[4] As we learned in Section 5.3.2, this terminal balance is equivalent to the net future worth of the investment. If the net future worth of the investment is zero, its NPW should also be zero.

[5] You will always have N rates of return. The issue is whether they are real or imaginary. If they are real, the question "Are they in the $(-100\%, \infty)$ interval?" should be asked. A **negative rate of return** implies that you never recover your initial investment.

5.6.2 Return on Invested Capital

Investment projects can be viewed as analogous to bank loans. We will now introduce the concept of rate of return based on the return on invested capital in terms of a project investment. A project's return is referred to as the internal rate of return (IRR) or the **yield** promised by an **investment project** over its **useful life**.

Definition 3

The internal rate of return is the interest rate charged on the unrecovered project balance of the investment such that, when the project terminates, the unrecovered project balance will be zero.

Suppose a company invests $10,000 in a computer with a three-year useful life and equivalent annual labour savings of $4021. Here, we may view the investing firm as the lender and the project as the borrower. The cash flow transaction between them would be identical to the amortized loan transaction described under Definition 1:

n	Beginning Project Balance	Return on Invested Capital	Ending Cash Payment	Project Balance
0	$0	$0	−$10,000	−$10,000
1	−10,000	−1,000	4,021	6,979
2	−6,979	−697	4,021	3,656
3	−3,656	−365	4,021	0

In our project balance calculation, we see that 10% is earned (or charged) on $10,000 during year 1, 10% is earned on $6979 during year 2, and 10% is earned on $3656 during year 3. This indicates that the firm earns a 10% rate of return on funds that remain *internally* invested in the project. Since it is a return that is *internal* to the project, we refer to it as the **internal rate of return**, or IRR. This means that the computer project under consideration brings in enough cash to pay for itself in three years and also to provide the firm with a return of 10% on its invested capital. Put differently, if the computer is financed with funds costing 10% annually, the cash generated by the investment will be exactly sufficient to repay the principal and the annual interest charge on the fund in three years.

Internal rate of return: This is the return that a company would earn if it invested in itself, rather than investing that money elsewhere.

Notice that only one cash outflow occurs at time 0, and the present worth of this outflow is simply $10,000. There are three equal receipts, and the present worth of these inflows is $4021(P/A, 10\%, 3) = \$10,000$. Since the NPW $= PW_{\text{Inflow}} - PW_{\text{Outflow}} = \$10,000 - \$10,000 = 0$, 10% also satisfies Definition 2 of the rate of return. Even though the preceding simple example implies that i^* coincides with IRR, only Definitions 1 and 3 correctly describe the true meaning of the internal rate of return. As we will see later, if the cash expenditures of an investment are not restricted to the initial period, several break-even interest rates (i^*s) may exist that satisfy Eq. (5.8). However, there may not be a rate of return that is *internal* to the project.

5.6.3 Simple versus Nonsimple Investments

We can classify an investment project by counting the number of sign changes in its net cash flow sequence. A change from either "+" to "−" or "−" to "+" is counted as one sign change. (We ignore a zero cash flow.) Then,

Simple investment: The project with only one sign change in the net cash flow series.

- A **simple investment** is an investment in which the initial cash flows are negative and only one sign change occurs in the remaining net cash flow series. If the initial flows are positive and only one sign change occurs in the subsequent net cash flows, they are referred to as **simple borrowing** cash flows.
- A **nonsimple investment** is an investment in which more than one sign change occurs in the cash flow series.

Multiple $i*$s, as we will see later, occur only in nonsimple investments. Three different types of investment possibilities are illustrated in Example 5.16.

EXAMPLE 5.16 Investment Classification

Consider the following three cash flow series and classify them into either simple or nonsimple investments:

Period	Net Cash Flow		
n	Project A	Project B	Project C
0	−$1000	−$1000	$1000
1	−500	3900	−450
2	800	−5030	−450
3	1500	2145	−450
4	2000		

SOLUTION

Given: Preceding cash flow sequences.

Find: Classify the investments shown into either simple or nonsimple investments.

- Project A represents a common simple investment. This situation reveals the NPW profile shown in Figure 5.16(a). The curve crosses the i-axis only once.
- Project B represents a nonsimple investment. The NPW profile for this investment has the shape shown in Figure 5.16(b). The i-axis is crossed at 10%, 30%, and 50%.
- Project C represents neither a simple nor a nonsimple investment, even though only one sign change occurs in the cash flow sequence. Since the first cash flow is positive, this is a **simple borrowing** cash flow, not an investment flow. Figure 5.16(c) depicts the NPW profile for this type of investment.

COMMENTS: Not all NPW profiles for nonsimple investments have multiple crossings of the i-axis. Clearly, then, we should place a high priority on discovering this situation early in our analysis of a project's cash flows. The quickest way to predict

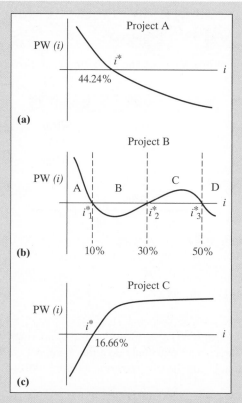

Figure 5.16 Present-worth profiles: (a) Simple investment, (b) nonsimple investment with multiple rates of return, and (c) simple borrowing cash flows.

multiple i^*s is to generate an NPW profile on a computer and check whether it crosses the horizontal axis more than once. In Appendix 5A, we will illustrate when to expect such multiple crossings by examining types of cash flows.

5.6.4 Methods of Finding i^*

Once we identify the type of an investment cash flow, several ways to determine its rate of return are available. Some of the most practical methods are as follows:

- Direct solution method,
- Trial-and-error method, and
- Computer solution method.

Direct Solution Method

For the special case of a project with only a two-flow transaction (an investment followed by a single future payment) or a project with a service life of two years of return, we can seek a direct mathematical solution for determining the rate of return. These two cases are examined in Example 5.17.

EXAMPLE 5.17 Finding $i*$ by Direct Solution: Two Flows and Two Periods

Consider two investment projects with the following cash flow transactions:

n	Project 1	Project 2
0	-$2000	-$2000
1	0	1300
2	0	1500
3	0	
4	3500	

Compute the rate of return for each project.

SOLUTION

Given: Cash flows for two projects.
Find: $i*$ for each project.

Project 1: Solving for $i*$ in $PW(i*) = 0$ is identical to solving $FW(i*) = 0$, because FW equals PW times a constant. We could do either here, but we will set $FW(i*) = 0$ to demonstrate the latter. Using the single-payment future-worth relationship, we obtain

$$FW(i*) = -\$2000(F/P, i*, 4) + \$3500 = 0,$$

$$\$3500 = \$2000(F/P, i*, 4) = \$2000(1 + i*)^4,$$

$$1.75 = (1 + i*)^4.$$

Solving for $i*$ yields

$$i* = \sqrt[4]{1.75} - 1$$
$$= 0.1502 \text{ or } 15.02\%.$$

Project 2: We may write the NPW expression for this project as

$$PW(i) = -\$2000 + \frac{\$1300}{(1 + i)} + \frac{\$1500}{(1 + i)^2} = 0.$$

Let $X = 1/(1 + i)$. We may then rewrite $PW(i)$ as a function of X as follows:

$$PW(X) = -\$2000 + \$1300X + \$1500X^2 = 0.$$

This is a quadratic equation that has the following solution:[6]

$$X = \frac{-1300 \pm \sqrt{1300^2 - 4(1500)(-2000)}}{2(1500)}$$

$$= \frac{-1300 \pm 3700}{3000}$$

$$= 0.8 \text{ or } -1.667.$$

Replacing X values and solving for i gives us

$$0.8 = \frac{1}{(1+i)} \rightarrow i = 25\%,$$

$$-1.667 = \frac{1}{(1+i)} \rightarrow i = -160\%.$$

Since an interest rate less than -100% has no economic significance, we find that the project's i^* is 25%.

COMMENTS: In both projects, one sign change occurred in the net cash flow series, so we expected a unique i^*. Also, these projects had very simple cash flows. When cash flows are more complicated, generally we use a trial-and-error method or a computer to find i^*.

Trial-and-Error Method

The first step in the trial-and-error method is to make an estimated **guess**[7] at the value of i^*. For a simple investment, we use the "guessed" interest rate to compute the present worth of net cash flows and observe whether it is positive, negative, or zero. Suppose, then, that PW(i) is negative.

Since we are aiming for a value of i that makes PW(i) = 0, we must raise the present worth of the cash flow. To do this, we lower the interest rate and repeat the process. If PW(i) is positive, however, we raise the interest rate in order to lower PW(i). The process is continued until PW(i) is approximately equal to zero. Whenever we reach the point where PW(i) is bounded by one negative and one positive value, we use **linear interpolation** to approximate i^*. This process is somewhat tedious and inefficient. (The trial-and-error method does not work for nonsimple investments in which the NPW function is not, in general, a monotonically decreasing function of the interest rate.)

EXAMPLE 5.18 Finding i^* by Trial and Error

The Imperial Chemical Company is considering purchasing a chemical analysis machine worth $13,000. Although the purchase of this machine will not produce any increase in sales revenues, it will result in a reduction of labour costs. In order to operate the machine properly, it must be calibrated each year. The machine has an

---

[6] The solution of the quadratic equation $aX^2 + bX + c = 0$ is $X = \dfrac{-b \pm \sqrt{b^2 - 4ac}}{2a}$.

[7] As we shall see later in this chapter, the ultimate objective of finding i^* is to compare it against the MARR. Therefore, it is a good idea to use the MARR as the initial guess.

expected life of six years, after which it will have no salvage value. The following table summarizes the annual savings in labour cost and the annual maintenance cost in calibration over six years:

Year (n)	Costs ($)	Savings ($)	Net Cash Flow ($)
0	13,000		−13,000
1	2,300	6,000	3,700
2	2,300	7,000	4,700
3	2,300	9,000	6,700
4	2,300	9,000	6,700
5	2,300	9,000	6,700
6	2,300	9,000	6,700

Find the rate of return for this project.

SOLUTION

Given: Cash flows over six years as shown in Figure 5.17.
Find: i^*.

We start with a guessed interest rate of 25%. The present worth of the cash flows is

$$PW(25\%) = -\$13,000 + \$3700(P/F, 25\%, 1) + \$4700(P/F, 25\%, 2)$$
$$+ \$6700(P/A, 25\%, 4)(P/F, 25\%, 2)$$
$$= \$3095.$$

Since this present worth is positive, we must raise the interest rate to bring PW toward zero. When we use an interest rate of 35%, we find that

$$PW(35\%) = -\$13,000 + \$3700(P/F, 35\%, 1) + \$4700(P/F, 35\%, 2)$$
$$+ \$6700(P/A, 35\%, 4)(P/F, 35\%, 2)$$
$$= -\$339.$$

We have now bracketed the solution: $PW(i)$ will be zero at i somewhere between 25% and 35%. Using straight-line interpolation, we approximate

$$i^* \cong 25\% + (35\% - 25\%)\left[\frac{3095 - 0}{3095 - (-339)}\right]$$
$$= 25\% + 10\%(0.9013)$$
$$= 34.01\%.$$

Now we will check to see how close this value is to the precise value of i^*. If we compute the present worth at this interpolated value, we obtain

$$PW(34\%) = -\$13,000 + \$3700(P/F, 34\%, 1) + \$4700(P/F, 34\%, 2)$$
$$+ \$6700(P/A, 34\%, 4)(P/F, 34\%, 2)$$
$$= -\$50.58.$$

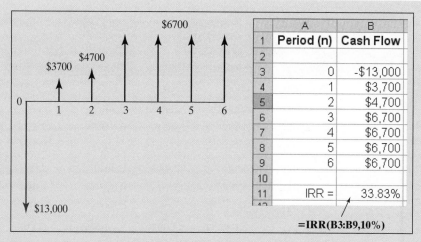

Figure 5.17 Cash flow diagram for a simple investment (Example 5.18).

As this is not zero, we may recompute i^* at a lower interest rate, say, 33%:

$$PW(33\%) = -\$13,000 + \$3700(P/F, 33\%, 1) + \$4700(P/F, 33\%, 2)$$
$$+ \$6700(P/A, 33\%, 4)(P/F, 33\%, 2)$$
$$= \$248.56.$$

With another round of linear interpolation, we approximate

$$i^* \cong 33\% + (34\% - 33\%)\left[\frac{248.56 - 0}{248.56 - (-50.58)}\right]$$
$$= 33\% + 1\%(0.8309)$$
$$= 33.83\%.$$

At this interest rate,

$$PW(33.83\%) = -\$13,000 + \$3700(P/F, 33.83\%, 1)$$
$$+ \$4700(P/F, 33.83\%, 2)$$
$$+ \$6700(P/A, 33.83\%, 4)(P/F, 33.83\%, 2)$$
$$= -\$0.49,$$

which is practically zero, so we may stop here. In fact, there is no need to be more precise about these interpolations, because the final result can be no more accurate than the basic data, which ordinarily are only rough estimates.

COMMENT: With Excel, you can evaluate the IRR for the project as =IRR (range,guess), where you specify the cell range for the cash flow (e.g., B3:B9) and the initial guess, such as 25%. Computing i^* for this problem in Excel, incidentally, gives us 33.8283%. Instead of using the factor notations, you may attempt to use a tabular approach as follows:

		Internal Rate of Return: What It Looks Like			
		Discount Rate: 25%		Discount Rate: 35%	
Year	Cash Flow	Factor	Amount	Factor	Amount
0	−$13,000	1.0000	−$13,000	1.0000	−$13,000
1	3,700	0.8000	2,960	0.7407	2,741
2	4,700	0.6400	3,008	0.5487	2,579
3	6,700	0.5120	3,430	0.4064	2,723
4	6,700	0.4096	2,744	0.3011	2,017
5	6,700	0.3277	2,196	0.2230	1,494
6	6,700	0.2621	1,756	0.1652	1,107
Total	+$22,200	NPW =	+$3,095	NPW =	−$339
				IRR = close to 34%	

Graphical Method

We don't need to do laborious manual calculations to find i^*. Many financial calculators have built-in functions for calculating i^*. It is worth noting that many online financial calculators or spreadsheet packages have i^* functions, which solve Eq. (5.8) very rapidly,[8] usually with the user entering the cash flows via a computer keyboard or by reading a cash flow data file. (For example, Microsoft Excel has an IRR financial function that analyzes investment cash flows, as illustrated in Example 5.18.)

The most easily generated and understandable graphic method of solving for i^* is to create the **NPW profile** on a computer. On the graph, the horizontal axis indicates the interest rate and the vertical axis indicates the NPW. For a given project's cash flows, the NPW is calculated at an interest rate of zero (which gives the vertical-axis intercept) and several other interest rates. Points are plotted and a curve is sketched. Since i^* is defined as the interest rate at which $PW(i^*) = 0$, the point at which the curve crosses the horizontal axis closely approximates i^*. The graphical approach works for both simple and nonsimple investments.

EXAMPLE 5.19 Graphical Approach to Estimate i^*

Consider the cash flow series shown in Figure 5.18. Estimate the rate of return by generating the NPW profile on a computer.

SOLUTION

Given: Cash flow series in Figure 5.18.

Find: (a) i^* by plotting the NPW profile and (b) i^* by using Excel.

---

[8] An alternative method of solving for i^* is to use a computer-aided economic analysis program. Cash Flow Analyzer (CFA) finds i^* visually by specifying the lower and upper bounds of the interest search limit and generates NPW profiles when given a cash flow series. In addition to the savings in calculation time, the advantage of computer-generated profiles is their precision. CFA can be found on the book's Companion Website.

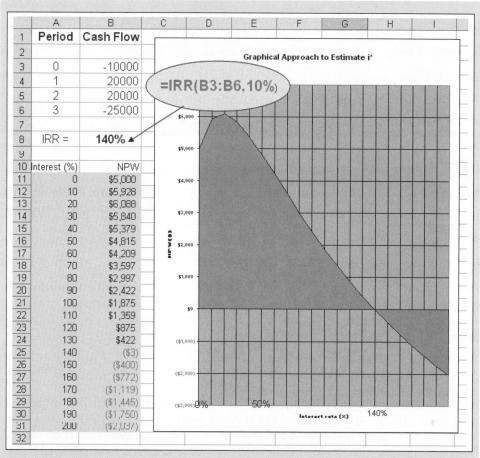

	A	B	C	D	E	F	G	H	I
1	Period	Cash Flow							
2									
3	0	-10000							
4	1	20000							
5	2	20000							
6	3	-25000							
7									
8	IRR =	140%							
9									
10	Interest (%)	NPW							
11	0	$5,000							
12	10	$5,928							
13	20	$6,088							
14	30	$5,840							
15	40	$5,379							
16	50	$4,815							
17	60	$4,209							
18	70	$3,597							
19	80	$2,997							
20	90	$2,422							
21	100	$1,875							
22	110	$1,359							
23	120	$875							
24	130	$422							
25	140	($3)							
26	150	($400)							
27	160	($772)							
28	170	($1,119)							
29	180	($1,445)							
30	190	($1,750)							
31	200	($2,037)							
32									

Graphical Approach to Estimate i*

=IRR(B3:B6,10%)

Figure 5.18 Graphical solution to rate-of-return problem for a typical nonsimple investment (Example 5.19).

(a) The present-worth function for the project cash flow series is

$$PW(i) = -\$10,000 + \$20,000(P/A, i, 2) - \$25,000(P/F, i, 3)$$

First we use $i = 0$ in this equation to obtain NPW = $5000, which is the vertical-axis intercept. Then we substitute several other interest rates—10%, 20%, ..., 140%—and plot these values of $PW(i)$ as well. The result is Figure 5.18, which shows the curve crossing the horizontal axis at roughly 140%. This value can be verified by other methods if we desire to do so. Note that, in addition to establishing the interest rate that makes NPW = 0, the NPW profile indicates where positive and negative NPW values fall, thus giving us a broad picture of those interest rates for which the project is acceptable or unacceptable. (Note also that a trial-and-error method would lead to some confusion: As you increase the interest rate from 0% to 20%, the NPW value also keeps increasing, instead of decreasing.) Even though the project is a nonsimple investment, the curve crosses the horizontal axis only once. As mentioned in the previous section, however, most nonsimple projects have more than one value of i^* that makes NPW = 0 (i.e., more than one i^* per project). In such a case, the NPW profile would cross the horizontal axis more than once.[9]

(b) With Excel, you can evaluate the IRR for the project with the function

$$= IRR(range, guess)$$

in which you specify the cell range for the cash flow and the initial guess, such as 10%.

5.7 Internal-Rate-of-Return Criterion

Now that we have classified investment projects and learned methods for determining the i^* value for a given project's cash flows, our objective is to develop an accept–reject decision rule that gives results consistent with those obtained from NPW analysis.

5.7.1 Relationship to PW Analysis

As we already observed earlier in this chapter, NPW analysis depends on the rate of interest used for the computation of NPW. A different rate may change a project from being considered acceptable to being unacceptable, or it may change the ranking of several projects:

- Consider again the NPW profile as drawn for the simple project in Figure 5.16(a). For interest rates below i^*, this project should be accepted because NPW > 0; for interest rates above i^*, it should be rejected.
- By contrast, for certain nonsimple projects, the NPW may look like the one shown in Figure 5.16(b). NPW analysis would lead you to accept the projects in regions **A** and **C**, but reject those in regions **B** and **D**. Of course, this result goes against intuition: A higher interest rate would change an unacceptable project into an acceptable one. The situation graphed in Figure 5.16(b) is one of the cases of multiple i^*s mentioned in Definition 2.

Therefore, for the simple investment situation in Figure 5.16(a), i^* can serve as an appropriate index for either accepting or rejecting the investment. However, for the nonsimple investment of Figure 5.16(b), it is not clear which i^* to use to make an accept–reject decision. Therefore, the i^* value fails to provide an appropriate measure of profitability for an investment project with multiple rates of return.

5.7.2 Decision Rule for Simple Investments

Suppose we have a simple investment. Why are we interested in finding the particular interest rate that equates a project's cost with the present worth of its receipts? Again, we may easily answer this question by examining Figure 5.16(a). In this figure, we notice two important characteristics of the NPW profile. First, as we compute the project's PW(i) at a varying interest rate i, we see that the NPW is positive for $i < i^*$, indicating that the project would be acceptable under PW analysis for those values of i. Second, the

---

[9] In Appendix 5A, we discuss methods of predicting the number of i^* values by looking at cash flows. However, generating an NPW profile to discover multiple i^*s is as practical and informative as any other method.

NPW is negative for $i > i^*$, indicating that the project is unacceptable for those values of i. Therefore, i^* serves as a **benchmark** interest rate, knowledge of which will enable us to make an accept–reject decision consistent with NPW analysis.

Note that, for a pure investment, i^* is indeed the IRR of the investment. (See Section 5.6.2.) Merely knowing i^*, however, is not enough to apply this method. Because firms typically wish to do better than break even (recall that at NPW = 0 we were indifferent to the project), a minimum acceptable rate of return (MARR) is indicated by company policy, management, or the project decision maker. If the IRR exceeds this MARR, we are assured that the company will more than break even. Thus, the IRR becomes a useful gauge against which to judge a project's acceptability, and the decision rule for a simple project is as follows:

If IRR > MARR, accept the project.

If IRR = MARR, remain indifferent.

If IRR < MARR, reject the project.

Note that this decision rule is designed to be applied for a single project evaluation. When we have to compare mutually exclusive investment projects, we need to apply the **incremental analysis approach**, as we shall see in Chapter 6.

EXAMPLE 5.20 Investment Decision for a Simple Investment

Merco Inc., a machinery builder in Hamilton, Ontario, is considering investing $1,250,000 in a complete structural beam-fabrication system. The increased productivity resulting from the installation of the drilling system is central to the project's justification. Merco estimates the following figures as a basis for calculating productivity:

- Increased fabricated steel production: 2000 tonnes/year.
- Average sales price of fabricated steel: $2566.50/tonne.
- Labour rate: $10.50/hour.
- Steel produced in a year: 15,000 tonnes.
- Cost of steel: $1950/tonne.
- Number of workers on layout, hole making, sawing, and material handling: 17.
- Additional maintenance cost: $128,500/year.

The cost of producing a tonne of fabricated steel is about $2170.50. With a selling price of $2566.50 per tonne, the resulting contribution to overhead and profit becomes $396 per tonne. Assuming that Merco will be able to sustain an increased production of 2000 tonnes per year by purchasing the system, engineers have estimated the projected additional revenue to be 2000 tonnes × $396 = $792,000.

Since the drilling system has the capacity to fabricate the full range of structural steel, two workers can run the system, one on the saw and the other on the drill. A third operator is required to operate a crane for loading and unloading materials. Merco estimates that, to do the equivalent work of these three workers with conventional manufacturing techniques would require, on the average, an additional 14 people for center punching, hole making with a radial or magnetic drill, and material handling. This translates into a labour savings in the amount of $294,000 per year

($14 \times 10.50×40 hours/week $\times$ 50 weeks/year). The system can last for 15 years, with an estimated after-tax salvage value of $80,000. However, after an annual deduction of $226,000 in corporate income taxes, the net investment costs, as well as savings, are as follows:

- Project investment cost: $1,250,000.
- Projected annual net savings:
  ($792,000 + $294,000) − $128,500 − $226,000 = $731,500.
- Projected after-tax salvage value at the end of year 15: $80,000.

   (a) What is the projected IRR on this fabrication investment?
   (b) If Merco's MARR is known to be 18%, is this investment justifiable?

SOLUTION

Given: Projected cash flows as shown in Figure 5.19 and MARR = 18%.
Find: (a) The IRR and (b) whether to accept or reject the investment.

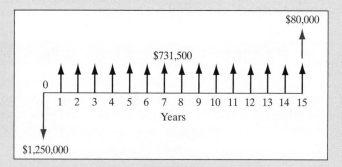

Figure 5.19 Cash flow diagram (Example 5.20).

(a) Since only one sign change occurs in the net cash flow series, the fabrication project is a simple investment. This indicates that there will be a unique rate of return that is internal to the project:

$$PW(i) = -$1,250,000 + $731,500(P/A, i, 15)$$
$$+ $80,000(P/F, i, 15)$$
$$= 0$$
$$i^* = 58.71\%.$$

(b) The IRR figure far exceeds Merco's MARR, indicating that the fabrication system project is an economically attractive one. Merco's management believes that, over a broad base of structural products, there is no doubt that the installation of the fabricating system would result in a significant savings, even after considering some potential deviations from the estimates used in the analysis.

5.7.3 Decision Rule for Nonsimple Investments

When applied to simple investment projects, the i^* provides an unambiguous criterion for measuring profitability. However, when multiple rates of return occur, none of them is an accurate portrayal of project acceptability or profitability. Clearly, then, we should place a high priority on discovering this situation early in our analysis of a project's cash flows. The most reliable way to predict multiple i^*s is to generate a NPW profile on computer and to check if it crosses the horizontal axis more than once.

In addition to the NPW profile, there are good—although somewhat more complex—analytical methods for predicting multiple i^*s. Perhaps more importantly, there is a good method, which uses an external interest rate, for refining our analysis when we do discover multiple i^*s. An external rate of return allows us to calculate a single valid internal rate of return; this will be discussed in Appendix 5A.

The flow chart in Figure 5.20 summarizes how you may proceed to apply the net cash flow sign test, the accumulated cash flow sign test, and net-investment test to calculate an IRR and make an accept-reject decision for an independent project. Some techniques referred to in this flow chart are covered in Appendix 5A. Given the complications involved in using IRR analysis to evaluate nonsimple projects, it is usually more desirable to use one of the other equivalence techniques like NPW, FW, and AE for this purpose. As an engineering manager, you should keep in mind the intuitive appeal of the rate-of-return measure. Once you have accepted a project on the basis of NPW or AE analysis, you may also wish to express its worth as an internal rate of return, for the benefits of your associates.

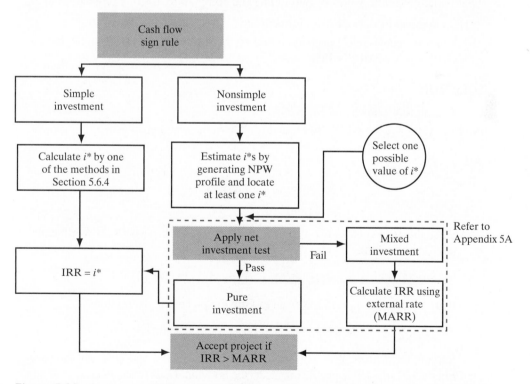

Figure 5.20 Summary of IRR criterion: A flow chart that summarizes how you may proceed to apply the net cash flow sign rule and net-investment test to calculate IRR for a pure as well as a mixed investment.

EXAMPLE 5.21 Analysis for a Nonsimple Investment

By outbidding its competitors, Trane Image Processing (TIP) has received a government contract worth $7,300,000 to build flight simulators for pilot training over two years. For some contracts, the government makes an advance payment when the contract is signed, but in this case the government will make two progressive payments: $4,300,000 at the end of the first year and the $3,000,000 balance at the end of the second year. The expected cash outflows required to produce the simulators are estimated to be $1,000,000 now, $2,000,000 during the first year, and $4,320,000 during the second year. The expected net cash flows from this project are summarized as follows:

Year	Cash Inflow	Cash Outflow	Net Cash Flow
0		$1,000,000	−$1,000,000
1	$4,300,000	2,000,000	2,300,000
2	3,000,000	4,320,000	−1,320,000

In normal situations, TIP would not even consider a marginal project such as this one. However, hoping that the company can establish itself as a technology leader in the field, management felt that it was worth outbidding its competitors. Financially, what is the economic worth of outbidding the competitors for this project? That is,

(a) Compute the values of $i*$s for this project.

(b) Make an accept–reject decision based on the results in part (a). Assume that the contractor's MARR is 15%.

SOLUTION

Given: Cash flow shown and MARR = 15%.

Find: (a) Compute the NPW, (b) $i*$, and (c) determine whether to accept the project.

(a)
$$PW(15\%) = -\$1,000,000 + \$2,300,000(P/F, 15\%, 1)$$
$$= -\$1,320,000(P/F, 15\%, 2)$$
$$= \$1890 > 0.$$

(b) Since the project has a two-year life, we may solve the net-present-worth equation directly via the quadratic formula:

$$-\$1,000,000 + \$2,300,000/(1 + i*) - \$1,320,000/(1 + i*)^2 = 0.$$

If we let $X = 1/(1 + i*)$, we can rewrite the preceding expression as

$$-1,000,000 + 2,300,000X - 1,320,000X^2 = 0.$$

Solving for X gives $X = 10/11$ and $10/12$, or $i* = 10\%$ and 20%. As shown in Figure 5.21, the NPW profile intersects the horizontal axis twice, once at 10% and again at 20%. The investment is obviously not a simple one; thus, neither 10% nor 20% represents the true internal rate of return of this government project.

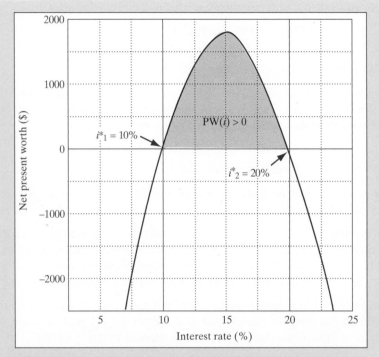

Figure 5.21 NPW plot for a nonsimple investment with multiple rates of return (Example 5.21).

(c) As shown in (b), this project is a nonsimple project. Instead of going through the more complicated IRR analysis for nonsimple projects, we can use the PW criterion to make the accept-reject decision. Since PW = $1890 > 0, as calculated in (a), the project meets the minimum requirement, and it is not as bad as initially believed.

COMMENTS: Example 5A.4 in Appendix 5A will illustrate how you go about finding the true internal rate of return for this nonsimple investment project.

SUMMARY

In this chapter, we have presented four methods for measuring the economic attractiveness or profitability of independent projects. We observed the following important results:

- Present, future, and annual worth are three equivalence methods of analysis in which a project's cash flows are converted to a present worth (PW), a future worth (FW), or an annual equivalent value (AE).

- The interest rate used in these equivalence calculations is MARR or the minimum attractive rate of return. It is generally dictated by management and is the interest rate at which a firm can always earn or borrow money under normal operating environment.

■ The three equivalent worth criteria are related in the following way:

$$FW(i) = PW(i)(F/P, i, N)$$
$$AE(i) = PW(i)(A/P, i, N) = FW(i)(A/F, i, N),$$

when FW is referenced to end of the project, N.

■ The capital recovery cost factor, or CR(i), is an important application of AE analysis in that it allows managers to calculate an annual equivalent of capital costs for ease of itemization with annual operating costs. The equation for CR(i) is

$$CR(i) = (P - S)(A/P, i, N) + iS,$$

where P = initial cost and S = salvage value.

■ Rate-of-return analysis measures project profitability in terms of a calculated interest rate or yield for the project.

■ **Rate of return** (ROR) is the interest rate earned on unrecovered project balance such that an investment's cash receipts make the terminal project balance equal to zero. The rate of return is an intuitively familiar and understandable measure of project profitability that many managers prefer to NPW or other equivalence measures.

■ Mathematically we can determine the rate of return for a given project cash flow series by locating a real positive interest rate that equates the net present worth of its cash flows to zero. This break-even interest rate is denoted by the symbol $i*$.

■ **Internal rate of return** (IRR) is a specific form of ROR that stresses the fact that we are concerned with the interest earned on the portion of the project balance that is internally invested, not those portions that are released by (borrowed from) the project.

■ To apply rate of return analysis correctly, we need to classify an investment into either a simple or a nonsimple investment. A **simple investment** is defined as one in which the initial cash flows are negative and only one sign change in the net cash flow occurs, whereas a nonsimple investment is one for which more than one sign change in the cash flow series occurs. Multiple $i*$s occur only in nonsimple investments. However, not all nonsimple investments will have multiple $i*$s.

■ For simple investments, ROR = IRR = $i*$.

■ For a nonsimple investment, because of the possibility of having multiple rates of return, it is recommended that the IRR analysis be abandoned and one of the equivalent worth methods (PW, FW, or AE) be used to make an accept/reject decision. Procedures are outlined in Appendix 5A for determining the rate of return internal to nonsimple investments. Once you find the IRR (or return on invested capital) for a nonsimple investment, you can use the same decision rule used for simple investments.

■ The three equivalent worth methods and the rate of return method result in identical conclusions with respect to the attractiveness or profitability of a project. Table 5.4 summarizes the decision rules for each of the methods.

■ The choice of one method over another is dependent on the specifics of the project being evaluated and analyst preference. Because the rate of return is an intuitively familiar and understandable measure of project profitability, many managers prefer it to the equivalent worth methods.

TABLE 5.4 Summary of Project Analysis Methods for Independent Projects

Analysis Method/ Equation Form(s)	Description	Comments/ Decision Rule
Payback period	A method for determing when in a project's history it breaks even.	Should be used as a screening method only—reflects liquidity, not profitablility of project.
Present equivalent $$PE(i) = \sum_{n=0}^{N} A_n(P/F, i, n)$$	An equivalence method that translates a project's cash flows into a net present worth value	If $PE(\text{MARR}) > 0$, accept If $PE(\text{MARR}) = 0$, remain indifferent If $PE(\text{MARR}) < 0$, reject
Future equivalent $$FE(i) = \sum_{n=0}^{N} A_n(F/P, i, N-n)$$	An equivalence method that translates a project's cash flows into a net future worth value	If $FE(\text{MARR}) > 0$, accept If $FE(\text{MARR}) = 0$, remain indifferent If $FE(\text{MARR}) < 0$, reject
Annual equivalent $$AE(i) = \left[\sum_{n=0}^{N} A_n(P/F, i, n) \right](A/P, i, N)$$	An equivalence method that translates a project's cash flows into a net annual worth value	If $AE(\text{MARR}) > 0$, accept If $AE(\text{MARR}) = 0$, remain indifferent If $AE(\text{MARR}) < 0$, reject
Rate of return $PE(i^*) = 0$ where $i^* = IRR$ (For simple investments only; see Appendix 5A for nonsimple investment case)	A method that calculates the interest rate earned on funds internal to the project	For both simple and nonsimple investments, If $IRR > \text{MARR}$, accept If $IRR = \text{MARR}$, remain indifferent If $IRR < \text{MARR}$, reject

PROBLEMS

Note: *Unless otherwise stated, all cash flows are after-tax cash flows. The interest rate (MARR) is also given on an after-tax basis.*

5.1 Camptown Togs Inc., a children's clothing manufacturer, has always found payroll processing to be costly because it must be done by a clerk so that the number of piece-goods coupons received by each employee can be collected and the types of tasks performed by each employee can be calculated. Not long ago, an industrial engineer designed a system that partially automates the process by means of a scanner that reads the piece-goods coupons. Management is enthusiastic about this system because it utilizes some personal computer systems that were purchased recently. It is expected that this new automated system will save $45,000 per year in labour. The new system will cost about $30,000 to build and test prior to operation. It is expected that operating costs, including income taxes, will be about $5000 per year. The system will have a five-year useful life. The expected net salvage value of the system is estimated to be $3000.

(a) Identify the cash inflows over the life of the project.

(b) Identify the cash outflows over the life of the project.

(c) Determine the net cash flows over the life of the project.

5.2 Refer to Problem 5.1, and answer the following questions:

(a) How long does it take to recover the investment?

(b) If the firm's interest rate is 15% after taxes, what would be the discounted payback period for this project?

5.3 Consider the following cash flows:

n	A	B	C	D
		Project's Cash Flow ($)		
0	−$2500	−$3000	−$5500	−$4000
1	300	2000	2000	5000
2	300	1500	2000	−3000
3	300	1500	2000	−2500
4	300	500	5000	1000
5	300	500	5000	1000
6	300	1500		2000
7	300			3000
8	300			

(a) Calculate the payback period for each project.

(b) Determine whether it is meaningful to calculate a payback period for project D.

(c) Assuming that $i = 10\%$, calculate the discounted payback period for each project.

5.4 Consider the following sets of investment projects, all of which have a three-year investment life:

| | Project's Cash Flow ($) | | | |
n	A	B	C	D
0	−$1500	−$1200	−$1600	−$3100
1	0	600	−1800	800
2	0	800	800	1900
3	3000	1500	2500	2300

(a) Compute the net present worth of each project at $i = 10\%$.

(b) Plot the present worth as a function of the interest rate (from 0% to 30%) for project B.

5.5 You need to know whether the building of a new warehouse is justified under the following conditions:

The proposal is for a warehouse costing $200,000. The warehouse has an expected useful life of 35 years and a net salvage value (net proceeds from sale after tax adjustments) of $35,000. Annual receipts of $37,000 are expected, annual maintenance and administrative costs will be $8000/year, and annual income taxes are $5000.

Given the foregoing data, which of the following statements are correct?

(a) The proposal is justified for a MARR of 9%.

(b) The proposal has a net present worth of $152,512 when 6% is used as the interest rate.

(c) The proposal is acceptable, as long as MARR $\leq 11.81\%$.

(d) All of the preceding are correct.

5.6 Your firm is considering purchasing an old office building with an estimated remaining service life of 25 years. Recently, the tenants signed long-term leases, which leads you to believe that the current rental income of $150,000 per year will remain constant for the first 5 years. Then the rental income will increase by 10% for every 5-year interval over the remaining life of the asset. For example, the annual rental income would be $165,000 for years 6 through 10, $181,500 for years 11 through 15, $199,650 for years 16 through 20, and $219,615 for years 21 through 25. You estimate that operating expenses, including income taxes, will be $45,000 for the first year and that they will increase by $3000 each year thereafter. You also estimate that razing the building and selling the lot on which it stands will realize a net amount of $50,000 at the end of the 25-year period. If you had the opportunity to invest your money elsewhere and thereby earn interest at the rate of 12% per annum, what would be the maximum amount you would be willing to pay for the building and lot at the present time?

5.7 Consider the following investment project:

n	A_n	i
0	−$42,000	10%
1	32,400	11
2	33,400	13
3	32,500	15
4	32,500	12
5	33,000	10

Suppose the company's reinvestment opportunities change over the life of the project as shown in the preceding table (i.e., the firm's MARR changes over the life of the project). For example, the company can invest funds available now at 10% for the first year, 11% for the second year, and so forth. Calculate the net present worth of this investment and determine the acceptability of the investment.

5.8 Cable television companies and their equipment suppliers are on the verge of installing new technology that will pack many more channels into cable networks, thereby creating a potential programming revolution with implications for broadcasters, telephone companies, and the consumer electronics industry.

Digital compression uses computer techniques to squeeze 3 to 10 programs into a single channel. A cable system fully using digital compression technology would be able to offer well over 100 channels, compared with about 35 for the traditional cable television system. If the new technology is combined with the increased use of optical fibers, it might be possible to offer as many as 300 channels.

A cable company is considering installing this new technology to increase subscription sales and save on satellite time. The company estimates that the installation will take place over two years. The system is expected to have an eight-year service life and produce the following savings and expenditures:

Digital Compression	
Investment	
Now	$500,000
First year	$3,200,000
Second year	$4,000,000
Annual savings in satellite time	$2,000,000
Incremental annual revenues due to new subscriptions	$4,000,000
Incremental annual expenses	$1,500,000
Incremental annual income taxes	$1,300,000
Economic service life	8 years
Net salvage value	$1,200,000

Note that the project has a 2-year investment period, followed by an 8-year service life (a total 10-year life for the project). This implies that the first annual savings will occur at the end of year 3 and the last will occur at the end of year 10. If the firm's MARR is 15%, use the NPW method to justify the economic worth of the project.

5.9 A large food-processing corporation is considering using laser technology to speed up and eliminate waste in the potato-peeling process. To implement the system, the company anticipates needing $3 million to purchase the industrial-strength lasers. The system will save $1,550,000 per year in labour and materials. However, it will require an additional operating and maintenance cost of $350,000. Annual income taxes will also increase by $150,000. The system is expected to have a 10-year service life and will have a salvage value of about $200,000. If the company's MARR is 18%, use the NPW method to justify the project.

5.10 Consider the following sets of investment projects, all of which have a three-year investment life:

Period	Project's Cash Flow			
(n)	A	B	C	D
0	-$12,500	-11,500	12,500	-13,000
1	5,400	-3,000	-7,000	5,500
2	14,400	21,000	-2,000	5,500
3	7,200	13,000	4,000	8,500

(a) Compute the net present worth of each project at $i = 15\%$.
(b) Compute the net future worth of each project at $i = 15\%$.

Which project or projects are acceptable?

5.11 Consider the following project balances for a typical investment project with a service life of four years:

n	A_n	Project Balance
0	-$1000	-$1000
1	( )	-1100
2	( )	-800
3	460	-500
4	( )	0

(a) Construct the original cash flows of the project.
(b) Determine the interest rate used in computing the project balance.
(c) Would the project be acceptable at $i = 15\%$?

5.12 Your R&D group has developed and tested a computer software package that assists engineers to control the proper chemical mix for the various process-manufacturing industries. If you decide to market the software, your first-year operating net cash flow is estimated to be $1,000,000. Because of market competition, product life will be about 4 years, and the product's market share will decrease by 25% each year over the previous year's share. You are approached by a big software house that wants to purchase the right to manufacture and distribute the product. Assuming that your interest rate is 15%, for what minimum price would you be willing to sell the software?

5.13 Consider the accompanying project balance diagram for a typical investment project with a service life of five years. The numbers in the figure indicate the beginning project balances.

(a) From the project balance diagram, construct the project's original cash flows.

(b) What is the project's conventional payback period (without interest)?

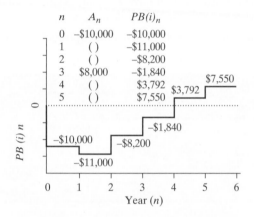

n	A_n	$PB(i)_n$
0	-$10,000	-$10,000
1	( )	-$11,000
2	( )	-$8,200
3	$8,000	-$1,840
4	( )	$3,792
5	( )	$7,550

5.14 Consider the following cash flows and present-worth profile:

	Net Cash Flows ($)	
Year	**Project 1**	**Project 2**
0	-$1000	-$1000
1	400	300
2	800	Y
3	X	800

(a) Determine the values of X and Y.

(b) Calculate the terminal project balance of project 1 at MARR $= 24\%$.

(c) Find the values of a, b, and c in the NPW plot.

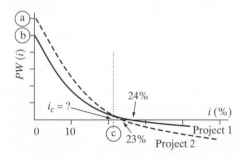

5.15 Consider the project balances for a typical investment project with a service life of five years, as shown in Table P5.15.

(a) Construct the original cash flows of the project and the terminal balance, and fill in the blanks in the table.

(b) Determine the interest rate used in the project balance calculation, and compute the present worth of this project at the computed interest rate.

TABLE P5.15 Investment Project Balances

n	A_n	Project Balance
0	−$3000	−$3000
1		−2700
2	1470	−1500
3		0
4		−300
5	600	

5.16 Refer to Problem 5.3, and answer the following questions:

(a) Graph the project balances (at $i = 10\%$) of each project as a function of n.

(b) By examining the graphical results in part (a), determine which project appears to be the safest to undertake if there is some possibility of premature termination of the projects at the end of year 2.

5.17 Consider the following investment projects:

			Project's Cash Flow		
n	A	B	C	D	E
0	−$1,800	−$5,200	−$3,800	−$4,000	−$6,500
1	−500	2,500	0	500	1,000
2	900	−4,000	0	2,000	3,600
3	1,300	5,000	4,000	3,000	2,400
4	2,200	6,000	7,000	4,000	
5	−700	3,000	12,000	1,250	

(a) Compute the future worth at the end of life for each project at $i = 12\%$.

(b) Determine the acceptability of each project.

5.18 Refer to Problem 5.17, and answer the following questions:

(a) Plot the future worth for each project as a function of the interest rate (0%–50%).

(b) Compute the project balance of each project at $i = 12\%$.

(c) Compare the terminal project balances calculated in (b) with the results obtained in Problem 5.17(a). Without using the interest factor tables, compute the future worth on the basis of the project balance concept.

5.19 Covington Corporation purchased a vibratory finishing machine for $20,000 in year 0. The useful life of the machine is 10 years, at the end of which the machine is estimated to have a salvage value of zero. The machine generates net annual revenues of $6000. The annual operating and maintenance expenses are estimated to be $1000. If Covington's MARR is 15%, how many years will it take before this machine becomes profitable?

5.20 Gene Research Inc. just finished a four-year program of R&D and clinical trial. It expects a quick approval from Health Canada. If Gene markets the product on its own, the company will require $30 million immediately ($n = 0$) to build a new manufacturing facility, and it is expected to have a 10-year product life. The R&D expenditure in the previous years and the anticipated revenues that the company can generate over the next 10 years are summarized as follows:

Period (n)	Cash Flow (Unit: $ million)
−3	−10
−2	−10
−1	−10
0	−10 − 30
1–10	100

Merck, a large drug company, is interested, in purchasing the R&D project and the right to commercialize the product from Gene Research Inc.; it wants to do so immediately ($n = 0$). What would be a starting negotiating price for the project from Merck? Assume that Gene's MARR $= 20\%$.

5.21 Consider the following independent investment projects:

n	Project Cash Flows		
	A	B	C
0	−$400	−$300	$100
1	150	140	−40
2	150	140	−40
3	350	140	−40
4	−200	110	
5	400	110	
6	300		

Assume that MARR $= 10\%$, and answer the following questions:

(a) Compute the net present worth for each project, and determine the acceptability of each.

(b) Compute the net future worth of each project at the end of each project period, and determine the acceptability of each project.

(c) Compute the present worth of each project with variable MARRs as follows: 10% for $n = 0$ to $n = 3$ and 15% for $n = 4$ to $n = 6$.

5.22 Consider the project balance profiles shown in Table P5.22 for proposed investment projects. Project balance figures are rounded to nearest dollars.

TABLE P5.22 **Profiles for Proposed Investment Projects**

	Project Balances		
n	A	B	C
0	−$1000	−$1000	−$1000
1	−1000	−650	−1200
2	−900	−348	−1440
3	−690	−100	−1328
4	−359	85	−1194
5	105	198	−1000
Interest rate used	10%	?	20%
NPW	?	$79.57	?

(a) Compute the net present worth of projects A and C.
(b) Determine the cash flows for project A.
(c) Identify the net future worth of project C.
(d) What interest rate would be used in the project balance calculations for project B?

5.23 Consider the following project balance profiles for proposed investment projects:

	Project Balances		
n	A	B	C
0	−$1000	−$1000	−$1000
1	−800	−680	−530
2	−600	−302	X
3	−400	−57	−211
4	−200	233	−89
5	0	575	0
Interest rate used	10%	18%	12%

Project balance figures are rounded to the nearest dollar.
(a) Compute the net present worth of each investment.
(b) Determine the project balance for project C at the end of period 2 if $A_2 = \$500$.
(c) Determine the cash flows for each project.
(d) Identify the net future worth of each project.

5.24 Maintenance money for a new building has been sought. Mr. Kendall would like to make a donation to cover all future expected maintenance costs for the building. These maintenance costs are expected to be $50,000 each year for the first five years, $70,000 each year for years 6 through 10, and $90,000 each year after that. (The building has an indefinite service life.)

(a) If the money is placed in an account that will pay 13% interest compounded annually, how large should the gift be?

(b) What is the equivalent annual maintenance cost over the infinite service life of the building?

5.25 Consider an investment project, the cash flow pattern of which repeats itself every five years forever as shown in the following diagram. At an interest rate of 14%, compute the capitalized equivalent amount for this project.

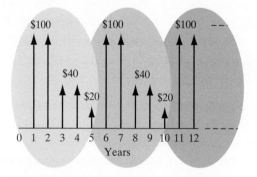

5.26 A group of concerned citizens has established a trust fund that pays 6% interest, compounded monthly, to preserve a historical building by providing annual maintenance funds of $30,000 forever. Compute the capitalized equivalent amount for these building maintenance expenses.

5.27 A newly constructed bridge costs $5,000,000. The same bridge is estimated to need renovation every 15 years at a cost of $1,000,000. Annual repairs and maintenance are estimated to be $100,000 per year.

(a) If the interest rate is 5%, determine the capitalized cost of the bridge.

(b) Suppose that, in (a), the bridge must be renovated every 20 years, not every 15 years. What is the capitalized cost of the bridge?

(c) Repeat (a) and (b) with an interest rate of 10%. What have you to say about the effect of interest on the results?

5.28 To decrease the costs of operating a lock in a large river, a new system of operation is proposed. The system will cost $650,000 to design and build. It is estimated that it will have to be reworked every 10 years at a cost of $100,000. In addition, an expenditure of $50,000 will have to be made at the end of the fifth year for a new type of gear that will not be available until then. Annual operating costs are expected to be $30,000 for the first 15 years and $35,000 a year thereafter. Compute the capitalized cost of perpetual service at $i = 8\%$.

5.29 Consider the following cash flows and compute the equivalent annual worth at $i = 10\%$:

n	Investment	A_n Revenue
0	−$5000	
1		$2000
2		2000
3		3000
4		3000
5		1000
6	2000	500

5.30 Consider the accompanying cash flow diagram. Compute the equivalent annual worth at $i = 12\%$.

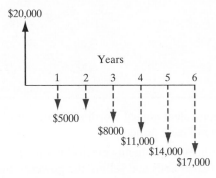

5.31 Consider the accompanying cash flow diagram. Compute the equivalent annual worth at $i = 10\%$.

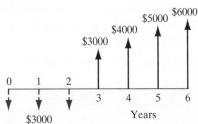

5.32 Consider the accompanying cash flow diagram. Compute the equivalent annual worth at $i = 13\%$.

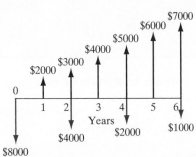

5.33 Consider the accompanying cash flow diagram. Compute the equivalent annual worth at $i = 8\%$.

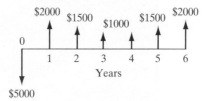

5.34 Consider the following sets of investment projects:

	Project's Cash Flow ($)			
n	A	B	C	D
0	−$2500	−$4500	−$8000	−$12,000
1	400	3000	−2000	2000
2	500	2000	6000	4000
3	600	1000	2000	8000
4	700	500	4000	8000
5	800	500	2000	4000

Compute the equivalent annual worth of each project at $i = 10\%$, and determine the acceptability of each project.

5.35 Sun-Devil Company is producing electricity directly from a solar source by using a large array of solar cells and selling the power to the local utility company. Because these cells degrade over time, thereby resulting in lower conversion efficiency and power output, the cells must be replaced every four years, which results in a particular cash flow pattern that repeats itself as follows: $n = 0$, −$500,000; $n = 1$, $600,000; $n = 2$, $400,000; $n = 3$, $300,000; and $n = 4$, $200,000. Determine the annual equivalent cash flows at $i = 12\%$.

5.36 Consider the following sets of investment projects:

	Project's Cash Flow			
n	A	B	C	D
0	−$7500	−$4000	−$5000	−$6600
1	0	1500	4000	3800
2	0	1800	3000	3800
3	15,500	2100	2000	3800

Compute the equivalent annual worth of each project at $i = 13\%$, and determine the acceptability of each project.

5.37 The cash flows for a certain project are as follows:

n	Investment	Operating Income	
		Net Cash Flow	
0	−$800		1st cycle
1		$900	
2		700	
3	−800	500	2nd cycle
4		900	
5		700	
6	−800	500	3rd cycle
7		900	
8		700	
9		500	

Find the equivalent annual worth for this project at $i = 10\%$, and determine the acceptability of the project.

5.38 Beginning next year, a foundation will support an annual seminar on campus by the earnings of a $100,000 gift it received this year. It is felt that 8% interest will be realized for the first 10 years, but that plans should be made to anticipate an interest rate of 6% after that time. What amount should be added to the foundation now to fund the seminar at the $10,000 level into infinity?

5.39 The owner of a business is considering investing $55,000 in new equipment. He estimates that the net cash flows will be $5000 during the first year, but will increase by $2500 per year the next year and each year thereafter. The equipment is estimated to have a 10-year service life and a net salvage value of $6000 at that time. The firm's interest rate is 12%.
(a) Determine the annual capital cost (ownership cost) for the equipment.
(b) Determine the equivalent annual savings (revenues).
(c) Determine whether this is a wise investment.

5.40 You are considering purchasing a dump truck. The truck will cost $45,000 and have an operating and maintenance cost that starts at $15,000 the first year and increases by $2000 per year. Assume that the salvage value at the end of five years is $9000 and the interest rate is 12%. What is the equivalent annual cost of owning and operating the truck?

5.41 Emerson Electronics Company just purchased a soldering machine to be used in its assembly cell for flexible disk drives. The soldering machine cost $250,000. Because of the specialized function it performs, its useful life is estimated to be five years. It is also estimated that at that time its salvage value will be $40,000. What is the capital cost for this investment if the firm's interest rate is 18%?

5.42 The present price (year 0) of kerosene is $0.50 per litre, and its cost is expected to increase by $0.04 per year. (At the end of year 1, kerosene will cost $0.54 per litre.) Mr. Garcia uses about 4000 litres of kerosene for space heating during a winter season. He has an opportunity to buy a storage tank for $600, and at the end of four years he can sell the storage tank for $100. The tank has a capacity to supply four years of Mr. Garcia's heating needs, so he can buy four years' worth of kerosene at its present price ($0.50), or he can invest his money elsewhere at 8%. Should he purchase the storage tank? Assume that kerosene purchased on a pay-as-you-go basis is paid for at the end of the year. (However, kerosene purchased for the storage tank is purchased now.)

5.43 Consider the following advertisement, which appeared in a local paper:

Pools-Spas-Hot Tubs—Pure Water without Toxic Chemicals: The comparative costs between the conventional chemical system (chlorine) and the IONETICS systems are as follows:

Item	Conventional System	IONETICS System
Annual costs		
Chemical	$471	
IONETICS		$ 85
Pump ($0.667/kWh)	$576	$ 100
Capital investment		$1200

Note that the IONETICS system pays for itself in fewer than two years.

Assume that the IONETICS system has a 12-year service life and the interest rate is 6%. What is the equivalent annual cost of operating the IONETICS system?

5.44 The cash flows for two investment projects are as follows:

	Project's Cash Flow	
n	A	B
0	−$4000	$5500
1	1000	−1400
2	X	−1400
3	1000	−1400
4	1000	−1400

(a) For project A, find the value of X that makes the equivalent annual receipts equal the equivalent annual disbursement at $i = 13\%$.

(b) Would you accept project B at $i = 15\%$, based on an AE criterion?

5.45 An industrial firm can purchase a special machine for $50,000. A down payment of $5000 is required, and the unpaid balance can be paid off in five equal year-end installments at 7% interest. As an alternative, the machine can be purchased for $46,000 in cash. If the firm's MARR is 10%, use the annual equivalent method to determine which alternative should be accepted.

5.46 An industrial firm is considering purchasing several programmable controllers and automating the company's manufacturing operations. It is estimated that the equipment will initially cost $100,000 and the labour to install it will cost $35,000. A service contract to maintain the equipment will cost $5000 per year. Trained service personnel will have to be hired at an annual salary of $30,000. Also estimated is an approximate $10,000 annual income-tax savings (cash inflow). How much will this investment in equipment and services have to increase the annual revenues after taxes in order to break even? The equipment is estimated to have an operating life of 10 years, with no salvage value because of obsolescence. The firm's MARR is 10%.

5.47 A construction firm is considering establishing an engineering computing centre. The centre will be equipped with three engineering workstations that cost $35,000 each, and each has a service life of five years. The expected salvage value of each workstation is $2,000. The annual operating and maintenance cost would be $15,000 for each workstation. At a MARR of 15%, determine the equivalent annual cost for operating the engineering centre.

5.48 You have purchased a machine costing $20,000. The machine will be used for two years, at the end of which time its salvage value is expected to be $10,000. The machine will be used 6000 hours during the first year and 8000 hours during the second year. The expected annual net savings will be $30,000 during the first year and $40,000 during the second year. If your interest rate is 10%, what would be the equivalent net savings per machine hour?

5.49 The engineering department of a large firm is overly crowded. In many cases, several engineers share one office. It is evident that the distraction caused by the crowded conditions reduces the productive capacity of the engineers considerably. Management is considering the possibility of providing new facilities for the department, which could result in fewer engineers per office and a private office for some. For an office presently occupied by five engineers, what minimum individual increase in effectiveness must result to warrant the assignment of only three engineers to an office if the following data apply?

- The office size is 4 × 8 metres.
- The average annual salary of each engineer is $80,000.
- The cost of the building is $1000 per square metre.
- The estimated life of the building is 25 years.
- The estimated salvage value of the building is 10% of the initial cost.
- The annual taxes, insurance, and maintenance are 6% of the initial cost.
- The cost of janitorial service, heating, and illumination is $50 per square metre per year.
- The interest rate is 12%.

Assume that engineers reassigned to other office space will maintain their present productive capability as a minimum.

5.50 Sam Tucker is a sales engineer at Buford Chemical Engineering Company. Sam owns two vehicles, and one of them is entirely dedicated to business use. His business car is a used small pickup truck, which he purchased with $11,000 of personal savings. On the basis of his own records and with data compiled by the Canadian Automobile Association (CAA), Sam has estimated the costs of owning and operating his business vehicle for the first three years as follows:

	First Year	Second Year	Third Year
Depreciation	$2879	$1776	$1545
Scheduled maintenance	100	153	220
Insurance	635	635	635
Registration and taxes	78	57	50
Total ownership cost	$3692	$2621	$2450
Nonscheduled repairs	35	85	200
Replacement tires	35	30	27
Accessories	15	13	12
Gasoline and taxes	688	650	522
Oil	80	100	100
Parking and tolls	135	125	110
Total operating costs	$988	$1003	$971
Total of all costs	$4680	$3624	$3421
Expected kilometres driven	14,500	13,000	11,500

If his interest rate is 6%, what should be Sam's reimbursement rate per kilometre so that he can break even?

5.51 A company is currently paying its employees $0.38 per kilometre to drive their own cars on company business. The company is considering supplying employees with cars, which would involve purchasing at $25,000, with an estimated three-year life, a net salvage value of $8000, taxes and insurance at a cost of $900 per year, and operating and maintenance expenses of $0.22 per kilometre. If the interest rate is 10% and the company anticipates an employee's annual travel to be 22,000 kilometres, what is the equivalent cost per kilometre (neglecting income taxes)?

5.52 An automobile that runs on electricity can be purchased for $30,000. The automobile is estimated to have a life of 12 years with annual travel of 20,000 kilometres. Every three years, a new set of batteries will have to be purchased at a cost of $3000. Annual maintenance of the vehicle is estimated to cost $700. The cost of recharging the batteries is estimated at $0.015 per kilometre. The salvage value of the batteries and the vehicle at the end of 12 years is estimated to be $2000. Suppose the MARR is 7%. What is the cost per kilometre to own and operate this vehicle, based on the preceding estimates? The $3000 cost of the batteries is a net value, with the old batteries traded in for the new ones.

5.53 The estimated cost of a completely installed and ready-to-operate 40-kilowatt generator is $30,000. Its annual maintenance costs are estimated at $500. The energy that can be generated annually at full load is estimated to be 100,000 kilowatt-hours. If the value of the energy generated is $0.08 per kilowatt-hour, how long will it take before this machine becomes profitable? Take the MARR to be 9% and the salvage value of the machine to be $2000 at the end of its estimated life of 15 years.

5.54 A large university that is currently facing severe parking problems on its campus is considering constructing parking decks off campus. A shuttle service could pick up students at the off-campus parking deck and transport them to various locations on campus. The university would charge a small fee for each shuttle ride, and the students could be quickly and economically transported to their classes. The funds raised by the shuttle would be used to pay for trolleys, which cost about $150,000 each. Each trolley has a 12-year service life, with an estimated salvage value of $3000. To operate each trolley, the following additional expenses will be incurred:

Item	Annual Expenses ($)
Driver	$50,000
Maintenance	10,000
Insurance	3,000

If students pay 10 cents for each ride, determine the annual ridership per trolley (number of shuttle rides per year) required to justify the shuttle project, assuming an interest rate of 6%.

5.55 Eradicator Food Prep Inc. has invested $7 million to construct a food irradiation plant. This technology destroys organisms that cause spoilage and disease, thus extending the shelf life of fresh foods and the distances over which they can be shipped. The plant can handle about 200,000 kilograms of produce in an hour, and it will be operated for 3600 hours a year. The net expected operating and maintenance costs (taking into account income-tax effects) would be $4 million per year. The plant is expected to have a useful life of 15 years, with a net salvage value of $900,000. The firm's interest rate is 15%.

(a) If investors in the company want to recover the plant investment within six years of operation (rather than 15 years), what would be the equivalent after-tax annual revenues that must be generated?

(b) To generate annual revenues determined in part (a), what minimum processing fee per kilogram should the company charge to its producers?

5.56 The local government of Grand Manan Island is completing plans to build a desalination plant to help ease a critical drought on the island. The drought has combined with new construction on Grand Manan to leave the island with an urgent need for a new water source. A modern desalination plant could produce fresh water from seawater for $5 per cubic metre. The cost of acquiring water from natural sources is about the same as that for desalting. The $3 million plant can produce 160 cubic metres of fresh water a day (enough to supply 295 households daily), more than a quarter of the island's total needs. The desalination plant has an estimated service life of 20 years, with no appreciable salvage value. The annual operating and maintenance costs would be about $250,000. Assuming an interest rate of 10%, what should be the minimum monthly water bill for each household?

5.57 A utility firm is considering building a 50-megawatt geothermal plant that generates electricity from naturally occurring underground heat. The binary geothermal system will cost $85 million to build and $6 million (including any income-tax effect) to operate per year. (Virtually no fuel costs will accrue compared with fuel costs related to a conventional fossil-fuel plant.) The geothermal plant is to last for

25 years. At that time, its expected salvage value will be about the same as the cost to remove the plant. The plant will be in operation for 70% (plant utilization factor) of the year (or 70% of 8760 hours per year). If the firm's MARR is 14% per year, determine the cost per kilowatt-hour of generating electricity.

5.58 A corporate executive jet with a seating capacity of 20 has the following cost factors:

Item	Cost
Initial cost	$12,000,000
Service life	15 years
Salvage value	$2,000,000
Crew costs per year	$225,000
Fuel cost per kilometre	$1.80
Landing fee	$250
Maintenance per year	$237,500
Insurance cost per year	$166,000
Catering per passenger trip	$75

The company flies three round trips from Halifax to Vancouver per week, a distance of 4,500 kilometres one way. How many passengers must be carried on an average trip in order to justify the use of the jet if the first-class round-trip fare is $3400? The firm's MARR is 15%. (Ignore income-tax consequences.)

5.59 You are going to buy a new car worth $14,500. The dealer computes your monthly payment to be $267 for 72 months' financing. What is the dealer's rate of return on this loan transaction?

5.60 Johnson Controls spent more than $2.5 million retrofitting a government complex and installing a computerized energy-management system for the Province of Saskatchewan. As a result, the province's energy bill dropped from an average of $6 million a year to $3.5 million. Moreover, both parties will benefit from the 10-year life of the contract. Johnson recovers half the money it saved in reduced utility costs (about $1.2 million a year over 10 years); Saskatchewan has its half to spend on other things. What is the rate of return realized by Johnson Controls in this energy-control system?

5.61 Pablo Picasso's 1905 portrait *Boy with a Pipe* sold for $104.2 million in an auction at Sotheby's Holdings Inc. on June 24, 2004, shattering the existing record for art and ushering in a new era in pricing for 20th-century paintings. The Picasso, sold by the philanthropic Greentree Foundation, cost Mr. Whitney about $30,000 in 1950. Determine the annual rate of appreciation of the artwork over 54 years.

5.62 Consider four investments with the following sequences of cash flows:
   (a) Identify all the simple investments.
   (b) Identify all the nonsimple investments.
   (c) Compute i^* for each investment.
   (d) Which project has no rate of return?

n	Net Cash Flow Project A	Project B	Project C	Project D
0	−$18,000	−$30,000	$34,578	−$56,500
1	30,000	32,000	−18,000	2,500
2	20,000	32,000	−18,000	6,459
3	10,000	−22,000	−18,000	−78,345

5.63 Consider the following infinite cash flow series with repeated cash flow patterns:

n	A_n
0	−$1000
1	400
2	800
3	500
4	500
5	400
6	800
7	500
8	500
⋮	⋮

Determine i^* for this infinite cash flow series.

5.64 Consider the following investment projects:

n	Project Cash Flow A	B	C	D	E
0	−$100	−$100	−$200	−$50	−$50
1	60	70	$20	120	−100
2	150	70	10	40	−50
3		40	5	40	0
4		40	−180	−20	150
5			60	40	150
6			50	30	100
7			400		100

(a) Classify each project as either simple or nonsimple.
(b) Use the quadratic equation to compute i^* for project A.
(c) Obtain the rate(s) of return for each project by plotting the NPW as a function of the interest rate.

5.65 Consider the projects in Table P5.65.

TABLE P5.65 Net Cash Flow for Four Projects

	Net Cash Flow			
n	A	B	C	D
0	−$2000	−$1500	−$1800	−$1500
1	500	800	5600	−360
2	100	600	4900	4675
3	100	500	−3500	2288
4	2000	700	7000	
5			−1400	
6			2100	
7			900	

(a) Classify each project as either simple or nonsimple.
(b) Identify all positive i^*s for each project.
(c) For each project, plot the present worth as a function of the interest rate (i).

5.66 Consider the following financial data for a project:

Initial investment	$50,000
Project life	8 years
Salvage value	$10,000
Annual revenue	$25,000
Annual expenses (including income taxes)	$ 9000

(a) What is i^* for this project?
(b) If the annual expense increases at a 7% rate over the previous year's expenses, but the annual income is unchanged, what is the new i^*?
(c) In part (b), at what annual rate will the annual income have to increase to maintain the same i^* obtained in part (a)?

5.67 Consider two investments, A and B, with the following sequences of cash flows:

	Net Cash Flow	
n	Project A	Project B
0	−$25,000	−$25,000
1	2,000	10,000
2	6,000	10,000
3	12,000	10,000
4	24,000	10,000
5	28,000	5000

(a) Compute $i*$ for each investment.

(b) Plot the present-worth curve for each project on the same chart, and find the interest rate that makes the two projects equivalent.

5.68 Consider an investment project with the following cash flows:

n	Net Cash Flow
0	−$120,000
1	94,000
2	144,000
3	72,000

(a) Find the IRR for this investment.

(b) Plot the present worth of the cash flow as a function of i.

(c) On the basis of the IRR criterion, should the project be accepted at MARR = 15%?

5.69 Agdist Corporation distributes agricultural equipment. The board of directors is considering a proposal to establish a facility to manufacture an electronically controlled "intelligent" crop sprayer invented by a professor at a local university. This crop sprayer project would require an investment of $10 million in assets and would produce an annual after-tax net benefit of $1.8 million over a service life of eight years. All costs and benefits are included in these figures. When the project terminates, the net proceeds from the sale of the assets will be $1 million. Compute the rate of return of this project. Is this a good project at MARR = 10%?

5.70 Consider an investment project with the following cash flows:

n	Cash Flow
0	−$5000
1	0
2	4840
3	1331

Compute the IRR for this investment. Is the project acceptable at MARR = 10%?

5.71 Consider the following cash flow of a certain project:

n	Net Cash Flow
0	−$2000
1	800
2	900
3	X

If the project's IRR is 10%,

(a) Find the value of X.

(b) Is this project acceptable at MARR = 8%?

5.72 You are considering a luxury apartment building project that requires an investment of $12,500,000. The building has 50 units. You expect the maintenance cost for the apartment building to be $250,000 the first year and $300,000 the second year. The maintenance cost will continue to increase by $50,000 in subsequent years. The cost to hire a manager for the building is estimated to be $80,000 per year. After five years of operation, the apartment building can be sold for $14,000,000. What is the annual rent per apartment unit that will provide a return on investment of 15%? Assume that the building will remain fully occupied during its five years of operation.

5.73 A machine costing $25,000 to buy and $3000 per year to operate will save mainly labour expenses in packaging over six years. The anticipated salvage value of the machine at the end of the six years is $5000. To receive a 10% return on investment (rate of return), what is the minimum required annual savings in labour from this machine?

5.74 Champion Chemical Corporation is planning to expand one of its propylene-manufacturing facilities. At $n = 0$, a piece of property costing $1.5 million must be purchased to build a plant. The building, which needs to be expanded during the first year, costs $3 million. At the end of the first year, the company needs to spend about $4 million on equipment and other start-up costs. Once the building becomes operational, it will generate revenue in the amount of $3.5 million during the first operating year. This will increase at the annual rate of 5% over the previous year's revenue for the next 9 years. After 10 years, the sales revenue will stay constant for another 3 years before the operation is phased out. (It will have a project life of 13 years after construction.) The expected salvage value of the land at the end of the project's life would be about $2 million, the building about $1.4 million, and the equipment about $500,000. The annual operating and maintenance costs are estimated to be approximately 40% of the sales revenue each year. What is the IRR for this investment? If the company's MARR is 15%, determine whether the investment is a good one. (Assume that all figures represent the effect of the income tax.)

5.75 Recent technology has made possible a computerized vending machine that can grind coffee beans and brew fresh coffee on demand. The computer also makes possible such complicated functions as changing $5 and $10 bills, tracking the age of an item, and moving the oldest stock to the front of the line, thus cutting down on spoilage. With a price tag of $4500 for each unit, Easy Snack has estimated the cash flows in millions of dollars over the product's six-year useful life, including the initial investment, as follows:

n	Net Cash Flow
0	−$20
1	8
2	17
3	19
4	18
5	10
6	3

(a) On the basis of the IRR criterion, if the firm's MARR is 18%, is this product worth marketing?

(b) If the required investment remains unchanged, but the future cash flows are expected to be 10% higher than the original estimates, how much of an increase in IRR do you expect?

(c) If the required investment has increased from $20 million to $22 million, but the expected future cash flows are projected to be 10% smaller than the original estimates, how much of a decrease in IRR do you expect?

Short Case Studies

ST5.1 Critics have charged that, in carrying out an economic analysis, the commercial nuclear power industry does not consider the cost of decommissioning, or "mothballing," a nuclear power plant and that the analysis is therefore unduly optimistic. As an example, consider a nuclear generating facility under construction: The initial cost is $1.5 billion (present worth at the start of operations), the estimated life is 40 years, the annual operating and maintenance costs the first year are assumed to be 4.6% of the initial cost and are expected to increase at the fixed rate of 0.05% each year, and annual revenues are estimated to be three times the annual operating and maintenance costs throughout the life of the plant.

(a) The criticism that the economic analysis is overoptimistic because it omits "mothballing" costs, since the addition of a cost of 50% of the initial cost to "mothball" the plant decreases the 10% rate of return to approximately 9.9%.

(b) If the estimated life of the plant is more realistically taken to be 25 years instead of 40 years, then the criticism is justified. By reducing the life to 25 years, the rate of return of approximately 9% without a "mothballing" cost drops to approximately 7.7% when a cost to "mothball" the plant (equal to 50% of the initial cost) is added to the analysis.

Comment on these statements.

ST5.2 The following is a letter that Professor Chan Park received from a local city engineer:

Dear Professor Park:

Thank you for taking the time to assist with this problem. I'm really embarrassed at not being able to solve it myself, since it seems rather straightforward. The situation is as follows:

A citizen of Opelika paid for concrete drainage pipe approximately 20 years ago to be installed on his property. (We have a policy that if drainage trouble exists on private property and the owner agrees to pay for the material, city crews will install it.) That was the case in this problem. Therefore, we are dealing with only material costs, disregarding labour.

However, this past year, we removed the pipe purchased by the citizen, due to a larger area drainage project. He thinks, and we agree, that he is due some refund for salvage value of the pipe due to its remaining life.

Problem:

- Known: 80' of 48" pipe purchased 20 years ago. Current quoted price of 48" pipe = $52.60/foot, times 80 feet = $4208 total cost in today's dollars.
- Unknown: Original purchase price.

- Assumptions: 50-year life; therefore, assume 30 years of life remaining at removal after 20 years. A 4% price increase per year, average, over 20 years.

Thus, we wish to calculate the cost of the pipe 20 years' ago. Then we will calculate, in today's dollars, the present salvage value after 20 years', use with 30 years of life remaining. Thank you again for your help. We look forward to your reply.

Charlie Thomas, P.E.
Director of Engineering
City of Opelika

After reading this letter, recommend a reasonable amount of compensation to the citizen for the replaced drainage pipe.

On the Companion Website that accompanies this text, you will find Excel templates and exercises, as well as the following analysis tools: Cash Flow Analyzer, Depreciation Analysis, Loan Analysis, and Interest Tables.

APPENDIX 5A

Computing IRR for Nonsimple (Mixed) Investments

To comprehend the nature of multiple $i*$s, we need to understand the investment situation represented by a cash flow series. The net investment test will indicate whether the $i*$ computed represents the true rate of return earned on the money invested in a project while it is actually in the project. As we shall see, the phenomenon of multiple $i*$s occurs only when the net investment test fails. When multiple positive rates of return for a cash flow series are found, in general none is suitable as a measure of project profitability, and we must proceed to the next analysis step: introducing an external rate of return.

5A.1 Predicting Multiple $i*$s

As hinted at in Example 5.16, for certain series of project cash flows, we may uncover the complication of multiple $i*$ values that satisfy Eq. (5.8). By analyzing and classifying cash flows, we may anticipate this difficulty and adjust our approach later. Here we will focus on the initial problem of whether we can predict a unique $i*$ for a project by examining its cash flow pattern. Two useful rules allow us to focus on sign changes (1) in net cash flows and (2) in accounting net profit (accumulated net cash flows).

Net Cash Flow Rule of Signs

One useful method for predicting an upper limit on the number of positive $i*$s of a cash flow stream is to apply the rule of signs: *The number of real $i*$s that are greater than −100% for a project with N periods is never greater than the number of sign changes in the sequence of the A_ns. A zero cash flow is ignored.*

An example is

Period	A_n	Sign Change
0	−$100	
1	−20	
2	+50	1
3	0	
4	+60	
5	−30	1
6	+100	1

Three sign changes occur in the cash flow sequence, so three or fewer real positive $i*$s exist.

It must be emphasized that the rule of signs provides an indication only of the *possibility* of multiple rates of return: The rule predicts only the *maximum* number of possible $i*$s. Many projects have multiple sign changes in their cash flow sequence, but still possess a unique real $i*$ in the $(-100\%, \infty)$ range.

Accumulated Cash Flow Sign Test

The accumulated cash flow is the sum of the net cash flows up to and including a given time. If the rule of cash flow signs indicates possible multiple $i*$s, we should proceed to the **accumulated cash flow sign test** to eliminate some possibility of multiple rates of return.

If we let A_n represent the net cash flow in period n and S_n represent the accumulated cash flow (the accounting sum) up to period n, we have the following:

Period (n)	Cash Flow (A_n)	Accumulated Cash Flow (S_n)
0	A_0	$S_0 = A_0$
1	A_1	$S_1 = S_0 + A_1$
2	A_2	$S_2 = S_1 + A_2$
$\vdots$	$\vdots$	$\vdots$
N	A_N	$S_N = S_{N-1} + A_N$

We then examine the sequence of accumulated cash flows $(S_0, S_1, S_2, S_3, \ldots, S_N)$ to determine the number of sign changes. *If the series S_n starts negatively and changes sign only once, then a unique positive $i*$ exists.* This cumulative cash flow sign rule is a more discriminating test for identifying the uniqueness of $i*$ than the previously described method.

EXAMPLE 5A.1 Predicting the Number of $i*$s

Predict the number of real positive rates of return for each of the following cash flow series:

Period	A	B	C	D
0	−$100	−$100	$0	−$100
1	−200	+50	−50	+50
2	+200	−100	+115	0
3	+200	+60	−66	+200
4	+200	−100		−50

SOLUTION

Given: Four cash flow series and cumulative flow series.
Find: The upper limit on number of $i*$s for each series.

The cash flow rule of signs indicates the following possibilities for the positive values of $i*$:

Project	Number of Sign Changes in Net Cash Flows	Possible Number of Positive Values of $i*$
A	1	1 or 0
B	4	4, 3, 2, 1, or 0
C	2	2, 1, or 0
D	2	2, 1, or 0

For cash flows B, C, and D, we would like to apply the more discriminating cumulative cash flow test to see if we can specify a smaller number of possible values of $i*$. Accordingly, we write

	Project B		Project C		Project D	
n	A_n	S_n	A_n	S_n	A_n	S_n
0	$-\$100$	$-\$100$	$\$0$	$\$0$	$-\$100$	$-\$100$
1	$+50$	-50	-50	-50	$+50$	-50
2	-100	-150	$+115$	$+65$	0	-50
3	$+60$	-90	-66	-1	$+200$	$+150$
4	-100	-190			-50	$+100$

Recall the test: If the series starts *negatively* and changes sign only once, a unique positive $i*$ exists.

- Only project D begins negatively and passes the test; therefore, we may predict a unique $i*$ value, rather than 2, 1, or 0 as predicted by the cash flow rule of signs. ($i_1^* = -75.16\%$ and $i_2^* = 35.05\%$)
- Project B, with no sign change in the cumulative cash flow series, has no rate of return.
- Project C fails the test, and we cannot eliminate the possibility of multiple $i*$s. ($i_1^* = 10\%$ and $i_2^* = 20\%$)

5A.2 Net-Investment Test: Pure versus Mixed Investments

To develop a consistent accept–reject decision rule with the IRR, we need to further classify a project into either a pure or a mixed investment:

- A project is said to be a **net investment** when the project balances computed at the project's $i*$ values, $\text{PB}(i^*)_n$, are either less than or equal to zero throughout the life of the investment, with the first cash flow being negative ($A_0 < 0$). The investment is *net* in the sense that the firm does not overdraw on its return at any point and hence the firm is *not indebted* to the project. This type of project is called a **pure investment**. In contrast, **pure borrowing** is defined as the situation in which $\text{PB}(i^*)_n$ values are positive or zero throughout the life of the loan, with $A_0 > 0$. *Simple investments will always be pure investments.*

Net investment test: A process to determine whether or not a firm borrows money from a project during the investment period.

Pure investment: An investment in which a firm never borrows money from the project.

• If any of the project balances calculated at the project's $i*$ is positive, the project is not a pure investment. A positive project balance indicates that, at some time during the project life, the firm acts as a borrower [$PB(i*)_n > 0$] rather than an investor in the project [$PB(i*)_n < 0$]. This type of investment is called a **mixed investment**.

Mixed investment: An investment in which a firm borrows money from the project during the investment period.

EXAMPLE 5A.2 Pure versus Mixed Investments

Consider the following four investment projects with known $i*$ values:

	Project Cash Flows			
n	**A**	**B**	**C**	**D**
0	−$1000	−$1000	−$1000	−$1000
1	−1000	1600	500	3900
2	2000	−300	−500	−5030
3	1500	−200	2000	2145
$i*$	33.64%	21.95%	29.95%	(10%, 30%, 50%)

Determine which projects are pure investments.

SOLUTION

Given: Four projects with cash flows and $i*$s as shown.
Find: Which projects are pure investments?

We will first compute the project balances at the projects' respective $i*$s. If multiple rates of return exist, we may use the largest value of $i*$ greater than zero.[1]

Project A:

$$PB(33.64\%)_0 = -\$1000$$
$$PB(33.64\%)_1 = -\$1000(1 + 0.3364) + (-\$1000) = -\$2336.40,$$
$$PB(33.64\%)_2 = -\$2336.40(1 + 0.3364) + \$2000 = -\$1122.36,$$
$$PB(33.64\%)_3 = -\$1122.36(1 + 0.3364) + \$1500 = 0.$$

$(-, -, -, 0)$: passes the net-investment test (pure investment).
Project B:

$$PB(21.95\%)_0 = -\$1000$$
$$PB(21.95\%)_1 = -\$1000(1 + 0.2195) + \$1600 = \$380.50,$$
$$PB(21.95\%)_2 = +\$380.50(1 + 0.2195) - \$300 = \$164.02,$$
$$PB(21.95\%)_3 = +\$164.02(1 + 0.2195) - \$200 = 0.$$

$(-, +, +, 0)$: fails the net-investment test (mixed investment).

[1] In fact, it does not matter which rate we use in applying the net-investment test. If one value passes the test, they will all pass. If one value fails, they will all fail.

Project C:

$$PB(29.95\%)_0 = -\$1000$$

$$PB(29.95\%)_1 = -\$1000(1 + 0.2995) + \$500 = -\$799.50,$$

$$PB(29.95\%)_2 = -\$799.50(1 + 0.2995) - \$500 = -\$1538.95,$$

$$PB(29.95\%)_3 = -\$1538.95(1 + 0.2995) + \$2000 = 0.$$

$(-, -, -, 0)$: passes the net-investment test (pure investment).

Project D: There are three rates of return. We can use any of them for the net investment test. Thus,

$$PB(50\%)_0 = -\$1000$$

$$PB(50\%)_1 = -\$1000(1 + 0.50) + \$3900 = \$2400,$$

$$PB(50\%)_2 = +\$2400(1 + 0.50) - \$5030 = -\$1430,$$

$$PB(50\%)_3 = -\$1430(1 + 0.50) + \$2145 = 0.$$

$(-, +, -, 0)$: fails the net-investment test (mixed investment).

COMMENTS: As shown in Figure 5A.1, projects A and C are the only pure investments. Project B demonstrates that the existence of a unique i^* is a necessary but not
sufficient condition for a pure investment.

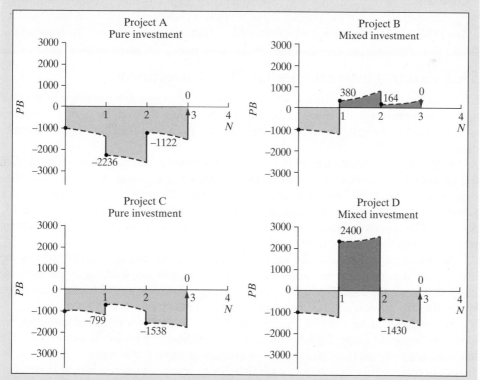

Figure 5A.1   Net-investment test (Example 5A.2).

EXAMPLE 5A.3 IRR for Nonsimple Project: Pure Investment

Consider project C in Example 5A.2. Assume that all costs and benefits are as stated for this independent project. Apply the net investment test and determine the acceptability of the project using the IRR criterion.

SOLUTION

Given: Cash flows, $i* = 29.95\%$, MARR = 15%.
Find: IRR, and determine whether the project is acceptable.

The net investment test was already applied in Example 5A.2, and it indicated that project C is a pure investment. In other words, the project balances were all less than or equal to zero, and the final balance was zero, as it must be if the interest rate used was $i*$. This proves that 29.95% is the true internal rate of return for the cash flows, that is, IRR = 29.95%. At MARR = 15%, IRR > MARR, thus, the project is acceptable. If we compute the PW of this project at $i = 15\%$, we obtain

$$PW(15\%) = -\$1000 + \$500(P/F, 15\%, 1) - \$500(P/F, 15\%, 2) + \$2000(P/F, 15\%, 3)$$
$$= \$371.74.$$

Since PW(15%) > 0, the project is also acceptable under the PW criterion. The IRR and the PW criteria will always produce the same decision if the criteria are applied correctly.

COMMENTS: In the case of a pure investment, the firm has funds committed to the project over the life of the project and at no time withdraws money from the project. The IRR is the return earned on the funds that remain internally invested in the project.

5A.3 External Interest Rate for Mixed Investments

In the case of a mixed investment, we can extend the economic interpretation of the IRR to the return-on-invested-capital measure if we are willing to make an assumption about what happens to the extra cash that the investor gets from the project during the intermediate years.

Project balance (PB): The amount of money committed to a project at a specific period.

First, the **project balance (PB)**, or investment balance, can also be interpreted from the viewpoint of a financial institution that borrows money from an investor and then pays interest on the PB. Thus, a negative PB means that the investor has money in a bank account; a positive PB means that the investor has borrowed money from the bank. Negative PBs represent interest paid by the bank to the investor; positive PBs represent interest paid by the investor to the bank.

Now, can we assume that the interest paid by the bank and the interest paid by the investor are the same for the same amount of balance? In our banking experience, we know that is not the case. Normally, the borrowing rate (interest paid by the investor) is higher than the interest rate on your deposit (interest paid by the bank).

However, when we calculate the project balance at an $i*$ for mixed investments, we notice an important point: Cash borrowed (released) from the project is assumed to earn the same interest rate through external investment as money that remains internally invested. In other words, in solving a cash flow for an unknown interest rate, it is assumed that money released from a project can be reinvested to yield a rate of return equal to that received from the project. In fact, we have been making this assumption regardless of whether a cash flow produces a unique positive i^*. Note that money is borrowed only when $\text{PB}(i^*) > 0$, and the magnitude of the borrowed amount is the project balance. When $\text{PB}(i^*) < 0$, no money is borrowed, even though the cash flow may be positive at that time.

In reality, it is not always possible for cash borrowed (released) from a project to be reinvested to yield a rate of return equal to that received from the project. Instead, it is likely that the rate of return available on a capital investment in the business is much different—usually higher—from the rate of return available on other external investments. Thus, it may be necessary to compute the project balances for a project's cash flow at two rates of interest—one on the internal investment and one on the external investments. As we will see later, by separating the interest rates, we can measure the **true rate of return** of any internal portion of an investment project.

5A.4 Calculation of Return on Invested Capital for Mixed Investments

For a mixed investment, we must calculate a rate of return on the portion of capital that remains invested internally. This rate is defined as the **true IRR** for the mixed investment and is commonly known as the **return on invested capital (RIC)**. Then, what interest rate should we assume for the portion of external investment? Insofar as a project is not a net investment, one or more periods when the project has a net outflow of money (a positive project balance) must later be returned to the project. This money can be put into the firm's investment pool until such time as it is needed in the project. The interest rate of this investment pool is the interest rate at which the money can in fact be invested outside the project.

> **Return on invested capital (RIC):** The amount that a company earns on the total investment it has made in its project.

Recall that the NPW method assumed that the interest rate charged to any funds withdrawn from a firm's investment pool would be equal to the MARR. In this book, *we will use the MARR as an established external interest rate* (i.e., the rate earned by money invested outside of the project). We can then compute the RIC as a function of the MARR by finding the value of the RIC that will make the terminal project balance equal to zero. (This implies that the firm wants to fully recover any investment made in the project and pays off any borrowed funds at the end of the project life.) This way of computing the rate of return is an accurate measure of the profitability of the project as represented by the cash flow. The following procedure outlines the steps for determining the IRR for a mixed investment:

Step 1. Identify the MARR (or external interest rate).

Step 2. Calculate $\text{PB}(i, \text{MARR})_n$ (or simply PB_n) according to the rule

$$\text{PB}(i, \text{MARR})_0 = A_0.$$

$$\text{PB}(i, \text{MARR})_1 = \begin{cases} \text{PB}_0(1 + i) + A_1, & \text{if } \text{PB}_0 < 0 \\ \text{PB}_0(1 + \text{MARR}) + A_1, & \text{if } \text{PB}_0 > 0 \end{cases}$$

$$\vdots$$

$$\text{PB}(i, \text{MARR})_n = \begin{cases} \text{PB}_{n-1}(1 + i) + A_n, & \text{if } \text{PB}_{n-1} < 0 \\ \text{PB}_{n-1}(1 + \text{MARR}) + A_n, & \text{if } \text{PB}_{n-1} > 0 \end{cases}$$

(As defined in the text, A_n stands for the net cash flow at the end of period n. Note that the terminal project balance must be zero.)

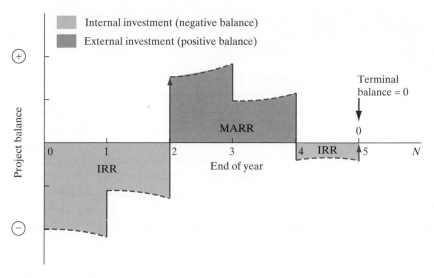

Figure 5A.2 Computational logic for IRR (mixed investment).

Step 3. Determine the value of i by solving the terminal project balance equation

$$PB(i, \text{MARR})_N = 0$$

The interest rate i is the RIC (or IRR) for the mixed investment.

Using the MARR as an external interest rate, we may accept a project if the IRR exceeds the MARR, and we should reject the project otherwise. Figure 5A.2 summarizes the IRR computation for a mixed investment.

EXAMPLE 5A.4 IRR for a Mixed Investment

Reconsider the contractor's flight-simulator project in Example 5.21. The project was a nonsimple and mixed investment. In the solution to Example 5.21, we abandoned the IRR criterion and used the PW criterion to make an accept/reject decision to simplify the decision-making process. We will now apply the procedures outlined to find the true IRR, or return on invested capital, for this mixed investment.

(a) Compute the IRR (RIC) for this project, assuming MARR = 15%.
(b) Make an accept/reject decision based on the results in part (a).

SOLUTION

Given: Cash flows shown in Example 5.21 and MARR = 15%.
Find: (a) Compute i^* and (b) RIC at MARR = 15%, and determine whether to accept the project.

(a) As calculated in Example 5.21, the project has multiple rates of return. This is obviously not a net investment, as the following table shows:

Net Investment Test N	Using $i^* = 10\%$			Using $i^* = 20\%$		
	0	1	2	0	1	2
Beginning balance	$0	−$1000	$1200	$0	−$1000	$1100
Return on investment	0	−100	120	0	−200	220
Payment	−1000	2300	−1320	−1000	2300	−1320
Ending balance	−$1000	$1200	0	−$1000	$1100	0

(Unit: $1000)

At $n = 0$, there is a net investment to the firm, so the project balance expression becomes

$$PB(i, 15\%)_0 = -\$1,000,000.$$

The net investment of $1,000,000 that remains invested internally grows at the interest rate i for the next period. With the receipt of $2,300,000 in year 1, the project balance becomes

$$PB(i, 15\%)_1 = -\$1,000,000(1 + i) + \$2,300,000$$
$$= \$1,300,000 - \$1,000,000i$$
$$= \$1,000,000(1.3 - i).$$

At this point, we do not know whether $PB(i, 15\%)_1$ is positive or negative; we want to know this in order to test for net investment and the presence of a unique i^*. It depends on the value of i, which we want to determine. Therefore, we need to consider two situations: (1) $i < 1.3$ and (2) $i > 1.3$.

- **Case 1:** $i < 1.3 \rightarrow PB(i, 15\%)_1 > 0$.
  Since this indicates a positive balance, the cash released from the project would be returned to the firm's investment pool to grow at the MARR until it is required back in the project. By the end of year 2, the cash placed in the investment pool would have grown at the rate of 15% [to $1,000,000(1.3 - i)$ $(1 + 0.15)$] and must equal the investment into the project of $1,320,000 required at that time. Then the terminal balance must be

$$PB(i, 15\%)_2 = \$1,000,000(1.3 - i)(1 + 0.15) - \$1,320,000$$
$$= \$175,000 - \$1,150,000i$$
$$= 0.$$

Solving for i yields

$$RIC = IRR = 0.1522, \text{ or } 15.22\% > 15\%.$$

The computational process is shown graphically in Figure 5A.3.

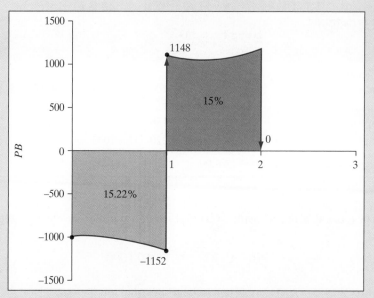

Figure 5A.3 Calculation of the IRR for a mixed investment (Example 5A.4).

- **Case 2:** $i > 1.3 \rightarrow \mathrm{PB}(i, 15\%)_1 < 0$.

   The firm is still in an investment mode. Therefore, the balance at the end of year 1 that remains invested will grow at the rate i for the next period. With the investment of \$1,320,000 required in year 2 and the fact that the net investment must be zero at the end of the project life, the balance at the end of year 2 should be

$$\mathrm{PB}(i, 15\%)_2 = \$1,000,000(1.3 - i)(1 + i) - \$1,320,000$$
$$= -\$20,000 + \$300,000i - \$1,000,000i^2$$
$$= 0$$

Solving for i gives

$$\mathrm{IRR} = 0.1 \text{ or } 0.2 < 1.3,$$

which violates the initial assumption that $i > 1.3$. Therefore, Case 1 is the only correct situation. Since it indicates that IRR > MARR, the project is acceptable, resulting in the same decision as obtained in Example 5.21 by applying the NPW criterion.

COMMENTS: In this example, we could have seen by inspection that Case 1 was correct. Since the project required an investment as the final cash flow, the project balance at the end of the previous period (year 1) had to be positive in order for the final balance to equal zero. Inspection does not generally work with more complicated cash flows.

5A.5 Trial-and-Error Method for Computing IRR for Mixed Investments

The trial-and-error approach to finding the IRR (RIC) for a mixed investment is similar to the trial-and-error approach to finding i^*. We begin with a given MARR and a guess for IRR and solve for the project balance. (A value of IRR close to the MARR is a good starting point for most problems.) Since we desire the project balance to approach zero, we can adjust the value of IRR as needed after seeing the result of the initial guess. For example, for a given pair of interest rates (IRR guess, MARR), if the terminal project balance is positive, the IRR guess value is too low, so we raise it and recalculate. We can continue adjusting our IRR guesses in this way until we obtain a project balance equal or close to zero.

EXAMPLE 5A.5 IRR for a Mixed Investment by Trial and Error

Consider project B in Example 5A.2. The project has the following cash flow:

n	A_n
0	−$1000
1	3900
2	−5030
3	2145

We know from an earlier calculation that this is a mixed investment. Compute the IRR for this project. Assume that MARR = 6%.

SOLUTION

Given: Cash flow as stated for mixed investment and MARR = 6%.
Find: IRR.

For MARR = 6%, we must compute i by trial and error. Suppose we guess $i = 8\%$:

$$\text{PB}(8\%, 6\%)_0 = -\$1000$$
$$\text{PB}(8\%, 6\%)_1 = -\$1000(1 + 0.08) + \$3900 = \$2820$$
$$\text{PB}(8\%, 6\%)_2 = +\$2820(1 + 0.06) - \$5030 = -\$2040.80$$
$$\text{PB}(8\%, 6\%)_3 = -\$2040.80(1 + 0.08) + \$2145 = -\$59.06$$

The net investment is negative at the end of the project, indicating that our trial $i = 8\%$ is too large. After several trials, we conclude that, for MARR = 6%, the IRR is approximately 6.13%. To verify the results, we write

$$\text{PB}(6.13\%, 6\%)_0 = -\$1000,$$
$$\text{PB}(6.13\%, 6\%)_1 = -\$1000.00(1 + 0.0613) + \$3900 = \$2838.66.$$
$$\text{PB}(6.13\%, 6\%)_2 = +\$2838.66(1 + 0.0600) - \$5030 = -\$2021.02,$$
$$\text{PB}(6.13\%, 6\%)_3 = -\$2021.02(1 + 00613) + \$2145 = 0.$$

The positive balance at the end of year 1 indicates the need to borrow from the project during year 2. However, note that the net investment becomes zero at the end of the project life, confirming that 6.13% is the IRR for the cash flow. Since IRR > MARR, the investment is acceptable.

COMMENTS: On the basis of the NPW criterion, the investment would be acceptable if the MARR was between zero and 10% or between 30% and 50%. The rejection region is $10\% < i < 30\%$ and $i > 50\%$. This can be verified in Figure 5.16(b). Note that the project also would be marginally accepted under the NPW analysis at MARR $= i = 6\%$:

$$
\begin{aligned}
\text{PW}(6\%) &= -\$1000 + 3900(P/F, 6\%, 1) \\
&= -\$5030(P/F, 6\%, 2) + 2145(P/F, 6\%, 3) \\
&= \$3.55 > 0
\end{aligned}
$$

SUMMARY

1. The possible presence of multiple i^*s (rates of return) can be predicted by
   - The net cash flow sign test
   - The cumulative cash flow sign test

   When multiple rates of return cannot be ruled out by the two methods, it is useful to generate a PW profile to approximate the values of i^*.

2. For all nonsimple investments, an i^* value should be exposed to the net investment test. Passing the net investment test indicates that the i^* is an internal rate of return and is therefore a suitable measure of project profitability. Failure to pass the test indicates project borrowing, a situation that requires further analysis by use of an external interest rate.

3. Return on invested capital analysis uses one rate (the firm's MARR) on externally invested balances and solves for another rate (IRR) on internally invested balances.

PROBLEMS

5A.1 Consider the following investment projects:

N	A	B	C	D	E	F
			Project Cash Flows			
0	−$100	−$100	−$100	−$100	−$100	−$100
1	200	470	−200	0	300	300
2	300	720	200	0	250	100
3	400	360	250	500	−40	400

(a) For each project, apply the sign rule to predict the number of possible $i*$s.
(b) For each project, plot the NPW profile as a function of i between 0 and 200%.
(c) For each project, compute the value(s) of $i*$.

5A.2 Consider the following investment projects:

n	Project 1	Project 2	Project 3
		Net Cash Flow	
0	−$1000	−$2000	−$1000
1	500	1560	1400
2	840	944	−100
IRR	?	?	?

Assume that MARR = 12% in the following questions:
(a) Compute $i*$ for each investment. If the problem has more than one $i*$, identify all of them.
(b) Compute IRR(true) for each project.
(c) Determine the acceptability of each investment.

5A.3 Consider the following investment projects:

n	A	B	C	D	E
			Project Cash Flow		
0	−$100	−$100	−$5	−$100	$200
1	100	30	10	30	100
2	24	30	30	30	−500
3		70	−40	30	−500
4		70		30	200
5				30	600

(a) Use the quadratic equation to compute $i*$ for A.

(b) Classify each project as either simple or nonsimple.

(c) Apply the cash flow sign rules to each project, and determine the number of possible positive $i*$s. Identify all projects having a unique $i*$.

(d) Compute the IRRs for projects B through E. Assume an external ROR of 10%.

(e) Apply the net-investment test to each project.

5A.4 Consider the following investment projects:

n	Net Cash Flow		
	Project 1	**Project 2**	**Project 3**
0	−$1,600	−$5,000	−$1,000
1	10,000	10,000	4,000
2	10,000	30,000	−4,000
3		−40,000	

Assume that MARR = 12% in the following questions:

(a) Identify the $i*$s for each investment. If the project has more than one $i*$, identify all of them.

(b) Which project(s) is (are) a mixed investment?

(c) Compute the IRR for each project.

(d) Determine the acceptability of each project.

5A.5 Consider the following investment projects:

n	Net Cash Flow		
	Project A	**Project B**	**Project C**
0	−$100	−$150	−$100
1	30	50	410
2	50	50	−558
3	80	50	252
4		100	
IRR	(23.24%)	(21.11%)	(20%, 40%, 50%)

Assume that MARR = 12% for the following questions:

(a) Identify the pure investment(s).

(b) Identify the mixed investment(s).

(c) Determine the IRR for each investment.

(d) Which project would be acceptable?

5A.6 The Boeing Company has received a NASA contract worth $460 million to build rocket boosters for future space missions. NASA will pay $50 million when the contract is signed, another $360 million at the end of the first year, and the $50 million balance at the end of second year. The expected cash outflows required to produce these rocket boosters are estimated to be $150 million now, $100 million during the first year, and $218 million during the second year. The firm's MARR is 12%. The cash flow is as follows:

n	Outflow	Inflow	Net Cash Flow
0	$150	$50	−$100
1	100	360	260
2	218	50	−168

(a) Show whether this project is or is not a mixed investment.
(b) Compute the IRR for this investment.
(c) Should Boeing accept the project?

5A.7 Consider the following investment projects:

	Net Cash Flow		
n	Project A	Project B	Project C
0	−$100		−$100
1	216	−150	50
2	−116	100	−50
3		50	200
4		40	
$i*$	?	15.51%	29.95%

(a) Compute $i*$ for project A. If there is more than one $i*$, identify all of them.
(b) Identify the mixed investment(s).
(c) Assuming that MARR = 10%, determine the acceptability of each project on the basis of the IRR criterion.

5A.8 Consider the following investment projects:

	Net Cash Flow				
n	A	B	C	D	E
0	−$1,000	−$5,000	−$2,000	−$2,000	−$1,000
1	3,100	20,000	1,560	2,800	3,600
2	−2,200	12,000	944	−200	−5,700
3		−3,000			3,600
$i*$	?	?	18%	32.45%	35.39%

Assume that MARR = 12% in the following questions:
(a) Compute $i*$ for projects A and B. If the project has more than one $i*$, identify all of them.

(b) Classify each project as either a pure or a mixed investment.

(c) Compute the IRR for each investment.

(d) Determine the acceptability of each project.

5A.9 Consider an investment project whose cash flows are as follows:

n	Net Cash Flow
0	−$5,000
1	10,000
2	30,000
3	−40,000

(a) Plot the present-worth curve by varying i from 0% to 250%.

(b) Is this a mixed investment?

(c) Should the investment be accepted at MARR $= 18\%$?

5A.10 Consider the following cash flows of a certain project:

n	Net Cash Flow
0	−$100,000
1	310,000
2	−220,000

The project's $i*$s are computed as 10% and 100%, respectively. The firm's MARR is 8%.

(a) Show why this investment project fails the net-investment test.

(b) Compute the IRR, and determine the acceptability of this project.

5A.11 Consider the following investment projects:

	Net Cash Flow		
n	Project 1	Project 2	Project 3
0	−$1,000	−$1,000	−$1,000
1	−1,000	1,600	1,500
2	2,000	−300	−500
3	3,000	−200	2,000

Which of the following statements is correct?

(a) All projects are nonsimple investments.

(b) Project 3 should have three real rates of return.

(c) All projects will have a unique positive real rate of return.

(d) None of the above.

On the Companion Website that accompanies this text, you will find Excel templates and exercises, as well as the following analysis tools: Cash Flow Analyzer, Depreciation Analysis, Loan Analysis, and Interest Tables.

SIX

Comparing Mutually Exclusive Alternatives

The proliferation of computers into all aspects of business has created an ever-increasing need for data capture systems that are fast, reliable, and cost-effective. One technology that has been adopted by many manufacturers, distributors, and retailers is a bar-coding system. Ultra Electronics, a leading manufacturer of underwater surveillance equipment, evaluated the economic benefits of installing an automated data acquisition system in their Halifax plant. They could use the system on a limited scale, such as for tracking parts and assemblies for inventory management, or opt for a broader implementation by recording information useful for quality control, operator efficiency, attendance, and other functions. All of these aspects are currently monitored, but although computers are used to manage the information, the recording is primarily conducted manually. The advantages of an automated data collection system, which include faster and more accurate data capture, quicker analysis and response to production changes, and savings due to tighter control over operations, could easily outweigh the cost of the new system.

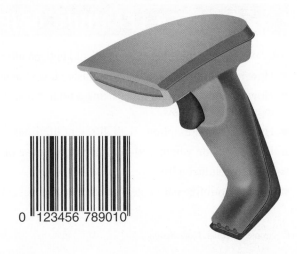

0 123456 789010

Two alternative systems from competing suppliers are under consideration. System 1 relies on handheld bar-code scanners that transmit their data to the host server on the local area network (LAN) using radio frequencies (RF). The hub of this wireless network can then be connected to their existing LAN and integrated with their current MRP II system and other management software. System 2 consists primarily of specialized data terminals installed at every collection point, with connected bar-code scanners where required. This system is configured in such a way to facilitate either phasing in the components over two stages or installing the system all at once. The phase-in option would allow Ultra to defer some of the capital investment while becoming thoroughly familiar with the functions introduced in the first stage.[1]

Either of these systems would satisfy Ultra's data collection needs. They each have some unique elements, and the company needs to compare the relative benefits of the features offered by each system. From the point of view of engineering economics, the two systems have different capital costs, and their operating and maintenance costs are not identical. One system may also be rated to last longer than the other system before replacement is required, particularly if the option of acquiring System 2 in phases is selected. There are many issues to be considered before making the best choice.

CHAPTER LEARNING OBJECTIVES

After completing this chapter, you should understand the following concepts:

- How firms compare mutually exclusive investment opportunities.
- How to compare alternatives using a total investment approach or an incremental analysis.
- How to compare options with different lifetimes.
- How to decide whether to make a part in-house or to buy it.
- How to conduct a life-cycle cost analysis.
- How to optimize parameters in engineering design.

[1] Ultra Electronics, Halifax, Nova Scotia.

Until now, we have considered situations involving either a single project alone or projects that were independent of one another. In both cases, we made the decision to reject or accept each project individually according to whether it met the MARR requirements, evaluated with either the PW, FW, AE, or IRR criterion.

In the real world of engineering practice, however, it is typical for us to have two or more choices of projects that are not independent of one another in seeking to accomplish a business objective. (As we shall see, even when it appears that we have only one project to consider, the implicit "do-nothing" alternative must be factored into the decision-making process.) In this chapter, we extend our evaluation techniques to multiple projects that are mutually exclusive. Other dependencies between projects will be considered in Chapter 12.

Often, various projects or investments under consideration do not have the same duration or do not match the desired study period. Adjustments must then be made to account for the differences. In this chapter, we explain the concept of an analysis period and the process of accommodating for different lifetimes, two important considerations that apply in selecting among several alternatives. In the first few sections of the chapter, all available options in a decision problem are assumed to have equal lifetimes. In Section 6.4 this restriction is relaxed.

6.1 Basic Elements of Mutually Exclusive Analysis

Mutually exclusive means that any one of several alternatives will fulfill the same need and that the selection of one alternative implies that the others will be excluded.

Take, for example, buying versus leasing an automobile for business use; when one alternative is accepted, the other is excluded. We use the terms **alternative** and **project** interchangeably to mean "decision option."

6.1.1 The "Do Nothing" Decision Option

When considering an investment, we are in one of two situations: Either the project is aimed at replacing an existing asset or system, or it is a new endeavour. In either case, a do-nothing alternative may exist. On the one hand, if a process or system already in place to accomplish our business objectives is still adequate, then we must determine which, if any, new proposals are economical replacements. If none are feasible, then we do nothing. On the other hand, if the existing system has failed, then the choice among proposed alternatives is mandatory (i.e., "do nothing" is not an option).

New endeavours occur as alternatives to the "green fields" do-nothing situation, which has zero revenues and zero costs (i.e., nothing currently exists). For most new endeavours, do nothing is generally an alternative, as we won't proceed unless at least one of the proposed alternatives is economically sound. In fact, undertaking even a single project entails making a decision between two alternatives because the do-nothing alternative is implicitly included. Occasionally, a new initiative must be undertaken, cost notwithstanding, and in this case the goal is to choose the most economical alternative, since "do nothing" is not an option.

When the option of retaining an existing asset or system is available, there are two ways to incorporate it into the evaluation of the new proposals. One way is to treat the do-nothing option as a distinct alternative; we cover this approach primarily in Chapter 11, where methodologies specific to replacement analysis are presented. The second approach, used mostly in this chapter, is to generate the cash flows of the new proposals relative to that of the do-nothing alternative. That is, for each new alternative, the **incremental costs** (and incremental savings or revenues if applicable) relative to "do nothing" are used in the economic evaluation. For a replacement-type problem, these costs are calculated by subtracting the do-nothing cash flows from those of each new alternative. For new endeavours, the incremental cash flows are the same as the absolute amounts associated with each alternative, since the do-nothing values are all zero.

Because the main purpose of this chapter is to illustrate how to choose among mutually exclusive alternatives, most of the problems are structured so that one of the options presented must be selected. Therefore, unless otherwise stated, it is assumed that "do nothing" is not an option, and costs and revenues can be viewed as incremental to "do nothing."

6.1.2 Service Projects versus Revenue Projects

When comparing mutually exclusive alternatives, we need to classify investment projects into either service or revenue projects. **Service projects** are projects whose revenues do not depend on the choice of project; rather, such projects *must produce the same amount of output (revenue).* In this situation, we certainly want to choose an alternative with the least input (or cost). For example, suppose an electric utility is considering building a new power plant to meet the peak-load demand during either hot summer or cold winter days. Two alternative service projects could meet this demand: a combustion turbine plant and a fuel-cell power plant. No matter which type of plant is selected, the firm will collect the same amount of revenue from its customers. The only difference is how much it will cost to generate electricity from each plant. If we were to compare these service projects, we would be interested in knowing which plant could provide the cheaper power (lower production cost). Further, if we were to use the NPW criterion to compare alternatives so as to minimize expenditures, we would choose the alternative with the **lower present-value** production cost over the service life of the plant.

Revenue projects, by contrast, are projects whose revenues depend on the choice of alternative. With revenue projects, we are not limiting the amount of input going into the project or the amount of output that the project would generate. Then our decision is to select the alternative with the largest net gains (output – input). For example, a TV manufacturer is considering marketing two types of high-resolution monitors. With its present production capacity, the firm can market only one of them. Distinct production processes for the two models could incur very different manufacturing costs, so the revenues from each model would be expected to differ due to divergent market prices and potentially different sales volumes. In this situation, if we were to use the NPW criterion, we would select the model that promises to bring in the higher net present worth.

6.1.3 Scale of Investment

Frequently, mutually exclusive investment projects may require different levels of investments. At first, it seems unfair to compare a project requiring a smaller investment with one requiring a larger investment. However, the disparity in scale of investment should not be of concern in comparing mutually exclusive alternatives, as long as you understand the basic assumption: *Funds not invested in the project will continue to earn interest at the MARR.* We will look at mutually exclusive alternatives that require different levels of investments for both service projects and revenue projects:

- **Service projects.** Typically, what you are asking yourself to do here is to decide whether the higher initial investment can be justified by additional savings that will occur in the future. More efficient machines are usually more expensive to acquire initially, but they will reduce future operating costs, thereby generating more savings.
- **Revenue projects.** If two mutually exclusive revenue projects require different levels of investments with varying future revenue streams, then your question is what to do with the difference in investment funds if you decide to go with the project that requires the smaller investment. To illustrate, consider the two mutually exclusive revenue projects illustrated in Figure 6.1. Our objective is to compare these two projects at a MARR of 10%.

Suppose you have exactly $4000 to invest. If you choose Project B, you do not have any leftover funds. However, if you go with Project A, you will have $3000 in unused funds. Our assumption is that these unused funds will continue to earn an interest rate that is the MARR. Therefore, the unused funds will grow at 10%, or $3933, at the end of the project term, or three years from now. Consequently, selecting Project A is equivalent to having a modified project cash flow as shown in Figure 6.1.

Let's calculate the net present worth for each option at 10%:

- **Project A:**

$$\text{PW}(10\%)_A = -\$1000 + \$450(P/F, 10\%, 1) + \$600(P/F, 10\%, 2)$$
$$+ \$500(P/F, 10\%, 3)$$
$$= \$283.$$

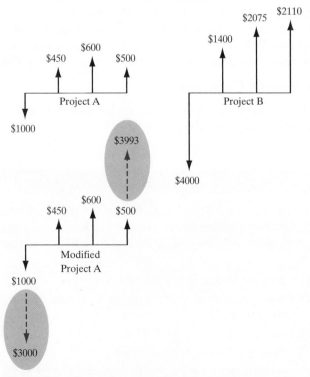

Figure 6.1 Comparing mutually exclusive revenue projects requiring different levels of investment.

- **Project B:**

$$\begin{aligned} \text{PW}(10\%)_B &= -\$4000 + \$1400(P/A, 10\%, 1) \\ &\quad + \$2075(P/F, 10\%, 2) + \$2110(P/F, 10\%, 3) \\ &= \$579. \end{aligned}$$

Clearly, Project B is the better choice. But how about the modified Project A? If we calculate the present worth for the modified Project A, we have the following:

- **Modified Project A:**

$$\begin{aligned} \text{PW}(10\%)_A &= -\$4000 + \$450(P/A, 10\%, 1) + \$600(P/F, 10\%, 2) \\ &\quad + \$4493(P/F, 10\%, 3) \\ &= \$283. \end{aligned}$$

This is exactly the same as the net present worth without including the investment consequence of the unused funds. It is not a surprising result, as the return on investment in the unused funds will be exactly 10%. If we also discount the funds at 10%, there will be no surplus. So, what is the conclusion? It is this: If there is any disparity in investment scale for mutually exclusive revenue projects, go ahead and calculate the net present worth for each option without worrying about the investment differentials.

6.2 Total-Investment Approach

Applying an evaluation criterion to each mutually exclusive alternative individually and then comparing the results to make a decision is referred to as the **total-investment approach**. Note that this approach guarantees valid results only when PW, FW, and AE criteria are used. (As shown later in this chapter, and in Chapter 13, the total-investment approach does not work for any decision criterion that is based on either a percentage (rate of return) or a ratio (e.g., a benefit–cost ratio). With percentages or ratios, you need to use the incremental investment approach, which also works with any decision criterion, including PW, FW, and AE.)

6.2.1 Present-Worth Comparison

When the project lives equal the analysis period, we compute the NPW for each project and select the one with the highest NPW as we did in Chapter 5. Example 6.1 illustrates this point.

EXAMPLE 6.1 Present-Worth Comparison: Three Alternatives

Bullard Company (BC) is considering expanding its range of industrial machinery products by manufacturing machine tables, saddles, machine bases, and other similar parts. Several combinations of new equipment and personnel could serve to fulfill this new function:

- **Method 1 (M1):** new machining centre with three operators.
- **Method 2 (M2):** new machining centre with an automatic pallet changer and three operators.
- **Method 3 (M3):** new machining centre with an automatic pallet changer and two task-sharing operators.

Each of these arrangements incurs different costs and revenues. The time taken to load and unload parts is reduced in the pallet-changer cases. Certainly, it costs more to acquire, install, and tool-fit a pallet changer, but because the device is more efficient and versatile, it can generate larger annual revenues. Although saving on labour costs, task-sharing operators take longer to train and are more inefficient initially. As the operators become more experienced at their tasks and get used to collaborating with one another, it is expected that the annual benefits will increase by 13% per year over the five-year study period. BC has estimated the investment costs and additional revenues as follows:

| | Machining Centre Methods | | |
	M1	M2	M3
Investment:			
Machine tool purchase	$121,000	$121,000	$121,000
Automatic pallet changer		$ 66,600	$ 66,600
Installation	$ 30,000	$ 42,000	$ 42,000
Tooling expense	$ 58,000	$ 65,000	$ 65,000
Total investment	$209,000	$294,600	$294,600
Annual benefits: Year 1			
Additional revenues	$ 55,000	$ 69,300	$ 36,000
Direct labour savings			$ 17,300
Setup savings		$ 4,700	$ 4,700
Year 1: Net revenues	$ 55,000	$ 74,000	$ 58,000
Years 2–5: Net revenues	constant	constant	g = 13%/year
Salvage value in year 5	$ 80,000	$120,000	$120,000

All cash flows include all tax effects. "Do nothing" is obviously an option, since BC will not undertake this expansion if none of the proposed methods is economically viable. If a method is chosen, BC expects to operate the machining centre over the next five years. On the basis of the use of the PW measure at i = 12%, which option would be selected?

SOLUTION

Given: Cash flows for three revenue projects and i = 12% per year.
Find: NPW for each project and which project to select.

For these revenue projects, the net-present-worth figures at i = 12% would be as follows:

- For Option M1,

$$PW(12\%)_{M1} = -\$209,000 + \$55,000(P/A, 12\%, 5)$$
$$+ \$80,000(P/F, 12\%, 5)$$
$$= \$34,657.$$

- For Option M2,

$$PW(12\%)_{M2} = -\$294{,}600 + \$74{,}000(P/A, 12\%, 5)$$
$$+ \$120{,}000(P/F, 12\%, 5)$$
$$= \$40{,}245.$$

- For Option M3,

$$PW(12\%)_{M3} = -\$294{,}600 + \$58{,}000(P/A_1, 13\%, 12\%, 5)$$
$$+ \$120{,}000(P/F, 12\%, 5)$$
$$= \$37{,}085.$$

Clearly, Option M2 is the most profitable. Given the nature of BC parts and shop orders, management decides that the best way to expand would be with an automatic pallet changer, but without task sharing.

COMMENTS: Of course, calculating the **future equivalent** of these mutually exclusive options will give the same optimal choice as the present equivalent analysis.

$$FW(12\%)_{M1} = \$34{,}657(F/P, 12\%, 5) = \$61{,}077.$$
$$FW(12\%)_{M2} = \$40{,}245(F/P, 12\%, 5) = \$70{,}925.$$
$$FW(12\%)_{M3} = \$37{,}085(F/P, 12\%, 5) = \$65{,}356.$$

Since the alternatives are compared over a fixed five-year period, the ratio of the future equivalent to the present equivalent is the same for all options:

$$\frac{FW_{M1}}{PW_{M1}} = \frac{FW_{M2}}{PW_{M2}} = \frac{FW_{M3}}{PW_{M3}} = (F/P, i, N)$$

This also implies that the ratio of future equivalents of any two options with identical life times is equal to the ratio of their present equivalents, as you can easily check:

$$\frac{FW_1}{FW_2} = \frac{PW_1}{PW_2}$$

6.2.2 Annual Equivalent Comparison

Mutually exclusive projects can also be compared by calculating the annual equivalent for each option, as shown in the following continuation of Example 6.1. We choose the project with the largest AE value.

EXAMPLE 6.2 Annual Equivalent Comparison: Three Alternatives

Consider again Example 6.1, in which we based our selection of three mutually exclusive options using the Present Worth and Future Worth criteria. Apply the annual equivalent approach to make the most economical choice.

SOLUTION

Given: Cash flows for three Machining Centre projects, $i = 12\%$ per year.
Find: AE of each project, which to select.

- For Option M1:

$$AE(12\%)_{M1} = -\$209,000(A/P, 12\%, 5) + \$55,000 + \$80,000(A/F, 12\%, 5)$$
$$= \$9614.$$

- For Option M2:

$$AE(12\%)_{M2} = -\$294,600(A/P, 12\%, 5) + \$74,000 + \$120,000(A/F, 12\%, 5)$$
$$= \$11,164.$$

- For Option M3:

$$AE(12\%)_{M3} = -\$294,600(A/P, 12\%, 5) +$$
$$\$58,000(P/A_1, 13\%, 12\%, 5)(A/P, 12\%, 5) +$$
$$\$120,000(A/F, 12\%, 5)$$
$$= \$10,288.$$

The best choice is option M2 because it has the highest AE value. As we expect, this result is consistent with the PW analysis. In fact, the annual equivalents calculated directly from the PW values in Example 6.1 would have yielded the same results as above. When two options' lifetimes are equal, the ratios of their AEs and PWs are equal, as follows:

$$\frac{AE_1}{AE_2} = \frac{PW_1(A/P,i,N)}{PW_2(A/P,i,N)} = \frac{PW_1}{PW_2}$$

So, the option with the higher AE always has a higher PW as well. This congruity of assessment criteria was noted in Chapter 5 for independent projects. Although the required consistency of analysis methods seems self-evident, particular care must be taken when applying the IRR criterion to mutually exclusive options, as shown in the following section.

6.3 Incremental Investment Analysis

In this section, we will present a method known as **incremental analysis**, a decision procedure that must be used when comparing two or more mutually exclusive projects based on the rate of return measure. This method may also be used when assessing alternatives using the PW, FW, or AE criteria.

6.3.1 Flaws in Project Ranking by IRR

As shown in the previous section, under the Total Investment Approach, the mutually exclusive project with the highest PW, FW, or AE is preferred. Unfortunately, the analogy does not carry over to IRR analysis: The project with the highest IRR may *not* be the preferred alternative. To illustrate the flaws inherent in comparing IRRs in order to choose from mutually exclusive projects, suppose you have two mutually exclusive alternatives, each with a one-year service life: One requires an investment of $1000 with a return of $2000, and the other requires $5000 with a return of $7000. You already obtained the IRRs and NPWs at MARR $= 10\%$ as follows:

n	A1	A2
0	$-\$1000$	$-\$5000$
1	2000	7000
IRR	100%	40%
PW(10%)	$818	$1364

Assuming that you have enough money in your investment pool to select either alternative, would you prefer the first project simply because you expect a higher rate of return?

On the one hand, we can see that A2 is preferred over A1 by the NPW measure. On the other hand, the IRR measure gives a numerically higher rating for A1. This inconsistency in ranking occurs because the NPW, NFW, and AE are **absolute (dollar)** measures of investment worth, whereas the IRR is a **relative (percentage)** measure and cannot be applied in the same way. That is, the IRR measure ignores the **scale** of the investment. Therefore, the answer to our question in the previous paragraph is no; instead, you would prefer the second project, with the lower rate of return but higher NPW. Either the NPW, FW, or the AE measure would lead to that choice, but a comparison of IRRs would rank the smaller project higher. Another approach, referred to as **incremental analysis**, is needed.

6.3.2 Incremental Investment Analysis

In the previous example, the more costly option requires an incremental investment of $4000 at an incremental return of $5000. Let's assume that you have exactly $5000 in your investment pool.

- If you decide to invest in option A1, you will need to withdraw only $1000 from your investment pool. The remaining $4000 will continue to earn 10% interest. One year later, you will have $2000 from the outside investment and $4400 from the investment pool. With an investment of $5000, in one year you will have $6400. The equivalent present worth of this change in wealth is PW(10%) $= -\$5000 + \$6400(P/F, 10\%, 1) = \$818$.

- If you decide to invest in option A2, you will need to withdraw $5000 from your investment pool, leaving no money in the pool, but you will have $7000 from your outside investment. Your total wealth changes from $5000 to $7000 in a year. The equivalent present worth of this change in wealth is PW(10%) = $-$5000 + $7000(P/F, 10\%, 1)$ = $1364.

In other words, if you decide to take the more costly option, certainly you would be interested in knowing that this additional investment can be justified at the MARR. The 10%-MARR value implies that you can always earn that rate from other investment sources (i.e., $4400 at the end of one year for a $4000 investment). However, in the second option, by investing the additional $4000, you would make an additional $5000, which is equivalent to earning at the rate of 25%. Therefore, the incremental investment can be justified.

Now we can generalize the decision rule for comparing mutually exclusive projects. For a pair of mutually exclusive projects (A and B, with B defined as the more costly option), we may rewrite B as

$$B = A + (B - A).$$

In other words, B has two cash flow components: (1) the same cash flow as A and (2) the incremental component $(B - A)$. Therefore, the only situation in which B is preferred to A is when the rate of return on the incremental component $(B - A)$ exceeds the MARR. Therefore, for two mutually exclusive projects, rate-of-return analysis is done by computing the *internal rate of return on the incremental investment* $(IRR\Delta)$ between the projects. Since we want to consider increments of investment, we compute the cash flow for the difference between the projects by subtracting the cash flow for the lower investment-cost project (A) from that of the higher investment-cost project (B). Then the decision rule is

Incremental IRR: IRR on the incremental investment from choosing a large project instead of a smaller project.

$$\text{If IRR}_{B-A} > \text{MARR, select B,}$$
$$\text{If IRR}_{B-A} = \text{MARR, select either project,}$$
$$\text{If IRR}_{B-A} < \text{MARR, select A,}$$

where $B - A$ is an investment increment (negative cash flow). *If a do-nothing alternative is allowed, the smaller cost option must be profitable (its IRR must be greater than the MARR) at first.* This means that you compute the rate of return for each alternative in the mutually exclusive group and then eliminate the alternatives whose IRRs are less than the MARR before applying the incremental analysis.

Since the incremental analysis method may also be used for PW, FW, and AE analyses, comparable decision rules apply:

	> 0	= 0	< 0
PW_{B-A} (MARR)	select B	select either one	select A
FW_{B-A} (MARR)	select B	select either one	select A
AE_{B-A} (MARR)	select B	select either one	select A

It may seem odd to you how this simple rule allows us to select the right project. Example 6.3 illustrates the incremental investment decision rule.

EXAMPLE 6.3 IRR on Incremental Investment: Two Alternatives

John Covington, a university student, wants to start a small-scale painting business during his off-school hours. To economize the start-up business, he decides to purchase some used painting equipment. He has two mutually exclusive options: Do most of the painting by himself by limiting his business to only residential painting jobs (B1) or purchase more painting equipment and hire some helpers to do both residential and commercial painting jobs that he expects will have a higher equipment cost, but provide higher revenues as well (B2). In either case, John expects to fold up the business in three years, when he graduates from university.

The cash flows for the two mutually exclusive alternatives are as follows:

n	B1	B2	B2 – B1
0	–$3,000	–$12,000	–$9,000
1	1,350	4,200	2,850
2	1,800	6,225	4,425
3	1,500	6,330	4,830
IRR	25.00%	17.43%	15.00%
PW (10%)	$841	$1,718	$877
AE (10%)	$338	$691	$353
FW (10%)	$1,120	$2,287	$1,167

Knowing that both alternatives are revenue projects, which project would John select at MARR $= 10\%$? (Note that both projects are also profitable at 10%.)

SOLUTION

Given: Incremental cash flow between two alternatives, MARR=10%

Find: Which alternative is preferable using: (a) IRR criterion, (b) PW criterion, (c) AE criterion, and (d) FW criterion

Given that $IRR_{B1} >$ MARR, $PW_{B1} > 0$, $AE_{B1} > 0$, and $FW_{B1} > 0$, by any criterion the cheapest option B1 is economically acceptable. The question is whether to make the additional investment, and opt for B2 instead. To choose the best project, we first calculate the incremental cash flow B2 − B1.

a) We compute the IRR on this increment of investment by solving

$$-\$9000 + \$2850(P/F, i, 1) + \$4425(P/F, i, 2) + \$4830(P/F, i, 3) = 0.$$

We obtain $i^*_{B2-B1} = 15\%$ as shown in the table above. By inspection of the incremental cash flows, we know it is a simple investment, so $IRR_{B2-B1} = i^*_{B2-B1}$. Since $IRR_{B2-B1} >$ MARR, we select B2, which is consistent with choosing the option with the largest PW. This result is illustrated in Figure 6.2. Notice that the IRR_{B2-B1} for the incremental cash flow curve B2 − B1 occurs

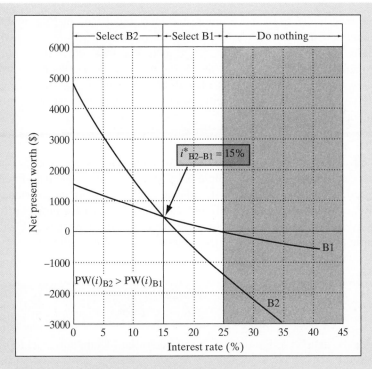

Figure 6.2 NPW profiles for B1 and B2 (Example 6. 3).

at the same interest rate at which the B1 and B2 curves intersect each other. This makes sense, as $PW_{B2 - B1} = PW_{B2} - PW_{B1} = 0$ at this interest rate of 15%. Note however that $IRR_{B2 - B1} \neq IRR_{B2} - IRR_{B1}$, which is why we cannot compare the IRRs for the individual options to arrive at a valid conclusion.

(b) $PW_{B2 - B1}(10\%) = \$877$, which is greater than zero, so the incremental investment is worthwhile, and we choose B2. Since $PW_{B2 - B1} = PW_{B2} - PW_{B1}$, stating that $PW_{B2 - B1} > 0$ is the same as saying that $PW_{B2} - PW_{B1} > 0$, or $PW_{B2} > PW_{B1}$. This is why we can directly compare the PWs of all the mutually exclusive options, and simply pick the best, rather than performing an incremental PW analysis, which takes longer.

(c) $AE_{B2 - B1}(10\%) = \$353 > 0$ so choose B2, which is consistent with the PW evaluation. Again, since $AE_{B2 - B1} = AE_{B2} - AE_{B1}$, the AE of each option may be compared directly, which is faster than applying this criterion to the incremental cash flow.

(d) $FW_{B2 - B1}(10\%) = \$1167 > 0$, which leads to the same optimal choice of B2, and this result could have been obtained directly from the difference of the future equivalents of the two options $FW_{B2} - FW_{B1}$.

COMMENTS: Why did we choose to look at the increment B2 − B1 instead of B1 − B2? Because we want the first flow of the incremental cash flow series to be negative (an investment flow), so that we can calculate an IRR. By subtracting the lower initial investment project from the higher, we guarantee that the first increment will be an investment flow. If we ignore the investment ranking, we might end up with an increment that involves *borrowing cash flow* and has no internal rate of return. This is indeed the case for B1 − B2. (i^*_{B1-B2} is also 15%, not −15%, but it has a different meaning: it is a borrowing rate, not a rate of return on your investment.) If, erroneously, we had compared this i^* with the MARR, we might have accepted project B1 over B2. This undoubtedly would have damaged our credibility with management!

When you have more than two mutually exclusive alternatives, they can be compared in pairs by successive examination. Example 6.4 illustrates how to compare three mutually exclusive alternatives. This example will indicate that the ranking inconsistency between PW and IRR can also occur when differences in the timing of a project's future cash flows exist, even if their initial investments are the same. (In Chapter 12, we will examine some multiple-alternative problems in the context of capital budgeting.)

EXAMPLE 6.4 IRR on Incremental Investment: Three Alternatives

Reconsider the data from Example 6.1, for the Bullard Company machining centre evaluation:

	Machining Centre Methods		
n	M1	M2	M3
0	−$209,000	−$294,600	−$294,600
1	55,000	74,000	58,000
2	55,000	74,000	65,540
3	55,000	74,000	74,060
4	55,000	74,000	83,688
5	$135,000	$194,000	$214,567
IRR=	17.62%	16.60%	15.99%

Which project would you select based on rate of return on incremental investment, assuming that MARR = 12%?

SOLUTION

Given: Cash flows given above, MARR = 12%.
Find: IRR on incremental investment and which alternative is preferable.

Step 1: Examine the IRR for each alternative. When "do nothing" is an option, we can eliminate any alternative that fails to meet the MARR.[2] In this example, all three alternatives exceed the MARR of 12%.

---

[2] When it is mandatory to pick one of the "new options" (i.e., if "do-nothing" is not allowed), then individual options cannot be rejected based on their IRR. Rather, their rate of return on incremental investment compared to any cheaper option must then be used as a screening criterion.

Step 2: Compare M1 and M2 in pairs.[3] M1, being the cheapest option, becomes the default choice and is termed the defender. To see whether the challenger, M2, is worth the extra investment, calculate the IRR on the incremental cash flow:

n	M2 − M1
0	−$85,600
1	19,000
2	19,000
3	19,000
4	19,000
5	59,000
IRR$_{M2\text{-}M1}$	14.16%

The incremental cash flow represents a simple investment, so the break-even interest rate calculated in the table is the valid internal rate of return. Since IRR$_{M2-M1}$ exceeds the MARR of 12%, M2 is preferred over M1, and M2 becomes the defender.

Step 3: Compare M3 and M2. When initial investments are equal, we progress through the cash flows until we find the first difference, then set up the increment so that this first non-zero flow is negative (i.e., an investment). In this problem, the existing order of M2 and M3 satisfies this requirement as shown in the following incremental investment table:

n	M3 − M2
0	0
1	−$16,000
2	−8,460
3	60
4	9688
5	20,567
IRR$_{M3\text{-}M2}$	6.66%

The incremental cash flows show a simple investment pattern. The (M3 − M2) increment has an unsatisfactory 6.66% rate of return; therefore, M3 is not preferred over M2. In summary, we conclude that M2 is the best alternative, consistent with our prior PW, AE, and FW analyses of this problem.

COMMENTS: It may be emphasized again that the IRR for each individual option, shown in the data table at the start of this problem, can be used for screening the alternatives for acceptability, but cannot be used to select the best option. Figure 6.3 demonstrates again that the option with the highest IRR is not necessarily the one with the highest PE at MARR.

---

[3] When faced with many alternatives, you may arrange them in order of increasing initial cost. This is not a required step, but it makes the comparison more tractable.

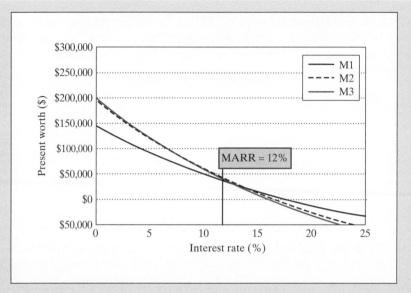

Figure 6.3 PW profiles for the three machining centre options (Example 6.4).

EXAMPLE 6.5 Incremental Analysis for Cost-Only Projects

Falk Corporation is considering two types of manufacturing systems to produce its shaft couplings over six years: (1) a cellular manufacturing system (CMS) and (2) a flexible manufacturing system (FMS). The average number of pieces to be produced with either system would be 544,000 per year. The operating cost, initial investment, and salvage value for each alternative are estimated as follows:

Items	CMS Option	FMS Option
Annual O&M costs:		
Annual labour cost	$1,169,600	$707,200
Annual material cost	832,320	598,400
Annual overhead cost	3,150,000	1,950,000
Annual tooling cost	470,000	300,000
Annual inventory cost	141,000	31,500
Annual income taxes	<u>1,650,000</u>	<u>1,917,000</u>
Total annual costs	$7,412,920	$5,504,100
Investment	$4,500,000	$12,500,000
Net salvage value	$500,000	$1,000,000

Figure 6.4 illustrates the cash flows associated with each alternative. The firm's MARR is 15%. On the basis of the IRR criterion, which alternative would be a better choice?

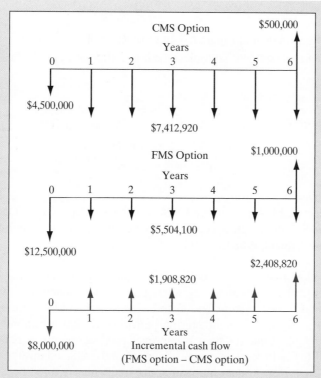

Figure 6.4 Comparison of mutually exclusive alternatives with equal revenues (cost only) (Example 6.5).

DISCUSSION: Since we can assume that both manufacturing systems would provide the same level of revenues over the analysis period, we can compare the two alternatives on the basis of cost only. (These are service projects.) Although we cannot compute the IRR for each option without knowing the revenue figures, we can still calculate the IRR on incremental cash flows. Since the FMS option requires a higher initial investment than that for the CMS, the incremental cash flow is the difference (FMS − CMS).

n	CMS Option	FMS Option	Incremental (FMS − CMS)
0	−$4,500,000	−$12,500,000	−$8,000,000
1	−7,412,920	−5,504,100	1,908,820
2	−7,412,920	−5,504,100	1,908,820
3	−7,412,920	−5,504,100	1,908,820
4	−7,412,920	−5,504,100	1,908,820
5	−7,412,920	−5,504,100	1,908,820
6	−7,412,920	−5,504,100	
	+ $500,000	+ $1,000,000	2,408,820

SOLUTION

Given: Cash flows shown in Figure 6.4 and MARR $= 15\%$ per year.

Find: Incremental cash flows, and select the better alternative on the basis of the IRR criterion.

First, we have

$$
\begin{aligned}
\text{PW}(i)_{\text{FMS}-\text{CMS}} &= -\$8{,}000{,}000 + \$1{,}908{,}820(P/A, i, 5) \\
&= +\$2{,}408{,}820(P/F, i, 6) \\
&= 0.
\end{aligned}
$$

Solving for i by trial and error yields 12.43%. Since $\text{IRR}_{\text{FMS}-\text{CMS}} = 12.43\% < 15\%$, we would select CMS. Although the FMS would provide an incremental annual savings of \$1,908,820 in operating costs, the savings are not large enough to justify the incremental investment of \$8,000,000.

COMMENTS: Note that the CMS option is marginally preferred to the FMS option. However, there are dangers in relying solely on the easily quantified savings in input factors—such as labour, energy, and materials—from the FMS and in not considering gains from improved manufacturing performance that are more difficult and subjective to quantify. Factors such as improved product quality, increased manufacturing flexibility (rapid response to customer demand), reduced inventory levels, and increased capacity for product innovation are frequently ignored because we have inadequate means for quantifying their benefits. If these intangible benefits were considered, the FMS option could come out better than the CMS option.

6.4 Analysis Period

The **analysis period** is the time span over which the economic effects of an investment will be evaluated. The analysis period may also be called the **study period** or **planning horizon**. The length of the analysis period may be determined in several ways: It may be a predetermined amount of time set by company policy, or it may be either implied or explicit in the need the company is trying to fulfill. (For example, a diaper manufacturer sees the need to dramatically increase production over a 10-year period in response to an anticipated "baby boom.") In either of these situations, we consider the analysis period to be the **required service period**.

When the required service period is not stated at the outset, the analyst must choose an appropriate analysis period over which to study the alternative investment projects. In such a case, one convenient choice of analysis period is the period of the useful life of the investment project.

When the useful life of an investment project does not match the analysis or required service period, we must make adjustments in our analysis. A further complication in a consideration of two or more mutually exclusive projects is that the investments themselves may have different useful lives. Accordingly, we must compare projects with different useful lives over an **equal time span**, which may require further adjustments in our analysis. (Figure 6.5 is a flowchart showing the possible combinations of the analysis period and the useful life of an investment.) In the preceding examples (6.1 to 6.5), the project lives

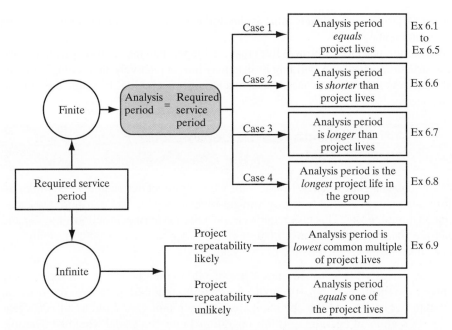

Figure 6.5   Analysis period implied in comparing mutually exclusive alternatives.

were all assumed to be equal, and the same as the study period. In the sections that follow, we will explore in more detail how to handle situations in which project lives differ from the analysis period and from one another. But we begin with the most straightforward situation: when the project lives and the analysis period coincide.

6.4.1 Analysis Period Differs From Project Lives

In Examples 6.1 to 6.5, we assumed the simplest scenario possible when analyzing mutually exclusive projects: The projects had useful lives equal to one another and to the required service period. In practice, this is seldom the case. Often, project lives do not match the required analysis period or do not match one another (or both). For example, two machines may perform exactly the same function, but one lasts longer than the other, and both of them last longer than the analysis period over which they are being considered. In the sections and examples that follow, we will develop some techniques for dealing with these complications.

Project's Life Is Longer Than Analysis Period

Project lives rarely conveniently coincide with a firm's predetermined required analysis period; they are often too long or too short. The case of project lives that are too long is the easier one to address.

Consider the case of a firm that undertakes a five-year production project when all of the alternative equipment choices have useful lives of seven years. In such a case, we analyze each project for only as long as the required service period (in this case, five years). We are then left with some unused portion of the equipment (in this case, two years' worth), which we include as salvage value in our analysis. **Salvage value** is the amount of money for which the equipment could be sold after its service to the project has been

Salvage value: The estimated value asset will realize upon its sale at the end of its useful life.

rendered. Alternatively, salvage value is the dollar measure of the remaining usefulness of the equipment.

A common instance of project lives that are longer than the analysis period occurs in the construction industry: A building project may have a relatively short completion time, but the equipment that is purchased has a much longer useful life. Example 6.6 illustrates this scenario.

EXAMPLE 6.6 Present-Worth Comparison: Project Lives Longer Than the Analysis Period

Waste Management Company (WMC) has won a contract that requires the firm to re-move radioactive material from government-owned property and transport it to a des-ignated dumping site. This task requires a specially made ripper–bulldozer to dig and load the material onto a transportation vehicle. Approximately 400,000 tonnes of waste must be moved in a period of two years.

- Model A costs $150,000 and has a life of 6000 hours before it will require any major overhaul. Two units of model A would be required to remove the material within two years, and the operating cost for each unit would run to $40,000/year for 2000 hours of operation. At this operational rate, the model would be operable for three years, at the end of which time it is estimated that the salvage value will be $25,000 for each machine.

- A more efficient model B costs $240,000 each, has a life of 12,000 hours without any major overhaul, and costs $22,500 to operate for 2000 hours per year to com-plete the job within two years. The estimated salvage value of model B at the end of six years is $30,000. Once again, two units of model B would be required to re-move the material within two years.

Since the lifetime of either model exceeds the required service period of two years (Figure 6.6), WMC has to assume some things about the used equipment at the end of that time. Therefore, the engineers at WMC estimate that, after two years, the model A units could be sold for $45,000 each and the model B units for $125,000 each. After considering all tax effects, WMC summarized the resulting cash flows (in thousand of dollars) for each project as follows:

Period	Model A		Model B	
0	−$300		−$480	
1	−80		−45	
2	−80	+90	−45	+250
3	−80	+50	−45	
4			−45	
5			−45	
6			−45	+60

Here, the figures in the boxes represent the estimated salvage values at the end of the analysis period (the end of year 2). Assuming that the firm's MARR is 15%, which option would be acceptable?

SOLUTION

Given: Cash flows for two alternatives as shown in Figure 6.6 and $i = 15\%$ per year. Find: NPW for each alternative and which alternative is preferred.

First, note that these are service projects, so we can assume the same revenues for both configurations. Since the firm explicitly estimated the market values of the assets at the end of the analysis period (two years), we can compare the two models directly. Because the benefits (removal of the waste) are equal, we can concentrate on the costs:

$$PW(15\%)_A = -\$300 - \$80(P/A, 15\%, 2) + \$90(P/F, 15\%, 2)$$
$$= -\$362;$$
$$PW(15\%)_B = -\$480 - \$45(P/A, 15\%, 2) + \$250(P/F, 15\%, 2)$$
$$= -\$364.$$

Model A has the least negative PW costs and thus would be preferred.

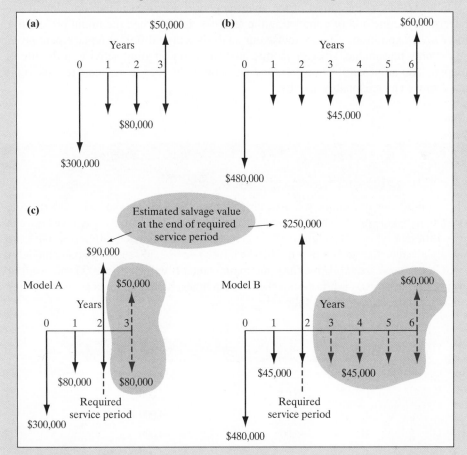

Figure 6.6 (a) Cash flow for model A; (b) cash flow for model B; (c) comparison of service projects with unequal lives when the required service period is shorter than the individual project life (Example 6.6).

Project's Life Is Shorter Than Analysis Period

When project lives are shorter than the required service period, we must consider how, at the end of the project lives, we will satisfy the rest of the required service period. Replacement projects—additional projects to be implemented when the initial project has reached the limits of its useful life—are needed in such a case. A sufficient number of replacement projects that match or exceed the required service period must be analyzed.

To simplify our analysis, we could assume that the replacement project will be exactly the same as the initial project, with the same costs and benefits. However, this assumption is not necessary. For example, depending on our forecasting skills, we may decide that a different kind of technology—in the form of equipment, materials, or processes—is a preferable replacement. Whether we select exactly the same alternative or a new technology as the replacement project, we are ultimately likely to have some unused portion of the equipment to consider as salvage value, just as in the case when the project lives are longer than the analysis period. Of course, we may instead decide to lease the necessary equipment or subcontract the remaining work for the duration of the analysis period. In this case, we can probably match our analysis period and not worry about salvage values.

In any event, at the outset of the analysis period, we must make some initial guess concerning the method of completing the analysis. Later, when the initial project life is closer to its expiration, we may revise our analysis with a different replacement project. This is only reasonable, since economic analysis is an ongoing activity in the life of a company and an investment project, and we should always use the most reliable, up-to-date data we can reasonably acquire.

EXAMPLE 6.7 Present-Worth Comparison: Project Lives Shorter Than the Analysis Period

The Smith Novelty Company, a mail-order firm, wants to install an automatic mailing system to handle product announcements and invoices. The firm has a choice between two different types of machines. The two machines are designed differently, but have identical capacities and do exactly the same job. The $12,500 semiautomatic model A will last three years, while the fully automatic model B will cost $15,000 and last four years. The expected cash flows for the two machines, including maintenance, salvage value, and tax effects, are as follows:

n	Model A	Model B
0	−$12,500	−$15,000
1	−5,000	−4,000
2	−5,000	−4,000
3	−5,000 + 2,000	−4,000
4		−4,000 + 1,500
5		

As business grows to a certain level, neither of the models may be able to handle the expanded volume at the end of year 5. If that happens, a fully computerized mail-order system will need to be installed to handle the increased business volume. In the scenario just presented, which model should the firm select at MARR = 15%?

SOLUTION

Given: Cash flows for two alternatives as shown in Figure 6.7, analysis period of five years, and $i = 15\%$.

Find: NPW of each alternative and which alternative to select.

Since both models have a shorter life than the required service period of five years, we need to make an explicit assumption of how the service requirement is to be met. Suppose that the company considers leasing equipment comparable to model A at an annual payment of $6000 (after taxes) and with an annual operating cost of $5000 for

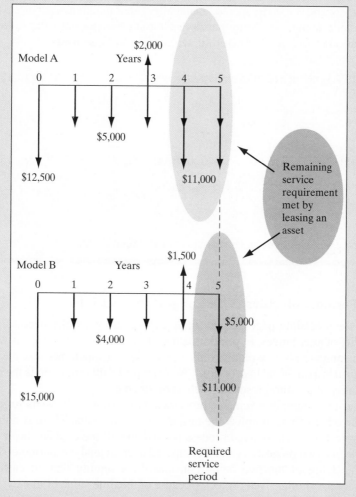

Figure 6.7 Comparison for service projects with unequal lives when the required service period is longer than the individual project life (Example 6.7).

the remaining required service period. In this case, the cash flow would look like that shown in Figure 6.7:

n	Model A	Model B
0	−$12,500	−$15,000
1	−5,000	−4,000
2	−5,000	−4,000
3	−5,000 + 2,000	−4,000
4	−5,000 − 6,000	−4,000 + 1,500
5	−5,000 − 6,000	−5,000 − 6,000

Here, the boxed figures represent the annual lease payments. (It costs $6000 to lease the equipment and $5000 to operate it annually. Other maintenance costs will be paid by the leasing company.) Note that both alternatives now have the same required service period of five years. Therefore, we can use NPW analysis:

$$PW(15\%)_A = -\$12,500 - \$5000(P/A, 15\%, 2) - \$3000(P/F, 15\%, 3)$$
$$- \$11,000(P/A, 15\%, 2)(P/F, 15\%, 3)$$
$$= -\$34,359.$$

$$PW(15\%)_B = -\$15,000 - \$4000(P/A, 15\%, 3) - \$2500(P/F, 15\%, 4)$$
$$- \$11,000(P/F, 15\%, 5)$$
$$= -\$31,031.$$

Since these are service projects, model B is the better choice.

Analysis Period Coincides With Longest Project Life

As seen in the preceding pages, equal future periods are generally necessary to achieve comparability of alternatives. In some situations, however, revenue projects with different lives can be compared if they require only a one-time investment because the task or need within the firm is a one-time task or need. An example of this situation is the extraction of a fixed quantity of a natural resource such as oil or coal.

Consider two mutually exclusive processes: One requires 10 years to recover some coal, and the other can accomplish the task in only eight years. There is no need to continue the project if the short-lived process is used and all the coal has been retrieved. In this example, the two processes can be compared over an analysis period of 10 years (the longest project life of the two being considered), assuming that no cash flows after eight years for the shorter lived project. Because of the time value of money, the revenues must be included in the analysis even if the price of coal is constant. Even if the total (undiscounted) revenue is equal for either process, that for the faster process has a larger present worth. Therefore, the two projects could be compared by using the NPW of each over its own life. Note that in this case the analysis period is determined by, and coincides with, the longest project life in the group. (Here we are still, in effect, assuming an analysis period of 10 years.)

EXAMPLE 6.8 Present-Worth Comparison: A Case Where the Analysis Period Coincides With the Project With the Longest Life in the Mutually Exclusive Group

The family-operated Foothills Ranching Company (FRC) owns the mineral rights to land used for growing grain and grazing cattle. Recently, oil was discovered on this property. The family has decided to extract the oil, sell the land, and retire. The company can lease the necessary equipment and extract and sell the oil itself, or it can lease the land to an oil-drilling company:

- **Drill option.** If the company chooses to drill, it will require $300,000 leasing expenses up front, but the net annual cash flow after taxes from drilling operations will be $600,000 at the end of each year for the next five years. The company can sell the land for a net cash flow of $1,000,000 in five years, when the oil is depleted.
- **Lease option.** If the company chooses to lease, the drilling company can extract all the oil in only three years, and FRC can sell the land for a net cash flow of $800,000 at that time. (The difference in resale value of the land is due to the increasing rate of land appreciation anticipated for this property.) The net cash flow from the lease payments to FRC will be $630,000 at the *beginning* of each of the next three years.

All benefits and costs associated with the two alternatives have been accounted for in the figures listed. Which option should the firm select at $i = 15\%$?

SOLUTION

Given: Cash flows shown in Figure 6.8 and $i = 15\%$ per year.

Find: NPW of each alternative and which alternative to select.

As illustrated in Figure 6.8, the cash flows associated with each option look like this:

n	Drill	Lease
0	−$300,000	$630,000
1	600,000	630,000
2	600,000	630,000
3	600,000	800,000
4	600,000	
5	1,600,000	

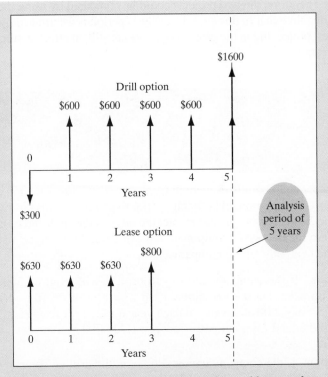

Figure 6.8 Comparison of revenue projects with unequal lives when the analysis period coincides with the project with the longest life in the mutually exclusive group (Example 6.8). In our example, the analysis period is five years, assuming no cash flow in years 4 and 5 for the lease option.

After depletion of the oil, the project will terminate. The present worth of each of the two options is as follows:

$$PW(15\%)_{Drill} = -\$300,000 + \$600,000(P/A, 15\%, 4)$$
$$+ \$1,600,000(P/F, 15\%, 5)$$
$$= \$2,208,470.$$

$$PW(15\%)_{Lease} = \$630,000 + \$630,000(P/A, 15\%, 2)$$
$$+ \$800,000(P/F, 15\%, 3)$$
$$= \$2,180,210.$$

Note that these are revenue projects; therefore, the drill option appears to be the marginally better option.

COMMENTS: The relatively small difference between the two NPW amounts ($28,260) suggests that the actual decision between drilling and leasing might be based on noneconomic issues. Even if the drilling option were slightly better, the company might prefer to forgo the small amount of additional income and select the lease option, rather than undertake an entirely new business venture and do its own drilling. A variable that might also have a critical effect on this decision is the sales value of the land in each alternative. The value of land is often difficult to forecast over any long period, and the firm may feel some uncertainty about the accuracy of its guesses. In Chapter 15, we will discuss sensitivity analysis, a method by which we can factor uncertainty about the accuracy of project cash flows into our analysis.

6.4.2 Analysis Period Is Not Specified

Our coverage so far has focused on situations in which an analysis period is known. When an analysis period is not specified, either explicitly by company policy or practice or implicitly by the projected life of the investment project, it is up to the analyst to choose an appropriate one. In such a case, the most convenient procedure is to choose an analysis on the basis of the useful lives of the alternatives. When the alternatives have equal lives, this is an easy selection. When the lives of at least some of the alternatives differ, we must select an analysis period that allows us to compare projects with different lifetimes on an equal time basis—that is, a **common service period**.

Lowest Common Multiple of Project Lives

A required service period of infinity may be assumed if we anticipate that an investment project will be proceeding at roughly the same level of production for some indefinite period. It is certainly possible to make this assumption mathematically, although the analysis is likely to be complicated and tedious. Therefore, in the case of an indefinitely ongoing investment project, we typically select a finite analysis period by using the **lowest common multiple** of project lives. For example, if alternative A has a three-year useful life and alternative B has a four-year life, we may select 12 years as the analysis or common service period. We would consider alternative A through four life cycles and alternative B through three life cycles; in each case, we would use the alternatives completely. We then accept the finite model's results as a good prediction of what will be the economically wisest course of action for the foreseeable future. The next example is a case in which we conveniently use the lowest common multiple of project lives as our analysis period.

Lowest common multiple of two numbers is the lowest number that can be divided by *both*.

EXAMPLE 6.9 Present-Worth Comparison: Unequal Lives, Lowest-Common-Multiple Method

Consider Example 6.7. Suppose that models A and B can each handle the increased future volume and that the system is not going to be phased out at the end of five years. Instead, the current mode of operation is expected to continue indefinitely. Suppose also that the two models will be available in the future without significant changes in price and operating costs. At MARR = 15%, which model should the firm select?

SOLUTION

Given: Cash flows for two alternatives as shown in Figure 6.9, $i = 15\%$ per year, and an indefinite period of need.

Find: NPW of each alternative and which alternative to select.

Recall that the two mutually exclusive alternatives have different lives, but provide identical annual benefits. In such a case, we ignore the common benefits and can make the decision solely on the basis of costs, as long as a common analysis period is used for both alternatives.

To make the two projects comparable, let's assume that, after either the three- or four-year period, the system would be reinstalled repeatedly, using the same model, and that the same costs would apply. The lowest common multiple of 3 and 4 is 12, so we will use 12 years as the common analysis period. Note that any cash flow difference between the alternatives will be revealed during the first 12 years. After that, the same cash flow pattern repeats every 12 years for an indefinite period. The replacement cycles and cash flows are shown in Figure 6.9. Here is our analysis:

- **Model A.** Four replacements occur in a 12-year period. The PW for the first investment cycle is

$$PW(15\%) = -\$12,500 - \$5000(P/A, 15\%, 3)$$
$$+ \$2000(P/F, 15\%, 3)$$
$$= -\$22,601.$$

With four replacement cycles, the total PW is

$$PW(15\%) = -\$22,601[1 + (P/F, 15\%, 3)$$
$$+ (P/F, 15\%, 6) + (P/F, 15\%, 9)]$$
$$= -\$53,657.$$

- **Model B.** Three replacements occur in a 12-year period. The PW for the first investment cycle is

$$PW(15\%) = -\$15,000 - \$4000(P/A, 15\%, 4)$$
$$+ \$1500(P/F, 15\%, 4)$$
$$= -\$25,562.$$

With three replacement cycles in 12 years, the total PW is

$$PW(15\%) = -\$25,562[1 + (P/F, 15\%, 4) + (P/F, 15\%, 8)]$$
$$= -\$48,534.$$

Clearly, model B is a better choice, as before.

COMMENTS: In Example 6.9, an analysis period of 12 years seems reasonable. The number of actual reinvestment cycles needed with each type of system will depend on the technology of the future system, so we may or may not actually need the four reinvestment cycles (model A) or three (model B) we used in our analysis. The validity of the analysis also depends on the costs of the system and labour remaining

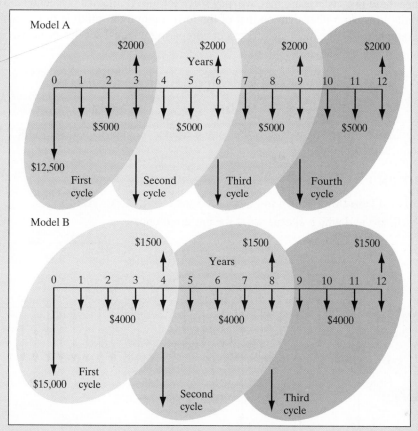

Figure 6.9 Comparison of projects with unequal lives when the required service period is infinite and the project is likely to be repeatable with the same investment and operations and maintenance costs in all future replacement cycles (Example 6.9).

constant. If we assume **constant-dollar prices** (see Chapter 14), this analysis would provide us with a reasonable result. (As you will see in Example 6.10, the annual-worth approach makes it mathematically easier to solve this type of comparison.) If we cannot assume constant-dollar prices in future replacements, we need to estimate the costs for each replacement over the analysis period. This will certainly complicate the problem significantly.

AE Analysis for Unequal Project Lives

Annual equivalent analysis also requires establishing common analysis periods, but AE offers some computational advantages as opposed to present worth analysis, provided the following criteria are met:

1. The service of the selected alternative is required on a continuous basis.
2. Each alternative will be replaced by an *identical* asset that has the same costs and performance.

When these two criteria are met, we may solve for the AE of each project on the basis of its initial lifespan, rather than on that of the lowest common multiple of the projects' lives. Example 6.10 illustrates the process of comparing unequal service projects.

EXAMPLE 6.10 Annual Equivalent Cost Comparison: Unequal Project Lives

Continuing Example 6.9, we apply the annual equivalent approach to select the most economical equipment.

SOLUTION

Given: Cost cash flows shown in Figure 6.10 and $i = 15\%$ per year.
Find: AE cost and which alternative is the preferred one.

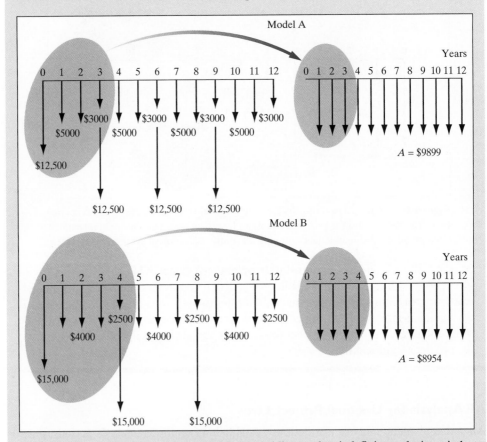

Figure 6.10 Comparison of projects with unequal lives and an indefinite analysis period using the annual equivalent-worth criterion (Example 6.10).

An alternative procedure for solving Example 6.10 is to compute the annual equivalent cost of an outlay of $12,500 for model A every three years and the annual equivalent cost of an outlay of $15,000 for model B every four years. Notice that the AE of each 12-year cash flow is the same as that of the corresponding three- or four-year cash flow (Figure 6.10). From Example 6.9, we calculate

- **Model A:** For a 3-year life,

$$PW(15\%) = \$22,601$$

$$AE(15\%) = 22,601(A/P, 15\%, 3)$$

$$= \$9899.$$

For a 12-year period (the entire analysis period),

$$PW(15\%) = \$53,657$$

$$AE(15\%) = 53,657(A/P, 15\%, 12)$$

$$= \$9899.$$

- **Model B:** For a 4-year life,

$$PW(15\%) = \$25,562$$

$$AE(15\%) = \$25,562(A/P, 15\%, 4)$$

$$= \$8954.$$

For a 12-year period (the entire analysis period),

$$PW(15\%) = \$48,534$$

$$AE(15\%) = \$48,534(A/P, 15\%, 12)$$

$$= \$8954.$$

Notice that the annual equivalent values that were calculated on the basis of the common service period are the same as those that were obtained over their initial lifespans. Thus, for alternatives with unequal lives, we will obtain the same selection by comparing the NPW over a common service period using repeated projects or by comparing the AE for initial lives.

IRR Analysis for Unequal Project Lives

Above, we discussed the use of the PW and AE criteria as bases for comparing projects with unequal lives. The IRR measure can also be used to compare projects with unequal lives, as long as we can establish a common analysis period. The decision procedure is then exactly the same as for projects with equal lives. It is likely, however, that we will have a multiple-root problem, which creates a substantial computational burden. For example, suppose we apply the IRR measure to a case in which one project has a five-year life and the other project has an eight-year life, resulting in a lowest common multiple of 40 years. When we determine the incremental cash flows over the analysis period, we are bound to observe many sign changes. This leads to the possibility of having many

$i*$s. Example 6.11 uses $i*$ to compare mutually exclusive projects, in which the incremental cash flows reveal several sign changes. (Our purpose is not to encourage you to use the IRR approach to compare projects with unequal lives. Rather, it is to show the correct way to compare them if the IRR approach must be used.)

EXAMPLE 6.11 IRR Analysis for Projects With Different Lives in Which the Increment Is a Nonsimple Investment

Consider Example 6.9, in which a mail-order firm wants to install an automatic mailing system to handle product announcements and invoices. The firm had a choice between two different types of machines. Using the IRR as a decision criterion, select the best machine. Assume a MARR of 15%, as before.

SOLUTION

Given: Cash flows for two projects with unequal lives, as shown in Figure 6.9, and MARR = 15%.

Find: The alternative that is preferable.

Since the analysis period is equal to the least common multiple of 12 years, we may compute the incremental cash flow over this 12-year period. As shown in Figure 6.11, we subtract the cash flows of model A from those of model B to form the increment of investment. (Recall that we want the first cash flow difference to be a negative value.) We can then compute the IRR on this incremental cash flow.

Five sign changes occur in the incremental cash flows, indicating a nonsimple incremental investment:

n	Model A		Model B		Model B − Model A
0	−$12,500		−$15,000		−$2,500
1		−5,000		−4,000	1,000
2		−5,000		−4,000	1,000
3	−12,500	−3,000		−4,000	11,500
4		−5,000	−15,000	−2,500	−12,500
5		−5,000		−4,000	1,000
6	−12,500	−3,000		−4,000	11,500
7		−5,000		−4,000	1,000
8		−5,000	−15,000	−2,500	−12,500
9	−12,500	−3,000		−4,000	11,500
10		−5,000		−4,000	1,000
11		−5,000		−4,000	1,000
12		−3,000		−2,500	500
	Four replacement cycles		Three replacement cycles		Incremental cash flows

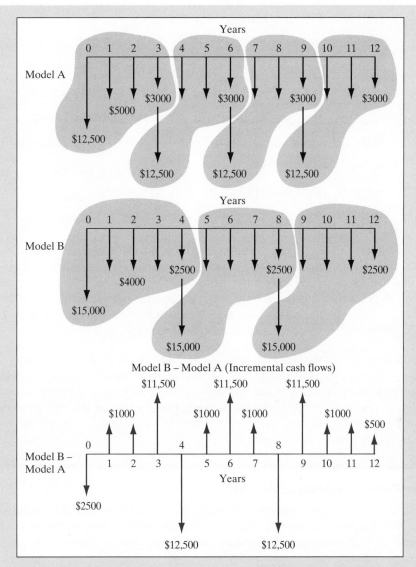

Figure 6.11 Comparison of projects with unequal lives (Example 6.11).

Even though there are five sign changes in the cash flow, there is only one positive i^* for this problem: 63.12%. Unfortunately, however, the investment is not a pure investment. We need to employ an external rate to compute the IRR in order to make a proper accept–reject decision (see Appendix 5A.5). Assuming that the firm's MARR is 15%, we will use a trial-and-error approach. Try $i = 20\%$:

$$PB(20\%, 15\%)_0 = -\$2500,$$

$$PB(20\%, 15\%)_1 = -\$2500(1.20) + \$1000 = -\$2000.$$

$$PB(20\%, 15\%)_2 = -\$2000(1.20) + \$1000 = -\$1400,$$

$$PB(20\%, 15\%)_3 = -\$1400(1.20) + \$11{,}500 = \$9820.$$

$$PB(20\%, 15\%)_4 = \$9820(1.15) - \$12,500 = -\$1207$$

$$PB(20\%, 15\%)_5 = -\$1207(1.20) + \$1000 = -\$448.40$$

$$PB(20\%, 15\%)_6 = -\$448.40(1.20) + \$11,500 = \$10,961.92$$

$$PB(20\%, 15\%)_7 = \$10,961.92(1.15) + \$1000 = \$13,606.21$$

$$PB(20\%, 15\%)_8 = \$13,606.21(1.15) - \$12,500 = \$3147.14$$

$$PB(20\%, 15\%)_9 = \$3147.14(1.15) + \$11,500 = \$15,119.21$$

$$PB(20\%, 15\%)_{10} = \$15,119.21(1.15) + \$1000 = \$18,387.09$$

$$PB(20\%, 15\%)_{11} = \$18,387.09(1.15) + \$1000 = \$22,145.16$$

$$PB(20\%, 15\%)_{12} = \$22,145.16(1.15) + \$500 = \$25,966.93$$

Since $PB(20\%, 15\%)_{12} > 0$, the guessed 20% is not the return on invested capital (RIC). We may increase the value of i and repeat the calculations. After several trials, we find that the RIC is 50.68%. Since $IRR_{B-A} > MARR$, model B would be selected, which is consistent with the NPW analysis. In other words, the additional investment over the years to obtain model B ($-\$2500$ at $n = 0$, $-\$12,500$ at $n = 4$, and $-\$12,500$ at $n = 8$) yields a satisfactory rate of return.

COMMENTS: Given the complications inherent in IRR analysis in comparing alternative projects, it is usually more desirable to employ one of the other equivalence techniques for this purpose. As an engineering manager, you should keep in mind the intuitive appeal of the rate-of-return measure. Once you have selected a project on the basis of NPW or AE analysis, you may also wish to express its worth as a rate of return, for the benefit of your associates.

Other Common Analysis Periods

In some cases, the lowest common multiple of project lives is an unwieldy analysis period to consider. Suppose, for example, that you were considering two alternatives with lives of 7 and 12 years, respectively. Besides making for tedious calculations, an 84-year analysis period may lead to inaccuracies, since, over such a long time, we can be less and less confident about the ability to install identical replacement projects with identical costs and benefits. In a case like this, it would be reasonable to use the useful life of one of the alternatives by either factoring in a replacement project or salvaging the remaining useful life, as the case may be. The important idea is that we must compare both projects on the same time basis.

6.5 Make-or-Buy Decision

Make-or-buy problems are among the most common of business decisions, and are therefore one of the most general applications of mutually exclusive economic evaluations. At any given time, a firm may have the option of either buying an item or producing it. Unlike the make-or-buy situation we will consider in Chapter 7, *if either the "make" or the "buy" alternative requires the acquisition of machinery or equipment, then it becomes an investment decision.* Since the cost of an outside service (the "buy" alternative) is usually quoted in terms of dollars per unit, it is easier to compare the two alternatives if the differential costs of the "make" alternative are also given in dollars per unit. This unit cost comparison requires the use of annual-worth analysis (see Section 5.5.5). The specific procedure is as follows:

Step 1. Determine the time span (planning horizon) for which the part (or product) will be needed.

Step 2. Determine the annual quantity of the part (or product).

Step 3. Obtain the unit cost of purchasing the part (or product) from the outside firm.

Step 4. Determine the equipment, manpower, and all other resources required to make the part (or product) in-house.

Step 5. Estimate the net cash flows associated with the "make" option over the planning horizon.

Step 6. Compute the annual equivalent cost of producing the part (or product).

Step 7. Compute the unit cost of making the part (or product) by dividing the annual equivalent cost by the required annual volume.

Step 8. Choose the option with the minimum unit cost.

EXAMPLE 6.12 Equivalent Worth: Outsourcing the Manufacture of Hard Drive Cases

Seagate Ltd. currently produces both the electronic storage media for hard drives and the cases for commercial use. An increased demand for hard drives is projected, and Seagate is deciding between increasing the internal production of hard drive cases or purchasing empty cases from an outside vendor. If Seagate purchases the cases from a vendor, the company must also buy specialized equipment to insert the electronic media, since its current loading machine is not compatible with the cases produced by the vendor under consideration. The projected production rate of hard drives is 79,815 units per week for 48 weeks of operation per year. The planning horizon is seven years. The accounting department has itemized the annual costs associated with each option as follows:

• **Make option (annual costs):**

Labour	$1,445,633
Materials	$2,048,511
Incremental overhead	$1,088,110
Total annual cost	$4,582,254

• **Buy option:**

Capital expenditure	
Acquisition of a new loading machine	$ 405,000
Salvage value at end of seven years	$ 45,000
Annual Operating Costs	
Labour	$ 251,956
Purchasing empty hard drive cases ($0.85/unit)	$3,256,452
Incremental overhead	$ 822,719
Total annual operating costs	$4,331,127

(Note the conventional assumption that cash flows occur in discrete lumps at the ends of years, as shown in Figure 6.12.) Assuming that Seagate's MARR is 14%, calculate the unit cost under each option.

SOLUTION

Given: Cash flows for two options and $i = 14\%$.
Find: Unit cost for each option and which option is preferred.

The required annual production volume is

$$79{,}815 \text{ units/week} \times 48 \text{ weeks} = 3{,}831{,}120 \text{ units per year.}$$

We now need to calculate the annual equivalent cost under each option:

• **Make option.** Since the "make option" is already given on an annual basis, the equivalent annual cost will be

$$\text{AEC}(14\%)_{\text{Make}} = \$4{,}582{,}254$$

• **Buy option.** The two cost components are capital cost and operating cost.
Capital cost:

$$\text{CR}(14\%) = (\$405{,}000 - \$45{,}000)(A/P, 14\%, 7)$$
$$+ (0.14)(\$45{,}000)$$
$$= \$90{,}249$$

$$\text{AEC}(14\%)_1 = CR(14\%) = \$90{,}249$$

Operating cost:

$$\text{AEC}(14\%)_2 = \$4{,}331{,}127$$

Total annual equivalent cost:

$$\text{AEC}(14\%)_{\text{Buy}} = \text{AEC}(14\%)_1 + \text{AEC}(14\%)_2 = \$4{,}421{,}376$$

Obviously, this annual equivalent calculation indicates that Seagate would be better off buying cases from the outside vendor. However, Seagate wants to know the unit costs in

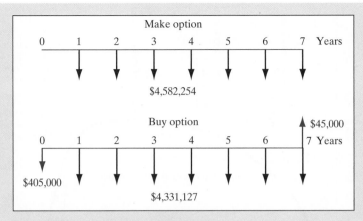

Figure 6.12 Make-or-buy analysis.

order to set a price for the product. In such a situation, we need to calculate the unit cost of producing the hard drives under each option. We do this by dividing the magnitude of the annual equivalent cost for each option by the annual quantity required:

- **Make option:**

$$\text{Unit cost} = \$4,582,254/3,831,120 = \$1.20/\text{unit}.$$

- **Buy option:**

$$\text{Unit cost} = \$4,421,376/3,831,120 = \$1.15/\text{unit}.$$

Buying the empty hard drive cases from the outside vendor and loading the electronic media in-house will save Seagate 5 cents per hard drive before any consideration of taxes.

COMMENTS: Two important noneconomic factors should also be considered. The first is the question of whether the quality of the supplier's component is better than, equal to, or worse than the component the firm is presently manufacturing. The second is the reliability of the supplier in terms of providing the needed quantities of the hard drive cases on a timely basis. A reduction in quality or reliability should virtually always rule out buying.

6.6 Life-Cycle Cost Analysis

Because 80% of the total life-cycle cost of a system occurs after the system has entered service, the best long-term system acquisition and support decisions are based on a full understanding of the total cost of acquiring, operating, and supporting the system (Figure 6.13). **Life-cycle cost analysis** (LCCA) enables the analyst to make sure that the selection of a design alternative is not based solely on the lowest initial costs, but also takes into account all the future costs over the project's usable life. Some of the unique features of LCCA are as follows:

- LCCA is used appropriately only to select from among design alternatives that would yield the same level of performance or benefits to the project's users during normal operations. If benefits vary among the design alternatives, then the alternatives cannot

Life-cycle cost analysis is useful when project alternatives fulfill the same performance requirements, but differ with respect to initial costs and operating costs.

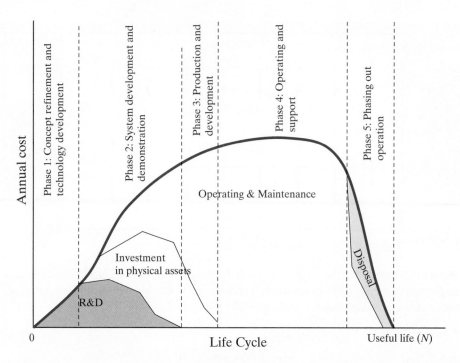

Figure 6.13 Stages of life-cycle cost. These include the concept refinement and technology development phase, the system development and demonstration phase, the production and deployment phase, the operating phase, and the disposal phase.

be compared solely on the basis of cost. Rather, the analyst would need to employ present-worth analysis or benefit–cost analysis (BCA), which measures the monetary value of life-cycle benefits as well as costs. BCA is discussed in Chapter 13.

- LCCA is a way to predict the most cost-effective solution; it does not guarantee a particular result, but allows the plant designer or manager to make a reasonable comparison among alternative solutions within the limits of the available data.

- To make a fair comparison, the plant designer or manager might need to consider the measure used. For example, the same process output volume should be considered, and if the two items being examined cannot give the same output volume, it may be appropriate to express the figures in cost per unit of output (e.g., $/tonne, or euro/kg), which requires a calculation of annual equivalent dollars generated. This calculation is based on the annual output.

In many situations, we need to compare a set of different design alternatives, each of which would produce the same number of units (constant revenues), but would require different amounts of investment and operating costs (because of different degrees of mechanization). Example 6.13 illustrates the use of the annual equivalent-cost concept to compare the cost of operating an existing pumping system with that of an improved system.

EXAMPLE 6.13 Pumping System With a Problem Valve[4]

Consider a single-pump circuit that transports a process fluid containing some solids from a storage tank to a pressurized tank. A heat exchanger heats the fluid, and a control valve regulates the rate of flow into the pressurized tank to 80 cubic metres per hour (m^3/h). The process is depicted in Figure 6.14.

The plant engineer is experiencing problems with a fluid control valve (FCV) that fails due to erosion caused by cavitations. The valve fails every 10 to 12 months at a cost of $4000 per repair. A change in the control valve is being considered: Replace the existing valve with one that can resist cavitations. Before the control valve is repaired again, the project engineer wants to look at other options and perform an LCCA on alternative solutions.

Engineering Solution Alternatives

The first step is to determine how the system is currently operating and why the control valve fails. Then the engineer can see what can be done to correct the problem. The control valve currently operates between 15% and 20% open and with considerable cavitation noise from the valve. It appears that the valve was not sized properly for the application. After reviewing the original design calculations, it was discovered that the pump was oversized: 110 m^3/h instead of 80 m^3/h. This resulted in a larger pressure drop across the control valve than was originally intended.

As a result of the large differential pressure at the operating rate of flow, and because the valve is showing cavitation damage at regular intervals, the engineer determines that the control valve is not suitable for this process.

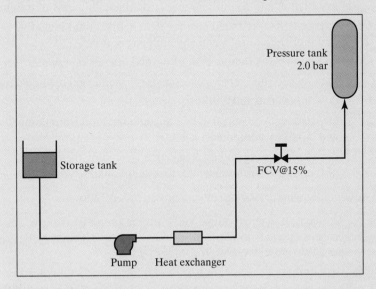

Figure 6.14 Sketch of a pumping system in which the control valve fails.

[4] *Pump Life Cycle Costs: A Guide to LCC Analysis for Pumping Systems.* DOE/GO-102001-1190. December 2000, Office of Industrial Technologies, U.S. Department of Energy and Hydraulic Institute.

The following four options are suggested:

- **Option A.** A new control valve can be installed to accommodate the high pressure differential.
- **Option B.** The pump impeller can be trimmed so that the pump does not develop as much head, resulting in a lower pressure drop across the current valve.
- **Option C.** A variable-frequency drive (VFD) can be installed and the flow control valve removed. The VFD can vary the pump speed and thus achieve the desired process flow.
- **Option D.** The system can be left as it is, with a yearly repair of the flow control valve to be expected.

Cost Summary

Pumping systems often have a lifespan of 15 to 20 years. Some cost elements will be incurred at the outset, and others will be incurred at different times throughout the lives of the different solutions evaluated. Therefore, you need to calculate a present value of the LCC to accurately assess the different solutions.

Some of the major LCC elements related to a typical pumping system are summarized as follows:

$$LCC = C_{ic} + C_{in} + C_e + C_o + C_m + C_s + C_{env} + C_d$$

LCC = life-cycle cost

C_{ic} = initial costs, purchase price (costs of pump, system, pipe, auxiliary services)

C_{in} = installation and commissioning costs (including cost of training)

C_e = energy costs (predicted cost for system operation, including cost of pump drive, controls, and any auxiliary service)

C_o = operation costs (labour cost of normal system supervision)

C_m = maintenance and repair costs (costs of routine and predicted repairs)

C_s = downtime costs (cost of loss of production)

C_{env} = environmental costs (cost due to contamination from pumped liquid and auxiliary equipment)

C_d = decommissioning and disposal costs (including the cost of restoration of the local environment and disposal of auxiliary services)

For each option, the major cost elements identified are as follows:

- **Option A.** The cost of a new control valve that is properly sized is $5000. The process operates at 80 m³/h for 6000 h/year. The energy cost is $0.08 per kWh and the motor efficiency is 90%.
- **Option B.** By trimming the impeller to 375 mm, the pump's total head is reduced to 42.0 m at 80 m³/h. This drop in pressure reduces the differential pressure across the control valve to less than 10 m, which better matches the valve's original design intent. The resulting annual energy cost with the smaller impeller is $6720 per year. It costs $2250 to trim the impeller. This cost includes the machining cost as well as the cost to disassemble and reassemble the pump.

TABLE 6.1 Cost Comparison for Options A Through D in the System With a Failing Control Valve

Cost	Change Control Valve (A)	Trim Impeller (B)	VFD (C)	Repair Control Valve (D)
Pump Cost Data				
Impeller diameter	430 mm	375 mm	430 mm	430 mm
Pump head	71.7 m	42.0 m	34.5 m	71.7 m
Pump efficiency	75.1%	72.1%	77%	75.1%
Rate of flow	80 m^3/h	80 m^3/h	80 m^3/h	80 m^3/h
Power consumed	23.1 kW	14.0 kW	11.6 kW	23.1 kW
Energy cost/year	$11,088	$6,720	$ 5,568	$11,088
New valve	$ 5,000	0	0	0
Modify impeller	0	$2,250	0	0
VFD	0	0	$20,000	0
Installation of VFD	0	0	$ 1,500	
Valve repair/year	0	0	0	$ 4,000

- **Option C.** A 30-kW VFD costs $20,000 and an additional $1500 to install. The VFD will cost $500 to maintain each year. It is assumed that it will not need any repairs over the project's eight-year life.
- **Option D.** The option to leave the system unchanged will result in a yearly cost of $4000 for repairs to the cavitating flow control value.

Table 6.1 summarizes financial as well as technical data related to the various options.

SOLUTION

Given: Financial data as summarized in Table 6.1.

Find: Which design option to choose.

Assumptions:

- The current energy price is $0.08/kWh.
- The process is operated for 6000 hours/year.
- The company has a cost of $500 per year for routine maintenance of pumps of this size, with a repair cost of $2500 every second year.
- There is no decommissioning cost or environmental disposal cost associated with this project.
- The project has an eight-year life.
- The interest rate for new capital projects is 8%, and an inflation rate of 4% is expected.

A sample LCC calculation for Option A is shown in Table 6.2. Note that the energy cost and other cost data are escalated at the annual rate of 4%. For example, the

TABLE 6.2 LCC Calculation for Option A

	A	B	C	D	E	F	G	H	I	J
1	**Calculation Chart for LCC: Option A**									
2										
3	Year No	0	1	2	3	4	5	6	7	8
4	Initial investment cost:	$5,000								
5	Installation and commissioning cost:	$0								
6										
7	Energy cost/year (calculated)		$11,532	$11,993	$12,472	$12,971	$13,490	$14,030	$14,591	$15,175
8	Routine maintenance		$520	$541	$562	$585	$608	$633	$658	$684
9	Repair cost every 2nd year			$2,704		$2,925		$3,163		$3,421
10	Operating costs		$0	$0	$0	$0	$0	$0	$0	$0
11	Other costs		$0	$0	$0	$0	$0	$0	$0	$0
12										
13	Downtime costs		$0	$0	$0	$0	$0	$0	$0	$0
14										
15	Decommissioning costs							$0		
16	Environmental & disposal costs							$0		
17										
18	**Sum of costs**	$5,000	$12,052	$15,238	$13,035	$16,481	$14,099	$17,826	$15,249	$19,280
19										
20	**Present costs**	$5,000	$11,159	$13,064	$10,348	$12,114	$9,595	$11,233	$8,898	$10,417
21	Present costs, energy		$10,677	$10,282	$9,901	$9,534	$9,181	$8,841	$8,514	$8,198
22	Present costs, routine maintenance		$481	$464	$446	$430	$414	$399	$384	$370
23										
24	**Sum of present costs**	**$91,827**								
25	Sum of energy costs	$75,129								
26	Sum of routine maintenance costs	$3,388								
27										
28	**Annual equivalent cost (8%)**	$15,979								
29	Cost per operating hour	**$2.66**								

current estimate of energy cost is $11,088. To find the cost at the end of year 1, we multiply $11,088 by $(1 + 0.04)$ yielding $11,532. For year 2, we multiply $11,532 by $(1 + 0.04)$ to obtain $11,993. Once we calculate the LCC in present worth ($91,827), we find the equivalent annual value ($15,979) at 8% interest. Finally, to calculate the cost per hour, we divide $15,979 by 6000 hours, resulting in $2.66. We can calculate the unit costs for other options in a similar fashion. (See Table 6.3.)

Option B, trimming the impeller, has the lowest life-cycle cost ($1.73 per hour) and is the preferred option for this example.

TABLE 6.3 Comparison of LCC for Options A Through D

	Option A Change Control Valve	Option B Trim Impeller	Option C VFD and Remove Control Valve	Option D Repair Control Valve
Input				
Initial investment cost:	$5,000	$2,250	$21,500	0
Energy price (present) per kWh:	0.080	0.080	0.080	0.080
Weighted average power of equipment in kW:	23.1	14.0	11.6	23.1
Average operating hours/year:	6,000	6,000	6,000	6,000
Energy cost/year (calculated) + energy price × weighted average power × average operating hours/year:	11,088	6,720	5,568	11,088
Maintenance cost (routine maintenance/year):	500	500	1,000	500
Repair every second year:	2,500	2,500	2,500	2,500
Other yearly costs:	0	0	0	4,000
Downtime cost/year:	0	0	0	0
Environmental cost:	0	0	0	0
Decommissioning/disposal (salvage) cost:	0	0	0	0
Lifetime in years:	8	8	8	8
Interest rate (%):	8.0%	8.0%	8.0%	8.0%
Inflation rate (%):	4.0%	4.0%	4.0%	4.0%
Output				
Present LCC value:	$91,827	$59,481	$74,313	$113,930
Cost per operating hour	**$2.66**	**$1.73**	**$2.16**	**$3.30**

6.7 Design Economics

Engineers are frequently involved in making design decisions that provide the required functional quality at the lowest cost. Another valuable extension of AE analysis is minimum-cost analysis, which is used to determine optimal engineering designs. The AE analysis method is useful when two or more cost components are affected differently by the same design element (i.e., for a single design variable, some costs may increase while others decrease). When the equivalent annual total cost of a design variable is a function of increasing and decreasing cost components, we can usually find the optimal value that will minimize the cost of the design with the formula

$$\text{AEC}(i) = a + bx + \frac{c}{x}, \tag{6.1}$$

where x is a common design variable and a, b, and c are constants.

To find the value of the common design variable that minimizes AE(i), we need to take the first derivative, equate the result to zero, and solve for x:

$$\frac{d\text{AEC}(i)}{dx} = b - \frac{c}{x^2}$$
$$= 0$$

$$x = x^* = \sqrt{\frac{c}{b}}. \tag{6.2}$$

(It is common engineering practice to denote the optimal solution with an asterisk.)

The logic of the first-order requirement is that an activity should, if possible, be carried to a point where its **marginal yield** $d\text{AEC}(i)/dx$ is zero. However, to be sure whether we have found a maximum or a minimum when we have located a point whose marginal yield is zero, we must examine it in terms of what are called the **second-order conditions**. Second-order conditions are equivalent to the usual requirements that the second derivative be negative in the case of a maximum and positive for a minimum. In our situation,

$$\frac{d^2\text{AEC}(i)}{dx^2} = \frac{2C}{x^3}.$$

As long as $C > 0$, the second derivative will be positive, indicating that the value x^* is the minimum-cost point for the design alternative. To illustrate the optimization concept, two examples are given, the first having to do with designing the optimal cross-sectional area of a conductor and the second dealing with selecting an optimal size for a pipe.

EXAMPLE 6.14 Optimal Cross-Sectional Area

A constant electric current of 5000 amps is to be transmitted a distance of 300 metres from a power plant to a substation for 24 hours a day, 365 days a year. A copper conductor can be installed for $16.50 per kilogram. The conductor will have an estimated life of 25 years and a salvage value of $1.65 per kilogram. The power loss from a conductor is inversely proportional to the cross-sectional area A of the conductor.

The resistance of a one square centimetre cross-sectional area conductor is 1.7241×10^{-4} Ω per metre. The cost of energy is \$0.0375 per kilowatt-hour, the interest rate is 9%, and the density of copper is 8894 kilograms per cubic metre. For the data given, calculate the optimum cross-sectional area (A) of the conductor.

DISCUSSION: The resistance of the conductor is the most important cause of power loss in a transmission line. The resistance of an electrical conductor varies directly with its length and inversely with its cross-sectional area according to the equation

$$R = \rho\left(\frac{L}{A}\right), \tag{6.3}$$

where R is the resistance of the conductor, L is the length of the conductor, A is the cross-sectional area of the conductor, and ρ is the resistivity of the conductor material.

A consistent set of units must be used. In SI units (the official designation for the Système International d'Unités system of units), L is in metres, A in square metres, and ρ in ohm-metres. In terms of the SI units, the copper conductor has a ρ value of 1.7241×10^{-8} Ω-metre, or 1.7241×10^{-4} Ω-cm²/m.

When current flows through a circuit, power is used to overcome resistance. The unit of electrical work is the kilowatt hour (kWh), which is equal to the power in kilowatts, multiplied by the hours during which work is performed. If the current (I) is steady, the energy loss through a conductor in time T can be expressed as

$$\text{Energy loss} = \frac{I^2RT}{1000} \text{ kWh} \tag{6.4}$$

where I is the current in amperes, R is the resistance in ohms (Ω), and T is the duration in hours during which work is performed.

SOLUTION

Given: Cost components as a function of cross-sectional area (A), $N = 25$ years, and $i = 9\%$.

Find: Optimal value of A.

Step 1: This classic minimum-cost example, the design of the cross-sectional area of an electrical conductor, involves increasing and decreasing cost components. Since the resistance of a conductor is inversely proportional to the size of the conductor, energy loss will decrease as the size of the conductor increases. More specifically, the annual energy loss in kilowatt-hours (kWh) in a conductor due to resistance is equal to

$$
\begin{aligned}
\text{Energy loss in kilowatt-hours} &= \frac{I^2RT}{1{,}000} = \frac{I^2T}{1{,}000} \times \frac{\rho L}{A} \\
&= \frac{5000^2 \times 24 \times 365}{1000} \times \frac{1.7241 \times 10^{-4} \times 300}{A} \\
&= \frac{11{,}327{,}337}{A} \text{ kWh}
\end{aligned}
$$

where A = Cross-sectional area in cm².

Step 2: The total cost of energy loss in dollars per year for the specified conductor material is

$$\text{Annual energy loss cost} = \frac{11{,}327{,}337}{A}(\phi)$$

$$= \frac{11{,}327{,}337}{A}(\$0.0375)$$

$$= \frac{\$424{,}775}{A},$$

where ϕ = cost of energy in dollars per kWh.

Step 3: As we increase the size of the conductor, however, it will cost more to build. First, we need to calculate the total amount of conductor material in kilograms. Since the cross-sectional area is given in square centimetres, we need to convert it to square metres before finding the material mass.

$$\text{Material mass in kilograms} = \frac{(300)(8894)A}{100^2}$$

$$= 267A$$

$$\text{Total material cost} = 267(A)(\$16.50)$$

$$= \$4406A$$

Here, we are looking for the trade-off between the cost of installation and the cost of energy loss.

Step 4: Since the copper material will be salvaged at the rate of $1.65 per kilogram at the end of 25 years, we can compute the capital recovery cost as follows:

$$\text{CR}(9\%) = [\$4406A - \$1.65(267A)](A/P, 9\%, 25) + \$1.65(276A)(0.09)$$

$$= 404A + 40A$$

$$= 444A.$$

Step 5: Using Eq. (6.1), we express the total annual equivalent cost (AEC) as a function of a design variable (A) as follows:

$$\text{AEC}(9\%) = +\underbrace{444A}_{\text{Capital cost}} + \underbrace{\frac{424{,}775}{A}}_{\text{Operating cost}}.$$

To find the minimum annual equivalent cost, we use the result of Eq. (6.2):

$$\frac{d\text{AEC}(9\%)}{dA} = +444 - \frac{424{,}755}{A^2} = 0,$$

$$A^* = \sqrt{\frac{424{,}775}{444}}$$

$$= 31 \text{ cm}^2$$

The minimum annual equivalent total cost is

$$\text{AEC}(9\%) = 444(31) + \frac{424{,}775}{31}$$

$$= \$27{,}466.$$

Figure 6.15 illustrates the nature of this design trade-off problem.

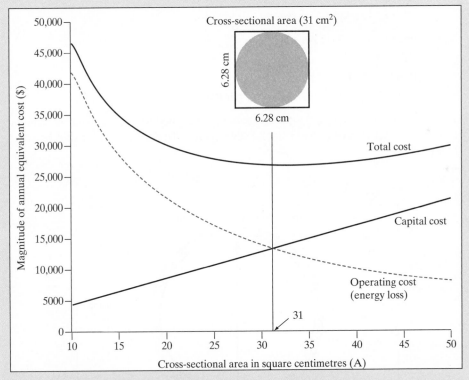

Figure 6.15 Optimal cross-sectional areas for a copper conductor. Note that the minimum point almost coincides with the crossing point of the capital-cost and operating-cost lines. This is not always true. Since the cost components can have a variety of cost patterns, the minimum point does not in general occur at the crossing point (Example 6.14).

EXAMPLE 6.15 Economical Pipe Size

As a result of the 1990 conflict in the Persian Gulf, Kuwait examined the feasibility of running a steel pipeline across the Arabian Peninsula to the Red Sea. The pipeline would have to be designed to handle 3 million barrels of crude oil per day under optimum conditions. The length of the line must be 1000 kilometres. Calculate the optimum pipeline diameter required for 20 years for the following data at $i = 10\%$:

$$\text{Pumping power} = \frac{Q\Delta P}{1000} \text{kW}$$

$$Q = \text{volume flow rate, metres}^3/\text{second}$$

$$\Delta P = \frac{128Q\mu L}{\pi D^4}, \text{pressure drop, pascals}$$

L = pipe length, metres

D = inside pipe diameter, metres

$t = 0.01D$, pipeline wall thickness, metres

$\mu = 3.5138$ kg/metre-second, oil viscosity

Power cost, $0.02 per kWh

Oil cost, $50 per barrel

Pipeline cost, $2.2 per kg of steel

Pump and motor costs, $261.50 per kW.

The salvage value of the steel after 20 years was assumed to be zero because removal costs exhaust scrap profits from steel. (See Figure 6.16 for the relationship between D and t.)

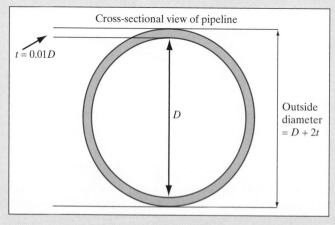

Figure 6.16 Designing economical pipe size to handle 3 million barrels of crude oil per day (Example 6.15).

DISCUSSION: In general, when a progressively larger size of pipe is used to carry a given fluid at a given volume flow rate, the energy required to move the fluid will progressively decrease. However, as we increase the pipe size, the cost of its construction will increase. In practice, to obtain the best pipe size for a particular situation, you may choose a reasonable, but small, starting size. Compute the energy cost of pumping fluid through this size of pipe and the total construction cost, and then compare the difference in energy cost with the difference in construction cost. When the savings in energy cost exceeds the added construction cost, the process may be repeated with progressively larger pipe sizes until the added construction cost exceeds the savings in energy cost. As soon as this happens, the best pipe size to use in the particular application is identified. However, this search process can be simplified by using the minimum-cost concept encompassed by Eqs. (6.1) and (6.2).

SOLUTION

Given: Cost components as a function of pipe diameter (D), $N = 20$ years, and $i = 10\%$.

Find: Optimal pipe size (D).

Step 1: Several different units are introduced; however, we need to work with common units. We will assume the following conversions:

$$1000 \text{ km} = 10^6 \text{ metres}$$
$$1 \text{ barrel} = 0.159 \text{ m}^3$$
$$\text{Density of steel} = 7861 \text{ kg/m}^3$$

Step 2: Power required to pump oil:

It is well known that, for any given set of operating conditions involving the flow of a noncompressible fluid, such as oil, through a pipe of constant diameter, a small-diameter pipe will have a high fluid velocity and a high fluid pressure drop through the pipe. This will require a pump that will deliver a high discharge pressure and a motor with large energy consumption. To determine the pumping power required to boost the pressure drop, we need to determine both the volume flow rate and the amount of pressure drop. Then we can calculate the cost of the power required to pump oil.

- Volume flow rate per hour:

$$
\begin{aligned}
Q &= 3{,}000{,}000 \text{ barrels/day} \times 0.159 \text{ m}^3/\text{barrel} \\
&= 477{,}000 \text{ m}^3/\text{day} \\
&= 19{,}875 \text{ m}^3/\text{hour} \\
&= 5.52 \text{ m}^3/\text{second.}
\end{aligned}
$$

- Pressure drop:

$$
\begin{aligned}
\Delta P &= \frac{128 Q \mu L}{\pi D^4} \\
&= \frac{128 \times 5.52 \times 3.5138 \times 10^6}{3.14159 \times D^4} \\
&= \frac{790{,}271{,}973}{3.14159 \times D^4} \text{Pascals.}
\end{aligned}
$$

• Pumping power required to boost the pressure drop:

$$\text{Power} = \frac{Q\Delta P}{1000}$$

$$= 1/1000 \times 5.52 \times 790{,}271{,}973/D^4$$

$$= \frac{4{,}362{,}301}{D^4} \text{kW.}$$

• Power cost to pump oil:

$$\text{Power cost} = \frac{4{,}362{,}301}{D^4} \text{kW} \times \$0.02/\text{kWh}$$

$$\times 24 \text{ hours/day} \times 365 \text{ days/year}$$

$$= \frac{\$764{,}275{,}135}{D^4}/\text{year.}$$

Step 3: Pump and motor cost calculation:

Once we determine the required pumping power, we can find the size of the pump and the motor costs. This is because the capacity of the pump and motor is proportional to the required pumping power. Thus,

$$\text{Pump and motor cost} = \frac{4{,}362{,}301}{D^4} \times \$261.50/\text{kW}$$

$$= \frac{\$1{,}140{,}741{,}712}{D^4}.$$

Step 4: Required amount and cost of steel:

The pumping cost will be counterbalanced by the lower costs for the smaller pipe, valves, and fittings. If the pipe diameter is made larger, the fluid velocity drops markedly and the pumping costs become substantially lower. Conversely, the capital cost for the larger pipe, fittings, and valves becomes greater. For a given cross-sectional area of the pipe, we can determine the total volume of the pipe as well as the weight. Once the total weight of the pipe is determined, we can easily convert it into the required investment cost. The calculations are as follows:

$$\text{Cross-sectional area} = 3.14159[(0.51D)^2 - (0.50D)^2]$$
$$= 0.032D^2 \text{ m}^2$$

$$\text{Total volume of pipe} = 0.032D^2 \text{ m}^2 \times 1{,}000{,}000 \text{ m}$$
$$= 32{,}000D^2 \text{ m}^3$$

$$\text{Total weight of steel} = 32{,}000D^2 \text{ m}^3 \times 7861 \text{ kg/m}^3$$
$$= 251{,}552{,}000D^2 \text{ kg}$$

$$\text{Total pipeline cost} = \$2.20/\text{kg} \times 251,552,000D^2 \text{ kg}$$
$$= \$553,414,000D^2.$$

Step 5: Annual equivalent cost calculation:

For a given total pipeline cost and its salvage value at the end of 20 years of service life, we can find the equivalent annual capital cost by using the formula for the capital recovery factor with return:

$$\text{Capital cost} = \left(\$553,414,000D^2 + \frac{\$1,140,741,712}{D^4} \right)(A/P, 10\%, 20)$$

$$= 65,026,145D^2 + \frac{134,037,151}{D^4}$$

$$\text{Annual power cost} = \frac{\$764,275,135}{D^4}.$$

Step 6: Economical pipe size:

Now that we have determined the annual pumping and motor costs and the equivalent annual capital cost, we can express the total equivalent annual cost as a fraction of the pipe diameter (D):

$$\text{AEC}(10\%) = 65,026,145D^2 + \frac{134,037,151}{D^4} + \frac{764,275,135}{D^4}.$$

To find the optimal pipe size (D) that results in the minimum annual equivalent cost, we take the first derivative of AEC(10%) with respect to D, equate the result to zero, and solve for D:

$$\frac{d\text{AEC}(10\%)}{dD} = 130,052,290D - \frac{3,593,249,144}{D^5}$$

$$= 0;$$

$$130,052,290D^6 = 3,593,249,144$$

$$D^6 = 27.6293,$$

$$D^* = 1.7387 \text{ m}.$$

Note that, ideally, the velocity in a pipe should be no more than approximately 3 m/sec, because friction wears the pipe. To check whether the preceding answer is reasonable, we may compute

$$Q = \text{velocity} \times \text{pipe inner area},$$

$$5.52 \text{ m}^3/\text{s} = V \frac{3.14159(1.7387)^2}{4},$$

$$V = 2.32 \text{ m/sec},$$

which is less than 3 m/sec. Therefore, the optimal answer as calculated is practical.

Step 7: Equivalent annual cost at optimal pipe size:

$$\text{Capital cost} = \left[\$553{,}414{,}000 \times 1.7387^2 + \frac{\$1{,}140{,}741{,}712}{1.7387^4} \right] (A/P, 10\%, 20)$$

$$= \left[\$65{,}026{,}145 \times 1.7387^2 + \frac{\$134{,}037{,}151}{1.7387^4} \right]$$

$$= \$211{,}245{,}591;$$

$$\text{Annual power cost} = \left(\frac{\$764{,}275{,}135}{1.7387^4} \right)$$

$$= \$83{,}627{,}885.$$

$$\text{Total annual cost} = \$211{,}245{,}591 + \$83{,}627{,}985$$

$$= \$294{,}873{,}476.$$

Step 8: Total annual oil revenue:

$$\text{Annual oil revenue} = \$50/\text{bbl} \times 3{,}000{,}000 \text{ bbl/day} \times 365 \text{ days/year}$$

$$= \$54{,}750{,}000{,}000/\text{year}.$$

Enough revenues are available to offset the capital cost as well as the operating cost.

COMMENTS: A number of other criteria exist for choosing the pipe size for a particular fluid transfer application. For example, low velocity may be required when erosion or corrosion concerns must be considered. Alternatively, higher velocities may be desirable for slurries when settling is a concern. Ease of construction may also weigh significantly in choosing a pipe size. On the one hand, a small pipe size may not accommodate the head and flow requirements efficiently; on the other hand, space limitations may prohibit the selection of large pipe sizes.

SUMMARY

- **Revenue projects** are those for which the income generated depends on the choice of project. **Service projects** are those for which income remains the same, regardless of which project is selected.

- Several alternatives that meet the same need are **mutually exclusive** if, whenever one of them is selected, the others will be rejected.

- In many cases, a **do-nothing alternative** is available, either because the need is being met by an existing system that could be replaced, or because we are assessing the viability of a new endeavour. When "do nothing" is not an option, one of the mutually exclusive alternatives under consideration must be chosen.

■ In **total investment analysis**, the PW (or FW or AE) of each mutually exclusive alternative is calculated separately, and the best choice is the one with the highest PW (or FW or AE).

■ In **incremental investment analysis**, the differences in cash flows between pairs of mutually exclusive alternatives are used as the basis for analysis. We always subtract the lower cost investment from the highest cost one. Basically we want to know that the extra investment required can be justified on the basis of the extra benefits generated in the future.

■ When using the IRR criterion, incremental investment analysis must be used to arrive at the correct optimal choice. This approach may also be used in conjunction with the PW, FW, or AE criteria. Given that B − A is an investment increment, the decision rules are as follows:

$$\text{If } IRR_{B-A} > MARR, \text{ select B,}$$
$$\text{If } IRR_{B-A} = MARR, \text{ select either one,}$$
$$\text{If } IRR_{B-A} < MARR, \text{ select A,}$$

or

	> 0	= 0	< 0
PW_{B-A} (MARR)	select B	select either one	select A
FW_{B-A} (MARR)	select B	select either one	select A
AE_{B-A} (MARR)	select B	select either one	select A

■ Mutually exclusive alternatives must be compared over the same **analysis period**.

■ When not specified by management or company policy, the analysis period to use in a comparison of mutually exclusive projects may be chosen by an individual analyst, perhaps to achieve more efficient computations. In general, the analysis period should be chosen to cover the required service period as highlighted in Fig. 6.5.

■ AE analysis is recommended for many situations involving the comparison of options with differing lives. If we can use a **lowest common multiple** duration, or **infinite planning horizon** with project repeatability, then the AE criterion involves less computation than other methods.

■ When comparing options with different lifetimes, the incremental cash flows often exhibit multiple sign changes, hence the IRR method becomes impractical to use.

■ **Life-Cycle Cost Analysis (LCCA)** is a way to predict the most cost-effective solution by allowing engineers to make a reasonable comparison among alternatives within the limit of the available data.

■ Design economics can be used to determine a design parameter that minimizes annual equivalent costs.

PROBLEMS

6.1 Consider the following cash flow data for two competing investment projects:

	Cash Flow Data (Unit: $ thousand)	
n	Project A	Project B
0	−$800	−$2635
1	−1500	−565
2	−435	820
3	775	820
4	775	1080
5	1275	1880
6	1275	1500
7	975	980
8	675	580
9	375	380
10	660	840

At $i = 12\%$, which of the two projects would be a better choice?

6.2 Consider the cash flows for the following investment projects.

	Project's Cash Flow				
n	A	B	C	D	E
0	−$1500	−$1500	−$3000	$1500	−$1800
1	1350	1000	1000	−450	600
2	800	800	X	−450	600
3	200	800	1500	−450	600
4	100	150	X	−450	600

(a) Suppose projects A and B are mutually exclusive. On the basis of the NPW criterion, which project would be selected? Assume that MARR = 15%.

(b) Repeat (a), using the NFW criterion.

(c) Find the minimum value of X that makes project C acceptable.

(d) Would you accept project D at $i = 18\%$?

(e) Assume that projects D and E are mutually exclusive. On the basis of the NPW criterion, which project would you select?

6.3 Consider two mutually exclusive investment projects, each with MARR = 12%, as shown in Table P6.3.

(a) On the basis of the NPW criterion, which alternative would be selected?

(b) On the basis of the FW criterion, which alternative would be selected?

TABLE P6.3 Two Mutually Exclusive Investment Projects

	Project's Cash Flow	
n	A	B
0	−$14,500	−$12,900
1	12,610	11,210
2	12,930	11,720
3	12,300	11,500

6.4 Consider the following two mutually exclusive investment projects, each with MARR = 15%:

	Project's Cash Flow	
n	A	B
0	−$6,000	−$8,000
1	800	11,500
2	14,000	400

(a) On the basis of the NPW criterion, which project would be selected?

(b) Sketch the PW(i) function for each alternative on the same chart between 0% and 50%. For what range of i would you prefer project B?

6.5 Two methods of carrying away surface runoff water from a new subdivision are being evaluated:

Method A. Dig a ditch. The first cost would be $60,000, and $25,000 of redigging and shaping would be required at five-year intervals forever.

Method B. Lay concrete pipe. The first cost would be $150,000, and a replacement would be required at 50-year intervals at a net cost of $180,000 forever.

At i = 12%, which method is the better one?

6.6 A local car dealer is advertising a standard 24-month lease of $1150 per month for its new XT 3000 series sports car. The standard lease requires a down payment of $4500, plus a $1000 refundable initial deposit *now*. The first lease payment is due at the end of month 1. In addition, the company offers a 24-month lease plan that has a single up-front payment of $30,500, plus a refundable initial deposit of $1000. Under both options, the initial deposit will be refunded at the end of month 24. Assume an interest rate of 6% compounded monthly. With the present-worth criterion, which option is preferred?

6.7 Two machines are being considered for a manufacturing process. Machine A has a first cost of $75,200, and its salvage value at the end of six years of estimated service life is $21,000. The operating costs of this machine are estimated to be $6800 per year. Extra income taxes are estimated at $2400 per year. Machine B has a first cost of $44,000, and its salvage value at the end of six years' service is estimated to be negligible. The annual operating costs will be $11,500. Compare these two mutually exclusive alternatives by the present-worth method at $i = 13\%$.

6.8 An electric motor is rated at 10 horsepower (HP) and costs $800. Its full load efficiency is specified to be 85%. A newly designed high-efficiency motor of the same size has an efficiency of 90%, but costs $1200. It is estimated that the motors will operate at a rated 10 HP output for 1500 hours a year, and the cost of energy will be $0.07 per kilowatt-hour. Each motor is expected to have a 15-year life. At the end of 15 years, the first motor will have a salvage value of $50 and the second motor will have a salvage value of $100. Consider the MARR to be 8%. (Note: 1 HP = 0.7457 kW.)

(a) Use the NPW criterion to determine which motor should be installed.

(b) In (a), what if the motors operated 2500 hours a year instead of 1500 hours a year? Would the motor you chose in (a) still be the choice?

6.9 Consider the following two mutually exclusive investment projects:

	Project's Cash Flow	
n	A	B
0	−$20,000	−$25,000
1	17,500	25,500
2	17,000	18,000
3	15,000	

On the basis of the NPW criterion, which project would be selected if you use an infinite planning horizon with project repeatability (the same costs and benefits) likely? Assume that $i = 12\%$.

6.10 Consider the following two mutually exclusive investment projects, which have unequal service lives:

	Project's Cash Flow	
n	A1	A2
0	−$900	−$1800
1	−400	−300
2	−400	−300
3	−400 + 200	−300
4		−300
5		−300
6		−300
7		−300
8		−300 + 500

(a) What assumption(s) do you need in order to compare a set of mutually exclusive investments with unequal service lives?

(b) With the assumption(s) defined in (a) and using $i = 10\%$, determine which project should be selected.

(c) If your analysis period (study period) is just three years, what should be the salvage value of project A2 at the end of year 3 to make the two alternatives economically indifferent?

6.11 Consider the following two mutually exclusive projects:

	B1		B2	
n	Cash Flow	Salvage Value	Cash Flow	Salvage Value
0	-$18,000		-$15,000	
1	-2,000	6,000	-2,100	6,000
2	-2,000	4,000	-2,100	3,000
3	-2,000	3,000	-2,100	1,000
4	-2,000	2,000		
5	-2,000	2,000		

Salvage values represent the net proceeds (after tax) from disposal of the assets if they are sold at the end of each year. Both B1 and B2 will be available (or can be repeated) with the same costs and salvage values for an indefinite period.

(a) Assuming an infinite planning horizon, which project is a better choice at MARR = 12%?

(b) With a 10-year planning horizon, which project is a better choice at MARR = 12%?

6.12 Consider the following cash flows for two types of models:

	Project's Cash Flow	
n	Model A	Model B
0	-$6,000	-$15,000
1	3,500	10,000
2	3,500	10,000
3	3,500	

Both models will have no salvage value upon their disposal (at the end of their respective service lives). The firm's MARR is known to be 15%.

(a) Notice that the models have different service lives. However, model A will be available in the future with the same cash flows. Model B is available at one time only. If you select model B now, you will have to replace it with model A at the end of year 2. If your firm uses the present worth as a decision criterion, which model should be selected, assuming that the firm will need either model for an indefinite period?

(b) Suppose that your firm will need either model for only two years. Determine the salvage value of model A at the end of year 2 that makes both models indifferent (equally likely).

6.13 An electric utility is taking bids on the purchase, installation, and operation of microwave towers. Following are some details associated with the two bids that were received. Which one is preferred?

	Cost per Tower	
	Bid A	**Bid B**
Equipment cost	$112,000	$98,000
Installation cost	$25,000	$30,000
Annual maintenance and inspection fee	$2,000	$2,500
Annual extra income taxes		$800
Life	40 years	35 years
Salvage value	$0	$0

6.14 Consider the following two investment alternatives:

	Project's Cash Flow	
n	**A1**	**A2**
0	−$15,000	−$25,000
1	9,500	0
2	12,500	X
3	7,500	X
PW(15%)	?	9,300

The firm's MARR is known to be 15%.

(a) Compute PW(15%) for A1.

(b) Compute the unknown cash flow X in years 2 and 3 for A2.

(c) Compute the project balance (at 15%) of A1 at the end of period 3.

(d) If these two projects are mutually exclusive alternatives, which one would you select?

6.15 Consider each of the after-tax cash flows shown in Table P6.15.

(a) Compute the project balances for projects A and D as a function of the project year at $i = 10\%$.

(b) Compute the net future-worth values for projects A and D at $i = 10\%$.

(c) Suppose that projects B and C are mutually exclusive. Suppose also that the required service period is eight years and that the company is considering leasing comparable equipment with an annual lease expense of $3000 for the remaining years of the required service period. Which project is a better choice?

TABLE P6.15 After-Tax Cash Flows

	Project's Cash Flow			
n	A	B	C	D
0	−$2500	−$7000	−$5000	−$5000
1	650	−2500	−2000	−500
2	650	−2000	−2000	−500
3	650	−1500	−2000	4000
4	600	−1500	−2000	3000
5	600	−1500	−2000	3000
6	600	−1500	−2000	2000
7	300		−2000	3000
8	300			

6.16 A mall with two levels is under construction. The plan is to install only nine escalators at the start, although the ultimate design calls for 16. The question arises as to whether to provide necessary facilities (stair supports, wiring conduits, motor foundations, etc.) that would permit the installation of the additional escalators at the mere cost of their purchase and installation or to defer investment in these facilities until the escalators need to be installed.

- **Option 1.** Provide these facilities now for all seven future escalators at $200,000.
- **Option 2.** Defer the investment in the facility as needed. Install two more escalators in two years, three more in five years, and the last two in eight years. The installation of these facilities at the time they are required is estimated to cost $100,000 in year 2, $160,000 in year 5, and $140,000 in year 8.

   Additional annual expenses are estimated at $3000 for each escalator facility installed. At an interest rate of 12%, compare the net present worth of each option over eight years.

6.17 An electrical utility is experiencing a sharp power demand that continues to grow at a high rate in a certain local area.

   Two alternatives are under consideration. Each is designed to provide enough capacity during the next 25 years, and both will consume the same amount of fuel, so fuel cost is not considered in the analysis.

- **Alternative A.** Increase the generating capacity now so that the ultimate demand can be met with additional expenditures later. An initial investment of $30 million would be required, and it is estimated that this plant facility would be in service for 25 years and have a salvage value of $0.85 million. The annual operating and maintenance costs (including income taxes) would be $0.4 million.
- **Alternative B.** Spend $10 million now and follow this expenditure with future additions during the 10th year and the 15th year. These additions would cost $18 million and $12 million, respectively. The facility would be sold 25 years from now, with a salvage value of $1.5 million. The annual operating and maintenance costs (including income taxes) initially will be $250,000 and will increase to $0.35 million after the second addition (from the 11th year to the 15th year) and to $0.45 million during the final 10 years. (Assume that these costs begin one year subsequent to the actual addition.)

On the basis of the present-worth criterion, if the firm uses 15% as a MARR, which alternative should be undertaken?

6.18 A large refinery–petrochemical complex is to manufacture caustic soda, which will use feedwater of 40,000 litres per day. Two types of feedwater storage installation are being considered over the 40 years of their useful life.

- **Option 1.** Build an 80,000-litre tank on a tower. The cost of installing the tank and tower is estimated to be $164,000. The salvage value is estimated to be negligible.
- **Option 2.** Place a tank of 80,000-litre capacity on a hill, which is 150 metres away from the refinery. The cost of installing the tank on the hill, including the extra length of service lines, is estimated to be $120,000, with negligible salvage value. Because of the tank's location on the hill, an additional investment of $12,000 in pumping equipment is required. The pumping equipment is expected to have a service life of 20 years, with a salvage value of $1000 at the end of that time. The annual operating and maintenance cost (including any income tax effects) for the pumping operation is estimated at $1000.

If the firm's MARR is known to be 12%, which option is better, on the basis of the present-worth criterion?

6.19 Two 150-horsepower (HP) motors are being considered for installation at a municipal sewage-treatment plant. The first costs $4500 and has an operating efficiency of 83%. The second costs $3600 and has an efficiency of 80%. Both motors are projected to have zero salvage value after a life of 10 years. If all the annual charges, such as insurance, maintenance, etc., amount to a total of 15% of the original cost of each motor, and if power costs are a flat 5 cents per kilowatt-hour, how many minimum hours of full-load operation per year are necessary to justify purchasing the more expensive motor at $i = 6\%$? (A conversion factor you might find useful is 1 HP = 746 watts = 0.746 kilowatts.)

6.20 Danford Company, a manufacturer of farm equipment, currently produces 20,000 units of gas filters per year for use in its lawn-mower production. The costs, based on the previous year's production, are reported in Table P6.20.

It is anticipated that gas-filter production will last five years. If the company continues to produce the product in-house, annual direct material costs will increase at the rate of 5%. (For example, annual material costs during the first production year will be $63,000.) Direct labour will also increase at the rate of 6% per year. However, variable overhead costs will increase at the rate of 3%, but the fixed overhead will remain at its current level over the next five years. John Holland

TABLE P6.20 Production costs

Item	Expense ($)
Direct materials	$ 60,000
Direct labour	180,000
Variable overhead (power and water)	135,000
Fixed overhead (light and heat)	70,000
Total cost	$445,000

Company has offered to sell Danford 20,000 units of gas filters for $25 per unit. If Danford accepts the offer, some of the manufacturing facilities currently used to manufacture the filter could be rented to a third party for $35,000 per year. In addition, $3.5 per unit of the fixed overheard applied to the production of gas filters would be eliminated. The firm's interest rate is known to be 15%. What is the unit cost of buying the gas filter from the outside source? Should Danford accept John Holland's offer, and why?

6.21 Saskatchewan Environmental Consulting (SEC) Inc. designs plans and specifications for asbestos abatement (removal) projects in public, private, and governmental buildings. Currently, SEC must conduct an air test before allowing the reoccupancy of a building from which asbestos has been removed. SEC subcontracts air-test samples to a laboratory for analysis by transmission electron microscopy (TEM). To offset the cost of TEM analysis, SEC charges its clients $100 more than the subcontractor's fee. The only expenses in this system are the costs of shipping the air-test samples to the subcontractor and the labour involved in shipping the samples. With the growth of the business, SEC is having to consider either continuing to subcontract the TEM analysis to outside companies or developing its own TEM laboratory. With recent government regulation requiring the removal of asbestos, SEC expects about 1000 air-sample testings per year over eight years. The firm's MARR is known to be 15%.

- **Subcontract option.** The client is charged $400 per sample, which is $100 above the subcontracting fee of $300. Labour expenses are $1500 per year, and shipping expenses are estimated to be $0.50 per sample.

- **TEM purchase option.** The purchase and installation cost for the TEM is $415,000. The equipment would last for eight years, at which time it should have no salvage value. The design and renovation cost is estimated to be $9500. The client is charged $300 per sample, based on the current market price. One full-time manager and two part-time technicians are needed to operate the laboratory. Their combined annual salaries will be $50,000. Material required to operate the lab includes carbon rods, copper grids, filter equipment, and acetone. The costs of these materials are estimated at $6000 per year. Utility costs, operating and maintenance costs, and the indirect labour needed to maintain the lab are estimated at $18,000 per year. The extra income-tax expenses would be $20,000.

(a) Determine the cost of an air-sample test by the TEM (in-house).

(b) What is the required number of air samples per year to make the two options equivalent?

6.22 The following cash flows represent the potential annual savings associated with two different types of production processes, each of which requires an investment of $12,000:

n	Process A	Process B
0	−$12,000	−$12,000
1	9,120	6,350
2	6,840	6,350
3	4,560	6,350
4	2,280	6,350

Assuming an interest rate of 15%,

(a) Determine the equivalent annual savings for each process.

(b) Determine the hourly savings for each process if it is in operation 2000 hours per year.

(c) Which process should be selected?

6.23 Hamilton Steel Inc. is considering replacing 20 conventional 25-HP, 230-V, 60-Hz, 1800-rpm induction motors in its plant with modern premium efficiency (PE) motors. Both types of motors have a power output of 18.650 kW per motor (25 HP × 0.746 kW/HP). Conventional motors have a published efficiency of 89.5%, while the PE motors are 93% efficient. The initial cost of the conventional motors is $13,000, whereas the initial cost of the proposed PE motors is $15,600. The motors are operated 12 hours per day, five days per week, 52 weeks per year, with a local utility cost of $0.07 per kilowatt-hour (kWh). The motors are to be operated at 75% load, and the life cycle of both the conventional and PE motor is 20 years, with no appreciable salvage value.

(a) At an interest rate of 13%, what are the savings per kWh achieved by switching from the conventional motors to the PE motors?

(b) At what operating hours are the two motors equally economical?

6.24 A certain factory building has an old lighting system, and lighting the building costs, on average, $20,000 a year. A lighting consultant tells the factory supervisor that the lighting bill can be reduced to $8000 a year if $50,000 were invested in relighting the building. If the new lighting system is installed, an incremental maintenance cost of $3000 per year must be taken into account. The new lighting system has zero salvage value at the end of its life. If the old lighting system also has zero salvage value, and the new lighting system is estimated to have a life of 20 years, what is the net annual benefit for this investment in new lighting? Take the MARR to be 12%.

6.25 Travis Wenzel has $2000 to invest. Usually, he would deposit the money in his savings account, which earns 6% interest compounded monthly. However, he is considering three alternative investment opportunities:

- **Option 1.** Purchasing a bond for $2000. The bond has a face value of $2000 and pays $100 every six months for three years, after which time the bond matures.

- **Option 2.** Buying and holding a stock that grows 11% per year for three years.

- **Option 3.** Making a personal loan of $2000 to a friend and receiving $150 per year for three years.

Determine the equivalent annual cash flows for each option, and select the best option.

6.26 A chemical company is considering two types of incinerators to burn solid waste generated by a chemical operation. Both incinerators have a burning capacity of 20 tonnes per day. The following data have been compiled for comparison:

	Incinerator A	Incinerator B
Installed cost	$1,200,000	$750,000
Annual O&M costs	$ 50,000	$ 80,000
Service life	20 years	10 years
Salvage value	$ 60,000	$ 30,000
Income taxes	$ 40,000	$ 30,000

If the firm's MARR is known to be 13%, determine the processing cost per tonne of solid waste incurred by each incinerator. Assume that incinerator B will be available in the future at the same cost.

6.27 Consider the cash flows for the following investment projects (MARR = 15%):

	Project's Cash Flow		
n	A	B	C
0	−$2500	−$4000	−$5000
1	1000	1600	1800
2	1800	1500	1800
3	1000	1500	2000
4	400	1500	2000

(a) Suppose that projects A and B are mutually exclusive. Which project would you select, based on the AE criterion?

(b) Assume that projects B and C are mutually exclusive. Which project would you select, based on the AE criterion?

6.28 An airline is considering two types of engine systems for use in its planes. Each has the same life and the same maintenance and repair record.

• **System A** costs $100,000 and uses 160 kL per 1000 hours of operation at the average load encountered in passenger service.

• **System B** costs $200,000 and uses 128 kL per 1000 hours of operation at the same level.

Both engine systems have three-year lives before any major overhaul is required. On the basis of the initial investment, the systems have 10% salvage values. If jet fuel currently costs $1.50/L and fuel consumption is expected to increase at the rate of 6% per year because of degrading engine efficiency, which engine system should the firm install? Assume 2000 hours of operation per year and a MARR of 10%. Use the AE criterion. What is the equivalent operating cost per hour for each engine?

6.29 Magna is considering one of two forklift trucks for its assembly plant. Truck A costs $15,000 and requires $3000 annually in operating expenses. It will have a $5000 salvage value at the end of its three-year service life. Truck B costs $20,000, but requires only $2000 annually in operating expenses; its service life is four years, at which time its expected salvage value will be $8000. The firm's

MARR is 12%. Assuming that the trucks are needed for 12 years and that no significant changes are expected in the future price and functional capacity of each truck, select the most economical truck, on the basis of AE analysis.

6.30 A small manufacturing firm is considering purchasing a new machine to modernize one of its current production lines. Two types of machines are available on the market. The lives of machine A and machine B are four years and six years, respectively, but the firm does not expect to need the service of either machine for more than five years. The machines have the following expected receipts and disbursements:

Item	Machine A	Machine B
First cost	$6500	$8500
Service life	4 years	6 years
Estimated salvage value	$600	$1000
Annual O&M costs	$800	$520
Change oil filter every other year	$100	None
Engine overhaul	$200 (every 3 years)	$280 (every 4 years)

The firm always has another option: leasing a machine at $3000 per year, fully maintained by the leasing company. After four years of use, the salvage value for machine B will remain at $1000.

(a) How many decision alternatives are there?

(b) Which decision appears to be the best at $i = 10\%$?

6.31 A plastic-manufacturing company owns and operates a polypropylene production facility that converts the propylene from one of its cracking facilities to polypropylene plastics for outside sale. The polypropylene production facility is currently forced to operate at less than capacity due to an insufficiency of propylene production capacity in its hydrocarbon cracking facility. The chemical engineers are considering alternatives for supplying additional propylene to the polypropylene production facility. Two feasible alternatives are to build a pipeline to the nearest outside supply source and to provide additional propylene by truck from an outside source. The engineers also gathered the following projected cost estimates:

- Future costs for purchased propylene excluding delivery: $0.473/kg.
- Cost of pipeline construction: $125,000/km (pipeline).
- Estimated length of pipeline: 300 km.
- Transportation costs by tank truck: 10¢/kg, utilizing a common carrier.
- Pipeline operating costs: $0.01/kg, excluding capital costs.
- Projected additional propylene needs: 80 Gg per year.
- Projected project life: 20 years.
- Estimated salvage value of the pipeline: 8% of the installed costs.

Determine the propylene cost per kilogram under each option if the firm's MARR is 18%. Which option is more economical?

6.32 The City of Mississauga is comparing two plans for supplying water to a newly developed subdivision:

- **Plan A** will take care of requirements for the next 15 years, at the end of which time the initial cost of $400,000 will have to be duplicated to meet the requirements of subsequent years. The facilities installed at dates 0 and 15 may be considered permanent; however, certain supporting equipment will have to be replaced every 30 years from the installation dates, at a cost of $75,000. Operating costs are $31,000 a year for the first 15 years and $62,000 thereafter, although they are expected to increase by $1000 a year beginning in the 21st year.

- **Plan B** will supply all requirements for water indefinitely into the future, although it will be operated only at half capacity for the first 15 years. Annual costs over this period will be $35,000 and will increase to $55,000 beginning in the 16th year. The initial cost of Plan B is $550,000; the facilities can be considered permanent, although it will be necessary to replace $150,000 worth of equipment every 30 years after the initial installation.

The city will charge the subdivision for the use of water on the basis of the equivalent annual cost. At an interest rate of 10%, determine the equivalent annual cost for each plan, and make a recommendation to the city.

6.33 A continuous electric current of 2,000 amps is to be transmitted from a generator to a transformer located 60 metres away. A copper conductor can be installed for $13/kg, will have an estimated life of 25 years, and can be salvaged for $2/kg. Power loss from the conductor will be inversely proportional to the cross-sectional area of the conductor and may be expressed as $0.0042039/A$ kilowatt, where A is in square metres. The cost of energy is $0.0825 per kilowatt-hour, the interest rate is 11%, and the density of copper is 8894 kg/m^3.
(a) Calculate the optimum cross-sectional area of the conductor.
(b) Calculate the annual equivalent total cost for the value you obtained in part (a).
(c) Graph the capital cost, energy-loss cost, and the total cost as a function of the cross-sectional area A, and discuss the impact of increasing energy cost on the optimum obtained in part (a).

6.34 Consider the following two mutually exclusive investment projects:

	Net Cash Flow	
n	Project A	Project B
0	−$300	−$800
1	0	1,150
2	690	40
$i*$	51.66%	46.31%

Assume that MARR = 15%.
(a) According to the IRR criterion, which project would be selected?
(b) Sketch the PW(i) function on the incremental investment $(B - A)$.

6.35 Consider two investments A and B with the following sequences of cash flows:

	Net Cash Flow	
n	**Project A**	**Project B**
0	−$120,000	−$100,000
1	20,000	15,000
2	20,000	15,000
3	120,000	130,000

(a) Compute the IRR for each investment.

(b) At MARR = 15%, determine the acceptability of each project.

(c) If A and B are mutually exclusive projects, which project would you select, based on the rate of return on incremental investment?

6.36 With $10,000 available, you have two investment options. The first is to buy a Guaranteed Investment Certificate (GIC) from a bank at an interest rate of 10% annually for five years. The second choice is to purchase a bond for $10,000 and invest the bond's interest in the bank at an interest rate of 9%. The bond pays 10% interest annually and will mature to its face value of $10,000 in five years. Which option is better? Assume that your MARR is 9% per year.

6.37 A manufacturing firm is considering the following mutually exclusive alternatives:

	Net Cash Flow	
n	**Project A1**	**Project A2**
0	−$2000	−$3000
1	1400	2400
2	1640	2000

Determine which project is a better choice at a MARR = 15%, based on the IRR criterion.

6.38 Consider the following two mutually exclusive alternatives:

	Net Cash Flow	
n	**Project A1**	**Project A2**
0	−$10,000	−$12,000
1	5,000	6,100
2	5,000	6,100
3	5,000	6,100

(a) Determine the IRR on the incremental investment in the amount of $2000.

(b) If the firm's MARR is 10%, which alternative is the better choice?

6.39 Consider the following two mutually exclusive investment alternatives:

	Net Cash Flow	
n	Project A1	Project A2
0	−$15,000	−$20,000
1	7,500	8,000
2	7,500	15,000
3	7,500	5,000
IRR	23.5%	20%

(a) Determine the IRR on the incremental investment in the amount of $5000. (Assume that MARR = 10%.)

(b) If the firm's MARR is 10%, which alternative is the better choice?

6.40 You are considering two types of automobiles. Model A costs $18,000 and model B costs $15,624. Although the two models are essentially the same, after four years of use model A can be sold for $9000, while model B can be sold for $6500. Model A commands a better resale value because its styling is popular among young university students. Determine the rate of return on the incremental investment of $2376. For what range of values of your MARR is model A preferable?

6.41 A plant engineer is considering two types of solar water heating system:

	Model A	Model B
Initial cost	$7,000	$10,000
Annual savings	$700	$1,000
Annual maintenance	$100	$50
Expected life	20 years	20 years
Salvage value	$400	$500

The firm's MARR is 12%. On the basis of the IRR criterion, which system is the better choice?

6.42 Consider the following investment projects:

	Net Cash Flow					
n	A	B	C	D	E	F
0	−$100	−$200	−$4000	−$2000	−$2000	−$3000
1	60	120	2410	1400	3700	2500
2	50	150	2930	1720	1640	1500
3	50					
*i**	28.89%	21.65%	21.86%	31.10%	121.95%	23.74%

Assume that MARR = 15%.

(a) Projects A and B are mutually exclusive. Assuming that both projects can be repeated for an indefinite period, which one would you select on the basis of the IRR criterion?

(b) Suppose projects C and D are mutually exclusive. According to the IRR criterion, which project would be selected?

(c) Suppose projects E and F are mutually exclusive. Which project is better according to the IRR criterion?

6.43 Toronto General Hospital is reviewing ways of cutting the cost of stocking medical supplies. Two new stockless systems are being considered, to lower the hospital's holding and handling costs. The hospital's industrial engineer has compiled the relevant financial data for each system as follows (dollar values are in millions):

	Current Practice	Just-in-Time System	Stockless Supply System
Start-up cost	$0	$2.5	$5
Annual stock holding cost	$3	$1.4	$0.2
Annual operating cost	$2	$1.5	$1.2
System life	8 years	8 years	8 years

The system life of eight years represents the period that the contract with the medical suppliers is in force. If the hospital's MARR is 10%, which system is more economical?

6.44 Consider the cash flows for the following investment projects:

			Project Cash Flow		
n	A	B	C	D	E
0	−$1000	−$1000	−$2000	$1000	−$1200
1	900	600	900	−300	400
2	500	500	900	−300	400
3	100	500	900	−300	400
4	50	100	900	−300	400

Assume that the MARR = 12%.

(a) Suppose A, B, and C are mutually exclusive projects. Which project would be selected on the basis of the IRR criterion?

(b) What is the borrowing rate of return (BRR) for project D?

(c) Would you accept project D at MARR = 20%?

(d) Assume that projects C and E are mutually exclusive. Using the IRR criterion, which project would you select?

6.45 Consider the following investment projects:

	Net Cash Flow		
n	**Project 1**	**Project 2**	**Project 3**
0	−$1000	−$5000	−$2000
1	500	7500	1500
2	2500	600	2000

Assume that MARR = 15%.
(a) Compute the IRR for each project.
(b) On the basis of the IRR criterion, if the three projects are mutually exclusive investments, which project should be selected?

6.46 Consider the following two investment alternatives:

	Net Cash Flow	
N	**Project A**	**Project B**
0	−$10,000	−$20,000
1	5,500	0
2	5,500	0
3	5,500	40,000
IRR	30%	?
PW(15%)	?	6300

The firm's MARR is known to be 15%.
(a) Compute the IRR of project B.
(b) Compute the NPW of project A.
(c) Suppose that projects A and B are mutually exclusive. Using the IRR, which project would you select?

6.47 Navtek Enterprises is considering acquiring an automatic screwing machine for its assembly operation of a personal computer. Three different models with varying automatic features are under consideration. The required investments are $360,000 for model A, $380,000 for model B, and $405,000 for model C. All three models are expected to have the same service life of eight years. The following financial information, in which model (B − A) represents the incremental cash flow determined by subtracting model A's cash flow from model B's, is available:

Model	IRR (%)
A	30%
B	15
C	25

Model	Incremental IRR (%)
(B − A)	5%
(C − B)	40
(C − A)	15

If the firm's MARR is known to be 12%, which model should be selected?

6.48 The GeoStar Company, a leading manufacturer of wireless communication devices, is considering three cost-reduction proposals in its batch job-shop manufacturing operations. The company has already calculated rates of return for the three projects, along with some incremental rates of return:

Incremental Investment	Incremental Rate of Return (%)
$A_1 - A_0$	18%
$A_2 - A_0$	20
$A_3 - A_0$	25
$A_2 - A_1$	10
$A_3 - A_1$	18
$A_3 - A_2$	23

A_0 denotes the do-nothing alternative. The required investments are $420,000 for A_1, $550,000 for A_2, and $720,000 for A_3. If the MARR is 15%, what system should be selected?

6.49 A manufacturer of electronic circuit boards is considering six mutually exclusive cost-reduction projects for its PC-board manufacturing plant. All have lives of 10 years and zero salvage values. The required investment and the estimated after-tax reduction in annual disbursements for each alternative are as follows, along with computed rates of return on incremental investments:

Proposal A_j	Required After-Tax Investment	Savings	Rate of Return (%)
A_1	$60,000	$22,000	35.0%
A_2	100,000	28,200	25.2
A_3	110,000	32,600	27.0
A_4	120,000	33,600	25.0
A_5	140,000	38,400	24.0
A_6	150,000	42,200	25.1

Incremental Investment	Incremental Rate of Return (%)
$A_2 - A_1$	9.0%
$A_3 - A_2$	42.8
$A_4 - A_3$	0.0
$A_5 - A_4$	20.2
$A_6 - A_5$	36.3

If the MARR is 15%, which project would you select, based on the rate of return on incremental investment?

6.50 Baby Doll Shop manufactures wooden parts for dollhouses. The worker is paid $8.10 an hour and, using a handsaw, can produce a year's required production (1600 parts) in just eight 40-hour weeks. That is, the worker averages five parts per hour when working by hand. The shop is considering purchasing a power band saw with associated fixtures, to improve the productivity of this operation. Three models of power saw could be purchased: Model A (the economy version), model B (the high-powered version), and model C (the deluxe high-end version). The major operating difference between these models is their speed of operation. The investment costs, including the required fixtures and other operating characteristics, are summarized as follows:

Category	By Hand	Model A	Model B	Model C
Production rate (parts/hour)	5	10	15	20
Labour hours required (hours/year)	320	160	107	80
Annual labour cost (@ $8.10/hour)	$2592	$1296	$867	$648
Annual power cost		$400	$420	$480
Initial investment		$4000	$6000	$7000
Salvage value		$400	$600	$700
Service life (years)		20	20	20

Assume that MARR = 10%. Are there enough savings to purchase any of the power band saws? Which model is most economical, based on the rate-of-return principle? (Assume that any effect of income tax has been already considered in the dollar estimates.) (*Source*: This problem is adapted with the permission of Professor Peter Jackson of Cornell University.)

6.51 Consider the following two mutually exclusive investment projects for which MARR = 15%:

	Net Cash Flow	
n	Project A	Project B
0	−$100	−$200
1	60	120
2	50	150
3	50	
IRR	28.89%	21.65%

On the basis of the IRR criterion, which project would be selected under an infinite planning horizon with project repeatability likely?

6.52 Consider the following two mutually exclusive investment projects:

	Net Cash Flow	
n	Project A1	Project A2
0	−$10,000	−$15,000
1	5,000	20,000
2	5,000	
3	5,000	

(a) To use the IRR criterion, what assumption must be made in comparing a set of mutually exclusive investments with unequal service lives?

(b) With the assumption defined in part (a), determine the range of MARRs that will indicate that project A1 should be selected.

Short Case Studies

ST6.1 Apex Corporation requires a chemical finishing process for a product under contract for a period of six years. Three options are available. Neither Option 1 nor Option 2 can be repeated after its process life. However, Option 3 will always be available from Regina Chemical at the same cost during the period that the contract is operative. Here are the options:

- **Option 1.** Process device A, which costs $100,000, has annual operating and labour costs of $60,000 and a useful service life of four years with an estimated salvage value of $10,000.
- **Option 2.** Process device B, which costs $150,000, has annual operating and labour costs of $50,000 and a useful service life of six years with an estimated salvage value of $30,000.
- **Option 3.** Subcontract out the process at a cost of $100,000 per year.

According to the present-worth criterion, which option would you recommend at $i = 12\%$?

ST6.2 Northern Electric was faced with providing electricity to a newly developed industrial park complex. The distribution engineering department needs to develop guidelines for the design of the distribution circuit. The "main feeder," which is the backbone of each 13-kV distribution circuit, represents a substantial investment by the company.[5]

Northern Electric has four approved main feeder construction configurations—triangular, as illustrated in the accompanying figure. The width of the easement sought depends on the planned construction configuration. If cross-arm construction is planned, a 5-metre easement is sought. A 3-metre wide easement is sought for vertical and triangular configurations.

[5] Example provided by Andrew Hanson.

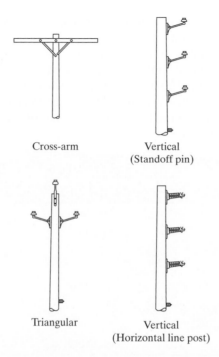

Cross-arm

Vertical
(Standoff pin)

Triangular

Vertical
(Horizontal line post)

Once the required easements are obtained, the line clearance department clears any foliage that would impede the construction of the line. The clearance cost is dictated by the typical tree densities along road rights-of-way. The average cost to trim one tree is estimated at $20, and the average tree density in the service area is estimated to be 50 trees per kilometre. The costs of each type of construction are as follows:

	Design Configurations			
Factors	**Cross-arm**	**Triangular**	**Horizontal Line**	**Standoff**
Easements	$487,000	$388,000	$388,000	$388,000
Line clearance	$613	$1,188	$1,188	$1,188
Line construction	$7,630	$7,625	$12,828	$8,812

Additional factors to consider in selecting the best main feeder configuration are as follows: In certain sections of Northern Electric's service territory, osprey often nest on transmission and distribution poles. The nests reduce the structural and electrical integrity of the poles. Cross-arm construction is most vulnerable to osprey nesting, since the cross-arm and braces provide a secure area for construction of the nest. Vertical and triangular construction do not provide such spaces. Furthermore, in areas where osprey are known to nest, vertical and triangular configuration have added advantages. The insulation strength of a construction configuration may favourably or adversely affect the reliability of the line for which the configuration is used. A common measure of line insulation strength is the critical flashover (CFO) voltage. The greater the CFO, the less susceptible the line is to nuisance flashovers from lightning and other electrical phenomena.

The utility's existing inventory of cross-arms is used primarily for main feeder construction and maintenance. The use of another configuration for main feeder construction would result in a substantial reduction in the inventory of cross-arms. The line crews complain that line spacing on vertical and triangular construction is too restrictive for safe live line work. Each accident would cost $65,000 in lost work and other medical expenses. The average cost of each flashover repair would be $3000. The following table lists the values of the factors involved in the four design configurations:

	Design Configurations			
Factors	Cross-arm	Triangular	Horizontal Line	Standoff
Nesting	Severe	None	None	None
Insulation strength, CFO (kV)	387	474	476	462
Annual flashover occurrence (n)	2	1	1	1
Annual inventory savings		$4521	$4521	$4521
Safety	OK	Problem	Problem	Problem

All configurations would last about 20 years, with no salvage value. It appears that noncross-arm designs are better, but engineers need to consider other design factors, such as safety, rather than just monetary factors when implementing the project. It is true that the line spacing on triangular construction is restrictive. However, with a better clearance design between phases for vertical construction, the hazard would be minimized. In the utility industry, the typical opposition to new types of construction is caused by the confidence acquired from constructing lines in the cross-arm configuration for many years. As more vertical and triangular lines are built, opposition to these configurations should decrease. Which of the four designs described in the preceding table would you recommend to the management? Assume Northern Electric's interest rate to be 12%.

ST6.3 Automotive engineers at Ford are considering the laser blank welding (LBW) technique to produce a windshield frame rail blank. The engineers believe that, compared with the conventional sheet metal blanks, LBW would result in a significant savings as follows:

1. Scrap reduction through more efficient blank nesting on coil.

2. Scrap reclamation (weld scrap offal into a larger usable blank).

The use of a laser welded blank provides a reduction in engineered scrap for the production of a window frame rail blank.

On the basis of an annual volume of 3000 blanks, Ford engineers have estimated the following financial data:

Description	Blanking Method	
	Conventional	Laser Blank Welding
Weight per blank (kg/part)	28.923	15.817
Steel cost/part	$ 14.98	$ 8.19
Transportation/part	$ 0.67	$ 0.42
Blanking/part	$ 0.50	$ 0.40
Die investment	$106,480	$83,000

The LBW technique appears to achieve significant savings, so Ford's engineers are leaning toward adopting it. Since the engineers have had no previous experience with LBW, they are not sure whether producing the windshield frames in-house at this time is a good strategy. For this windshield frame, it may be cheaper to use the services of a supplier that has both the experience with, and the machinery for, laser blanking. Ford's lack of skill in laser blanking may mean that it will take six months to get up to the required production volume. If, however, Ford relies on a supplier, it can only assume that supplier labour problems will not halt the production of Ford's parts. The make-or-buy decision depends on two factors: the amount of new investment that is required in laser welding and whether additional machinery will be required for future products. Assuming a lifetime of 10 years and an interest rate of 16%, recommend the best course of action. Assume also that the salvage value at the end of 10 years is estimated to be insignificant for either system. If Ford considers the subcontracting option, what would be the acceptable range of contract bid (unit cost per part)?

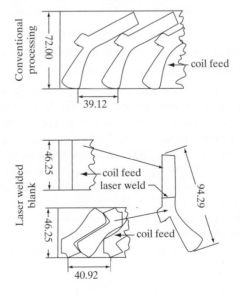

ST6.4 A Credit Valley hospital is to decide which type of boiler fuel system will most efficiently provide the required steam energy output for heating, laundry, and sterilization purposes. The current boilers were installed in the early 1950s and are now obsolete. Much of the auxiliary equipment is also old and in need of repair. Because of these general conditions, an engineering recommendation was made to replace the entire plant with a new boiler plant building that would house modern equipment. The cost of demolishing the old boiler plant would be almost a complete loss, as the salvage value of the scrap steel and used brick was estimated to be only about $1000. The hospital's engineer finally selected two alternative proposals as being worthy of more intensive analysis. The hospital's annual energy requirement, measured in terms of steam output, is approximately 66 Gg of steam. As a rule of thumb for analysis, 1 kg of steam is approximately 2.32 MJ, and 26.8m^3 of natural gas is also approximately 1 GJ. The two alternatives are as follows:

- **Proposal 1.** Replace the old plant with a new coal-fired boiler plant that costs $1,770,300. To meet the requirements for particulate emission as set by Environment Canada, this coal-fired boiler, even if it burned low-sulfur coal, would need an electrostatic precipitator, which would cost approximately $100,000. The plant would last for 20 years. One kilogram of dry coal yields about 33.24 MJ. To find the heat input requirements, it is necessary to divide by the relative boiler efficiency for the type of fuel. The boiler efficiency for coal is 0.75. The price of coal is estimated to be $125/tonne.

- **Proposal 2.** Build a gas-fired boiler plant with heat oil, and use the new plant as a standby. This system would cost $889,200 and have an expected service life of 20 years. Since small household or commercial gas users that are entirely dependent on gas have priority, large plants must have an oil switchover capability. It has been estimated that 6% of 66 Gg of steam energy (or 3.96 Gg) would come about as a result of the switch to oil. The boiler efficiency with each fuel would be 0.78 for gas and 0.81 for oil, respectively. The heat value of natural gas is approximately 39 MJ/m^3, and for heating oil it is 38.6 MJ/L. The estimated gas price is 40¢/m^3, and the price of heating oil is 86¢/L.

(a) Calculate the annual fuel costs for each proposal.
(b) Determine the unit cost per steam kg for each proposal. Assume that $i = 10\%$.
(c) Which proposal is the more economical?

ST6.5 Toxic Toy Company is facing an uncertain but impending deadline to phase out a phthalate called DINP from its infant toy products. Used to "soften" the vinyl, it is found in a wide range of consumer products, including perfumes, nail polish, vinyl floors, detergents, lubricants, food packaging, soap, paint, shampoo, toys, air fresheners, and plastic bags. Studies suggest that certain phthalates are hazardous to reproduction and development and may cause health effects such as liver and kidney failure in young children when products are sucked or chewed for extended periods. On June 19, 2009, Health Minister Leona Aglukkaq announced that the government is moving to ban phthalates in soft vinyl toys and children's products; however, the timeline is still unknown. Toxic has been pursuing other means of creating similar plastics using safer compounds, and they are considering two options (over an eight-year time frame) for modifying their manufacturing process:

- **Option 1:** Retrofitting the plant now to adapt the chemical process to be DNIP-free, and remain a market leader in infant toys. Because implementing the new processes on a large scale is untested, it may cost more to operate the facility for some time while learning the new system.

- **Option 2:** Deferring the retrofitting until federal regulations are introduced would be cheaper due to expected improvement in plastics processing technology and know-how, but there will be tough market competition.

The financial data for the two options, given below, remain constant after the retrofit:

	Option 1	Option 2
Investment timing	Now	n months from now
Initial investment	$10 million	$10 million $-$ $100,000/month
System life	8 years	8 years
Salvage value	$1 million	$1 million $+$ $50,000/month
Annual revenue	$15 million	$15 million $-$ $150,000/month
Annual O&M costs	$6 million	$6 million $+$ $600,000/year

(a) What assumptions must be made when comparing these options?

(b) If Toxic's interest rate is 1%/month (compounded monthly), when are the best and worst times to retrofit the plant?

ST6.6 The Chiller Cooling Technology Company, a maker of automobile air-conditioners, faces an uncertain, but impending, deadline to phase out the traditional chilling technique, which uses chlorofluorocarbons (CFCs), a family of refrigerant chemicals believed to attack the earth's protective ozone layer. Chiller has been pursuing other means of cooling and refrigeration. As a near-term solution, the engineers recommend a cold technology known as absorption chiller, which uses plain water as a refrigerant and semiconductors that cool down when charged with electricity. Chiller is considering two options:

- **Option 1.** Retrofit the plant now to adapt the absorption chiller and continue to be a market leader in cooling technology. Because of untested technology on a large scale, it may cost more to operate the new facility while personnel are learning the new system.

- **Option 2.** Defer the retrofitting until the federal deadline, which is three years away. With expected improvement in cooling technology and technical know-how, the retrofitting cost will be cheaper, but there will be tough market competition, and the revenue would be less than that of Option 1.

The financial data for the two options are as follows:

	Option 1	Option 2
Investment timing	Now	3 years from now
Initial investment	$6 million	$5 million
System life	8 years	8 years
Salvage value	$1 million	$2 million
Annual revenue	$15 million	$11million
Annual O&M costs	$6 million	$7 million

(a) What assumptions must be made when comparing these two options?
(b) If Chiller's MARR is 15%, which option is the better choice, based on the IRR criterion?

ST6.7 An oil company is considering changing the size of a small pump that is currently operational in wells in an oil field. If this pump is kept, it will extract 50% of the known crude-oil reserve in the first year of its operation and the remaining 50% in the second year. A pump larger than the current pump will cost $1.6 million, but it will extract 100% of the known reserve in the first year. The total oil revenues over the two years are the same for both pumps, namely, $20 million. The advantage of the large pump is that it allows 50% of the revenues to be realized a year earlier than with the small pump.

	Current Pump	Larger Pump
Investment (year 0)	0	$1.6 million
Revenue (year 1)	$10 million	$20 million
Revenue (year 2)	$10 million	0

If the firm's MARR is known to be 20%, what do you recommend, based on the IRR criterion?

ST6.8 You have been asked by the president of the company you work for to evaluate the proposed acquisition of a new injection moulding machine for the firm's manufacturing plant. Two types of injection moulding machines have been identified, with the following estimated cash flows:

	Net Cash Flow	
n	Project 1	Project 2
0	−$30,000	−$40,000
1	20,000	43,000
2	18,200	5,000
IRR	18.1%	18.1%

You return to your office, quickly retrieve your old engineering economics text, and then begin to smile: Aha—this is a classic rate-of-return problem! Now, using a calculator, you find out that both projects have about the same rate of return: 18.1%. This figure seems to be high enough to justify accepting the project, but you recall that the ultimate justification should be done with reference to the firm's MARR. You call the accounting department to find out the current MARR the firm should use in justifying a project. "Oh boy, I wish I could tell you, but my boss will be back next week, and he can tell you what to use," says the accounting clerk.

A fellow engineer approaches you and says, "I couldn't help overhearing you talking to the clerk. I think I can help you. You see, both projects have the same IRR, and on top of that, project 1 requires less investment, but returns more cash flows ($-\$30,000 + \$20,000 + \$18,200 = \$8200$, and $-\$40,000 + \$43,000 + \$5,200 = \$8,000$); thus, project 1 dominates project 2. For this type of decision problem, you don't need to know a MARR!"

(a) Comment on your fellow engineer's statement.

(b) At what range of MARRs would you recommend the selection of project 2?

On the Companion Website that accompanies this text, you will find Excel templates and exercises, as well as the following analysis tools: Cash Flow Analyzer, Depreciation Analysis, Loan Analysis, and Interest Tables.

Analysis of Project Cash Flows

CHAPTER SEVEN

Cost Concepts Relevant to Decision Making

***High Hopes for Beer Bottles*[1]** Three hundred billion beer bottles a year worldwide is a mighty tempting target for the plastics industry. Brewers generally say they need a bottle that provides shelf life of over 120 days with less than 15% loss of CO_2 and admittance of no more than 1 ppm of oxygen. Internal or external coatings, and three- or five-layer polyethylene terephthalate (PET) structures using barrier materials are being evaluated to reach that performance.

What is the least expensive way to make a 0.5L PET barrier bottle? Summit International LLC, of Smyrna, Georgia, which specializes in preform and container development and market research, compared the manufacturing costs of five different barrier technologies against a standard monolayer PET bottle. Summit looked at three-layer and five-layer and at internally and externally coated containers, all of them reheat stretch blow moulded. It found the bottle with an external coating to be least costly, while the internally coated bottle was the most expensive.

The firm compared a five-layer structure with an oxygen-scavenger material, a three-layer bottle with a US$5.50/kg barrier material, and a second three-layer structure with a US$13.50/kg barrier. Also compared were a bottle coated inside using Sidel's new Actis plasma technology and a bottle coated on the outside.

Capital investment (preform and bottle machines, utilities, downstream equipment, quality control, spare parts, and installation) for producing 20,000 bottles/hour is US$10.8 million for the five-layer bottle, US$9.9 million for both three-layer structures, US$9.2 million for internal coating, US$7.5 million for external coating, and US$6.8 million with no barrier.

---

[1] "Blow Molding Close-Up: Prospects Brighten for PET Beer Bottles," Mikell Knights, Plastic Technology Online, © copyright 2005, Gardner Publications, Inc.

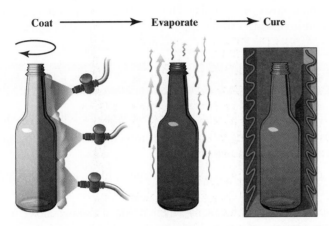

Coat ⟶ Evaporate ⟶ Cure

Spray coating of external PET bottles.

Direct manufacturing cost per 1000 (materials, energy, labour, maintenance, and scrap) amounts to US$66.57 for three layers with the expensive barrier, US$59.35 for five layers, US$55.34 for the external coating, US$54.63 for three layers with the low-cost barrier, US$46.90 for internal coating, and US$44.63 without barrier.

You may be curious how all of those cost data were estimated in this opening vignette. Before we study the different kinds of engineering economic decision problems, we need to understand the concept of various costs. In a manufacturing company, engineers must make decisions involving materials, plant facilities, and the in-house capabilities of company personnel. Consider, for example, the manufacture of food processors. In terms of selecting materials, several of the parts could be made of plastic, whereas others must be made of metal. Once materials have been chosen, engineers must consider the production methods, the shipping weight, and the method of packaging necessary to protect the different types of materials. In terms of actual production, parts may be made in-house or purchased from an outside vendor, depending on the availability of machinery and labour.

Present economic studies: Various economic analyses for short operating decisions.

All these operational decisions (commonly known as **present economic studies** in traditional engineering economic texts) require estimating the costs associated with various production or manufacturing activities. Because these costs also provide the basis for developing successful business strategies and planning future operations, it is important to understand how various costs respond to changes in levels of business activity. In this chapter, we discuss many of the possible uses of cost data. We also discuss how to define, classify, and estimate costs for each use. Our ultimate task in doing so is to explain how costs are classified in making various engineering economic decisions.

CHAPTER LEARNING OBJECTIVES

After completing this chapter, you should understand the following concepts:

- Various cost terminologies that are common in cost accounting and engineering economic studies.

- How a cost item reacts or responds to changes in the level of production or business activities.

- The types of cost data that management needs in making choice between alternative courses of action.

- The types of present economic studies frequently performed by engineers in manufacturing and business environment.

- How to develop a production budget related to operating activities.

7.1 General Cost Terms

In engineering economics, the term **cost** is used in many different ways. Because there are many types of costs, each is classified differently according to the immediate needs of management. For example, engineers may want cost data to prepare external reports, to prepare planning budgets, or to make decisions. Also, each different use of cost data demands a different classification and definition of *cost*. For example, the preparation of external financial reports requires the use of historical cost data, whereas decision making may require current cost data or estimated future cost data.

Our initial focus in this chapter is on manufacturing companies because their basic activities (acquiring raw materials, producing finished goods, marketing, etc.) are commonly found in most other businesses. Therefore, the understanding of costs in a manufacturing company can be helpful in understanding costs in other types of business organizations.

7.1.1 Manufacturing Costs

Several types of manufacturing costs incurred by a typical manufacturer are illustrated in Figure 7.1. In converting raw materials into finished goods, a manufacturer incurs various costs associated with operating a factory. Most manufacturing companies divide manufacturing costs into three broad categories: direct raw material costs, direct labour costs, and manufacturing overhead.

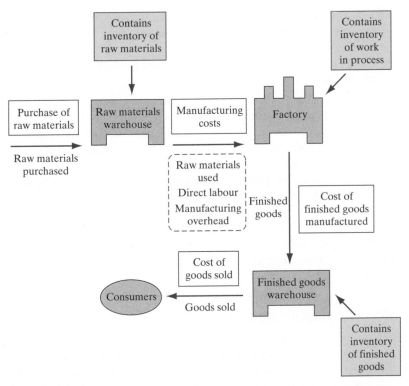

Figure 7.1   Various types of manufacturing costs incurred by a manufacturer.

Direct Raw Materials

Direct raw materials are any materials that are used in the final product and that can be easily traced to it. Some examples are wood in furniture, steel in bridge construction, paper in printing firms, and fabric for clothing manufacturers. It is important to note that the finished product of one company can become the raw materials of another company. For example, the computer chips produced by Intel are a raw material used by Dell in its personal computers.

Direct cost: A cost that can be directly traced to producing specific goods or services.

Direct Labour

Like direct raw materials, direct labour incurs costs that go into the production of a product. The labour costs of assembly-line workers, for example, would be direct labour costs, as would the labour costs of welders in metal-fabricating industries, carpenters or bricklayers in home building, and machine operators in various manufacturing operations.

Manufacturing Overhead

Manufacturing overhead, the third element of manufacturing cost, includes all costs of manufacturing except the costs of direct materials and direct labour. In particular, manufacturing overhead includes such items as the costs of indirect materials; indirect labour; maintenance and repairs on production equipment; heat and light, property taxes, depreciation, and insurance on manufacturing facilities; and overtime premiums. The most important thing to note about manufacturing overhead is the fact that, unlike direct materials and direct labour, it is not easily traceable to specific units of output. In addition,

Overhead: A reference in accounting to all costs not including or related to direct labour, materials, or administration costs.

many manufacturing overhead costs do not change as output changes, as long as the production volume stays within the capacity of the plant. For example, depreciation of factory buildings is unaffected by the amount of production during any particular period. If, however, a new building is required to meet any increased production, manufacturing overhead will certainly increase.

- Sometimes it may not be worth the effort to trace the costs of materials that are relatively insignificant in the finished products. Such minor items include the solder used to make electrical connections in a computer circuit board and the glue used to bind this textbook. Materials such as solder and glue are called *indirect materials* and are included as part of manufacturing overhead.
- Sometimes we may not be able to trace some of the labour costs to the creation of a product. We treat this type of labour cost as a part of manufacturing overhead, along with indirect materials. *Indirect labour* includes the wages of janitors, supervisors, material handlers, and night security guards. Although the efforts of these workers are essential to production, it would be either impractical or impossible to trace their costs to specific units of product. Therefore, we treat such labour costs as indirect labour costs.

7.1.2 Nonmanufacturing Costs

Two additional costs incurred in supporting any manufacturing operation are (1) marketing or selling costs and (2) administrative costs. Marketing or selling costs include all costs necessary to secure customer orders and get the finished product or service into the hands of the customer. Breakdowns of these types of costs provide data for control over selling and administrative functions in the same way that manufacturing cost breakdowns provide data for control over manufacturing functions. For example, a company incurs costs for

- **Overhead.** Heat and light, property taxes, and depreciation or similar items associated with the company's selling and administrative functions.
- **Marketing.** Advertising, shipping, sales travel, sales commissions, and sales salaries. Marketing costs include all executive, organizational, and clerical costs associated with sales activities.
- **Administrative functions.** Executive compensation, general accounting, public relations, and secretarial support, associated with the general management of an organization.

Matching principle: The accounting principle that requires the recognition of all costs that are associated with the generation of the revenue reported in the income statement.

7.2 Classifying Costs for Financial Statements

For purposes of preparing financial statements, we often classify costs as either period costs or product costs. To understand the difference between them, we must introduce the matching concept essential to any accounting studies. In financial accounting, the **matching principle** states that *the costs incurred in generating a certain amount of revenue should be recognized as expenses in the same period that the revenue is recognized.* This matching principle is the key to distinguishing between period costs and product costs. Some costs are matched against periods and become expenses immediately. Other costs are matched against products and do not become expenses until the products are sold, which may be in the next accounting period.

7.2.1 Period Costs

Period costs are costs charged to expenses in the period in which they are incurred. The underlying assumption is that the associated benefits are received in the same period the cost is incurred. Some specific examples are all general and administrative expenses, selling expenses, and insurance and income tax expenses. Therefore, advertising costs, executives' salaries, sales commissions, public-relations costs, and other nonmanufacturing costs discussed earlier would all be period costs. Such costs are not related to the production and flow of manufactured goods, but are deducted from revenue in the income statement. In other words, period costs will appear on the income statement as expenses during the time in which they occur.

7.2.2 Product Costs

Some costs are better matched against products than they are against periods. Costs of this type—called **product costs**—consist of the costs involved in the purchase or manufacture of goods. In the case of manufactured goods, product costs are the costs of direct materials, direct labour costs, and manufacturing overhead. Product costs are not viewed as expenses; rather, they are the cost of creating inventory. Thus, product costs are considered an asset until the associated goods are sold. At the time they are sold, the costs are released from inventory as expenses (typically called cost of goods sold) and matched against sales revenue. Since product costs are assigned to inventories, they are also known as *inventory costs*. In theory, product costs include all manufacturing costs—that is, all costs relating to the manufacturing process. As shown in Figure 7.2, product costs appear on financial statements when the inventory, or final goods, is sold, not when the product is manufactured.

   To understand product costs more fully, let us look briefly at the flow of costs in a manufacturing company. By doing so, we will be able to see how product costs move through the various accounts and affect the balance sheet and the income statement in the course of the manufacture and sale of goods. The flows of period costs and product costs through the financial statements are illustrated in Figure 7.3. All product costs filter through the balance-sheet statement as "inventory cost." If the product gets sold, the inventory costs in the balance-sheet statement are transferred to the income statement under the head "cost of goods sold."

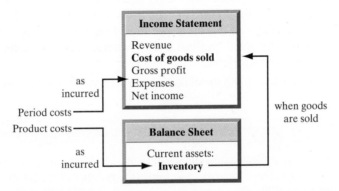

Figure 7.2   How the period costs and product costs flow through financial statements from the manufacturing floor to sales.

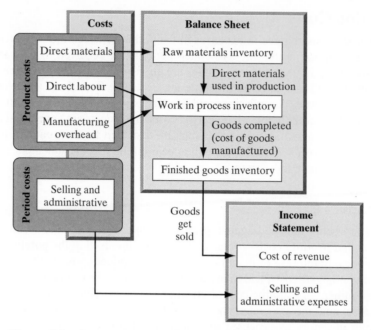

Figure 7.3   Cost flows and classifications in a manufacturing company.

- **Raw-materials inventory.** This account in the balance-sheet statement represents the unused portion of the raw materials on hand at the end of the fiscal year.
- **Work-in-process inventory.** This balance-sheet entry consists of the partially completed goods on hand in the factory at year-end. When raw materials are used in production, their costs are transferred to the work-in-process inventory account as direct materials. Note that direct labour cost and manufacturing overhead cost are also added directly to work-in-process, which can be viewed as the assembly line in a manufacturing plant, where workers are stationed and where products slowly take shape as they move from one end of the line to the other.
- **Finished-goods inventory.** This account shows the cost of finished goods that are on hand and awaiting sale to customers at year-end. As the goods are completed, accountants transfer the cost from the work-in-process account into the finished-goods account. At this stage, the goods await sale to a customer. As the goods are sold, their cost is transferred from the finished-goods account into the cost-of-goods-sold (or cost-of-revenue) account. At this point, we finally treat the various material, labour, and overhead costs that were involved in the manufacture of the units being sold as *expenses* in the income statement.

Example 7.1 serves to explain the classification scheme for financial statements.

EXAMPLE 7.1 Classifying Costs for Uptown Ice Cream Shop

Here is a look at why it costs $2.50 for a single-dip ice cream cone at a typical store in Edmonton, Alberta. The annual sales volume (the number of ice cream cones sold) averages around 185,000 cones, bringing in revenue of $462,500. This is equivalent to selling more than 500 cones a day, assuming a seven-day operation. The following table shows the unit price of an ice cream cone and the costs that go into producing the product:

Items	Total Cost	Unit Cost*	% of Price
Ice cream (cream, sugar, milk, and milk solids)	$120,250	$0.65	26%
Cone	9,250	0.05	2
Rent	132,275	0.71	29
Wages	46,250	0.25	10
Benefit	9,250	0.05	2
GST	23,125	0.13	5
Income taxes	14,800	0.08	3
Debt service	42,550	0.23	9
Supplies	16,650	0.09	4
Utilities	14,800	0.08	3
Other expenses (insurance, advertising, fees, and heating and lighting for shop)	9,250	0.05	2
Profit	24,050	0.13	5
Total	**$462,500**	**$2.50**	**100**

*Based on an annual volume of 185,000 cones.

If you were to classify the operating costs into either product costs or period costs, how would you do it?

SOLUTION:

Given: Financial data just described.

Find: Classify the cost elements into product costs and period costs.

The following is a breakdown of the two kinds of costs:

- **Product costs:** Costs incurred in preparing 185,000 ice cream cones per year.

Raw materials:	
Ice cream @ $0.65	$120,250
Cone @ $0.05	9,250
Labour:	
Wages @ $0.25	46,250
Benefits @ $0.05	9,250
Overhead:	
Supplies @ $0.09	16,650
Utilities @ $0.08	14,800
Total product cost	$216,450

- **Period costs:** Costs incurred in running the shop regardless of sales volume.

Business taxes:	
GST @ $0.13	$ 23,125
Income taxes @ $0.08	14,800
Operating expenses:	
Rent @ $0.71	132,275
Debt service @ $0.23	42,550
Other @ $0.05	9,250
Total period cost	$222,000

7.3 Cost Classification for Predicting Cost Behaviour

In engineering economic analysis, we need to predict how a certain cost will behave in response to a change in activity. For example, a manager will want to estimate the impact a 5% increase in production will have on the company's total wages before he or she decides whether to alter production. **Cost behaviour** describes how a cost item will react or respond to changes in the level of business activity.

7.3.1 Volume Index

In general, the operating costs of any company are likely to respond in some way to changes in the company's operating volume. In studying cost behaviour, we need to determine some measurable volume or activity that has a strong influence on the amount of cost incurred. The unit of measure used to define volume is called a **volume index**. A volume index may be based on production inputs, such as tonnes of coal processed, direct labour hours used, or machine-hours worked; or it may be based on production outputs, such as the number of kilowatt-hours generated. For a vehicle, the number of kilometres driven per year may be used as a volume index. Once we identify a volume index, we try to find out how costs change in response to changes in the index.

7.3.2 Cost Behaviours

Accounting systems typically record the cost of resources acquired and track their subsequent usage. Fixed costs and variable costs are the two most common cost behaviour patterns. An additional category known as "mixed costs" contains two parts, the first of which is fixed and the other of which varies with the volume of output.

Fixed cost: A cost that remains constant, regardless of any change in a company's activity.

Fixed Costs

The costs of providing a company's basic operating capacity are known as the company's **fixed cost** or **capacity cost**. For a cost item to be classified as fixed, it must have a relatively wide span of output over which costs are expected to remain constant. This span is called the **relevant range**. In other words, fixed costs do not change within a given period, although volume may change. In the case of an automobile, for example, the annual in-surance premium, and licence fee are fixed costs, since they are independent of the number of kilometres driven per year. Some other typical examples are building rents; depreciation of buildings, machinery, and equipment; and salaries of administrative and production personnel. In our Uptown Ice Cream Shop example, we may classify expenses such as rent, debt service, and other (insurance, advertising, professional fees) as fixed costs (costs that are fixed in total for a given period of time and for a wide range of production levels).

Variable Costs

In contrast to fixed operating costs, **variable operating costs** have a close relationship to the level of volume of a business. If, for example, volume increases 10%, a total variable cost will also increase by approximately 10%. Gasoline is a good example of a variable automobile cost because fuel consumption is directly related to kilometres driven. Similarly, the cost of replacing tires will increase as a vehicle is driven more.

In a typical manufacturing environment, direct labour and material costs are major variable costs. In our Uptown Shop example, the variable costs would include the cost of the ice cream and cone (direct materials), wages, benefits, income taxes, GST, and supplies. Both GST and business taxes are related to sales volume. In other words, if the store becomes busy, more servers are needed, which will increase the payroll as well as taxes.

> **Variable cost:** A cost that changes in proportion to a change in a company's activity or business.

Mixed Costs

Some costs do not fall precisely into either the fixed or the variable category, but contain elements of both. We refer to these as mixed costs (or **semivariable costs**). In our automobile example, **depreciation** (loss of value) is a mixed cost. On the one hand, some depreciation occurs simply from the passage of time, regardless of how many kilometres a car is driven, and this represents the fixed portion of depreciation. On the other hand, the more kilometres an automobile is driven a year, the faster it loses its market value, and this represents the variable portion of depreciation. A typical example of a mixed cost in manufacturing is the cost of electric power. Some components of power consumption, such as lighting, are independent of the operating volume, while other components (e.g., the number of machine-hours equipment is operated) may vary directly with volume. In our Uptown Shop example, the utility cost can be a mixed cost item: Some lighting and heating requirements might stay the same, but the use of power mixers will be in proportion to sales volume.

> **Mixed cost:** Costs are fixed for a set level of production or consumption, becoming variable after the level is exceeded.

Average Unit Cost

The foregoing description of fixed, variable, and mixed costs was expressed in terms of total volume over a given period. We often use the term **average cost** to express activity cost on a per unit basis. In terms of unit costs, the description of cost is quite different:

- The variable cost per unit of volume is a constant.
- Fixed cost per unit varies with changes in volume: As the volume increases, the fixed cost per unit decreases.
- The mixed cost per unit also changes as volume changes, but the amount of change is smaller than that for fixed costs.

To explain the behaviour of the fixed, variable, mixed, and average costs in relation to volume, we will use the popular *Driving Costs* brochure prepared annually by the

Canadian Automobile Association (CAA). CAA announced the release of its 2009 edition of this brochure on May 20, 2009. To access this brochure, please visit www.caa.ca/publicAffairs/public-affairs-reports-e.cfm. This brochure can help you calculate how much it costs to own and operate your private vehicle each year. The national averages and approximate driving costs provided will help you understand what factors affect the cost of driving.

The assumptions used in preparation of the *Driving Costs* brochure are summarized in Table 7.1. Based on these assumptions, the costs of owning and operating a 2009 Chevrolet Cobalt LT four-door sedan are estimated as follows:

Operating costs:

• Fuel	6.79 cents per kilometre
• Maintenance	2.36 cents per kilometre
• Tires	1.55 cents per kilometre
Total	10.70 cents per kilometre

Ownership costs:

• Comprehensive insurance ($250 deductible)	$1780
• Licence, registration, taxes	$111
• Depreciation (18,000 kilometres annually)	$3857
• Finance charge (10% down, loan @7.25%/4 years)	$768
Cost per year	$6516
Cost per day	$17.85
Added depreciation costs (per 1000 kilometres over 18,000 kilometres annually)	$24

TABLE 7.1 Assumptions Used in Calculating the Average Cost of Owning and Operating a New Vehicle

What's Covered	Costs Base
Fuel	National average self-service price 81.2 ¢ per litre for regular-grade gasolines.
Maintenance	Costs of retail parts and labour for routine maintenance, as specified by the vehicle manufacturer.
Tires	Costs are based on the price of one set of replacement tires of the same quality, size, and ratings as those that came with the vehicle.
Insurance	A full-coverage policy for a mature driver with a good driving record and commuting fewer than 16 kilometres per day to and from work.
Licence, Registration, and Taxes	All government taxes and fees payable at time of purchase, as well as fees due each year to keep the vehicle licensed and registered.
Depreciation	Based on the difference between the new-vehicle purchase price and the estimated trade-in value at the end of four years.
Finance	Based on a four-year loan at 7.25% interest with a 10% down payment.

Total cost per kilometre:

• Operating cost per kilometre × 18,000 kilometres	$1926
• Ownership cost per day × 365 days per year	$6516
Total cost per year	$8442
Average cost per kilometre ($8442/18,000)	46.9 cents

Now, if you drive the same vehicle for 24,000 kilometres instead of 18,000 kilometres, you may be interested in knowing how the average cost per kilometre would change. Example 7.2 illustrates how you determine the average cost as a function of mileage.

EXAMPLE 7.2 Calculating Average Cost per Kilometre as a Function of Mileage

Table 7.2 itemizes the operating and ownership costs associated with driving a passenger car by fixed, variable, and mixed classes. Note that the only change from the preceding list is in the depreciation amount. Using the given data, develop a cost–volume chart and calculate the average cost per kilometre as a function of the annual mileage.

DISCUSSION: First we may examine the effect of driving an additional 1000 kilometres over the allotted 18,000 kilometres. Since the loss in the car's value due to driving an additional 1000 kilometres over 18,000 kilometres is estimated to be $24, the total cost per year and the average cost per kilometre, based on 19,000 kilometres, can be recalculated as follows:

- Added depreciation cost: $24.
- Added operating cost: 1000 kilometres × 10.70 cents = $107.
- Total cost per year: $8442 + $24 + $107 = $8573.
- Average cost per kilometre ($8573/19,000 kilometres): 45.12 cents per kilometre.

Note that the average cost comes down as you drive more, as the ownership cost per kilometre is further reduced.

SOLUTION

Given: Financial data.

Find: The average cost per kilometre at an annual operating volume between 18,000 and 24,000 kilometres.

In Table 7.3, we summarize the costs of owning and operating the automobile at various annual operating volumes from 18,000 to 24,000 kilometres. Once the total cost figures are available at specific volumes, we can calculate the effect of volume on unit (per kilometre) costs by converting the total cost figures to average unit costs.

To estimate annual costs for any assumed mileage, we construct a **cost–volume diagram** as shown in Figure 7.4(a). Further, we can show the relation between volume (kilometres driven per year) and the three cost classes separately, as in Figure 7.4(b) through (d). We can use these cost–volume graphs to estimate both the separate and

TABLE 7.2 Cost Classification of Owning and Operating a Passenger Car

Cost Classification	Reference	Cost
Variable costs:		
Fuel per kilometre		$0.068
Maintenance per kilometre		$0.024
Tires per kilometre		$0.016
Annual fixed costs:		
Insurance (comprehensive)		$1780
Licence, registration, taxes		$111
Finance charge		$768
Mixed costs: Depreciation		
Fixed portion per year (18,000 kilometres)		$3857
Variable portion per kilometre (above 18,000 kilometres)		$0.024

combined costs of operating the car at other possible volumes. For example, an owner who expects to drive 24,000 kilometres in a given year may estimate the total cost at $9227.25, or 38.4 cents per kilometre. In Figure 7.4, all costs more than those necessary to operate at the zero level are known as variable costs. Since the fixed cost is $6516 a year, the remaining $2711.25 is variable cost. By combining all the fixed and variable elements of cost, we can state simply the following: if the total mileage

TABLE 7.3 Operating Costs as a Function of Mileage Driven

Volume Index (km)	18,000	19,000	20,000	21,000	22,000	23,000	24,000
Variable costs ($0.107 per km)	$1926	$2033	$2140	$2247	$2354	$2461	$2568
Fixed costs:	2659	2659	2659	2659	2659	2659	2659
Mixed costs (depreciation):							
Fixed portion:	3857	3857	3857	3857	3857	3857	3857
Variable portion ($0.024/km):	–	24	48	72	96	120	144
Total variable cost	1926	2057	2188	2319	2450	2581	2712
Total fixed cost	6516	6516	6516	6516	6516	6516	6516
Total costs	$8442	$8573	$8704	$8835	$8966	$9097	$9228
Cost per km	$0.4690	$0.4512	$0.4352	$0.4207	$0.4075	$0.3955	$0.3845

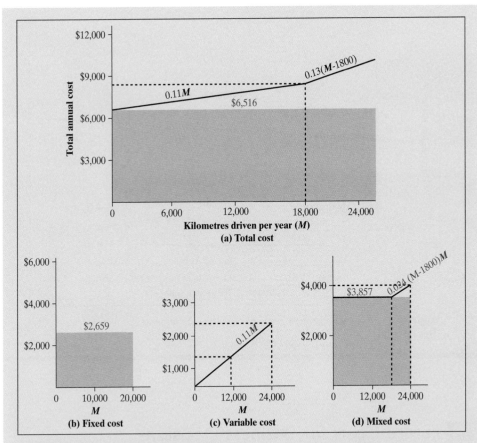

Figure 7.4 Cost–volume relationships pertaining to annual automobile costs (Example 7.2).

driven in a year is no more than 18,000 kilometres, the total cost of owning and operating the vehicle is simply $6536 plus 10.70 cents for each kilometre driven; if the total mileage driven in a year is more than 18,000 kilometres, the total cost of owning and operating the vehicle will be $8442 plus 13.10 cents for each kilometre driven beyond 18,000 kilometres. Note that the slope of the variable cost curve changes at 18,000 kilometres driven per year. This is because if the mileage driven in a year is lower than 18,000 kilometres, the total depreciation does not change. On the other hand, if the mileage driven in a year is more than 18,000 kilometres, there is a depreciation charge of $24 for each additional kilometre.

Figure 7.5 illustrates graphically the average unit cost of operating the automobile. The average fixed unit cost, represented by the height of the middle curve in the figure, declines steadily as the volume increases. The average unit cost is high when volume is low because the total fixed cost is spread over a relatively few units of volume. In other words, the total fixed costs remain the same regardless of the number of kilometres driven, but the average fixed cost decreases on a per kilometre basis as the number of kilometres driven increases.

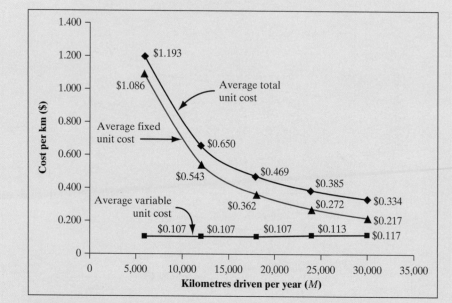

Figure 7.5 Average cost per kilometre of owning and operating a car (Example 7.2).

7.4 Future Costs for Business Decisions

In the previous sections, our focus has been on classifying cost data that serve management's need to control and evaluate the operations of a firm. However, these data are historical in nature; they may not be suitable for management's planning future business operations. We are not saying that historical cost data are of no use in making decisions about the future. In fact, they serve primarily as the first step in predicting the uncertain future. However, the types of cost data that management needs in making choices between alternative courses of action are different from historical cost data.

7.4.1 Differential Cost and Revenue

As we have seen throughout the text, decisions involve choosing among alternatives. In business decisions, each alternative has certain costs and benefits that must be compared with the costs and benefits of the other available alternatives. A difference in cost between any two alternatives is known as a **differential cost**. Similarly, a difference in revenue between any two alternatives is known as **differential revenue**. A differential cost is also known as an incremental cost, although, technically, an incremental cost should refer only to an increase in cost from one alternative to another.

Cost–volume relationships based on differential costs find many engineering applications. In particular, they are useful in making a variety of short-term operational decisions. Many short-run problems have the following characteristics:

- The base case is the status quo (the current operation or existing method), and we propose an alternative to the base case. If we find the alternative to have lower costs than the base case, we accept the alternative, assuming that nonquantitative factors do not offset the cost advantage. The **differential (incremental) cost** is the difference in total cost that results from selecting one alternative instead of another. If several alternatives are possible, we select the one with the maximum savings from the base. Problems of this type are often called trade-off problems because one type of cost is traded off for another.
- New investments in physical assets are not required.
- The planning horizon is relatively short (a week or a month—certainly less than a year).
- Relatively few cost items are subject to change by management decision.

> **Incremental cost** is the overall change that a company experiences by producing one additional unit of good.

Some common examples of short-run problems are method changes, operations planning, and make-or-buy decisions.

Method Changes

Often, we may derive the best information about future costs from an analysis of historical costs. Suppose the proposed alternative is to consider some new method of performing an activity. Then, as Example 7.3 shows, if the differential costs of the proposed method are significantly lower than the current method, we adopt the new method.

EXAMPLE 7.3 Differential Cost Associated With Adopting a New Production Method

The engineering department at an auto-parts manufacturer recommends that the current dies (the base case) be replaced with higher quality dies (the alternative), which would result in substantial savings in manufacturing one of the company's products. The higher cost of materials would be more than offset by the savings in machining time and electricity. If estimated monthly costs of the two alternatives are as shown in the accompanying table, what is the differential cost for going with better dies?

SOLUTION

Given: Financial data.
Find: Which production method is preferred.

In this problem, the differential cost is $-\$5000$ a month. The differential cost's being negative indicates a saving, rather than an addition to total cost. An important point to remember is that differential costs usually include variable costs, but fixed costs are affected only if the decision involves going outside of the relevant range. Although the production volume remains unchanged in our example, the slight increase in depreciation expense is due to the acquisition of new machine tools (dies). All other items of fixed cost remained within their relevant ranges.

	Current Dies	Better Dies	Differential Cost
Variable costs:			
Materials	$150,000	$170,000	+$20,000
Machining labour	85,000	64,000	−21,000
Electricity	73,000	66,000	−7,000
Fixed costs:			
Supervision	25,000	25,000	0
Taxes	16,000	16,000	0
Depreciation	40,000	43,000	+3,000
Total	$392,000	$387,000	−$5,000

Operations Planning

In a typical manufacturing environment, when demand is high, managers are interested in whether to use a one-shift-plus-overtime operation or to add a second shift. When demand is low, it is equally possible to explore whether to operate temporarily at very low volume or to shut down until operations at normal volume become economical. In a chemical plant, several routes exist for scheduling products through the plant. The problem is which route provides the lowest cost. Example 7.4 illustrates how engineers may use cost–volume relationships in a typical operational analysis.

Break-even analysis: An analysis of the level of sales at which a project would make zero profit.

EXAMPLE 7.4 Break-Even Volume Analysis

Sandstone Corporation has one of its manufacturing plants operating on a single-shift five-day week. The plant is operating at its full capacity (24,000 units of output per week) without the use of overtime or extra shifts. Fixed costs for single-shift operation amount to $90,000 per week. The average variable cost is a constant $30 per unit, at all output rates, up to 24,000 units per week. The company has received an order to produce an extra 4000 units per week beyond the current single-shift maximum capacity. Two options are being considered to fill the new order:

- **Option 1.** Increase the plant's output to 36,000 units a week by adding overtime, by adding Saturday operations, or both. No increase in fixed costs is entailed, but the variable cost is $36 per unit for any output in excess of 24,000 units per week, up to a 36,000-unit capacity.
- **Option 2.** Operate a second shift. The maximum capacity of the second shift is 21,000 units per week. The variable cost of the second shift is $31.50 per unit, and the operation of a second shift entails additional fixed costs of $13,500 per week.

Determine the range of operating volume that will make Option 2 profitable.

SOLUTION

Given: Financial data.

Find: The break-even volume that will make both options indifferent.

In this example, the operating costs related to the first-shift operation will remain unchanged if one alternative is chosen instead of another. Therefore, those costs are irrelevant to the current decision and can safely be left out of the analysis. Consequently, we need to examine only the increased total cost due to the additional operating volume under each option. (This kind of study is known as *incremental analysis*.) Let Q denote the additional operating volume. Then we have

- **Option 1.** Overtime and Saturday operation: $\$36Q$.
- **Option 2.** Second-shift operation: $\$13,500 + \$31.50Q$.

We can find the break-even volume (Q_b) by equating the incremental cost functions and solving for Q:

$$36Q = 13,500 + 31.5Q,$$
$$4.5Q = 13,500,$$
$$Q_b = 3000 \text{ units.}$$

If the additional volume exceeds 3000 units, the second-shift operation becomes more efficient than overtime or Saturday operation. A break-even (or cost–volume) graph based on the foregoing data is shown in Figure 7.6. The horizontal scale

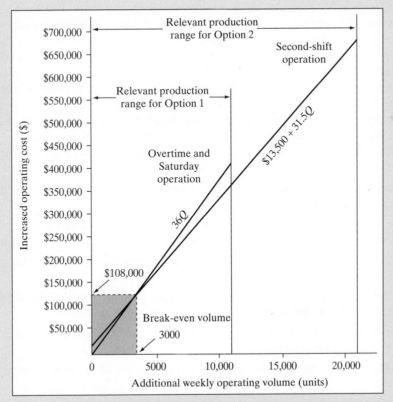

Figure 7.6 Cost–volume relationships of operating overtime and a Saturday operation versus second-shift operation beyond 24,000 units (Example 7.4).

represents additional volume per week. The upper limit of the relevant volume range for Option 1 is 12,000 units, whereas the upper limit for Option 2 is 21,000 units. The vertical scale is in dollars of cost. The operating savings expected at any volume may be read from the cost–volume graph. For example, the break-even point (zero savings) is 3000 units per week. Option 2 is a better choice, since the additional weekly volume from the new order exceeds 3000 units.

Make-or-Buy (Outsourcing) Decision

In business, the make-or-buy decision arises on a fairly frequent basis. Many firms perform certain activities using their own resources and pay outside firms to perform certain other activities. It is a good policy to constantly seek to improve the balance between these two types of activities by asking whether we should outsource some function that we are now performing ourselves or vice versa. Make-or-buy decisions are often grounded in the concept of *opportunity cost*.

7.4.2 Opportunity Cost

Opportunity cost: The benefits you could have received by taking an alternative action.

Opportunity cost may be defined as the potential benefit that is given up as you seek an alternative course of action. In fact, virtually every alternative has some opportunity cost associated with it. For example, suppose you have a part-time job while attending college that pays you $200 per week. You would like to spend a week at the beach during spring break, and your employer has agreed to give you the week off. What would be the opportunity cost of taking the time off to be at the beach? The $200 in lost wages would be an opportunity cost.

In an economic sense, opportunity cost could mean the contribution to income that is forgone by not using a limited resource in the best way possible. Or we may view opportunity costs as cash flows that could be generated from an asset the firm already owns, provided that such flows are not used for the alternative in question. In general, *accountants do not post opportunity cost in the accounting records of an organization. However, this cost must be explicitly considered in every decision.* In sum,

- An opportunity cost arises when a project uses a resource that may already have been paid for by the firm.
- When a resource that is already owned by a firm is being considered for use in a project, that resource has to be priced on its next-best alternative use, which may be

  1. A sale of the asset, in which case the opportunity cost is the expected proceeds from the sale, net of any disposal tax effect.
  2. Renting or leasing the asset out, in which case the opportunity cost is the expected present value of the after-tax revenue from the rental or lease.
  3. Some use elsewhere in the business, in which case the opportunity cost is the cost of replacing the resource.
  4. That the asset has been abandoned or is of no use. Then the opportunity cost is zero.

EXAMPLE 7.5 Opportunity Cost: Lost Rental Income (Opportunity Cost)

Benson Company is a farm equipment manufacturer that currently produces 20,000 units of gas filters annually for use in its lawn mower production. The expected annual production cost of the gas filters is summarized as follows:

Variable costs:	
Direct materials	$100,000
Direct labour	190,000
Power and water	35,000
Fixed costs:	
Heating and light	20,000
Depreciation	100,000
Total cost	**$445,000**

Tompkins Company has offered to sell Benson 20,000 units of gas filters for $17 per unit. If Benson accepts the offer, some of the manufacturing facilities currently used to manufacture the filters could be rented to a third party at an annual rent of $35,000. Should Benson accept Tompkins's offer, and why?

SOLUTION

Given: Financial data; production volume = 20,000 units.

Find: Whether Benson should outsource the gas filter operation.

	Make Option	Buy Option	Differential Cost (Make–Buy)
Variable costs:			
Direct materials	$100,000		$100,000
Direct labour	190,000		190,000
Power and water	35,000		35,000
Gas filters		340,000	−340,000
Fixed costs:			
Heating and light	20,000	20,000	0
Depreciation	100,000	100,000	0
Rental income lost	35,000		35,000
Total cost	$480,000	$460,000	$20,000
Unit cost	$24.00	$23.00	$1.00

This problem is unusual in the sense that the buy option would generate a rental fee of $35,000. In other words, Benson could rent out the current manufacturing facilities if it

> were to purchase the gas filters from Tompkins. To compare the two options, we need to examine the cost of each option.
>
> The buy option has a lower unit cost and saves $1 for each use of a gas filter. If the lost rental income (opportunity cost) were not considered, however, the decision would favor the make option.

7.4.3 Sunk Costs

Sunk cost: A cost that has been incurred and cannot be reversed.

A **sunk cost** is a cost that has already been incurred by past actions. Sunk costs are not relevant to decisions because they cannot be changed regardless of what decision is made now or in the future. The only costs relevant to a decision are costs that vary among the alternative courses of action being considered. To illustrate a sunk cost, suppose you have a very old car that requires frequent repairs. You want to sell the car, and you figure that the current market value would be about $1200 at best. While you are in the process of advertising the car, you find that the car's water pump is leaking. You decided to have the pump repaired, which cost you $200. A friend of yours is interested in buying your car and has offered $1300 for it. Would you take the offer, or would you decline it simply because you cannot recoup the repair cost with that offer? In this example, the $200 repair cost is a sunk cost. You cannot change this repair cost, regardless of whether you keep or sell the car. Since your friend's offer is $100 more than the best market value, it would be better to accept the offer.

7.4.4 Marginal Cost

We make decisions every day, as entrepreneurs, professionals, executives, investors, and consumers, with little thought as to where our motivations come from or how our assumptions of logic fit a particular academic regimen. As you have seen, the engineering economic decisions that we describe throughout this text owe much to English economist Alfred Marshall (1842–1924) and his concepts of **marginalism**. In our daily quest for material gain, rarely do we recall the precepts of microeconomics or marginal utility, yet the ideas articulated by Marshall remain some of the most useful core principles guiding economic decision making.

Definition

Marginal cost: The cost associated with one additional unit of production.

Another cost term useful in cost–volume analysis is marginal cost. We define **marginal cost** as the added cost that would result from increasing the rate of output by a single unit. The accountant's differential-cost concept can be compared to the economist's marginal-cost concept. In speaking of changes in cost and revenue, the economist employs the terms *marginal cost* and *marginal revenue*. The revenue that can be obtained from selling one more unit of product is called **marginal revenue**. The cost involved in producing one more unit of product is called **marginal cost**.

EXAMPLE 7.6 Marginal Costs versus Average Costs

Consider a company that has an available electric load of 37 horsepower and that purchases its electricity at the following rates:

kWh/Month	@$/kWh	Average Cost ($/kWh)
First 1500	$0.050	$0.050
Next 1250	0.035	$\dfrac{\$75 + 0.0350(X - 1500)}{X}$
Next 3000	0.020	$\dfrac{\$118.75 + 0.020(X - 2750)}{X}$
All over 5750	0.010	$\dfrac{\$178.25 + 0.010(X - 5750)}{X}$

According to this rate schedule, the unit variable cost in each rate class represents the marginal cost per kilowatt-hours (kWh). Alternatively, we may determine the average costs in the third column by finding the cumulative total cost and dividing it by the total number of kWh (X). Suppose that the current monthly consumption of electric power averages 3200 kWh. On the basis of this rate schedule, determine the marginal cost of adding one more kWh and, for a given operating volume (3200 kWh), the average cost per kWh.

SOLUTION

Given: Marginal cost schedule for electricity; operating volume $= 3200$ kWh.

Find: Marginal and average cost per kWh.

The marginal cost of adding one more kWh is $0.020. The average variable cost per kWh is calculated as follows:

kWh	Rate ($/kWh)	Cost
First 1500	0.050	$75.00
Next 1250	0.035	43.75
Remaining 450	0.020	9.00
Total		$127.75

The average variable cost per kWh is $127.75/3200 kWh $=$ $0.0399 kWh. Or we can find the value by using the formulas in the third column of the rate schedule:

$$\frac{\$118.75 + 0.020(3200 - 2750)}{3200} = \frac{\$127.75}{3200 \text{ kWh}} = \$0.0399 \text{ kWh}$$

Changes in the average variable cost per unit are the result of changes in the marginal cost. As shown in Figure 7.7, the average variable cost continues to fall because the marginal cost is lower than the average variable cost over the entire volume.

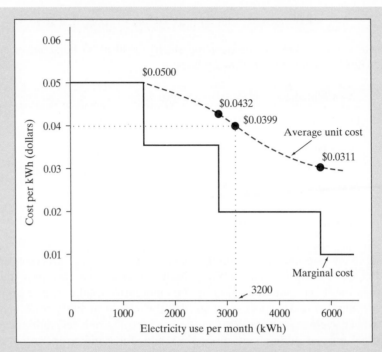

Figure 7.7 Marginal versus average cost per kWh (Example 7.6).

Marginal analysis: A technique used in microeconomics by which very small changes in specific variables are studied in terms of the effect on related variables and the system as a whole.

Marginal Analysis

The fundamental concept of marginal thinking is that you are where you are, and what is past is irrelevant. (This means that we should ignore the sunk cost.) The point is whether you move forward, and you will do so if the benefits outweigh the costs, even if they do so by a smaller margin than before. You will continue to produce a product or service until the cost of doing so equals the revenue derived. Microeconomics suggests that business persons and consumers measure progress not in great leaps, but in small incremental steps. Rational persons will reevaluate strategies and take new actions if benefits exceed costs on a marginal basis. If there is more than one alternative to choose, do so always with the alternative with greatest marginal benefit.

As we have mentioned, the economist's marginal-cost concept is the same as the accountant's differential-cost concept. For a business problem on maximization of profit, economists place a great deal of emphasis on the importance of **marginal analysis**: Rational individuals and institutions should perform the most cost-effective activities first. Specifically,

- When you examine marginal costs, you want to use the least expensive methods first and the most expensive last (if at all).
- Similarly, when you examine marginal revenue (or demand) the most revenue-enhancing methods should be used first, the least revenue enhancing last.

- If we need to consider the revenue along with the cost, the volume at which the total revenue and total cost are the same is known as the *break-even point*. If the cost and revenue function are assumed to be linear, this point may also be calculated, by means of the following formula:

$$\text{Break-even volume} = \frac{\text{Fixed costs}}{\text{Sales price per unit} - \text{Variable cost per unit}}$$

$$= \frac{\text{Fixed costs}}{\text{Marginal contribution per unit}} \qquad (7.1)$$

The difference between the unit sales price and the unit variable cost is the producer's **marginal contribution**, also known as **marginal income**. This means that each unit sold contributes toward absorbing the company's fixed costs.

EXAMPLE 7.7 Profit-Maximization Problem: Marginal Analysis[2]

Suppose you are a chief executive officer (CEO) of a small pharmaceutical company that manufactures generic aspirin. You want the company to maximize its profits. You can sell as many aspirins as you make at the prevailing market price. You have only one manufacturing plant, which is the constraint. You have the plant working at full capacity Monday through Saturday, but you close the plant on Sunday because on Sundays you have to pay workers overtime rates, and it is not worth it. The marginal costs of production are constant Monday through Saturday. Marginal costs are higher on Sundays, only because labour costs are higher.

Now you obtain a long-term contract to manufacture a brand-name aspirin. The costs of making the generic aspirin or the brand-name aspirin are identical. In fact, there is no cost or time involved in switching from the manufacture of one to the other. You will make much larger profits from the brand-name aspirin, but the demand is limited. One day of manufacturing each week will permit you to fulfill the contract. You can manufacture both the brand-name and the generic aspirin. Compared with the situation before you obtained the contract, your profits will be much higher if you now begin to manufacture on Sundays—even if you have to pay overtime wages.

- *Generic aspirin.* Each day, you can make 1000 cases of generic aspirin. You can sell as many as you make, for the market price of $10 per case. Every week you have fixed costs of $5000 (property tax and insurance). No matter how many cases you manufacture, the cost of materials and supplies is $2 per case; the cost of labour is $5 per case, except on Sundays, when it is $10 per case.
- *Brand-name aspirin.* Your order for the brand-name aspirin requires that you manufacture 1000 cases per week, which you sell for $30 per case. The cost for the brand-name aspirin is identical to the cost of the generic aspirin.

What do you do?

[2] "Profit Maximization Problem," by David Hemenway and Elon Kohlberg, *Economic Inquiry,* October 1, 1997, Page 862, Copyright 1997 Western Economic Association International.

SOLUTION

Given: Sales price, $10 per case for generic aspirin, $30 per case for brand aspirin; fixed cost, $5000; variable cost, $7 per case during weekdays, $12 per case on Sunday operation; weekly production, 6000 cases of generic aspirin, 1000 cases of brand-name aspirin.

Find: (a) Optimal production mix and (b) break-even volume.

(a) *Optimal production mix*: The marginal costs of manufacturing brand-name aspirin are constant Monday through Saturday; they rise substantially on Sunday and are above the marginal revenue from manufacturing generic aspirin. In other words, your company should manufacture the brand-name aspirin first. Your marginal revenue is the highest the "first" day, when you manufacture the brand-name aspirin. It then falls and remains constant for the rest of the week. On the seventh day (Sunday), the marginal revenue from manufacturing the generic aspirin is still below the marginal cost. You should manufacture brand-name aspirin one day a week and generic aspirin five days a week. On Sundays, the plant should close.

- The marginal revenue from manufacturing on Sunday is $10,000 (1000 cases times $10 per case).
- The marginal cost from manufacturing on Sunday is $12,000 (1000 cases times $12 per case—$10 labour + $2 materials).
- Profits will be $2000 lower than revenue if the plant operates on Sunday.

(b) *Break-even volume*: Let Q represent the total number of cases of aspirin produced in a week including brand name and generic. The total revenue and cost functions can be represented as follows:

$$\text{Total revenue function:} \begin{cases} 30Q & \text{for } 0 \le Q \le 1000 \\ 30{,}000 + 10(Q - 1000) & \text{for } 1000 < Q \le 7000, \end{cases}$$

$$\text{Total cost function:} \begin{cases} 5000 + 7Q & \text{for } 0 \le Q \le 6000 \\ 47{,}000 + 12(Q - 6000) & \text{for } 6000 \le Q \le 7000. \end{cases}$$

Table 7.4 shows the various factors involved in the production of the brand-name and the generic aspirin.

- If you produce the brand-name aspirin first, the break-even volume is

$$30Q - 7Q - 5000 = 0,$$
$$Q_b = 217.39.$$

- If you produce the generic aspirin first, the break-even volume is

$$10Q - 7Q - 5000 = 0,$$
$$Q_b = 1666.67.$$

Clearly, scheduling the production of the brand-name aspirin first is the better strategy, as you can recover the fixed cost ($5000) much faster by selling just 217.39 cases of brand-name aspirin. Also, Sunday operation is not economical, as the marginal cost exceeds the marginal revenue by $2000, as shown in Figure 7.8.

TABLE 7.4 Net Profit Calculation as a Function of Production Volume

	Total Production Volume (Q)	Product Mix	Incremental Revenue	Total Revenue	Variable Cost	Fixed Cost	Total Cost	Net Profit
	0		0	0	0	$5,000	$5,000	–$5,000
Mon	1,000	Brand name	$30,000	$30,000	$7,000	0	12,000	18,000
Tue	2,000	Generic	10,000	40,000	7,000	0	19,000	21,000
Wed	3,000	Generic	10,000	50,000	7,000	0	26,000	24,000
Thu	4,000	Generic	10,000	60,000	7,000	0	33,000	27,000
Fri	5,000	Generic	10,000	70,000	7,000	0	40,000	30,000
Sat	6,000	Generic	10,000	80,000	7,000	0	47,000	33,000
Sun	7,000	Generic	10,000	90,000	12,000	0	59,000	31,000

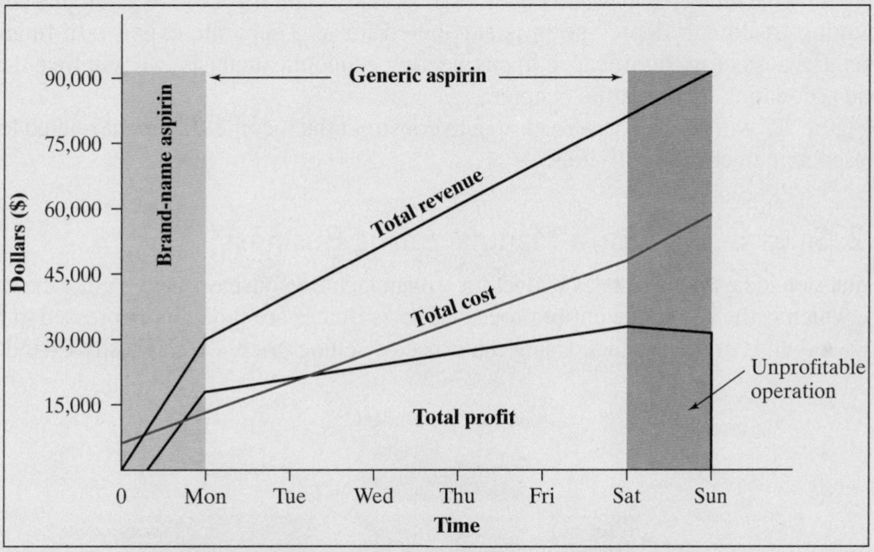

Figure 7.8 Weekly profits as a function of time. Sunday operation becomes unprofitable, because the marginal revenue stays at $10 per case whereas the marginal cost increases to $12 per case (Example 7.7).

7.5 Estimating Profit From Production

Up to this point, we have defined various cost elements and their behaviours in a manufacturing environment. Our ultimate objective is to develop a project's cash flows; we do so in Chapter 10, where the first step is to formulate the budget associated with the sales and production of the product. As we will explain in that chapter, the income statement developed in the current chapter will be the focal point for laying out the project's cash flows.

7.5.1 Calculation of Operating Income

Accountants measure the net income of a specified operating period by subtracting expenses from revenues for that period. These terms can be defined as follows:

1. The **project revenue** is the income earned[3] by a business as a result of providing products or services to customers. Revenue comes from sales of merchandise to customers and from fees earned by services performed for clients or others.

2. The **project expenses** that are incurred[4] are the cost of doing business to generate the revenues of the specified operating period. Some common expenses are the cost of the goods sold (labour, material, inventory, and supplies), depreciation, the cost of employees' salaries, the business's operating costs (such as the cost of renting buildings and the cost of insurance coverage), and income taxes.

The business expenses just listed are accounted for in a straightforward fashion on a company's income statement and balance sheet: The amount paid by the organization for each item would translate, dollar for dollar, into expenses in financial reports for the period. One additional category of expenses—the purchase of new assets—is treated by depreciating the total cost gradually over time. Because capital goods are given this unique accounting treatment, depreciation is accounted for as a separate expense in financial reports. Because of its significance in engineering economic analysis, we will treat depreciation accounting in a separate chapter.

Figure 7.9 will be used as a road map to construct the income statement related to the proposed manufacturing activities.

7.5.2 Sales Budget for a Manufacturing Business

The first step in laying out a sales budget for a manufacturing business is to predict the dollar sales, which is the key to the entire process. The estimates are typically expressed in both dollars and units of the product. Using the expected selling price, we can easily extend unit

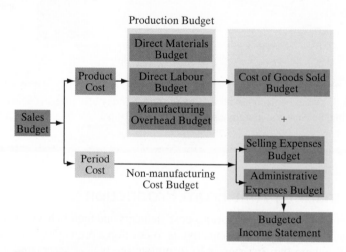

Figure 7.9 Process of creating a master production budget.

---

[3] Note that the cash may be received in a different accounting period.
[4] Note that the cash may be paid in a different accounting period.

sales to revenues. Often, the process is initiated by individual salespersons or managers predicting sales in their own area for the budget period. Of course, an aggregate sales budget is influenced by economic conditions, pricing decisions, competition, marketing programs, and so on. To illustrate, suppose that you want to create a sales budget schedule on a quarterly basis. Projected sales units are 1000 units during the first quarter, 1200 units during the second quarter, 1300 units during the third quarter, and 1500 units during the fourth quarter. The projected sales price is $15 per unit. Then an estimated sales schedule looks like the following:

Sales Budget Schedule (Year 2010)—Product X				
	1Q	2Q	3Q	4Q	Annual Total
Budgeted units	1,000	1,200	1,300	1,500	5,000
Sales price	$ 15	$ 15	$ 15	$ 15	$ 15
Estimated sales	$ 15,000	$ 18,000	$ 19,500	$ 22,500	$ 75,000

7.5.3 Preparing the Production Budget

Once the sales budget is known, we can prepare the production activities. Using the sales budget for the number of project units needed, we prepare an estimate of the number of units to be produced in the budget period. Note that the production budget is the basis for projecting the cost-of-goods-manufactured budget, which we will discuss subsequently. To illustrate, we may use the following steps:

1. From the sales budget, record the projected number of units to be sold.
2. Determine the desired number of units to carry in the ending inventory; usually, that number is a percentage of next quarter's needs. In our example, we will assume that 20% of the budgeted units will be the desirable ending inventory position in each quarter.
3. *Add* to determine the total number of units needed each period.
4. Determine the projected number of units in the beginning inventory. Note that the beginning inventory is last quarter's ending inventory. In our example, we will assume that the first quarter's beginning inventory is 100 units.
5. *Subtract* to determine the projected production budget.

Then a typical production budget may look like the following:

Production Budget (Year 2010)—Product X				
	1Q	2Q	3Q	4Q	Annual Total
Budgeted units to be sold	1000	1200	1300	1500	5000
Desired ending inventory	200	240	260	300	1000
Total units needed	1200	1440	1560	1800	6000
Less beginning inventory	100	200	240	260	800
Units to produce	1100	1240	1320	1540	5200

Materials Budgets

Once we know how many units we need to produce, we are ready to develop direct materials budgets. We may use the following steps:

1. From the production budget, copy the projected number of units to be produced.
2. *Multiply* by the amount of raw materials needed per unit to calculate the amount of materials needed. In our example, we will assume that each production unit consumes $4 of materials.
3. Calculate the desired ending inventory (the number of units required in the ending inventory × $4).
4. *Add* to calculate the total amount of materials needed.
5. *Subtract* the beginning inventory, which is last quarter's ending inventory, to calculate the amount of raw materials needed to be purchased.
6. Calculate the net cost of raw materials.

Then a typical direct materials budget may look like the following:

Direct Materials Budget (Year 2010)—Product X					
	1Q	2Q	3Q	4Q	Annual Total
Units to produce	1,100	1,240	1,320	1,540	5,200
Unit cost of materials	$ 4	$ 4	$ 4	$ 4	
Cost of materials for units to be produced	$ 4,400	$ 4,960	$ 5,280	$ 6,160	$ 20,800
Plus cost of materials in ending inventory	$ 800	$ 960	$ 1,040	$ 1,200	$ 4,000
Total cost of materials needed	$ 5,200	$ 5,920	$ 6,320	$ 7,360	$ 24,800
Less cost of materials in beginning inventory	$ 400	$ 800	$ 960	$ 1,040	$ 3,200
Cost of materials to purchase	$ 4,800	$ 5,120	$ 5,360	$ 6,320	$ 21,600

Direct Labour Budget for a Manufacturing Business

As with the materials budgets, once the production budget has been completed, we can easily prepare the direct labour budget. This budget allows the firm to estimate labour requirements—both labour hours and dollars—in advance. To illustrate, use the following steps:

1. From the production budget, copy the projected number of units to be produced.
2. *Multiply* by the direct labour cost per unit to calculate the total direct labour cost. In our example, we will assume that the unit direct labour cost is $1.27 in 1Q, $1.30 in 2Q, $1.32 in 3Q, and $1.35 in 4Q.

Then a typical direct labour budget may look like the following:

Direct Labour Budget (Year 2010)—Product X					
	1Q	2Q	3Q	4Q	Annual Total
Units to produce	1100	1240	1320	1540	5200
× Direct labour cost per unit	$ 1.27	$ 1.30	$ 1.32	$ 1.35	
Total direct labour cost ($)	$ 1379	$ 1612	$ 1742	$ 2079	$ 5244

Overhead Budget for a Manufacturing Business

The overhead budget should provide a schedule of all costs of production other than direct materials and direct labour. Typically, the overhead budget is expressed in dollars, based on a predetermined overhead rate. In preparing a manufacturing overhead budget, we may take the following steps:

1. From the production budget, copy the projected number of units to be produced.
2. *Multiply* by the variable overhead rate to calculate the budgeted variable overhead. In our example, we will assume the variable overhead rate to be $1.50 per unit. There are several ways to determine this overhead rate. One common approach (known as traditional standard costing) is to divide the expected total overhead cost by the budgeted number of direct labour hours (or units). Another approach is to adopt an **activity-based costing** concept, allocating indirect costs against the activities that caused them. We will not review this accounting method here, but it can more accurately reflect indirect cost improvement than traditional standard costing can.
3. *Add* any budgeted fixed overhead to calculate the total budgeted overhead. In our example, we will assume the fixed overhead to be $230 each quarter.

Activity-based costing (ABC) identifies opportunities to improve business process effectiveness and efficiency by determining the "true" cost of a product or service.

Then a typical manufacturing overhead budget may look like the following:

Manufacturing Overhead Budget (Year 2010)—Product X					
	1Q	2Q	3Q	4Q	Annual Total
Units to produce	1100	1240	1320	1540	5200
Variable mfg overhead rate per unit ($1.50)	$ 1650	$ 1860	$ 1980	$ 2310	$ 7800
Fixed mfg overhead	$ 230	$ 230	$ 230	$ 230	$ 920
Total overhead	$ 1880	$ 2090	$ 2210	$ 2540	$ 8720

7.5.4 Preparing the Cost-of-Goods-Sold Budget

The production budget developed in the previous section shows how much it would cost to produce the required production volume. Note that the number of units to be produced includes both the anticipated number of sales units and the number of units in inventory.

The cost-of-goods-sold budget is different from the production budget because we do not count the costs incurred to carry the inventory. Therefore, we need to prepare a budget that details the costs related to the sales, not the inventory. Typical steps in preparing a cost-of-goods-sold budget are as follows:

1. From the sales budget, and not the production budget, copy the budgeted number of sales units.
2. *Multiply* by the direct material cost per unit to estimate the amount of direct materials.
3. *Multiply* by the direct labour cost per unit to estimate the direct labour.
4. *Multiply* by the manufacturing overhead per unit to estimate the overhead.

Then a typical cost-of-goods-sold budget may look like the following:

Cost of goods sold: A figure reflecting the cost of the product or good that a company sells to generate revenue.

Cost of Goods Sold (Year 2010)—Product X					
	1Q	2Q	3Q	4Q	Annual Total
Budgeted sales units	1,000	1,200	1,300	1,500	5,000
Direct materials ($4/unit)	$ 4,000	$ 4,800	$ 5,200	$ 6,000	$ 20,000
Direct labour ($3/unit)	$ 1,270	$ 1,570	$ 1,720	$ 2,020	$ 6,580
Mfg overhead:					
Variable ($1.50 per unit)	$ 1,500	$ 1,800	$ 1,950	$ 2,250	$ 7,500
Fixed	$ 230	$ 230	$ 230	$ 230	$ 920
Cost of goods sold	$ 7,000	$ 8,400	$ 9,100	$ 10,500	$ 35,000

7.5.5 Preparing the Nonmanufacturing Cost Budget

To complete the entire production budget, we need to add two more items related to production: the selling expenses and the administrative expenses.

Selling Expenses Budget for a Manufacturing Business

Since we know the sales volume, we can develop a selling expenses budget by considering the budgets of various individuals involved in marketing the products. The budget includes both variable cost items (shipping, handling, and sales commission) and fixed items, such as advertising and salaries for marketing personnel. To prepare the selling expenses budget, we may take the following steps:

1. From the sales budget, copy the projected number of unit sales.
2. List the variable selling expenses, such as the sales commission. In our example, we will assume that the sales commission is calculated at 5% of unit sales.
3. List the fixed selling expenses, typically rent, depreciation expenses, advertising, and other office expenses. In our example, we will assume the following: rent, $500 per quarter; advertising, $300 per quarter; office expense, $200 per quarter.
4. *Add* to calculate the total budgeted selling expenses.

Then a typical selling expenses budget may look like the following:

Selling Expenses (Year 2010)—Product X					
	1Q	2Q	3Q	4Q	Annual Total
Budgeted unit sales ($)	$ 15,000	$ 18,000	$ 19,500	$ 22,500	$ 75,000
Variable expenses:					
Commission	$ 750	$ 900	$ 975	$ 1,125	$ 3,750
Fixed expenses:					
Rent	$ 500	$ 500	$ 500	$ 500	$ 2,000
Advertising	$ 300	$ 300	$ 300	$ 300	$ 1,200
Office expenses	$ 200	$ 200	$ 200	$ 200	$ 800
Total selling expenses	$ 1,750	$ 1,900	$ 1,975	$ 2,125	$ 7,750

Administrative Expenses Budget for a Manufacturing Business

Another nonmanufacturing expense category to consider is administrative expenses, a category that contains mostly fixed-cost items such as executive salaries and the depreciation of administrative buildings and office furniture. To prepare an administrative expense budget, we may take the following steps:

1. List the variable administrative expenses, if any.
2. List the fixed administrative expenses, which include salaries, insurance, office supplies, and utilities. We will assume the following quarterly expenses: salaries, $1400; insurance, $135; office supplies, $300; other office expenses, $150.
3. *Add* to calculate the total administrative expenses.

Then a typical administrative expenses budget may look like the following:

Administrative Expenses (Year 2010)—Product X					
	1Q	2Q	3Q	4Q	Annual Total
Variable expenses:					
Fixed expenses:					
Salaries	$ 1400	$ 1400	$ 1400	$ 1400	$ 5600
Insurance	$ 135	$ 135	$ 135	$ 135	$ 540
Office supplies	$ 300	$ 300	$ 300	$ 300	$ 1200
Utilities (phone, power, water, etc.)	$ 500	$ 500	$ 500	$ 500	$ 2000
Other office expenses	$ 150	$ 150	$ 150	$ 150	$ 600
Total administrative expenses	$ 2485	$ 2485	$ 2485	$ 2485	$ 9940

7.5.6 Putting It All Together: The Budgeted Income Statement

Now we are ready to develop a consolidated budget that details a company's projected sales, costs, expenses, and net income. To determine the net income, we need to include any other operating expenses, such as interest expenses and renting or leasing expenses. Once we calculate the taxable income, we can determine the federal taxes to pay. The procedure to follow is

1. Copy the net sales from the sales budget.
2. Copy the cost of goods sold.
3. Copy both the selling and administrative expenses.
4. Gross income = Sales − Cost of goods sold.
5. Operating income = Sales − (Cost of goods sold + Operating expenses).
6. Determine any interest expense incurred during the budget period.
7. Calculate the income tax at a percentage of taxable income. In our example, let's assume a 35% tax rate.
8. Net income = Income from operations − Income tax, which summarizes the projected profit from the budgeted sales units.

Then the budgeted income statement would look like the format shown in Table 7.5.

Once you have developed a budgeted income statement, you may be able to examine the profitability of the manufacturing operation. Three margin figures are commonly used to quickly assess the profitability of the operation: gross margin, operating margin, and net profit margin.

Gross Margin

The gross margin reveals how much a company earns, taking into consideration the costs that it incurs for producing its products or services. In other words, gross margin is equal to gross income divided by net sales and is expressed as a percentage:

$$\text{Gross margin} = \frac{\text{Gross income}}{\text{Net sales}}. \tag{7.2}$$

In our example, gross margin = \$40,000/\$75,000 = 53%. Gross margin indicates how profitable a company is at the most fundamental level. Companies with higher gross margins will have more money left over to spend on other business operations, such as research and development or marketing.

Operating Margin

Operating margin gives analysts an idea of how much a company makes (before interest and taxes) on each dollar of sales.

The **operating margin** is defined as the operating profit for a certain period, divided by revenues for that period:

$$\text{Operating margin} = \frac{\text{Operating income}}{\text{Net sales}}. \tag{7.3}$$

In our example, operating margin = \$22,310/\$75,000 = 30%. Operating *profit margin* indicates how effective a company is at controlling the costs and expenses associated with its normal business operations.

TABLE 7.5 Budgeted Income Statement

	1Q	2Q	3Q	4Q	Annual Total
		Budgeted Income Statement (Year 2010)—Product X			
Sales	$ 15,000	$ 18,000	$ 19,500	$ 22,500	$ 75,000
Cost of goods sold	$ 7,000	$ 8,400	$ 9,100	$ 10,500	$ 35,000
Gross income	$ 8,000	$ 9,600	$ 10,400	$ 12,000	$ 40,000
Operating expenses:					
Selling expenses	$ 1,750	$ 1,900	$ 1,975	$ 2,125	$ 7,750
Administrative expenses	$ 2,485	$ 2,485	$ 2,485	$ 2,485	$ 9,940
Operating income	$ 3,765	$ 5,215	$ 5,940	$ 7,390	$ 22,310
Interest expenses	$ —	$ —	$ —	$ —	$ —
Net income before taxes	$ 3,765	$ 5,215	$ 5,940	$ 7,390	$ 22,310
Income taxes (35%)	$ 1,318	$ 1,825	$ 2,079	$ 2,587	$ 7,809
Net income	$ 2,447	$ 3,390	$ 3,861	$ 4,804	$ 14,502

Net Profit Margin

As we defined the term in Chapter 2, the net profit margin is obtained by dividing net profit by net revenues, often expressed as a percentage:

$$\text{Net profit margin} = \frac{\text{Net income}}{\text{Net sales}}. \tag{7.4}$$

This number is an indication of how effective a company is at cost control. The higher the net profit margin, the more effective the company is at converting revenue into actual profit. In our example, the net profit margin = $14,502/$75,000 = 19%. The net profit margin is a good way of comparing companies in the same industry, since such companies are generally subject to similar business conditions.

7.5.7 Looking Ahead

The operational decision problem described in this chapter tends to have a relatively short-term horizon (weekly or monthly). That is, such decisions do not commit a firm to a certain course of action over a relatively long period. If operational decision problems significantly affect the amount of funds that must be invested in a firm, fixed costs will have to increase. In Example 7.7, if the daily aspirin production increases to 10,000 cases per day, exceeding the current production capacity, the firm must make new investments in plant and equipment. Since the benefits of the expansion will continue to occur over an extended period, we need to consider the economic effects of the fixed costs over the life of the assets. Also, we have not discussed how we actually calculate the depreciation amount related to production or the amount of income taxes to be paid from profits generated from the operation. Doing so requires an understanding of the concepts of **capital investment, depreciation, and income taxes**, which we will discuss in the following chapters.

SUMMARY

In this chapter, we examined several ways in which managers classify costs. How the costs will be used dictates how they will be classified:

■ Most manufacturing companies divide **manufacturing costs** into three broad categories: *direct materials*, *direct labour*, and *manufacturing overhead*. **Nonmanufacturing costs** are classified into two categories: *marketing* or *selling costs* and *administrative costs*.

■ For the purpose of valuing inventories and determining expenses for the balance sheet and income statement, costs are classified as either **product costs** or **period costs**.

■ For the purpose of predicting **cost behaviour**—how costs will react to changes in activity—managers commonly classify costs into two categories: variable costs and fixed costs.

■ An understanding of the following **cost–volume relationships** is essential to developing successful business strategies and to planning future operations:

Fixed operating costs: Costs that do not vary with production volume.

Variable operating costs: Costs that vary with the level of production or sales.

Average costs: Costs expressed in terms of units obtained by dividing total costs by total volumes.

Differential (incremental) costs: Costs that represent differences in total costs, which result from selecting one alternative instead of another.

Opportunity costs: Benefits that could have been obtained by taking an alternative action.

Sunk costs: Past costs not relevant to decisions because they cannot be changed no matter what actions are taken.

Marginal costs: Added costs that result from increasing rates of outputs, usually by single units.

Marginal analysis: In economic analysis, we need to answer the apparently trivial question, "Is it worthwhile?"—whether the action in question will add sufficiently to the benefits enjoyed by the decision maker to make performing the action worth the cost. This is the heart of marginal-decision making—the statement that an action merits performance if, and only if, as a result, we can expect to be better off than we were before.

■ At the level of plant operations, engineers must make decisions involving materials, production processes, and the in-house capabilities of company personnel. Most of these operating decisions do not require any investments in physical assets; therefore, they depend solely on the cost and volume of business activity, without any consideration of the time value of money.

■ Engineers are often asked to prepare the production budgets related to their operating division. Doing this requires a knowledge of budgeting scarce resources, such as labour and materials, and an understanding of the overhead cost. The same budgeting practice will be needed in preparing the estimates of costs and revenues associated with undertaking a new project.

PROBLEMS

7.1 Identify which of the following transactions and events are product costs and which are period costs:
- Storage and material handling costs for raw materials.
- Gains or losses on the disposal of factory equipment.
- Lubricants for machinery and equipment used in production.
- Depreciation of a factory building.
- Depreciation of manufacturing equipment.
- Depreciation of the company president's automobile.
- Leasehold costs for land on which factory buildings stand.
- Inspection costs of finished goods.
- Direct labour cost.
- Raw-materials cost.
- Advertising expenses.

7.2 Identify which of the following costs are fixed and which are variable:
- Wages paid to temporary workers.
- Property taxes on a factory building.
- Property taxes on an administrative building.
- Sales commission.
- Electricity for machinery and equipment in the plant.
- Heating and air-conditioning for the plant.
- Salaries paid to design engineers.
- Regular maintenance on machinery and equipment.
- Basic raw materials used in production.
- Factory fire insurance.

7.3 The accompanying figures are a number of cost behaviour patterns that might be found in a company's cost structure. The vertical axis on each graph represents total cost, and the horizontal axis on each graph represents level of activity (volume). For each of the situations that follow, identify the graph that illustrates the cost pattern involved. Any graph may be used more than once.[5]

(a) Electricity bill—a flat-rate fixed charge, plus a variable cost after a certain number of kilowatt-hours are used.

(b) City water bill, which is computed as follows:

First 1,000,000 litres $1000 flat or less rate

Next 10,000 litres $0.003 per litre used

Next 10,000 litres $0.006 per litre used

Next 10,000 litres $0.009 per litre used

And so on.

(c) Depreciation of equipment, where the amount is computed by the straight-line method. When the depreciation rate was established, it was anticipated that the obsolescence factor would be greater than the wear-and-tear factor.

---

[5] Adapted originally from a CPA exam, and the same materials are also found in R. H. Garrison and E. W. Noreen, *Managerial Accounting*, 8th ed. Irwin, 1997, copyright © Richard D. Irwin, p. 271.

(d) Rent on a factory building donated by the city, where the agreement calls for a fixed fee payment unless 200,000 labour-hours or more are worked, in which case no rent need be paid.

(e) Cost of raw materials, where the cost decreases by 5 cents per unit for each of the first 100 units purchased, after which it remains constant at $2.50 per unit.

(f) Salaries of maintenance workers, where one maintenance worker is needed for every 1000 machine-hours or fewer (that is, 0 to 1000 hours requires one maintenance worker, 1001 to 2000 hours requires two maintenance workers, etc.).

(g) Cost of raw materials used.

(h) Rent on a factory building donated by the county, where the agreement calls for rent of $100,000 less $1 for each direct labour-hour worked in excess of 200,000 hours, but a minimum rental payment of $20,000 must be paid.

(i) Use of a machine under a lease, where a minimum charge of $1000 must be paid for up to 400 hours of machine time. After 400 hours of machine time, an additional charge of $2 per hour is paid, up to a maximum charge of $2000 per period.

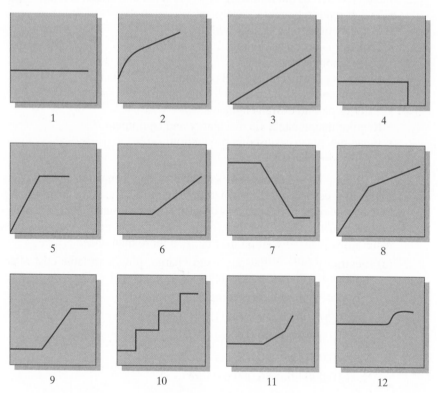

7.4 Harris Company manufactures a single product. Costs for 2008 for output levels of 1000 and 2000 units are as follows:

Units produced	1,000	2,000
Direct labour	$30,000	$30,000
Direct materials	20,000	40,000
Overhead:		
Variable portion	12,000	24,000
Fixed portion	36,000	36,000
Selling and administrative costs:		
Variable portion	5,000	10,000
Fixed portion	22,000	22,000

At each level of output, compute the following:
(a) Total manufacturing costs.
(b) Manufacturing costs per unit.
(c) Total variable costs.
(d) Total variable costs per unit.
(e) Total costs that have to be recovered if the firm is to make a profit.

7.5 Bragg & Stratton Company manufactures a specialized motor for chainsaws. The company expects to manufacture and sell 30,000 motors in 2010. It can manufacture an additional 10,000 motors without adding new machinery and equipment. Bragg & Stratton's projected total costs for the 30,000 units are as follows:

Direct materials	$150,000
Direct labour	300,000
Manufacturing overhead:	
Variable portion	100,000
Fixed portion	80,000
Selling and administrative costs:	
Variable portion	180,000
Fixed portion	70,000

The selling price for the motor is $80.
(a) What is the total manufacturing cost per unit if 30,000 motors are produced?
(b) What is the total manufacturing cost per unit if 40,000 motors are produced?
(c) What is the break-even price on the motors?

7.6 The accompanying chart shows the expected monthly profit or loss of Cypress Manufacturing Company within the range of its monthly practical operating capacity. Using the information provided in the chart, answer the following questions:
(a) What is the company's break-even sales volume?
(b) What is the company's marginal contribution rate?

(c) What effect would a 5% decrease in selling price have on the break-even point in (a)?

(d) What effect would a 10% increase in fixed costs have on the marginal contribution rate in (b)?

(e) What effect would a 6% increase in variable costs have on the break-even point in (a)?

(f) If the chart also reflects $20,000 monthly depreciation expenses, compute the sale at the break-even point for cash costs.

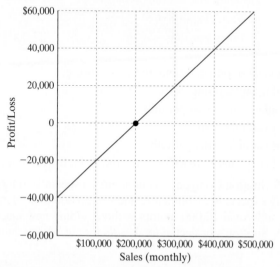

7.7 The accompanying graph is a cost–volume–profit graph. In the graph, identify the following line segments or points:

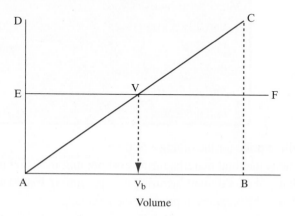

(a) Line EF represents _____.

(b) The horizontal axis AB represents _____, and the vertical axis AD represents _____.

(c) Point V represents _____.

(d) The distance CB divided by the distance AB is _____.

(e) The point V_b is a break-even _____.

7.8 A cost–volume–profit (CVP) graph is a useful technique for showing relationships between costs, volume, and profits in an organization.

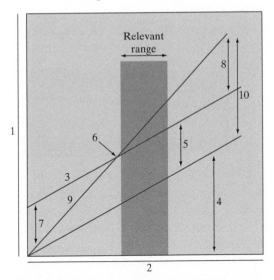

(a) Identify the numbered components in the accompanying CVP graph.

No.	Description	No.	Description
1		6	
2		7	
3		8	
4		9	
5		10	

(b) Using the typical CVP relationship shown, fill in the missing amounts in each
of the following four situations (each case is independent of the others):

Case	Units Sold	Sales	Variable Expenses	Contribution Margin per Unit	Fixed Expenses	Net Income (Loss)
A	9000	$270,000	$162,000		$ 90,000	
B		$350,000		$15	$170,000	$ 40,000
C	20,000		$280,000	$ 6		$ 35,000
D	5000	$100,000			$ 82,000	($12,000)

7.9 An executive from a large merchandising firm has called your vice-president for production to get a price quote for an additional 100 units of a given product. The vice-
president has asked you to prepare a cost estimate. The number of hours required
to produce a unit is five. The average labour rate is $12 per hour. The materials cost

$14 per unit. Overhead for an additional 100 units is estimated at 50% of the direct labour cost. If the company wants to have a 30% profit margin, what should be the unit price to quote?

7.10 The Morton Company produces and sells two products: A and B. Financial data related to producing these two products are summarized as follows:

	Product A	Product B
Selling price	$ 10.00	$12.00
Variable costs	$ 5.00	$10.00
Fixed costs	$ 2000	$ 600

(a) If these products are sold in the ratio of 4A for 3B, what is the break-even point?

(b) If the product mix has changed to 5A for 5B, what would happen to the break-even point?

(c) In order to maximize the profit, which product mix should be pushed?

(d) If both products must go through the same manufacturing machine and there are only 30,000 machine hours available per period, which product should be pushed? Assume that product A requires 0.5 hour per unit and B requires 0.25 hour per unit.

7.11 Pearson Company manufactures a variety of electronic printed circuit boards (PCBs) that go into cellular phones. The company has just received an offer from an outside supplier to provide the electrical soldering for Pearson's Motorola product line (Z-7 PCB, slimline). The quoted price is $4.80 per unit. Pearson is interested in this offer, since its own soldering operation of the PCB is at its peak capacity.

- **Outsourcing option.** The company estimates that if the supplier's offer was accepted, the direct labour and variable overhead costs of the Z-7 slimline would be reduced by 15% and the direct material cost would be reduced by 20%.

- **In-house production option.** Under the present operations, Pearson manufactures all of its own PCBs from start to finish. The Z-7 slimlines are sold through Motorola at $20 per unit. Fixed overhead charges to the Z-7 slimline total $20,000 each year. The further breakdown of producing one unit is as follows:

Direct materials	$ 7.50
Direct labour	5.00
Manufacturing overhead	4.00
Total cost	$16.50

The manufacturing overhead of $4.00 per unit includes both variable and fixed manufacturing overhead, based on a production of 100,000 units each year.

(a) Should Pearson Company accept the outside supplier's offer?

(b) What is the maximum unit price that Pearson Company should be willing to pay the outside supplier?

Short Case Study

ST7.1 The Hamilton Flour Company is currently operating its mill six days a week, 24 hours a day, on three shifts. At current prices, the company could easily obtain a sufficient volume of sales to take the entire output of a seventh day of operation each week. The mill's practical capacity is 270 tonnes of flour per day. Note that

- Flour sells for $275 per tonne and the price of wheat is $0.16 per kilogram. About 1.2 kilograms of wheat are required for each kilogram of flour. Fixed costs now average $4200 a day, or $15.56 per tonne. The average variable cost of mill operation, almost entirely wages, is $7.56 per tonne.
- With Sunday operation, wages would be doubled for Sunday work, which would bring the variable cost of Sunday operation to $15.12 per tonne. Total fixed costs per week would increase by $420 (or $29,820) if the mill was to operate on Sunday.

(a) Using the information provided, compute the break-even volumes for six-day and seven-day operation.

(b) What are the marginal contribution rates for six-day and seven-day operation?

(c) Compute the average total cost per tonne for six-day operation and the net profit margin per tonne before taxes.

(d) Would it be economical for the mill to operate on Sundays? (Justify your answer numerically.)

On the Companion Website that accompanies this text, you will find Excel templates and exercises, as well as the following analysis tools: Cash Flow Analyzer, Depreciation Analysis, Loan Analysis, and Interest Tables.

EIGHT

Depreciation

Source: Illustration based on photo courtesy of Park Thermal International.

Formed in 1970, IMP Aerospace provides depot-level aircraft engineering and maintenance support to the Canadian Department of National Defence (DND) and Norwegian Air Force, as well as other international customers. IMP Aerospace also specializes in repair and overhaul of aircraft components as well as manufacturing aircraft parts to support its core business.

As part of ongoing process improvement initiatives, IMP Aerospace's Continuous Improvement Department reviewed its internal manufacturing process with a goal of reducing its manufacturing lead time to better support its customers. As a result, the material processing step of *Heat Treat* (aluminum parts) and/or *CAD Plate* was identified as a bottleneck, adding 3 to 10 days to the manufacturing cycle and significantly more days when the part had to be sent to a subcontractor. *Heat Treat* involves a salt bath and an aging oven and *CAD Plate* involved the same aging oven. This oven is more than 25 years old, energy inefficient, inflexible (can only be used once a day), and not very reliable (incurring maintenance downtime). To all but eliminate the manufacturing bottleneck, IMP Aerospace chose to purchase a new air furnace to replace the salt bath and two new aging ovens from a Canadian company called Park Thermal to replace the current aging oven. The new equipment is more flexible in that it can be computer monitored and controlled, as well as having the ability to do larger parts and handle more than one batch a day. The cost of the new equipment is well over $300,000.

Now ask yourself, How does the cost of this system affect the financial position of IMP? In the long run, the system promises to create greater wealth for the organization in many ways, including improved throughput, shorter delivery lead time, energy savings, and less subcontracting. In the short run, however, the high cost of this system will have a negative impact on the organization's "bottom line" because it involves high initial costs that are only gradually recovered from the benefits of the system.

Another consideration should come to mind. This state-of-the-art equipment will inevitably wear out over time, and even if its productive service extends over many years, the cost of maintaining its high level of functioning will increase as the individual components wear out and need to be replaced. Of even greater concern is the question of how long this system will be "state of the art." When will the competitive advantage the firm has just acquired become a competitive disadvantage through obsolescence?

One of the facts of life that organizations must deal with and account for is that fixed assets lose their value—even as they continue to function and contribute to the engineering projects that use them. This loss of value, called **depreciation**, can involve deterioration and obsolescence.

The main function of **depreciation accounting** is to account for the cost of fixed assets in a pattern that matches their decline in value over time. The cost of the processing system we have just described, for example, will be allocated over several years in the firm's financial statements so that its pattern of costs roughly matches its pattern of service. In this way, as we shall see, depreciation accounting enables the firm to stabilize the statements of financial position that it distributes to stockholders and the outside world.

On a project level, engineers must be able to assess how the practice of depreciating fixed assets influences the investment value of a given project. To do this, they need to estimate the allocation of capital costs over the life of the project, which requires understanding the conventions and techniques that accountants use to depreciate assets. In this chapter, we will review the conventions and techniques of asset depreciation.

We begin by discussing the nature and significance of depreciation, distinguishing its general economic definition from the related, but different, accounting view of depreciation. We then focus our attention almost exclusively on the methods that accountants use to allocate depreciation expenses, and the rules and laws that govern asset depreciation under the Canadian income tax system. Knowledge of these rules will prepare you to apply them in assessing the depreciation of assets acquired in engineering projects. Finally, we turn our attention to tax depreciation in the U.S., which utilizes similar ideas.

CHAPTER LEARNING OBJECTIVES

After completing this chapter, you should understand the following concepts:

- How to account for the loss of value of an asset in business.
- The meaning and types of depreciation.
- The difference between book depreciation and tax depreciation.
- The effects of depreciation on net income calculation.
- The capital cost allowance (CCA) system for tax depreciation in Canada.
- The fundamentals of the U.S. MACRS tax depreciation method.

8.1 Asset Depreciation

Fixed assets such as equipment and real estate are economic resources that are acquired to provide future cash flows. Generally, **depreciation** can be defined as the gradual decrease in utility of fixed assets with use and time. While this general definition does not adequately capture the subtleties inherent in a more specific definition of depreciation, it does provide us with a starting point for examining the variety of underlying ideas and practices that are discussed in this chapter. Figure 8.1 will serve as a road map for understanding the distinctions inherent in the meaning of depreciation we will explore here.

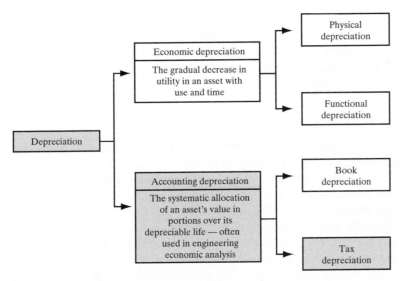

Figure 8.1 Classification of depreciation.

8.1.1 Economic Depreciation

This chapter is primarily concerned with accounting depreciation, which is the form of depreciation that provides an organization with the information needed to assess its financial position and calculate its income taxes. It would also be useful, however, to discuss briefly the economic ideas upon which accounting depreciation is based. In the course of the discussion, we will develop a precise definition of economic depreciation that will help us distinguish between various conceptions of depreciation.

If you have ever owned a car, you are probably familiar with the term *depreciation* as it is used to describe the decreasing value of your vehicle. Because a car's reliability and appearance usually decline with age, the vehicle is worth less with each passing year. You can calculate the economic depreciation accumulated for your car by subtracting the current market value or "black book" value of the car from the price you originally paid for it.[1] We can define **economic depreciation** as follows:

$$\text{Economic depreciation} = \text{Purchase price} - \text{market value}$$

Physical and functional depreciation are categories of economic depreciation.

Physical depreciation can be defined as a reduction in an asset's capacity to perform its intended service due to physical impairment. Physical depreciation can occur in any fixed asset in the form of (1) deterioration from interaction with the environment, including such agents as corrosion, rotting, and other chemical changes; and (2) wear and tear from use. Physical depreciation leads to a decline in performance and high maintenance costs.

Functional depreciation occurs as a result of changes in the organization or in technology that decrease or eliminate the need for an asset. Examples of functional depreciation include obsolescence attributable to advances in technology, a declining need for the services performed by an asset, or the inability to meet increased quantity or quality demands.

---

[1] See www.canadianblackbook.com.

The measurement of economic depreciation does not require that an asset be sold: the market value of the asset can be closely estimated without actually testing its value in the marketplace. The need to have a precise scheme for recording the on-going decline in the value of an asset as a part of the accounting process leads us to an exploration of how organizations account for depreciation.

8.1.2 Accounting Depreciation

The acquisition of fixed assets is an important activity for a business organization. This is true whether the organization is starting up or acquiring new assets to remain competitive. Like other disbursements, the cost of these fixed assets must be recorded as expenses on a firm's balance sheet and income statement. However, unlike costs such as maintenance, material, and labour, the costs of fixed assets are not treated simply as expenses to be accounted for in the year that they are acquired. Rather, these assets are **capitalized**; that is, their costs are distributed by subtracting them as expenses from gross income—one part at a time over a number of periods. The systematic allocation of the initial cost of an asset in parts over a time, known as the asset's **depreciable life**, is what we mean by **accounting depreciation**. Because accounting depreciation is the standard of the business world, we sometimes refer to it more generally as **asset depreciation**.

Accounting depreciation is based on the **matching concept**: A fraction of the cost of the asset is chargeable as an expense in each of the accounting periods in which the asset provides service to the firm, and each charge is meant to be a percentage of the whole cost which "matches" the percentage of the value utilized in the given period. The matching concept suggests that the accounting depreciation allowance generally reflects, at least to some extent, the actual economic depreciation of the asset.

In engineering economic analysis, we exclusively use the concept of accounting depreciation. This is because accounting depreciation provides a basis for determining the income taxes associated with any project undertaken. As we will see in Chapter 9, depreciation has a significant influence on the income and cash position of a firm. With a clear understanding of the concept of depreciation, we will be able to appreciate fully the importance of utilizing depreciation as a means to maximize the value both of engineering projects and of the organization as a whole.

8.2 Factors Inherent in Asset Depreciation

The process of depreciating an asset requires that we make several preliminary determinations: (1) what is the cost of the asset? (2) what is the asset's value at the end of its useful life? (3) what is the depreciable life of the asset? and, finally, (4) what method of depreciation do we choose? In this section, we will discuss each of these questions.

8.2.1 Depreciable Property

As a starting point, it is important to recognize what constitutes a **depreciable asset**, that is, a property for which a firm may take depreciation deductions against income. To see if an item qualifies as depreciable property for the purpose of Canadian taxes, detailed descriptions are provided in the Income Tax Act. In general, though, depreciable property has the following characteristics:

1. It must be used in business or held for production of income.
2. It must have a definite useful life, and that life must be longer than one year.
3. It must be something that wears out, decays, gets used up, becomes obsolete, or loses value from natural causes.

Depreciable property includes buildings, machinery, equipment, and vehicles. Inventories are not depreciable property because these are held primarily for sale to customers in the ordinary course of business. If an asset has no definite useful life, the asset cannot be depreciated. For example, *you cannot depreciate land.*[2]

As long as assets meet the conditions listed in the Income Tax Act, depreciation may be claimed by any business enterprise, whether a sole proprietorship, partnership, or corporation. For example, an individual who periodically uses her car for business, or part of her house as a work space, may claim depreciation on a prorated portion of the capital cost of the property.

8.2.2 Cost Basis

The **cost basis** of an asset represents the total cost that is claimed as an expense over an asset's life, i.e., the sum of the annual depreciation expenses. The cost basis generally includes the actual cost of an asset and all other incidental expenses, such as freight, site preparation, installation, and legal, accounting, and engineering fees. This total cost, rather than the cost of the asset only, must be the depreciation basis charged as an expense over an asset's life.

Besides being used in figuring depreciation deductions, an asset's cost basis is used in calculating the gain or loss to the firm if the asset is ever sold or salvaged. We will discuss these topics in Sections 9.1.2 and 9.5.

EXAMPLE 8.1 Cost Basis

Lanier Corporation purchased an automatic hole-punching machine priced at $62,500. The vendor's invoice included a sales tax of $9375. Lanier also paid the inbound transportation charges of $725 on the new machine as well as labour cost of $2150 to install the machine in the factory. Lanier also had to prepare the site at the cost of $3500 before installation. Determine the cost basis for the new machine for depreciation purposes.

SOLUTION

Given: Invoice price = $62,500, freight = $725, installation cost = $2150, and site preparation = $3500.

Find: The cost basis.

The cost of the machine that is applicable for depreciation is computed as follows:

---

[2] This also means that you cannot depreciate the cost of clearing, grading, planting, and landscaping. All such expenses are considered part of the cost of the land. (An exception is land acquired as part of a timber limit.)

Cost of new hole-punching machine	$62,500
Freight	725
Installation labour	2,150
Site preparation	3,500
Cost of machine (cost basis)	$68,875

COMMENTS: Why do we include all the incidental charges relating to the acquisition of a machine in its cost? Why not treat these incidental charges as expenses of the period in which the machine is acquired? The matching of costs and revenue is the basic accounting principle. Consequently, the total costs of the machine should be viewed as an asset and allocated against the future revenue that the machine will generate. All costs incurred in acquiring the machine are costs of the services to be received from using the machine.

If the asset is purchased by trading in a similar asset, the difference between the **book value** (cost basis minus the total accumulated depreciation) and the trade-in allowance must be considered in determining the cost basis for the new asset. If the trade-in allowance exceeds the book value, the difference (known as **unrecognized gain**) needs to be subtracted from the cost basis of the new asset. If the opposite is true (**unrecognized loss**), it should be added to the cost basis for the new asset.

EXAMPLE 8.2 Cost Basis With Trade-in Allowance

In Example 8.1, suppose Lanier purchased the hole-punching press by trading in a similar machine and paying cash for the remainder. The trade-in allowance was $5000 and the book value of the hole-punching machine that was traded in was $4000. Determine the cost basis for this hole-punching press.

SOLUTION

Given: Accounting data from Example 8.1; trade-in allowance = $5000.

Find: The revised cost basis.

Old hole-punching machine (book value)	$4,000
Less: Trade-in allowance	5,000
Unrecognized gains	$1,000
Cost of new hole-punching machine	$62,500
Less: Unrecognized gains	(1,000)
Freight	725
Installation labour	2,150
Site preparation	3,500
Cost of machine (cost basis)	$67,875

8.2.3 Useful Life and Salvage Value

The **useful life** is an estimate of the duration over which the asset is expected to fulfill its intended service. Referring to Figure 8.1, the useful life may be based on the asset's anticipated physical longevity, which depends on many factors, such as its durability, severity of planned usage, and harshness of its environment. For other types of property, functionality may be the limiting factor, particularly for rapidly changing technologies such as computers, electronics, and communication equipment. Predicting the useful life of an asset is required for some depreciation computations and most economic analyses.

The **salvage value** (also known as scrap value or residual value) is an asset's value at the end of its life; it is the amount eventually recovered through sale, trade-in, or salvage, net of all costs incurred for disposal and restoring the site to its original condition if required. Since there are many unknown factors that may affect this value, a rough estimate must be made in advance to perform an economic analysis. If the demolition or removal costs are expected to exceed receipts from selling the used asset, a negative salvage value is applied. For long-lived assets, a zero salvage value is often assumed for simplicity, as any small residual value may be negated by the disposal costs. Often, the estimated salvage value will not equal the projected book value at the end of the asset's useful life. Since accounting depreciation allocates the cost of the asset over time to match its anticipated usage, the book value is not intended to equal the salvage value. Adjustments must be made to account for differences between the salvage value (estimated, or actual at the time of disposal) and the book value, as shown in the following sections and Chapter 9.

8.2.4 Depreciation Methods: Book and Tax Depreciation

As mentioned earlier, most firms calculate depreciation in two different ways, depending on whether the calculation is (1) intended for financial reports (**book depreciation method**), such as for the balance sheet or income statement or (2) for the Canada Revenue Agency (CRA), for the purpose of calculating taxes (**capital cost allowance**, or **CCA**). In Canada, this distinction is totally legitimate under CRA regulations, as it is in many other countries. Calculating depreciation differently for financial reports and for tax purposes allows for the following benefits:

- It enables firms to report depreciation to stockholders and other significant outsiders based on the matching concept. Therefore the actual loss in value of the assets is generally reflected.
- It allows firms to benefit from the tax advantages of depreciating assets more quickly than would be possible using the matching concept. In many cases, capital cost allowance permits firms to defer paying income taxes. This does not mean that they pay less tax overall, because the total depreciation expense accounted for over time is the same in either case. However, because CCA methods generally result in greater depreciation in earlier years than do common book depreciation methods, the tax benefit of depreciation is enjoyed earlier, and firms generally pay lower taxes in the initial years of an investment project. Typically this leads to a better cash position in early years, and the added cash leads to greater future wealth because of the time value of the funds.

As we proceed through the chapter, we will make increasing use of the distinction between depreciation accounting for financial reporting and depreciation accounting used for income tax calculation. Now that we have established the context for our interest in both book depreciation and CCA for tax purposes, we can survey the different methods with an accurate perspective.

8.3 Book Depreciation Methods

Three different methods are commonly used to calculate the periodic depreciation allowances. These are (1) the straight-line method, (2) accelerated methods, and (3) the units-of-production method. In engineering economic analysis, we are primarily interested in depreciation in the context of income tax computation. Nonetheless, a number of reasons make the study of book depreciation methods useful. First, capital cost allowance is based largely on the same principles that are used in certain book depreciation methods. Second, firms continue to use book depreciation methods for financial reporting to stockholders and outside parties. Finally, some resource allowances discussed in Section 8.6 are based on some of these three book depreciation methods.

8.3.1 Straight-Line (SL) Method

The **straight-line (SL) method** of depreciation interprets a fixed asset as one that provides its services in a uniform fashion. The asset provides an equal amount of service in each year of its useful life. SL is the prevailing book depreciation method, due to its reasonable assumptions and simplicity.

The straight-line method charges, as an expense, an equal fraction of the net cost of the asset each year, as expressed by the relation

$$D_n = \frac{(P-S)}{N},$$ (8.1)

where D_n = Depreciation charge during year n
P = Cost of the asset including installation expenses
S = Salvage value at the end of the asset's useful life
N = Useful life.

The book value of the asset at the end of n years is then defined as

Book value in a given year = Cost basis − total depreciation charges made to date or

$$B_n = P - (D_1 + D_2 + D_3 + \ldots + D_n) = P - \frac{n(P-S)}{N}$$ (8.2)

EXAMPLE 8.3 Straight-Line Depreciation

Consider the following automobile data:

Cost basis of the asset, P = $10,000
Useful life, N = 5 years
Estimated salvage value, S = $2000

Compute the annual depreciation allowances and the resulting book values using the straight-line depreciation method.

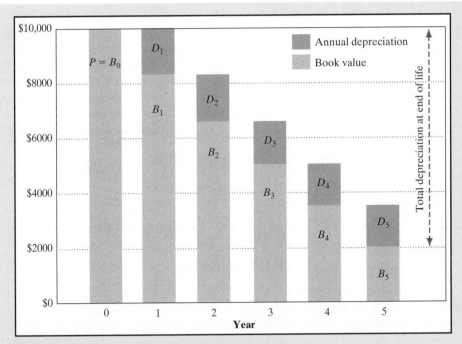

Figure 8.2   Straight-line depreciation method (Example 8.3).

SOLUTION

Given: $P = \$10,000$, $S = \$2000$, $N = 5$ years.
Find: D_n and B_n for $n = 1$ to 5.

The straight-line depreciation rate is 1/5 or 20%. Therefore, the annual depreciation charge is

$$D_n = (0.20)(\$10,000 - \$2000) = \$1600.$$

Then, the asset would have the following book values during its useful life:

n	B_{n-1}	D_n	B_n
1	$10,000	$1600	$8400
2	8,400	1600	6800
3	6,800	1600	5200
4	5,200	1600	3600
5	3,600	1600	2000

where B_{n-1} represents the book value before the depreciation charge for year n. This situation is illustrated in Figure 8.2. The time zero book value, B_o, is the cost basis of the asset, P.

8.3.2 Accelerated Methods

The second depreciation concept recognizes that the stream of services provided by a fixed asset may decrease over time; in other words, the stream may be greatest in the first year of an asset's service life and least in its last year. This pattern may occur because the mechanical efficiency of an asset tends to decline with age, because maintenance costs tend to increase with age, or because of the increasing likelihood that better equipment will become available and make the original asset obsolete. This reasoning leads to a method that charges a larger fraction of the cost as an expense of the early years than of the later years. Any such method is called an **accelerated method**. The two most widely used accelerated methods are the **declining balance method** and the **sum-of-years'-digits method**.

Declining Balance (DB) Method

The **declining balance (DB) method** of calculating depreciation allocates a fixed fraction of the beginning book balance each year. The fraction, d, is obtained as follows:

$$d = (1/N)(\text{multiplier}). \tag{8.3}$$

When the multiplier equals 1.0, the method is called single declining balance (or simply DB). This is the basis for most of the CCA rates for income tax purposes in Canada, although in that case the rate is specified by the government for different types of assets, so the taxpayer does not set the N value themselves. Other common mulipliers for book depreciation include 1.5 (called 150% DB) and 2.0 (called 200%, or double declining balance, DDB). As N increases, d decreases. This method results in a situation in which depreciation is highest in the first year and decreases over the asset's depreciable life.

The factor, d, can be utilized to determine depreciation charges for a given year, D_n, as follows:

$$D_1 = dP,$$
$$D_2 = d(P - D_1) = dP(1 - d),$$
$$D_3 = d(P - D_1 - D_2) = dP(1 - d)^2,$$

and thus for any year, n, we have a depreciation charge, D_n, of

$$D_n = dP(1 - d)^{n-1} \quad n \geq 1. \tag{8.4}$$

We can also compute the total declining balance depreciation (TDB) at the end of n years as follows:

$$
\begin{aligned}
TDB_n &= D_1 + D_2 + \cdots + D_n \\
&= dP + dP(1 - d) + dP(1 - d)^2 + \cdots + dP(1 - d)^{n-1} \\
&= dP[1 + (1 - d) + (1 - d)^2 + \cdots + (1 - d)^{n-1}].
\end{aligned} \tag{8.5}
$$

Multiplying Eq. (8.5) by $(1 - d)$, we obtain

$$TDB_n(1 - d) = dP[(1 - d) + (1 - d)^2 + (1 - d)^3 + \cdots + (1 - d)^n]. \tag{8.6}$$

Subtracting Eq. (8.5) from Eq. (8.6) and dividing by d gives

$$TDB_n = P[1 - (1 - d)^n]. \tag{8.7}$$

The book value, B_n, at the end of n years will be the cost of the asset P minus the total depreciation at the end of n years:

$$B_n = P - TDB_n$$
$$= P - P[1 - (1 - d)^n]$$
$$B_n = P(1 - d)^n. \qquad (8.8)$$

EXAMPLE 8.4 Declining Balance Depreciation

Consider the following accounting information for a photocopier:

$$\text{Cost basis of the asset, } P = \$10,000$$
$$\text{Useful life, } N = 5 \text{ years}$$
$$\text{Estimated salvage value, } S = \$3277.$$

Compute the annual depreciation charges and the resulting book values using the declining balance depreciation method (Figure 8.3).

SOLUTION

Given: $P = \$10,000$, $S = \$3277$, $N = 5$ years.
Find: D_n and B_n for $n = 1$ to 5.

The book value at the beginning of the first year is \$10,000, and the declining balance rate (d) is $(1/5) = 20\%$. Then, the depreciation deduction for the first year will be \$2000 ($20\% \times \$10,000 = \$2000$). To figure the depreciation deduction in the second year, we must first adjust the book value for the amount of depreciation we deducted in the first year. The first year's depreciation is subtracted from the beginning book

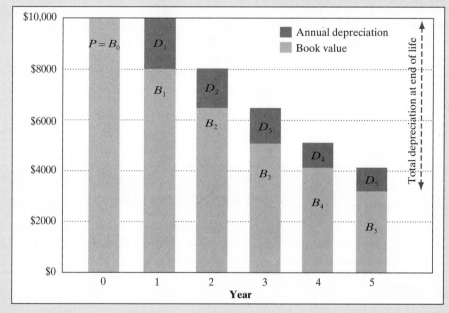

Figure 8.3   Declining balance method (Example 8.4).

value ($10,000 − $2000 = $8000). This amount is multiplied by the rate of depreciation ($8000 × 20% = $1600). By continuing the process, we obtain:

n	B_{n-1}	D_n	B_n
1	$10,000	$2,000	$8,000
2	$8,000	$1,600	$6,400
3	$6,400	$1,280	$5,120
4	$5,120	$1,024	$4,096
5	$4,096	$819	$3,277

Alternatively, we could use Eqs. (8.4) and (8.8) to calculate D_n and B_n directly.

The declining balance is illustrated in terms of the book value each year in Figure 8.3.

Salvage value (S) must be estimated independently of depreciation analysis. In Example 8.4, the final book value (B_N) conveniently equals the estimated salvage value of $3277, an occurrence that is rather unusual in the real world. When $B_N \neq S$, the following adjustments may be made in our calculations for book depreciation. The approaches below do not apply for tax purposes, however, as the CRA has specific procedures to follow when the book and salvage values differ (see Section 8.4.3 and Chapter 9).

Case 1: $B_N > S$

When $B_N > S$, we are faced with a situation in which we have not depreciated the entire cost of the asset. If you would prefer to reduce the book value of an asset to its salvage value as quickly as possible, it can be done by switching from DB to SL whenever SL depreciation would result in larger depreciation charges and therefore a more rapid reduction in the book value of the asset. The switch from DB to SL depreciation can take place in any of the N years, the objective being to identify the optimal year to switch. The switching rule is as follows: If depreciation by DB in any year is less than (or equal to) what it would be by SL, then we should switch to and remain with the SL method for the remaining duration of the project's depreciable life. Switch in year n when the following condition holds:

$$dB_{n-1} \leq \frac{B_{n-1} - S}{N - n + 1}. \tag{8.9}$$

EXAMPLE 8.5 Declining Balance With Conversion to Straight-Line Depreciation ($B_N > S$)

Suppose the asset given in Example 8.4 has a salvage value of $2700 instead of $3277. Determine the optimal time to switch from DB to SL depreciation and the resulting depreciation schedule.

SOLUTION

Given: $P = \$10,000$, $S = \$2700$, $N = 5$ years, $d = 20\%$.
Find: Optimal conversion time, D_n and B_n for $n = 1$ to 5.

Computing the SL depreciation from the beginning of each year n, to the final year N, and comparing to the DB calculated in Example 8.4, we use the decision rule given in Eq. (8.9) to determine when to switch:

If Switch to SL at Beginning of Year	SL Depreciation	DB Depreciation	Decision
2	($8000 − 2700)/4 = $1325	< $1600	Do not switch
3	($6400 − 2700)/3 = $1233	< $1280	Do not switch
4	($5120 − 2700)/2 = $1210	> $1024	Switch to SL

The optimal time (year 4) in this situation corresponds to n' in Figure 8.4(a). The resulting depreciation schedule is:

Year	DB with Switching to SL	End-of-Year Book Value
1	$2000	$8000
2	1600	6400
3	1280	5120
4	1210	3910
5	1210	2700
Total	$7300	

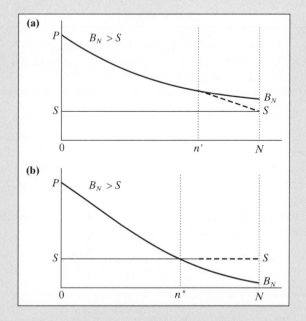

Figure 8.4 Adjustments to the declining balance method: (a) Switch from DB to the SL after n'; (b) no further depreciation allowances are available after n'' (Examples 8.5 and 8.6).

Case 2: $B_N < S$

With a relatively high salvage value, it is possible that the book value of the asset could decline below the estimated salvage value. To avoid deducting depreciation charges that would drop the book value below the salvage value, you simply stop depreciating the asset whenever you get down to $B_n = S$.

EXAMPLE 8.6 Declining Balance, $B_N < S$

Suppose the asset described in Example 8.4 has a salvage value of $4500 instead of $3277. Determine the declining balance schedule which culminates in a final book value equal to the estimated salvage value.

SOLUTION

Given: $P = \$10{,}000$, $S = \$4500$, $N = 5$ years, $d = 20\%$.
Find: D_n and B_n for $n = 1$ to 5.

End of Year	D_n	B_n
1	$0.2(\$10{,}000) = \$2{,}000$	$\$10{,}000 - \$2{,}000 = \$8{,}000$
2	$0.2(8{,}000) = 1{,}600$	$\$8{,}000 - 1{,}600 = 6{,}400$
3	$0.2(6{,}400) = 1{,}280$	$6{,}400 - 1280 = 5{,}120$
4	$0.2(5{,}120) = 1{,}024 > \boxed{620}$	$5{,}120 - 620 = 4{,}500$
5	0	$4{,}500 - 0 = 4{,}500$
	Total $= \$8{,}000$	

Note that B_4 would be less than $S = \$4500$, if the full deduction ($1024) had been taken. Therefore, we adjust D_4 to 620, making $B_4 = \$4500$. D_5 is zero and B_5 remains at $4500. Year 4 is equivalent to n'' in Figure 8.4(b).

Sum-of-Years'-Digits (SOYD) Method

Another accelerated method for allocating the cost of an asset is called **sum-of-years'-digits** (SOYD) depreciation. Compared with SL depreciation, SOYD results in larger depreciation charges during the early years of an asset's life and smaller charges as the asset reaches the end of its estimated useful life. Unlike the DB method, SOYD guarantees that $B_n = S$ automatically.

In the SOYD method, the numbers 1, 2, 3, ..., N are summed, where N is the estimated years of useful life. We find this sum by the equation[3]

$$SOYD = 1 + 2 + 3 + \ldots + N = \frac{N(N+1)}{2}. \tag{8.10}$$

[3] You may derive this sum equation by writing the SOYD expression in two ways:
$SOYD = 1 + 2 + 3 + \ldots + N$
$SOYD = N + (N-1) + \ldots + 1.$
Then you add these two expressions and solve for SOYD
$2SOYD = (N+1) + (N+1) + \ldots + (N+1)$
$= N(N+1)$
$SOYD = N(N+1)/2.$

The depreciation rate each year is a fraction in which the denominator is the SOYD and the numerator is, for the first year, N; for the second year, $N-1$; for the third year, $N-2$; and so on. Each year the depreciation charge is computed by dividing the remaining useful life by the SOYD and by multiplying this ratio by the total amount to be depreciated $(P-S)$.

$$D_n = \frac{N-n+1}{SOYD}(P-S). \tag{8.11}$$

EXAMPLE 8.7 SOYD Depreciation

Compute the SOYD depreciation schedule for Example 8.3.

$$\begin{aligned} \text{Cost basis of the asset, } P &= \$10,000 \\ \text{Useful life, } N &= 5 \text{ years} \\ \text{Salvage value, } S &= \$2000. \end{aligned}$$

SOLUTION

Given: $P = \$10,000$, $S = \$2000$, $N = 5$ years.
Find: D_n and B_n for $n = 1$ to 5.

We first compute the sum-of-years' digits:

$$SOYD = 1 + 2 + 3 + 4 + 5 = 5(5+1)/2 = 15$$

End of Year	D_n			B_n
1	(5/15)	($10,000 – $2,000) =	$2,667	$7,333
2	(4/15)	(10,000 – 2,000) =	2,133	5,200
3	(3/15)	(10,000 – 2,000) =	1,600	3,600
4	(2/15)	(10,000 – 2,000) =	1,067	2,533
5	(1/15)	(10,000 – 2,000) =	533	2,000

This situation is illustrated in Figure 8.5.

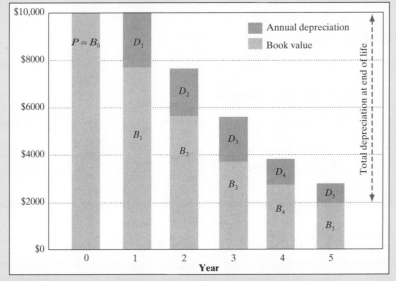

Figure 8.5 Sum-of-years' digits method (Example 8.7).

8.3.3 Units-of-Production (UP) Method

Straight-line depreciation is most applicable if the machine is used for the same amount of time each year. What happens when a punch press machine is run 1670 hours one year and 780 the next, or when some of its output is shifted to a new machining center? This leads us to a consideration of another depreciation concept that views the asset as consisting of a bundle of service units; unlike the SL and accelerated methods, however, this concept does not assume that the service units will be consumed in a time-phased pattern. Rather the cost of each service unit is the net cost of the asset divided by the total number of such units. The depreciation charge for a period is then related to the number of service units consumed in that period. This leads to the **units-of-production (UP) method,** according to which the depreciation in any year is given by

$$D_n = \frac{\text{Service units consumed during year } n}{\text{Total service units}}(P - S) \tag{8.12}$$

When using the units-of-production method, depreciation charges are made proportional to the ratio of actual output to the total expected output, and usually this ratio is figured in machine hours. The advantage of using this method is that depreciation varies with production volume, and therefore the method gives a more accurate picture of machine usage. A disadvantage of the units-of-production method is that collecting data on machine usage is somewhat tedious, as are the accounting methods. This method can be useful for depreciating equipment used to exploit natural resources, if the resources will be depleted before the equipment wears out. It is not, however, considered a practical method for general use in depreciating industrial equipment.

EXAMPLE 8.8 Units-of-Production Depreciation

A truck for hauling coal has an estimated net initial cost of $55,000 and is expected to give service for 250,000 kilometres, resulting in a $5000 salvage value. Compute the allowed depreciation amount for the year in which the truck usage was 30,000 kilometres.

SOLUTION

Given: $P = \$55,000$, $S = \$5000$, total service units = 250,000 kilometres, usage for this year = 30,000 kilometres.

Find: Depreciation amount in this year.

The depreciation expense in a year in which the truck travelled 30,000 kilometres would be

$$\frac{30,000 \text{ kilometres}}{250,000 \text{ kilometres}}(\$55,000 - \$5000) = (\frac{3}{25})(\$50,000)$$

$$= \$6000.$$

8.4 Capital Cost Allowance for Income Tax

The federal and provincial governments in Canada collect income taxes from businesses, as they do from individuals. Since income taxes often have a major impact on the economic viability of engineering projects, we will study this topic in some detail in Chapters 9, 10, and 11. The Income Tax Act, administered by the Canada Revenue Agency, also recognizes that depreciable assets lose value over time. Although the purchase cost of depreciable assets may not be deducted as an expense for tax purposes in the year of acquisition, a portion of the acquisition cost of such assets may be deducted from income each year, which may lower the taxes payable. This deduction is referred to as capital cost allowance (CCA).

Although many features of book depreciation also apply to capital cost allowance calculations, the Income Tax Act defines its own nomenclature specifically for the CCA context. The following table shows the correspondence between terms, although the interpretations are slightly different in some cases.

Book Depreciation Term	Tax Depreciation Term
asset	property
depreciation	capital cost allowance
cost basis	capital cost
book value	undepreciated capital cost
salvage value	proceeds of disposition

8.4.1 CCA System

The capital cost allowance system differs from book depreciation practices in two significant ways. First of all, the Income Tax Act specifies exactly how the CCA must be calculated, so there is no leeway as there is for financial reporting. This ensures fairness and consistency across all business taxpayers. The key provisions for calculating CCA are presented in the following sections.

Secondly, most property items are not depreciated individually, but are grouped into classes (also known as asset pools) as decreed by the tax regulations. The frequently used CCA classes are displayed in Table 8.1. The property in most classes must be depreciated using the declining-balance method, but a few classes follow more complex procedures, usually variants of the straight-line method. To limit the maximum allowance, the tax rules specify a percentage rate (CCA rate) for depreciation of each of the DB classes, and guidelines for determining the depreciable life for all property in the straight-line classes (e.g., Classes 13 and 14 in Table 8.1).

For example, the milling machines, saws, and drill presses owned by a machine shop would be combined into Class 43 (if they were all acquired after February 25, 1992). At the end of each year, the undepreciated capital cost (i.e., the income tax equivalent of current book value) of all this equipment may be depreciated by 30%, and the resulting capital cost allowance deducted as an expense before calculating income taxes. A company may claim up to the maximum allowance in each class in each year. This is the general practice, but for a variety of reasons, such as an unprofitable year, the full CCA is not always claimed. In the examples in this text, it is assumed that the maximum allowable CCA is claimed unless stated otherwise.

You may wonder what the rationale is for pooling property in this way. Three main principles apply. First, similar assets are generally grouped together, as they would have comparable usage characteristics and lifetimes, and the CCA rates are set accordingly.

TABLE 8.1 Capital Cost Allowance Rates and Classes

Class Number	Description	CCA Rate
1	Most buildings made of brick, stone, or cement acquired after 1987, including their component parts such as electric wiring, lighting fixtures, plumbing, heating and cooling equipment, elevators, and escalators (additional allowance of 6% for buildings used for manufacturing and processing in Canada and 2% for buildings used for other non-residential purposes, for buildings acquired after March 18, 2007)	4%
3	Most buildings made of brick, stone, or cement acquired before 1988, including their component parts as listed in Class 1 above	5%
6	Buildings made of frame, log, stucco on frame, galvanized iron, or corrugated metal that are used in the business of farming or fishing, or that have no footings below-ground; fences and most greenhouses	10%
7	Canoes, boats, and most other vessels, including their furniture, fittings, or equipment	15%
8	Property that is not included in any other class such as furniture, calculators and cash registers (that do not record multiple sales taxes), photocopy and fax machines, printers, display fixtures, refrigeration equipment, machinery, tools costing $500 or more, and outdoor advertising billboards and greenhouses with rigid frames and plastic covers	20%
9	Aircraft, including furniture, fittings, or equipment attached, and their spare parts	25%
10	Automobiles (except taxis and others used for lease or rent), vans, wagons, trucks, buses, tractors, trailers, drive-in theatres, general-purpose electronic data-processing equipment (e.g., personal computers) and systems software, and timber-cutting and removing equipment	30%
10.1	Passenger vehicles costing more than $30,000 if acquired after 2000	30%
12	Chinaware, cutlery, linen, uniforms, dies, jigs, moulds or lasts, computer software (except systems software), cutting or shaping parts of a machine, certain property used for earning rental income such as apparel or costumes, and videotape cassettes; certain property costing less than $500 such as kitchen utensils, tools, and medical or dental equipment acquired after May 1, 2006	100%
13	Property that is leasehold interest (the maximum CCA rate depends on the type of leasehold and the terms of the lease)	N/A
14	Patents, franchises, concessions, and licences for a limited period – the CCA is limited to whichever is less: • the capital cost of the property spread out over the life of the property; or • the undepreciated capital cost of the property at the end of the tax year Class 14 also includes patents, and licences to use patents for a limited period, that you elect not to include in Class 44	N/A
16	Automobiles for lease or rent, taxicabs, and coin-operated video games or pinball machines; certain tractors and large trucks acquired after December 6, 1991, that are used to haul freight and that weigh more than 11,788 kilograms	40%
17	Roads, sidewalks, parking-lot or storage areas, telephone, telegraph, or non-electronic data communication switching equipment	8%
38	Most power-operated movable equipment acquired after 1987 used for moving, excavating, placing, or compacting earth, rock, concrete, or asphalt	30%
39	Machinery and equipment acquired after 1987 that is used in Canada mainly to manufacture and process goods for sale or lease	25%
43	Manufacturing and processing machinery and equipment acquired after February 25, 1992, described in Class 39 above	30%
44	Patents and licences to use patents for a limited or unlimited period that the corporation acquired after April 26, 1993 – However, you can elect not to include such property in Class 44 by attaching a letter to the return for the year the corporation acquired the property. In the letter, indicate the property you do not want to include in Class 44	25%
45	Computer equipment that is "general-purpose electronic data processing equipment and system software" included in paragraph f of Class 10 acquired after March 22, 2004. Also see class 50 and 52.	45%
46	Data network infrastructure equipment that supports advanced telecommunication applications, acquired after March 22, 2004 – it includes assets such as switches, multiplexers, routers, hubs, modems, and domain name servers that are used to control, transfer, modulate and direct data, but does not include office equipment such as telephones, cell phones or fax machines, or property such as wires, cables or structures	30%
50	General-purpose computer equipment and systems software acquired after March 18, 2007, and before January 28, 2009, that is not used principally as electronic process control, communications control, or monitor equipment, and the systems software related to such equipment, and data handling equipment that is not ancillary to general-purpose computer equipment.	55%
52	General-purpose computer equipment and systems software acquired after January 27, 2009, and before February 2011.	100%

Source: Canada Revenue Agency (CRA), T2 Corporation—Income Tax Guide 2009 (www.cra-arc.gc.ca/E/pub/tg/t4012/t4012-09e.pdf). Reproduced with permission of the Minister of Public Works and Government Services Canada, 2010.

For example, buildings of sturdier materials (Classes 1 and 3) normally last longer than those in Class 6, and so are depreciated more slowly under the lower CCA rate. Machinery and vehicles wear out faster than buildings, and fall into classes with higher rates. Secondly, some combinations seem incompatible, such as the taxicabs and pinball machines in Class 16. However, the government has decided that they can be assigned to the same high CCA rate class to reflect the degree of wear-and-tear on such assets. Thirdly, since the capital cost allowance can act as an economic incentive for businesses, the CCA rates may reflect government policies or special circumstances. As an example of the latter, notice that "general-purpose electronic data processing equipment" acquired up to March 22, 2004, falls into Class 10 (CCA rate = 30%), whereas if it were bought between March 23, 2004, and March 18, 2007, it belongs in Class 45 (CCA rate = 45%), and if purchased after that it is assigned to Class 50 (CCA rate = 55%). This distinction was introduced to allow a faster depreciation of computers and application software to ease the tax burden on companies that have to upgrade or replace such assets fairly frequently because of the rapidly evolving technology.

Table 8.1 is included in the Income Tax Guide for Corporations and Businesses because it covers the majority of property owned by most businesses. Class numbers not listed here either relate to highly specialized assets, e.g., powered industrial lift track such as Class 42 for fibre-optic cable (CCA rate = 12%), or are applicable to property acquired during limited periods, such as Class 40, which includes specific assets (e.g., powered industrial lift trucks) purchased in 1988 or 1989. Several of the classes retained in Table 8.1 are also time-delimited. On the surface, it is not apparent why automobiles in Class 10.1 are not simply included in Class 10, since the same CCA rate applies in both cases. The reason for the varied classifications is that certain depreciation details for more expensive cars (Class 10.1) differ, such as how to calculate the tax effect upon disposition and a ceiling on the cost base (i.e., $30,000 before sales tax from 2001–2008) even if the car costs more than that. Since tax laws and regulations are changed regularly, it is important to note that we must use whatever rates, classifications, and adjustments are mandated *at the time an asset is acquired*.

These comments illustrate that the CCA rules can be very complicated, and recourse must often be made to accountants and other tax experts. However, sufficient information is provided in this text to deal with simple situations, or to develop after-tax cash flow estimates robust enough to make engineering economic decisions. The student interested in studying the CCA system or general tax rules in more detail can refer to the following sources (in increasing order of comprehensiveness):

- T2 Corporation—Income Tax Guide, Canada Revenue Agency.
- Interpretation Bulletins (IT) and Circulars (IC) available from CRA, or on the Web (www.cra_arc.gc.ca). In particular, Bulletin IT285R2 Capital Cost Allowance— General Comments discusses many aspects of the CCA system.
- Income Tax Act and Regulations. Several publishers compile this information with annual or bi-annual editions, such as *The Practitioner's Income Tax Act*, David M. Sherman (ed.), Carswell Publishing; *Canadian Master Tax Guide*, CCH Canadian Ltd.; *Federal Income Tax Act*, The Canadian Institute of Chartered Accountants, Ernst & Young Electronic Publishing Services Inc.

8.4.2 Available-for-Use Rule

In some cases, the capital cost allowance for an asset may not be claimed beginning in the year that it is acquired. The earliest taxation year in which CCA may be claimed for property is when it becomes **available-for-use**. This term has a specific meaning within the context of the Income Tax Act, and the restrictions differ for building versus non-building property, as described below.

A building is considered available for use on the earliest of several dates. The following are some examples of these dates:

- when the business uses all or substantially all of the building for its intended purpose;
- when the construction of the building is completed;
- the second taxation year after the date the property was acquired;
- immediately before disposal of the property.

Property other than a building is considered available for use on the earliest of several dates. The following are some examples of these dates:

- when the property is first used to produce income;
- when the business can use the property to either produce a saleable product or perform a saleable service;
- the second taxation year after the date the property was acquired;
- immediately before disposal of the property.

EXAMPLE 8.9 Available-for-Use Rule for CCA

Kasement Windows Ltd. decided to expand their production capacity and automate more of their operation in anticipation of a forecasted housing construction boom. On May 20, 2008, they purchased a vacant lot adjacent to their current plant and signed a contract to have a new facility built. On August 3, 2008, they received a new conveyor/assembly system that they temporarily housed in one corner of the new facility, which was not yet finished. The conveyor system required custom jigs to hold the window components, and these jigs had not yet arrived. On December 1, 2008, they acquired a used frame welder and an extruder at a good price 1 and stored them in the new facility. At the same time, they also purchased a forklift truck for the new production plant, but inventory was piling up so much at the existing plant that they started using the truck immediately. Construction of the new facility was completed on February 12, 2009. In the meantime, in January 2009 the company decided to invest in a state-of-the-art welder instead and traded in the used one they had bought for a slight loss. They were ready for full production at the new facility on June 1, 2009. They initially had to rely on manual material handling and assembly, and by the end of 2009 they still hadn't used the conveyor system since the custom jigs never fit properly. Determine when each new piece of property became available for use. The company's taxation year matches the calendar year.

SOLUTION

Given: Timing of acquisitions and usage of several pieces of property.

Find: When each piece of property is considered available-for-use for CCA calculations.

The building is available for use in 2009 because construction was completed then. The truck is available for use in 2008 because it was used immediately in December as part of their production (i.e., to generate income). Although the extruder was acquired in 2008, it was not installed or used until June 2009, which is the taxation year when it became available-for-use. The used frame welder was never installed, so it becomes available-for-use and added into the pool of depreciable property just before it was sold, hence in 2009. The state-of-the-art welder was first applied to production in the year 2009, which is when it became available-for-use. The conveyor still had not been used in the year 2009. The second taxation year following the acquisition date is 2010, so this equipment is considered available-for-use in 2010, even if Kasement has not used the conveyor by then. (*Note:* The jigs are in a separate CCA class from the conveyor, as shown in following section.)

Comments: The gist of this rule is that no CCA can be claimed for property in the year it is acquired, or in the following year, unless it is available-for-use. Thus there is a maximum two-year limit on this restriction.

8.4.3 Calculating the Capital Cost Allowance

To claim a capital cost allowance, a corporation must fill out a Schedule 8 form according to the T2 Corporation Income Tax Guide (T4012). Self-employed professionals, sole proprietors, or business partnerships perform equivalent calculations on their T2125 tax form. We will first present the CCA form in Figure 8.6, and accompanying guidelines for filling out each of the fields. As this is a summary only, and tax rules change periodically, you should refer to the current tax guides when performing calculations for an actual business. We will then proceed to provide examples to explain and illustrate most of the key elements in the CCA calculation. Each declining balance CCA class must be entered on a separate line of the form. Property belonging to one of the straight-line classes must have a separate calculation of CCA performed for each asset, which depends on many factors including the nature of the asset and its duration. The total allowable capital cost allowance for the firm is the sum of the CCAs across all classes.

- **Column 1—Class number** Each class of property is identified with the assigned class number. Generally, all depreciable property of the same class is grouped together. However, for some property types a separate record has to be maintained for each asset even though they are in the same class. The CCA is then calculated based on the undepreciated capital cost of all the property in that class.
- **Column 2—Undepreciated capital cost (UCC) at the beginning of the year** This is the value of all the assets in the class at the start of the year. It is set equal to the UCC value of all the assets in the class at the end of the preceding taxation year.
- **Column 3—Cost of acquisitions during the year** The total cost (including shipping, installation, and related fees) of depreciable property acquired by the business and available for use in the taxation year is entered by class of assets. *Land is excluded*, as it is not a depreciable property.
- **Column 4—Net adjustments** Sometimes, amounts must be entered that either increase or decrease the capital cost of a property. A partial list of such quantities is

included here. The capital cost of a property must be *reduced* by the following amounts:
- any GST/HST input tax credit or rebate received in the year;
- any federal **investment tax credit** (ITC) used to reduce taxes in the preceding taxation year (note: the government provides such incentives to encourage investment in certain activities or regions of the country; these transactions will be reviewed in Chapter 9).
- any provincial or territorial ITCs received in the current year;
- any government assistance received in the current year.

The capital cost of a property must be increased by the following amounts:
- any refund of GST/HST input tax credit previously deducted;
- any depreciated property transferred from an amalgamated or wound-up subsidiary;
- any government assistance the business repaid which was previously used to reduce the capital cost.

- **Column 5—Proceeds of dispositions during the year** The net proceeds of disposition (after deducting costs of removal of the asset or restoring the site) are entered for each class. If these proceeds exceed the capital cost of the property, enter the capital cost. (Note: the amount by which disposition proceeds exceed the original capital cost results in a capital gain—see Chapter 9—which is taxed at a different rate than other income.)

- **Column 6—Undepreciated capital cost (UCC)** The value in this cell is calculated by adding columns 2 and 3, subtracting or adding the amount in column 4 depending on the type of adjustment, and subtracting the amount in column 5. For each class, we must now distinguish between three possible states:
  (1) *the UCC in column 6 is positive, and property remains in the class at the end of the year:* Proceed with the calculations in column 7;
  (2) *the UCC in column 6 is negative:* This generally results when more CCA has been claimed than the actual loss of value on assets bought and sold in the class. Not only are you unable to claim CCA for the property in that class in the current year, but this recapture of CCA must be added to the firm's income, which is then taxed. The recapture of CCA is entered in column 10, and no further calculations are performed for that class.
  (3) *the UCC in column 6 is positive, but no property remains in the class at the end of the year:* This positive quantity results when the last asset(s) is sold for less than the undepreciated capital of all the property that was owned in the class. Since no assets remain in the class, this amount or terminal loss may be deducted directly from the business' income in the taxation year. The terminal loss is entered in column 11, and no further calculations are performed for that class.

Note that recapture and terminal loss rules do not apply to a few specific classes, including passenger vehicles in Class 10.1.

- **Column 7—50% rule** Most new property that is available for use during the taxation year is only eligible for 50% of the normal maximum CCA for the year (see Section 8.4.4). A reduced UCC must be calculated accordingly before applying the CCA rate. The adjustment equals one-half of the net amount of additions to the class (the net cost of acquisitions minus the proceeds of dispositions). This amount must be entered into column 7. Sometimes, adjustments must be made to this total based on amounts in column 4.

Canada Revenue Agency | Agence du revenu du Canada

CAPITAL COST ALLOWANCE (CCA) (2006 and later tax years)

SCHEDULE 8
Code 0601

Name of corporation

Business Number

Tax year-end
Year | Month | Day

For more information, see the section called "Capital Cost Allowance" in the T2 Corporation Income Tax Guide.

Is the corporation electing under Regulation 1101(5q)? 101 1 Yes ☐ 2 No ☐

1 Class number	2 Undepreciated capital cost at the beginning of the year (undepreciated capital cost at the end of the year from column 13 of last year's CCA schedule)	3 Cost of acquisitions during the year (new property must be available for use) See note 1 below	4 Net adjustments (show negative amounts in brackets) See note 2 below	5 Proceeds of dispositions during the year (amount not to exceed the capital cost)	6 Undepreciated capital cost (column 2 plus column 3 plus or minus column 4 minus column 5)	7 50% rule (1/2 of the amount, if any, by which the net cost of acquisitions exceeds column 5) See note 3 below	8 Reduced undepreciated capital cost (column 6 minus column 7)	9 CCA rate %	10 Recapture of capital cost allowance	11 Terminal loss	12 Capital cost allowance (column 8 multiplied by column 9; or a lower amount) See note 4 below	13 Undepreciated capital cost at the end of the year (column 6 minus column 12)
200	201	203	205	207		211		212	213	215	217	220
1.												
2.												
3.												
4.												
5.												
6.												
7.												
8.												
9.												
10.												

Totals

Note 1. Include any property acquired in previous years that has now become available for use. This property would have been previously excluded from column 3. List separately any acquisitions that are not subject to the 50% rule, see Regulation 1100(2) and (2.2).

Note 2. Include amounts transferred under section 85, or on amalgamation and winding-up of a subsidiary. See the T2 Corporation Income Tax Guide for other examples of adjustments to include in column 4.

Note 3. The net cost of acquisitions is the cost of acquisitions (column 3) plus or minus certain adjustments from column 4. For exceptions to the 50% rule, see Interpretation Bulletin IT-285, Capital Cost Allowance - General Comments.

Note 4. If the tax year is shorter than 365 days, prorate the CCA claim. Some classes of property do not have to be prorated. See the T2 Corporation Income Tax Guide for more information.

Enter the total of column 10 on line 107 of Schedule 1.
Enter the total of column 11 on line 404 of Schedule 1.
Enter the total of column 12 on line 403 of Schedule 1.

T2 SCH 8 (06)

(Vous pouvez obtenir ce formulaire en français à www.arc.gc.ca ou au 1-800-959-3376)

Canada

Figure 8.6 Capital cost allowance form.

Source: Canada Revenue Agency (CRA), Capital Cost Allowance (CCA) (2006 and later tax years) (www.cra-arc.gc.ca/E/pbg/tf/t2sch8/t2sch8-06-e.pdf). Reproduced with permission of the Minister of Public Works and Government Services Canada, 2010.

- **Column 8—Reduced undepreciated capital cost** A reduced UCC is required because of the 50% rule, by subtracting the amount in column 7 from column 6.
- **Column 9—CCA rate** The prescribed CCA rate for the class is entered here. For a class where the rate is not specified in the Income Tax Act, enter N/A.
- **Column 10—Recapture of capital cost allowance** The recapture amount calculated in column 6 is entered here (if applicable).
- **Column 11—Terminal loss** The terminal loss calculated in column 6 is entered here (if applicable).
- **Column 12—Capital cost allowance** The maximum allowable CCA for each class is calculated as the product of the reduced UCC from column 8 multiplied by the CCA rate in column 9. The business may claim any amount up to the maximum. If the taxation year is shorter than 365 days for a start-up business, the CCA for most property must be prorated (see tax guide for details).
- **Column 13—Undepreciated capital cost at the end of the year** The UCC from the end of the previous year, adjusted for acquisitions, dispositions, and government assistance (plus a few other adjustments as described in column 4) is reduced by the CCA claimed in the current year, resulting in the final UCC at the end of the current taxation year. This result is calculated by subtracting the amount in column 12 from the amount in column 6. For any class with a terminal loss or recapture of CCA, the UCC at the end of the year is always nil.

8.4.4 The 50% Rule

The 50% rule, also known as the half-year convention, was introduced into the income tax system on November 12, 1981. For most classes, it limits the maximum capital cost allowance of new property acquired during the year to 50% of what it would otherwise be. This rule reflects an "average" amount of depreciation for the year of acquisition across all businesses' new assets, since property purchased early in the taxation year has been subject to almost an entire year of usage, while property acquired near the end is practically brand new. More precisely, the 50% rule is applied to the **net acquisitions** in each class, equal to the amount by which purchases exceed dispositions during the year.

For a single asset, let us compare the pattern of CCA and UCC over time with and without the 50% rule.

$$
\begin{aligned}
\text{Let:} \quad P &= \text{the capital cost of the property} \\
U_n &= \text{the undepreciated capital cost at the end of year } n \\
CCA_n &= \text{the maximum claimable capital cost allowance in year } n \\
d &= \text{the prescribed CCA rate (declining-balance).}
\end{aligned}
$$

Year	Without 50% Rule		With 50% Rule	
	CCA	UCC	CCA	UCC
0		$U_0 = P$		$U_0 = P$
1	$CCA_1 = Pd$	$U_1 = P(1-d)$	$CCA_1 = Pd/2$	$U^1 = P(1-d/2)$
2	$CCA_2 = Pd(1-d)$	$U_2 = P(1-d)^2$	$CCA_2 = Pd(1-d/2)$	$U_2 = P(1-d/2)(1-d)$
$n\,(\geq 2)$	$CCA_n = Pd(1-d)^{n-1}$	$U_n = P(1-d)^n$	$CCA_n = Pd(1-d/2)(1-d)^{n-2}$	$U_n = P(1-d/2)(1-d)^{n-1}$

Most, but not all, depreciable property is subject to the 50% rule. Exemptions are listed below:

- Most Class 10 property must *comply* with the 50% rule, *except for*: television cable boxes or decoders; a Canadian film or video production.
- Most Class 12 property is *exempt* from the 50% rule *except for:* a die, jig, pattern, mould or last; the cutting or shaping part in a machine; the film or videotape comprising a television commercial message; a certified feature film or production; computer software (non-systems).
- All property in Classes 13, 14, 15, 23, 24, 27, 29 and 34 is *exempt* from the 50% rule (note: some provisions in classes 13, 24, 27, 29 and 34 require a reduction in their first-year CCA by 50%, but not using the same procedures as the 50% rule).
- Any property whose earliest available-for-use date is 358 days after acquisition (see section 8.4.2) is *exempt* from the 50% rule.

Take special note of the capital cost allowance for Class 12 items ($d = 100\%$). Those assets which are exempt from the 50% rule may be fully depreciated in the year that they are available for use. The remaining items in Class 12 are fully depreciated over two years. Most Class 12 items that are of interest for engineering economic decisions (machine parts, software) fall under the 50% rule.

EXAMPLE 8.10 CCA Calculations for Multiple Property Classes

Reconsider Kasement Windows' situation from Example 8.9. Assume that their undepreciated capital costs at the end of 2007 by CCA class are given by:

CCA Class	UCC_{2007}
1	$1,766,419
3	194,980
8	63,971
10	0
12	17,518
43	483,602

At that time, Class 1 included their existing plant (which was constructed after 1987); the only asset in Class 3 was an old warehouse that they used for excess inventory; Class 8 contained various office equipment; Class 12 encompassed certain machine parts; and Class 43 included all of their manufacturing equipment. Aside from the property transactions described in Example 8.9, Kasement finally got the proper jigs installed and working on their conveyor/assembly system in the year 2010. In that year, they sold the forklift truck, deciding that the conveyor system satisfied their material handling needs. They also sold their old warehouse in the year 2010, since the new facility has ample storage capacity. Given the following depreciable property transaction data, calculate the maximum CCA that Kasement can claim for the years 2008, 2009, and 2010.

Depreciable Property Transactions 2008–2010

Asset	CCA Class	Year Available for Use	Capital Cost	Disposition Year	Proceeds of Disposition
Forklift truck	10	2008	$18,000	2010	$5,000
New facility*	1	2009	$2,600,000		
Extruder	43	2009	$41,000		
Used welder	43	2009	$29,000	2009	$28,000
New welder	43	2009	$54,000		
Conveyor/assembler	43	2010	$112,000		
Conveyor jigs	12	2010	$13,000		
Warehouse	3			2010	$200,000

*This includes the new building costs only, excluding the land, as land is not depreciable.

SOLUTION

Given: UCC_{2007} for CCA classes, and depreciable property transactions for 2008–2010. Find: CCA for all classes for 2008, 2009, and 2010.

By following the procedure for calculating CCA, a Schedule 8 form is completed for each of the three years, as shown in Table 8.2.

Comments on 2008 Schedule 8: The UCC at the start of 2008 (column 2) must equal the UCC at the end of 2007. Under the 50% rule, only half of the value of the truck may be claimed in the first year ($CCA_{2008} = 1/2Pd = 1/2*18,000*30\% = \$2700$). Also, the remaining Class 12 property depreciates to $0 under the 100% rate.

Comments on 2009 Schedule 8: As with most CCA calculations, all of the new assets in Class 43 are pooled together. The 50% rule only applies to the *net* new acquisitions in each class (i.e., column 3 minus 5 in Class 43). Even though the used welder was never installed, it must still be recorded on Schedule 8 before it is sold. Therefore, the $1000 loss on this machine does not become a write-off but must be depreciated annually. Although the used welder became available-for-use the year after it was acquired, it is still subject to the 50% rule. As opposed to accounting depreciation, where an explicit adjustment to the cost basis must be made when the trade-in value does not equal the book value (see Example 8.2), the CCA system takes care of this automatically.

Comments on 2010 Schedule 8: In Class 3, because the sale price of the warehouse is higher than the UCC at the start of the year, the intermediate UCC result in column 6 is negative. Since Kasement has actually sold the property for more than the UCC of this class, Canada Revenue Agency taxes this difference by requiring the company to add this "recaptured CCA" to their income for 2010. This amount is transferred to column 10, and no further calculations are performed for this class. When the forklift truck in Class 10 is sold, it is the last property in the class. Since it is not fully depreciated, the balance in column 6 is copied to column 11, and this "terminal loss" is deducted as an expense in 2010. The jigs are one type of Class 12 assets governed by

TABLE 8.2 Capital Cost Allowance for Kasement Windows (Example 8.10).

Kasement Windows—Schedule 8 for 2008 (see Figure 8.6 for column headings)

1	2	3	4	5	6	7	8	9	10	11	12	13
1	$1,766,419				$1,766,419		$1,766,419	4%			$70,656	$1,695,762
3	$194,980				$194,980		$194,980	5%			$9,749	$185,231
8	$63,971				$63,971		$63,971	20%			$12,794	$51,176
10	$0	$18,000			$18,000	$9,000	$9,000	30%			$2,700	$15,300
12	$17,518				$17,518		$17,518	100%			$17,518	$0
43	$483,602				$483,602		$483,602	30%			$145,080	$338,521
										Total CCA	$258,497	

Kasement Windows—Schedule 8 for 2009 (see Figure 8.6 for column headings)

1	2	3	4	5	6	7	8	9	10	11	12	13
1	$1,695,762	$2,600,000			$4,294,762		$4,294,762	4%			$171,790	$4,122,972
3	$185,231				$185,231		$185,231	5%			$9,262	$175,969
8	$51,176				$51,176		$51,176	20%			$10,235	$40,941
10	$15,300				$15,300		$15,300	30%			$4,590	$10,710
12	$0				$0		$0	100%			$0	$0
43	$338,521	$124,000		$28,000	$434,521	$48,000	$386,521	30%			$115,956	$318,565
										Total CCA	$311,833	

Kasement Windows—Schedule 8 for 2010 (see Figure 8.6 for column headings)

1	2	3	4	5	6	7	8	9	10	11	12	13
1	$4,122,972				$4,122,972		$4,122,972	4%			$164,919	$3,958,053
3	$175,969			$200,000	−$24,031			5%	$24,031			
8	$40,941				$40,941		$40,941	20%			$8,188	$32,753
10	$10,710			$5,000	$5,710			30%		$5,710		
12	$0	$13,000			$13,000	$6,500	$6,500	100%			$6,500	$6,500
43	$318,565	$112,000			$430,565		$430,565	30%			$129,170	$301,396
										Total CCA	$308,777	

the 50% rule. The new conveyor system is *not* subject to the 50% rule because it became available for use in the second taxation year after acquisition.

Answer: $CCA_{2008} = \$258{,}497$ $CCA_{2009} = \$311{,}833$ $CCA_{2010} = \$308{,}777$

8.4.5 CCA for Individual Projects

Since a principal aim of engineering economic analysis is to evaluate the viability of one or more independent projects, the functioning of the CCA system presents a unique challenge. Because most depreciable property is pooled into classes, the tax implications of purchasing new assets for the project being evaluated depend on activity in the class (acquisitions and dispositions) unrelated to the current project. For example, we may apply the 50% rule for the first-year CCA of a new Class 43 machine for a proposed project, but the effect would be incorrect if any Class 43 property from elsewhere in the company is disposed of in the same year (because the 50% rule applies to the *net* acquisitions).

Another difficulty arises when predicting the tax implications upon disposal of an asset in our project evaluation. If the asset is the only property in its class, we may foresee a terminal loss or recapture upon its disposal, depending on the assumed salvage value. However, in the much more common situation where:

(1) the asset for the proposed project is outlived by other property in the class, and

(2) the asset is sold for more, or less, than its undepreciated capital cost,

then the CCA effects of this asset continue indefinitely after its disposal. That is, although the UCC of a class accounts for the sale of property (column 5 in Figure 8.6), there are residual effects on the pool in future years if the salvage value S of an asset does not equal its undepreciated capital cost U. This complicates the calculation of equivalence measures for project evaluation, particularly when using the IRR method. The details of disposal tax effects are presented in Chapter 9.

For the sake of consistency and simplicity, the following assumptions are used in this text (unless noted otherwise) for calculating the CCA of new assets on a project basis:

- Ignore any other activity in the class from outside the project. Implications of this include applying the 50% rule (if applicable) to a new asset in the project, and allowing the asset to fully depreciate every year of its useful life (which otherwise may not happen if there is a recapture of CCA in the class due to assets outside of the project).
- The **disposal tax effect** is based solely on the difference $U - S$ in the year of disposal. That is, a gain relative to its UCC on disposal $(S > U)$ gets taxed, while a relative loss from the sale $(S < U)$ results in tax savings. This approximation is generally acceptable since the error compared with the actual CCA effect is often small, and the disposal often occurs years into the future with little impact on the project viability (see Chapter 9 examples).
- Property is disposed of at the end of the year. This allows the disposal tax effect to coincide with the salvage value transaction.

EXAMPLE 8.11 CCA for a Project

Brigitte's Bakery in Port-aux-Basques, Newfoundland, is considering expanding to a second location in Corner Brook. They would take out a lease on January 1, renewable every four years, in a vacant store. The lease costs $10,000 per year. They would install $13,000 worth of equipment for selling and storing the baked goods: a cash register, display counter, refrigeration unit for pastries, and so on. If sales go as well as anticipated, after two years they will spend $6000 on capital improvements to the store and install an oven for $3000 so that they can produce some goods locally rather than ship everything from Port-aux-Basques. To evaluate this proposal over an eight-year time horizon, Brigitte assumes that the salvage value of all equipment combined in eight years would be $2000. What CCA amounts need to be included in this project evaluation?

SOLUTION

Given: Capital costs for an assortment of depreciable properties, and project life.
Find: CCA over project life.

As a tenant, Brigitte's Bakery has a **leasehold interest** in the store property, so any capital improvements to the building are eligible for a CCA allowance under Class 13. The capital cost allowance for this straight-line class is calculated as follows:[4]

- A *prorated portion* of the capital cost equals the lesser of:
  (a) one-fifth of the capital cost
  (b) the capital cost divided by the number of 12-month periods from the start of the year when the improvement was made to the end of the original lease plus first succeeding renewal (if applicable)
- Only half of the prorated portion may be claimed in the first year, with the balance claimable in the year following the end of the eight-year amortization period.

There are six years from the time of the capital improvements until the end of the lease's first renewal, so the prorated portion equals $6000/6 = $1000 (half in the first year).

The equipment and oven fall into Class 8 (CCA rate = 20%).

End-of-Year	Equipment CCA	Oven CCA	Leasehold CCA	Total CCA
1	$1300			$1300
2	$2340			$2340
3	$1872	$300	$500	$2672
4	$1498	$540	$1000	$3038
5	$1198	$432	$1000	$2630
6	$958	$346	$1000	$2304
7	$767	$276	$1000	$2043
8	$614	$221	$1000	$1835

[4]See CRA Interpretation Bulletin IT464R. Several other conditions can apply for this class of property, with respect to multiple concurrent leases, multiple renovations, etc.

Comments: This example illustrates some features of evaluating a project in isolation, while also demonstrating calculations for a CCA class which is not based on declining balance. By treating the project independently of other assets in the company, we have avoided several complications. The CCA claimed for the equipment in year 1 is only accurate if no other equipment in that class was disposed of in that year. The same holds for the acquisition of the oven. The CCA in any year for the equipment or the leasehold improvement could be disallowed if the UCC were driven to zero by the disposition of property outside of the project. By using the disposal tax effect simplification described above, where $U_8 = \$3338$ and $S_8 = \$2000$, then a terminal loss on the equipment of $1338 is assumed at the end eight years, independently of any other equipment in that class owned by Brigitte's. A similar adjustment may be made for the leasehold which has a UCC of $500 after eight years, and an assumed salvage value of zero.

8.5 Additions or Alterations to Depreciable Assets

If any major alterations or repairs (engine overhaul) or additions (improvements) are made during the life of the asset, we need to determine whether these actions will extend the life of the asset or will increase the originally estimated salvage value. When either of these situations arises, a revised estimate of useful life should be made, and the periodic depreciation expense should be updated accordingly. We will examine how repairs, alterations, or improvements affect both book and tax depreciations.

8.5.1 Revision of Book Depreciation

Recall that book depreciation rates are based on estimates of the useful life of assets. Estimates of useful life are seldom precise. Therefore, after a few years of use you may find that the asset could last for a considerably longer or shorter period than was originally estimated. If this happens, the annual depreciation expense, based on the estimated useful life, may be either excessive or inadequate. (If the repairs or improvements do not extend the life or increase the salvage value of the asset, these costs may be treated as maintenance expenses during that year.) The procedure for correcting the book depreciation schedule is to revise the current book value and to allocate this cost over the remaining years of useful life.

8.5.2 Revision of Tax Depreciation

Under the capital cost allowance system, capital expenditures (additions or alterations) on existing depreciable property are treated as new depreciable property acquisitions. The CCA class of the capital expenditure on additions or alterations is the class that would apply to the original property if it were made available-for-use at the same time as the addition or alteration.

Aside from the general prerequisites provided above regarding increased life or value of the property, Canada Revenue Agency publishes guidelines on how to distinguish between current expenditures on maintenance and repairs, versus capital upgrades (IT-128R bulletin). The principles include a test of **enduring benefit**. A part, added to a depreciable property, which requires regular replacement is not considered a capital expenditure.

A distinction is also made between **maintenance** and **betterment**. Bringing a property back to its original condition is one indication that the repair should be treated as a current expenditure, unless it involves a material improvement beyond the property's original condition. For example, replacing a roof is normally a current expense, unless the new roof is clearly of better quality or durability than the original one. Other criteria involve the relative value of the addition or improvement compared with the capital cost of the existing property, and whether the upgrade is an **integral part** of the original asset, or a separate marketable asset. If the relative value of the addition is small, or the new part is an integral part of the original property, the cost is more likely treated as a current expense. If in doubt, these and related criteria may be reviewed with a qualified accountant or other tax expert.

EXAMPLE 8.12 Depreciation Adjustment for an Overhauled Asset

In January 2000, Kendall Manufacturing Company purchased a new numerical control machine at a cost of $60,000. The machine had an expected life of 10 years at the time of purchase and a zero expected salvage value at the end of the 10 years. For book depreciation purposes, no major overhauls had been planned for that period and the machine was being depreciated using the straight-line method toward a zero salvage value, or $6000 per year. For tax purposes, the machine was Class 43 property (CCA rate = 30%). In January, 2003, however, the machine was thoroughly overhauled and rebuilt at a cost of $15,000. It was estimated that the overhaul would extend the machine's useful life by five years (see Figure 8.7).

(a) Calculate the book value and depreciation for the year 2005 on a straight-line basis.

(b) Calculate the CCA and UCC for the year 2005 for this machine.

SOLUTION

Given: $P = \$60,000$, $S = \$0$, $N = 10$ years, machine overhaul = $15,000, extended life = 15 years from the original purchase date.

Find: B_6 and D_6 for book depreciation, U_6 and CCA_6 for tax depreciation.

(a) Since an improvement was made at the beginning of 2003, the book value of the asset at that time consists of the existing book value plus the cost added to the asset. First, the existing book value at the end of 2002 is calculated:

$$B_3 \text{ (before improvement)} = \$60,000 - 3(\$6000) = \$42,000.$$

After adding the improvement cost of $15,000, the revised book value is

$$B_3 \text{ (after improvement)} => \$42,000 + \$15,000 = \$57,000.$$

To calculate the book depreciation in the year 2005, which is three years after the improvement, we need to calculate the annual straight-line depreciation amount with the extended useful life. The remaining useful life before the improvement was made was seven years. Therefore, the revised remaining useful life should be 12 years. The

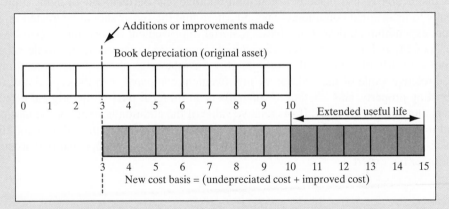

Figure 8.7   Revision of book depreciation as additions or improvements are made as described in Example 8.12

revised annual depreciation is then $57,000/12 = $4750. Using the straight-line depreciation method, we compute the book value and depreciation for year 2005 as follows:

$$B_6 = \$42,750 \qquad D_6 = \$4750.$$

(b) For tax depreciation, overhaul expenditure is treated as new property. The amount is added to Class 43 because *if* the original machine had been bought in 2003, it would have been placed in Class 43 at that time. To determine the total CCA and UCC in year 6, we calculate them separately for the two properties (the original asset and the "improvement") and then add them.

| | Original Machine | | Overhaul | |
Year	CCA (d = 30%)	UCC (year-end)	CCA (d = 30%)	UCC (year-end)
2000	$9,000*	$51,000		
2001	$15,300	$35,700		
2002	$10,710	$24,990		
2003	$7,497	$17,493	$2,250*	$12,750
2004	$5,248	$12,245	$3,825	$8,925
2005	$3,674	$8,571	$2,678	$6,247

*The 50% rule applies to both assets in their respective first years.

The total CCA for this machine and its overhaul in the year 2005 is $6352 (3674 + 2678) resulting in an end-of-year UCC of $14,818 (8571 + 6247).

These results could also be obtained directly by using the CCA formulas from section 8.4.4 as follows:

For the original machine, $n = 6$ years (from acquisition to the end of year 2005):

$$\text{CCA}_{2005} = Pd(1 - d/2)(1 - d)^{n-2} = \$60,000(0.30)(1 - 0.15)(1 - 0.30)^4 = \$3674$$
$$U_{2005} = P(1 - d/2)(1 - d)^{n-1} = \$60,000(1 - 0.15)(1 - 0.30)^5 = \$8571$$

For the overhaul, $n = 3$ years (from improvement completed to the end of year 2005):

$$CCA_{2005} = Pd(1 - d/2)(1 - d)^{n-2} = \$15{,}000(0.30)(1 - 0.15)(1 - 0.30)^1 = \$2678$$
$$U_{2005} = P(1 - d/2)(1 - d)^{n-1} = \$15{,}000(1 - 0.15)(1 - 0.30)^2 = \$6247$$

8.6 Tax Depreciation in the U.S. (Optional)

Canada's largest trading partner is the United States. With many companies having established operations and plants on both sides of the border, it is useful to get a basic appreciation of the U.S. system of tax depreciation. In 1981, the Internal Revenue Service (IRS) adopted the **Modified Accelerated Cost Recovery System (MACRS)**, which superseded previous methods.

8.6.1 MACRS Depreciation

Under the MACRS scheme, simple guidelines were established that created several classes of assets, each with a more or less arbitrary life called a **recovery period**. (*Note:* Recovery periods do not necessarily bear any relationship to expected useful lives.) There are eight categories of assets, as shown in Table 8.3. They are all depreciated using either the straight-line method as indicated or specified declining balance (DB) rate with a switch

TABLE 8.3 MACRS Property Classification

Recovery Period	Applicable Property	DB Rate	SL*
3 years	Special tools for the manufacture of plastic products, fabricated metal products, and motor vehicles	200%	switch
5 years	Automobiles, light trucks, high-tech equipment, equipment used for research and development, computerized telephone switching systems	200%	switch
7 years	Manufacturing equipment, office furniture, fixtures	200%	switch
10 years	Vessels, barges, tugs, railroad cars	200%	switch
15 years	Wastewater plants, telephone distribution plants, similar utility property	150%	switch
20 years	Municipal sewers, electrical power plants	150%	switch
$27\frac{1}{2}$ years	Residential rental property	—	Y
39 years	Nonresidential real property (commercial buildings), including elevators and escalators	—	Y

* "Y" indicates that SL is used; "switch" means that a switch is made from DB to SL when warranted, as explained in Section 8.3.2.

to the straight-line method when it becomes beneficial (see Section 8.3.2). Under the MACRS, *the salvage value of a property is always treated as zero.*

As with the Canadian guidelines presented in the preceding sections, these rules are deceptively simple, as there are many exceptions, and the rules and rates are periodically updated by the government. Therefore, current documentation must be consulted when performing important after-tax analyses.

8.6.2 MACRS Depreciation Rules

Applying the methods described in the preceding section, the MACRS establishes prescribed annual depreciation rates, called **recovery allowance percentages**, for all assets within each class. These rates, as set forth in 1986 and 1993, are shown in Table 8.4. The yearly recovery, or depreciation expense, is determined by multiplying the asset's depreciation base by the applicable recovery allowance percentage.

TABLE 8.4 MACRS Depreciation Schedules for Personal Properties with Half-Year Convention, Declining-Balance Method

Year	Class Depreciation Rate	3 200%	5 200%	7 200%	10 200%	15 150%	20 150%
1		33.33	20.00	14.29	10.00	5.00	3.750
2		44.45	32.00	24.49	18.00	9.50	7.219
3		14.81*	19.20	17.49	14.40	8.55	6.677
4		7.41	11.52*	12.49	11.52	7.70	6.177
5			11.52	8.93*	9.22	6.93	5.713
6			5.76	8.92	7.37	6.23	5.285
7				8.93	6.55*	5.90*	4.888
8				4.46	6.55	5.90	4.522
9					6.56	5.91	4.462*
10					6.55	5.90	4.461
11					3.28	5.91	4.462
12						5.90	4.461
13						5.91	4.462
14						5.90	4.461
15						5.91	4.462
16						2.95	4.461
17							4.462
18							4.461
19							4.462
20							4.461
21							2.231

* Year to switch from declining balance to straight line. (*Source*: IRS Publication 534. *Depreciation*. Washington, D.C.: U.S. Government Printing Office, December 2005.)

Half-Year Convention

The MACRS recovery percentages shown in Table 8.4 use the half-year convention; that is, it is assumed that all assets are placed in service at midyear and that they will have *zero* salvage value. As a result, only a half-year of depreciation is allowed for the first year that a property is placed in service. With half of one year's depreciation being taken in the first year, a full year's depreciation is allowed in each of the remaining years of the asset's recovery period, and the remaining half-year's depreciation is taken in the year following the end of the recovery period. A half-year depreciation is also allowed for the year in which the property is disposed of, or is otherwise retired from service, any time before the end of the recovery period. To demonstrate how the MACRS depreciation percentages were calculated by the IRS with use of the half-year convention, consider Example 8.13.

EXAMPLE 8.13 MACRS Depreciation: Personal Property

A taxpayer wants to place in service a $10,000 asset that is assigned to the five-year class. Compute the MACRS percentages and the depreciation amounts for the asset.

SOLUTION

Given: Five-year asset, half-year convention, $\alpha = 40\%$, and $S = 0$.
Find: MACRS depreciation percentages D_n for $10,000 asset.

For this problem, we use the following equations:

$$\text{Straight-line rate} = \tfrac{1}{5} = 0.20,$$

$$200\% \text{ declining balance rate} = 2(0.20) = 40\%,$$

$$\text{Under MACRS, salvage } (S) = 0.$$

Then, beginning with the first taxable year and ending with the sixth year, MACRS deduction percentages are computed as follows:

Year	Calculation (%)		MACRS Percentage
1	$\tfrac{1}{2}$-year DDB depreciation $= 0.5(0.40)(100\%)$	=	20%
2	DDB depreciation $= (0.40)(100\% - 20\%)$	=	32%
	SL depreciation $= (1/4.5)(100\% - 20\%)$	=	17.78%
3	DDB depreciation $= (0.40)(100\% - 52\%)$	=	19.20%
	SL depreciation $= (1/3.5)(100\% - 52\%)$	=	13.71%
4	DDB depreciation $= (0.40)(100\% - 71.20\%)$	=	11.52%
	SL depreciation $= (1/2.5)(100\% - 71.20\%)$	=	11.52%
5	SL depreciation $= (1/1.5)(100\% - 82.72\%)$	=	11.52%
6	$\tfrac{1}{2}$-year SL depreciation $= (0.5)(11.52\%)$	=	5.76%

In year 2, we check to see what the SL depreciation would be. Since 4.5 years are left to depreciate, SL depreciation = $(1/4.5)(100\% - 20\%) = 17.78\%$. The DDB depreciation is greater than the SL depreciation, so DDB still applies. Note that SL depreciation $\geq$ DDB depreciation in year 4, so we switch to SL then.

We can calculate the depreciation amounts from the percentages we just determined. In practice, the percentages are taken directly from Table 8.4, supplied by the IRS. The results are as follows and are also shown in Figure 8.8.

Year n	MACRS Percentage (%)		Depreciation Basis		Depreciation Amount (D_n)
1	20	×	$10,000	=	$2,000
2	32	×	10,000	=	3,200
3	19.20	×	10,000	=	1,920
4	11.52	×	10,000	=	1,152
5	11.52	×	10,000	=	1,152
6	5.76	×	10,000	=	576

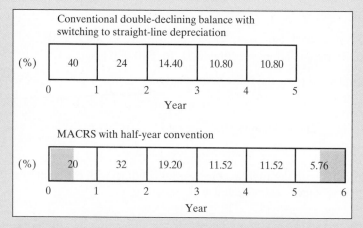

Figure 8.8 MACRS with a five-year recovery period (Example 8.13).

Note that when an asset is disposed of before the end of a recovery period, only half of the normal depreciation is allowed. If, for example, the $10,000 asset were to be disposed of in year 2, the MACRS deduction for that year would be $1600.

Improvements and Repairs

For U.S. tax depreciation purposes, repairs or improvements made to any property are treated as *separate* property items. The recovery period for a repair or improvement to the initial property normally begins on the date the repaired or improved property is placed in service. The recovery class of the repair or improvement is the recovery class that would apply to the property if it were placed in service at the same time as the repair or improvement. In Example 8.14, we redo Example 8.12 using U.S. rules to illustrate the procedure for correcting the depreciation schedule for an asset that had repairs or improvements made to it during its depreciable life.

EXAMPLE 8.14 Depreciation Adjustment for an Overhauled Asset in the U.S.

Reconsider Kendall Manufacturing Company from Example 8.12 with the following changes:

- Assume it is based in the U.S.
- For tax purposes, the machine was classified as a seven-year MACRS property.
- The major overhaul took place in December 2002.

  (a) Calculate the book depreciation for the year 2005 on a straight-line basis.
  (b) Calculate the tax depreciation for the year 2005 for this machine.

SOLUTION

Given: $I = \$60,000$, $S = \$0$, $N = 10$ years, machine overhaul $= \$15,000$, and extended life $= 15$ years from the original purchase.

Find: D_6 for book depreciation and D_6 for tax depreciation.

(a) Since an improvement was made at the end of the year 2002, the book value of the asset at that time consisted of the original book value plus the cost added to the asset. First, the original book value at the end of 2002 is calculated:

$$B_3 \text{ (before improvement)} = \$60,000 - 3(\$6000) = \$42,000.$$

After the improvement cost of \$15,000 is added, the revised book value is

$$B_3 \text{ (after improvement)} = \$42,000 + \$15,000 = \$57,000.$$

To calculate the book depreciation in the year 2005, which is three years after the improvement, we need to calculate the annual straight-line depreciation amount with the extended useful life. The remaining useful life before the improvement was made was seven years. Therefore, the revised remaining useful life should be 12 years. The revised annual depreciation is then $\$57,000/12 = \$4750$. Using the straight-line depreciation method, we compute the depreciation amount for 2005 as follows:

$$D_6 = \$4750.$$

(b) For tax depreciation purposes, the improvement made is viewed as a separate property with the same recovery period as that of the initial asset. Thus, we need to calculate both the tax depreciation under the original asset and that of the new asset. For the seven-year MACRS property, the sixth-year depreciation allowance is 8.92% of \$60,000, or \$5352. The third-year depreciation for the improved asset is 17.49% of \$15,000, or \$2623. Therefore, the total tax depreciation in 2005 is

$$D_6 = \$5352 + \$2623 = \$7975.$$

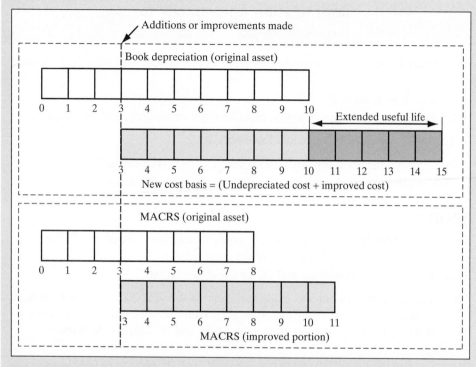

Figure 8.9 Revision of depreciation amount as additions or improvements are made as described in Example 8.14.

Figure 8.9 illustrates how additions or improvements are treated in the U.S. in revising depreciation amounts for book and tax purposes.

SUMMARY

■ Machine tools and other manufacturing equipment, and even the factory buildings themselves, are subject to wear over time. This **economic depreciation** generally results in increased maintenance and lower reliability; as assets age, periodic replacement of **capital property** is required.

■ The entire acquisition cost of a machine cannot be properly charged to any one year's production; rather, the cost should be spread (or capitalized) over the years in which the machine is in service, in accordance with the **matching principle**. The cost charged to operations during a particular year is called **depreciation**. Several different meanings and applications of depreciation have been presented in this chapter. From an engineering economics point of view, our primary concern is with **accounting depreciation:** The systematic allocation of an asset's value over its depreciable life.

- Accounting depreciation can be broken into two categories:
  1. **Book depreciation**—the method of depreciation used for financial reports and pricing products;
  2. **Tax depreciation**—the method of depreciation used for calculating taxable income and income taxes; it is governed by tax legislation.
- The four components of information required to calculate depreciation are:
  1. The cost basis of the asset,
  2. The salvage value of the asset,
  3. The depreciable life of the asset, and
  4. The method of its depreciation.

  Table 8.5 summarizes the differences in the way these components are treated for purposes of book and tax depreciation.
- Many firms select straight-line depreciation for book depreciation because of its relative ease of calculation.
- Many **capital cost allowance (CCA)** classes employ accelerated methods of depreciation, which gives taxpayers a break: It allows them to take earlier and faster advantage of the tax-deferring benefits of depreciation.

TABLE 8.5 Summary of Book versus Tax Depreciation

Components of Depreciation	Book Depreciation	Tax Depreciation (CCA)
Cost basis (Capital cost)	Based on the actual cost of the asset, plus all incidental costs such as freight, site preparation, installation, etc.	Same as book depreciation
Salvage value (Proceeds of disposition)	Estimated at the outset of depreciation analysis. If the final book value does not equal the estimated salvage value, we may need to make adjustments in our depreciation calculations.	Estimated in order to calculate the terminal loss or recapture of CCA. Also needed for most SL classes.
Depreciable life	Firms may select their own estimated useful lives, following generally acceptable accounting practices.	Not applicable to declining balance classes. Estimates required for other classes.
Method of depreciation	Firms may select from the following: • straight-line • accelerated methods (declining balance, double declining balance, and sum-of-years'-digits) • units-of-production	Depreciation method is specified by class; primarily declining balance and straight-line.

- Among the many regulations specific to tax depreciation, two rules are prevalent. The **available-for-use rule** dictates when property may be included in the capital cost base for its class. The **50% rule**, which applies to most CCA classes, specifies that only half of the amount of net acquisitions may be used for calculation of their first year's CCA.

- If the **undepreciated capital cost (UCC)** in a class becomes negative, a **recapture of CCA** is triggered. If the UCC is greater than zero in the year that the final asset in a class is disposed of, a **terminal loss** may be claimed.

- For tax depreciation in the U.S., the **Modified Accelerated Cost Recovery System (MACRS)** is used, with all assets falling into one of eight categories with differing recovery periods.

- Given the frequently changing nature of depreciation and tax law, we must use whatever rates, classifications, and adjustments, mandated *at the time an asset is acquired.*

PROBLEMS

Assumptions:
- Tax year equals the calendar year (unless specified otherwise).
- Property is available-for-use when acquired (unless specified otherwise).
- 50% rule for CCA applies as described in the text.

8.1 A machine, purchased for $45,000, had a depreciable life of four years. It will have an expected salvage value of $5000 at the end of the depreciable life. Using the straight-line method, what is the book value at the end of year 2?

8.2 Consider Problem 8.1. If the double declining balance (200% DB) method is used, what is the depreciation amount for year 2?

8.3 Consider Problem 8.1. If the sum-of-the years'-digits (SOYD) method is used, what is the depreciation amount for year 2 and the book value at the end of that year?

8.4 Your accounting records indicate that an asset in use has an undepreciated capital cost of $11,059. The asset cost $30,000 when it was purchased, and it has been depreciated under CCA Class 8 (rate = 20%). Based on the information available, determine how many years the asset has been in service.

8.5 A machine now in use was purchased four years ago at a cost of $5000. It has a book value of $1300. It can be sold now for $2300, but could be used for three more years, at the end of which time it would have no salvage value. What is the amount of economic depreciation now for this asset?

8.6 To automate one of their production processes, Waterloo Corporation bought three flexible manufacturing cells at a price of $500,000 each. When they were delivered, Waterloo paid freight charges of $25,000 and handling fees of $12,000. Site preparation for these cells cost $35,000. Six foremen, each earning $15 an hour, worked five 40-hour weeks to set up and test the manufacturing cells. Special wiring and other materials applicable to the new manufacturing cells cost $1500. Determine the cost basis (amount to be capitalized) for these cells.

8.7 Leo Smith Inc. paid $495,000 in 2007 for a used office building. This purchase price represents $375,000 for the building and $120,000 for the land. Calculate the capital cost allowance over the first five years of ownership.

8.8 A new drill press was purchased for $95,000 by trading in a similar machine that had a book value of $25,000. Assuming that the trade-in allowance is $20,000 and that $75,000 cash is to be paid for the new asset, what is the cost basis of the new asset for book depreciation purposes?

8.9 A lift truck priced at $35,000 is acquired by trading in a similar lift truck and paying cash for the remaining balance. Assuming that the trade-in allowance is $10,000, and the book value of the asset traded in is $6000, what is the cost basis of the new asset for the computation of book depreciation?

8.10 A firm is trying to decide whether to keep an item of construction equipment which typically lasts for eight years. The firm is using DDB for book depreciation purposes, and this is the fourth year of ownership. The item cost $150,000 new. What was the depreciation in year 3?

8.11 Compute the double declining balance (DDB) depreciation schedule for the following asset with switching to straight-line:

Cost of the asset, P	$60,000
Useful life, N	8 years
Salvage value, S	$5000

8.12 Compute the SOYD depreciation schedule for the following asset:

Cost of the asset, P	$12,000
Useful life, N	5 years
Salvage value, S	$2000

(a) What is the denominator of the depreciation fraction?
(b) What is the amount of depreciation for the first full year of use?
(c) What is the book value of the asset at the end of the fourth year?

8.13 A truck for hauling coal has an estimated net cost of $85,000 and is expected to give service for 250,000 kilometres. Its salvage value will be $5000 and depreciation will be charged at a rate of 32 cents per kilometre. Compute the allowed depreciation amount for the truck usage amounting to 55,000 kilometres per year.

8.14 A diesel-powered generator with a cost of $60,000 is expected to have a useful operating life of 50,000 hours. The expected salvage value of this generator is $8000. In its first operating year, the generator was operated 5000 hours. Determine the depreciation for the year.

8.15 The Harris Foundry Company purchased new casting equipment in 2004 at a cost of $180,000. Harris also paid $35,000 to have the equipment delivered and installed. The casting machine has an estimated useful life of 12 years, and it will be depreciated as a Class 43 asset (CCA rate = 30%).
(a) What is the capital cost of the casting equipment?
(b) What will be the CCA each year for the life of the casting equipment?

8.16 Suppose that a taxpayer in the U.S. places in service a $20,000 asset that is assigned to the six-year class (say, a new property class) with the half-year convention. Develop the MACRS deductions, assuming a 200% declining balance rate followed by switching to straight line.

8.17 A piece of machinery purchased at a cost of $68,000 has an estimated salvage value of $9000 and an estimated useful life of five years. It was placed in service on May 1 of the current fiscal year, which ends on December 31. The asset falls into CCA Class 43. Determine the capital cost allowances over the useful life.

8.18 Suppose that a taxpayer acquires new drafting software for $10,000. Calculate the CCA and UCC over the next four years.

8.19 General Service Contractor Company paid $100,000 for a house and lot. The value of the land was appraised at $65,000, and the value of the house at $35,000. The house was then torn down at an additional cost of $5000 so that a warehouse could be built on the lot at a cost of $50,000. What is the total value of the property with the warehouse? For book depreciation purposes, what is the cost basis for the warehouse?

8.20 Consider the following data on an asset:

Cost of the asset, P	$100,000
Useful life, N	5 years
Salvage value, S	$10,000

Compute and graph the annual depreciation allowances and the resulting book values, using
(a) The straight-line depreciation method,
(b) Declining balance method (with switching to SL if required), and
(c) Sum-of-the-years' digits method.

8.21 Consider the following data on an asset:

Cost of the asset, P	$30,000
Useful life, N	7 years
Salvage value, S	$8000

Compute the annual depreciation allowances and the resulting book values using the DB with switching to SL if required.

8.22 The double declining balance method is to be used for an asset with a cost of $80,000, estimated salvage value of $22,000, and estimated useful life of six years.
(a) What is the depreciation for the first three fiscal years, assuming that the asset was placed in service at the beginning of the year?
(b) If switching to the straight-line method is allowed, when is the optimal time to switch?

8.23 Upjohn Company purchased new packaging equipment with an estimated useful life of five years. Cost of the equipment was $20,000 and the salvage value was estimated to be $3000 at the end of year 5. Compute the annual depreciation expenses through the five-year life of the equipment under each of the following methods of book depreciation:

(a) Straight-line

(b) Double declining balance method (limit the depreciation expense in the fifth year to an amount that will cause the book value of the equipment at year end to equal the $3000 estimated salvage value).

(c) Sum-of-years'-digits method.

8.24 A second-hand bulldozer acquired at the beginning of the fiscal year at a cost of $58,000 has an estimated useful life of 12 years, when its salvage value is assumed to be $8000. Determine the following:

(a) The amount of annual depreciation by the straight-line method,

(b) The amount of depreciation for the third year computed by the declining balance method,

(c) The amount of depreciation for the second year computed by the sum-of-years'-digits method.

8.25 Ingot Land Company owned four trucks dedicated primarily for its landfill business. The company's accounting record indicates the following:

	Truck Type			
Description	A	B	C	D
Purchase cost ($)	50,000	25,000	18,500	35,600
Salvage value ($)	5,000	2,500	1,500	3,500
Useful life (kilometres)	200,000	120,000	100,000	200,000
Accumulated depreciation as year begins ($)	0	1,500	8,925	24,075
Kilometres driven during year	25,000	12,000	15,000	20,000

Determine the amount of depreciation for each truck during the year using units-of-production method.

8.26 Zerex Paving Company purchased a hauling truck on January 1, 2009, at a cost of $32,000. The truck has a useful life of eight years with an estimated salvage value of $5000. The straight-line method is used for book purposes. For tax purposes the truck would be depreciated as Class 10 property with a declining balance rate of 30%. Determine the annual depreciation amount to be taken over the useful life of the hauling truck for both book and tax purposes.

8.27 On October 1, you purchased a residential home in which to locate your professional office for $150,000. The appraisal is divided into $30,000 for the land and $120,000 for the building.

(a) In your first year of ownership, how much CCA can you deduct? (Assume that the entire house is used for business.)

(b) Suppose that the property was sold at $187,000 at the end of the fourth year of ownership. What is the undepreciated capital cost of the building?

8.28 On July 9, 2006, Twovens Ltd. purchased a used spindle machine at a bankruptcy sale for $34,000. They finally installed it in their factory on January 12, 2009.

Calculate the CCA and UCC for this Class 43 machine for five years, starting in the year when it first became available for use.

8.29 Armour Construction Company has just acquired a new Structural-Testing Sledge Hammer for $7500. By hitting a bridge or other structure, the feedback signals from the piezoelectric load cell embedded in the hammer head gives the engineer valuable information on structural integrity and resonant frequencies. They expect the equipment to last for six years, with a zero salvage value at that time. For book depreciation, Armour will use declining balance, with the rate determined by the useful life (no 50% rule). For tax purposes, the hammer falls into Class 8 property. Using the depreciation formulas presented in Section 8.4.4, calculate the book depreciation and capital cost allowance in the sixth year of ownership, and the UCC at the end of that time.

8.30 A manufacturing company has purchased four assets:

Item	Asset Type			
	Lathe	**Truck**	**Building**	**Photocopier**
Initial cost ($)	45,000	25,000	800,000	40,000
Book life	12 yr	200,000 km	50 yr	5 yr
CCA class	43	10	1	8
Salvage value ($)	3,000	2,000	100,000	0
Book depreciation	DDB	UP	SL	SOYD

For book depreciation, the units-of-production (UP) method was used for the truck. Usage of the truck was 22,000 kilometres and 25,000 kilometres during the first two years, respectively.

(a) Calculate the book depreciation for each asset in the first two years.

(b) Calculate the capital cost allowance for each asset in the first two years.

(c) If the lathe is to be depreciated over the early portion of its life using DDB and then by switching to SL for the remainder of the asset's life, when should the switch occur?

8.31 For each of four assets in the following table, determine the missing amounts for the year indicated. For asset type IV, annual usage is 15,000 kilometres.

Types of Asset	I	II	III	IV
Depreciating methods	SL	DDB	SOYD	UP
End of year	—	4	3	3
Initial cost ($)	10,000	18,000	—	30,000
Salvage value($)	2,000	2,000	7,000	0
Book value ($)	3,000	2,333	—	—
Depreciable life	8 yrs	5 yr	5 yr	90,000 km
Depreciating amount ($)	—	—	16,600	—
Accumulated depreciation ($)	—	15,680	66,400	—

8.32 Flint Metal Shop purchased a stamping machine for $147,000 on March 1, 2009. It is expected to have a useful life of 10 years, salvage value of $27,000, production of 250,000 units, and working hours of 30,000. During 2009 Flint used the stamping machine for 2450 hours to produce 23,450 units. From the information given, compute the book depreciation expense for 2009 under each of the following methods:

(a) Straight-line
(b) Units of production method
(c) Working hours
(d) Sum-of-years'-digits
(e) Declining balance
(f) Double declining balance.

8.33 Three assets were purchased in February 2005 and placed in service according to the following table.

Property Type	Date Placed in Service	Capital Cost	CCA Class
Car	Feb. 12, 2005	$15,000	10
Arc Welder	Nov. 11, 2005	$12,000	43
Freezer	Jan. 6, 2006	$8,000	8

Compute the CCA by year for each asset up to, and including, the year 2009.

8.34 Otto-Rentals Ltd. is setting up a new car rental operation. In the first year they purchase 12 vehicles for a total of $296,000. In year 2, they buy two more vehicles for $54,000. In year 3, they sell four vehicles for $58,000, and buy five new cars for $142,000. In year 4, they sell three vehicles for $32,000. These vehicles belong in CCA Class 16. Using columns similar to Schedule 8 (Figure 8.6), calculate the CCA and UCC for the four years for this pool of vehicles.

8.35 Given the data below, identify the depreciation method used for each depreciation schedule as one of the following:

• Double declining balance (DDB) depreciation
• Sum-of-years'-digits depreciation
• DDB with conversion to straight-line, assuming a zero salvage value
• CCA with 50% rule.

First cost	$80,000
Book depreciation life	7 years
Salvage value	$24,000
CCA Class	8

	Depreciation Schedule			
n	A	B	C	D
1	$14,000	22,857	8,000	22,857
2	12,000	16,327	14,400	16,327
3	10,000	11,661	11,520	11,661
4	8,000	5,154	9,216	8,330
5	6,000	0	7,373	6,942
6	4,000	0	5,898	6,942
7	2,000	0	4,719	6,942
8	0	0	3,775	0

8.36 A machine bought by Maine Machine Shoppe is classified as a seven-year MACRS property. Compute the book value for tax purposes for the U.S. Internal Revenue Service (IRS) at the end of three years. The cost basis is $145,000.

8.37 Eldridge Inc., a large environmental consulting company, wishes to set up an office in Winnipeg. They sign a seven-year lease for commercial space, with no option for renewal. Occupancy begins next January 1. In January they undertake major renovations, including new walls to partition the offices, an air conditioning system, and replacement of the drafty casement windows with new vinyl-clad sliding models. It is ready for the employees to move in on February 1. The capital cost of these improvements totals $22,000, and since Eldridge has a leasehold interest, they may claim CCA under Class 13. Calculate the maximum CCA claimable in years 1 to 3.

8.38 Perkins Construction Company bought a building for $1,200,000 25 years ago; it is to be used as a warehouse. A number of major structural repairs, completed at the beginning of the current year at a cost of $125,000, are expected to extend the life of the building 10 years beyond the original estimate. The building has been depreciated by the straight-line method. Salvage value is expected to be negligible and has been ignored. The book value of the building before the structural repairs is $400,000.

(a) What has the amount of annual depreciation been in past years?

(b) What is the book value of the building after the repairs have been recorded?

(c) What is the amount of depreciation for the current year, using the straight-line method?

8.39 The Dow Ceramic Company purchased a glass moulding machine in 2006 for $140,000. The company has been depreciating the machine over an estimated useful life of 10 years, assuming no salvage value, by the straight-line method of depreciation. For tax purposes, the machine has been depreciated as a Class 43 property. At the beginning of 2009, Dow overhauled the machine at a cost of $25,000. As a result of the overhaul, Dow estimated that the useful life of the machine would extend five years beyond the original estimate.

(a) Calculate the book depreciation for year 2011.

(b) Calculate the CCA for year 2011.

Short Case Studies

ST8.1 In 2008, three recent Industrial Engineering graduates are considering setting up a new consulting company, ErgoTech Ltd., to provide services on ergonomic designs and evaluations. Initially, they expect to invest $7000 on furniture, photocopier, and testing equipment (Class 8), $3500 on a computer including systems software (Class 50), $2400 on general and specialized software (Class 12), and $13,000 for a small car (Class 10) to visit clients. If the business grows as well as expected, they foresee the following property transactions over the coming years:

Year 1: purchase additional ergonomics software for $800

Year 2: purchase a second computer for $4000, and specialized software for $1000

Year 3: sell the car for $6000 and purchase a van for $20,000 for ease of transporting testing equipment; sell the old photocopier for $1000, and buy a better model costing $2500

Year 4: sell the original computer for $500 (uninstalling all software except the systems software); spend $2000 to upgrade the motherboard, RAM– and video card on the second computer, which will significantly improve its performance

Fill out a table similar to Schedule 8 shown in Figure 8.6 to show the capital cost allowances for each of these classes, and an annual total, for ErgoTech's first four years of business.

ST8.2 On January 2, 2005, Hines Food Processing Company purchased a machine priced at $75,000 that dispenses a premeasured amount of tomato juice into a can. The estimated useful life of the machine is estimated at 12 years with a salvage value of $4500. At the time of purchase, Hines incurred the following additional expenses:

Freight-in	$800
Installation cost	2500
Testing costs prior to regular operation	1200

Book depreciation was calculated by the straight-line method but, for tax purposes, the machine belonged in Class 43. In January 2007, accessories costing $2000 were added to the machine to reduce its operating costs. These accessories need to be replaced every year as they wear out.

(a) Calculate the book depreciation expense for 2008.

(b) Calculate the tax depreciation expense for 2008.

ST8.3 On January 2, 2004, Allen Flour Company purchased a new machine at a cost of $63,000. Installation costs for the machine were $2000. The machine was expected to have a useful life of 10 years with a salvage value of $4000. The company uses straight-line depreciation for financial reporting. On January 3, 2006, the machine broke down, and an extraordinary repair had to be made to the machine at a cost of $6000. The repair resulted in extending the machine's life to 13 years, but left the salvage value unchanged. On January 2, 2007, an improvement was made to the machine in the amount of $3000, which increases the machine's productivity and increases the salvage value to $6000, but does not affect the remaining useful life. Determine book depreciation expenses for the years December 31, 2004, 2005, 2006, and 2007.

ST8.4 At the beginning of the fiscal year, Borland Company acquired new equipment at a cost of $65,000. The equipment has an estimated life of five years and an estimated salvage value of $5000.

(a) Determine the annual depreciation (for financial reporting) for each of the five years of estimated useful life of the equipment, the accumulated depreciation at the end of each year, and the book value of the equipment at the end of each year by (1) straight-line method, (2) the double declining balance method, and (3) the sum-of-years'-digits method.

(b) Determine the annual depreciation for tax purposes assuming that the equipment falls into CCA class 43.

(c) Assume that the equipment was depreciated under CCA class 43. In the first month of the fourth year, the equipment was traded in for similar equipment priced at $82,000. The trade-in allowance on the old equipment was $10,000, and cash was paid for the balance. Calculate the CCA for Class 43 in the fourth year, and the UCC at the end of that year.

ST8.5 There are aspects of the capital cost allowance system that frequently arise during engineering economics evaluations, but are not described in detail in the T2 Tax Guide. For each of the following, look up the associated Interpretation Bulletin on the Canada Revenue Agency website *www.cra-arc.gc.ca*, and provide a short summary of the key points:

(a) IT128R Capital Cost Allowance—Depreciable Property

(b) IT422 Definition of Tools

(c) IT472 Capital Cost Allowance—Class 8 property

On the Companion Website that accompanies this text, you will find Excel templates and exercises, as well as the following analysis tools: Cash Flow Analyzer, Depreciation Analysis, Loan Analysis, and Interest Tables.

CHAPTER NINE

Corporate Income Taxes

Source: Illustration based on photo © Cornelius20 | Dreamstime.com.

The total revenues and income taxes paid by five well-known companies for the tax year ending in 2008 are summarized below (dollars in millions).

Company	Revenues	Earnings Before Income Taxes	Income Taxes	Net Income[*]	Effective (Average) Tax Rate (%)
BCE Inc.**	$17,698	$1,820	$469	$1,351	25.8%
Bombardier Inc.	$17,506	$439	$122	$317	27.8%
Encana Corp.	$30,064	$8,577	$2,633	$5,944	30.7%
SNC-Lavalin Group Inc.	$7,107	$403	$85	$318	21.1%
RBC Capital Trust	$21,582	$6,005	$1,369	$4,636	22.8%

[*] Before extraordinary items.
** All amounts in U.S. dollars.

What do these companies have in common? All are among the largest and most successful corporations operating in Canada, from a wide variety of sectors. But how do we explain the apparent discrepancy in the rates of taxation, which range from 21.1% in the case of SNC-Lavalin to 30.7% for Encana? The fact is that although each company has a statutory combined federal/provincial tax rate, the unique nature of business activities in any given year can result in taxes being paid at effective rates that may be higher or lower than the statutory rate. For example, the non-deductibility of certain major investments may raise the total tax payable significantly. Conversely, governments provide incentives for encouraging specified activities through tax credits and other allowances that may decrease the tax owing.

In this chapter, we will focus on combined federal plus provincial/ territorial income taxes. When you operate a business, any profits or losses you incur are subject to income tax consequences. Therefore, we cannot ignore the impact of income taxes in project evaluation.

Federal and provincial/territorial tax laws are exceptionally complex. While the overall conceptual framework for income tax calculation changes little from year to year, the specific details are subject to frequent changes. The approach presented in this chapter is a simplistic, but reasonably accurate, method of calculating corporate income taxes, which reflects the basic structure of the Canadian tax system. This treatment can be readily adapted to reflect future changes in the tax laws. The discussion and examples are based on 2008 regulations and rates.[1]

The chapter begins with a description of the general approach used in the calculation of corporate income taxes. This is followed by sections that describe the calculation of corporate income taxes and, more specifically, tax rates to be used in the evaluation of a new project.

It must be emphasized that corporations may have unique tax considerations that are well beyond the scope of the treatment presented. These situations should be referred to an accountant or a lawyer who specializes in tax matters.

---

[1] On the Companion Website that accompanies this text, you will find the current tax rates and revised versions of Tables 9.1 and 9.2.

CHAPTER LEARNING OBJECTIVES

After completing this chapter, you should understand the following concepts:

■ The general scheme of Canadian corporate taxes.

■ How to determine ordinary gains and capital gains.

■ How to determine the appropriate tax rate to use in project analysis.

■ The relationship between net income and net cash flow.

■ The availability of investment tax credits (ITC) for certain types of expenditures.

■ How to calculate the disposal tax effect when selling off assets.

9.1 Income Tax Fundamentals

There are a number of concepts that apply to corporate income tax calculations. However, in its most simple form, an income tax calculation can be represented as follows:

$$\text{Income taxes} = (\text{tax rate}) \times (\text{taxable income}).$$

In the following sections, these terms are defined and descriptions are given of the various elements needed to assess their values.

9.1.1. Tax Rate Definitions

Three different tax rates can be defined for corporate tax situations.

The **average or effective tax rate** is that value which gives the total income taxes payable by a corporation divided by the total taxable income, i.e., income from **all** sources:

$$\text{Average tax rate} = (\text{total income taxes}) / (\text{total taxable income}).$$

The determination of an average tax rate occurs only after the detailed tax calculations have been completed. This calculation gives a general sense of the tax burden for a specific business after accounting for its specific circumstances and deductions. However, the average tax rate is of little interest for the purpose of project evaluation.

The **marginal tax rate** represents the tax rate that is applicable to the next dollar of taxable income. Given the fact that a corporation has an existing level of taxable income and associated income taxes, how much additional income tax is payable if the taxable income increases by one dollar? Thus:

$$\frac{\text{Income tax on the next}}{\text{dollar of taxable income}} = (\text{marginal tax rate}) \times (\$1).$$

As we will see in later sections, the marginal tax rate depends on a number of factors, including the current level of taxable income and the type of additional taxable income.

The **incremental tax rate** is the tax rate that applies to a specified increment of taxable income over and above the existing level of taxable income and results in a corresponding increment of income tax. When this increment of taxable income is the result of a new project or investment, we need the incremental tax rate, i.e., an average tax rate over the increment of new taxable income, to quantify the tax effects in an economic evaluation. Hence, the incremental tax rate is of the most interest for our purposes:

$$\begin{matrix} \text{Income taxes due to a} \\ \text{new project or investment} \end{matrix} = \begin{matrix} \text{(incremental} \\ \text{tax rate)} \end{matrix} \times \begin{matrix} \text{(Taxable income due to a} \\ \text{new project investment).} \end{matrix}$$

To illustrate, let's consider a corporation having an existing level of annual taxable income of $70,000, which is taxable at an average rate of 35%. A new investment is under consideration that will increase annual taxable income by $25,000, of which the first $10,000 is taxable at 40% and the next $15,000 is taxable at 45%.

	Before Investment	**After Investment**
Taxable income	$70,000	$95,000
Income taxes	$70,000 × 35% = $24,500	$70,000 × 35% = $24,500
		+
		$10,000 × 40% = $ 4,000
		+
		$15,000 × 45% = $ 6,750
		Total $35,250

Following the investment, the total taxes will be $35,250 on taxable income of $95,000, which corresponds to an average tax rate for the corporation of 37.1% ($35,250/$95,000). The increase in taxes as a result of the investment is $10,750 ($35,250 − $24,500) from the additional $25,000 in taxable income. Therefore, the incremental tax rate for this investment is 43.0% ($10,750/$25,000). There are two marginal tax rates. In the interval of taxable income between $70,000 and $80,000, the marginal rate is 40%, and between $80,000 and $95,000, it is 45%. With this knowledge of the marginal rates, the incremental rate for the investment could have been calculated by pro-rating these marginal rates in relation to the total increase in taxable income:

$$\text{Incremental tax rate for the investment} = \frac{\$10,000}{\$25,000} \times 40\% + \frac{\$15,000}{\$25,000} \times 45\%$$

$$= 43.0\%.$$

In certain situations, the average, marginal, and incremental tax rates can be identical. We will demonstrate this in the sections on corporate income tax calculations.

9.1.2 Capital Gains (Losses)

Capital assets of corporations include all depreciable assets (buildings, equipment, etc.) and nondepreciable assets (land, stock, bonds, etc.). When capital assets are sold for more than the purchase price a profit or **capital gain** may be realized. Conversely, if the selling price is less than the purchase price a **capital loss** may result. The determination of gain or loss is as follows:

$$\text{Capital gain (loss)} = \text{selling price} - \text{cost base.}$$

The selling price represents the sale proceeds minus any selling expense and the **cost base** usually includes the purchase price plus the cost of improvements and expenses incurred in acquiring the capital asset. For depreciable assets, the cost base equals the original capital cost. Depreciable assets and personal-use property (i.e., property owned mainly for the personal use or enjoyment of an individual related to the corporation) do not incur capital losses.

Capital gains (losses) must be considered in the calculation of taxable income. Any gains will increase taxable income while losses will decrease taxable income. However, losses can only be counted to the extent that there are offsetting capital gains. So, if there are a series of capital gains, CG_1, CG_2, CG_3..., and capital losses, CL_1, CL_2, CL_3..., the net capital gain, which must be greater than or equal to zero, is calculated as

$$\text{Net capital gain} = \sum_{n=1}^{X} CG_n + \sum_{n=1}^{Y} CL_n \geq 0$$

where X and Y represent the number of capital gains and losses respectively with the losses expressed as negative values. Unused capital losses in any given year can be used to offset capital gains reported in the three previous years or retained and used to offset capital gains in future years.

For our purposes, whenever a project or investment incurs a capital loss, we will assume there are sufficient capital gains from other sources such that the capital loss can be used to reduce the taxable income from the project or investment in the same year.

The specific details on the inclusion of capital gains (losses) in corporate tax calculations are described in Sections 9.5.

EXAMPLE 9.1 Capital Gain on Land Transactions

Senstech is a Winnipeg-based manufacturer of electronic sensor and alarm systems. In 2005, it acquired land in three locations within the province to construct supply/distribution facilities. By 2009, only one of these facilities had been constructed and the company decided to sell the other two pieces of property. One of these had been purchased for $65,000 and there were associated acquisition costs of $5000. When it was sold in 2009, the actual price was $95,000 and there were legal costs of $3500. The second property had been purchased for $45,000 and sold for $35,000. The associated acquisition and selling costs were $3000 and $1500 respectively. Determine the net capital gain (loss) in 2009 for the land transactions.

SOLUTION

Given: Land purchase and selling price and associated costs.

Find: Net capital gain (loss) for 2009.

First, we have to determine the selling price adjusted for associated costs.

Capital	Actual Selling Price	−	Associated Costs of Sale	=	Adjusted Selling Price for Gains Purposes
Property 1	$95,000		$3,500		$91,500
Property 2	$35,000		$1,500		$33,500

The cost base is determined as follows:

	Purchase Price	+	Associated Costs of Acquisition	=	Cost Base
Property 1	$65,000		$5,000		$70,000
Property 2	$45,000		$3,000		$48,000

The capital gain (loss) is the difference between the adjusted selling price and the cost base.

$$\text{Property 1 capital gain} = \$91,500 - \$70,000 = \$21,500$$
$$\text{Property 2 capital loss} = \$33,500 - \$48,000 = (\$14,500)$$
$$\text{Net capital gain} = \$7000$$

So Senstech would have a net capital gain in 2009 of $7000 as a result of the two land sales.

9.2 Net Income

Firms invest in a project because they expect it to increase their wealth. If the project does this—if project revenues exceed project costs—we say it has generated a **profit**, or **income**. If the project reduces a firm's wealth—if project costs exceed project revenues—we say that the project has resulted in a **loss**. One of the most important roles of the accounting function within an organization is to measure the amount of profit or loss a project generates each year, or in any other relevant time period. Any profit generated will be taxed. The accounting measure of a project's after-tax profit during a particular time period is known as **net income**.

9.2.1 Calculation of Net Income

Accountants measure the net income of a specified operating period by subtracting expenses from revenues for that period. These terms can be defined as follows:

1. The **project revenue** is the income earned[2] by a business as a result of providing products or services to customers. Revenue comes from sales of merchandise to customers and from fees earned by services performed for clients.

---

[2] Note that the cash may be received in a different accounting period.

2. The **project expenses** that are incurred[3] are the cost of doing business to generate the revenues of the specified operating period. Common expenses include labour, raw materials, supplies, supervision, sales, depreciation; and allocated portions of utilities, rent, insurance, property taxes, and corporate functions such as marketing, engineering, administrative, and management; as well as interest on borrowed money and income taxes.

The business expenses listed above are accounted for in a straightforward fashion on a company's income statement and balance sheet. The amount paid by an organization for each item would translate dollar for dollar into expenses in financial reports for the period. For manufacturing organizations, accountants treat interest and income taxes separately but place the other expenses in two broad categories—**cost of goods sold and operating expenses**. The **operating expenses** include all the general, administrative, and selling expenses. The **cost of goods sold** includes all other expenses, including depreciation, which are reported in the subcategories of direct labour, direct materials, and manufacturing overhead.

As described in Chapter 8, the depreciation expense arises from the systematic allocation of the cost of an asset over time. In the following section, we will discuss how tax depreciation accounting is used in the calculation of taxable income and the resulting net income. It should be noted that this net income calculation may differ from that published in corporate financial reports. As noted in Section 8.2.4, these reports are based on book depreciation.

9.2.2 Treatment of Depreciation Expenses

Whether you are starting or maintaining a business, you will probably need to acquire assets (such as buildings and equipment). The cost of this property becomes part of your business expenses. The accounting treatment of capital expenditures differs from the treatment of manufacturing and operating expenses, such as cost of goods sold and business operating expenses. As you recall from Chapter 8, **capital expenditures must be capitalized**, i.e., they must be systematically allocated as expenses over their depreciable lives. Therefore, when you acquire a piece of property that has a productive life extending over several years, you cannot deduct the total costs from profits in the year the asset was purchased. Instead, a capital cost allowance[4] schedule is established over the life of the asset, and an appropriate allowance is included in the company's deductions from profit each year. Because it plays a role in reducing taxable income, depreciation accounting is of special concern to a company. In the next section, we will investigate the relationship between tax depreciation and net income.

9.2.3 Taxable Income and Income Taxes

Corporate taxable income is defined as follows:

$$\text{Taxable income} = \text{gross income (revenues)} - \text{expenses}.$$

---

[3] Note that the cash may be paid in a different accounting period.
[4] This allowance is based on the total cost basis of the property.

As noted in Section 9.1, income taxes are calculated as:

$$\text{Income taxes} = (\text{tax rate}) \times (\text{taxable income}).$$

(We will discuss how we determine the applicable tax rate in Section 9.3.) We then calculate net income as follows:

$$\text{Net income} = \text{taxable income} - \text{income taxes}.$$

A more common format is to present the net income in the following tabular income statement:

Item
Gross income (revenues)
Expenses
Cost of goods sold[5]
Capital cost allowance (CCA)
Operating expenses
Interest[6]
Taxable income
Income taxes
Net income

Our next example illustrates this relationship using numerical values.

EXAMPLE 9.2 Net Income Within a Year

A company buys a numerically controlled (NC) machine for $40,000 (year 0) and uses it for five years, after which it is scrapped. This equipment falls into CCA Class 43. Therefore, its declining balance CCA rate is 30% and the CCA claimable in the first year is $6,000, or 15% of the initial cost. (The 50% rule applies in the first year.) The cost of the goods produced by this NC machine should include a charge for the depreciation of the machine. Suppose the company estimates the following revenues and expenses including the depreciation for the first operating year.

$$\text{Sales revenue} = \$52,000$$
$$\text{Cost of goods sold} = \$20,000$$
$$\text{CCA on NC machine} = \$6,000$$
$$\text{Operating expenses} = \$5,000$$

If the company pays taxes at an incremental rate of 40% on the taxable income from the project, what is the net project income during the first year?

---

[5] The formal accounting definition of this item includes depreciation expenses. Here, depreciation in the form of CCA is represented as a separate item in the income statement and the definition of cost of goods sold is modified accordingly.

[6] Interest expense is included for completeness but is ignored elsewhere in the chapter. It is dealt with in Chapter 10.

SOLUTION

Given: Gross income and expenses as stated, income tax rate = 40%.
Find: Net income.

At this point, we will defer the discussion of how the tax rate (40%) is determined and treat it as given. We consider the purchase of the machine to have been made at time 0, which is also the beginning of year 1. (Note that our example explicitly assumes that the only CCA claimed for year 1 is that for the NC machine, a situation that may not be typical.)

Item	Amount
Gross income (revenues)	$52,000
Expenses	
Cost of goods sold	20,000
CCA	6,000
Operating expenses	$5,000
Taxable income	$21,000
Taxes (40%)	$8,400
Net income	$12,600

COMMENTS: In this example, the inclusion of a tax depreciation or CCA expense reflects the true cost of doing business. This expense is meant to match the amount of the $40,000 total cost of the machine that has been put to use or "used up" during the first year. This example also highlights some of the reasons why income tax laws govern the depreciation of assets for tax calculation purposes. If the company were allowed to claim the entire $40,000 as a year 1 expense, a discrepancy would exist between the one-time cash outlay for the machine's cost and the gradual benefits of its productive use. This discrepancy would lead to dramatic variations in the firm's income taxes and net income. Net income would become a less accurate measure of the organization's performance. On the other hand, failing to account for this cost would lead to increased reported profit during the accounting period. In this situation, the profit would be a "false profit" in that it would not accurately account for the usage of the machine. Depreciating the cost over time allows the company a logical distribution of costs that matches the utilization of the machine's value.

Identical considerations apply to net income values based on book depreciation, which appear in financial reports. Unless the basis for book depreciation calculations is reasonable in terms of distributing an asset's cost over its life, the reported net income values will be misleading.

9.2.4 Net Income versus Cash Flow

Traditional accounting stresses net income as a means of measuring a firm's profitability, but we will explain why cash flows are the relevant data to be used in project evaluation. As seen in Section 9.2.1, net income is an accounting measure based, in part, on the matching concept. Costs become expenses as they are matched against revenue. The actual timing of cash inflows and outflows is ignored.

Over the life of a firm, net incomes and net cash inflows will usually be the same. However, the timing of incomes and cash inflows can differ substantially. Given the time value of money, it is better to receive cash now rather than later, because cash can be invested to earn more cash. (You cannot invest net income.) For example, consider two firms and their income and cash flow schedules over two years as follows:

		Company A	Company B
Year 1	Net income	$1,000,000	$1,000,000
	Cash flow	1,000,000	0
Year 2	Net income	1,000,000	1,000,000
	Cash flow	1,000,000	2,000,000

Both companies have the same amount of net income and cash sum over two years, but Company A returns $1 million cash yearly, while Company B returns $2 million at the end of the second year. If you received $1 million at the end of the first year from Company A, you could invest it at 10%, for example. While you would receive only $2 million in total from Company B at the end of the second year, you would receive in total $2.1 million from Company A.

Apart from the concept of the time value of money, certain expenses do not even require a cash outflow. Depreciation is the best example of this type of expense. Even though capital cost allowance is deducted from revenue, no cash is paid to anyone.

In Example 9.2, we have just seen that the annual capital cost allowance has an important impact on both taxable and net income. However, although capital cost allowance has a direct impact on net income, it is not a cash outlay; as such, it is important to distinguish between annual income in the presence of depreciation and annual cash flow.

The situation described in Example 9.2 serves as a good vehicle to demonstrate the difference between tax depreciation costs as expenses and the cash flow generated by the purchase of a fixed asset. In this example, cash in the amount of $40,000 was expended in year 0, but the $6000 of CCA charged against the income in year 1 is not a cash outlay. Figure 9.1 summarizes the difference.

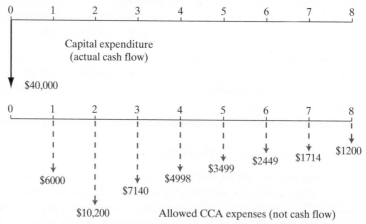

Figure 9.1   Cash flow versus CCA expenses for an asset with a cost basis of $40,000.

Net income (**accounting profit**) is important for accounting purposes, but **cash flows** are more important for project evaluation purposes. As we will now demonstrate, net income can provide us with a starting point to estimate the cash flow of a project.

The procedure for calculating net income is identical to that used for obtaining net cash flow (after-tax) from operations, with the exception of CCA, which is excluded from the net cash flow computation (it is needed only for computing income taxes). **Assuming that revenues are received and expenses are paid in cash**, we can obtain the net cash flow by adding the **non-cash expense** (i.e., CCA) to net income, which cancels the operation of subtracting it from revenues.

$$\text{Cash flows} = \text{net income} + \text{non-cash expense (CCA)}.$$

Example 9.3 illustrates this relationship.

EXAMPLE 9.3 Cash Flow versus Net Income

Using the situation described in Example 9.2, assume that (1) all sales were cash sales, and (2) all expenses except depreciation were paid during year 1. How much cash would be generated from operations?

SOLUTION

Given: Net income components.
Find: Cash flow.

We can generate a cash flow statement by simply examining each item in the income statement and determining which items actually represent receipts or disbursements. Some of the assumptions listed in the problem statement make this process simpler.

Item	Income Statement	Cash Flow
Gross income (revenues)	$52,000	$52,000
Expenses		
Cost of goods sold	20,000	–20,000
CCA	6,000	
Operating expenses	5,000	–5,000
Taxable income	21,000	
Taxes (40%)	8,400	–8,400
Net income	$12,600	
Net cash flow		$18,600

Column 2 shows the income statement, while Column 3 shows the statement on a cash flow basis. The sales of $52,000 are all cash sales. Costs other than depreciation were $25,000; these were paid in cash, leaving $27,000. CCA is not a cash flow—the firm did not pay out $6000 in CCA expenses. Taxes, however, are paid in cash, so the $8400 for taxes must also be deducted from the $27,000, leaving a net cash flow from operations of $18,600.

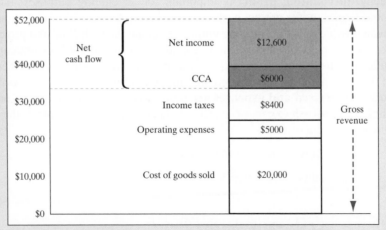

Figure 9.2 Net income versus net cash flow (Example 9.3).

Check: As shown in Figure 9.2, this $18,600 is exactly equal to net income plus CCA: $12,600 + $6000 = $18,600.

As we've just seen, CCA has an important impact on annual cash flow in its role as an accounting expense that reduces taxable income and thus taxes. (Although capital cost allowance expenses are not actual cash flows, depreciation has a positive impact on the after-tax cash flow of the firm.) Of course, during the year in which an asset is actually acquired, the cash disbursed to purchase it creates a significant negative cash flow and, during the depreciable life of the asset, the depreciation charges will affect the taxes paid and, therefore, cash flows.

As shown in Example 9.3, we can see clearly that depreciation, through its influence on taxes, plays a critical role in project cash flow analysis, which we will explore further in Chapter 10.

9.3 Corporate Income Tax Calculation

Now that we have learned what elements constitute taxable income, we turn our attention to the process of computing income taxes. The corporate tax rate is applied to the taxable income of a corporation, which is defined as its gross income minus allowable deductions. As we briefly discussed in Section 9.2, the allowable deductions include the cost of goods sold, CCA, operating expenses, and interest.

The cash expenses attributable to manufacturing a product or delivering a service are often referred to as operating and maintenance, or O&M, costs and may be shortened further to just operating costs (see Section 10.2.1). Examples and problems throughout the book use both the more rigorous cost of goods sold and operating expense categories and the less well defined O&M or operating cost designations.

9.3.1 Combined Corporate Tax Rates on Operating Income

The combined corporate tax rate consists of a federal rate plus a provincial or territorial rate. A corporation's federal tax rate is determined by corporate size (as measured by taxable income), ownership (private versus public and Canadian versus non-Canadian), type of business (manufacturing versus nonmanufacturing), the types of income (operating versus investment), and the source of income (inside versus outside Canada). In general, provincial/territorial rates depend on the same considerations.

If we restrict the discussion of federal rates to operating income of Canadian-controlled corporations with all such income earned in Canada, there are essentially only three corporate tax rates—small business, nonmanufacturing, and manufacturing. The calculation of these rates for 2008 is shown in Table 9.1. Many of these rates have continued to drop in recent years, as have some conditions for their application, so it is imperative to consult up-to-date tax tables when conducting an actual engineering economics evaluation.

In all cases, the starting point is the basic federal tax rate of 38%, which is reduced by the 10% federal tax abatement (which is intended to compensate corporations for provincial tax obligations). Private Canadian corporations (as opposed to public corporations listed on the stock exchange) whose taxable capital[7] is less than $10 million qualify for the small business deduction (SBD) of 17% on the first $400,000 of taxable income, referred to as the **business limit**. For corporations with taxable capital exceeding $10 million, the SBD percentage is reduced linearly until it reaches zero when the taxable capital is $15 million.

Corporations that derive at least 10% of their gross revenue from manufacturing and processing goods in Canada qualify for the 8.5% manufacturing and processing profits deduction (MPPD). The tax laws dictate what types of activities qualify for the MPPD, and there are numerous exclusions, notably natural resource enterprises, that are governed by distinct rules. Furthermore, the MPPD cannot be applied to income eligible for the small business deduction (SBD). There is also a federal tax "rate reduction" of 8.5% that applies to most nonmanufacturing income for Canadian corporations. As shown in Table 9.1, this reduction cannot be combined with either the SBD or the MPPD.

The **combined corporate tax** rate is obtained by adding the federal value for a given type of corporation from Table 9.1 to the corresponding provincial/territorial value from Table 9.2.

Since corporate tax rates are considered **non-progressive**, i.e., independent of the level of taxable income, the average combined tax rate is the same as the marginal combined

TABLE 9.1   Federal Corporate Tax Structure for 2008[1]

	Type of Corporation		
	Small Business[2]	Manufacturing	Nonmanufacturing
Basic federal tax	38.0%	38.0%	38.0%
Less: federal abatement	10.0%	10.0%	10.0%
Less: SBD	17.0%	—	—
Less: MPPD	—	8.5%	—
Less: rate reduction[3]	—	—	8.5%
Total federal tax rate	11.0%	19.5%	19.5%

[1] Adapted from KPMG (www.kpmg.ca/en/services/tax/taxrates), reprinted with permission.
[2] Only Canadian-Controlled Private Corporations (CCPC) are eligible for SBD; SBD business limit for federal taxation is $400,000.
[3] Applicable to most nonmanufacturing income for all Canadian-controlled corporations.

[7]Taxable capital includes the company's capital stock, retained earnings, long-term debt, advances received, surpluses, reserves, and so on.

TABLE 9.2 Provincial/Territorial Corporate Tax Structure for 2008[1]

	Small Business[2]	Manufacturing	Nonmanufacturing
Province/Territory			
British Columbia	3.91%[3]	11.5%[3]	11.5%
Alberta	3.0%	10.0%	10.0%
Saskatchewan	4.5%	10.0%	12.5%[3]
Manitoba	2.0%	13.5%[3]	13.5%[3]
Ontario	5.5%	12.0%	14.0%
Quebec	8.0%	11.4%	11.4%
New Brunswick	5.0%	13.0%	13.0%
Nova Scotia	5.0%	16.0%	16.0%
Prince Edward Island	3.47%[3]	16.0%	16.0%
Newfoundland & Labrador	5.0%	5.0%	14.0%
Yukon	2.5/4.0%[4]	2.5%	15.0%
Northwest Territories	4.0%	11.5%	11.5%
Nunavut	4.0%	12.0%	12.0%

[1] Adapted from KPMG (www.kpmg.ca/en/services/tax/taxrates), reprinted with permission.
[2] Only Canadian-Controlled Private Corporations (CCPC) are eligible for SBD; business limit varies from $400,000 to $500,000 depending on the province/territory.
[3] Average for the year, as these provinces dropped their rates partway through the year.
[4] The first number is for small business manufacturing and the second for nonmanufacturing small businesses.

tax rate and any incremental combined tax rate. This statement must be qualified to the extent that the new project does not place the corporation in a different tax category. For example, a manufacturing firm whose entire taxable income qualifies for the small business deduction may become a manufacturing type of business as a result of a project that increases production and sales. These considerations are illustrated in Example 9.4.

9.3.2 Income Taxes on Operating Income

The corporate tax rates from Section 9.3.1 can be applied directly to determine income taxes on operating income as illustrated in the following example.

EXAMPLE 9.4 Corporate Income Taxes

A Canadian-controlled and private mail-order computer company in Truro, Nova Scotia, sells computer supplies and peripherals. The company leased showroom space and a warehouse for $20,000 a year and installed $100,000 worth of inventory checking and packaging equipment. At a 30% CCA rate, the allowed capital cost allowance (CCA) for this capital expenditure within the first year will amount to $15,000. The store was completed and operations began on January 1. The company had a gross income of $1,450,000 for the calendar year. Supplies and all operating expenses other than the lease expense were itemized as follows:

Cost of merchandise sold in the year	$500,000
Employee salaries and benefits	$250,000
Other supplies and expenses	$ 90,000
Total expenses	$840,000

Compute the taxable income for the company. How much will the company pay in federal and provincial income taxes for the year?

SOLUTION

Given: Income, cost information, and CCA.

Find: Taxable income, federal and provincial income taxes.

First we compute the taxable income as follows:

Gross revenues	$1,450,000
Expenses	$ 840,000
Lease expense	$ 20,000
CCA	$ 15,000
Taxable income	$ 575,000

Now we must calculate the federal and provincial income tax.

Federal Income Tax		
Taxable income		$575,000
The first $400,000 is taxable at the rate of 11% (Small business federal corporate tax rate)		
Federal tax	= $400,000 × 11%	$ 44,000
The remaining $175,000 is taxable at the rate of 19.5% (Nonmanufacturing corporate tax rate)		
Federal tax	= $175,000 × 19.5%	$ 34,125
Total federal tax payable		$ 78,125

Provincial Income Tax (Based upon Nova Scotia tax rates)		
Taxable income		$575,000
The first $400,000 is taxable at the rate of 5% (Small business provincial corporate tax rate)[1]		
Provincial tax	= $400,000 × 5%	$ 20,000
The remaining $175,000 is taxable at the rate of 16% (Standard provincial corporate tax rate)		
Provincial tax	= $175,000 × 16%	$ 28,000
Total provincial tax payable		$ 48,000
Total combined corporate income tax payable		$126,125

[1] NS business limit also equals $400,000.

COMMENTS: Because the mail order computer company is private and Canadian controlled, it qualifies for the small business deduction (SBD) on the first $400,000 of taxable income. Therefore, the combined corporate tax rate on the $400,000 is 16% (11% federal + 5% Nova Scotia). The combined rate on the taxable income above $400,000 is the nonmanufacturing rate of 35.5% (19.5% federal + 16.0% Nova Scotia). The computer company's combined average or effective tax rate is 21.93% (total income taxes of $126,125 divided by taxable income of $575,000).

Projects that this company undertakes to increase revenue or reduce expenses will keep taxable income well above the $400,000 small business maximum. Above this maximum all additional taxable income is taxed at a combined rate of 35.5%, until the company's taxable capital exceeds $10 million, at which point the SBD rate is affected. Therefore, this is both the marginal rate and the incremental rate that should be used in the economic evaluation of any project this company may be considering.

It should also be noted that if this company had been a public rather than a private company, the SBD would not apply. In such a case all the company's taxable income would have been taxed at a combined rate of 35.5%. The average, marginal, and incremental rates would be identical and equal to 35.5%.

9.3.3 Corporate Operating Losses

Although corporations strive to be profitable, there may be years in which a corporation operates at a loss and pays no taxes. Ordinary corporate operating losses can be carried back to each of the preceding three years and forward for the following 20 years and can be used to offset taxable income in those years. The loss must be applied first to the earliest year, then to the next earliest year, and so forth. For example, an operating loss in 2009 can be used to reduce taxable income for 2006, 2007, or 2008, resulting in tax refunds or credits; any remaining losses can then be carried forward and used in 2010, 2011, and so forth, to the year 2029.

9.3.4 Income Taxes on Non-operating Income

Non-operating income is derived from activities outside a company's mainstream business. For example, a manufacturer of oil- and gas-processing equipment may have income that is not directly related to the sale of its equipment. Such income could include interest payments from investments, rent, and capital gains. The tax rates applicable to such income differ from those discussed in Section 9.3.1. We will only consider the tax considerations around capital gains.

Capital gains are taxable at half of their value. The applicable federal and provincial/territorial tax rates are the same as the nonmanufacturing rates in Table 9.1 and 9.2. The income tax on capital gains or the **capital gains tax** is calculated as follows:

Tax on capital gain = 1/2 × capital gain × (federal rate + provincial/territorial rate).

Frequently, the 1/2 factor is combined with the tax rate term to give a capital gains combined tax rate, t_{CG}, and the tax calculation becomes

Tax on capital gain = t_{CG} × capital gain.

You should recall from the discussion in Section 9.1.2 that capital losses must be offset against capital gains and we assumed that these were always available. Such a loss actually reduces the taxes payable. The tax saving resulting from a capital loss is:

$$\text{Tax saving from capital loss} = t_{CG} \times \text{capital loss.}$$

9.3.5 Investment Tax Credits

The federal government may use the tax regulations to provide incentives that encourage investments in certain types of activities within specific areas of Canada. The **investment tax credit** (ITC) is such an incentive. Income taxes can be reduced by a percentage of the expenditures that are eligible for an ITC. The investment tax credits earned during the year are added to an ITC pool of accumulated credits, where the balance of the pool has been reduced by the amount of ITCs claimed in any preceding year. The qualifying activities and conditions for claiming ITCs are altered frequently by the government, but as of 2008 the following types of expenditures are eligible:

- Qualified Property—Expenditures on buildings, machinery, and equipment for use in designated activities (manufacturing, exploration, and development of natural resources, natural resource processing, logging, farming, and fishing) within specific geographical areas (Newfoundland and Labrador, Prince Edward Island, Nova Scotia, New Brunswick, Gaspé Peninsula, and certain offshore regions) earn investment tax credits at a rate of 10% for 2008.
- Scientific Research and Experimental Development (SR&ED)—Corporations earn investment tax credits at a rate of 20% of the amount of investment in scientific research and development activities (with certain restrictions on the types of expenditures) conducted in Canada. Small Canadian-controlled private corporations (CCPC) get a preferential ITC rate of 35% on the first $3 million of SR&ED investment—called "expenditure limit" (EL). If the CCPC's taxable income (Y) for the preceding year exceeds $400,000, the limit on SR&ED expenditures eligible for the higher 35% rate will be reduced as follows:

$$EL = \max\{0, \$3,000,000 - 10 \times (Y - \$400,000)\}$$

so that the EL drops to $0 if Y is $700,000 or more.
- Pre-production mining expenditures—Expenditures on exploration and pre-production preparation for mining qualifying minerals (e.g., diamonds, and base or precious metals) can be claimed at an ITC rate of 10%.
- Child-care spaces—Companies (other than those whose main business is child care) may claim ITCs on expenditures for building and equipping new child-care facilities, at a rate of 25% per child-care space. The limit on eligible expenditure is $40,000 per child-care space.
- Apprenticeship job creation—Businesses may claim an ITC equal to 10% of the wages of a qualified apprentice during their first 24 months of apprenticeship. The apprentice must work in a qualified trade, of which there are about 50 in Canada (www.red-seal.ca/Site/trades/analist_e.htm).

The investment tax credit (ITC), calculated as follows, can be deducted directly against taxes owing by the corporation:

$$\text{investment tax credit} = \text{value of eligible expenditures} \times \text{rate of ITC.}$$

Table 9.3 summarizes the ITC rates that apply, depending on various factors. Some investments that qualify for ITC are capital property acquisitions, while others are current (i.e., operating) expenditures. When the investment tax credits are earned from expenditures toward the capital cost of qualified depreciable assets, the UCC in that class must be reduced by the ITC amount in the year *following* the corresponding ITC claim for that investment. An ITC on depreciable property cannot be claimed until the asset is available for use. Furthermore, any ITC claimed is deducted from the adjusted cost base of capital property for the purpose of calculating a capital gain (see Section 9.1.2).

When the ITCs derive from qualified current expenditures being deducted, the deductible expenses for the following year are reduced by the amount of the credit. The net effect of this is that after-tax dollars are saved now at the expense of future before-tax deductions, which is an overall benefit to the taxpayer. For example, a company with a marginal tax rate of 40% that claims an ITC of $5000 in year 1 must deduct $5000 from similar qualified expenditures the following year before calculating taxes, thus "losing" $2000 (= 40% × 5000) of tax benefit then.

A corporation does not always deduct ITCs in the year they are earned, particularly when they have no taxable income in that year. There are three other alternatives to claiming the ITCs:

1. ITCs can be "carried back" three years, to be deducted from taxes paid in those years;
2. ITCs can be "carried forward" 20 years, to be deducted from taxes owing in a later year;
3. For ITCs earned on investments by CCPCs in qualified property or SR&ED, eligible taxpayers may choose to claim a **refundable investment tax credit**, in which case they receive the ITC in cash as a tax refund, even if they have no taxable income in that year.

Example 9.5 illustrates how ITCs provide a tax break for an eligible company that expands its manufacturing capacity.

TABLE 9.3 Investment Tax Credit Rates (2008) and Key Conditions

Expenditure category	Rate (%)	Region	Refundable tax credit	Limits
SR&ED	20 or 35[1]	Canada	Y	$3,000,000[2]
Qualified property	10	Atlantic provinces and offshore;		
		Gaspé Peninsula	Y	n/a
Pre-production mining	10	Canada	N	n/a
Child-care spaces	25	Canada	N	$10,000 per child-care space
Apprenticeship job creation	10	Canada	N	$2,000/year/ employee

[1] 35% is available only for smaller CCPCs (see description in this section).
[2] Applies only as a limit to the amount of expenditures eligible for the 35% ITC rate for qualified CCPCs; this expenditure limit is reduced depending on the CCPC's preceding-year taxable income and/or taxable capital.

EXAMPLE 9.5 Investment Tax Credits

During 2009, Medea Metals, near Moncton, New Brunswick, invested $2,000,000 in precision processing machinery to expand their operation (Class 43) and also hired seven apprentices to assist with the increased workload. Their combined salaries eligible for ITCs amounted to $140,000 for the year. Calculate the impact of the investment tax credits for 2009 and 2010.

SOLUTION

Given: Expenditures on Qualified Property and on Eligible Apprentices.
Find: The ITCs and their impact on taxes and on CCA for two years.

For the current expenditures on apprentices the ITC rate is 10%, so there will be a $14,000 credit against the 2009 tax payable. This amount of $14,000 must be included as income in 2010.

The ITC for the capital investment is 10% in the Atlantic Provinces, thus there will be a credit of $200,000 against Tax Payable in 2009. The Class 43 depreciation rate is 30%. The 50% rule will be applied to the total capital investment of $2,000,000 to calculate the capital cost allowance for 2009 as follows:

$$CCA_{2009} = {}^1/_2 * 0.30*\$2,000,000 = \$300,000$$

Therefore, this allowance will reduce the UCC at the end of 2009 to $1,700,000.

In 2010, the $200,000 ITC must be deducted from the January 1, 2010, UCC, leaving a balance of $1,500,000. So the 2010 CCA will be:

$$CCA_{2010} = 0.30*\$1,500,000 = \$450,000$$

The rules governing ITCs are specific and complicated, so tax experts and guidelines should be consulted as required. Current information on available investment tax credits can be found on the Canada Revenue Agency (CRA) website (www.cra-arc.gc.ca), Information Circular IC78-4R3SR.

9.4 Practical Issues Around Corporate Income Tax

The detailed material in Section 9.3 provides a basis for calculating the correct corporate tax rate. However, in practice (and most of the problems in this book are a case in point) the incremental corporate tax appropriate to a particular project has already been determined and this value is given. But do we know whether the company can actually use all the CCA available? How often are taxes paid? These questions need resolution before we can proceed to build income taxes into a project evaluation.

9.4.1 Incremental Corporate Tax Rate Specified

When the incremental tax rate is known, the calculation of taxes can proceed without reference to the type of corporation or the province/territory in which it operates.

Since the Capital Gains Inclusion Rate is altered periodically by the government, for many problems we will use the rate of 50%, which is current at the time of writing. *This will be the default value in this text when not explicitly stated otherwise.*

EXAMPLE 9.6 Capital Gains Tax on Land Transactions

Calculate the capital gains tax on the Senstech land transactions in Example 9.1. Senstech is a private Canadian-controlled manufacturing corporation with annual taxable income of several million dollars.

SOLUTION

Given: Net capital gain from Example 9.1 of $7000.
Find: Capital gains tax.

Senstech's incremental combined tax rate is the manufacturing federal rate of 19.5% plus the Manitoba rate of 13.5%, for 33.0% total. The capital gains tax rate is taken as $1/2$ of this value, or 16.5%. The capital gains tax associated with the land transaction is

$$\text{capital gains tax} = 16.5\% \times \$7000$$
$$= \$1155.$$

9.4.2 Claiming Maximum Deductions

The level of taxable income is dependent upon the gross income or revenues and the deductions that that can be claimed against revenues. These deductions include all the expenses associated with delivering a good or service as well as debt interest and CCA. Revenue Canada does not permit the total deductions to exceed the revenues. In other words, the taxable income must be greater than or equal to zero.

The constraint of non-negative taxable income can limit the deductions actually used. For example, it may not be possible to claim all of the CCA in a particular year because there is insufficient income. In this case, the unused CCA is claimed in subsequent years.

Such a limitation adds considerable complexity to the calculation of taxes. *However for our purposes, we will always assume that for any investment all available deductions can be claimed at their maximum level in any given year.* This means that negative taxable income will arise if the available deductions exceed the revenue. The income taxes for such an investment then become a positive cash flow. The underlying assumption is that there are other business operations in the company that already pay taxes. The total taxes payable on these other business operations will be reduced by the "positive" income taxes for the new investment. These tax savings are a credit or cash inflow to the investment under consideration. This concept will be explained by way of example in Chapter 10.

9.4.3 Timing of Corporate Income Tax Payments

The taxation year for a corporation is its fiscal or business year. During its fiscal year, a corporation must make monthly payments of federal and provincial income tax based on an estimate of the taxes actually owed by the corporation. This estimate may be nothing more than one-twelfth of taxes paid in the preceding fiscal year.

Although a corporation has up to six months after the end of its fiscal year to file federal and provincial tax returns, it must make a final tax payment within approximately two months of the end of its fiscal year. This payment represents the difference between total income tax payable and the amount paid in monthly instalments. The actual deadline is dependent on various factors such as corporate size and type of industry, and these types of considerations can add considerable complexity to an engineering economic analysis. *For our purposes, corporate investments will be assumed to occur at the beginning of a fiscal year, and the associated income taxes will be paid as a lump sum at the end of each fiscal year unless indicated otherwise.*

9.5 Depreciable Assets: Disposal Tax Effects

As in the disposal of capital assets, there are generally gains or losses on the sale (or exchange) of depreciable assets. To calculate a gain or loss, we first need to determine undepreciated capital cost (UCC) of the depreciable asset at the time of disposal. When a depreciable asset used in business is sold for an amount different from its UCC, this gain or loss has an effect on income taxes. The gain or loss is found as follows:

$$\text{Gains (losses)} = \text{salvage value} - \text{UCC}.$$

where the salvage value represents the proceeds from the sale minus any selling expense or removal cost.

These gains, known as **recaptured depreciation** or **recaptured CCA**, are taxable. In the unlikely event that an asset is sold for an amount greater than its capital cost, the gains (salvage value – UCC) are divided into two parts for tax purposes:

$$\begin{aligned}
\text{Gains} &= \text{salvage value} - \text{UCC} \\
&= (\text{salvage value} - \text{cost base}) + (\text{cost base} - \text{UCC}) \\
&= \text{capital gains} + \text{recaptured CCA}.
\end{aligned}$$

Recall from Section 8.2.2 that cost base means the purchase cost of an asset plus any incidental costs, such as freight and installation. As illustrated in Figure 9.3, in this case

$$\begin{aligned}
\text{Capital gains} &= \text{salvage value} - \text{cost base} \\
\text{Recaptured CCA} &= \text{cost base} - \text{UCC}.
\end{aligned}$$

This distinction is necessary because capital gains are taxed at a capital gain tax rate and recaptured CCA is taxed at the ordinary income tax rate. As described previously, current tax law allows a lower rate of taxation for capital gains. When disposal of a depreciable asset results in a loss, the loss reduces taxes payable.

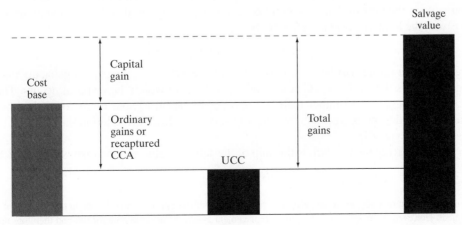

Figure 9.3 Capital gains and recaptured CCA.

Section 9.5.1 describes an approximate representation of disposal tax effects for the declining balance and the straight-line depreciation methods. The Companion Website that accompanies this text provides a more rigorous analysis of disposal tax effects for the declining balance and straight-line depreciation methods. Unless otherwise stated, all examples and problems throughout this book will follow the approach described in Section 9.5.1.

9.5.1 Calculation of Disposal Tax Effects

The manner in which gains or losses from an asset disposal affects taxes depends upon the depreciation method used, the assumed timing of the disposal, and whether or not the asset disposal depletes the asset class. As was discussed in Chapter 8, Section 8.4.5, the disposal of an asset does not normally deplete the asset class. The situation is further complicated if there are new acquisitions added to the asset class in the year in which disposal of an asset occurs. The loss or recapture of CCA from the disposal is built into the undepreciated capital cost base for the entire asset class as a result of the asset pool accounting procedure. The tax effect attributable to the disposal will then be spread over a number of future years, which makes the calculation of the total disposal tax effect very tedious. Because of this difficulty and because disposal tax effects are normally not an important factor in determining the acceptability of a project, we assume that the tax implications of the disposal are completely realized in the year of disposal and there are no new acquisitions to the asset class in that year.[8] *Specifically, we assume that the disposal occurs just prior to the end of the last year of service and that the CCA in the year of asset disposal is calculated in the usual manner and without reference to the disposal.* The disposal tax effect, G, then represents the tax implications over and above the tax savings realized in the year of disposal from CCA. The effects of these assumptions and variations to them are analyzed in more detail on the Companion Website that accompanies this text.

Because of the above assumptions, any gains on disposal are added to income, and any losses incurred are treated as expenses for assets depreciated by any depreciation method. This greatly simplifies the calculation of the disposal tax effect because G is now merely the extra tax payment or the tax savings realized in the year of disposal, i.e.,

$$G = t(U_{\text{Disposal}} - S),$$

where $U_{\text{Disposal}} = UCC_N = UCC_{N-1} - CCA_N$. UCC_{N-1} is the undepreciated capital cost of the asset at the end of the year prior to disposal, and CCA_N is the capital cost allowance claimable in the year of disposal with the actual disposal ignored. When capital gains are involved, the total disposal tax effect G becomes

$$G = t(U_{\text{Disposal}} - P) - t_{CG}(S - P)$$

where P is the initial cost base (time zero installed cost), t_{CG} is the capital gains tax rate, and U_{Disposal} is the asset's UCC at the end of the year in which the disposal occurs. The choice of $(U_{\text{Disposal}} - S)$ rather than $(S - U_{\text{Disposal}})$ is arbitrary, but more convenient because it provides the correct sign for the after-tax cash flow correction in the last year of the asset's service life.

The net salvage value, NS, is the sum of the salvage value and the disposal tax effect:

$$NS = S + G$$

---

[8] This assumption may appear overly restrictive in situations which require asset replacement over the life of a project. However, it can be argued that these replacement assets are not available for use and therefore not added to the asset class until the beginning of the year following the disposal of the original asset.

EXAMPLE 9.7 Disposal Tax Effects on Depreciable Assets

A company purchased a drill press costing $250,000. The drill press is classified as a CCA Class 43 property with a declining balance rate of 30%. If it is sold at the end of three years, compute the gains (losses) for the following four salvage values: (a) $150,000, (b) $104,125, (c) $90,000, and (d) $280,000. Assume that the company's combined federal and provincial tax rate is 40% and that capital gains are taxed at $1/2$ of their value, i.e., effectively taxed at 30% in this example.

SOLUTION

Given: a CCA Class 43 asset, cost base = $250,000, sold three years after purchase.

Find: Disposal tax effects, and net salvage value from the sale if sold for $150,000, $104,125, $90,000, or $280,000

Year	Capital Cost Allowance (30% CCA rate)	UCC
0		$250,000
1	$37,500	212,500
2	63,750	148,750
3	44,625	104,125

The CCA amount in year 1 is reduced by 50% because of the 50% rule. The UCC at the end of year 3 is $104,125.

(a) Case 1: UCC < salvage value < cost base

$$\text{Disposal tax effects} = G = t(U_{\text{Disposal}} - S)$$
$$= 0.4 \times (\$104,125 - \$150,000)$$
$$= -\$18,350$$

$$\text{Net salvage value} = \text{salvage value} + \text{disposal tax effects}$$
$$= \$150,000 - \$18,350$$
$$= \$131,650.$$

This situation (salvage value exceeds UCC) is denoted as Case 1 in Figure 9.4.

(b) Case 2: Salvage value = UCC

In Case 2, the UCC is again $104,125. Thus, if the drill press's salvage value is $104,125, equal to the UCC, no taxes are calculated on that salvage value. Therefore, the net proceeds equal the salvage value.

(c) Case 3: Salvage value < UCC

Case 3 illustrates a loss when the salvage value (say, $90,000) is less than the UCC. We compute the net salvage value after tax as follows:

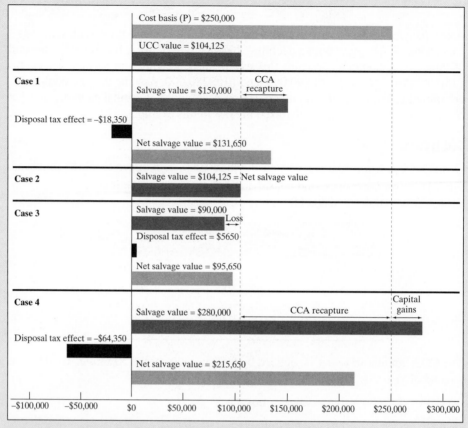

Figure 9.4 Calculations of gains or losses (Example 9.7).

$$\text{Disposal tax effects} = G = t(U_{\text{Disposal}} - S)$$
$$= 0.4 \times (\$104,125 - \$90,000)$$
$$= \$5,650$$

$$\text{Net salvage value} = \$90,000 + \$5,650 = \$95,650$$

(d) Case 4: Salvage value > cost base

This situation is not likely for most depreciable assets (except for real property). Nevertheless, the tax treatment on this gain is as follows:

$$\text{Capital gains} = \text{salvage value } (S) - \text{cost base } (P)$$
$$= \$280,000 - \$250,000$$
$$= \$30,000.$$

$$\text{Capital gains tax} = \$30,000 \times 1/2 \times 0.4$$
$$= \$6,000$$

$$\text{Disposal tax effects} = -\text{capital gains tax} + t(U_{Disposal} - P)$$
$$= -\$6,000 + 0.4 \times (\$104,125 - \$250,000)$$
$$= -\$6,000 - \$58,350$$
$$= -\$64,350$$
$$\text{Net salvage value} = \$280,000 - \$64,350$$
$$= \$215,650$$

COMMENTS: Note that in (c) the disposal tax effects are positive, which represents a reduction in tax due to the loss and increases the net proceeds. The corporation would still be paying tax, but less than if the asset had not been sold at a loss.

SUMMARY

■ Explicit consideration of taxes is a necessary aspect of any complete economic study of an investment project.

■ Income taxes equal a taxable income amount multiplied by an appropriate tax rate.

■ Three tax rates were defined in this chapter: **marginal tax rate**, which is the rate applied to the next dollar of income earned; **effective (average) tax rate**, which is the ratio of total income tax paid to total taxable income; and **incremental tax rate**, which is the average rate applied to the incremental income generated by a new investment project.

■ **Capital gains** are taxed at half of their value (2009 regulations). **Capital losses** are deducted from capital gains and net remaining losses may be carried backward and forward for consideration in other tax years.

■ Since we are interested primarily in the measurable financial aspects of depreciation, we consider the effects of depreciation or **capital cost allowance (CCA)** on two important measures of an organization's financial position, **net income** and **cash flow**. Once we understand that depreciation has a significant influence on the income and cash position of a firm, we will be able to appreciate fully the importance of utilizing CCA as a means to maximize the value of engineering projects and of the organization as a whole.

■ For corporations, the Canadian tax system has the following characteristics:

1. Tax rates are not progressive. The rates are constant and do not increase with earnings.
2. Tax rates depend upon the type and characteristics of the corporation.
3. The combined federal/provincial (territorial) tax rate is the sum of the federal and provincial/territorial rates, which are set independently.

■ An **investment tax credit** is a direct reduction of income taxes payable, arising from the acquisition of depreciable assets. Government uses the investment tax credit to stimulate investments in specific assets, for specific initiatives, in specific industries, or in specific areas of the country.

■ Disposal of depreciable assets—as with disposal of capital assets—may result in gains (**recaptured depreciation**) or losses, which must be considered in calculating taxes.

■ Net salvage value of an asset is the actual salvage value adjusted for the **disposal tax effect.**

PROBLEMS

Note: *The following considerations apply to the problems herein.*

1. *All tax rates (given or to be determined) are combined federal plus provincial/ territorial values.*
2. *The tax year is the calendar year.*
3. *Unless stated otherwise, use the tax rates provided in the text. For some problems, you may be requested to use current tax rates, which are available on the Companion Website that accompanies this text or from CRA.*
4. *Assume tax rates are unchanged from year to year, when multiple years are to be considered.*
5. *Unless indicated otherwise, corporations are private and Canadian controlled, i.e., eligible for the small business deduction.*
6. *The maximum CCA can be used in each year.*
7. *The 50% rule applies to new assets in declining balance CCA classes.*

9.1 Which of the following statements is correct?
   (a) Over a project's life, a typical business will generate a greater amount of total project cash flows (undiscounted) if CCA is claimed more quickly.
   (b) No matter how quickly CCA is claimed, total tax obligations over a project's life remain unchanged.
   (c) CCA recapture equals the cost base minus an asset's UCC at the time of disposal.
   (d) Cash flows normally include CCA since it represents a cost of doing business.

9.2 You purchased a box-making machine that cost $50,000 five years ago. At that time, the system was estimated to have a service life of five years with salvage value of $5,000. These estimates are still good. The property has been depreciated at a declining balance CCA rate of 30%. Now (at the end of year 5 from purchase) you are considering selling the machine for $10,000. What UCC should you use in determining the disposal tax effect?

9.3 Omar Shipping Company bought a tugboat for $75,000 (year 0) and expected to use it for five years, after which it will be sold for $12,000. Suppose the company estimates the following revenues and expenses for the first operating year.

Operating revenue	$200,000
Operating expenses	$84,000
CCA	$4,000

If the company pays taxes at the rate of 30% on its taxable income, what is the net income during the first year?

9.4 In Problem 9.3, assume for the moment that (1) all sales are for cash, and (2) all costs, except depreciation, were paid during year 1. How much cash would have been generated from operations?

9.5 Gilbert Corporation had a gross income of $500,000 in tax year 1, $150,000 in salaries, $50,000 in wages, and $60,000 in CCA for an asset purchased three years ago. Ajax Corporation has a gross income of $500,000 in tax year 1, $150,000 in salaries, and $110,000 in wages. Gilbert and Ajax are both Quebec-based public manufacturing companies. Apply the tax rates in Tables 9.1 and 9.2 and determine which of the following statements is correct.

(a) Both corporations will pay the same amount of income taxes in year 1.

(b) Both corporations will have the same amount of net cash flows in year 1.

(c) Ajax Corporation will have a larger net cash flow than Gilbert in year 1.

(d) Gilbert Corporation will have a larger taxable income than Ajax Corporation in year 1.

9.6 A consumer electronics company was formed in Saskatoon, Saskatchewan, to manufacture and sell GPS units for cars. The company purchased a warehouse for $500,000 and converted it into a manufacturing plant. It completed installation of assembly equipment worth $1,500,000 on January 31. The plant began operation on February 1. The company had a gross income of $2,500,000 for the calendar year. Manufacturing costs and all operating expenses, excluding the capital expenditures, were $1,280,000. The CCA for capital expenditures amounted to $128,000.

(a) Compute the taxable income of this company.

(b) How much will the company pay in income taxes for the year?

9.7 Consider a CCA Class 43 asset ($d = 30\%$), which was purchased at $60,000.

The applicable salvage values would be $20,000 in year 3, $10,000 in year 5, and $5000 in year 6, respectively. Compute the gain or loss amounts when the asset is disposed of

(a) in year 3,

(b) in year 5, and

(c) in year 6.

9.8 An electrical appliance company purchased an industrial robot costing $300,000 in year 0. The industrial robot to be used for welding operations is depreciated at a 30% declining balance rate for tax purposes. If the robot is to be sold after five years, compute the amounts of gains (losses) and the disposal tax effect for the following three salvage values with a corporate tax rate of 34%.

(a) $10,000

(b) $125,460

(c) $200,000

9.9 LaserMaster Inc., an Edmonton, Alberta–based CCPC laser-printing service company, had a sales revenue of $1,250,000 during the current tax year. The following represents the other financial information relating to the tax year.

Labour expenses	$350,000
Material costs	$185,000
CCA	$32,500
Rental expenses	$57,200
Proceeds from the sale of old printers	$23,000

(The printers had a combined UCC of $20,000 at the time of sale.)

(a) Determine the taxable income for the tax year.

(b) Determine the taxable gains for the tax year.

(c) Determine the amount of income taxes.

(d) Determine the disposal tax effect.

9.10 CanWest Inc. bought a machine for $50,000 on January 2, 2009. Management expects to use the machine for 10 years, at which time it will have a $1000 salvage value. Consider the following questions independently. Parts (a) through (d) relate to book depreciation.

(a) If CanWest uses straight-line depreciation, what is the book value of the machine on December 31, 2011?

(b) If CanWest uses double declining balance depreciation, what is the depreciation expense for 2011?

(c) If CanWest uses double declining balance depreciation switching to straight-line depreciation, when is the optimal time to switch?

(d) If CanWest uses sum-of-years'-digits depreciation, what is the total depreciation on December 31, 2014?

(e) If CanWest sold the machine on April 1, 2012, at a price of $30,000, what was the disposal tax effect? The corporate tax rate is 35% and the machine is depreciated at a 30% declining balance rate for tax purposes.

9.11 Buffalo Ecology Corporation is an Ontario-based company that expects to generate a taxable income of $450,000 from its regular business in the current tax year. The company is considering a new venture: cleaning oil spills made by fishing boats in lakes. This new venture is expected to generate additional taxable income of $150,000.

(a) Determine the firm's marginal tax rates before and after the venture.

(b) Determine the firm's average tax rates before and after the venture.

9.12 Repeat Problem 9.11 assuming Buffalo Ecology is not eligible for the small business deduction.

9.13 Precision Machining, a New Brunswick company, expects to have annual taxable income of $470,000 from its regular business over the next six years. The company is considering the proposed acquisition of a new milling machine. The machine's installed price is $200,000. The machine falls into a 30% declining balance CCA class, and it will have an estimated salvage value of $30,000 at the end of six years. The machine is expected to generate an additional before-tax revenue of $80,000 per year.

(a) What is the total amount of economic depreciation for the milling machine, if the asset is sold at $30,000 at the end of six years?

(b) Determine the company's marginal tax rates over the next six years without the machine.

(c) Determine the company's average tax rates over the next six years with the machine.

9.14 Major Electrical Company is located in Red Deer, Alberta. It expects to have an annual taxable income of $350,000 from its residential accounts over the next two years. This CPCC is bidding on a two-year wiring service for a large apartment complex. This commercial service requires the purchase of a new truck equipped with wire-pulling tools at a cost of $50,000. The equipment falls into CCA Class 10 and will be retained for future use (instead of selling) after two years, indicating no gain or loss on this property. The project will bring in an additional annual revenue of $200,000, but it is expected to incur additional annual operating costs of $100,000. Compute the incremental tax rate applicable to the project's operating profits for the next two years.

9.15 Okanagan Juice Corporation is a public, manufacturing-type company in British Columbia. It estimates its taxable income for next year at $2,000,000. The company is considering expanding its product line by introducing apple-peach juice for the next year. The market responses could be (1) good, (2) fair, or (3) poor. Depending on the market response, the expected additional taxable incomes are (1) $2,000,000 for a good response, (2) $500,000 for fair response, and (3) a loss of $100,000 for a poor response.

(a) Determine the incremental tax rate applicable to each situation.

(b) Determine the average tax rate that results from each situation.

(c) What is the marginal tax rate prior to the expansion and after the expansion with the three market scenarios?

9.16 A small manufacturing company in Aylmer, Quebec, has an estimated annual taxable income of $95,000. Owing to an increase in business, the company is considering purchasing a new machine that will generate an additional (before-tax) net annual revenue of $50,000 over the next five years. The new machine requires an investment of $100,000. It will be depreciated at a declining balance CCA rate of 30%.

(a) What is the increment in income tax due to the purchase of the new machine in tax year 1?

(b) What is the incremental tax rate due to the purchase of the new equipment in year 1?

9.17 Simon Machine Tools Company is a Canadian-controlled manufacturing company located in London, Ontario. It is considering the purchase of a new grinding machine to process special orders. The following financial information is available.

- Without the project: The company expects to have taxable income of $500,000 each year from its regular business over the next three years.
- With the project: This three-year project requires the purchase of a new grinder at a cost of $50,000. The equipment falls into CCA Class 43. The grinder will be sold at the end of project life for $10,000. The project will be bringing in an additional annual revenue of $80,000, but it is expected to incur additional annual operating costs of $20,000.

(a) What are the additional taxable incomes (due to undertaking the project) during years 1 through 3, respectively?

(b) What are the additional income taxes (due to undertaking the new orders) during years 1 through 3, respectively?

(c) Compute the disposal tax effect when the asset is disposed of at the end of year 3.

9.18 A company purchased a new forging machine to manufacture disks for airplane turbine engines. The new press cost $3,500,000, and is considered a CCA Class 43 asset. The company has to pay annual city property taxes for ownership of this forging machine at a rate of 1.2% on the beginning-of-year undepreciated capital cost.

(a) Determine the UCC of the asset at the beginning of each tax year.

(b) Determine the total amount of property taxes over the machine's depreciable life.

9.19 Genoasis Inc., a CCPC located in Charlottetown, PEI, conducts pharmaceutical research through the use of animal genes. Based on recent promising results, they have decided to invest $5,000,000 next year in new facilities, comprising a building costing $3.5M (Class 1) and equipment totalling another $1.5M (which falls into Class 8). These expenditures qualify for federal investment tax credits under the SR&ED program. Their taxable income for this year amounts to $525,000. Calculate the impact of these ITCs and the CCA for these new investments over the next two years.

9.20 Nutrival Ltd, an agri-chemical producer in Brandon, Manitoba, has decided to invest in an on-site daycare facility since most of its workers are young and the current demand for this service is expected to grow. The costs for adding an extension to their existing building to accommodate 20 youngsters is $900,000 (Class 1), with an additional $75,000 spent on furnishings for the new centre. The furnishings fall into Class 8 for tax depreciation. Calculate the ITCs that Nutrival is eligible to receive for this investment in child-care spaces and show the CCA and UCC for the year the facility is built, and the following year.

9.21 Quick Printing Company of Halifax, Nova Scotia, had sales revenue of $1,450,000 from operations during tax year 1. Here are some operating data on the company:

Labour expenses	$550,000
Materials costs	185,000
CCA	32,500
Rental expenses	45,000
Proceeds from sale of old equipment that had a UCC of $20,000	23,000

(a) What is Quick's taxable income?

(b) What are Quick's taxable gains?

(c) What are Quick's marginal and effective (average) tax rates?

(d) What is Quick's net cash flow after tax?

9.22 Valdez Corporation commenced operations on January 1, 2007. The company's financial performance during its first year of operation was as follows:

- Sales revenues, $1,500,000.
- Labour, material, and overhead costs, $600,000.
- The company purchased a warehouse worth $500,000 in February. To finance this warehouse, the company issued $500,000 of long-term bonds on January 1, which carry an interest rate of 10%. The first interest payment occurred on December 31. The bond interest is tax-deductible.

- For depreciation purposes, the purchase cost of the warehouse is divided into $100,000 in land and $400,000 in building. The building is a CCA Class 1 asset and is depreciated accordingly.
- On January 5, the company purchased $200,000 of equipment, which falls into CCA Class 43.
- The corporate tax rate is 40%.

(a) Determine the total CCA allowed in 2007.

(b) Determine Valdez's income taxes for 2007.

Short Case Studies

ST9.1 A machine now in use that was purchased three years ago at a cost of $4000 has an undepreciated capital cost of $1800. It can be sold for $2500, but could be used for three more years, at the end of which time it would have no salvage value. The annual O&M costs amount to $10,000 for the old machine. A new machine can be purchased at an invoice price of $14,000 to replace the present equipment. Freight-in will amount to $800, and the installation cost will be $200. The new machine has an expected service life of five years and will have no salvage value at the end of that time. With the new machine, the expected direct cash savings amount to $8000 in the first year and $7000 for each of the next two years. Corporate income taxes are at an annual rate of 40%. The present machine has been depreciated at a declining balance CCA rate of 30% and the proposed machine would be depreciated at the same rate. (*Note:* Each question should be considered independently.)

(a) If the old asset is to be sold now, what would be the amount of its equivalent economic depreciation?

(b) For depreciation purposes, what would be the first cost of the new machine (cost base)?

(c) If the old machine is to be sold now, what would be the disposal tax effect?

(d) If the old machine can be sold for $5,000 now instead of $2,500, what would be the disposal tax?
  The following questions relate to book depreciation:

(e) If the old machine had been depreciated using 175% DB and then switching to SL depreciation, what would be the current book value?

(f) If the machine were not replaced by the new one, when would be the time to switch from DB to SL?

ST9.2 Van-Line Company, a small electronics repair firm located in Sydney, Nova Scotia, expects an annual taxable income of $70,000 from its regular business. The company is considering expanding its repair business by buying some new precision tools (Class 8). This expansion would bring in an additional annual income of $30,000, but will require an additional expense of $10,000 each year over the next three years. Using applicable current tax rates, answer the following:

(a) What is the marginal tax rate in tax year 1?

(b) What is the average tax rate in tax year 1?

(c) Suppose that the new business expansion requires a capital investment of $20,000 (CCA calculated at a 20% declining balance rate). What is the PW of the total income taxes to be paid over the project life, at $i = 10\%$?

ST9.3 Electronic Measurement and Control Company (EMCC) has developed a laser speed detector that emits infrared light invisible to humans and radar detectors alike. For full-scale commercial marketing, EMCC needs to invest $5 million in new manufacturing facilities. The system is priced at $3000 per unit. The company expects to sell 5000 units annually over the next five years. The new manufacturing facilities will be depreciated at a declining balance CCA rate of 30%. The expected salvage value of the manufacturing facilities at the end of five years is $1.6 million. The manufacturing cost for the detector is $1200 per unit, excluding depreciation expenses. The operating and maintenance costs are expected to run to $1.2 million per year. EMCC has a combined income tax rate of 35%, and undertaking this project will not change the current marginal tax rate.

(a) Determine the incremental taxable income, income taxes, and net income due to undertaking this new product for the next five years.

(b) Determine the gains or losses associated with the disposal of the manufacturing facilities at the end of five years.

ST9.4 Diamonid is a start-up diamond coating public corporation planning to manufacture a microwave plasma reactor that synthesizes diamonds. Diamonid anticipates that the industry demand for diamonds will skyrocket over the next decade for use in industrial drills, high-performance microchips, and artificial human joints, among other things. Diamonid has decided to raise $50 million through issuing common stocks for investment in plant ($10 million) and equipment ($40 million) to establish a facility at Edmonton, Alberta. Each reactor can be sold at a price of $100,000 per unit. Diamonid can expect to sell 300 units per year during the next eight years. The unit manufacturing cost is estimated at $30,000, excluding CCA. The operating and maintenance cost for the plant is estimated at $12 million per year. Diamonid expects to phase out the operation at the end of eight years and revamp the plant and equipment to adopt a new diamond manufacturing technology. At that time, Diamonid estimates that the salvage values for the plant and equipment will be about 60% and 10% of the original investments, respectively. The plant and equipment will be depreciated at declining balance CCA rates of 20% and 30% respectively.

(a) If the 2008 corporate tax rates apply over the project life, determine the combined income tax rate each year.

(b) Determine the gains or losses at the time of plant revamping.

(c) Determine the net income each year over the plant life.

On the Companion Website that accompanies this text, you will find Excel templates and exercises, as well as the following analysis tools: Cash Flow Analyzer, Depreciation Analysis, Loan Analysis, and Interest Tables.

CHAPTER — TEN

Developing Project Cash Flows

Imperial Gives Kearl Green Light: Battered Oil Patch Welcomes Decision to Proceed with Mine's $8 Billion Phase I[1] Imperial Oil Ltd.'s approval of a major new $8 billion project is sparking optimism for a revival in the oil sands after delays of several projects due to low crude prices and high costs. Kearl is an oil sands mining and extraction project being developed by Imperial Oil (as operator) and ExxonMobil Canada. The project is located 70 kilometres northeast of Fort McMurray, Alberta, at Kearl Lake. Kearl Oil Sands Project's total recoverable bitumen is estimated to be 4.6 billion barrels. It is located near the surface and will be recovered using open-pit mining methods similar to those used at existing oil sands mines in the region. Imperial is planning this project in three phases, with selective clearing, mining, and reclamation continuing over the 50-year life of the project. The start-up phase is expected to be in operation in 2012, with a production capacity of about 100,000 barrels per day. Each of the two additional phases will increase the production capacity by about 100,000 barrels per day. At its peak, Kearl could produce up to 345,000 barrels per day. Kearl's production based on full development plans is expected to average about 300,000 barrels of bitumen per day.

The consultation process for Kearl began in 1997 when Mobil Oil Canada first announced the Kearl Oil Sands Project. Since that time, Imperial has been involved in many discussions, meetings, presentations, and workshops to consult with stakeholders in an effort to address their concerns in its project plans. As early as 2004, Imperial estimated the capital investment required to develop the initial phase of this project to be $5 billion to $8 billion. Last year, Imperial Oil

---

[1] Nathan Vanderklippe, "Imperial Gives Kearl Green Light; Battered Oil Patch Welcomes Decision to Proceed With Mine's $8-billion Phase 1," *Globe and Mail* May 26, 2009, page B1.

delayed the go-ahead decision of the project, negotiating with service providers and waiting for better deals before ordering equipment in hopes of reducing total investment required. Today, with the final nod from management, the total estimated investment required is around $8 billion.

Given today's global economic downturn, tight financial markets, low crude oil prices, and the threat of costly new greenhouse gas regulations, what has made Imperial Oil believe that this project will generate the financial returns required by its owners on the estimated investment of $8 billion?

The total investment required is not as high as last year, which improves the viability of the project. Construction of the first phase will employ up to 5000 people. The province's construction workforce has fallen from 160,000 to about 120,000 due to the recent economic downturn. The costs of labour, equipment, and materials for the construction project are much lower than if started last year.

A key factor influencing the profitability of the project is the estimated crude oil price. Today, crude oil trades around US$62 per barrel. Imperial's management believes that crude will rebound to above US$80 a barrel, the price analysts say Kearl will need to turn a 10% after-tax profit. The Kearl project will offset declines in conventional oil production and help meet North America's energy demands for many years to come.

Other influential factors include the labour costs when the project is in operation in 2012 (Kearl will employ between 1000 and 1200 workers during Phase 1), the costs of equipment such as shovels, trucks, crushers, and so on, and the tightening emissions regulations.

The key issue again is how Imperial justified the investment of $8 billion when it finally approved this product. Clearly, Imperial's first step would be to estimate the magnitude of the revenue stream over the 50-year life of the project. The crude oil price and the production level would be the basis for possible revenue projections. The production quantity depends on the internal operation of the project while the crude oil price depends on external factors such as the production level of conventional oil, the world economic growth and energy demand, the production level of other oil sands projects, and environmental regulations. Costs of running the project over its life should also be estimated. Thus, any project's justification depends upon the ability to estimate potential cash flows.

Kearl Process Overview

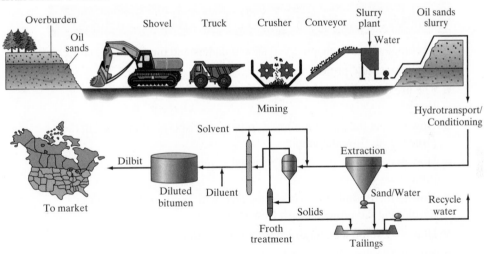

Projecting cash flows is the most important—and the most difficult—step in the analysis of a capital project. Typically, a capital project initially requires investment outlays and only later produces annual net cash inflows. A great many variables are involved in forecasting cash flows, and many individuals, ranging from engineers to cost accountants and marketing people, participate in the process. This chapter provides the general principles on which the determination of a project's cash flows are based.

To help us imagine the range of activities that are typically initiated by project proposals, we begin the chapter with an overview in Section 10.1 of how firms classify projects. A variety of projects exists, each having its own characteristic set of economic concerns. Section 10.2 then defines the typical cash flow elements of engineering projects. Once we have defined these elements, we will examine how to develop cash flow statements that are used to analyze the economic value of projects. In Section 10.3, we use several examples to demonstrate the development of after-tax project cash flow statements. Then, Section 10.4 presents an alternative technique for developing after-tax cash flows; this approach is called the generalized cash flow approach. Another alternative called the after-tax cash flow diagram approach is given in Section 10.5. By the time you have finished this chapter, not only will you understand the format and significance of after-tax cash flow statements, but you will also know how to develop them yourself.

CHAPTER LEARNING OBJECTIVES

After completing this chapter, you should understand the following concepts:

■ What constitute project cash flow elements.

■ The use of the income statement approach in developing project cash flows.

■ How to treat the gains and losses related to disposal of an asset in the project cash flow statement.

■ How to determine the working capital requirement and its impact on project cash flows.

■ How to incorporate the costs associated with financing a project in developing the project's cash flow statement.

■ How to develop a generalized cash flow model.

■ How to use the after-tax cash flow diagram approach to develop cash flows.

■ The analysis of a lease-or-buy decision on an after-tax basis.

10.1 Cost–Benefit Estimation for Engineering Projects

Before the economics of an engineering project can be evaluated, it is necessary to estimate the various cost and revenue components that describe the project. Engineering projects may range from something as simple as the purchase of a new milling machine to the design and construction of a multibillion-dollar process or resource recovery complex.

The engineering projects appearing in this book as examples and problems already include the necessary cost and revenue estimates. Developing adequate estimates for these quantities is extremely important and can be a time-consuming activity. Although cost-estimating techniques are not the focus of this book, it is worthwhile to mention some of the approaches that are taken. Some are used in the context of simple projects that are straightforward and involve little or no engineering design; others are employed on complex projects that tend to be large and may involve many thousands of hours of engineering design. Obviously, there are also projects that fall between these extremes, and some combination of approaches may be appropriate in these cases.

10.1.1 Simple Projects

Projects in this category usually involve a single "off-the-shelf" component or a series of such components that are integrated together in a simple manner. The acquisition of a new milling machine is an example.

The installed cost is the price of the equipment as determined from catalogues or supplier quotations, shipping and handling charges, and the cost of building modifications and changes to utility requirements. The latter may call for some design effort to define the scope of the work, which is the basis for contractor quotations.

Project benefits are in the form of new revenue and/or cost reduction. Estimating new revenue requires agreement on the total number of units produced and the selling price per unit. These quantities are related through supply and demand considerations in the marketplace. In highly competitive product markets, sophisticated marketing studies are often required to establish price–volume relationships. These studies are undertaken as one of the first steps in an effort to define the appropriate scale for the project.

The ongoing costs required to operate and maintain equipment can be estimated at various levels of detail and accuracy. Familiarity with the cost–volume relationships for similar facilities allows the engineer to establish a "ballpark" cost. Some of this information may be used in conjunction with more detailed estimates in other areas. For example, maintenance costs may be estimated as a percentage of the installed cost. Such percentages are derived from historical data and are frequently available from equipment suppliers. Other costs, such as the cost of manpower and energy, may be estimated in detail to reflect specific local considerations. The most comprehensive and time-consuming type of estimate is a set of detailed estimates around each type of cost associated with the project.

10.1.2 Complex Projects

The estimates developed for complex projects involve the same general considerations as those that apply to simple projects. However, complex projects usually include specialized equipment that is not "off the shelf" and must be fabricated from detailed engineering drawings. For certain projects, such drawings are not even available until after a commitment has been made to proceed with the project. The typical phases of a project are as follows:

- Development.
- Conceptual design.
- Preliminary design.
- Detailed design.

Depending upon the project, some phases may be combined. Project economics is performed during each of these phases to confirm the attractiveness of the project and the incentive to continue working on it. The types of estimates and estimating techniques are a function of the stage of the project.

When the project involves the total or partial replacement of an existing facility, the benefits of the project are usually well known at the outset in terms of pricing–volume relationships or cost reduction potential. In the case of natural-resource projects, however, oil, gas, and mineral price forecasts are subject to considerable uncertainty.

At the development phase, work is undertaken to identify technologies that may be needed and to confirm their technical viability. Installed cost estimates and operating estimates are based on evaluations of the cost of similar existing facilities or their parts.

Conceptual design examines the issues of the scale of the project and alternative technologies. Again, estimates tend to be based on large pieces of, or processes within, the overall project that correspond to similar facilities already in operation elsewhere.

Preliminary design takes the most attractive alternative from the conceptual design phase to a level of detail that yields specific sizing and layout estimates for the actual equipment and associated infrastructure. Estimates at this stage tend to be based on individual pieces of equipment. Once more, the estimating basis is similar pieces of equipment in use elsewhere.

For very large projects, the cost of undertaking detailed engineering is prohibitive unless the project is going forward. At this stage, detailed fabrication and construction drawings that provide the basis for an actual vendor quotation become available.

The accuracy of the estimates improves with each phase as the project becomes defined in greater detail. Normally, the installed cost estimate from each phase includes a contingency that is some fraction of the actual estimate calculated. The contingency at the development phase can be 50% to 100%; after preliminary design, it decreases to something in the order of 10%.

More information on cost-estimating techniques can be found in reference books on project management and cost engineering. Industry-specific data books are also available for some sectors, for which costs are summarized on some normalized basis, such as dollars per square metre of building space or dollars per tonne of material moved in operating mines. In these cases, the data are categorized by type of building and type of mine. Engineering design companies maintain extensive databases of such cost information.

10.2 Incremental Cash Flows

When a company purchases a fixed asset such as equipment, it makes an investment. The company commits funds today in the expectation of earning a return on those funds in the future. Such investments are similar to those made by a bank when it lends money. In the case of a bank loan, the future cash flow consists of interest plus repayment of the principal. For a fixed asset, the future return is in the form of cash flows generated by the profitable use of the asset. In evaluating a capital investment, we are concerned only with those cash flows that result directly from the investment. These cash flows, called **differential** or **incremental cash flows**, represent the change in the firm's total cash flow that occurs as a direct result of the investment. In this section, we will look into some of the cash flow elements that are common to most investments.

10.2.1 Elements of Cash Outflows

We first consider the potential uses of cash in undertaking an investment project:

- **Purchase of New Equipment.** A typical project usually involves a cash outflow in the form of an initial investment in equipment. The relevant investment costs are incremental ones, such as the cost of the asset, shipping and installation costs, and the cost of training employees to use the new asset. If the purchase of a new asset results in the sale of an existing asset, the net proceeds from this sale reduce the amount of the incremental investment. In other words, the incremental investment represents the total amount of additional funds that must be committed to the investment project. When existing equipment is sold, the transaction results in an accounting gain or loss, which is dependent on whether the amount realized from the sale is greater or less than the equipment's book value. In either case, when existing assets are disposed of, the relevant amount by which the new investment is reduced consists of the proceeds of the sale, adjusted for tax effects.

- **Investments in Working Capital.** Some projects require an investment in nondepreciable assets. If, for example, a project increases a firm's revenues, then more funds will be needed to support the higher level of operation. Investment in nondepreciable assets is often called **investment in working capital**. In accounting, **working capital** is the amount carried in cash, accounts receivable, and inventory that is available to meet day-to-day operating needs. For example, additional working capital may be needed to meet the greater volume of business that will be generated by a project. Part of this increase in current assets may be supplied from increased accounts payable, but the remainder must come from permanent capital. This additional working capital is as much a part of the initial investment as the equipment itself. (We explain the amount of working capital required for a typical investment project in Section 10.3.2.)

> **Working capital** measures how much in liquid assets a company has available to build its business.

- **Manufacturing, Operating, and Maintenance Costs.** The costs associated with manufacturing a new product need to be determined. Typical manufacturing costs include labour, materials, and overhead costs, the last of which cover items such as power, water, and indirect labour. Investments in fixed assets normally require periodic outlays for repairs and maintenance and for additional operating costs, all of which must be considered in investment analysis.

- **Leasing Expenses.** When a piece of equipment or a building is leased (instead of purchased) for business use, leasing expenses become cash outflows. Many firms lease computers, automobiles, industrial equipment, and other items that are subject to technological obsolescence.

- **Interest and Repayment of Borrowed Funds.** When we borrow money to finance a project, we need to make interest payments as well as payments on the principal. Proceeds from both short-term borrowing (bank loans) and long-term borrowing (bonds) are treated as cash inflows, but repayments of debts (both principal and interest) are classified as cash outflows.

- **Income Taxes and Tax Credits.** Any income tax payments following profitable operations should be treated as cash outflows for capital budgeting purposes. As we learned in Chapter 9, when an investment is made in depreciable assets, depreciation is an expense that offsets part of what would otherwise be additional taxable income. This strategy is called a **tax shield**, or **tax savings**, and we must take it into account when calculating income taxes. If any investment **tax credit** is allowed, it will directly reduce income taxes, resulting in cash inflows.

10.2.2 Elements of Cash Inflows

The following are potential sources of cash inflow over the project life:

- **Borrowed Funds.** If you finance your investment by borrowing, the borrowed funds will appear as cash inflow to the project at the time they are borrowed. From these funds, the purchase of new equipment or any other investment will be paid out.
- **Operating Revenues.** If the primary purpose of undertaking the project is to increase production or service capacity to meet increased demand, the new project will be bringing in additional revenues.
- **Cost Savings (or Cost Reduction).** If the primary purpose of undertaking a new project is to reduce operating costs, the amount involved should be treated as a cash inflow for capital budgeting purposes. A reduction in costs is equivalent to an increase in revenues, even though the actual sales revenues may remain unchanged.
- **Salvage Value.** In many cases, the estimated salvage value of a proposed asset is so small and occurs so far in the future that it may have no significant effect on the decision to undertake a project. Furthermore, any salvage value that is realized may be offset by removal and dismantling costs. In situations where the estimated salvage value is significant, the net salvage value is viewed as a cash inflow at the time of disposal. The **net salvage value** of the existing asset is the selling price of the asset, minus any costs incurred in selling, dismantling, and removing it, and this value is subject to taxable gain or loss.
- **Working-Capital Release.** As a project approaches termination, inventories are sold off and receivables are collected. That is, at the end of the project, these items can be liquidated at their cost. As this occurs, the company experiences an end-of-project cash flow that is equal to the net working-capital investment that was made when the project began. This recovery of working capital is not taxable income, since it merely represents a return of investment funds to the company.

Figure 10.1 sums up the various types of cash flows that are common in engineering investment projects.

10.2.3 Classification of Cash Flow Elements

Once the cash flow elements (both inflows and outflows) are determined, we may group them into three categories: (1) cash flow elements associated with operations, (2) cash flow elements associated with investment activities (capital expenditures), and (3) cash flow elements associated with project financing (such as borrowing). The main purpose of grouping cash flows this way is to provide information about the operating, investing, and financing activities of a project. You probably have noticed that these three groups of cash flows are used in the cash flow statement prepared by a company as discussed in Chapter 2. The difference is that a company's cash flow statement includes all cash flows generated by all projects of the company. In this chapter, we focus on the cash flows of an individual project. Another difference is that a company's cash flow statement reports the past cash flows that were incurred in a fiscal year, while the cash flow statement that is discussed in this chapter refers to future cash flows of the project if it is carried out.

Operating Activities

In general, cash flows from operations include current sales revenues, the cost of goods sold, operating expenses, and income taxes. Cash flows from operations should generally

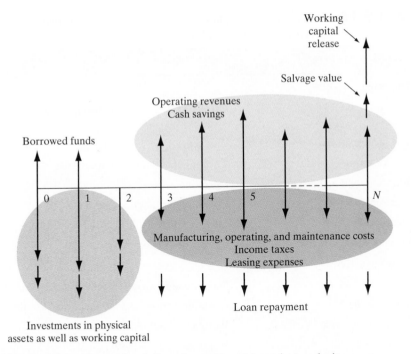

Figure 10.1 Types of cash flow elements used in project analysis.

reflect the cash effects of transactions entering into the determination of net income. The interest portion of a loan repayment is a deductible expense that is allowed in determining net income, and it is included in operating activities. Since we usually look only at yearly flows, it is logical to express all cash flows on a yearly basis.

As mentioned in Chapter 9, we can determine the net cash flow from operations on the basis of either (1) net income or (2) cash flow as determined by computing income taxes in a separate step. When we use net income as the starting point for cash flow determination, we should add any noncash expenses (mainly depreciation and amortization expenses) to net income to estimate the net cash flow from the operation. Recall that depreciation (or amortization) is not a cash flow, but is deducted, along with operating expenses and lease costs, from gross income to find taxable income and therefore taxes. Accountants calculate net income by subtracting taxes from taxable income, but depreciation—which is not a cash flow—was already subtracted to find taxable income, so it must be added back to taxable income if we wish to use the net-income figure as an intermediate step along the path to after-tax cash flow. Mathematically, it is easy to show that the two approaches are identical. Thus,

$$\text{Cash flow from operation} = \text{Net income} + (\text{Depreciation or amortization}).$$

Investing Activities

In general, three types of investment cash flows are associated with buying a piece of equipment: the original investment, the salvage value at the end of the useful life of the equipment, and the working-capital investment or recovery. We will assume that our outflow for both capital investment and working-capital investment take place in year 0. It is possible, however, that both investments will occur, not instantaneously, but rather, over a

TABLE 10.1 Classifying Cash Flow Elements and Their Equivalent Terms as Practised in Business

Cash Flow Element	Other Terms Used in Business
Operating activities:	
Gross income	Gross revenue, sales revenue, gross profit, operating revenue
Cost savings	Cost reduction
Manufacturing expenses	Cost of goods sold, cost of revenue
Operations and maintenance cost	Operating expenses
Operating income	Operating profit, gross margin
Interest expenses	Interest payments, debt cost
Income taxes	Income taxes owed
Investing activities:	
Capital investment	Purchase of new equipment, capital expenditure
Salvage value	Net selling price, disposal value, resale value
Investment in working capital	Working-capital requirement
Release of working capital	Working-capital recovery
Gains taxes	Capital gains taxes, ordinary gains taxes
Financing activities:	
Borrowed funds	Borrowed amounts, loan amount
Repayment of principal	Loan repayment

few months as the project gets into gear; we could then use year 1 as an investment year. (Capital expenditures may occur over several years before a large investment project becomes fully operational. In this case, we should enter all expenditures as they occur.) For a small project, either method of timing these flows is satisfactory, because the numerical differences are likely to be insignificant.

Financing Activities

Cash flows classified as financing activities include (1) the amount of borrowing and (2) the repayment of principal. Recall that interest payments are tax-deductible expenses, so they are classified as operating, not financing, activities.

The net cash flow for a given year is simply the sum of the net cash flows from operating, investing, and financing activities. Table 10.1 can be used as a checklist when you set up a cash flow statement, because it classifies each type of cash flow element as an operating, investing, or financing activity.

10.3 Developing Cash Flow Statements

In this section, we will illustrate, through a series of numerical examples, how we actually prepare a project's cash flow statement; a generic version is shown in Figure 10.2, in which we first determine the net income from operations and then adjust the net

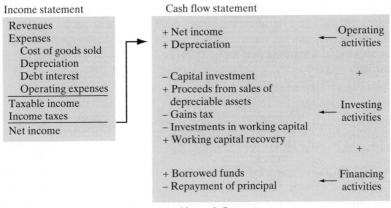

Figure 10.2   A popular format used for presenting a cash flow statement.

income by adding any noncash expenses, mainly depreciation (or amortization). We will also consider a case in which a project generates a negative taxable income for an operating year.

In general accounting practice, negative numbers are often presented as numbers in parentheses. For example, -200 is written as (200). We have seen this practice in the financial statements of Research In Motion Ltd. presented in Chapter 2. In this chapter, when we introduce the income statement approach for generating net cash flows for a project, we will follow this practice. As you will see, to generate the net cash flows of a project, we use both the income statement and the cash flow statement. In both statements, negative numbers will be expressed as numbers in parentheses. One difference between these two statements is described below. In the income statement, we don't use positive and negative signs to differentiate cash inflows from cash outflows. For example, in the income statement, revenue of $3000 is simply written as $3000 and cost of $1000 is simply presented as $1000, even though revenue is a cash inflow and cost is a cash outflow. However, in the cash flow statement, we explicitly use positive and negative signs to differentiate cash inflows from cash outflows. For example, in the cash flow statement, the purchase cost of an asset (an investment) in the amount of $20,000 is written as $(20,000), a payment of the principal portion of a loan in the amount of $1200 will be written as $(1200), and a disposal tax effect equal to -2000 will be written as $(2000) (this disposal tax effect indicates that it is a tax payment, which is a cash outflow).

As we noted earlier, the income statement and the cash flow statement that are used in this chapter are for individual projects, not for the total operation of the whole company. To calculate the taxable income attributed to the project in question, we simply subtract cost from revenue. If this results in a negative number, we record it as a negative number in the income statement. The income tax attributed to this project is calculated as the product of the tax rate and the negative taxable income, which will result in a negative tax amount. The negative taxable income generated by this project will reduce the total taxable income of the company. The negative income tax due to this project indicates that it will reduce the total tax payment of the whole company. The underlying assumption for this process is that the company always has enough revenues from its other operations to cover the losses of the project being analyzed. If one is considering starting a company with a single project and losses are projected during the first few years of its operation, these losses can be carried forward by a maximum of seven years. In this case, the tax savings will be realized in years later than the years of the actual losses. Such a scenario can be analyzed similarly using concepts and approaches we are introducing in this book. However, we will focus on evaluation of projects assuming that there is no need to carry forward losses, that is, there are always other money-making projects so that any losses of the project being analyzed will generate immediate tax savings.

10.3.1 When Projects Require Only Operating and Investing Activities

We will start with the simple case of generating after-tax cash flows for an investment project with only operating and investment activities. In the sections ahead, we will add complexities to this problem by including working-capital investments (Section 10.3.2) and borrowing activities (Section 10.3.3).

EXAMPLE 10.1 Cash Flow Statement: Operating and Investing Activities for an Expansion Project

A computerized machining centre has been proposed for a small tool-manufacturing company. If the new system, which costs $125,000, is installed, it will generate annual revenues of $100,000 and will require $20,000 in annual labour, $12,000 in annual material expenses, and another $8000 in annual overhead (power and utility) expenses. The automation facility would be classified as a Class 43 property with a CCA rate of 30%.

The company expects to phase out the facility at the end of five years, at which time it will be sold for $50,000. Find the year-by-year after-tax net cash flow for the project at a 40% marginal tax rate based on the net income and determine the after-tax net present worth of the project at the company's MARR of 15%.

DISCUSSION: We can approach the problem in two steps by using the format shown in Figure 10.2 to generate an income statement and then a cash flow statement. We will follow this form in our subsequent listing of givens and unknowns. In year 0 (that is, at present) we have an investment cost of $125,000 for the equipment.[2] Capital cost allowances (CCA) will be calculated for years 1 to 5. The revenues and costs are uniform annual flows in years 1 to 5. We can see that once we find depreciation allowances for each year, we can easily compute the results for years 1 to 4, which have fixed revenue and expense entries along with the variable CCA. In year 5, we will need to incorporate the salvage value of the asset and any tax effect from its disposal.

We will use the business convention that no signs (positive or negative) be used in preparing the income statement, except in the situation where we have a negative taxable income or tax savings. In that situation, we will use parentheses to denote a negative entry. However, in preparing the cash flow statement, we will explicitly observe the convention that parentheses indicate a cash outflow.

---

[2] We will assume that the asset is purchased and placed in service at the beginning of year 1 (or the end of year 0), and the first year's capital cost allowance will be claimed at the end of year 1.

SOLUTION

Given: Preceding cash flow information.

Find: After-tax cash flow.

Before presenting the cash flow table, we need to do some preliminary calculations. The following notes explain the essential items in Table 10.2:

- Calculation of capital cost allowance

  The initial purchase cost of the system is $P = \$125,000$. This system is a Class 43 property. According to Table 8.1, a Class 43 property has a capital cost allowance rate, or CCA rate, of 30%, that is, $d = 30\%$. We should also apply the 50% rule for the first year of a newly acquired asset. As summarized in Section 8.4.4, we can use the following equations for CCA calculations:

  $$CCA_1 = \frac{1}{2}Pd, \text{ for year 1}$$

  $$CCA_n = \left(1 - \frac{1}{2}\right)(1 - d)^{n-2}, \text{ for year } n \geq 2.$$

  We find that the capital cost allowances for years 1 through 5 are $18,750, $31,875, $22,313, $15,619, and $10,933, respectively. These calculated CCA values are used in Table 10.2 to complete the income statement.

TABLE 10.2 Cash Flow Statement for the Automated Machining Centre Project Using the Income Statement Approach (Example 10.1)

Year	0	1	2	3	4	5
Income Statement						
Revenues		$100,000	$100,000	$100,000	$100,000	$100,000
Expenses						
Labour		20,000	20,000	20,000	20,000	20,000
Material		12,000	12,000	12,000	12,000	12,000
Overhead		8,000	8,000	8,000	8,000	8,000
CCA (30%)		18,750	31,875	22,313	15,619	10,933
Taxable income		$41,250	$28,125	$37,688	$44,381	$49,067
Income taxes (40%)		16,500	11,250	15,075	17,753	19,627
Net income		$24,750	$16,875	$22,613	$26,629	$29,440
Cash Flow Statement						
Operating activities						
Net income		$24,750	$16,875	$22,613	$26,629	$29,440
CCA		18,750	31,875	22,313	15,619	10,933
Investment activities						
Investment	$(125,000)					
Salvage						50,000
Disposal tax effect						(9,796)
Net cash flow	$(125,000)	$43,500	$48,750	$44,925	$42,248	$80,578

- Salvage value and disposal tax effect

    At the end of year 5, the asset is disposed of. A salvage value in the amount of $50,000 is received. We need to calculate the disposal tax effect as we may either have to pay some additional income tax or realize a tax saving.

    1. The total capital cost allowance that has been claimed over the five-year period is equal to $18,750 + $31,875 + $22,313 + 15,619 + $10,933 = $99,489. The undepreciated capital cost at the end of year 5 is equal to the initial purchase price minus the total CCA that has been claimed, that is, $U = \$125,000 - \$99,489 = \$25,511$.

    2. An alternative way of finding the undepreciated capital cost at the time of disposal is to use the following formula taken from Section 8.4.4:

    $$U_n = P\left(1 - \frac{d}{2}\right)(1 - d)^{n-1}.$$

    Substituting in the values of $P = \$125,000$, $d = 30\%$, $n = 5$, we have

    $$U_5 = \$25,511.$$

    3. The disposal tax effect G is

    $$G = t(U5 - S) = 40\%(\$25,511 - \$50,000) = -\$9,796.$$

    Since the disposal tax effect is negative, it represents a negative cash flow. This means that additional tax in the amount of $9,796 has to be paid due to the disposal.

    Table 10.2 presents a summary of the cash flow profile.[3]

- Investment analysis

    Once we obtain the project's after-tax net cash flows, we can determine their equivalent present worth at the firm's interest rate. The after-tax cash flow series

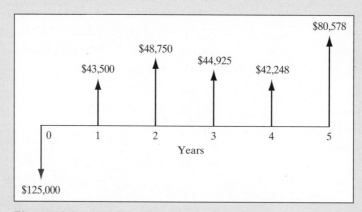

Figure 10.3   Cash flow diagram (Example 10.1).

---

[3] Even though disposal tax effects have an effect on income tax calculations, they should not be viewed as ordinary operating income. Therefore, in preparing the income statement, capital expenditures and related items such as capital gains tax, disposal tax effects, and salvage value are not included. Nevertheless, these items represent actual cash flows in the year they occur and must be shown in the cash flow statement.

from the cash flow statement is shown in Figure 10.3. Since this series does not contain any patterns to simplify our calculations, we must find the net present worth of each cash flow. Using $i = 15\%$, we have

$$PW(15\%) = -\$125,000 + \$43,500(P/F, 15\%, 1)$$
$$+ \$48,750(P/F, 15\%, 2) + \$44,925(P/F, 15\%, 3)$$
$$+ \$42,248(P/F, 15\%, 4) + \$80,578(P/F, 15\%, 5)$$
$$= \$43,443.$$

This means that investing \$125,000 in this automated facility would bring in enough revenue to recover the initial investment and the cost of funds, with a surplus of \$43,443.

10.3.2 When Projects Require Working-Capital Investments

In many cases, changing a production process by replacing old equipment or by adding a new product line will have an impact on cash balances, accounts receivable, inventory, and accounts payable. For example, if a company is going to market a new product, inventories of the product and larger inventories of raw materials will be needed. Accounts receivable from sales will increase, and management might also decide to carry more cash because of the higher volume of activities. These investments in working capital are investments just like those in depreciable assets, except that they have no tax effects: The flows always sum to zero over the life of a project, but the inflows and outflows are shifted in time, so they do affect net present worth.

Consider the case of a firm that is planning a new product line. The new product will require a two-month's supply of raw materials at a cost of \$40,000. The firm could provide \$40,000 in cash on hand to pay them. Alternatively, the firm could finance these raw materials via a \$30,000 increase in accounts payable (60-day purchases) by buying on credit. The balance of \$10,000 represents the amount of net working capital that must be invested.

Working-capital requirements differ according to the nature of the investment project. For example, larger projects may require greater average investments in inventories and accounts receivable than would smaller ones. Projects involving the acquisition of improved equipment entail different considerations. If the new equipment produces more rapidly than the old equipment, the firm may be able to decrease its average inventory holdings because new orders can be filled faster as a result of using the new equipment. (One of the main advantages cited in installing advanced manufacturing systems, such as flexible manufacturing systems, is the reduction in inventory made possible by the ability to respond to market demand more quickly.) Therefore, it is also possible for working-capital needs to decrease because of an investment. If inventory levels were to decrease at the start of a project, the decrease would be considered a cash inflow, since the cash freed up from inventory could be put to use in other places. (See Example 10.5.)

Two examples illustrate the effects of working capital on a project's cash flows. Example 10.2 shows how the net working-capital requirement is computed, and Example 10.3 examines the effects of working capital on the automated machining centre project discussed in Example 10.1.

EXAMPLE 10.2 **Working-Capital Requirements**

Suppose that in Example 10.1 the tool-manufacturing company's annual revenue projection of $100,000 is based on an annual volume of 10,000 units (or 833 units per month). Assume the following accounting information:

Price (revenue) per unit	$10
Unit variable manufacturing costs:	
Labour	$2
Material	$1.20
Overhead	$0.80
Monthly volume	833 units
Finished-goods inventory to maintain	2-month supply
Raw-materials inventory to maintain	1-month supply
Accounts payable	30 days
Accounts receivable	60 days

The accounts receivable period of 60 days means that revenues from the current month's sales will be collected two months later. Similarly, accounts payable of 30 days indicates that payment for materials will be made approximately one month after the materials are received. Determine the working-capital requirement for this operation.

SOLUTION

Given: Preceding information.

Find: Working-capital requirement.

Figure 10.4 illustrates the working-capital requirements for the first 12-month period. In year 1, suppose the company sells 10,000 units. Because the accounts payable period is 60 days, the company will receive payments during year 1 from its customers only for the 8333 units sold in the first 10 months of the year. The revenue for the sales

	During year 1		
	Income/Expense reported	Actual cash received/paid	Difference
Sales	$100,000 (10,000 units)	$83,333	–$16,666
Expenses	$40,000 (10,000 units)	$46,665 (11,667 units)	+$6,665
Income taxes	$16,500	$16,500	0
Net amount	$43,500	$20,168	–$23,331

This differential amount must be invested at the beginning of the year

Figure 10.4 Illustration of working-capital requirements (Example 10.2).

made in the last two months of year 1 will not be received until year 2. Accounts receivable of $16,666 (two months' sales) means that in year 1 the company will have cash inflows of $83,333, less than the projected sales of $100,000 ($8333 × 12). In years 2 to 5, collections will be $100,000, equal to sales, because the beginning and ending accounts receivable will be $16,666, with sales of $100,000. Collection of the final accounts receivable of $16,666 would occur in the first two months of year 6, but can be added to the year-5 revenue to simplify the calculations. The important point is that cash inflow lags sales by $16,666 in the first year.

Assuming that the company wishes to build up two months' inventory of finished goods during the first year, it must produce 833 × 2 = 1666 more units than are sold the first year. The extra cost of these goods in the first year is 1666 units ($4 variable cost per unit), or $6665. The finished-goods inventory of $6665 represents the variable cost incurred to produce 1,666 more units than are sold in the first year. In years 2 to 4, the company will produce and sell 10,000 units per year, while maintaining its 1,666-unit supply of finished goods. In the final year of operations, the company will produce only 8,334 units (for 10 months) and will use up the finished-goods inventory. As 1,666 units of the finished-goods inventory get liquidated during the last year (exactly, in the first two months of year 6), a working capital in the amount of $6,665 will be released. Along with the collection of the final accounts receivable of $16,666, a total working-capital release of $23,331 will be realized when the project terminates. Now we can calculate the working-capital requirements as follows:

Accounts receivable	
(833 units/month × 2 months × $10)	$16,666
Finished-goods inventory	
(833 units/month × 2 months × $4)	6,665
Raw-materials inventory	
(833 units/month × 1 month × $1.20)	1,000
Accounts payable (purchase of raw materials)	
(833 units/month × 1 month × $1.20)	(1,000)
Net change in working capital	$23,331

COMMENTS: During the first year, the company produces 11,666 units to maintain two months' finished-goods inventory, but it sells only 10,000 units. On what basis should the company calculate the net income during the first year (i.e., use 10,000 or 11,666 units)? *Any* increases in inventory expenses will reduce the taxable income; therefore, this calculation is based on 10,000 units. The reason is that the accounting measure of net income is based on the **matching concept**. If we report revenue when it is *earned* (whether it is actually received or not), and we report expenses when they are *incurred* (whether they are paid or not), we are using the *accrual method* of accounting. By tax law, this method *must* be used for purchases and sales whenever business transactions involve an inventory. Therefore, most manufacturing and merchandising businesses use the accrual basis in recording revenues and expenses. Any cash inventory expenses not accounted for in the net-income calculation will be reflected in changes in working capital.

EXAMPLE 10.3 Cash Flow Statement, Including Working Capital

Update the after-tax cash flows for the automated machining centre project of Example 10.1 by including a working-capital requirement of $23,331 in year 0 and full recovery of the working capital at the end of year 5.

SOLUTION

Given: Flows as in Example 10.1, with the addition of a $23,331 working-capital requirement.

Find: Net after-tax cash flows with working capital and present worth.

Using the procedure just outlined, we group the net after-tax cash flows for this machining centre project as shown in Table 10.3. As the table indicates, investments in working capital are cash outflows when they are expected to occur, and recoveries are treated as cash inflows at the times they are expected to materialize. In this example, we assume that the investment in working capital made at period 0 will be

TABLE 10.3 Cash Flow Statement for Automated Machining Centre Project With Working Capital Requirement (Example 10.3)

Year	0	1	2	3	4	5
Income Statement						
Revenues		$100,000	$100,000	$100,000	$100,000	$100,000
Expenses						
Labour		20,000	20,000	20,000	20,000	20,000
Materials		12,000	12,000	12,000	12,000	12,000
Overhead		8,000	8,000	8,000	8,000	8,000
CCA (30%)		18,750	31,875	22,313	15,619	10,933
Taxable income		$41,250	$28,125	$37,688	$44,381	$49,067
Income taxes (40%)		16,500	11,250	15,075	17,753	19,627
Net income		$24,750	$16,875	$22,613	$26,629	$29,440
Cash Flow Statement						
Operating activities						
Net income		$24,750	$16,875	$22,613	$26,629	$29,440
CCA		18,750	31,875	22,313	15,619	10,933
Investment activities						
Investment	$(125,000)					
Salvage						50,000
Disposal tax effects						(9,796)
Working capital	(23,331)					23,331
Net cash flow	$(148,331)	$43,500	$48,750	$44,925	$42,248	$103,909

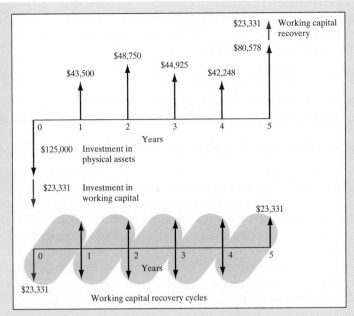

Figure 10.5   Cash flow diagram (Example 10.3).

recovered at the end of the project's life. (See Figure 10.5.)[4] Moreover, we also assume a full recovery of the initial working capital. However, in many situations, the investment in working capital may not be fully recovered (e.g., inventories may deteriorate in value or become obsolete). The equivalent net present worth of the after-tax cash flows, including the effects of working capital, is calculated as

$$PW(15\%) = -\$148,331 + \$43,500(P/F, 15\%, 1) + \cdots$$
$$+ \ \$103,909(P/F, 15\%, 5)$$
$$= \$31,712.$$

This present-worth value is $11,731 less than that in the situation with no working-capital requirement (Example 10.1), demonstrating that working-capital requirements play a critical role in assessing a project's worth.

---

[4] In fact, we could assume that the investment in working capital would be recovered at the end of the first operating cycle (say, year 1). However, the same amount of investment in working capital has to be made again at the beginning of year 2 for the second operating cycle, and the process repeats until the project terminates. Therefore, the net cash flow transaction looks as though the initial working capital will be recovered at the end of the project's life:

Period	0	1	2	3	4	5
Investment	−$23,331	−$23,331	−$23,331	−$23,331	−$23,331	0
Recovery	0	23,331	23,331	23,331	23,331	23,331
Net flow	−$23,331	0	0	0	0	$23,331

COMMENT: The $11,731 reduction in present worth is just the present worth of an annual series of 15% interest payments on the working capital, which is borrowed by the project at time 0 and repaid at the end of year 5:

$$\$23,331(15\%)(P/A, 15\%, 5) = \$11,731.$$

The investment tied up in working capital results in lost earnings.

10.3.3 When Projects Are Financed With Borrowed Funds

Debt ratio:
Debt capital divided by total assets. This will tell you how much the company relies on debt to finance assets.

Many companies use a mixture of debt and equity to finance their physical plant and equipment. The ratio of total debt to total investment, generally called the **debt ratio**, represents the percentage of the total initial investment provided by borrowed funds. For example, a debt ratio of 0.3 indicates that 30% of the initial investment is borrowed and the rest is provided from the company's earnings (also known as **equity**). Since interest is a tax-deductible expense, companies with a high tax rate may incur lower after-tax financing costs by financing through debt. (Along with the effect of debt on taxes, the method of repaying the loan can have a significant impact. We will discuss the issue of project financing in Chapter 12.)

EXAMPLE 10.4 Cash Flow Statement With Financing (Borrowing)

Rework Example 10.3, assuming that $62,500 of the $125,000 paid for the investment is obtained through debt financing (debt ratio = 0.5). The loan is to be repaid in equal annual installments at 10% interest over five years. The remaining $62,500 will be provided by equity (e.g., from retained earnings).

SOLUTION

Given: Same as in Example 10.3, but $62,500 is borrowed and then repaid in equal installments over five years at 10% interest.
Find: Net after-tax cash flows in each year.

We first need to compute the size of the annual loan repayment installments:

$$\$62,500(A/P, 10\%, 5) = \$16,487.$$

Next, we determine the repayment schedule of the loan by itemizing both the interest and principal represented in each annual repayment:

Year	Beginning Balance	Interest Payment	Principal Payment	Ending Balance
1	$62,500	$6,250	$10,237	$52,263
2	52,263	5,226	11,261	41,002
3	41,002	4,100	12,387	28,615
4	28,615	2,861	13,626	14,989
5	14,989	1,499	14,988	0

The resulting after-tax cash flow is detailed in Table 10.4. The present-value equivalent of the after-tax cash flow series is

$$PW(15\%) = -\$85,831 + \$29,513(P/F, 15\%, 1) + \cdots$$
$$+ \$88,021(P/F, 15\%, 5)$$
$$= \$44,729.$$

When this amount is compared with the amount found in the case when there was no borrowing ($31,712), we see that debt financing actually increases the present worth by $13,017. This surprising result is largely caused by the firm's being able to borrow the funds at a cheaper rate (10%) than its MARR (opportunity cost rate) of 15%. We should be careful in interpreting the result, however: It is true, to some extent, that firms can usually borrow money at lower rates than their MARR, but if the firm can borrow money at a significantly lower rate, that also affects its MARR, because the borrowing rate is one of the elements determining the MARR. Therefore, a significant difference in present values between "with borrowings" and "without borrowings" is not expected in practice. We will also address this important issue in Chapter 12.

TABLE 10.4 Cash Flow Statement for Automated Machining Centre Project With Debt Financing (Example 10.4)

Year	0	1	2	3	4	5
Income Statement						
Revenues		$100,000	$100,000	$100,000	$100,000	$100,000
Expenses						
Labour		20,000	20,000	20,000	20,000	20,000
Materials		12,000	12,000	12,000	12,000	12,000
Overhead		8,000	8,000	8,000	8,000	8,000
Debt interest		6,250	5,226	4,100	2,861	1,499
CCA (30%)		18,750	31,875	22,313	15,619	10,933
Taxable income		$35,000	$22,899	$33,587	$41,520	$47,568
Income tax (40%)		14,000	9,159	13,435	16,608	19,027
Net income		$21,000	$13,739	$20,152	$24,912	$28,541
Cash Flow Statement						
Operating activities						
Net income		$21,000	$13,739	$20,152	$24,912	$28,541
CCA		18,750	31,875	22,313	15,619	10,933
Investment activities						
Investment	$(125,000)					
Salvage						50,000
Disposal tax effect						(9,796)
Working capital	(23,331)					23,331
Financing activities						
Principal repayment	$62,500	(10,237)	(11,261)	(12,387)	(13,626)	(14,988)
Net cash flow	$(85,831)	$29,513	$34,353	$30,078	$26,905	$88,021

10.3.4 When Projects Result in Negative Taxable Income

In a typical project year, revenues may not be large enough to offset expenses, thereby resulting in a negative taxable income. Now, a negative taxable income does *not* mean that a firm does not need to pay income tax; rather, the negative figure can be used to reduce the taxable incomes generated by other business operations.[5] Therefore, a negative taxable income usually results in a **tax savings**. When we evaluate an investment project with the use of an incremental tax rate, we also assume that the firm has sufficient taxable income from other activities, so that changes due to the project under consideration will not change the incremental tax rate.

When we compare **cost-only** mutually exclusive projects (service projects), we have no revenues to consider in their cash flow analysis. In this situation, we typically assume no revenue (zero), but proceed as before in constructing the after-tax cash flow statement. With no revenue to match expenses, we have a negative taxable income, again resulting in tax savings. Example 10.5 illustrates how we may develop an after-tax cash flow statement for this type of project.

Tex savings (shield): The reduction in income taxes that results from taking an allowable deduction from taxable income.

EXAMPLE 10.5 Project Cash Flows for a Cost-Only Project

An Aluminum fabricator in Ontario produces aluminum coils, sheets, and plates. Its annual production runs at 400 million kilograms. In an effort to improve its current production system, an engineering team, led by the divisional vice-president, went on a fact-finding tour of Japanese aluminum and steel companies to observe their production systems and methods. Cited among the team's observations were large fans that the Japanese companies used to reduce the time that coils need to cool down after various processing operations. Cooling the hot process coils with the fans was estimated to significantly reduce the queue or work-in-process (WIP) inventory buildup allowed for cooling. The approach also reduced production lead time and improved delivery performance. The possibility of reducing production time and, as a consequence, the WIP inventory, excited the team members, particularly the vice-president. After the trip, Neal Thompson, the plant engineer, was asked to investigate the economic feasibility of installing cooling fans at the plant. Neal's job is to justify the purchase of cooling fans for his plant. He was given one week to prove that the idea was a good one. Essentially, all he knew was the number of fans, their locations, and the project cost. Everything else was left to Neal's devices. Suppose that he compiled the following financial data:

- The project will require an investment of $563,000 in cooling fans now.
- The cooling fans would provide 16 years of service with no appreciable salvage values, considering the removal costs.
- It is expected that the amount of time required between hot rolling and the next operation would be reduced from five days to two days. Cold-rolling queue time also would be reduced, from two days to one day for each cold-roll pass. The net effect of these changes would be a reduction in the WIP inventory at a value of

---

[5] Even if the firm does not have any other taxable income to offset in the current tax year, the operating loss can be carried back to each of the preceding three years and forward for the next 20 years to offset taxable income in those years.

$2,121,000. Because of the lead time involved in installing the fans, as well as the consumption of stockpiled WIP inventory, this working-capital release will be realized one year after the fans are installed. We also assume that the value of WIP will go back to the current level at the end of the project.

- The cooling fans have a CCA rate of 20%.
- Annual electricity costs are estimated to rise by $86,000.
- The firm's after-tax required rate of return is known to be 20% for this type of cost reduction project.
- The company's incremental tax rate is 40%.

Develop the project cash flows over the service period of the fans, and determine whether the investment is a wise one, based on 20% interest.

SOLUTION

Given: Required investment = $563,000; service period = 16 years; salvage value = $0; CCA rate for cooling fans = 20%; working-capital release = $2,121,000 1 year later; annual operating cost (electricity) = $86,000.

Find: (a) annual after-tax cash flows; (b) make the investment decision on the basis of the NPW.

(a) Because we can assume that the annual revenues would stay the same as they were before and after the installation of the cooling fans, we can treat these unknown revenue figures as zero. Table 10.5 summarizes the cash flow statement for the cooling-fan project. With no revenue to offset the expenses, the taxable income will be negative, resulting in tax savings. Note that the working-capital recovery (as opposed to working-capital investment for a typical investment project) is shown in year 1. Note also that the disposal tax effect is not zero even though the salvage value of the cooling fans is zero.

(b) At $i = 20\%$, the NPW of this investment would be

$$PW(20\%) = -\$563,000 + \$2,091,920(P/F, 20\%, 1)$$
$$- \$11,064(P/F, 20\%, 2) + \ldots$$
$$- \$2,163,686(P/F, 20\%, 16)$$
$$= \$949,144$$

Even with only one-time savings in WIP, this cooling-fan project is economically justifiable.

COMMENTS: As the cooling fans reach the end of their service lives, we need to add the working-capital investment ($2,120,000) at the end of year 16, working under the assumption that the plant will return to the former system without the cooling fans and thus will require the additional investment in working capital.

  If the new system has proven to be effective, and the plant will remain in service, we need to make another investment to purchase a new set of cooling fans at the end of year 16. As a consequence of this new investment, there will be a working-capital release in the amount of $2,120,000. However, this investment should bring benefits

TABLE 10.5 Cash Flow Statement for the Cooling Fan Project without Revenue (Example 10.5)

Year	0	1	2	3	4	5	6	7	8	9	10	11	12	13	14	15	16
Income Statement																	
Revenues																	
Expenses																	
CCA		$56,300	$101,340	$81,072	$64,858	$51,886	$41,509	$33,207	$26,566	$21,253	$17,002	$13,602	$10,881	$8,705	$6,964	$5,571	$4,457
Electricity cost		86,000	86,000	86,000	86,000	86,000	86,000	86,000	86,000	86,000	86,000	86,000	86,000	86,000	86,000	86,000	86,000
Taxable income		(142300)	(187340)	(167072)	(150858)	(137886)	(127509)	(119207)	(112566)	(107253)	(103002)	(99602)	(96881)	(94705)	(92964)	(91571)	(90457)
Income taxes (40%)		(56920)	(74936)	(66829)	(60343)	(55154)	(51004)	(47683)	(45026)	(42901)	(41201)	(39841)	(38753)	(37882)	(37186)	(36628)	(36183)
Net income		(85380)	(112404)	(100243)	(90515)	(82732)	(76505)	(71524)	(67539)	(64352)	(61801)	(59761)	(58129)	(56823)	(55778)	(54943)	(54274)
Cash Flow Statement																	
Operating activities																	
Net income		(85380)	(112404)	(100243)	(90515)	(82732)	(76505)	(71524)	(67539)	(64352)	(61801)	(59761)	(58129)	(56823)	(55778)	(54943)	(54274)
CCA		56,300	101,340	81,072	64,858	51,886	41,509	33,207	26,566	21,253	17,002	13,602	10,881	8,705	6,964	5,571	4,457
Investment activities																	
Cooling fans	$(563,000)																
Salvage value																	0
Disposal tax effects																	7131
Working capital	2,121,000																(2,121,000)
Net cash flow	$(563,000)	$2,091,920	$(11,064)	$(19,171)	$(25,657)	$(30,846)	$(34,996)	$(38,317)	$(40,974)	$(43,099)	$(44,799)	$(46,159)	$(47,247)	$(48,118)	$(48,814)	$(49,372)	$(2,163,686)

Note: The working capital release attributable to reduction in work-in-process inventories will be materialized at the end of year 1.

to the second cycle of the operation, so that it should be charged against the cash flows for the second cycle, not the first cycle.

10.3.5 When Projects Require Multiple Assets

Up to this point, our examples have been limited to situations in which only one asset was employed in a project. In many situations, however, a project may require the purchase of multiple assets with different property classes. For example, a typical engineering project may involve more than just the purchase of equipment: It may need a building in which to carry out manufacturing operations. The various assets may even be placed in service at different points in time. What we then have to do is itemize the timing of the investment requirement and the depreciation allowances according to the placement of the assets. Example 10.6 illustrates the development of project cash flows that require multiple assets.

EXAMPLE 10.6 A Project Requiring Multiple Assets

Langley Manufacturing Company (LMC), a manufacturer of fabricated metal products, is considering purchasing a new computer-controlled milling machine to produce a custom-ordered metal product. The following summarizes the relevant financial data related to the project:

- The machine costs $90,000. The costs for its installation, site preparation, and wiring are expected to be $10,000. The machine also needs special jigs and dies, which will cost $12,000. The milling machine is expected to last 10 years, the jigs and dies five years. The machine will have a $10,000 salvage value at the end of its life. The special jigs and dies are worth only $1,000 as scrap metal at any time in their lives. From Table 8.1, we find that the milling machine has a CCA rate of 30% while the jigs and dies have a CCA rate of 100%.

- LMC needs to either purchase or build an 8,000 m^2 warehouse in which to store the product before it is shipped to the customer. LMC has decided to purchase a building near the plant at a cost of $160,000. For tax depreciation purposes, the warehouse cost of $160,000 is divided into $120,000 for the building with a CCA rate of 4% and $40,000 for land. At the end of 10 years, the building will have a salvage value of $80,000, but the value of the land will have appreciated to $110,000.

- The revenue from increased production is expected to be $150,000 per year. The additional annual production costs are estimated as follows: materials, $22,000; labour, $32,000; energy $3500; and other miscellaneous costs, $2500.

- For the analysis, a 10-year life will be used. LMC has a marginal tax rate of 40% and a MARR of 18%. No money is borrowed to finance the project. Capital gains will be taxed at 20%.[6]

DISCUSSION: Three types of assets are to be considered in this problem: the milling machine, the jigs and dies, and the warehouse. The first two assets are equipment and

---

[6] Capital gains for corporations are taxed via the so-called capital gains inclusion rate, according to Canada Revenue Agency. This capital gains inclusion rate is revised from time to time by Canada Revenue Agency. As of 2009, this inclusion rate is 50%. This means that the tax rate on capital gains is 50% of the company's marginal tax rate.

the last is a real property. The cost basis for each asset has to be determined separately. For the milling machine, we need to add the site-preparation expense to the cost basis, whereas we need to subtract the land cost from the warehouse cost to establish the correct cost basis for the real property. The various cost bases are as follows:

- The milling machine: $90,000 + $10,000 = $100,000.
- Jigs and dies: $12,000.
- Warehouse (building): $120,000.
- Warehouse (land): $40,000.

Since the jigs and dies last only five years, we need to make a specific assumption regarding the replacement cost at the end of that time. In this problem, we will assume that the replacement cost would be approximately equal to the cost of the initial purchase. Note that the 50% rule needs to be used to calculate the CCA for the first year's use of each depreciable asset.

SOLUTION

Given: Preceding cash flow elements, $t_m = 40\%$, and MARR $= 18\%$.
Find: Net after-tax cash flow and NPW.

Table 10.6 and Figure 10.6 summarize the net after-tax cash flows associated with the multiple-asset investment. Capital cost allowances need to be calculated for the building part of the warehouse (4%), the milling machine (30%), and the jigs and dies (100%). The 50% rule must be used for the first year's use of each depreciable asset. Take the jigs and dies as an example. Even though its CCA rate is 100%, it cannot be fully depreciated in the first year due to the 50% rule.

$$CCA_1 = \$12,000 \, (100\%)/2 = \$6000$$
$$CCA_2 = \$12,000 \, (1 - 100\%/2)(100\%) = \$6,000$$

These values will be entered in years 1 and 2 and years 6 and 7 as shown in Table 10.6. The CCA values for other depreciable assets are also shown in Table 10.6.

For each asset, depreciable or not, we need to calculate its disposal tax effect when it is sold. For the building part of the warehouse, the land part of the warehouse, and the milling machine, we need to calculate their disposal tax effects at the end of year 10. For the jigs and dies, we need to calculate their disposal tax effects at the end of year 5 and at the end of year 10. Take the building part of the warehouse as an example. The undepreciated capital cost and the disposal tax effect at the end of year 10 are calculated as follows:

$$U_{10} = \$120,000(1 - 2\%)(1 - 4\%)^9 = \$81,442$$
$$G_{10} = t(U_{10} - S_{10}) = 40\% \, (\$81,442 - \$80,000) = \$577$$

This disposal tax effect of $577 is a cash inflow that occurs at the end of year 10 in addition to the salvage value of the building part of the warehouse in the amount of $80,000. The land part of the warehouse is not depreciable. There is a capital gain when the land is sold. Since the tax rate on capital gains is 20%, we find the disposal tax effect for the land as follows:

$$G = -20\% \, (\$110,000 - \$40,000) = -\$14,000.$$

TABLE 10.6 Cash Flow Statement for LMC's Machining Centre Project With Multiple Assets (Example 10.6)

Year	0	1	2	3	4	5	6	7	8	9	10
Income Statement											
Revenues		$150,000	$150,000	$150,000	$150,000	$150,000	$150,000	$150,000	$150,000	$150,000	$150,000
Expenses											
Materials		22,000	22,000	22,000	22,000	22,000	22,000	22,000	22,000	22,000	22,000
Labour		32,000	32,000	32,000	32,000	32,000	32,000	32,000	32,000	32,000	32,000
Energy		3,500	3,500	3,500	3,500	3,500	3,500	3,500	3,500	3,500	3,500
Others		2,500	2,500	2,500	2,500	2,500	2,500	2,500	2,500	2,500	2,500
CCA											
Building (4%)		2,400	4,704	4,516	4,335	4,162	3,995	3,836	3,682	3,535	3,393
Machines (30%)		15,000	25,500	17,850	12,495	8,747	6,123	4,286	3,000	2,100	1,470
Tools (100%)		6,000	6,000	-	-	-	6,000	6,000	-	-	-
Taxable income		66,600	53,796	67,634	73,170	77,092	73,882	75,879	83,318	84,365	85,137
Income taxes (40%)		26,640	21,518	27,054	29,268	30,837	29,553	30,351	33,327	33,746	34,055
Net income		$39,960	$32,278	$40,580	$43,902	$46,255	$44,329	$45,527	$49,991	$50,619	$51,082
Cash Flow Statement											
Operating activities											
Net income		$39,960	$32,278	$40,580	$43,902	$46,255	$44,329	$45,527	$49,991	$50,619	$51,082
CCA		23,400	36,204	22,366	16,830	12,908	16,118	14,121	6,682	5,635	4,863
Investment activities											
Land	$(40,000)										110,000
Building	(120,000)										80,000
Machines	(100,000)										10,000
Tools (1st cycle)	(12,000)					1000					1,000
Tools (2nd cycle)						(12,000)					
Disposal tax effects											
Land											(14,000)
Building											577
Machines											(2,628)
Tools						(400)					(400)
Net cash flow	$(272,000)	$63,360	$68,482	$62,946	$60,732	$47,763	$60,447	$59,649	$56,673	$56,254	$240,494

Note: Investment in tools (jigs and dies) will be repeated at the end of year 5 with the same initial purchase costs.

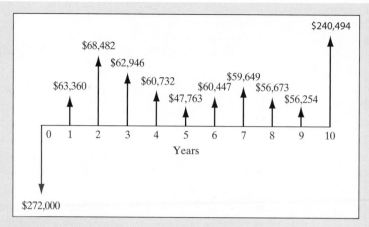

Figure 10.6 Cash flow diagram (Example 10.6).

The negative sign indicates that this is extra tax that the company has to pay due to disposal of the land. The disposal tax effects associated with disposal of each asset are tabulated below:

Property (Asset)	Cost Base	Salvage Value (S)	Undepreciated Capital Cost (U)	Gain/Losses (U–S)	Disposal Tax Effects
Land	$40,000	$110,000	$40,000	–$70,000	–$14,000
Building	120,000	80,000	81,442	1,442	577
Milling machine	100,000	10,000	3,430	–6,570	–2,628
Jigs and dies	12,000	1,000	0	–1,000	–400

Using the net cash flows given in Table 10.6, we can find the present worth of the project to be

$$PW(18\%) = -\$272,000 + \$63,360(P/F, 18\%, 1) + \$68,482(P/F, 18\%, 2)$$
$$+ \ldots + \$240,494(P/F, 18\%, 10)$$
$$= \$36,218 > 0,$$

and the IRR for this investment is about 21%, which exceeds the MARR. Therefore, this project is acceptable.

COMMENT: Note that the losses (gains) when assets are disposed of can be classified into two types: CCA losses (gains) and capital gains. Only the $70,000 for land represents capital gains, whereas others represent CCA losses (gains) upon disposal.

10.3.6 When Investment Tax Credits Are Allowed

As discussed in Section 9.3.5, the investment tax credit (ITC) has been used as a vehicle to encourage investments of certain types or in certain areas of Canada. A corporation may earn investment tax credits by acquiring certain property or making certain expenditures in certain geographical areas.

How do we adjust our after-tax cash flow analysis to consider the effects of a tax credit? The current tax regulation indicates the following:[7] The investment tax credit reduces the company's income tax payment by that amount at the end of the first year the asset is put in service. At the same time, the cost base for calculation of capital cost allowances has to be reduced by the amount of ITC. Thus, this will affect the capital cost allowances to be claimed for this asset and the disposal tax effect of this asset.

EXAMPLE 10.7 Considering Investment Tax Credits

Suppose in Example 10.6 that a 35% ITC was allowed on the purchase of the milling machine and the jigs and dies. How does the ITC affect the profitability of the investment?

SOLUTION

Given: Net cash flows in Table 10.6, ITC = 35%.
Find: The present worth of the project.

The ITC amount for each asset is

Milling machine: $0.35 \times \$100,000 = \$35,000$
Jigs and dies: $0.35 \times \$12,000 = \$4,200$

The ITC for the milling machine occurs in year 1 while the ITC for jigs and dies occurs in both year 1 and year 6 because a new set of jigs and dies is to be purchased at the beginning of year 1 and at the beginning of year 6. The new cost base for each asset for CCA calculations is:

Milling machine: $\$100,000 - \$35,000 = \$65,000$
Jigs and dies: $\$12,000 - \$4,200 = \$7,800.$

The capital cost allowances for these two assets will be changed and the disposal tax effects of the two assets will be changed too. The updated net cash flows are given in Table 10.7. We can then find the new PW with ITC considered as follows:

New PW = $-\$272,000 + \$99,620(P/F, 18\%, 1)$
$\qquad +\$64,072(P/F, 18\%, 2) + \dots + \$239,808(P/F, 18\%, 10)$
$\qquad = \$60,987$

New IRR = 24%.

Note that the ITC increases the profitability of the project. The present worth is increased from $36,218 (calculated in Example 10.6) to $60,987 and the IRR is increased from 21% (calculated in Example 10.6) to 24% because of the investment credits. (For a certain investment situation, the ITC could increase the profitability of the project from somewhat marginal to moderate.) This increase will certainly encourage the firm to undertake the investment.

---

[7] Check with Canada Revenue Agency for details on ITC. The website of Canada Revenue Agency is at www.cra-arc.gc.ca.

TABLE 10.7 Cash Flow Statement for LMC's Machining Centre Project With Multiple Assets Considering Investment Tax Credit (Example 10.7)

Year	0	1	2	3	4	5	6	7	8	9	10
Income Statement											
Revenues		$150,000	$150,000	$150,000	$150,000	$150,000	$150,000	$150,000	$150,000	$150,000	$150,000
Expenses											
Materials		22,000	22,000	22,000	22,000	22,000	22,000	22,000	22,000	22,000	22,000
Labour		32,000	32,000	32,000	32,000	32,000	32,000	32,000	32,000	32,000	32,000
Energy		3,500	3,500	3,500	3,500	3,500	3,500	3,500	3,500	3,500	3,500
Others		2,500	2,500	2,500	2,500	2,500	2,500	2,500	2,500	2,500	2,500
CCA											
Building (4%)		2,400	4,704	4,516	4,335	4,162	3,995	3,836	3,682	3,535	3,393
Machines (30%)		9,750	16,575	11,603	8,122	5,685	3,980	2,786	1,950	1,365	956
Tools (100%)		3,900	3,900				3,900	3,900			
Taxable income		73,950	64,821	73,882	77,543	80,153	78,125	79,479	84,368	85,100	85,651
Income taxes (40%)		29,580	25,928	29,553	31,017	32,061	31,250	31,791	33,747	34,040	34,260
Net income		$44,370	$38,893	$44,329	$46,526	$48,092	$46,875	$47,687	$50,621	$51,060	$51,391
Cash Flow Statement											
Operating activities											
Net income		$44,370	$38,893	$44,329	$46,526	$48,092	$46,875	$47,687	$50,621	$51,060	$51,391
CCA		16,050	25,179	16,118	12,457	9,847	11,875	10,521	5,632	4,900	4,349
Investment activities											
Land	(40,000)										110,000
Building	(120,000)										80,000
Machines	(100,000)										10,000
Tools (1st cycle)	(12,000)					1,000					
Tools (2nd cycle)						(12,000)					1,000
ITC											
Machines		35,000									
Tools		4,200					4,200				
Disposal tax effects											
Land											(14,000)
Building											577
Machines											(3,108)
Tools						(400)					(400)
Net cash flow	$(272,000)	$99,620	$64,072	$60,447	$58,983	$46,539	$62,950	$58,209	$56,253	$55,960	$239,808

Note: Investment in tools (jigs and dies) will be repeated at the end of year 5 with the same initial purchase costs.

10.4 Generalized Cash Flow Approach

As we mentioned in Chapter 9, corporate tax rates are non-progressive. For a company in a given business category, its tax rate on new projects is usually the same as the tax rate on its existing projects. This means that we can apply the same tax rate to each individual taxable item. By aggregating individual items, we obtain the project's net cash flows. This approach is referred to as the **generalized cash flow approach**. As we shall see in later chapters, it affords several analytical advantages. In particular, when we compare service projects, the generalized cash flow method is computationally efficient.

10.4.1 Setting Up Net Cash Flow Equations

To produce a generalized cash flow table, we first examine each cash flow element. We can do this as follows, using the scheme for classifying cash flows that we have just developed:

$$
\begin{aligned}
A_n = \ & + \text{Revenues at time } n, \ (R_n) \\
& - \text{Expenses excluding CCA and} \\
& \quad \text{debt interest at time } n, \ (E_n) \\
& - \text{Interest portion of debt payment at time } n, \ (I_n) \\
& - \text{Income taxes at time } n, \ (T_n)
\end{aligned}
\left.\vphantom{\begin{aligned}1\\2\\3\\4\end{aligned}}\right\} \text{Operating activities}
$$

$$
\begin{aligned}
& - \ \text{Investment at time } n, \ (P_n) \\
& + \ \text{Net proceeds from sale at time } n, \ (S_n + G_n) \\
& - \ \text{Working capital investment at time } n, \ (W_n) \\
& + \ \text{Working capital recovery at time } n, \ (W_n')
\end{aligned}
\left.\vphantom{\begin{aligned}1\\2\\3\\4\end{aligned}}\right\} \text{Investing activities}
$$

$$
\begin{aligned}
& + \ \text{Proceeds from loan at time } n, \ (L_n) \\
& - \ \text{Principal portion of debt payment at time } n, (PP_n),
\end{aligned}
\left.\vphantom{\begin{aligned}1\\2\end{aligned}}\right\} \text{Financing activities.}
$$

where A_n is the net after-tax cash flow at the end of period n.

Capital cost allowance is *not* a cash flow and is therefore excluded from E_n (although it must be considered in calculating income taxes). Note also that $(S_n + G_n)$ represents the net salvage value after adjustments for disposal tax effect (G_n). Not all terms are relevant in calculating a cash flow in every year; for example, the term $(S_n + G_n)$ appears only when the asset is disposed of.

In terms of symbols, we can express A_n as

$$
\begin{aligned}
A_n = \ & R_n - E_n - I_n - T_n && \longleftarrow && \text{Operating activities} \\
& - P_n + (S_n + G_n) - W_n + W_n' && \longleftarrow && \text{Investing activities} \\
& + L_n - PP_n && \longleftarrow && \text{Financing activities.}
\end{aligned}
\tag{10.1}
$$

If we designate T_n as the total income taxes paid at time n and t as the marginal tax rate, income taxes on this project are

$$
\begin{aligned}
T_n &= (\text{Taxable income})(\text{marginal tax rate}) \\
&= (R_n - E_n - I_n - CCA_n)t \\
&= (R_n - E_n)t - (I_n + CCA_n)t.
\end{aligned}
\tag{10.2}
$$

The term $(I_n + CCA_n)t$ is known as the tax shield (or tax savings) from financing and capital cost allowance. Now substituting the result of Eq. (10.2) into Eq. (10.1), we obtain

$$A_n = (R_n - E_n - I_n)(1 - t) + t \times CCA,$$
$$- P_n + (S_n + G_n) - W_n + W_n'$$
$$+ L_n - PP_n. \tag{10.3}$$

10.4.2 Presenting Cash Flows in Compact Tabular Formats

After-tax cash flow components over the project life can be grouped by type of activity in a compact tabular format as follows:

	End of Period			
Cash Flow Elements	**0**	**1**	**2**	**...N**
Operating activities:				
$+(1 - t)(R_n)$				
$-(1 - t)(E_n)$				
$-(1 - t)(I_n)$				
$+t \times CCA_n$				
Investment activities:				
$-P_n$				
$+(S_n + G_n)$				
$-W_n$				
$+W_n'$				
Financing activities:				
$+L_n$				
$-PP_n$				
Net cash flow				
A_n				

However, in preparing their after-tax cash flow, most business firms adopt the income statement approach presented in previous sections, because they want to know the accounting income along with the cash flow statement.

EXAMPLE 10.8 Generalized Cash Flow Approach

Consider again Example 10.4. Use the generalized cash flow approach to obtain the after-tax cash flows:

SOLUTION

Given:

Investment cost $(P_0) = \$125,000$,

Investment in working capital $(W_0) = \$23,331$,

Annual revenues $(R_n) = \$100,000$, $n = 1, 2, ..., 5$

Annual expenses other than CCA and debt interest $(E_n) = \$40,000$, $n = 1, 2, ..., 5$

Debt interest (I_n), years 1 to 5 = \$6250, \$5226, \$4100, \$2861, \$1499, respectively,

Principal repayment (PP_n), years 1 to 5 = \$10,237, \$11,261, \$12,387, \$13,626, \$14,988, respectively,

Capital cost allowance (CCA_n), years 1 to 5 = \$18,750, \$31,875, \$22,313, \$15,619, \$10,933 respectively,

Marginal tax rate $(t) = 40\%$, and

Salvage value $(S_5) = \$50,000$.

Find: Annual after-tax cash flows (A_n).

Step 1: Find the cash flow at year 0:

1. Investment in depreciable asset $(P_0) = -\$125,000$.
2. Investment in working capital $(W_0) = -\$23,331$.
3. Borrowed funds $(L_0) = \$62,500$.
4. Net cash flow $(A_0) = -\$125,000 - \$23,331 + \$62,500 = -\$85,831$.

Step 2: Find the cash flow in years 1 to 4:

$$A_n = (R_n - E_n - I_n)(1 - t) + t \times CCA_n - PP_n.$$

n	Net Operating Cash Flow (\$)
1	$(100,000 - 40,000 - 6,250)(0.60) + 18,750(0.40) - 10,237 = \$29,513$
2	$(100,000 - 40,000 - 5,226)(0.60) + 31,875(0.40) - 11,261 = \$34,353$
3	$(100,000 - 40,000 - 4,100)(0.60) + 22,313(0.40) - 12,387 = \$30,078$
4	$(100,000 - 40,000 - 2,861)(0.60) + 15,619(0.40) - 13,626 = \$26,905$

Step 3: Find the cash flow for year 5:

1. Operating cash flow:
   $(100,000 - 40,000 - 1,499)(0.60) + 10,933(0.40) = \$39,474$.
2. Net salvage value
   (refer to Example 10.1) $= S + G = \$50,000 - \$9,796 = \$40,204$.
3. Recovery of working capital, $W_5' = \$23,331$.
4. Net cash flow in year 5, $A_5 = \$39,474 + \$40,204 + \$23,331 - \$14,988 = \$88,021$.

Our results and overall calculations are summarized in Table 10.8. Checking them against the results we obtained in Table 10.4 confirms them.

TABLE 10.8 Cash Flow Statement for Example 10.4 Using the Generalized Cash Flow Approach (Example 10.8)

	0	1	2	3	4	5
(1 – 0.40) (Revenue)		$ 60,000	$ 60,000	$ 60,000	$ 60,000	$ 60,000
–(1 – 0.40) (Expenses)		(24,000)	(24,000)	(24,000)	(24,000)	(24,000)
–(1 – 0.40) (Debt interest)		(3,750)	(3,136)	(2,460)	(1,717)	(899)
+(0.40) (CCA)		7,500	12,750	8,925	6,248	4,373
Investment	$(125,000)					
Net proceeds from sale						$40,204
Investment in working capital	(23,331)					
Recovery of working capital						23,331
Borrowed funds	62,500					
Principal repayment		(10,237)	(11,261)	(12,387)	(13,626)	(14,988)
Net cash flow	$(85,831)	$ 29,513	$ 34,353	$ 30,078	$ 26,905	$88,021

10.4.3 Lease-or-Buy Decision

A lease-or-buy decision begins only after a company has decided that the acquisition of a piece of equipment is necessary to carry out an investment project. Having made this decision, the company may be faced with several alternative methods of financing the acquisition: cash purchase, debt purchase, or acquisition via a lease.

Before comparing these different ways of acquiring an asset, we must define the terms *lessee* and *lessor*. The lessee is the party leasing the property, and the lessor is the owner of property that is leased. Leasing takes several different forms, the most important of which are operating (or service) leases and financial (or capital) leases.

Operating Leases

Automobiles and trucks are frequently acquired under operating leases, and the operating lease contract for office equipment, such as computers and copying machines, is becoming common too. These operating leases have the following three characteristics.

1. The lessor maintains and services the leased equipment, and the cost of this maintenance is figured into the lease payments.
2. Operating leases are not fully amortized; in other words, the payments required under the lease contract are not sufficient to recover the full cost of the equipment.
3. Operating leases frequently contain a cancellation clause, which gives the lessee the right to cancel the lease before the expiration of the basic agreement.

These operating lease contracts are written for a period considerably shorter than the expected economic life of the leased equipment, and the lessor expects to recover all investment costs through subsequent renewal payments, through subsequent leases to other lessees, or by sale of the leased equipment. The cancellation clause is an important consideration to the lessee, because the equipment can be returned if it is rendered obsolete by technological developments or is no longer needed because of a decline in business.

Financial Leases

In contrast to operating leases, financial leases are usually drawn up as net leases; that is, the lessee assumes responsibility for paying most of the operating and maintenance costs of the equipment, and payment to the lessor is primarily for principal and interest. The lessee's obligations under a net financial lease are, therefore, similar to obligations incurred under a debt financial instrument. However, the difference is that leasing does not increase the debt ratio of the company and the lessee is not the owner of the asset. Companies utilizing financial leases treat them as a cheap way of obtaining fixed assets for long-term use.

Now we are ready to compare these three methods—cash purchase, debt purchase, and leasing—of obtaining a capital asset. With a debt purchase, the present-worth expression is similar to that for a cash purchase, except that it has additional items: loan repayments and a tax shield on interest payments. The only way that the lessee can evaluate the cost of a lease is to compare it against the best available estimate of what the cost would be if the lessee owned the equipment.

To lay the groundwork for a more general analysis, we shall first consider how to analyze the lease-or-buy decision for a project with a single fixed asset for which the company expects a service life of N periods. Since the revenue is the same for both alternatives, we need only consider the incremental cost of owning the asset and the incremental cost of leasing. Using the generalized cash flow approach, we may express the present equivalent incremental cost of owning the asset by borrowing, PEC, as

$$\text{PEC}(i)_{\text{Buy}} = \quad + \text{ PW of loan payment including principal and interest}$$
$$+ \text{ PW of after-tax O\&M costs}$$
$$- \text{ PW of tax credit on CCA and interest}$$
$$- \text{ PW of net proceeds from sale.} \qquad (10.4)$$

Note that the acquisition (investment) cost of the asset is offset by the same amount of borrowing at time 0, so that we need to consider only the loan repayment series.

If the asset is to be purchased without using borrowed money at all, the present equivalent incremental cost of owning the asset can be expressed as:

$$\text{PEC}(i)_{\text{buy}} = \text{Initial purchase cost}$$
$$+ \text{ PW of after-tax O\&M costs}$$
$$- \text{ PW of tax credits on CCA}$$
$$- \text{ PW of net proceeds from sale.}$$

Suppose that the firm can lease the asset at a constant amount per period. Then the project's incremental cost of leasing becomes

$$\text{PEC}(i)_{\text{Lease}} = \text{PW of after-tax lease expenses.}$$

If the lease does not provide for the maintenance of the equipment leased, then the lessee must assume this responsibility. In that situation, the maintenance term in Eq. (10.4) can be dropped in the calculation of the incremental cost of owning the asset.

The criterion for the decision to lease as opposed to purchasing the asset thus reduces to a comparison between $\text{PEC}(i)_{\text{Buy}}$ and $\text{PEC}(i)_{\text{Lease}}$. Purchasing is preferred if the combined present equivalent cost of owning the asset is smaller than the present equivalent cost of leasing it.

EXAMPLE 10.9 Lease-or-Buy Decision

The Montreal Electronics Company is considering replacing an old, half-tonne-capacity industrial forklift truck. The truck has been used primarily to move goods from production machines into storage. The company is working nearly at capacity and is operating on a two-shift basis, 300 days per year. Montreal management is considering the possibility of either purchasing or leasing a new truck. The plant engineer has compiled the following data for the management:

- The capital cost of a gas-powered truck is $20,000. The new truck would use about 30 litres of gasoline (per eight-hour shift) at a cost of 90¢ per litre. If the truck is operated 16 hours per day, its expected life will be four years, and an engine overhaul at a cost of $1500 will be required at the end of two years.

- The Windsor Industrial Truck Company was servicing the old forklift truck, and Montreal would buy the new truck through Windsor. Windsor offers a service agreement to users of its trucks that provides for a monthly visit by an experienced service representative to lubricate and tune the trucks and costs $120 per month. Insurance cost for the truck is $650 per year.

- The truck is classified as Class 10 property with a capital cost allowance rate of 30%. Montreal has a 40% tax rate. The estimated resale value of the truck at the end of four years will be 15% of the original cost.

- Windsor also has offered to lease a truck to Montreal. Windsor will maintain the equipment and guarantee to keep the truck in serviceable condition at all times. In the event of a major breakdown, Windsor will provide a replacement truck, at its expense. The cost of the operating lease plan is $10,200 per year. The contract term is three years' minimum, with the option to cancel on 30 days' notice thereafter.

- The company can secure short-term debt at 10% interest rate.

- Based on recent experience, the company expects that funds committed to new investments should earn at least a 12% rate of return after taxes.

Compare the cost of owning versus leasing the truck.

DISCUSSION: We may calculate the fuel costs for each day as

$$(30 \text{ litres/shift})(2 \text{ shifts/day})(90\text{¢/litre}) = \$54 \text{ per day}.$$

The truck will operate 300 days per year, so the annual fuel cost will be $16,200. However, both alternatives require the company to supply its own fuel, so the fuel cost is not relevant for our decision making. Therefore, we may drop this common cost item from our calculation.

SOLUTION

Given: Cost information as given, MARR = 12%, marginal tax rate = 40%.
Find: Incremental cost of owning versus leasing the truck.

(a) Owning the truck

To compare the incremental cost of ownership with the incremental cost of leasing, we make the following additional estimates and assumptions.

- **Step 1:** The preventive-maintenance contract, which costs $120 per month (or $1,440 per year) will be adopted. With an annual insurance cost of $650, the present worth of the after-tax O&M cost is

$$P_1 = -(\$1440 + \$650)(1 - 0.40)(P/A, 12\%, 4) = -\$3809.$$

- **Step 2:** The engine overhaul is not expected to increase either the salvage value or the service life. Therefore, the overhaul cost ($1,500) will be expensed all at once rather than capitalized. The present equivalent worth of this after-tax overhaul expense is

$$P_2 = -\$1500(1 - 0.40)(P/F, 12\%, 2) = -\$717.$$

- **Step 3:** If Montreal decided to purchase the truck through debt financing, the first step in determining financing costs would be to compute the annual installment of the debt-repayment schedule. Assuming that the entire investment of $20,000 is financed at a 10% interest rate, the annual payment would be

$$A = \$20,000(A/P, 10\%, 4) = \$6309.$$

The present equivalent worth of this loan-payment series is

$$P_3 = -\$6309(P/A, 12\%, 4) = -\$19,163.$$

- **Step 4:** The interest payment each year (10% of the beginning balance in the year) is calculated as follows:

Year	Beginning Balance	Interest Charged	Annual Payment	Ending Balance
1	$20,000	$2,000	−$6,309	$15,691
2	15,691	1,569	−6,309	10,951
3	10,951	1,095	−6,309	5,737
4	5,737	573	−6,309	

- **Step 5:** With a 30% CCA rate, the combined tax savings due to capital cost allowances and interest payments can be calculated as follows:

n	CCA_n	I_n	Combined Tax Savings
1	$3,000	$2,000	$5,000 × 0.40 = $2,000
2	5,100	1,569	6,669 × 0.40 = 2,668
3	3,570	1,095	4,665 × 0.40 = 1,866
4	2,499	573	3,072 × 0.40 = 1,229

Therefore, the present equivalent of the combined tax credit is

$$P_4 = \$2000(P/F, 12\%, 1) + \$2668(P/F, 12\%, 2)$$
$$+ \$1866(P/F, 12\%, 3) + \$1229(P/F, 12\%, 4)$$
$$= \$6022.$$

- **Step 6:** With the estimated salvage value of 15% of the initial investment ($3000) and the undepreciated capital cost, we compute the net proceeds from the sale of the truck at the end of four years as follows:

$$\text{Undepreciated capital cost} = \$20,000(1 - 15\%)(1 - 30\%)^3 = \$5831$$
$$\text{Losses} = U - S = \$5831 - \$3000 = \$2831$$
$$\text{Tax savings} = 0.40 \times \$2831 = \$1132$$
$$\text{Net proceeds from sale} = \$3000 + \$1132 = \$4132.$$

The present equivalent amount of the net salvage value is

$$P_5 = \$4132(P/F, 12\%, 4) = \$2626.$$

- **Step 7:** Therefore, the net present worth of owning the truck through 100% debt financing is

$$PW(12\%)_{\text{Buy}} = P_1 + P_2 + P_3 + P_4 + P_5$$
$$= -\$15,041.$$

This means that the present equivalent cost (PEC) of owning the truck is $15,041.

(b) Leasing the truck

How does the cost of acquiring a forklift truck under the lease compare with the cost of owning the truck?

- **Step 1:** The lease payment is also a tax-deductible expense. The net cost of leasing has to be computed explicitly on an after-tax basis. Assume that the lease payments are made at the end of each year. The calculation of the annual incremental leasing costs is as follows.

$$\text{Annual lease payments (12 months)} = \$10,200$$
$$\text{Less 40\% taxes} = 4,080$$
$$\text{Annual net costs after taxes} = \$6,120.$$

- **Step 2:** The total present worth of leasing is

$$PW(12\%)_{\text{Lease}} = -\$6120(P/A, 12\%, 4)$$
$$= -\$18,589.$$

This means that the present equivalent cost (PEC) of leasing the truck is $18,589. Comparing this number with the present equivalent cost of owning the truck, purchasing the truck with debt financing would save Montreal $3742 in the present equivalent costs.

COMMENTS: In our example, leasing the truck appears to be more expensive than purchasing the truck with debt financing. You should not conclude, however, that leasing is always more expensive than owning. In many cases, analysis favours a lease option. The interest rate, salvage value, lease-payment schedule, and debt financing all have an important effect on decision making.

In Example 10.9, we have assumed that the lease payments occur at the end of year. However, many leasing contracts require the payments to be made at the beginning of every year. As we have stated before, tax payments or tax savings are assumed to be realized only at the end of year. In this case, we can construct the cash flow diagram as shown in Figure 10.7 to describe the after-tax cash flows of the leasing option.

Using the after-tax cash flow diagram in Figure 10.7, we can find the present equivalent cost of lease financing when lease payment occurs at the beginning of the year as follows:

$$PEC(i)_{\text{Lease}} = LC + LC(P/A, i, N-1) - LC \times t\,(P/A, i, N). \qquad (10.5)$$

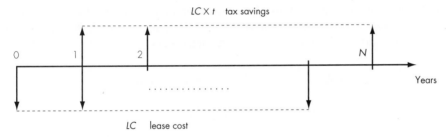

Figure 10.7   After tax cash flow diagram for lease option when lease payment occurs at beginning of year.

10.5 The After-Tax Cash Flow Diagram Approach

The income statement approach given in Section 10.3 and the generalized cash flow approach given in Section 10.4 require a year-by-year estimate of each cash flow element. These estimates are then combined in a manner that yields a net after-tax cash flow for each year. The resulting net cash flows are the basis for the subsequent discounted cash flow analysis.

The advantage of these two approaches is that they detail the magnitude of each annual net cash flow and indicate whether each value is positive or negative. Prior to the availability of spreadsheet software packages, problems that had a large number of cash flow elements and a long service life were very time-consuming. An alternative approach utilizing the after-tax cash flow diagram can help simplify numerical calculations in these cases. An added advantage of the after-tax cash flow diagram approach is that it depicts graphically the time location of each type of cash flow. In this section, we introduce this after-tax cash flow diagram approach for evaluating projects on the after-tax basis.

The after-tax cash flow diagram approach is quite similar to the generalized cash flow approach introduced in Section 10.4. Eq. (10.3) depicts the possible elements of the after-tax net cash flows that may appear on the cash flow diagram. For any specific problem, the relevant cash flow elements that comprise the net cash flows can be portrayed on the cash flow diagram.

For simplicity, we will consider investments requiring only one capital asset in the following discussions. When there is more than one asset, one needs to include the initial investment, the final net salvage value, and the tax savings due to CCA for each asset in the after-tax cash flow diagram. In the following, we will consider three investment scenarios: (1) when there is no debt financing involved; (2) when the debt financing is in the form of an amortized loan; and (3) when debt financing is through a bond series.

10.5.1 When There Is No Debt Financing

Consider a typical project financed with equity only with a life of N years. We also assume that there is no working capital requirement in this project. The initial investment P occurs at time 0 and the disposal with a salvage value S at year N. The net salvage value is $S + G$, where G is the disposal tax effect. The revenue in year n is R_n and the cost in year n is E_n. CCA_n denotes the capital cost allowance in year n. With these cash flow elements, based on Eq. (10.3), we know that $R_n(1 - t)$, $E_n(1 - t)$, and $t \times CCA_n$ are in years 1 through N.

These cash flow elements can now be represented on a cash flow diagram as shown in Figure 10.8. Here the wavy lines indicate an arbitrary pattern of cash flows.

The present worth (PW) of cash flows in the diagram is

$$
\begin{aligned}
PW = -P &+ (1 - t)\sum_{n=1}^{N} R_n(P/F, i, n) \\
&- (1 - t)\sum_{n=1}^{N} E_n(P/F, i, n) \\
&+ t\sum_{n=1}^{N} CCA_n(P/F, i, n) \\
&+ (S + G)(P/F, i, N).
\end{aligned}
\tag{10.6}
$$

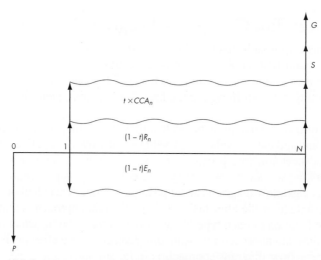

Figure 10.8 The after-tax cash flow diagram with one capital asset and no debt financing (wavy-line cash flows indicate that they may not be equal cash flow series).

If R_n and E_n follow a standard pattern (uniform series, linear gradient series, or geometric series) or a combination thereof, as they often do, we will be able to replace the summation terms for R_n and E_n with interest factors. However, the $t \times CCA_n$ terms do not exhibit a nice cash flow pattern because of the 50% rule for the first year. If the 50% rule did not apply, we would have a geometric gradient series with $g = -d$ in these terms. Even with the 50% rule, these $t \times CCA_n$ terms for $n > 1$ still exhibit such a geometric gradient pattern with $g = -d$. By adding and then subtracting a cash flow in the amount of

$$\frac{tPd}{2(1 - d)}$$

to the first year's cash flow, we can calculate the present equivalent of all $t \times CCA_n$ terms for $n \ge 1$ using the following equation:

$$PW_{t \times CCA_n} = tPd \, \frac{1 - \dfrac{d}{2}}{1 - d} (P/A_1, -d, i, N) - \frac{tPd}{2(1 - d)}(P/F, i, 1), \qquad (10.7)$$

where d is the CCA rate.

For example, if $R_n = R$ and $E_n = E$ are both constant, equation (10.6) can be written as:

$$PW = -P + (1 - t)R(P/A, i, N) - (1 - t)E(P/A, i, N)$$

$$+ (S + G)(P/F, i, N)$$

$$+ tPd\frac{1 - \dfrac{d}{2}}{1 - d} (P/A_1, -d, i, N)$$

$$- \frac{tPd}{2(1 - d)}(P/F, i, 1). \qquad (10.8)$$

10.5.2 When There Is Debt Financing in the Form of an Amortized Loan

If an amortized loan at an interest rate of i_d is included in project financing, the project cash flow diagram is as shown in Figure 10.9. In Figure 10.9, P_d is the amount of borrowed money, $A = P_d \times (A/P, i_d, N)$ represents the constant annual debt payment including both principal and interest payments, and I_n is the interest portion in the nth annual payment. I_n can be calculated with Eq. (4.14) in Chapter 4. With the after-tax cash flow diagram given in Figure 10.9, one can find the present worth of the project with this kind of debt financing.

10.5.3 When There Is Debt Financing in the Form of Bond-Type Loan

If the amount of debt, P_d, can be paid off with a single lump sum at the end of the project and only interest charges are paid from year to year during the project life, the interest payment, I, is constant in each year. This type of loan can be realized in the form of issuing bonds. It may also be negotiated directly with a lending institution. In this case, the constant interest payment in every year can be calculated as:

$$I = P_d \times i_d.$$

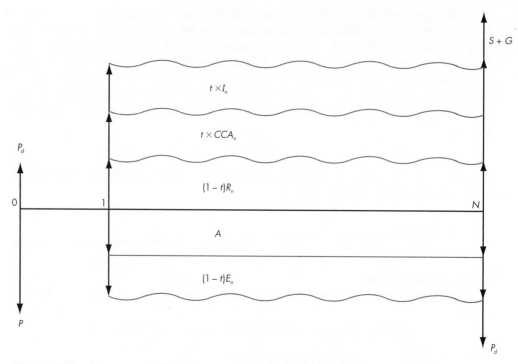

Figure 10.9 After-tax cash flow diagram with one capital asset and an amortized loan (wavy-line cash flows indicate that they may not be equal cash flow series).

The after-tax cash flow diagram of the project with a single capital asset and with this kind of debt financing is shown in Figure 10.10. One can then find the present worth of the project with this kind of debt financing using Figure 10.10.

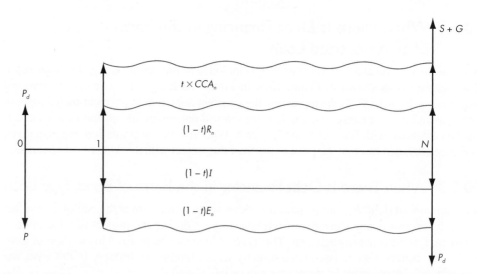

Figure 10.10 After-tax cash flow diagram with one capital asset and the bond-type debt financing (wavy-line cash flows indicate that they may not be equal cash flow series).

10.5.4 Other Considerations

As we discussed earlier in Section 10.3, there may be other considerations in a project such as multiple capital assets, working capital requirement, and multiple types of costs like labour, materials, and utilities.

When the working capital requirement is constant, we simply add the working capital investment cash flow at time 0 and its recovery at the end of the project in year N. The present worth evaluation can be conducted based on the corresponding after-tax cash flow diagram.

Some projects require investment in multiple capital assets. For a non-depreciable asset, we simply add the initial investment at time 0 and the net salvage value at the end of year N. For each class of depreciable assets (i.e., with a given CCA rate), we can treat all assets in the same class like a single asset. For each asset with a different CCA rate, we need to include its initial investment at time zero, the tax savings due to CCA at the end of each year $(1 - N)$, and then the final net salvage value at the end of year N.

In Section 10.4.3, we have used this after-tax cash flow diagram approach in our analysis of the equipment leasing option when the lease payments occur at the beginning of each year. This same approach can be used to analyze the leasing option when the lease payments are occurring at the end of each year.

In the following, we will use the after-tax cash flow diagram approach to evaluate a project that has a single asset and requirement for working-capital investment.

EXAMPLE 10.10 The After-Tax Cash Flow Diagram Approach Involving a Single Asset and Working Capital Requirement

Reconsider the problem solved in Example 10.3 where there is a working-capital requirement of $23,331 to be invested at the beginning of the project and fully recovered at the end of the project. Other financial data on this project are given in Example 10.1.

SOLUTION

Given: Cash flows in Example 10.1, plus a working capital requirement of $23,331.
Find: Whether the project is acceptable with a MARR of 15%.

Using the notation introduced in Section 10.4 and used earlier in this section, the after-tax cash flow diagram for this project is shown in Figure 10.11. As shown in Figure 10.11, the initial investment in the machining centre is $125,000; the annual revenue is constant at $100,000; the annual operating costs including labour, materials, and overhead are also constant at $40,000; and the working capital in the amount of $23,331 is invested at time 0 and fully recovered at the end of year 5. Other quantities that are referred to in Figure 10.11 are listed below:

$$t = 40\%;$$
$$d = 30\%;$$
$$S = \$50,000;$$
$$G = -\$9,796 \text{ as calculated in Example 10.1;}$$

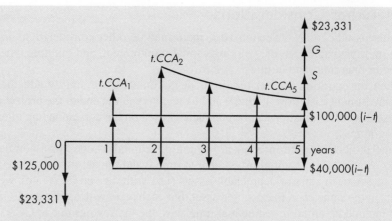

Figure 10.11 Cash flow diagram (Example 10.10).

We can use Eq. (10.7) to calculate the present worth of all CCA tax credits, so there is no need to calculate the CCA value for each year explicitly.

Using Eq. (10.8), considering working-capital investment and recovery, and/or referring to the cash flow diagram in Figure 10.11, we can find the present worth of this project as

$$
\begin{aligned}
PW = \ & -\$125,000 - \$23,331 - \$40,000\,(1 - 40\%)\,(P/A, 15\%, 5) \\
& + \$100,000\,(P/A, 15\%, 5) + (\$23,331 + \$50,000 - \$9,796)(P/F, 15\%, 5) \\
& + 40\% \times \$125,000 \times 30\%\,(1 - 15\%)\,(P/A_1, -30\%, 15\%, 5)\,/(1 - 30\%) \\
& - 40\% \times \$125,000 \times 30\%\,(P/F, 15\%, 1)/2/(1 - 30\%) \\
= \ & \$31,712.
\end{aligned}
$$

Since this PW value is positive at MARR = 15%, the project is acceptable. This conclusion is the same as that reached in Example 10.3.

SUMMARY

■ Identifying and estimating relevant project cash flows is perhaps the most challenging aspect of engineering economic analysis. All cash flows can be organized into one of the following three categories:

1. Operating activities.

2. Investing activities.

3. Financing activities.

■ The following types of cash flow are the most common flows a project may generate:

1. New investment and disposal of existing assets.

2. Salvage value (or net selling price).

3. Working capital.

4. Working-capital release.

5. Cash revenues or savings.

6. Manufacturing, operating, and maintenance costs.

7. Interest and loan payments.

8. Taxes and tax credits.

In addition, although not cash flows, the following elements may exist in a project analysis, and they must be accounted for:

1. Depreciation expenses.

2. Amortization expenses.

Table 10.1 summarizes these elements and organizes them as investment, financing, or operating elements.

■ The **income statement approach** is typically used in organizing project cash flows. This approach groups cash flows according to whether they are operating, investing, or financing functions. This approach is especially suitable when a spreadsheet software is used.

■ The **generalized cash flow approach** (shown in Table 10.7) to organizing cash flows can be used when a project does not change a company's marginal tax rate. The cash flows can be generated more quickly and the formatting of the results is less elaborate than with the income statement approach. There are also analytical advantages, which we will discover in later chapters. However, the generalized cash flow approach is less intuitive and not commonly understood by business people.

■ The **after-tax cash flow diagram approach** provides a graphical representation of the cash flows involved in a project. It enables users to take advantage of the interest factors and regular cash flow patterns covered in Chapter 3.

PROBLEMS

10.1   You are considering a luxury apartment building project that requires an investment of $12,500,000. The building has 50 units. You expect the maintenance cost for the apartment building to be $250,000 the first year and $300,000 the second year, after which it will continue to increase by $50,000 in subsequent years. The cost to hire a manager for the building is estimated to be $80,000 per year. After five years of operation, the apartment building can be sold for $14,000,000. What is the annual rent per apartment unit that will provide a return on investment of 15% after tax? Assume that the building will remain fully occupied during the five years. Assume also that your regular tax rate is 35% and the capital gains tax rate is 17.5%. The building has a CCA rate of 4% and will be sold at the end of the fifth year of ownership.

10.2   An automobile-manufacturing company is considering purchasing an industrial robot to do spot welding, which is currently done by skilled labour. The initial cost of the robot is $185,000, and the annual labour savings are projected to be $120,000. The robot is a Class 43 property with a CCA rate of 30%. The robot will be used for seven years, at the end of which time the firm expects to sell it for $40,000. The company's marginal tax rate is 35% over the project period. Determine the net after-tax cash flows for each period over the project life.

10.3   A firm is considering purchasing a machine that costs $55,000. It will be used for six years, and the salvage value at that time is expected to be zero. The machine will save $25,000 per year in labour, but it will incur $7000 in operating and maintenance costs each year. The machine is a Class 43 property with a CCA rate of 30%. The firm's tax rate is 40% and its after-tax MARR is 15%. Should the machine be bought?

10.4 A Calgary company is planning to market an answering device for people working alone who want the prestige that comes with having a secretary, but who cannot afford one. The device, called Tele-Receptionist, is similar to a voice-mail system. It uses digital recording technology to create the illusion that a person is operating the switchboard at a busy office. The company purchased a 4,000 m² building and converted it to an assembly plant for $600,000 ($100,000 worth of land and $500,000 worth of building). Assembly equipment worth $500,000 was installed. The company expects to have a gross annual income of $2,200,000 over the next five years. Annual manufacturing costs and all other operating expenses (excluding depreciation) are projected to be $1,280,000. For tax depreciation purposes, the assembly plant building has a CCA rate of 4% and the assembly equipment has a CCA rate of 30%. The property value of the land and the building at the end of year 5 would both appreciate as much as 15% over the initial purchase cost. The residual value of the assembly equipment is estimated to be about $50,000 at the end of year 5. The firm's marginal tax rate is expected to be about 40% over the project period. The capital gains tax rate is 20%. Determine the project's after-tax cash flows over the period of five years.

10.5 A highway contractor is considering buying a new trench excavator that costs $180,000 and can dig a 1-metre-wide trench at the rate of 5 metres per hour. With the machine adequately maintained, its production rate will remain constant for the first 1200 hours of operation and then decrease by 0.5 metres per hour each year. The excavator is expected to dig 2 kilometres each year. Maintenance and operating costs will be $15 per hour. The excavator has a CCA rate of 30%. At the end of five years, the excavator will be sold for $40,000. Assuming that the contractor's marginal tax rate is 34% per year, determine the annual after-tax cash flow.

10.6 A small children's clothing manufacturer is considering an investment to computerize its management information system for material requirement planning, piece-goods coupon printing, and invoice and payroll. An outside consultant has been retained to estimate the initial hardware requirement and installation costs. He suggests the following:

PC systems	
(15 PCs, four printers)	$85,000
Local area networking system	15,000
System installation and testing	4,000

The expected life of the computer system is five years, with no expected salvage value. The proposed system is classified as a Class 45 property with a CCA rate of 45%. A group of computer consultants needs to be hired to develop various customized software packages to run on the system. Software development costs will be $20,000 and can be expensed during the first tax year. The new system will eliminate two clerks, whose combined annual payroll expenses are $52,000. Additional annual expenses to run this computerized system are expected to be

$12,000. Borrowing is not considered an option for this investment, nor is a tax credit available for the system. The firm's expected marginal tax rate over the next six years will be 35%. The firm's interest rate is 13%. Compute the after-tax cash flows over the life of the investment.

10.7   A firm has been paying a print shop $18,000 annually to print the company's monthly newsletter. The agreement with this print shop has now expired, but it could be renewed for a further five years. The new subcontracting charges are expected to be 12% higher than they were under the previous contract. The company is also considering the purchase of a desktop publishing system with a high-quality laser printer driven by a microcomputer. With appropriate text and graphics software, the newsletter can be composed and printed in near-typeset quality. A special device is also required to print photos in the newsletter. The following estimates have been quoted by a computer vendor:

Personal computer	$4,500
Colour laser printer	6,500
Photo device/scanner	5,000
Software	2,500
Total cost basis	$18,500
Annual O&M costs	10,000

The salvage value of each piece of equipment at the end of five years is expected to be only 10% of the original cost. The company's marginal tax rate is 40%, and the desktop publishing system can be considered a Class 45 property with a CCA rate of 45%.

(a) Determine the projected net after-tax cash flows for the investment.

(b) Compute the IRR for this project.

(c) Is the project justifiable at MARR = 12%?

10.8   An asset in Class 8 with a CCA rate of 20% costs $120,000 and has a zero estimated salvage value after six years of use. The asset will generate annual revenues of $300,000 and will require $80,000 in annual labour and $50,000 in annual material expenses. There are no other revenues and expenses. Assume a tax rate of 40%.

(a) Compute the after-tax cash flows over the project life.

(b) Compute the NPW at MARR = 12%. Is the investment acceptable?

10.9   An automaker is considering installing a three-dimensional (3-D) computerized car-styling system at a cost of $300,000 (including hardware and software). With the 3-D computer modelling system, designers will have the ability to view their design from many angles and to fully account for the space required for the engine and passengers. The digital information used to create the computer model can be revised in consultation with engineers, and the data can be used to run milling machines that make physical models quickly and precisely. The automaker expects to decrease the turnaround time for designing a new automobile model (from configuration to final design) by 22%. The expected savings in dollars is $250,000 per year. The training and operating maintenance cost for the new system is expected to be $50,000 per year. The system has a five-year useful life and can be depreciated with a CCA rate of 30% for tax purposes. The system will have an estimated salvage value of $5,000. The automaker's marginal tax rate is 40%. Determine the annual cash flows for this investment. What is the return on investment for the project?

10.10 A facilities engineer is considering a $50,000 investment in an energy management system (EMS). The system is expected to save $10,000 annually in utility bills for N years. After N years, the EMS will have a zero salvage value. In an after-tax analysis, what would N need to be in order for the investment to earn a 10% return? Assume a 35% CCA rate and a 35% tax rate.

10.11 A corporation is considering purchasing a machine that will save $130,000 per year before taxes. The cost of operating the machine, including maintenance, is $20,000 per year. The machine will be needed for five years, after which it will have a zero salvage value. Its CCA rate is 30%. The marginal income tax rate is 40%. If the firm wants 15% IRR after taxes, how much can it afford to pay for this machine?

10.12 The Doraville Machinery Company is planning to expand its current spindle product line. The required machinery would cost $500,000. The building that will house the new production facility would cost $1.5 million. The land would cost $250,000, and $150,000 working capital would be required. The product is expected to result in additional sales of $875,000 per year for 10 years, at which time the land can be sold for $500,000, the building for $700,000, and the equipment for $50,000. All of the working capital will be recovered. The annual disbursements for labour, materials, and all other expenses are estimated to be $425,000. The firm's income tax rate is 40%, and any capital gains will be taxed at 20%. The CCA rates for the building and the equipment are 4% and 30%, respectively. The firm's MARR is known to be 15% after taxes.

(a) Determine the projected net after-tax cash flows from this investment. Is the expansion justified?

(b) Compare the IRR of this project with that of a situation with no working capital.

10.13 An industrial engineer proposed the purchase of scanning equipment for the company's warehouse and weave rooms. The engineer felt that the purchase would provide a better system of locating cartons in the warehouse by recording the locations of the cartons and storing the data in the computer.

The estimated investment, annual operating and maintenance costs, and expected annual savings are as follows:

- Cost of equipment and installation: $65,500.
- Project life: 6 years.
- Expected salvage value: $3,000.
- Investment in working capital (fully recoverable at the end of the project life): $10,000.
- Expected annual savings on labour and materials: $55,800.
- Expected annual expenses: $8,120.
- CCA rate: 30%.

The firm's marginal tax rate is 35%.

(a) Determine the net after-tax cash flows over the project life.

(b) Compute the IRR for this investment.

(c) At MARR = 18%, is the project acceptable?

10.14 Sarnia Chemical Corporation is considering investing in a new composite material. R&D engineers are investigating exotic metal–ceramic and ceramic–ceramic composites to develop materials that will withstand high temperatures, such as those to be encountered in the next generation of jet fighter engines. The company expects a three-year R&D period before these new materials can be applied to commercial products.

The following financial information is presented for management review:

- **R&D cost.** $5 million over a three-year period. Annual R&D expenditure of $0.5 million in year 1, $2.5 million in year 2, and $2 million in year 3. For tax purposes, these R&D expenditures will be expensed rather than amortized.

- **Capital investment.** $5 million at the beginning of year 4. This investment consists of $2 million in a building and $3 million in plant equipment. The company already owns a piece of land as the building site.

- **Capital cost allowance.** The building's CCA rate is 4% and the plant equipment has a CCA rate of 30%.

- **Project life.** 10 years after a three-year R&D period.

- **Salvage value.** 10% of the initial capital investment for the equipment and 50% for the building (at the end of the project life).

- **Total sales.** $45 million (at the end of year 4), with an annual sales growth rate of 10% per year (compound growth) during the next five years (year 5 through year 9) and −10% (negative compound growth) per year for the remaining project life.

- **Out-of-pocket expenditures.** 80% of annual sales.

- **Working capital.** 10% of annual sales (considered as an investment at the beginning of each production year and investments fully recovered at the end of the project life).

- **Marginal tax rate.** 40%.

    (a) Determine the net after-tax cash flows over the project life.

    (b) Determine the IRR for this investment.

    (c) Determine the equivalent annual worth for the investment at MARR = 20%.

10.15 Refer to the data in Problem 10.1. If the firm expects to borrow the initial investment ($12,500,000) at 10% over five years (paying back the loan in equal annual payments of $3,297,469), determine the project's net cash flows.

10.16 In Problem 10.2, to finance the industrial robot, the company will borrow the entire amount from a local bank, and the loan will be paid off at the rate of $37,000 per year, plus 10% on the unpaid balance. Determine the net after-tax cash flows over the project life.

10.17 Refer to the financial data in Problem 10.9. Suppose that 50% of the initial investment of $300,000 will be borrowed from a local bank at an interest rate of 11% over five years (to be paid off in five equal annual payments). Recompute the after-tax cash flow.

10.18 A special-purpose machine tool set would cost $20,000. The tool set will be financed by a $10,000 bank loan repayable in two equal annual installments at 10% compounded annually. The tool is expected to provide annual (material) savings of $30,000 for two years and has a CCA rate of 100%. The tool will require annual O&M costs in the amount of $5,000. The salvage value at the end of the two years is expected to be $8,000. Assuming a marginal tax rate of 40% and MARR of 15%, what is the net present worth of this project? You may use the following worksheet in your calculation:

Cash Flow Statement	0	1	2
Operating activities			
Net income		8,400	8,686
CCA		10,000	10,000
Investment activities			
Investment	−20,000		
Salvage			8,000
Gains tax (40%)			
Financial activities			
Borrowed funds	10,000		
Principal repayment	0		
Net cash flow	−$10,000		

10.19 A.M.I. Company is considering installing a new process machine for the firm's manufacturing facility. The machine costs $200,000 installed, will generate additional revenues of $80,000 per year, and will save $55,000 per year in labour and material costs. The machine will be financed by a $150,000 bank loan repayable in three equal annual principal installments, plus 9% interest on the outstanding balance. The machine has a CCA rate of 30%. The useful life of the machine is 10 years, after which it will be sold for $20,000. The combined marginal tax rate is 40%.

(a) Find the year-by-year after-tax cash flow for the project.

(b) Compute the IRR for this investment.

(c) At MARR = 18%, is the project economically justifiable?

10.20 Consider the following financial information about a retooling project at a computer manufacturing company:

- The project costs $2.1 million and has a five-year service life.
- The retooling project has a CCA rate of 30%.
- At the end of the fifth year, any assets held for the project will be sold. The expected salvage value will be about 10% of the initial project cost.
- The firm will finance 40% of the project money from an outside financial institution at an interest rate of 10%. The firm is required to repay the loan with five equal annual payments.
- The firm's incremental (marginal) tax rate on the investment is 35%.
- The firm's MARR is 18%.
- With the preceding financial information,
    (a) Determine the after-tax cash flows.
    (b) Compute the annual equivalent worth for this project.

10.21 A fully automatic chucker and bar machine is to be purchased for $40,000, to be borrowed with the stipulation that it be repaid with six equal end-of-year payments at 12% compounded annually. The machine is expected to provide an annual revenue of $10,000 for six years and has a CCA rate of 30%. The salvage value at the end of six years is expected to be $3,000. Assume a marginal tax rate of 36% and a MARR of 15%.

(a) Determine the after-tax cash flow for this asset over six years.

(b) Determine whether the project is acceptable on the basis of the IRR criterion.

10.22 A manufacturing company is considering acquiring a new injection moulding machine at a cost of $110,000. Because of a rapid change in product mix, the need for this particular machine is expected to last only eight years, after which time the machine is expected to have a salvage value of $10,000. The annual operating cost is estimated to be $5,000. The addition of the machine to the current production facility is expected to generate an annual revenue of $40,000. The firm has only $70,000 available from its equity funds, so it must borrow the additional $40,000 required at an interest rate of 10% per year, with repayment of principal and interest in eight equal annual amounts. The applicable marginal income tax rate for the firm is 40%. Assume that the asset qualifies for a Class 43 property with a CCA rate of 30%.

(a) Determine the after-tax cash flows.

(b) Determine the NPW of this project at MARR $= 14\%$.

10.23 Suppose an asset has a first cost of $6,000, a life of five years, a salvage value of $2,000 at the end of five years, and a net annual before-tax revenue of $1,500. The firm's marginal tax rate is 35%. The asset has a CCA rate of 15%.

(a) Using the generalized cash flow approach, determine the cash flow after taxes.

(b) Rework part (a), assuming that the entire investment would be financed by a bank loan at an interest rate of 9%.

(c) Given a choice between the financing methods of parts (a) and (b), show calculations to justify your choice of which is the better one at an interest rate of 9%.

10.24 A construction company is considering acquiring a new earthmover. The purchase price is $125,000, and an additional $25,000 is required to modify the equipment for special use by the company. The equipment has a CCA rate of 30%, and it will be sold after five years (the project life) for $50,000. The purchase of the earthmover will have no effect on revenues, but the machine is expected to save the firm $60,000 per year in before-tax operating costs, mainly labour. The firm's marginal tax rate is 40%. Assume that the initial investment is to be financed by a bank loan at an interest rate of 10%, payable annually. Determine the after-tax cash flows by using the generalized cash flow approach and the present worth of the investment for this project if the firm's MARR is known to be 12%.

10.25 Air West, a leading regional airline that is now carrying 54% of all the passengers flying within Canada, is considering adding a new long-range aircraft to its fleet. The aircraft being considered for purchase is the McDonnell Douglas DC-9-532 "Funjet," which is quoted at $62 million per unit. McDonnell Douglas requires a 10% down payment at the time of delivery, and the balance is to be paid over a 10-year period at an interest rate of 12% compounded annually. The actual payment schedule calls for only interest payments over the 10-year period, with the original principal amount to be paid off at the end of the 10th year. Air West expects to generate $35 million per year by adding this aircraft to its current fleet, but also estimates an operating and maintenance cost of $20 million per year. The aircraft is expected to have a 15-year service life with a salvage value of 15% of the original purchase price. The aircraft is a Class 9 property with a CCA rate of 25%. The firm's combined tax rate is 38%, and its required minimum attractive rate of return is 18%.

(a) Use the generalized cash flow approach to determine the cash flow associated with the debt financing.

(b) Is this project acceptable?

10.26 The Toronto Division of Vermont Machinery Inc. manufactures drill bits. One of the production processes of a drill bit requires tipping, whereby carbide tips are inserted into the bit to make it stronger and more durable. The tipping process usually requires four or five operators, depending on the weekly workload. The same operators are assigned to the stamping operation, in which the size of the drill bit and the company's logo are imprinted into the bit. Vermont is considering acquiring three automatic tipping machines to replace the manual tipping and stamping operations. If the tipping process is automated, Vermont engineers will have to redesign the shapes of the carbide tips to be used in the machines. The new design requires less carbide, resulting in a savings of material. The following financial data have been compiled:

- Project life: six years.
- Expected annual savings: reduced labour, $56,000; reduced material, $75,000; other benefits (reduction in carpal tunnel syndrome and related problems), $28,000; reduced overhead, $15,000.
- Expected annual O&M costs: $22,000.
- Tipping machines and site preparation: equipment costs (three machines), including delivery, $200,000; site preparation, $20,000.
- Salvage value: $30,000 (three machines) at the end of six years.
- CCA rate: 30%.
- Investment in working capital: $25,000 at the beginning of the project year, and that same amount will be fully recovered at the end of the project year.
- Other accounting data: marginal tax rate of 39% and MARR of 18%.

To raise $220,000, Vermont is considering the following financing options:

- **Option 1.** Use the retained earnings of the tipping machines to finance them.
- **Option 2.** Secure a 12% term loan over six years (to be paid off in six equal annual installments).
- **Option 3.** Lease the tipping machines. Vermont can obtain a six-year financial lease on the equipment including site preparation (with, however, no maintenance service included in the lease) for payments of $55,000 at the *beginning* of each year.

(a) Determine the net after-tax cash flows for each financing option.

(b) What is Vermont's present-value of owning the equipment by borrowing?

(c) What is Vermont's present-value of leasing the equipment?

(d) Recommend the best course of action for Vermont.

10.27 The headquarters building owned by a rapidly growing company is not large enough for the company's current needs. A search for larger quarters revealed two new alternatives that would provide sufficient room, enough parking, and the desired appearance and location. The company now has three options:

- **Option 1.** Lease the new quarters for $144,000 per year to be paid at the beginning of each year.
- **Option 2.** Purchase the new quarters for $850,000, including a $150,000 cost for land.
- **Option 3.** Remodel the current headquarters building.

It is believed that land values will not change over the ownership period, but the value of all structures will decline to 10% of the purchase price in 30 years. Annual property tax payments are expected to be 5% of the purchase price. The present headquarters building is already paid for and is now valued at $300,000. The land it is on is appraised at $60,000 and this price will not change in 30 years. The structure can be remodelled at a cost of $300,000 to make it comparable to other alternatives. The remodelled building can be sold for $60,000 in 30 years. However, the remodelling will occupy part of the existing parking lot. An adjacent, privately owned parking lot can be leased for 30 years under an agreement that the first year's rental of $9,000 will increase by $500 each year to be paid at the beginning of each year. The annual property taxes on the remodelled property will again be 5% of the present valuation, plus the cost of remodelling. The new quarters are expected to have a service life of 30 years, and the desired rate of return on investments is 12%. Assume that the firm's marginal tax rate is 40% and that the new building and remodelled structure are Class 3 property with a CCA rate of 4%. It is further assumed that if either option 1 or option 2 is selected, the old headquarter can be leased for $60,000, to be received at the begining of each year. If the annual upkeep costs are the same for all three alternatives, which one is preferable? Ignore the CCA effects of the current building.

10.28 An international manufacturer of prepared food items needs 50,000,000 kWh of electrical energy a year, with a maximum demand of 10,000 kW. The local utility currently charges $0.085 per kWh, a rate considered high throughout the industry. Because the firm's power consumption is so large, its engineers are considering installing a 10,000-kW steam-turbine plant. Three types of plant have been proposed (units in thousands of dollars):

	Plant A	Plant B	Plant C
Average station heat rate (BTU/kWh)	16,500	14,500	13,000
Total investment (boiler/turbine/electrical/structures)	$8,530	$9,498	$10,546
Annual operating cost:			
Fuel	1,128	930	828
Labour	616	616	616
O&M	150	126	114
Supplies	60	60	60
Insurance and property taxes	10	12	14

The service life of each plant is expected to be 20 years. The plant investment has a CCA rate of 30%. The expected salvage value of the plant at the end of its useful life is about 10% of its original investment. The firm's MARR is known to be 12%. The firm's marginal income tax rate is 39%.

(a) Determine the unit power cost ($/kWh) for each plant.

(b) Which plant would provide the most economical power?

10.29 The Jacob Company needs to acquire a new lift truck for transporting its final product to the warehouse. One alternative is to purchase the truck for $41,000, which will be financed by the bank at an interest rate of 12%. The loan must be repaid in four equal installments, payable at the end of each year. Under the borrow-to-purchase arrangement, Jacob would have to maintain the truck at a cost of $1,200, payable at year-end. Alternatively, Jacob could lease the truck under a four-year contract for a lease payment of $11,000 per year. Each annual lease payment must be made at the beginning of each year. The truck would be maintained by the lessor. The truck has a CCA rate of 30%, and it has a salvage value of $10,000, which is the expected market value after four years, at which time Jacob plans to replace the truck irrespective of whether it leases or buys. Jacob has a marginal tax rate of 40% and a MARR of 15%.

(a) What is Jacob's cost of leasing, in present worth?

(b) What is Jacob's cost of owning, in present worth?

(c) Should the truck be leased or purchased?

This is an operating lease, so the truck would be maintained by the lessor.

10.30 Janet Wang, an electrical engineer for Instrument Control Inc. (ICI), has been asked to perform a lease–buy analysis of a new pin-inserting machine for ICI's PC-board manufacturing.

- **Buy Option.** The equipment costs $130,000. To purchase it, ICI could obtain a term loan for the full amount at 10% interest, payable in four equal end-of-year annual installments. The machine has a CCA rate of 30%. Annual revenues of $200,000 and operating costs of $40,000 are anticipated. The machine requires annual maintenance at a cost of $10,000. Because technology is changing rapidly in pin-inserting machinery, the salvage value of the machine is expected to be only $20,000.

- **Lease Option.** Business Leasing Inc. (BLI) is willing to write a four-year operating lease on the equipment for payments of $44,000 at the beginning of each year. Under this arrangement, BLI will maintain the asset, so that the annual maintenance cost of $10,000 will be saved. ICI's marginal tax rate is 40%, and its MARR is 15% during the analysis period.

(a) What is ICI's present-value (incremental) cost of owning the equipment?

(b) What is ICI's present-value (incremental) cost of leasing the equipment?

(c) Should ICI buy or lease the equipment?

10.31 Consider the following lease-versus-borrow-and-purchase problem:

- **Borrow-and-purchase option:**

  1. Jensen Manufacturing Company plans to acquire sets of special industrial tools with a four-year life and a cost of $200,000, delivered and installed. The tools have a CCA rate of 30%.

2. Jensen can borrow the required $200,000 at a rate of 10% over four years. Four equal end-of-year annual payments would be made in the amount of $63,094 = $200,000(A/P, 10%, 4). The annual interest and principal payment schedule, along with the equivalent present worth of these payments, is as follows:

End of Year	Interest	Principal
1	$20,000	$43,094
2	15,961	47,403
3	10,950	52,144
4	5,736	57,358

3. The estimated salvage value for the tool sets at the end of four years is $20,000.

4. If Jensen borrows and buys, it will have to bear the cost of maintenance, which will be performed by the tool manufacturer at a fixed contract rate of $12,000 per year.

- **Lease option:**

   1. Jensen can lease the tools for four years at an annual rental charge of $70,000, payable at the end of each year.

   2. The lease contract specifies that the lessor will maintain the tools at no additional charge to Jensen.

Jensen's tax rate is 40%. Any gains will also be taxed at 40%.

(a) What is Jensen's PW of after-tax cash flow of leasing at $i = 15\%$?

(b) What is Jensen's PW of after-tax cash flow of owning at $i = 15\%$?

10.32 Tom Torbiak has decided to acquire a new car for his business. One alternative is to purchase the car outright for $18,000, financing the car with a bank loan for the net purchase price. The bank loan has a 12.6% interest rate with equal annual payments for three years. The terms of each alternative are as follows:

Buy	Lease
$18,000	$5100 per year
	3-year open-end lease
	Annual mileage allowed = 20,000 km

If Tom takes the lease option, he is required to pay $500 for a security deposit, refundable at the end of the lease, and $5100 a year at the beginning of each year for three years. The car is a Class 10 property with a CCA rate of 30%. The car has a salvage value of $5,800, which is the expected market value after three years, at which time Tom plans to replace the car irrespective of whether he leases or buys. Tom's marginal tax rate is 35%. His MARR is known to be 13% per year.

(a) Determine the annual cash flows for each option.

(b) Which option is better?

10.33 The Boggs Machine Tool Company has decided to acquire a pressing machine. One alternative is to lease the machine under a three-year contract for a lease payment of $15,000 per year, with payments to be made at the beginning of each year. The lease would include maintenance. The second alternative is to purchase the machine outright for $80,000, financing the machine with a bank loan for the net purchase price and amortizing the loan over a three-year period at an interest rate of 12% per year.

Under the borrow-to-purchase arrangement, the company would have to maintain the machine at an annual cost of $5,000, payable at year-end. The machine falls into Class 43 property with a CCA rate of 30% and has a salvage value of $50,000, which is the expected market value at the end of year 3, at which time the company plans to replace the machine irrespective of whether it leases or buys. Boggs has a tax rate of 40% and a MARR of 15%.

(a) What is Boggs' PW cost of leasing?

(b) What is Boggs' PW cost of owning?

(c) From the financing analysis in (a) and (b), what are the advantages and disadvantages of leasing and owning?

10.34 An asset is to be purchased for $25,000. The asset is expected to provide revenue of $10,000 a year and have operating costs of $2500 a year. The asset is considered to be a Class 8 property with a CCA rate of 20%. The company is planning to sell the asset at the end of year 5 for $5,000. Given that the company's marginal tax rate is 30% and that it has a MARR of 10% for any project undertaken, answer the following questions:

(a) What is the net cash flow for each year, given that the asset is purchased with borrowed funds at an interest rate of 12%, with repayment in five equal end-of-year payments?

(b) What is the net cash flow for each year, given that the asset is leased at a rate of $3500 paid at the beginning of each year (a financial lease)?

(c) Which method (if either) should be used to obtain the new asset?

10.35 Enterprise Capital Leasing Company is in the business of leasing tractors to construction companies. The firm wants to set a three-year lease payment schedule for a tractor purchased at $53,000 from the equipment manufacturer. The asset is classified as a Class 16 property with a CCA rate of 40%. The tractor is expected to have a salvage value of $22,000 at the end of three years' rental. Enterprise will require a lessee to make a security deposit in the amount of $1500, which is refundable at the end of the lease term. Enterprise's marginal tax rate is 35%. If Enterprise wants an after-tax return of 10%, what lease payment schedule should be set?

Short Case Studies

ST10.1 Global Aluminum's Laval plant is considering making a major investment of $140 million ($5 million for land, $40 million for buildings, and $95 million for manufacturing equipment and facilities) to develop a stronger, lighter material, called aluminum lithium, that will make aircraft sturdier and more fuel efficient. Aluminum lithium, which has been sold commercially for only a few years as an alternative to composite materials, will likely be the material of choice for the next generation of commercial and military aircraft, because it is so much lighter than conventional aluminum alloys, which use a combination of copper, nickel, and magnesium to harden aluminum. Another advantage of aluminum lithium is that it is cheaper than composites. The firm predicts that aluminum lithium will account for about 5% of the structural weight of the average commercial aircraft within five years and 10% within 10 years. The proposed plant, which has an estimated service life of 12 years, would have a capacity of about 10 million kilograms of aluminum lithium, although domestic consumption of the material is expected to be only 3 million kilograms during the first four years, 5 million for the next three years, and 8 million for the remaining life of the plant. Aluminum lithium costs $12 a kilogram to produce, and the firm would expect to sell it at $17 a kilogram. The building has a CCA rate of 4%. All manufacturing equipment and facilities have a CCA rate of 30%. At the end of the project life, the land will be worth $8 million, the building $30 million, and the equipment $10 million. Assuming that the firm's marginal tax rate is 40% and its capital gains tax rate is 20%,

(a) Determine the net after-tax cash flows.

(b) Determine the IRR for this investment.

(c) Determine whether the project is acceptable if the firm's MARR is 15%.

ST10.2 Morgantown Mining Company is considering a new mining method at its Blacksville mine. The method, called longwall mining, is carried out by a robot. Coal is removed by the robot, not by tunnelling like a worm through an apple, which leaves more of the target coal than is removed, but rather by methodically shuttling back and forth across the width of the deposit and devouring nearly everything. The method can extract about 75% of the available coal, compared to 50% for conventional mining, which is done largely with machines that dig tunnels. Moreover, the coal can be recovered far more inexpensively. Currently, at Blacksville alone, the company mines 5 million tonnes a year with 2200 workers. By installing two longwall robot machines, the company can mine 5 million tonnes with only 860 workers. (A robot miner can dig more than 6 tonnes a minute.) Despite the loss of employment, the United Mine Workers union generally favours longwall mines, for two reasons: The union officials are quoted as saying, (1) "It would be far better to have highly productive operations that were able to pay our folks good wages and benefits than to have 2,200 shovellers living in poverty," and (2) "Longwall mines are inherently safer in their design." The company projects the following financial data upon installation of the longwall mining:

Robot installation (2 units)	$19.3 million
Total amount of coal deposit	50 million tonnes
Annual mining capacity	5 million tonnes
Project life	10 years
Estimated salvage value	$0.5 million
Working capital requirement	$2.5 million
Expected additional revenues:	
Labour savings	$6.5 million
Accident prevention	$0.5 million
Productivity gain	$2.5 million
Expected additional expenses:	
O&M costs	$2.4 million

(a) Estimate the firm's net after-tax cash flows over the project life if the firm uses the unit-production method to depreciate assets. The firm's marginal tax rate is 40%.

(b) Estimate the firm's net after-tax cash flows if the firm chooses to depreciate the robots on the basis of MACRS (seven-year property classification).

ST10.3 National Parts Inc., an auto-parts manufacturer, is considering purchasing a rapid prototyping system to reduce prototyping time for form, fit, and function applications in automobile-parts manufacturing. An outside consultant has been called in to estimate the initial hardware requirement and installation costs. He suggests the following:

- Prototyping equipment: $215,000.
- Posturing apparatus: $10,000.
- Software: $15,000.
- Maintenance: $36,000 per year by the equipment manufacturer.
- Resin: Annual liquid polymer consumption of 1600 litres at $87.5 per litre.

The expected life of the system is six years, with an estimated salvage value of $30,000. The proposed system is classified as a Class 43 property with a CCA rate of 30%. A group of computer consultants must be hired to develop customized software to run on the system. Software development costs will be $20,000 and can be expensed during the first tax year. The new system will reduce prototype development time by 75% and material waste (resin) by 25%. This reduction in development time and material waste will save the firm $314,000 and $35,000 annually, respectively. The firm's expected marginal tax rate over the next six years will be 40%. The firm's interest rate is 20%.

(a) Assuming that the entire initial investment will be financed from the firm's retained earnings (equity financing), determine the after-tax cash flows over the life of the investment. Compute the NPW of this investment.

(b) Assuming that the entire initial investment will be financed through a local bank at an interest rate of 13% compounded annually, determine the net after-tax cash flows for the project. Compute the NPW of the investment.

(c) Suppose that a financial lease is available for the prototype system at $62,560 per year, payable at the beginning of each year. Compute the NPW of the investment with lease financing.

(d) Select the best financing option.

ST10.4 National Office Automation Inc. (NOAI) is a leading developer of imaging systems, controllers, and related accessories. The company's product line consists of systems for desktop publishing, automatic identification, advanced imaging, and office information markets. The firm's manufacturing plant in Windsor, Ontario, consists of eight different functions: cable assembly, board assembly, mechanical assembly, controller integration, printer integration, production repair, customer repair, and shipping. The process to be considered is the transportation of pallets loaded with eight packaged desktop printers from printer integration to the shipping department. Several alternatives for minimizing operating and maintenance costs have been examined. The two most feasible alternatives are the following:

- **Option 1.** Use gas-powered lift trucks to transport pallets of packaged printers from printer integration to shipping. The truck also can be used to return printers that must be reworked. The trucks can be leased at a cost of $5465 per year. With a maintenance contract costing $6317 per year, the dealer will maintain the trucks. Both need to be paid at the beginning of each year. A fuel cost of $1660 per year is expected. The truck requires a driver for each of the three shifts, at a total cost of $58,653 per year for labour. It is estimated that transportation by truck would cause damages to material and equipment totalling $10,000 per year.

- **Option 2.** Install an automatic guided vehicle system (AGVS) to transport pallets of packaged printers from printer integration to shipping and to return products that require rework. The AGVS, using an electrical powered cart and embedded wire-guidance system, would do the same job that the truck currently does, but without drivers. The total investment costs, including installation, are itemized as follows:

Vehicle and system installation	$147,255
Staging conveyor	24,000
Power supply lines	5,000
Transformers	2,500
Floor surface repair	6,000
Batteries and charger	10,775
Shipping	6,500
Sales tax	6,970
Total AGVS system cost	$209,000

NOAI could obtain a term loan for the full investment amount ($159,000) at a 10% interest rate. The loan would be amortized over five years, with payments made at the end of each year. The AGVS has a CCA rate of 30%, and it has an estimated service life of 10 years and no salvage value. If the AGVS is installed, a maintenance contract would be obtained at a cost of $20,000, payable at the beginning of each year. The firm's marginal tax rate is 35% and its MARR is 15%.

(a) Determine the net cash flows for each alternative over 10 years.

(b) Determine the best course of action.

On the Companion Website that accompanies this text, you will find Excel templates and exercises, as well as the following analysis tools: Cash Flow Analyzer, Depreciation Analysis, Loan Analysis, and Interest Tables.

Special Topics in Engineering Economics

CHAPTER ELEVEN

Replacement Decisions

Options for the Existing Johnson Street Bridge in the City of Victoria, British Columbia[1] In April 2009, the results of an overall condition assessment of the Johnson Street Bridge identified extensive corrosion in steel structural beams, obsolete mechanical and electrical systems, and significant seismic vulnerability. This comprehensive assessment noted that substantial investment in the bridge would be required by 2012 to avoid further deterioration, operation cost increase, and possible closure. The existing bridge is of the Bascule type, in which one end may rise to allow larger marine vessels to go through. The bridge carries approximately 30,000 trips per day, including vehicles, pedestrians, and cyclists.

Based on the bridge condition assessment report, the city council ruled out the option of **Do-Nothing** due to risk, liability, and economic impact.

The **Replacement** option would adopt the Rolling Bascule design with an estimated investment of $63 million. This design is able to address all the concerns on the current bridge and provide excellent service to the City of Victoria, with an estimated life of 100 years.

Technical work on the **Rehabilitation** option is yet to be completed. Major questions to be answered include what major components and systems of the existing bridge are to be replaced and how the foundation is to be reinforced to meet the latest construction standards. The total investment required for this option is not yet available.

---

[1] The Johnson Street Bridge Project, City of Victoria, British Columbia, www.johnsonstreetbridge.com.

Sources: (Top) Image of Johnson Street Bridge in downtown Victoria, B.C. (Bottom) Artwork of possible future Johnson Street Bridge. Reproduction of architect design rendering. Copyright © 2009 City of Victoria. All rights reserved.

Currently, the only money in place for the Replacement option is up to one-third of the project cost to a maximum of $21 million provided by the federal government under the Build Canada Fund—Major Infrastructure Component. To use this money for the Rehabilitation option, the city has to demonstrate that the concerns identified in the condition assessment report can be satisfactorily addressed with this option. The city is planning to borrow the rest to get the project moving forward.

In Chapters 5 and 6, we presented methods that helped us choose the best of a number of investment alternatives. The problems we examined in those chapters concerned primarily profit-adding projects. However, economic analysis is also frequently performed on projects with existing facilities or profit-maintaining projects—those projects whose primary purpose is not to increase sales, but rather, simply to maintain ongoing operations. In practice, profit-maintaining projects less frequently involve the comparison of new machines; instead, the problem often facing management is whether to buy new and more efficient equipment or to continue to use existing equipment. This class of decision problems is known as the **replacement problem**. In this chapter, we examine the basic concepts and techniques related to replacement analysis.

CHAPTER LEARNING OBJECTIVES

After completing this chapter, you should understand the following concepts:

- What makes the replacement decision problems differ from the other capital investment decisions.
- What types of financial information should be collected to conduct a typical replacement decision problem.
- How to compare a defender with a challenger on the basis of opportunity cost concept.
- How to determine the economic service life for any given asset.
- How to determine the optimal time to replace a defender.
- How to consider the tax effects in replacement analysis.

11.1 Replacement Analysis Fundamentals

In this section and the next two, we examine three aspects of the replacement problem: (1) approaches to comparing defender and challenger, (2) the determination of economic service life, and (3) replacement analysis when the required service period is long. The impact of income tax regulations will be ignored; in Section 11.4, we revisit these replacement problems, taking income taxes into account.

11.1.1 Basic Concepts and Terminology

Replacement projects are decision problems involving the replacement of existing obsolete or worn-out assets. The continuation of operations is dependent on these assets. The failure to make an appropriate decision results in a slowdown or shutdown of the operations. The question is when existing equipment should be replaced with more efficient equipment. This situation has given rise to the use of the terms **defender** and **challenger**, terms commonly used in the boxing world. In every boxing class, the current defending champion is constantly faced with a new challenger. In replacement analysis, the defender is the existing machine (or system), and the challenger is the best available replacement equipment.

An existing piece of equipment will be removed at some future time, either when the task it performs is no longer necessary or when the task can be performed more efficiently by newer and better equipment. The question is not *whether* the existing piece of equipment will be removed, but *when* it will be removed. A variation of this question is why we should replace existing equipment at the current time, rather than postponing replacement of the equipment by repairing or overhauling it. Another aspect of the defender–challenger comparison concerns deciding exactly which equipment is the best challenger. If the defender is to be replaced by the challenger, we would generally want to install the very best of the possible alternatives.

Current Market Value

The most common problem encountered in considering the replacement of existing equipment is the determination of what financial information is actually relevant to the analysis. Often, a tendency to include irrelevant information in the analysis is apparent. To illustrate this type of decision problem, let us consider Example 11.1.

EXAMPLE 11.1 Information Relevant to Replacement Analysis

Macintosh Printing Inc. purchased a $20,000 printing machine two years ago. The company expected this machine to have a five-year life and a salvage value of $5000. The company spent $5000 last year on repairs, and current operating costs are running at the rate of $8000 per year. Furthermore, the anticipated salvage value of the machine has been reduced to $2500 at the end of the machine's remaining useful life. In addition, the company has found that the current machine has a market value of $10,000 today. The equipment vendor will allow the company this full amount as a trade-in on a new machine. What values for the defender are relevant to our analysis?

SOLUTION

In this example, several different dollar amounts relating to the defender are presented:

1. **Original cost.** The printing machine was purchased for $20,000.
2. **Current market value.** The company estimates the old machine's current market value at $10,000.
3. **Trade-in allowance.** This is the same as the market value. (In other problems, however, it could be different from the market value.)
4. **Repair cost.** $5000 was spent last year to repair the machine.
5. **Operating costs.** If kept, the machine costs $8000 per year to operate.
6. **Future salvage value.** The machine has a salvage value of $2500 three years from today.

COMMENTS: In this example and in all defender analyses, the **current market value** of the equipment is always a relevant cost. The original cost, past repair cost, and trade-in value are irrelevant. A common misconception is that the trade-in value is the same as the current market value of the equipment and thus could be used to assign a suitable current value to the equipment. This is not always true, however. For example, a car dealer typically offers a trade-in value on a customer's old car to reduce the price of a new car. Would the dealer offer the same value on the old car if he or she were not also selling the new one? The answer is no, generally. In many instances, the trade-in allowance is inflated to make the deal look good, and the price of the new car is also inflated to compensate for the dealer's trade-in cost. In this type of situation, the trade-in value does not represent the true value of the item, so we should not use it in economic analysis.[2]

Sunk Costs

Sunk costs are costs that have already been incurred and that cannot be recovered to any significant degree.

As mentioned in Section 7.4.3, a **sunk cost** is any past cost that is unaffected by any future investment decision. In Example 11.1, the company spent $20,000 to buy the machine two years ago. Last year, $5000 more was spent on the machine. The total accumulated expenditure is $25,000. If the machine were sold today, the company could get only $10,000 back (Figure 11.1). It is tempting to think that the company would lose $15,000 in addition to the cost of the new machine if the old machine were to be sold and replaced with a new one. This is an incorrect way of doing economic analysis, however. In a proper engineering economic analysis, only future costs should be considered; past or sunk costs should be ignored. Thus, the value of the defender that should be used in a replacement analysis should be its current market value, not what it cost when it was originally purchased and not the cost of repairs that have already been made to the machine.

Sunk costs are money that is gone, and no present action can recover them. They represent past actions—the results of decisions made in the past. In making economic decisions at the present time, one should consider only the possible outcomes of the various decisions

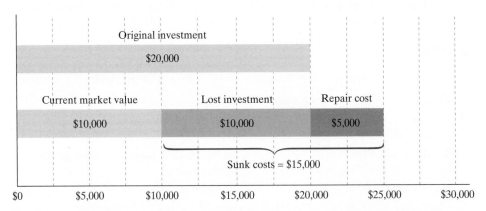

Figure II.I Sunk cost associated with an asset's disposal as described in Example 11.1.

[2] If we do make the trade, however, the actual net cash flow at the time of the trade, properly used, is certainly relevant.

and pick the one with the best possible future results. Using sunk costs in arguing one option over the other would only lead to more bad decisions.

Operating Costs

The driving force for replacing existing equipment is that it becomes more expensive to operate with time. The total cost of operating a piece of equipment may include repair and maintenance costs, wages for the operators, energy consumption costs, and costs of materials. Increases in any one or a combination of these cost items over a period of time may impel us to find a replacement for the existing asset. The challenger is usually newer than the defender and often incorporates improvements in design and newer technology. As a result, some or all of the cost items for the challenger are likely to be less expensive than those for the defender.

We will call the sum of the various cost items related to the operation of an asset the **operating costs**. As is illustrated in the sections that follow, keeping the defender involves a lower initial cost than purchasing the challenger, but higher annual operating costs. Usually, operating costs increase over time for both the defender and the challenger. In many instances, the labour costs, material costs, and energy costs are the same for the defender and the challenger and do not change with time. It is the repair and maintenance costs that increase and cause the operating costs to increase each year as an asset ages.

When repair and maintenance costs are the only cost items that differ between the defender and the challenger on a year-by-year basis, we need to include only those costs in the operating costs used in the analysis. Regardless of which cost items we choose to include in the operating costs, it is essential that the same items be included for both the defender and the challenger. For example, if energy costs are included in the operating costs of the defender, they should also be included in the operating costs of the challenger. A more comprehensive discussion of the various types of costs incurred in a complex manufacturing facility was provided in Chapter 7.

11.1.2 Opportunity Cost Approach to Comparing Defender and Challenger

Although replacement projects are a subcategory of the mutually exclusive categories of project decisions we studied in Chapter 6, they do possess unique characteristics that allow us to use specialized concepts and analysis techniques in their evaluation. We consider a basic approach to analyzing replacement problems commonly known as the **opportunity cost approach**.

The basic issue is how to treat the proceeds from the sale of the old equipment. In fact, if you decide to keep the old machine, this potential sales receipt is forgone. The opportunity cost approach views the net proceeds from sale as the opportunity cost of keeping the defender. In other words, we consider the current market value as a cash outflow for the defender (or an investment required in order to keep the defender).

EXAMPLE 11.2 Replacement Analysis Using the Opportunity Cost Approach

Consider again Example 11.1. The company has been offered a chance to purchase another printing machine for $15,000. Over its three-year useful life, the new machine will reduce the usage of labour and raw materials sufficiently to cut operating costs from $8000 a year to $6000. It is estimated that the new machine can be sold for $5500 at the end of year 3. If the new machine were purchased, the old machine would be sold to another company, rather than traded in for the new machine.

Suppose that the firm will need either machine (old or new) for only three years and that it does not expect a new, superior machine to become available on the market during this required service period. Assuming that the firm's interest rate is 12%, decide whether replacement is justified now.

SOLUTION

- **Option 1: Keep the defender.**
  If the decision is to keep the defender, the opportunity cost approach treats the $10,000 current market value of the defender as an investment cost. The annual operating cost for the next three years will be $8000 per year, and the defender's salvage value three years from today will be $2500. The cash flow diagram for the defender is shown in Figure 11.2(a).

Opportunity cost approach views the net proceeds from the sale of the old machine as an investment required to keep the old asset.

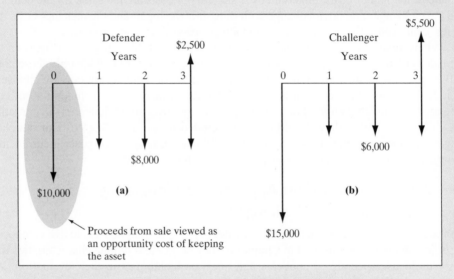

Figure 11.2 Comparison of defender and challenger based on the opportunity cost approach (Example 11.2).

- **Option 2: Replace the defender with the challenger.**
  The new machine costs $15,000. The annual operating cost of the challenger is $6000. The salvage value of the challenger three years later will be $5500. The cash flow diagram for this option is shown in Figure 11.2(b). We calculate the net present equivalent cost (PEC) and annual equivalent cost (AEC) for each of the two options as follows:

$$\text{PEC}(12\%)_D = \$10,000 + \$8000(P/A, 12\%, 3)$$
$$- \$2500(P/F, 12\%, 3)$$
$$= \$27,435,$$

$$\text{AEC}(12\%)_D = \text{PEC}(12\%)_D \ (A/P, 12\%, 3)$$
$$= \$11{,}423,$$

$$\text{PEC}(12\%)_C = \$15{,}000 + \$6000(P/A, 12\%, 3) - \$5500(P/F, 12\%, 3)$$
$$= \$25{,}496,$$

$$\text{AEC}(12\%)_C = \text{PEC}(12\%)_C \ (A/P, 12\%, 3)$$
$$= \$10{,}615.$$

Because of the annual difference of \$808 in favour of the challenger, the replacement should be made now.

COMMENTS: If our analysis showed instead that the defender should not be replaced now, we still need to address the question of whether the defender should be kept for one or two years and then replaced with the challenger. This is a valid question that requires more data on salvage values over time. We address the situation later, in Section 11.3. Recall that we assumed the same service life for both the defender and the challenger in Examples 11.1 and 11.2. In general, however, old equipment has a relatively short remaining life compared with new equipment, so this assumption is overly simplistic. In the next section, we discuss how to find the economic service life of equipment.

11.2 Economic Service Life

Perhaps you have seen a 50-year-old automobile that is still in service. Provided that it receives the proper repair and maintenance, almost anything can be kept operating for an extended period of time. If it's possible to keep a car operating for an almost indefinite period, why aren't more old cars spotted on the streets? Two reasons are that some people may get tired of driving the same old car, and other people may want to keep a car as long as it will last, but they realize that repair and maintenance costs will become excessive.

Economic service life is the remaining useful life of an asset that results in the minimum annual equivalent cost.

In general, we need to consider economically how long an asset should be held once it is placed in service. For instance, a truck-rental firm that frequently purchases fleets of identical trucks may wish to arrive at a policy decision on how long to keep each vehicle before replacing it. If an appropriate lifespan is computed, a firm could stagger a schedule of truck purchases and replacements to smooth out annual capital expenditures for its overall truck purchases.

The costs of owning and operating an asset can be divided into two categories: **capital costs** and **operating costs**. Capital costs have two components: the initial investment and the salvage value at the time of disposal of the asset. The initial investment for the challenger is simply its purchase price. For the defender, we should treat the opportunity cost as its initial investment. We will use N to represent the length of time in years the asset will be kept, P to denote the initial investment, and S_N to designate the salvage value at the end of the ownership period of N years.

The annual equivalent of capital costs, which is called the capital recovery cost (see Section 5.5.3), over the period of N years can be calculated with the following equation:

$$CR(i) = P(A/P, i, N) - S_N(A/F, i, N).$$ (11.1)

Generally speaking, as an asset becomes older, its salvage value becomes smaller. As long as the salvage value is less than the initial cost, the capital recovery cost is a decreasing function of N. In other words, the longer we keep an asset, the lower the capital recovery cost becomes. If the salvage value is equal to the initial cost no matter how long the asset is kept, the capital recovery cost is constant.

As described earlier, the operating costs of an asset include operating and maintenance (O&M) costs, labour costs, material costs, and energy consumption costs. Labour costs, material costs, and energy costs are often constant for the same equipment from year to year if the usage of the equipment remains constant. However, O&M costs tend to increase as a function of the age of the asset. Because of the increasing trend of the O&M costs, the total operating costs of an asset usually increase as long as the asset ages. We use OC_n to represent the total operating costs in year n of the ownership period and $OC(i)$ to represent the annual equivalent of the operating costs over a lifespan of N years at interest rate i. Then $OC(i)$ can be expressed as

$$OC(i) = \left(\sum_{n=1}^{N} OC_n \, (P/F, i, n) \right)(A/P, i, N).$$ (11.2)

As long as the annual operating costs increase with the age of the equipment, $OC(i)$ is an increasing function of the life of the asset. If the annual operating costs are the same from year to year, $OC(i)$ is constant and equal to the annual operating costs, no matter how long the asset is kept.

The total annual equivalent costs of owning and operating an asset ($AEC(i)$) are a summation of the capital recovery costs and the annual equivalent of operating costs of the asset:

$$AEC(i) = CR(i) + OC(i).$$ (11.3)

The economic service life of an asset is defined to be the period of useful life that minimizes the annual equivalent costs of owning and operating the asset. On the basis of the foregoing discussions, we need to find the value of N that minimizes AEC as expressed in Eq. (11.3). If $CR(i)$ is a decreasing function of N and $OC(i)$ is an increasing function of N, as is often the case, AEC will be a convex function of N with a unique minimum point. (See Figure 11.3.) In this book, we assume that AEC has a unique minimum point. If the salvage value of the asset is constant and equal to the initial cost, and if the annual operating cost increases with time, then AEC is an increasing function of N and attains its minimum at $N = 1$. In this case, we should try to replace the asset as soon as possible. If, however, the annual operating cost is constant and the salvage value is less than the initial cost and decreases with time, then AEC is a decreasing function of N. In this case, we would try to delay the replacement of the asset as much as possible. Finally, if the salvage

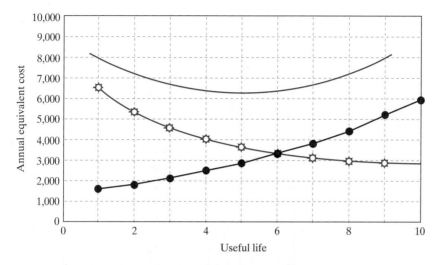

Figure 11.3 A schematic illustrating the trends of capital recovery cost (ownership cost), annual operating cost, and total annual equivalent cost.

value is constant and equal to the initial cost and the annual operating costs are constant, then AEC will also be constant. In this case, when to replace the asset does not make any economic difference.

If a new asset is purchased and operated for the length of its economic life, the annual equivalent cost is minimized. If we further assume that a new asset of identical price and features can be purchased repeatedly over an indefinite period, we would always replace this kind of asset at the end of its economic life. By replacing perpetually according to an asset's economic life, we obtain the minimum AEC stream over an indefinite period. However, if the identical-replacement assumption cannot be made, we will have to use the methods to be covered in Section 11.3 to carry out a replacement analysis. The next example explains the computational procedure for determining an asset's economic service life.

EXAMPLE 11.3 Economic Service Life of a Lift Truck

Suppose a company has a forklift, but is considering purchasing a new electric-lift truck that would cost $18,000, have operating costs of $1000 in the first year, and have a salvage value of $10,000 at the end of the first year. For the remaining years, operating costs increase each year by 15% over the previous year's operating costs. Similarly, the salvage value declines each year by 25% from the previous year's salvage value. The lift truck has a maximum life of seven years. An overhaul costing $3000 and $4500 will be required during the fifth and seventh years of service, respectively. The firm's required rate of return is 15%. Find the economic service life of this new machine.

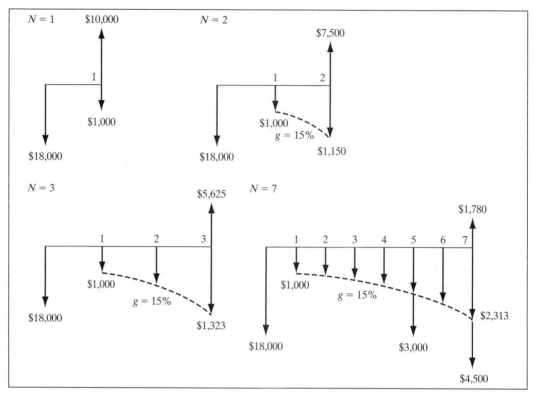

Figure 11.4 Cash flow diagrams for the options of keeping the asset for one year, two years, three years, and seven years (Example 11.3).

DISCUSSION: For an asset whose revenues are either unknown or irrelevant, we compute its economic life on the basis of the costs for the asset and its year-by-year salvage values. To determine an asset's economic service life, we need to compare the options of keeping the asset for one year, two years, three years, and so forth. The option that results in the lowest annual equivalent cost (AEC) gives the economic service life of the asset.

- **$N = 1$: One-year replacement cycle.** In this case, the machine is bought, used for one year, and sold at the end of year 1. The cash flow diagram for this option is shown in Figure 11.4. The annual equivalent cost for this option is

$$\text{AEC}(15\%) = \$18,000(A/P, 15\%, 1) + \$1000 - \$10,000$$
$$= \$11,700.$$

Note that $(A/P, 15\%, 1) = (F/P, 15\%, 1)$ and the annual equivalent cost is the equivalent cost at the end of year 1, since $N = 1$. Because we are calculating the annual equivalent cost in the computation of AEC(15%), we have treated cost items with a positive sign, while the salvage value has a negative sign.

- $N = 2$: **Two-year replacement cycle.** In this case, the truck will be used for two years and disposed of at the end of year 2. The operating cost in year 2 is 15% higher than that in year 1, and the salvage value at the end of year 2 is 25% lower than that at the end of year 1. The cash flow diagram for this option is also shown in Figure 11.4. The annual equivalent cost over the two-year period is

$$\text{AEC}(15\%) = [\$18{,}000 + \$1000(P/A_1, 15\%, 15\%, 2)](A/P, 15\%, 2)$$

$$-\$7500(A/F, 15\%, 2)$$

$$= \$8653.$$

- $N = 3$: **Three-year replacement cycle.** In this case, the truck will be used for three years and sold at the end of year 3. The salvage value at the end of year 3 is 25% lower than that at the end of year 2; that is, $\$7500(1 - 25\%) = \$5625$. The operating cost per year increases at a rate of 15%. The cash flow diagram for this option is also shown in Figure 11.4. The annual equivalent cost over the three-year period is

$$\text{AEC}(15\%) = [\$18{,}000 + \$1000(P/A_1, 15\%, 15\%, 3)](A/P, 15\%, 3)$$

$$- \$5625(A/F, 15\%, 3)$$

$$= \$7406.$$

- Similarly, we can find the annual equivalent costs for the options of keeping the asset for four, five, six, and seven years. One has to note that there is an additional cost of overhaul in year 5 and again in year 7. The cash flow diagram when $N = 7$ is shown in Figure 11.4. The computed annual equivalent costs for each of these options are

$$N = 4, \text{AEC}(15\%) = \$6678,$$

$$N = 5, \text{AEC}(15\%) = \$6642,$$

$$N = 6, \text{AEC}(15\%) = \$6258,$$

$$N = 7, \text{AEC}(15\%) = \$6394.$$

From the preceding calculated AEC values for $N = 1, \ldots, 7$, we find that AEC(15%) is smallest when $N = 6$. If the truck were to be sold after six years, it would have an annual cost of \$6258 per year. If it were to be used for a period other than six years, the annual equivalent costs would be higher than \$6258. Thus, a lifespan of six years for this truck results in the lowest annual cost. We conclude that the economic service life of the truck is six years. By replacing the assets perpetually according to an

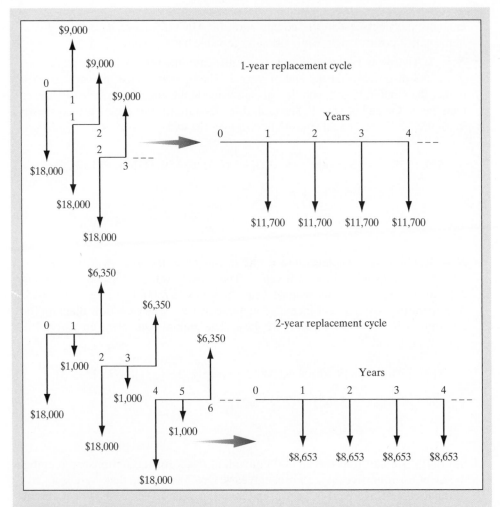

Figure 11.5 Conversion of an infinite number of replacement cycles to infinite AE cost streams (Example 11.3).

economic life of six years, we obtain the minimum annual equivalent cost stream. Figure 11.5 illustrates this concept. Of course, we should envision a long period of required service for this kind of asset.

11.3 Replacement Analysis When the Required Service Is Long

Now that we understand how the economic service life of an asset is determined, the next question is how to use these pieces of information to decide whether now is the time to replace the defender. If now is not the right time, when *is* the optimal time to replace the defender? Before presenting an analytical approach to answer this question, we consider several important assumptions.

11.3.1 Required Assumptions and Decision Frameworks

In deciding whether now is the time to replace the defender, we need to consider the following three factors:

- Planning horizon (study period).
- Technology.
- Relevant cash flow information.

Planning Horizon (Study Period)

By the planning horizon, we simply mean the service period required by the defender and a sequence of future challengers. The infinite planning horizon is used when we are simply unable to predict when the activity under consideration will be terminated. In other situations, it may be clear that the project will have a definite and predictable duration. In those cases, replacement policy should be formulated more realistically on the basis of a finite planning horizon.

Technology

Predictions of technological patterns over the planning horizon refer to the development of types of challengers that may replace those under study. A number of possibilities exist in predicting purchase cost, salvage value, and operating cost that are dictated by the efficiency of a new machine over the life of an existing asset. If we assume that all future machines will be the same as those now in service, we are implicitly saying that no technological progress in the area will occur. In other cases, we may explicitly recognize the possibility of machines becoming available in the future that will be significantly more efficient, reliable, or productive than those currently on the market. (Personal computers are a good example.) This situation leads to the recognition of technological change and obsolescence. Clearly, if the best available machine gets better and better over time, we should certainly investigate the possibility of delaying an asset's replacement for a couple of years—a viewpoint that contrasts with the situation in which technological change is unlikely.

Revenue and Cost Patterns Over the Life of an Asset

Many varieties of predictions can be used to estimate the patterns of revenue, cost, and salvage value over the life of an asset. Sometimes revenue is constant, but costs increase, while salvage value decreases, over the life of a machine. In other situations, a decline in revenue over the life of a piece of equipment can be expected. The specific situation will determine whether replacement analysis is directed toward cost minimization (with constant revenue) or profit maximization (with varying revenue). We formulate a replacement policy for an asset whose salvage value does not increase with age.

Decision Frameworks

To illustrate how a decision framework is developed, we indicate a replacement sequence of assets by the notation $(j_0, n_0), (j_1, n_1), (j_2, n_2), \ldots, (j_K, n_K)$. Each pair of numbers (j, n) indicates a type of asset and the lifetime over which that asset will be retained. The defender, asset 0, is listed first; if the defender is replaced now, $n_0 = 0$. A sequence of pairs may cover a finite period or an infinite period. For example, the sequence $(j_0, 2), (j_1, 5), (j_2, 3)$ indicates retaining the defender for two years, then replacing the defender with an asset of type j_1 and using it for five years, and then replacing j_1 with an asset of type j_2 and using it for three years. In this situation, the total planning horizon

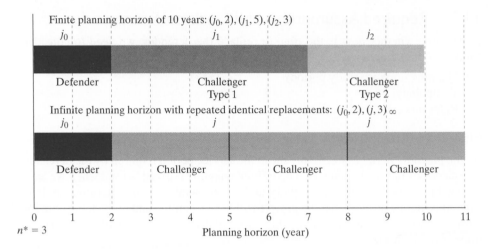

Figure II.6 Types of typical replacement decision frameworks.

covers 10 years $(2 + 5 + 3)$. The special case of keeping the defender for n_0 periods, followed by infinitely repeated purchases and the use of an asset of type j for n^* years, is represented by $(j_0, n_0), (j, n^*)_\infty$. This sequence covers an infinite period, and the relationship is illustrated in Figure 11.6.

Decision Criterion

Although the economic life of the defender is defined as the additional number of years of service that minimizes the annual equivalent cost (or maximizes the annual equivalent revenue), that is *not* necessarily the *optimal* time to replace the defender. The correct replacement time depends on data on the challenger, as well as on data on the defender.

As a decision criterion, the AE or AEC method provides a more direct solution when the planning horizon is infinite. When the planning horizon is finite, the PW or PEC method is more convenient to use. We will develop the replacement decision procedure for both situations. We begin by analyzing an infinite planning horizon without technological change. Even though a simplified situation such as this is not likely to occur in real life, the analysis of this replacement situation introduces methods that will be useful in analyzing infinite-horizon replacement problems with technological change.

II.3.2 Replacement Strategies Under the Infinite Planning Horizon

Consider a situation in which a firm has a machine that is in use in a process that is expected to continue for an indefinite period. Presently, a new machine will be on the market that is, in some ways, more effective for the application than the defender is. The problem is when, if at all, the defender should be replaced with the challenger.

Under the infinite planning horizon, the service is required for a very long time. Either we continue to use the defender to provide the service, or we replace the defender with the best available challenger for the same service requirement. In this case, we may apply the following procedure in replacement analysis:

1. Compute the economic lives of both the defender and the challenger. Let's use N_D^* and N_C^* to indicate the economic lives of the defender and the challenger, respectively. The annual equivalent costs for the defender and the challenger at their respective economic lives are indicated by AEC_D^* and AEC_C^*.

2. Compare AEC_D^* and AEC_C^*. If AEC_D^* is bigger than AEC_C^*, it is more costly to keep the defender than to replace it with the challenger. Thus, the challenger should replace the defender now. If AEC_D^* is smaller than AEC_C^*, it costs less to keep the defender than to replace it with the challenger. Thus, the defender should *not* be replaced now. The defender should continue to be used at least for the duration of its economic life if there are no technological changes over that life.

3. If the defender should not be replaced now, when should it be replaced? First we need to continue to use it until its economic life is over. Then we should calculate the cost of running the defender for one more year after its economic life. If this cost is greater than AEC_C^*, the defender should be replaced at the end of its economic life. Otherwise, we should calculate the cost of running the defender for the second year after its economic life. If this cost is bigger than AEC_C^*, the defender should be replaced one year after its economic life. The process should be continued until we find the optimal replacement time. This approach is called **marginal analysis**; that is, we calculate the incremental cost of operating the defender for just one more year. In other words, we want to see whether the cost of extending the use of the defender for an additional year exceeds the savings resulting from delaying the purchase of the challenger. Here, we have assumed that the best available challenger does not change.

Note that this procedure might be applied dynamically. For example, it may be performed annually for replacement analysis. Whenever there are updated data on the costs of the defender or new challengers available on the market, the new data should be used in the procedure. Example 11.4 illustrates the procedure.

EXAMPLE 11.4 Replacement Analysis Under an Infinite Planning Horizon

Advanced Electrical Insulator Company is considering replacing a broken inspection machine, which has been used to test the mechanical strength of electrical insulators, with a newer and more efficient one.

• If repaired, the old machine can be used for another five years, although the firm does not expect to realize any salvage value from scrapping it at that time. However, the firm can sell it now to another firm in the industry for $5000. If the machine is kept, it will require an immediate $1200 overhaul to restore it to operable condition. The overhaul will neither extend the service life originally estimated nor increase the value of the inspection machine. The operating costs are estimated at $2000 during the first year, and these are expected to increase by $1500 per year thereafter. Future market values are expected to decline by $1000 per year.

• The new machine costs $10,000 and will have operating costs of $2000 in the first year, increasing by $800 per year thereafter. The expected salvage value is $6000 after one year and will decline 15% each year. The company requires a rate of return of 15%. Find the economic life for each option, *and* determine when the defender should be replaced.

SOLUTION

1. **Economic service life:**
   • **Defender.** If the company retains the inspection machine, it is in effect deciding to overhaul the machine and invest the machine's current market value in that alternative. The opportunity cost of the machine is $5000. Because an overhaul costing $1200 is also needed to make the machine operational, the total initial investment in the machine is $5000 + $1200 = $6200. Other data for the defender are summarized as follows:

n	Overhaul	Forecasted Operating Cost	Market Value if Disposed of
0	$1200		$5000
1	0	$2000	$4000
2	0	$3500	$3000
3	0	$5000	$2000
4	0	$6500	$1000
5	0	$8000	0

We can calculate the annual equivalent costs if the defender is to be kept for one year, two years, three years, and so forth. For example, the cash flow diagram for $N = 4$ years is shown in Figure 11.7. The annual equivalent costs for four years are as follows:

$$N = 4 \text{ years: } \text{AEC}(15\%) = \$6200(A/P, 15\%, 4) + \$2000$$
$$+ \$1500(A/G, 15\%, 4) - \$1000(A/F, 15\%, 4)$$
$$= \$5961.$$

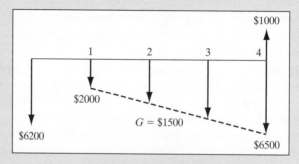

Figure 11.7 Cash flow diagram for defender when $N = 4$ years (Example 11.4).

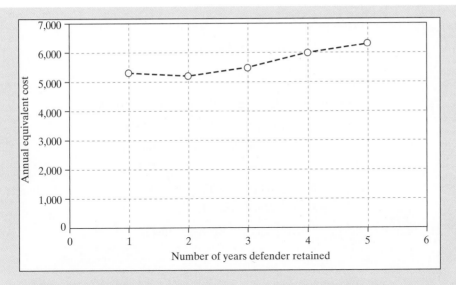

Figure 11.8 AEC as a function of the life of the defender (Example 11.4).

The other AEC figures can be calculated with the following equation:

$$\text{AEC}(15\%)_N = \$6200(A/P, 15\%, N) + \$2000 + \$1500(A/G, 15\%, N)$$
$$- \$1000(5 - N)(A/F, 15\%, N) \text{ for } N = 1, 2, 3, 4, 5;$$

$$N = 1: \text{AEC}(15\%) = \$5130,$$
$$N = 2: \text{AEC}(15\%) = \$5116,$$
$$N = 3: \text{AEC}(15\%) = \$5500,$$
$$N = 4: \text{AEC}(15\%) = \$5961,$$
$$N = 5: \text{AEC}(15\%) = \$6434.$$

When $N = 2$ years, we get the lowest AEC value. Thus, the defender's economic life is two years. Using the notation we defined in the procedure, we have

$$N_D{}^* = 2 \text{ years}$$
$$\text{AEC}_D{}^* = \$5116$$

The AEC values as a function of N are plotted in Figure 11.8. Actually, after computing AEC for $N = 1, 2,$ and 3, we can stop right there. There is no need to compute AEC for $N = 4$ and $N = 5$, because AEC is increasing when $N > 2$ and we have assumed that AEC has a unique minimum point.

• **Challenger.** The economic life of the challenger can be determined with the same procedure we used in this example for the defender and in Example 11.3. A summary of the general equation for calculating AEC for the challenger follows. You don't have to summarize such an equation when you need to determine the economic life of an asset, as long as you follow the procedure illustrated in Example 11.3. The equation is

$$\text{AEC}(15\%)_N = \$10{,}000(A/P, 15\%, N) + \$2000$$
$$+ \$800(A/G, 15\%, N)$$
$$- \$6000(1 - 15\%)^{N-1}(A/F, 15\%, N).$$

The results of "plugging in" the values and solving are as follows:

$$N = 1 \text{ year: AEC}(15\%) = \$7500,$$
$$N = 2 \text{ years: AEC}(15\%) = \$6151,$$
$$N = 3 \text{ years: AEC}(15\%) = \$5857,$$
$$N = 4 \text{ years: AEC}(15\%) = \$5826,$$
$$N = 5 \text{ years: AEC}(15\%) = \$5897.$$

The economic life of the challenger is four years; that is,

$$N_C{}^* = 4 \text{ years.}$$

Thus,

$$\text{AEC}_C{}^* = \$5826.$$

2. **Should the defender be replaced now?**
Since $\text{AEC}_D{}^* = \$5{,}116 < \text{AEC}_C{}^* = \$5826$, the defender should not be replaced now. If there are no technological advances in the next few years, the defender should be used for at least $N_D{}^* = 2$ more years. However, it is not necessarily best to replace the defender right at the high point of its economic life.

3. **When should the defender be replaced?**
If we need to find the answer to this question today, we have to calculate the cost of keeping and using the defender for the third year from today. That is, what is the cost of not selling the defender at the end of year 2, using it for the third year, and replacing it at the end of year 3? The following cash flows are related to this question:

(a) Opportunity cost at the end of year 2: equal to the market value then, or $3000.

(b) Operating cost for the third year: $5000.

(c) Salvage value of the defender at the end of year 3: $2000.

The following diagram represents these cash flows:

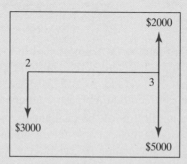

The cost of using the defender for one more year from the end of its economic life is

$$\$3000 \times 1.15 + \$5000 - \$2000 = \$6450.$$

Now compare this cost with the $AEC_C^* = \$5826$ of the challenger. It is greater than AEC_C^*. Thus, it is more expensive to keep the defender for the third year than to replace it with the challenger. Accordingly, we conclude that we should replace the defender at the end of year 2. If this one-year cost is still smaller than AEC_C^*, we need to calculate the cost of using the defender for the fourth year and then compare that cost with the AEC_C^* of the challenger.

In replacement analysis, it is common for a defender and its challenger to have different economic service lives. The annual-equivalent approach is frequently used, but it is important to know that we use the AEC method in replacement analysis, not because we have to deal with the problem of unequal service lives, but rather because the AEC approach provides some computational advantage for a special class of replacement problems.

In Chapter 6, we discussed the general principle for comparing mutually exclusive alternatives with unequal service lives. In particular, we pointed out that use of the AEC method relies on the concept of repeatability of projects and one of two assumptions: an infinite planning horizon or a common service period. In defender–challenger situations, however, repeatability of the defender cannot be assumed. In fact, by virtue of our definition of the problem, we are not repeating the defender, but replacing it with its challenger, an asset that in some way constitutes an improvement over the current equipment. Thus, the assumptions we made for using an annual cash flow analysis with unequal service life alternatives are not valid in the usual defender–challenger situation.

The complication—the unequal-life problem—can be resolved if we recall that the replacement problem at hand is not *whether* to replace the defender, but *when* to do so. When the defender is replaced, it will always be by the challenger—the best available equipment. An identical challenger can then replace the challenger repeatedly. In fact, we really are comparing the following two options in replacement analysis:

1. **Replace the defender now.** The cash flows of the challenger will be used from today and will be repeated because an identical challenger will be used if replacement becomes necessary again in the future. This stream of cash flows is equivalent to a cash flow of AEC_C^* each year for an infinite number of years.

2. **Replace the defender, say, x years later.** The cash flows of the defender will be used in the first x years. Starting in year $x + 1$, the cash flows of the challenger will be used indefinitely.

The annual-equivalent cash flows for the years beyond year x are the same for these two options. We need only to compare the annual-equivalent cash flows for the first x years to determine which option is better. This is why we can compare AEC_D^* with AEC_C^* to determine whether now is the time to replace the defender.

11.3.3 Replacement Strategies Under the Finite Planning Horizon

If the planning period is finite (for example, eight years), a comparison based on the AE method over a defender's economic service life does not generally apply. The procedure for solving such a problem with a finite planning horizon is to establish all "reasonable" replacement patterns and then use the PW or PEC value for the planning period to select the most economical pattern. To illustrate this procedure, consider Example 11.5.

EXAMPLE 11.5 Replacement Analysis Under the Finite Planning Horizon (PW Approach)

Consider again the defender and the challenger in Example 11.4. Suppose that the firm has a contract to perform a given service, using the current defender or the challenger for the next eight years. After the contract work, neither the defender nor the challenger will be retained. What is the best replacement strategy?

SOLUTION

Recall again the annual equivalent costs for the defender and challenger under the assumed holding periods (a boxed number denotes the minimum AEC value at $N_D^* = 2$ and $N_C^* = 4$, respectively):

	Annual Equivalent Cost ($)	
n	Defender	Challenger
1	5130	7500
2	5116	6151
3	5500	5857
4	5961	5826
5	6434	5897

Many ownership options would fulfill an eight-year planning horizon, as shown in Figure 11.9. Of these options, six appear to be the most likely by inspection. These options are listed, and the present equivalent cost for each option is calculated, as follows:

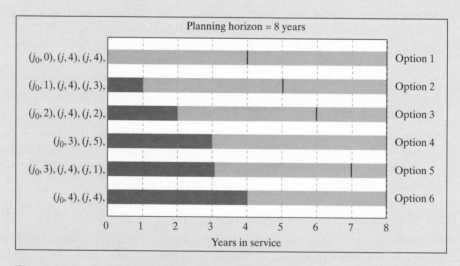

Figure 11.9 Some likely replacement patterns under a finite planning horizon of eight years (Example 11.5).

- **Option 1:** $(j_0, 0), (j, 4), (j, 4)$

$$PEC(15\%)_1 = \$5826(P/A, 15\%, 8)$$
$$= \$26,143.$$

- **Option 2:** $(j_0, 1), (j, 4), (j, 3)$

$$PEC(15\%)_2 = \$5130(P/F, 15\%, 1)$$
$$+ \$5826(P/A, 15\%, 4)(P/F, 15\%, 1)$$
$$+ \$5857(P/A, 15\%, 3)(P/F, 15\%, 5)$$
$$= \$25,573.$$

- **Option 3:** $(j_0, 2), (j, 4), (j, 2)$

$$PEC(15\%)_3 = \$5116(P/A, 15\%, 2)$$
$$+ \$5826(P/A, 15\%, 4)(P/F, 15\%, 2)$$
$$+ \$6151(P/A, 15\%, 2)(P/F, 15\%, 6)$$
$$= \$25,217 \leftarrow \text{minimum}.$$

- **Option 4:** $(j_0, 3), (j, 5)$

$$PEC(15\%)_4 = \$5500(P/A, 15\%, 3)$$
$$+ \$5897(P/A, 15\%, 5)(P/F, 15\%, 3)$$
$$= \$25,555.$$

- **Option 5:** $(j_0, 3), (j, 4), (j, 1)$

$$PEC(15\%)_5 = \$5500(P/A, 15\%, 3)$$
$$+ \$5826(P/A, 15\%, 4)(P/F, 15\%, 3)$$
$$+ \$7500(P/F, 15\%, 8)$$
$$= \$25,946.$$

- **Option 6:** $(j_0, 4), (j, 4)$

$$PEC(15\%)_6 = \$5961(P/A, 15\%, 4)$$
$$+ \$5826(P/A, 15\%, 4)(P/F, 15\%, 4)$$
$$= \$26,529.$$

An examination of the present equivalent cost of a planning horizon of eight years indicates that the least-cost solution appears to be Option 3: Retain the defender for two years, purchase the challenger and keep it for four years, and purchase another challenger and keep it for two years.

COMMENTS: In this example, we examined only six decision options that were likely to lead to the best solution, but it is important to note that several other possibilities have not been looked at. To explain, consider Figure 11.10, which shows a graphical representation of various replacement strategies under a finite planning horizon. For

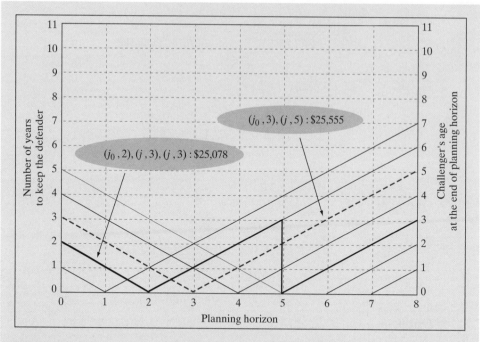

Figure 11.10 Graphical representations of replacement strategies under a finite planning horizon (Example 11.5).

example, the replacement strategy $[(j_0, 2), (j, 3), (j, 3)]$ (shown as a solid line in the figure) is certainly feasible, but we did not include it in the previous computation. Actually, this option gives an even lower PEC value. It's better than all six options listed earlier. Naturally, as we extend the planning horizon, the number of possible decision options can easily multiply. To make sure that we indeed find the optimal solution for such a problem, an optimization technique such as dynamic programming can be used.[3]

11.3.4 Consideration of Technological Change

Thus far, we have defined the challenger simply as the best available replacement for the defender. It is more realistic to recognize that the replacement decision often involves an asset now in use versus a candidate for replacement—that is, in some way, an improvement on the current asset. This, of course, reflects technological progress that is ongoing continually. Future models of a machine are likely to be more effective than a current model. In most areas, technological change appears as a combination of gradual advances in effectiveness; the occasional technological breakthrough, however, can revolutionize the character of a machine.

The prospect of improved future challengers makes a current challenger a less desirable alternative. By retaining the defender, we may have an opportunity to acquire an improved challenger later. If this is the case, the prospect of improved future challengers may affect a current decision between a defender and its challenger. It is difficult to forecast future technological trends in any precise fashion. However, in developing a long-term replacement policy, we need to take technological change into consideration.

[3] F. S. Hillier and G. S. Lieberman, *Introduction to Operations Research,* 8th ed. (New York: McGraw-Hill, 2005).

11.4 Replacement Analysis With Tax Considerations

Up to this point, we have covered various concepts and techniques that are useful in replacement analysis in general. In this section, we illustrate how to use those concepts and techniques to conduct replacement analysis on an after-tax basis.

To apply the concepts and methods covered in Sections 11.1 through 11.3 in an after-tax comparison of defender and challenger, we have to incorporate the tax effects (gains or losses) whenever an asset is disposed of. Whether the defender is kept or the challenger is purchased, we also need to incorporate the tax effects of capital cost allowances into our analysis.

Replacement studies require a knowledge of the depreciation schedule and of taxable gains or losses at disposal of the asset. Note that the depreciation schedule is determined at the time the asset is acquired, whereas the relevant tax law determines the gains from tax effects at the time of disposal. In this section, we will use the same examples (Example 11.1 through 11.4) to illustrate how to do the following analyses on an after-tax basis:

1. Calculate the net proceeds due to disposal of the defender (Example 11.6).
2. Use the opportunity cost approach in comparing defender and challenger (Example 11.7).
3. Calculate the economic life of the defender or the challenger (Example 11.8).
4. Conduct replacement analysis under the infinite planning horizon (Example 11.9).

EXAMPLE 11.6 Net Proceeds From the Disposal of an Old Machine

Suppose that, in Example 11.1, the $20,000 capital expenditure has a CCA rate of 20%. If the firm's marginal income tax rate is 40%, determine the taxable gains (or losses) and the net proceeds from disposal of the old printing machine.

SOLUTION

First we need to find the undepreciated capital cost of the old printing machine today. Applying the 50% rule to the original cost and using $d = 20\%$, we have

$$U = \$20,000 \left(1 - \frac{20\%}{2} \right)(1 - 20\%) = \$14,400,$$

so we compute the following:

Undepreciated capital cost	=	$14,400
Current market value	=	$10,000
Losses	=	$4,400
Tax savings = $4,400(0.40)	=	$1,760
Net proceeds from the sale	=	$10,000 + $1,760
	=	$11,760

This calculation is illustrated in Figure 11.11.

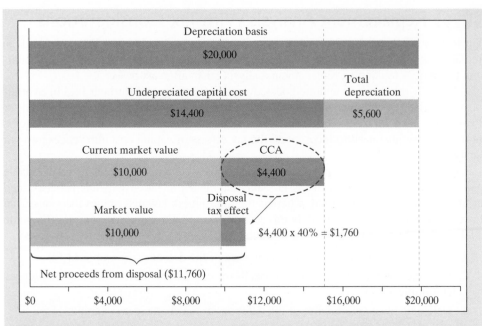

Figure II.II Net proceeds from the sale of the old printing machine—defender (Example 11.6).

EXAMPLE II.7 Replacement Analysis Using the Opportunity Cost Approach

Suppose, in Example 11.2, that the new machine has the same CCA rate of 20% as the old machine. The firm's after-tax interest rate (MARR) is 12%, and the marginal income tax rate is 40%. Use the opportunity cost approach to decide whether replacement of the old machine is justified now.

SOLUTION

Tables 11.1 and 11.2 show the worksheet formats the company uses to analyze a typical replacement project with the generalized cash flow approach. Each line is numbered, and a line-by-line description of the table follows:

- **Option 1: Keep the defender.**

    Lines 1–4: If the old machine is kept, the CCA schedule for the next three years would be $2880, $2304, and $1843. The undepreciated capital cost at the end of year 3 is $7373.

    Lines 5–6: Repair costs in the amount of $5000 were already incurred before the replacement decision. This is a sunk cost and should not be considered in the analysis. If a repair in the amount of $5000 is required to keep the defender in serviceable condition, it will show as an expense in year 0. If the old machine is

TABLE 11.1 Replacement Worksheet: Option 1—Keep the Defender (Example 11.7)

n	−2	−1	0	1	2	3
Financial data						
(cost information):						
(1) CCA		$2,000	$ 3,600	$2,880	$2,304	$1,843
(2) Undepreciated capital cost	$20,000	$18,000	$14,400	$11,520	$9,216	$7,373
(3) Salvage value						$2,500
(4) Loss from sale						−$4,873
(5) Repair cost		$5,000				
(6) O&M costs				$8,000	$8,000	$8,000
Cash flow statement:						
(7) Opportunity cost			−$11,760			
(8) Net proceeds from sale						$4,449
(9) −O&M cost × (0.6)				−$4,800	−$4,800	−$4,800
(10) +CCA × (0.4)				$1,152	$ 922	$ 737
(11) Net cash flow			−$11,760	−$3,648	−$3,878	$ 386

Note: The highlighted data represent sunk costs.

TABLE 11.2 Replacement Worksheet: Option 2—Replace the Defender (Example 11.7)

n	−2	−1	0	1	2	3
Financial data						
(cost information):						
(1) Cost of new printer			$15,000			
(2) CCA				$ 1,500	$2,700	$2,160
(3) Undepreciated capital cost				$13,500	$10,800	$8,640
(4) Salvage value						$5,500
(5) Loss from sale						$3,140
(6) O&M costs				$6,000	$6,000	$6,000
Cash flow statement:						
(7) Investment cost			−$15,000			
(8) Net proceeds from sale						$6,756
(9) −O&M cost × (0.6)				−$3,600	−$3,600	−$3,600
(10) +CCA × (0.4)				$600	$1,080	$864
(11) Net cash flow			−$15,000	−$3,000	−$2,520	$4,020

retained for the next three years, the before-tax annual O&M costs are as shown in Line 6.

Lines 7–11: Recall that the CCA allowances result in a tax reduction equal to the CCA amount multiplied by the tax rate. The operating expenses are multiplied by the factor of $(1 - $ the tax rate$)$ to obtain the after-tax O&M. For a situation in which the asset is retained for three years, Table 11.1 summarizes the cash flows obtained by using the generalized cash flow approach.

- **Option 2: Replace the defender.**

  Line 1: The purchase price of the new machine, including installation and freight charges, is listed in Table 11.2.

  Line 2: The CCA schedule, along with the undepreciated capital cost for the new machine $(d = 20\%)$, is shown.

  Line 5: With the salvage value estimated at $5500, we expect a loss ($3140 = $8640 - 5500) on the sale of the new machine at the end of year 3.

  Line 6: The O&M costs for the new machine are listed.

If the decision to keep the defender had been made, the opportunity cost approach would treat the $11,760 current net market value of the defender as an investment cost. Figure 11.12 illustrates the cash flows related to these decision options.

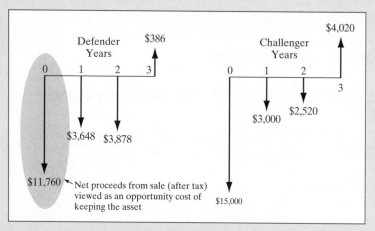

Figure II.12 Comparison of defender and challenger on the basis of the opportunity cost approach (Example 11.7).

Since the lifetimes of the defender and challenger are the same, we can use either PEC or AEC analysis as follows:

$$\text{PEC}(12\%)_{\text{Old}} = \$11,760 + \$3648(P/F, 12\%, 1) + \$3878(P/F, 12\%, 2)$$

$$- \$386(P/F, 12\%, 3)$$

$$= \$17,834;$$

$$\text{AEC}(12\%)_{\text{Old}} = \$17,834(A/P, 12\%, 3)$$

$$= \$7425;$$

$$PEC(12\%)_{New} = \$15,000 + \$3000(P/F, 12\%, 1)$$
$$+ \$2520(P/F, 12\%, 2) - \$4020(P/F, 12\%, 3)$$
$$= \$16,826;$$
$$AEC(12\%)_{New} = \$16,826(A/P, 12\%, 3)$$
$$= \$7006.$$

Because of the annual difference of $419 in favour of the challenger, the replacement should be made now.

COMMENTS: Recall that we assumed the same service life for both the defender and the challenger. In general, however, old equipment has a relatively short remaining life compared with new equipment, so that assumption is too simplistic. When the defender and challenger have unequal lifetimes, we must make an assumption in order to obtain a common analysis period. A typical assumption is that, after the initial decision, we make **perpetual replacements** with assets similar to the challenger. Certainly, we can still use PEC analysis with actual cash flows, but that would require evaluating *infinite* cash flow streams.

EXAMPLE 11.8 Economic Service Life of a Lift Truck

Consider again Example 11.3, but with the following additional data: The asset belongs to Class 43 property with a CCA rate of 30%. The firm's marginal tax rate is 40%, and its after-tax MARR is 15%. Find the economic service life of this new machine.

DISCUSSION: To determine an asset's economic service life, we first list the gains or losses that will be realized if the truck were to be disposed of at the end of each operating year. In doing so, we need to compute the undepreciated capital cost at the end of each operating year, assuming that the asset would be disposed of at that time. As summarized in Table 11.3, these values provide a basis for identifying the relevant after-tax cash flows at the end of an assumed operating period.

SOLUTION

Two approaches may be used to find the economic life of an asset: (a) the generalized cash flow approach and (b) the tabular approach.

(a) Generalized cash flow approach:
   Since we have only a few cash flow elements (O&M, CCA, and salvage value), an efficient way to obtain the after-tax cash flow is to use the generalized cash flow approach discussed in Section 10.4. Table 11.4 summarizes the cash flows for two-year ownership obtained by using the generalized cash flow approach.
   If we use the expected operating costs and the salvage values from Table 11.3, we can continue to generate yearly after-tax entries for the asset's remaining

TABLE 11.3 Forecasted Operating Costs and Net Proceeds From Sale as a Function of Holding Period (Example 11.8)

Holding Period	O&M	Permitted Annual Capital Cost Allowances over the Holding Period							Total CCA	Undepreciated Capital Cost	Expected Market Value	Disposal Tax Effect	Net A/T Salvage Value
		1	2	3	4	5	6	7					
1	$1,000	$2,700							$ 2,700	$15,300	$10,000	$2,120	$12,120
2	$1,150	$2,700	$4,590						$ 7,290	$10,710	$ 7,500	$1,284	$ 8,784
3	$1,323	$2,700	$4,590	$3,213					$10,503	$ 7,497	$ 5,625	$ 749	$ 6,374
4	$1,521	$2,700	$4,590	$3,213	$2,249				$12,752	$ 5,248	$ 4,219	$ 412	$ 4,630
5	$4,749	$2,700	$4,590	$3,213	$2,249	$1,574			$14,326	$ 3,674	$ 3,164	$ 204	$ 3,368
6	$2,011	$2,700	$4,590	$3,213	$2,249	$1,574	$1,102		$15,429	$ 2,571	$ 2,373	$ 79	$ 2,452
7	$6,813	$2,700	$4,590	$3,213	$2,249	$1,574	$1,102	$771	$16,200	$ 1,800	$ 1,780	$ 8	$ 1,788

Note: Asset price of $18,000, CCA rate = 30%; in year 5, normal operating expense ($1,749) + overhaul ($3,000); in year 7, normal operating expense ($2,313) + another engine overhaul ($4,500).

TABLE 11.4 **After-Tax Cash Flow Calculation for Owning and Operating the Asset for Two Years (Example 11.10)**

n	0	1	2
Financial data (cost information):			
(1) Cost of new printer	$18,000		
(2) CCA		$2,700	$4,590
(3) Undepreciated capital cost		$15,300	$10,710
(4) Salvage value			$7,500
(5) Gain (loss) from sale			−$3,210
(6) O&M costs		$1,000	$1,150
Cash flow statement:			
(7) Investment cost	−$18,000		
(8) Net proceeds from sale			$8,784
(9) −O&M cost × (0.6)		−$600	−$690
(10) +CCA × (0.4)		$1,080	$6,836
(11) Net cash flow	−$18,000	$480	$9,930

physical life. For the first two operating years, we compute the equivalent annual costs of owning and operating the asset as follows:

- $n = 1$, *one-year replacement cycle:*

$$\text{AEC}(15\%) = \left\{ \$18,000 + \begin{bmatrix} (0.6)(\$1,000) - (0.4)(\$2,700) \\ -\$12,120 \end{bmatrix} \right.$$

$$\left. (P/F, 15\%, 1) \right\} (A/P, 15\%, 1)$$

$$= \$7,043(1.15)$$
$$= \$8100.$$

- $n = 2$, *two-year replacement cycle:*

$$\text{AEC}(15\%) = \left\{ \begin{matrix} \$18,000 + [0.6(\$1,000) - 0.4(\$2,700)](P/F, 15\%, 1) \\ + [0.6(\$1,150) - 0.4(\$4,590) + \$8,784](P/F, 15\%, 2) \end{matrix} \right\}$$

$$(A/P, 15\%, 2)$$

$$= \$10,074(0.6151)$$
$$= \$6,197.$$

Similarly, the annual equivalent costs for the subsequent years can be computed as shown in Table 11.5 (column 16). If the truck were to be sold after six years, it would have a minimum annual cost of $4,432 per year, and this is the life that is most favourable for comparison purposes. That is, by replacing the asset perpetually according to an economic life of six years, we obtain the minimum infinite annual equivalent cost stream. Figure 11.13 illustrates this concept. Of course,

TABLE 11.5 Tabular Calculation of Economic Service Life (Example 11.8)

(1) Holding Period N	(2) Market Value	(3) Undeprec Capital Cost	(4) A/T Market Value	(5) PE of Market Value	(6) O&M Cost	(7) A/T O&M Cost	(8) PE of O&M Cost	(9) Cum. PE of O&M Cost	(10) Capital Cost Allowance	(11) CCA Credit	(12) Cum. PE of CCA Credit	(13) Total PE Cost	(14) Capital Recovery Cost	(15) AE A/T Operating Cost	(16) Total AE Cost
0	18,000	18000	$18,000	$18,000	0	$ —	$ —	$ —							
1	10,000	15300	$12,120	$10,539	1000	$ 600	$ 522	$ 522	2700	1080	$ 939	$ 7,043	$8,580	$(480)	$8,100
2	7,500	10710	$ 8,784	$ 6,642	1150	$ 690	$ 522	$1,043	4590	1836	$2,327	$10,074	$6,987	$(790)	$6,197
3	5,625	7497	$ 6,374	$ 4,191	1323	$ 794	$ 522	$1,565	3213	1285	$3,172	$12,202	$6,048	$(704)	$5,344
4	4,219	5248	$ 4,630	$ 2,647	1521	$ 913	$ 522	$2,087	2249	900	$3,687	$13,753	$5,377	$(560)	$4,817
5	3,164	3674	$ 3,368	$ 1,674	4749	$2,849	$1,417	$3,504	1574	630	$4,000	$15,829	$4,870	$(148)	$4,722
6	2,373	2571	$ 2,452	$ 1,060	2011	$1,207	$ 522	$4,025	1102	441	$4,190	$16,775	$4,476	$ (44)	$4,432
7	1,780	1800	$ 1,788	$ 672	6813	$4,088	$1,537	$5,562	771	309	$4,307	$18,583	$4,165	$ 302	$4,467

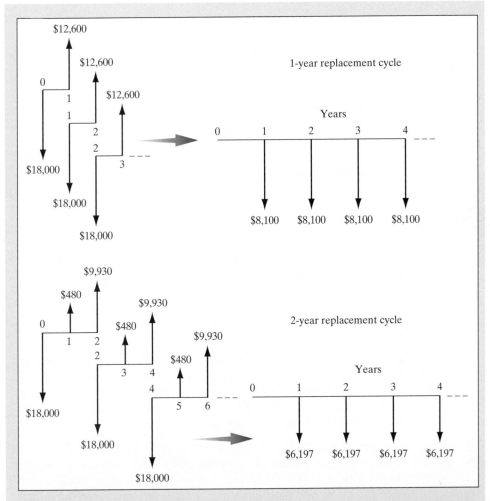

Figure 11.13 Conversion of an infinite number of replacement cycles to infinite AE cost streams (Example 11.8).

we should envision a long period of required service for the asset, its life no doubt being heavily influenced by market values, O&M costs, and CCA credits.

(b) Tabular approach:

The tabular approach separates the annual cost elements into two parts, one associated with the capital recovery of the asset and the other associated with operating the asset. In computing the capital recovery cost, we need to determine the after-tax salvage values at the end of each holding period, as calculated previously in Table 11.3. Then, we compute the total annual equivalent cost of the asset for any given year's operation using Eq. (11.3). The tabular approach for this example is shown in Table 11.5. The abbreviations in the

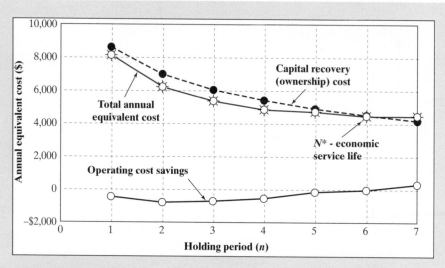

Figure 11.14 Economic service life obtained by finding the minimum AEC (Example 11.8). Note that we treat the cost items with a positive sign.

headings of the table include after-tax (A/T), undepreciated (Undeprec), present equivalent (PE), cumulative (Cum.), annual equivalent (AE), and capital cost allowance (CCA). A cumulative PE value gives the sum of the present equivalents of the cash flows from year 1 to the length of the holding period. A/T market value is also called net salvage value or net market value.

If we examine the annual equivalent costs itemized in Table 11.5 (columns 14 and 15), we see that, as the asset ages, the annual operating cost savings decrease except the first year due to the 50% rule. At the same time, capital recovery costs decrease with prolonged use of the asset. The combination of decreasing capital recovery costs and increasing annual operating costs results in the total annual equivalent cost taking on a form similar to that depicted in Figure 11.14. Even though an expensive overhaul is required during the fifth year of service, it is more economical to keep the equipment over a six-year life.

In this example, there is an equipment overhaul cost in year 5 and then in year 7. We have treated these overhaul costs as tax deductible expenses in the presented solutions. Whether such overhaul costs should be treated as capital expenditures depends on whether they extend the life of the equipment. If they do, they should be treated as capital expenditures, and capital cost allowances may be claimed during the useful life of these capital expenditures. Canada Revenue Agency regulations should be consulted to determine whether these overhaul costs are tax deductable expenses or must be treated as capital expenditures.

EXAMPLE 11.9 Replacement Analysis Under the Infinite Planning Horizon

Recall Example 11.4, in which Advanced Electrical Insulator Company is considering replacing a broken inspection machine. Let's assume the following additional data:

- The old machine was bought two years ago at a cost of $11,000. Its CCA rate is 30%. The machine could be used for another five years, but the firm does not expect to realize any salvage value from scrapping it in five years.
- The new machine has the same CCA rate as the old machine.

The marginal income tax rate is 40%, and the after-tax MARR is 15%. Find the useful life for each option presented in Example 11.4, *and* decide whether the defender should be replaced now or later.

SOLUTION

1. **Economic service life:**
   - **Defender.** The after-tax or net salvage values are calculated as follows:

Year	Salvage Value	Undepreciated Capital Cost	Disposal Tax Effect	Net Salvage Value
0	5000	6545	618	5618
1	4000	4582	233	4233
2	3000	3207	83	3083
3	2000	2245	98	2098
4	1000	1571	229	1229
5	–	1100	440	440

If the company retains the inspection machine, it is in effect deciding to overhaul the machine and invest the machine's current market value (after taxes) in that alternative. Again we assume that the overhaul costs are tax deductable right away at time 0. Although the company will make no physical cash flow transaction, it is withholding the market value of the inspection machine (the opportunity cost) from the investment. The after-tax O&M costs are as follows:

n	Overhaul	Forecasted O&M Cost	After-Tax O&M Cost
0	$1200		$1200(1 - 0.40) = $720
1	0	$2000	2000(1 - 0.40) = 1200
2	0	3500	3500(1 - 0.40) = 2100
3	0	5000	5000(1 - 0.40) = 3000
4	0	6500	6500(1 - 0.40) = 3900
5	0	8000	8000(1 - 0.40) = 4800

Using the current year's market value as the investment required to retain the defender, we obtain the results in Table 11.6, indicating that the remaining useful life of the defender is two more years with the lowest AEC value of $3408, *in the absence of future challengers.* The overhaul (repair) cost of $1200 in year 0 is treated as a deductible operating expense for tax purposes, as long as it does not add value

TABLE 11.6 Economics of Retaining the Defender for *N* More Years (Example 11.9)

(1) Holding Period N	(2) Market Value	(3) Undeprec Capital Cost	(4) A/T Market Value	(5) PE of Market Value	(6) O&M Cost	(7) A/T O&M Cost	(8) PE of O&M Cost	(9) Cum. PE of O&M Cost	(10) Capital Cost Allowance	(11) CCA Credit	(12) Cum. PE of CCA Credit	(13) Total PE Cost	(14) Capital Recovery Cost	(15) AE A/T Operating Cost	(16) Total AE A/T Cost
0	5,000	6545	$5,618	$5,618	1200	$ 720	$ 720	$ 720							
1	4,000	4582	$4,233	$3,681	2000	$1,200	$1,043	$1,763	$1,964	$785	$ 683	$ 3,018	$2,228	$1,243	$3,471
2	3,000	3207	$3,083	$2,331	3500	$2,100	$1,588	$3,351	$1,374	$550	$1,099	$ 5,540	$2,022	$1,386	$3,408
3	2,000	2245	$2,098	$1,379	5000	$3,000	$1,973	$5,324	$ 962	$385	$1,352	$ 8,211	$1,856	$1,740	$3,596
4	1,000	1571	$1,229	$ 702	6500	$3,900	$2,230	$7,554	$ 673	$269	$1,506	$10,964	$1,722	$2,118	$3,840
5	0	1100	$ 440	$ 219	8000	$4,800	$2,386	$9,940	$ 471	$189	$1,599	$13,740	$1,611	$2,488	$4,099

to the property. (Any repair or improvement expenses that increase the value of the property must be capitalized using the same CCA rate of the property.) In Table 11.6, we use only the net salvage value at time 0 and the net salvage at the end of year N to calculate the capital recovery cost for holding the asset for N years. The overhaul cost at time 0 is included in the annual equivalent after-tax operating cost calculations.

- **Challenger.** For the challenger, we also need to determine the undepreciated capital cost of the asset at the end of each period to compute the after-tax salvage value. This is done in Table 11.7. With the after-tax salvage values computed in that table, we are now ready to find the economic service life of the challenger by generating AEC value entries. These calculations are summarized in Table 11.8. The economic life of the challenger is four years, with an AEC(15%) value of $3946.

2. **Optimal time to replace the defender:** Since the AEC value for the defender's remaining useful life (two years) is $3408, which is less than $3946, the decision will be to keep the defender for now. Of course, the defender's remaining useful life of two years does not imply that the defender should actually be kept for two years before the company switches to the challenger. The reason for this is that the defender's remaining useful life of two years was calculated without considering what type of challenger would be available in the future. When a challenger's financial data are available, we need to enumerate all replacement timing possibilities. Since the defender can be used for another five years, six replacement strategies exist:

- Replace the defender with the challenger now.
- Replace the defender with the challenger in year 1.
- Replace the defender with the challenger in year 2.
- Replace the defender with the challenger in year 3.
- Replace the defender with the challenger in year 4.
- Replace the defender with the challenger in year 5.

The possible replacement cash patterns associated with each of these alternatives are shown in Figure 11.15, assuming that the costs and efficiency of the current challenger remain unchanged in future years. From the figure, we observe that, on an annual basis, the cash flows after the remaining physical life of the defender are the same.

Before we evaluate the economics of various replacement-decision options, recall the AEC values for the defender and the challenger under the assumed service lives (a boxed figure denotes the minimum AEC value at $n_0 = 2$ and $n^* = 4$):

	Annual Equivalent Cost	
n	Defender	Challenger
1	$3,471	$5,100
2	3,408	4,249
3	3,596	4,014
4	3,840	3,946
5	4,099	3,947
6		3,986
7		4,046

TABLE 11.7 Forecasted Operating Costs and Net Proceeds From Sale as a Function of Holding Period—Challenger (Example 11.9)

CCA over the Holding Periods							Total CCA	Undepreciated Capital Cost	Market Value	Tax Effect	Net Salvage Value
1	2	3	4	5	6	7					
$1,500							$1,500	$8,500	$6,000	$1,000	$7,000
$1,500	$2,550						$4,050	$5,950	$5,100	$ 340	$5,440
$1,500	$2,550	$1,785					$5,835	$4,165	$4,335	$ (68)	$4,267
$1,500	$2,550	$1,785	$1,250				$7,085	$2,916	$3,685	$ (308)	$3,377
$1,500	$2,550	$1,785	$1,250	$875			$7,959	$2,041	$3,132	$ (436)	$2,696
$1,500	$2,550	$1,785	$1,250	$875	$612		$8,571	$1,429	$2,662	$ (493)	$2,169
$1,500	$2,550	$1,785	$1,250	$875	$612	$429	$9,000	$1,000	$2,263	$ (505)	$1,758

Note: Asset price of $10,000 with a CCA rate of 30%.

TABLE 11.8 Economics of Owning and Operating the Challenger for N More Years (Example 11.9)

(1) Holding Period N	(2) Market Value	(3) Undeprec Capital Cost	(4) A/T Market Value	(5) PE of Market Value	(6) O&M Cost	(7) A/T O&M Cost	(8) PE of O&M Cost	(9) Cum. PE of O&M Cost	(10) Capital Cost Allowance	(11) CCA Credit	(12) Cum. PE of CCA Credit	(13) Total PE Cost	(14) Capital Recovery Cost	(15) AE A/T Operating Cost	(16) Total AE A/T Cost
0	**10,000**	10000	$10,000	$10,000	**0**	$ –	$ –	$ –							
1	**6,000**	8500	$ 7,000	$ 6,087	**2000**	$1,200	$1,043	$1,043	1500	600	$ 522	$ 4,435	$4,500	$ 600	$5,100
2	**5,100**	5950	$ 5,440	$ 4,113	**2800**	$1,680	$1,270	$2,314	2550	1020	$1,293	$ 6,907	$3,621	$ 628	$4,249
3	**4,335**	4165	$ 4,267	$ 2,806	**3600**	$2,160	$1,420	$3,734	1785	714	$1,762	$ 9,166	$3,151	$ 863	$4,014
4	**3,685**	2916	$ 3,377	$ 1,931	**4400**	$2,640	$1,509	$5,243	1250	500	$2,048	$11,264	$2,826	$1,119	$3,946
5	**3,132**	2041	$ 2,696	$ 1,340	**5200**	$3,120	$1,551	$6,795	875	350	$2,222	$13,232	$2,583	$1,364	$3,947
6	**2,662**	1429	$ 2,169	$ 938	**6000**	$3,600	$1,556	$8,351	612	245	$2,328	$15,085	$2,395	$1,591	$3,986
7	**2,263**	1000	$ 1,758	$ 661	**6800**	$4,080	$1,534	$9,885	429	171	$2,393	$16,832	$2,245	$1,801	$4,046

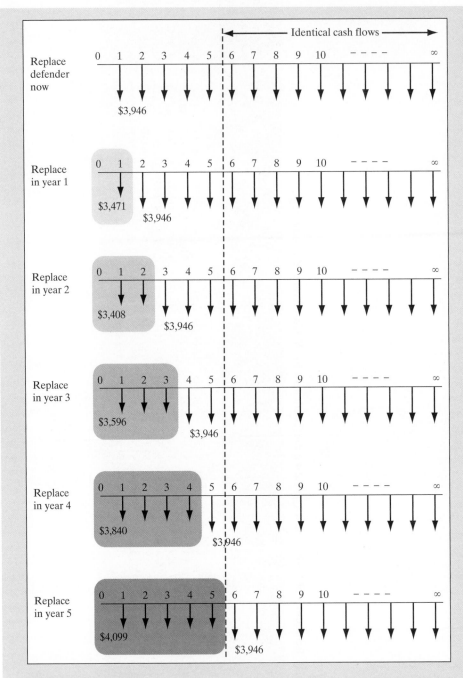

Figure 11.15 Equivalent annual cash flow streams when the defender is kept for *n* years followed by infinitely repeated purchases of the challenger every four years (Example 11.9).

Instead of using the marginal analysis in Example 11.4, we will use PEC analysis, which requires an evaluation of infinite cash flow streams. (The result is the same under both analyses.) Immediate replacement of the defender by the challenger is equivalent to computing the PEC for an infinite cost stream of $3946. If we use the capitalized equivalent-worth approach of Chapter 5 ($CE(i) = A/i$), we obtain

- $n = 0$:

$$\text{PEC}(15\%)_{n_0=0} = (1/0.15)(\$3946)$$
$$= \$26,307.$$

Suppose we retain the old machine n more years and then replace it with the new one. Now we will compute $\text{PEC}(i)_{n_0=n}$:

- $n = 1$:

$$\text{PEC}(15\%)_{n_0=1} = \$3471(P/A, 15\%, 1) + \$26,307(P/F, 15\%, 1)$$
$$= \$25,894.$$

- $n = 2$:

$$\text{PEC}(15\%)_{n_0=2} = \$3408(P/A, 15\%, 2) + \$26,307(P/F, 15\%, 2)$$
$$= \$25,432.$$

- $n = 3$:

$$\text{PEC}(15\%)_{n_0=3} = \$3596(P/A, 15\%, 3) + \$26,307(P/F, 15\%, 3)$$
$$= \$25,508.$$

- $n = 4$:

$$\text{PEC}(15\%)_{n_0=4} = \$3840(P/A, 15\%, 4) + \$26,307(P/F, 15\%, 4)$$
$$= \$26,004.$$

- $n = 5$:

$$\text{PEC}(15\%)_{n_0=5} = \$4099(P/A, 15\%, 5) + \$26,307(P/F, 15\%, 5)$$
$$= \$26,820.$$

This leads us to conclude that the defender should be kept for two more years. The PEC of $25,432 represents the net cost associated with retaining the defender for two years, replacing it with the challenger, and then replacing the challenger every four years for an indefinite period.

SUMMARY

■ In replacement analysis, the **defender** is an existing asset; the **challenger** is the best available replacement candidate.

■ The **current market value** is the value to use in preparing a defender's economic analysis. **Sunk costs**—past costs that cannot be changed by any future investment decision—should not be considered in a defender's economic analysis.

■ The basic approach to analyzing replacement problems is the **opportunity cost approach**. The opportunity cost approach views the net proceeds from the sale of the defender as an opportunity cost of keeping the defender. That is, instead of deducting the salvage value from the purchase cost of the challenger, we consider the net proceeds as an investment required to keep the asset.

■ The **economic service life** is the remaining useful life of a defender *or* a challenger that results in the minimum equivalent annual cost or maximum annual equivalent revenue. We should use the respective economic service lives of the defender and the challenger when conducting a replacement analysis.

■ Ultimately, in replacement analysis, the question is not *whether* to replace the defender, but *when* to do so. The AEC method provides a marginal basis on which to make a year-by-year decision about the best time to replace the defender. As a general decision criterion, the PEC method provides a more direct solution to a variety of replacement problems, with either an infinite or a finite planning horizon, or a technological change in a future challenger.

■ The role of **technological change** in improving assets should be evaluated in making long-term replacement plans: If a particular item is undergoing rapid, substantial technological improvements, it may be prudent to delay replacement (to the extent where the loss in production does not exceed any savings from improvements in future challengers) until a desired future model is available.

PROBLEMS

11.1 Columbus Electronics Company is considering replacing a half-tonne-capacity forklift truck that was purchased three years ago at a cost of $18,000. The diesel-operated forklift was originally expected to have a useful life of eight years and a zero estimated salvage value at the end of that period. The truck has not been dependable and is frequently out of service while awaiting repairs. The maintenance expenses of the truck have been rising steadily and currently amount to about $3000 per year. The truck could be sold for $7000. If retained, the truck will require an immediate $1500 overhaul to keep it in operating condition. This overhaul will neither extend the originally estimated service life nor increase the value of the truck. The updated annual operating costs, engine overhaul cost, and market values over the next five years are estimated as follows:

n	O&M	Engine Overhaul	Market Value
−3			
−2			
−1			
0		$1500	$7000
1	$3000		4000
2	3500		3000
3	3800		1500
4	4500		1000
5	4800	5000	0

A drastic increase in O&M costs during the fifth year is expected due to another overhaul, which will again be required to keep the truck in operating condition. The firm's MARR is 15%.

(a) If the truck is to be sold now, what will be its sunk cost?

(b) What is the opportunity cost of not replacing the truck now?

(c) What is the annual equivalent cost of owning and operating the truck for two more years?

(d) What is the annual equivalent cost of owning and operating the truck for five years?

11.2 Komatsu Cutting Technologies is considering replacing one of its CNC machines with one that is newer and more efficient. The firm purchased the CNC machine 10 years ago at a cost of $150,000. The machine had an expected economic life of 12 years at the time of purchase and an expected salvage value of $12,000 at the end of the 12 years. The original salvage estimate is still good, and the machine has a remaining useful life of 2 years. The firm can sell this old machine now to another firm in the industry for $35,000. A new machine can be purchased for $175,000, including installation costs. It has an estimated useful (economic) life of 8 years. The new machine is expected to reduce cash operating expenses by $30,000 per year over its 8-year life, at the end of which the machine is estimated to be worth only $5000. The company has a MARR of 12%.

(a) If you decided to retain the old machine, what is the opportunity (investment) cost of retaining the old asset?

(b) Compute the cash flows associated with retaining the old machine in years 1 to 2.

(c) Compute the cash flows associated with purchasing the new machine in years 1 to 8. (Use the opportunity cost concept.)

(d) If the firm needs the service of these machines for an indefinite period and no technology improvement is expected in future machines, what will be your decision?

11.3 Air Links, a commuter airline company, is considering replacing one of its baggage-handling machines with a newer and more efficient one. The old machine has a remaining useful life of five years. The salvage value expected from scrapping the old machine at the end of five years is zero, but the company can sell the machine now to another firm in the industry for $10,000. The new baggage-handling machine has a purchase price of $130,000 and an estimated useful life of seven years. It has an estimated salvage value of $30,000 and is expected to realize economic savings on electric power usage, labour, and repair costs and also to reduce the amount of damaged luggage. In total, an annual savings of $50,000 will be realized if the new machine is installed. The firm uses a 15% MARR. Using the opportunity cost approach,

(a) What is the initial cash outlay required for the new machine?

(b) What are the cash flows for the defender in years 0 to 5?

(c) Should the airline purchase the new machine?

11.4 Duluth Medico purchased a digital image-processing machine three years ago at a cost of $50,000. The machine had an expected life of eight years at the time of purchase and an expected salvage value of $5,000 at the end of the eight years. The old machine has been slow at handling the increased business volume, so management is considering replacing the machine. A new machine can be purchased for $85,000, including installation costs. Over its five-year life, the machine will reduce cash operating expenses by $30,000 per year. Sales are not expected to change. At the end of its useful life, the machine is estimated to be worthless. The old machine can be sold today for $10,000. The firm's interest rate for project justification is known to be 15%. The firm does not expect a better machine (other than the current challenger) to be available for the next five years. Assuming that the economic service life of the new machine, as well as the remaining useful life of the old machine, is five years,

(a) Determine the cash flows associated with each option (keeping the defender versus purchasing the challenger).

(b) Should the company replace the defender now?

11.5 The Northwest Manufacturing Company is currently manufacturing one of its products on a hydraulic stamping press machine. The unit cost of the product is $12, and 3000 units were produced and sold for $19 each during the past year. It is expected that both the future demand of the product and the unit price will remain steady at 3000 units per year and $19 per unit. The old machine has a remaining useful life of three years. The old machine could be sold on the open market now for $6500. Three years from now, the old machine is expected to have a salvage value of $1200. The new machine would cost $35,500, and the unit manufacturing cost on the new machine is projected to be $11. The new machine has an expected economic life of five years and an expected salvage value of $6300. The appropriate MARR is 12%. The firm does not expect a significant improvement in technology, and it needs the service of either machine for an indefinite period.

(a) Compute the cash flows over the remaining useful life of the old machine if the firm decides to retain it.

(b) Compute the cash flows over the economic service life if the firm decides to purchase the machine.

(c) Should the machine be acquired now?

11.6 A firm is considering replacing a machine that has been used to make a certain kind of packaging material. The new, improved machine will cost $32,000 installed and will have an estimated economic life of 10 years, with a salvage value of $2500. Operating costs are expected to be $1000 per year throughout the service life of the machine. The old machine (still in use) had an original cost of $25,000 four years ago, and at the time it was purchased, its service life (physical life) was estimated to be seven years, with a salvage value of $5000. The old machine has a current market value of $7900. If the firm retains the old machine, its updated market values and operating costs for the next four years will be as follows:

Year-End	Market Value	Operating Costs
0	$7900	
1	4300	$3200
2	3300	3700
3	1100	4800
4	0	5850

The firm's MARR is 12%.

(a) Working with the updated estimates of market values and operating costs over the next four years, determine the remaining economic life of the old machine.

(b) Determine whether it is economical to make the replacement now.

(c) If the firm's decision is to replace the old machine, when should it do so?

11.7 The University Resumé Service has just invested $7500 in a new desktop publishing system. From past experience, the owner of the company estimates its after-tax cash returns as

$$A_n = \$8000 - \$4000(1 + 0.15)^{n-1},$$

$$S_n = \$6000(1 - 0.3)^n,$$

where A_n stands for the net after-tax cash flows from operation of the system during period n and S_n stands for the after-tax salvage value at the end of period n.

(a) If the company's MARR is 12%, compute the economic service life of the system.

(b) Explain how the economic service life varies with the interest rate.

11.8 A special-purpose machine is to be purchased at a cost of $18,000. The following table shows the expected annual operating and maintenance cost and the salvage values for each year of the machine's service:

Year of Service	O&M Costs	Market Value
1	$2,500	$14,400
2	3,250	11,520
3	4,225	9,216
4	5,493	7,373
5	7,140	5,898

(a) If the interest rate is 10%, what is the economic service life for this machine?

(b) Repeat (a), using $i = 20\%$.

11.9 A special-purpose turnkey stamping machine was purchased four years ago for $21,000. It was estimated at that time that this machine would have a life of 10 years and a salvage value of $3000, with a removal cost of $1500. These estimates are still good. The machine has annual operating costs of $2000. A new machine that is more efficient will reduce the operating costs to $1000, but it will require an investment of $22,000, plus $1000 for installation. The life of the new machine is estimated to be 12 years, with a salvage of $2000 and a removal cost of $1500. An offer of $6000 has been made for the old machine, and the purchaser is willing to pay for removal of the machine. Find the economic advantage of replacing or of continuing with the present machine. State any assumptions that you make. (Assume that MARR = 8%).

11.10 A five-year-old defender purchased at $8000 has a current market value of $4000 and expected O&M costs of $3000 this year, increasing by $1500 per year. Future market values are expected to decline by $1000 per year. The machine can be used for another three years. The challenger costs $7000 and has O&M costs of $2000 per year, increasing by $1000 per year. The machine will be needed for only three years, and the salvage value at the end of that time is expected to be $2000. The MARR is 15%.

(a) Determine the annual cash flows for retaining the old machine for three years.

(b) Determine whether now is the time to replace the old machine. First show the annual cash flows for the challenger.

11.11 Greenleaf Company is considering purchasing a new set of air-electric quill units to replace an obsolete one. The machine currently being used for the operation has a market value of zero; however, it is in good working order, and it will last for at least an additional five years. The new quill units will perform the operation with so much more efficiency that the firm's engineers estimate that labour, material, and other direct costs will be reduced by $3000 a year if the units are installed. The new set of quill units costs $11,000, delivered and installed, and its economic life is estimated to be five years with zero salvage value. The firm's MARR is 10%.

(a) What investment is required to keep the old machine?

(b) Compute the cash flow to use in the analysis for each option.

(c) If the firm uses the internal-rate-of-return criterion, should the firm buy the new machine on that basis?

11.12 Wu Lighting Company is considering replacing an old, relatively inefficient vertical drill machine that was purchased 7 years ago at a cost of $11,000. The machine had an original expected life of 12 years and a zero estimated salvage value at the end of that period. The divisional manager reports that a new machine can be bought and installed for $15,000. Further, over its 5-year life, the machine will expand sales from $10,000 to $11,500 a year and will reduce the usage of labour and raw materials sufficiently to cut annual operating costs from $7000 to $5000. The new machine has an estimated salvage value of $2000 at the end of its 5-year life. The old machine's current market value is $2000; the firm's MARR is 15%.

(a) Should the new machine be purchased now?

(b) What current market value of the old machine would make the two options equal?

11.13 Advanced Robotics Company is faced with the prospect of replacing its old call-switching system, which has been used in the company's headquarters for 10 years. This particular system was installed at a cost of $100,000, and it was assumed that it would have a 15-year life with no appreciable salvage value. The current annual operating costs for the old system are $20,000, and these costs would be the same for the rest of its life. A sales representative from North Central Bell is trying to sell the company a computerized switching system that would require an investment of $220,000 for installation. The economic life of this computerized system is estimated to be 10 years, with a salvage value of $18,000, and the system will reduce annual operating costs to $5000. No detailed agreement has been made with the sales representative about the disposal of the old system. Determine the ranges of resale value associated with the old system that would justify installation of the new system at a MARR of 14%.

11.14 A company is currently producing chemical compounds by a process that was installed 10 years ago at a cost of $90,000. It was assumed that the process would have a 20-year life with a zero salvage value. The current market value of the process however, is $60,000, and the initial estimate of its economic life is still good. The annual operating costs associated with the process are $18,000. A sales representative from National Instrument Company is trying to sell a new chemical-compound-making process to the company. This new process will cost $180,000, have a service life of 10 years and a salvage value of $20,000, and reduce annual operating costs to $4000. Assuming that the company desires a return of 12% on all investments, should it invest in the new process?

11.15 Eight years ago, a lathe was purchased for $55,000. Its operating expenses were $8700 per year. An equipment vendor offers a new machine for $52,000. An allowance of $8500 would be made for the old machine when the new one is purchased. The old machine is expected to be scrapped at the end of five years. The new machine's economic service life is five years with a salvage value of $12,000. The new machine's O&M cost is estimated to be $4200 for the first year, increasing at an annual rate of $500 thereafter. The firm's MARR is 12%. What option would you recommend?

11.16 The Toronto Taxi Cab Company has just purchased a new fleet of models for the year 2010. Each brand-new cab cost $20,000. From past experience, the company estimates after-tax cash returns for each cab as

$$A_n = \$43,620 - 20,000(1 + 0.15)^{n-1},$$
$$S_n = \$20,000(1 - 0.15)^n,$$

where, again, A_n stands for net after-tax cash flows from the cab's operation during period n and S_n stands for the after-tax salvage value of the cab at the end of period n. The management views the replacement process as a constant and infinite chain.

(a) If the firm's MARR is 10% and it expects no major technological and functional change in future models, what is the optimal period (constant replacement cycle) to replace its cabs? (Ignore inflation.)

(b) What is the internal rate of return for a cab if it is retired at the end of its economic service life? What is the internal rate of return for a sequence of identical cabs if each cab in the sequence is replaced at the optimal time?

11.17 Four years ago, an industrial batch oven was purchased for $24,000. If sold now, the machine will bring $2000. If sold at the end of the year, it will bring $1500. Salvage value will decline by 20% each additional year. Annual operating costs for subsequent years are $3800. A new machine will cost $49,000 with a 12-year life and have a $3000 salvage value. The operating cost will be $3000 as of the end of each year, with the $6000-per-year savings due to better quality control. If the firm's MARR is 10%, should the machine be purchased now?

11.18 Vancouver Ceramic Company has an automatic glaze sprayer that has been used for the past 10 years. The sprayer can be used for another 10 years and will have a zero salvage value at that time. The annual operating and maintenance costs for the sprayer amount to $15,000 per year. Due to an increase in business, a new sprayer must be purchased. Vancouver Ceramic is faced with two options:

• **Option 1.** If the old sprayer is retained, a new smaller capacity sprayer will be purchased at a cost of $49,000, and it will have a $5000 salvage value in 10 years.

This new sprayer will have annual operating and maintenance costs of $12,000. The old sprayer has a current market value of $6000.

- **Option 2.** If the old sprayer is sold, a new sprayer of larger capacity will be purchased for $85,000. This sprayer will have a $9000 salvage value in 10 years and will have annual operating and maintenance costs of $24,000.

Which option should be selected at MARR = 12%?

11.19 The annual equivalent after-tax costs of retaining a defender machine over four years (physical life) or operating its challenger over six years (physical life) are as follows:

n	Defender	Challenger
1	$3250	$5850
2	2550	4280
3	2700	3250
4	3350	3550
5		4050
6		5550

If you need the service of either machine for only the next 10 years, what is the best replacement strategy? Assume a MARR of 12% and no improvements in technology in future challengers.

11.20 The after-tax annual equivalent worth of retaining a defender over four years (physical life) or operating its challenger over six years (physical life) are as follows:

n	Defender	Challenger
1	$13,450	$12,350
2	13,550	13,050
3	13,850	13,650
4	13,250	13,450
5		13,050
6		12,550

If you need the service of either machine for only the next eight years, what is the best replacement strategy? Assume a MARR of 12% and no improvements in technology in future challengers. Note that the numbers in the table are not AEC values. They are AE values.

11.21 An existing asset that cost $15,000 two years ago has a market value of $12,000 today, an expected salvage value of $2000 at the end of its remaining useful life of six more years, and annual operating costs of $4000. A new asset under consideration as a replacement has an initial cost of $13,000, an expected salvage value of $4000 at the end of its economic life of three years, and annual operating costs of $2000. It is assumed that this new asset could be replaced by another one identical in every respect after three years, if desired. Use a MARR of 11%, a six-year study period, and PEC calculations to decide whether the existing asset should be replaced by the new one.

11.22 Repeat Problem 11.21, using the AEC criterion.

11.23 Redo Problem 11.1, but with the following additional information: The asset is classified as a Class 43 property with a CCA rate of 30%. The firm's marginal tax rate is 40%.

11.24 Redo Problem 11.2, but with the following additional information: The asset is classified as a Class 43 property with a CCA rate of 30%. The firm's marginal tax rate is 40%.

11.25 Redo Problem 11.3, but with the following additional information:
  • The old machine was bought three years ago at $100,000.
  • Both the new and the old machines have a CCA rate of 30%.
  • The company's marginal tax rate is 40%.

11.26 Redo Problem 11.4, but with the following additional information:
  • Both the old and the new machines have a CCA rate of 20%.
  • The marginal tax rate is 35%.

11.27 Redo Problem 11.5, but with the following additional information:
  • The old stamping machine was bought four years ago at $50,000.
  • The CCA rate for both the new & the old machine is 30%.
  • The firm's marginal tax rate is 40%.

11.28 Redo Problem 11.6, but with the following additional information:
  • The CCA rate for both the new & the old machine is 30%.
  • The company's marginal tax rate is 35%.

11.29 A machine has a first cost of $10,000 with a CCA rate of 40%. End-of-year salvage values and annual O&M costs are provided over its useful life as follows:

Year End	Salvage Value	Operating Costs
1	$5300	$1500
2	3900	2100
3	2800	2900
4	1800	3800
5	1400	4800
6	600	5900

(a) Determine the economic life of the machine if the MARR is 15% and the marginal tax rate is 40%.

(b) Determine the economic life of the machine if the MARR is 5% and the marginal tax rate remains at 40%.

11.30 Given the data

$$P = \$20{,}000,$$

$$S_n = 12{,}000 - 2000n,$$

$$U_n = 20{,}000 - 2500n,$$

$$O\&M_n = 3{,}000 + 2000(n - 1), \text{ and}$$

$$t_m = 0.40,$$

where P = Asset purchase price,

S_n = Market value at the end of year n,

U_n = Undepreciated capital cost at the end of year n,

$O\&M_n$ = O&M cost during year n, and

t_m = Marginal tax rate,

(a) Determine the economic service life of the asset if $i = 10\%$.

(b) Determine the economic service life of the asset if $i = 25\%$.

(c) Assume that $i = 0$ and determine the economic service life of the asset mathematically (i.e., use the calculus technique for finding the minimum point, as described in Section 6.7 of Chapter 6).

11.31 Redo Problem 11.8, but with the following additional information:
  • For tax purposes, the CCA rate of the machine is 15%.
  • The firm's marginal tax rate is 40%.

11.32 Quintana Electronic Company is considering purchasing new robot-welding equipment to perform operations currently being performed by less efficient equipment. The new machine's purchase price is $160,000, delivered and installed. A Quintana industrial engineer estimates that the new equipment will produce savings of $30,000 in labour and other direct costs annually, compared to the current equipment. He estimates the proposed equipment's economic life at 10 years, with a zero salvage value. The current equipment, bought 7 years ago at $130,000 with a CCA rate of 30%, is in good working order and will last, physically, for at least 10 more years. Quintana Company expects to pay income taxes at 40%. Quintana uses a 10% discount rate for analysis performed on an after-tax basis. The new equipment has a CCA rate of 30%.

(a) Assuming that the current equipment has zero salvage value, should the company buy the proposed equipment?

(b) Assuming that the current equipment has a salvage value today of $45,000 and that, if the current equipment is retained for 10 more years, its salvage value will be zero, should the company buy the proposed equipment?

(c) Assume that the new equipment will save only $15,000 a year, but that its economic life is expected to be 12 years. If other conditions are as described in part (b), should the company buy the proposed equipment?

11.33 Quintana Company decided to purchase the equipment described in Problem 11.32 (hereafter called "Model A" equipment). Two years later, even better equipment (called "Model B") came onto the market, making Model A obsolete, with no re-sale value. The Model B equipment costs $400,000 delivered and installed, but it is expected to result in annual savings of $75,000 over the cost of operating the Model A equipment. The economic life of Model B is estimated to be 10 years, with a zero salvage value. (Model B also has a CCA rate of 30%.)

(a) What action should the company take?

(b) If the company decides to purchase the Model B equipment, a mistake must have been made because good equipment, bought only two years previously, is being scrapped. How did this mistake come about?

11.34 Redo Problem 11.9, but with the following additional information:
- The CCA rate for both the old machine and the new machine is 30%.
- The company's marginal tax rate is 30%.

11.35 Redo Problem 11.10, but with the following additional information:
- The CCA rate for both the defender and the challenger is 25%.
- The marginal tax rate is 40%.

11.36 Redo Problem 11.11, but with the following additional information:
- The current undepreciated capital cost of the old machine is $5,000.
- Both the old and new machines have a CCA rate of 30%.
- The company's marginal tax rate is 40%.

11.37 Redo Problem 11.12, but with the following additional information:
- Both machines have a CCA rate of 30%.
- The marginal tax rate is 40%.

11.38 Redo Problem 11.13, with the following additional information:
- The machines are Class 43 property with a CCA rate of 30%.
- The company's marginal tax rate is 40%.

11.39 Five years ago, a conveyor system was installed in a manufacturing plant at a cost of $35,000. It was estimated that the system, which is still in operating condition, would have a useful life of eight years, with a salvage value of $3000. If the firm continues to operate the system, its market values and operating costs for the next three years are updated as follows:

Year-End	Market Value	Operating Cost
0	$8,000	
1	5,200	$ 6,000
2	3,500	7,000
3	1,200	10,000

A new system can be installed for $42,500; it would have an estimated economic life of 10 years, with a salvage value of $3500. Operating costs are expected to be

$1,500 per year throughout the service life of the system. The firm's MARR is 18%. The conveyor systems are Class 43 property with a CCA rate of 30%. The firm's marginal tax rate is 35%.

(a) Decide whether to replace the existing system now.

(b) If the decision is to replace the existing system, when should replacement occur?

11.40 Redo Problem 11.14, but with the following additional information:
- Assume that the CCA rate for this chemical process is 20%.
- The marginal tax rate is 40%.

11.41 Redo Problem 11.15, but with the following additional information:
- The CCA rate is 30%.
- The marginal tax rate if 35%.

11.42 Redo Problem 11.17, but with the following additional information:
- The CCA rate is 30%.
- The marginal tax rate is 40%.

11.43 Redo Problem 11.18, but with the following additional information:
- **Option 1:** The old sprayer has an undepreciated capital cost of $4000. Both have a CCA rate of 30%.
- **Option 2:** The larger capacity sprayer also has a CCA rate of 30%.
- The company's marginal tax rate is 40%.

11.44 A six-year-old computer numerical control (CNC) machine originally cost $9000, and its current market value is $1500. If the machine is kept in service for the next five years, its O&M costs and salvage value are estimated as follows:

End of Year	O&M Costs		Salvage Value
	Operation and Repairs	Delays Due to Breakdowns	
1	$1300	$600	$1200
2	1500	800	1000
3	1700	1000	500
4	1900	1200	0
5	2000	1400	0

It is suggested that the machine be replaced by a new CNC machine of improved design at a cost of $5500. It is believed that this purchase will completely eliminate breakdowns and the resulting cost of delays and that operation and repair costs will be reduced $200 a year from what they would be with the old machine. Assume a five-year life for the challenger and a $1000 terminal salvage value. The CCA rate for both machines is 30%. The firm's MARR is 12%, and its marginal tax rate is 30%. Should the old machine be replaced now?

11.45 Redo Problem 11.21, but with the following additional information:
- The CCA rate for both assets is 25%.
- The marginal tax rate is 30%.

Short Case Studies

ST11.1 Chevron Overseas Petroleum Inc. entered into a 1993 joint venture with the Republic of Kazakhstan, a former republic of the old Soviet Union, to develop the huge Tengiz oil field.[4] Unfortunately, the climate in the region is harsh, making it difficult to keep oil flowing. The untreated oil comes out of the ground at 114°F. Even though the pipelines are insulated, as the oil gets further from the well on its way to be processed, hydrate salts begin to precipitate out of the liquid phase as the oil cools. These hydrate salts create a dangerous condition by forming plugs in the line.

The method for preventing this trap pressure condition is to inject methanol (MeOH) into the oil stream. This keeps the oil flowing and prevents hydrate salts from precipitating out of the liquid phase. The present methanol loading and storage facility is a completely manual controlled system, with no fire protection and with a rapidly deteriorating tank that causes leaks. The scope of repairs and upgrades is extensive. The storage tanks are rusting and are leaking at their riveted joints. The manual control system causes frequent tank overfills. There is no fire protection system, as water is not available at the site.

The present storage facility has been in service for five years with an initial installation cost of $250,000. Permit requirements mandate upgrades to achieve minimum acceptable Kazakhstan standards. Upgrades in the amount of $104,000 will extend the life of the current facility to about 10 years. However, upgrades will not completely stop the leaks. The expected spill and leak losses will amount to $5000 a year. The annual operating costs are expected to be $40,000.

As an alternative to the old facility, a new methanol storage facility can be designed on the basis of minimum acceptable international oil industry practices. The new facility, which would cost $325,000, would last about 12 years before a major upgrade would be required. However, it is believed that oil transfer technology will be such that methanol will not be necessary in 10 years. The pipeline heating and insulation systems will make methanol storage and use systems obsolete. With a lower risk of leaks, spills, and evaporation loss and a more closely monitored system, the expected annual operating cost would be $10,000. Assume a tax rate of 30%.

(a) Assume that the storage tanks (the new ones as well as the upgraded ones) will have no salvage value at the end of their useful lives (after considering the removal costs) and that the tanks will be depreciated by the straight-line method with $S = 0$ and $N = 10$ according to Kazakhstan's tax law. If Chevron's interest rate is 20% for foreign projects, which option is a better choice?

---

[4] This example was provided by Mr. Joel M. Haight of the Chevron Oil Company.

(b) How would the decision change as you consider the risk of spills (resulting in cleanup costs) and the evaporation of the product having an environmental impact?

ST11.2 National Woodwork Company, a manufacturer of window frames, is considering replacing a conventional manufacturing system with a flexible manufacturing system (FMS). The company cannot produce rapidly enough to meet demand. One manufacturing problem that has been identified is that the present system is expected to be useful for another five years, but will require an estimated $105,000 per year in maintenance, which will increase $10,000 each year as parts become more scarce. The current market value of the existing system is $140,000. The present system has an undepreciated capital cost of $110,000 with a CCA rate of 30%.

The proposed system will reduce or entirely eliminate setup times, and each window can be made as it is ordered by the customer, who phones the order into the head office, where details are fed into the company's main computer. These manufacturing details are then dispatched to computers on the manufacturing floor, which are, in turn, connected to a computer that controls the proposed FMS. This system eliminates the warehouse space and material-handling time that are needed when the conventional system is used.

Before the FMS is installed, the old equipment will be removed from the job shop floor at an estimated cost of $100,000. This cost includes needed electrical work on the new system. The proposed FMS will cost $2,200,000. The economic life of the machine is expected to be 10 years, and the salvage value is expected to be $120,000. The change in window styles has been minimal in the past few decades and is expected to continue to remain stable in the future. The proposed equipment has a CCA rate of 30%. The total annual savings will be $664,243: $12,000 attributed to a reduction in the number of defective windows, $511,043 from the elimination of 13 workers, $100,200 from the increase in productivity, and $41,000 from the near elimination of warehouse space and material handling. The O&M costs will be only $45,000, increasing by $2000 per year. The National Woodwork's MARR is about 15%, and the expected marginal tax rate over the project years is 40%.

(a) What assumptions are required to compare the conventional system with the FMS?

(b) With the assumptions defined in (a), should the FMS be installed now?

ST11.3 In 2 × 4 and 2 × 6 lumber production, significant amounts of wood are present in sideboards produced after the initial cutting of logs. Instead of processing the sideboards into wood chips for the paper mill, Union Camp Company uses an "edger" to reclaim additional lumber, resulting in savings for the company. An edger is capable of reclaiming lumber by any of the following three methods: (1) removing rough edges, (2) splitting large sideboards, and (3) salvaging 2 × 4 lumber from low-quality 4 × 4 boards. Union Camp Company's engineers have discovered that a significant reduction in production costs could be achieved simply by replacing the original edger with a newer, laser-controlled model.

The old edger was placed in service eight years ago at a cost of $200,000. Any machine scrap value would offset the removal cost of the equipment. No market exists for this obsolete equipment. The old edger needs two operators. During the cutting operation, the operator makes edger settings, using his or her judgment. The operator has no means of determining exactly what dimension of lumber could be recovered from a given sideboard and must guess at the proper setting to recover the highest grade of lumber. Furthermore, the old edger is not capable of salvaging good-quality 2 × 4's from poor-quality 4 × 4's. The defender can continue in service for another five years with proper maintenance. Following are the financial data for the old edger:

Current market value	$0
Annual maintenance cost	$2500 in year 1, increasing at a rate of 15% each year over the previous year's cost
Annual operating costs (labour and power)	$65,000

The new edger has numerous advantages over its defender, including laser beams that indicate where cuts should be made to obtain the maximum yield by the edger. The new edger requires a single operator, and labour savings will be reflected in lower operating and maintenance costs of $35,000 a year. The following gives the estimated costs and the depreciation methods associated with the new edger:

Estimated Cost	
Equipment	$ 55,700
Equipment installation	21,500
Building	47,200
Conveyor modification	14,500
Electrical (wiring)	16,500
Subtotal	$155,400
Engineering	7,000
Construction management	20,000
Contingency	16,200
Total	$198,600
Useful life of new edger	10 years
Salvage value	
Building (tear down)	$0
Equipment	10% of the original cost
Annual O&M costs	$35,000

	CCA Rates	
Building	10%	
Equipment and installation	30%	

Twenty-five percent of the total mill volume passes through the edger. A 12% improvement in yield is expected to be realized with the new edger, resulting in an improvement of $(0.25)(0.12) = 3\%$ in the total mill volume. The annual savings due to the improvement in productivity is thus expected to be $57,895.

(a) Should the defender be replaced now if the mill's MARR and marginal tax rate are 16% and 40%, respectively?

(b) If the defender will eventually be replaced by the current challenger, when is the optimal time to perform the replacement?

ST11.4 Rivera Industries, a manufacturer of home heating appliances, is considering purchasing an Amada Turret Punch Press, a more advanced piece of machinery, to replace its present system that uses four old presses. Currently, the four smaller presses are used (in varying sequences, depending on the product) to produce one component of a product until a scheduled time when all machines must retool to set up for a different component. Because the setup cost is high, production runs of individual components take a long time and result in large inventory buildups of one component. These buildups are necessary to prevent extended backlogging while other products are being manufactured.

The four presses in use now were purchased six years ago at a price of $120,000. The manufacturing engineer expects that these machines can be used for eight more years, but they will have no market value after that. The presses have a CCA rate of 30%. Their present market value is estimated to be $40,000. The average setup cost, which is determined by the number of labour hours required times the labour rate for the old presses, is $80 per hour, and the number of setups per year expected by the production control department is 200, yielding a yearly setup cost of $16,000. The expected operating and maintenance cost for each year in the remaining life of the old system is estimated as follows:

Year	Setup Costs	O&M Costs
1	$16,000	$15,986
2	16,000	16,785
3	16,000	17,663
4	16,000	18,630
5	16,000	19,692
6	16,000	20,861
7	16,000	22,147
8	16,000	23,562

These costs, which were estimated by the manufacturing engineer with the aid of data provided by the vendor, represent a reduction in efficiency and an increase in needed service and repair over time.

The price of the two-year-old Amada Turret Punch Press is $145,000 and would be paid for with cash from the company's capital fund. In addition, the company would incur installation costs totalling $1200. An expenditure of $12,000 would be required to recondition the press to its original condition. The reconditioning would extend the Amada's economic service life to eight years, after which time the machine would have no salvage value. The cash savings of the Amada over the present system are due to the reduced setup time. The average setup cost of the Amada is $15, and the machine would incur 1000 setups per year, yielding a yearly setup cost of $15,000. The savings due to the reduced setup time occur because of the reduction in carrying costs associated with that level of inventory at which the production run and ordering quantity are reduced. The Accounting Department has estimated that at least $26,000, and probably $36,000, per year could be saved by shortening production runs. The operating and maintenance costs of the Amada, as estimated by the manufacturing engineer, are similar to, but somewhat less, than the O&M costs of the present system.

Year	Setup Costs	O&M Costs
1	$15,000	$11,500
2	15,000	11,950
3	15,000	12,445
4	15,000	12,990
5	15,000	13,590
6	15,000	14,245
7	15,000	14,950
8	15,000	15,745

The reduction in the O&M costs is caused by the age difference of the machines and the reduced power requirements of the Amada.

If Rivera Industries delays the replacement of the current four presses for another year, the secondhand Amada machine will no longer be available, and the company will have to buy a brand-new machine at a price of $210,450, installed. The expected setup costs would be the same as those for the secondhand machine, but the annual operating and maintenance costs would be about 10% lower than the estimated O&M costs for the secondhand machine. The expected economic service life of the brand-new press would be eight years, with no salvage value. All these presses, new or old, have a CCA rate of 30%.

Rivera's MARR is 12% after taxes, and the marginal income tax rate is expected to be 40% over the life of the project.

(a) Assuming that the company would need the service of either press for an indefinite period, what would you recommend?

(b) Assuming that the company would need the press for only five more years, what would you recommend?

ST11.5 Tiger Construction Company purchased its current bulldozer (a Caterpillar D8H) and placed it in service six years ago at a cost of $375,000. Since the purchase of the Caterpillar, new technology has produced changes in machines resulting in an increase in productivity of approximately 20%. The Caterpillar worked in a system with a fixed (required) production level to maintain overall system productivity. As the Caterpillar aged and logged more downtime, more hours had to be scheduled to maintain the required production. Tiger is considering purchasing a new bulldozer (a Komatsu K80A) to replace the Caterpillar. The following data have been collected by Tiger's civil engineer:

	Defender (Caterpillar D8H)	Challenger (Komatsu K80A)
Useful life	Not known	Not known
Purchase price		$400,000
Salvage value if kept for		
0 years	$75,000	$400,000
1 year	60,000	300,000
2 years	50,000	240,000
3 years	30,000	190,000
4 years	30,000	150,000
5 years	10,000	115,000
Fuel use (litre/hour)	11.30	16
Maintenance costs:		
1	$46,800	$35,000
2	46,800	38,400
3	46,800	43,700
4	46,800	48,300
5	46,800	58,000

	Defender (Caterpillar D8H) (continued)	Challenger (Komatsu K80A) (continued)
Operating hours (hours/year):		
1	1,800	2,500
2	1,800	2,400
3	1,700	2,300
4	1,700	2,100
5	1,600	2,000
Productivity index	1.00	1.20
Other relevant information:		
Fuel cost ($/litre)		$1.20
Operator's wages ($/hour)		$23.40
Market interest rate (MARR)		15%
Marginal tax rate		40%
CCA rate:	30%	30%

(a) A civil engineer notices that both machines have different working hours and hourly production capacities. To compare the different units of capacity, the engineer needs to devise a combined index that reflects the machine's productivity as well as actual operating hours. Develop such a combined productivity index for each period.
(b) Adjust the operating and maintenance costs by the index you have come up with.
(c) Compare the two alternatives. Should the defender be replaced now?
(d) If the following price index were forecasted for the next five project years, should the defender be replaced now?

	Forecasted Price Index			
Year	General Inflation	Fuel	Wage	Maintenance
0	100	100	100	100
1	108	110	115	108
2	116	120	125	116
3	126	130	130	124
4	136	140	135	126
5	147	150	140	128

On the Companion Website that accompanies this text, you will find Excel templates and exercises, as well as the following analysis tools: Cash Flow Analyzer, Depreciation Analysis, Loan Analysis, and Interest Tables.

TWELVE

Capital-Budgeting Decisions

Hotels Go to the Mattresses[1] In 2005, Marriott International Inc. launched a major initiative to replace nearly every bed in seven of its chains. At the Marriott chain, which was slated for the most extensive upgrade, each king-size bed donned 300-thread-count 60% cotton sheets, seven pillows instead of five, a pillowy mattress cover, a white duvet, and a "bed scarf" draped along the bottom of the bed. The year-long project cost approximately $190 million, and the company says that the cost, along with the planned marketing efforts, made it its biggest initiative ever.

[1] Christina Binkley, "Hotels Go to the Mattresses," *The Wall Street Journal,* January 25, 2005.

The bed replacements raised some awkward issues, such as what to do with all the old beds. Some hotels were trying to give them to charity, but "homeless shelters don't really have a use for king-size beds," says one Marriott executive. Housekeepers complained that stuffing down comforters into all the new duvets was more time consuming than making a traditional bed with bedspread.

All the pressure to make hotel beds better finally forced the industry to reveal—and start to abandon—its dirtiest little secret: the fact that those colourful bedspreads on hotel-room beds sometimes get washed only a few times a year at most. Marriott hopes it will top rivals with a pledge to wash its white duvets between each guest visit, an initiative the company calls "Clean for You." When it comes to bedspread sanitation, hotel chains are typically loath to reveal how often they wash the bedcovers. Regular laundering is costly, in terms of both housekeeping and laundering, as well as in wear and tear on expensive linens.

Hotels are doing all this because their research shows that people are willing to pay more for luxurious beds. Bill Marriott says that once the new beds were installed, Marriott hotels were able to charge as much as $30 a night more.

Marriott purchased approximately 628,000 beds for hotels in seven chains. However, only the full-service Marriott and Renaissance chains received the new white duvets, which the company launders regularly as part of the Clean for You campaign. Less expensive Marriott chains like Courtyard and SpringHill Suites, along with Residence Inn, now triple-sheet their beds by putting the extra sheet on top of the (less frequently laundered) outermost bedspread.

All this came as the hotel industry experienced an economic boom, with room rates and occupancies rising faster than any time since the dot-com boom, leaving hotels extra cash to spend on improvements. Much of the increase in travel came from business travellers, who tend to be pickier and less price-sensitive than vacationers—meaning they want things like better beds and often don't mind paying for them since their expense accounts are picking up the tab.

Without budget limitations, the replacement problem would be more of a logistic issue: simply select the option with the most revenue-enhancing potential for each targeted hotel. However, to replace or upgrade all beds over a short period of time cost in excess of $190 million, and the company needed to find a way to finance this large-scale project. Because of the size of the financing involved, the firm's cost of capital increased during the project period. In this circumstance, the choice of an appropriate interest rate (MARR) for use in the project evaluation became a critical issue. Given these budget and other restrictions, the company aimed to determine the least-cost replacement/upgrade strategy.

Capital budgeting is the planning process used to determine a firm's long term investments.

In this chapter, we present the basic framework of **capital budgeting**, which involves investment decisions related to fixed assets. Here, the term **capital budget** includes planned expenditures on fixed assets; **capital budgeting** encompasses the entire process of analyzing projects and deciding whether they should be included in the capital budget. In previous chapters, we focused on how to evaluate and compare investment projects—the analysis aspect of capital budgeting. In this chapter, we focus on the budgeting aspect. Proper capital-budgeting decisions require a choice of the method of project financing, a schedule of investment opportunities, and an estimate of the minimum attractive rate of return (MARR).

CHAPTER LEARNING OBJECTIVES

After completing this chapter, you should understand the following concepts:

- How a corporation raises its capital to finance a project.
- How to determine the cost of debt.
- How to determine the cost of equity.
- How to determine the marginal cost of capital.
- How to determine the MARR in project evaluation.
- How to create an optimal project portfolio under capital rationing.

12.1 Methods of Financing

In previous chapters, we focused on problems relating to investment decisions. In reality, investment decisions are not always independent of the source of finance. For convenience, however, in economic analysis investment decisions are usually separated from finance decisions: First the investment project is selected, and then the source of financing is considered. After the source is chosen, appropriate modifications to the investment decision are made.

We have also assumed that the assets employed in an investment project are obtained with the firm's own capital (retained earnings) or from short-term borrowings. In practice, this arrangement is not always attractive or even possible. If the investment calls for a significant infusion of capital, the firm may raise the needed capital by issuing stock. Alternatively, the firm may borrow the funds by issuing bonds to finance such purchases. In this section, we will first discuss how a typical firm raises new capital from external sources. Then we will discuss how external financing affects after-tax cash flows and how the decision to borrow affects the investment decision.

The two broad choices a firm has for financing an investment project are **equity financing** and **debt financing**.[2] We will look briefly at these two options for obtaining

---

[2] A hybrid financing method, known as *lease financing*, was discussed in Section 10.4.3.

external investment funds and also examine their effects on after-tax cash flows. The basic terminology and definitions relating to stocks and bonds, to which we will refer in the following sections, were presented in detail in Chapter 2.

12.1.1 Equity Financing

Equity financing can take one of two forms: (1) the use of retained earnings otherwise paid to stockholders or (2) the issuance of stock. Both forms of equity financing use funds invested by the current or new owners of the company.

Until now, many of our economic analyses presumed that companies had cash on hand to make capital investments—implicitly, we were dealing with cases of financing by retained earnings. A simplified view of a company's cash flows in Figure 12.1 illustrates the source of **retained earnings**. Several deductions are applied to the revenue to arrive at the taxable income, including interest payments owed on borrowed money, *I*, and the capital cost allowance, CCA. After the income tax, *T*, has been calculated, the CCA is added back in, leaving the firm with available cash. Once the debts for the period are repaid, the firm may allocate a portion of the remaining money to the company's owners—the shareholders—in the form of a dividend and use the remaining retained earnings to reinvest in the company now or keep some on hand for future needs.

If a company does not have sufficient cash on hand to make an investment and does not wish to borrow in order to fund the investment, financing can be arranged by selling stock to raise the required funds. (Many small biotechnology and computer firms raise capital by going public and selling common stock.) To do this, the company has to decide how much money to raise, the type of securities to issue (common stock or preferred stock), and the basis for pricing the issue.

Once the company has decided to issue common stock, it must estimate **flotation costs**—the expenses it will incur in connection with the issue, such as investment bankers' fees, lawyers' fees, accountants' costs, and the cost of printing and engraving. Usually, an investment banker will buy the issue from the company at a discount, below the price at which the stock is to be offered to the public. (The discount usually represents the *flotation costs.*) If the company is already publicly owned, the offering price will commonly be based on the existing market price of the stock. If the company is going public for the first time, no established price will exist, so investment bankers have to estimate the expected market price at which the stock will sell after the stock issue. Example 12.1 illustrates how the flotation cost affects the cost of issuing common stock.

Flotation cost: The costs associated with the issuance of new securities.

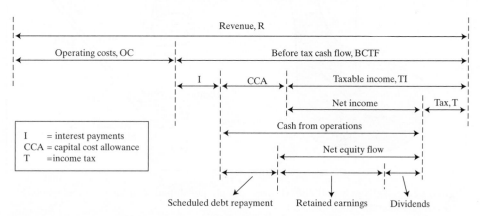

Figure 12.1   Retained earnings.

EXAMPLE 12.1 Issuing Common Stock

Scientific Sports Inc. (SSI), a golf club manufacturer, has developed a new metal club (Driver). The club is made out of titanium alloy, an extremely light and durable metal with good vibration-damping characteristics (Figure 12.2). The company expects to acquire considerable market penetration with this new product. To produce it, the company needs a new manufacturing facility, which will cost $10 million. The company decided to raise this $10 million by selling common stock. The firm's current stock price is $30 per share. Investment bankers have informed management that the new public issue must be priced at $28 per share because of decreasing demand, which will occur as more shares become available on the market. The flotation costs will be 6% of the issue price, so SSI will net $26.32 per share. How many shares must SSI sell to net $10 million after flotation expenses?

SOLUTION

Let X be the number of shares to be sold. Then total flotation cost will be

$$(0.06)(\$28)(X) = 1.68X.$$

To net $10 million, we must have

$$\text{Sales proceeds} - \text{flotation cost} = \text{Net proceeds},$$
$$28X - 1.68X = \$10,000,000,$$
$$26.32X = \$10,000,000,$$
$$X = 379,940 \text{ shares}.$$

Now we can figure out the flotation cost for issuing the common stock. The cost is

$$1.68(379,940) = \$638,300.$$

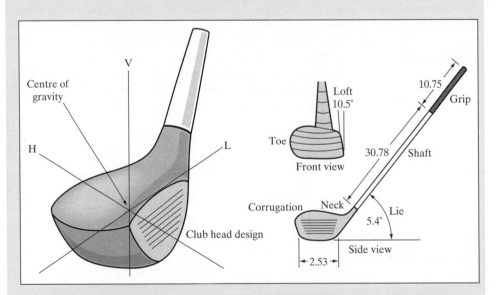

Figure 12.2 SSI's new golf club (Driver) design, developed with the use of advanced engineering materials (Example 12.1).

12.1.2 Debt Financing

The second major type of financing a company can select is **debt financing**, which includes both short-term borrowing from financial institutions and the sale of long-term bonds, wherein money is borrowed from investors for a fixed period. With debt financing, the interest paid on the loans or bonds is treated as an expense for income-tax purposes. Since interest is a tax-deductible expense, companies in high marginal tax jurisdictions may incur lower after-tax financing costs with a debt. In addition to influencing the borrowing interest rate and tax bracket, the loan-repayment method can affect financing costs.

When the debt-financing option is used, we need to separate the interest payments from the repayment of the loan for our analysis. The interest-payment schedule depends on the repayment schedule established at the time the money is borrowed. The two common debt-financing methods are as follows:

1. **Bond Financing.** This type of debt financing does not involve the partial payment of principal; only interest is paid each year (or semiannually). The principal is paid in a lump sum when the bond matures. (See Section 4.6 for bond terminologies and valuation.) Bond financing is similar to equity financing in that flotation costs are involved when bonds are issued.

2. **Term Loans.** Term loans involve an equal repayment arrangement according to which the sum of the interest payments and the principal payments is uniform; interest payments decrease, while principal payments increase, over the life of the loan. Term loans are usually negotiated directly between the borrowing company and a financial institution, generally a commercial bank.

Example 12.2 illustrates how these different methods can affect the cost of issuing bonds or term loans.

EXAMPLE 12.2 Debt Financing

Consider again Example 12.1. Suppose SSI has instead decided to raise the $10 million by debt financing. SSI could issue a mortgage bond or secure a term loan. Conditions for each option are as follows:

- **Bond financing.** The flotation cost is 1.8% of the $10 million issue. The company's investment bankers have indicated that a five-year bond issue with a face value of $1000 can be sold at $985 per share. The bond would require annual interest payments of 12%.
- **Term loan.** A $10 million bank loan can be secured at an annual interest rate of 11% for five years; it would require five equal annual installments.

 (a) How many $1000 par value bonds would SSI have to sell to raise the $10 million?
 (b) What are the annual payments (interest and principal) on the bond?
 (c) What are the annual payments (interest and principal) on the term loan?

SOLUTION

 (a) To net $10 million, SSI would have to sell

$$\frac{\$10,000,000}{(1 - 0.018)} = \$10,183,300$$

TABLE 12.1 Two Common Methods of Debt Financing (Example 12.2)

	0	1	2	3	4	5
1. Bond financing: No principal repayments until end of life						
Beginning balance	$10,338,380	$10,338,380	$10,338,380	$10,338,380	$10,338,380	$10,338,380
Interest owed		1,240,606	1,240,606	1,240,606	1,240,606	1,240,606
Repayment						
Interest payment		(1,240,606)	(1,240,606)	(1,240,606)	(1,240,606)	(1,240,606)
Principal payment						(10,338,380)
Ending balance	$10,338,380	$10,338,380	$10,338,380	$10,338,380	$10,338,380	0
2. Term loan: Equal annual repayments [$10,000,000(A/P, 11%, 5) = $2,705,703]						
Beginning balance	$10,000,000	$10,000,000	$ 8,394,297	$ 6,611,967	$ 4,633,580	$ 2,437,571
Interest owed		1,100,000	923,373	727,316	509,694	268,133
Repayment						
Interest payment		(1,100,000)	(923,373)	(727,316)	(509,694)	(268,133)
Principal payment		(1,605,703)	(1,782,330)	(1,978,387)	(2,196,009)	(2,437,570)
Ending balance	$10,000,000	$ 8,394,297	$ 6,611,967	$ 4,633,580	$ 2,437,571	0

worth of bonds and pay $183,300 in flotation costs. Since the $1000 bond will be sold at a 1.5% discount, the total number of bonds to be sold would be

$$\frac{\$10,183,300}{\$985} = 10,338.38.$$

(b) For the bond financing, the annual interest is equal to

$$\$10,338,380(0.12) = \$1,240,606.$$

Only the interest is paid each period; thus, the principal amount owed remains unchanged.

(c) For the term loan, the annual payments are

$$\$10,000,000(A/P, 11\%, 5) = \$2,705,703.$$

The principal and interest components of each annual payment are summarized in Table 12.1. Note that, although SSI would be responsible for a $10,338,380 debt with the bond issue, they only net $10,000,000 new capital because of the flotation cost and the discount sale price.

12.1.3 Capital Structure

Capital structure: The means by which a firm is financed.

The ratio of total debt to total capital, generally called the **debt ratio**, or **capital structure**, represents the percentage of the total capital provided by borrowed funds. For example, a debt ratio of 0.4 indicates that 40% of the capital is borrowed and the remaining funds are provided from the company's equity (retained earnings or stock offerings). This type of financing is called **mixed financing**.

Borrowing affects a firm's capital structure, and firms must determine the effects of a change in the debt ratio on their market value before making an ultimate financing decision. Even if debt financing is attractive, you should understand that companies do not simply borrow funds to finance projects. A firm usually establishes a **target capital structure**, or **target debt ratio**, after considering the effects of various financing methods. This target may change over time as business conditions vary, but a firm's management always strives to achieve the target whenever individual financing decisions are considered. On the one hand, when the actual debt ratio is below the target level, any new capital will probably be raised by issuing debt. On the other hand, if the debt ratio is currently above the target, expansion capital will be raised by issuing stock.

How does a typical firm set the target capital structure? This is a rather difficult question to answer, but we can list several factors that affect the capital-structure policy. First, capital-structure policy involves a trade-off between risk and return. As you take on more debt for business expansion, the inherent business risk[3] also increases, but investors view business expansion as a healthy indicator for a corporation with higher expected earnings. When investors perceive higher business risk, the firm's stock price tends to be depressed. By contrast, when investors perceive higher expected earnings, the firm's stock price tends to increase. The optimal capital structure is thus the one that strikes a balance between business risk and expected future earnings. The greater the firm's business risk, the lower is its optimal debt ratio.

Second, a major reason for using debt is that interest is a deductible expense for business operations, which lowers the effective cost of borrowing. Dividends paid to common stockholders, however, are not deductible. If a company uses debt, it must pay interest on this debt, whereas if it uses equity, it pays dividends to its equity investors (shareholders). A company needs $1 in before-tax income to pay $1 of interest, but if the company is in the 34% tax bracket, it needs $1/(1 - 0.34) = 1.52 of before-tax income to pay a $1 dividend.

Third, financial flexibility—the ability to raise capital on reasonable terms from the financial market—is an important consideration. Firms need a steady supply of capital for stable operations. When money is tight in the economy, investors prefer to advance funds to companies with a healthy capital structure (lower debt ratio). These three elements (business risk, taxes, and financial flexibility) are major factors that determine the firm's optimal capital structure. Example 12.3 illustrates how a typical firm finances a large-scale engineering project by maintaining the predetermined capital structure.

EXAMPLE 12.3 Project Financing Based on an Optimal Capital Structure

Consider again SSI's $10 million venture project in Example 12.1. Suppose that SSI's optimal capital structure calls for a debt ratio of 0.5. After reviewing SSI's capital structure, the investment banker convinced management that it would be better

---

[3] Unlike equity financing, in which dividends are optional, debt interest and principal (face value) must be repaid on time. Also, uncertainty is involved in making projections of future operating income as well as expenses. In bad times debt can be devastating, but in good times the tax deductibility of interest payments increases profits to owners.

off, in view of current market conditions, to limit the stock issue to $5 million and to raise the other $5 million as debt by issuing bonds. Because the amount of capital to be raised in each category is reduced by half, the flotation cost would also change. The flotation cost for common stock would be 8.1%, whereas the flotation cost for bonds would be 3.2%. As in Example 12.2, the five-year, 12% bond will have a par value of $1000 and will be sold for $985.

Assuming that the $10 million capital would be raised from the financial market, the engineering department has detailed the following financial information:

- The new venture will have a five-year project life.
- The $10 million capital will be used to purchase land for $1 million, a building for $3 million, and equipment for $6 million. The plant site and building are already available, and production can begin during the first year. The building and the equipment fall into CCA Class 1 ($d = 4\%$) and CCA Class 43 ($d = 30\%$), respectively. The 50% rule is applicable. At the end of year 5, the salvage value for each asset is as follows: the land $1.5 million, the building $2 million, and the equipment $2.5 million. The capital gains inclusion rate is 50%.
- For common stockholders, an annual cash dividend in the amount of $2 per share is planned over the project life. This steady cash dividend payment is deemed necessary to maintain the market value of the stock.
- The unit production cost is $50.31 (material, $22.70; labour and overhead (excluding depreciation), $10.57; and tooling, $17.04).
- The unit price is $250, and SSI expects an annual demand of 20,000 units.
- The operating and maintenance cost, including advertising expenses, would be $600,000 per year.
- An investment of $500,000 in working capital is required at the beginning of the project; the amount will be fully recovered when the project terminates.
- The firm's marginal tax rate is 40%, and this rate will remain constant throughout the project period.

(a) Determine the after-tax cash flows for this investment with external financing.

(b) Is this project justified at an interest rate of 20%?

DISCUSSION: As the amount of financing and flotation costs change, we need to re-calculate the number of shares (or bonds) to be sold in each category. For a $5 million common stock issue, the flotation cost increases to 8.1%.[4] The number of shares to be sold to net $5 million is $5,000,000/0.919/28 = 194,311$ shares (or $5,440,708$). For a $5 million bond issue, the flotation cost is 3.2%. Therefore, to net $5 million, SSI has to sell $5,000,000/0.968/985 = 5,243.95$ units of $1,000 par value. This implies that SSI is effectively borrowing $5,243,948, upon which figure the annual bond interest will be calculated. The annual bond interest payment is $5,243,948(0.12) = $629,274$.

---

[4] Flotation costs are higher for small issues than for large ones due to the existence of fixed costs: Certain costs must be incurred regardless of the size of the issue, so the percentage of flotation costs increases as the size of the issue gets smaller.

SOLUTION

(a) *After-tax cash flows.* Table 12.2 summarizes the after-tax cash flows for the new venture. The following calculations and assumptions were used in developing the table:

- Revenue: $250 \times 20,000 = \$5,000,000$ per year.
- Costs of goods: $50.31 \times 20,000 = \$1,006,200$ per year.
- Bond interest: $\$5,243,948 \times 0.12 = \$629,274$ per year.

TABLE 12.2 Effects of Project Financing on After-Tax Cash Flows (Example 12.3)

	0	1	2	3	4	5
Income Statement						
Revenue		$5,000,000	$5,000,000	$5,000,000	$5,000,000	$5,000,000
Expenses						
Cost of goods		1,006,200	1,006,200	1,006,200	1,006,200	1,006,200
O&M		600,000	600,000	600,000	600,000	600,000
Bond Interest		629,274	629,274	629,274	629,274	629,274
CCA						
Building		60,000	117,600	112,896	108,380	104,045
Equipment		900,000	1,530,000	1,071,000	749,700	524,790
Taxable income		1,804,526	1,116,926	1,580,630	1,906,446	2,135,691
Income taxes		721,810	446,770	632,252	762,578	854,276
Net Income		$1,082,716	$670,156	$948,378	$1,143,868	$1,281,415
Cash Flow Statement						
Operating						
Net income		$1,082,716	$670,156	$948,378	$1,143,868	$1,281,415
CCA		960,000	1,647,600	1,183,896	858,080	628,835
Investment						
Land	(1,000,000)					1,500,000
Building	(3,000,000)					2,000,000
Equipment	(6,000,000)					2,500,000
Working Capital	(500,000)					500,000
Disposal tax effects						(411,365)
Financing						
Common stock	5,000,000					(5,440,708)
Bond	5,000,000					(5,243,948)
Cash dividend		(388,622)	(388,622)	(388,622)	(388,622)	(388,622)
Net cash flow	($500,000)	$1,654,094	$1,929,134	$1,743,652	$1,613,326	($3,074,393)

- Capital cost allowance: Assuming that the building is placed in service in January, the first year's CCA percentage is 2.0%. Therefore, the allowed CCA amount is $3,000,000 × 0.02 = $60,000. The CCA in each subsequent year would be 4.0% of the undepreciated capital cost at the beginning of that year. For the second year, the CCA would be

$$($3,000,000 - $60,000) \times 0.04 = $117,600.$$

CCA on equipment is calculated in a similar manner using a 30% rate (Year 1 = 15%).

- Disposal tax effect:

Capital gain on land = $1,500,000 − $1,000,000 = $500,000

Tax on capital gain = −0.4 × 1/2 × $500,000 = −$100,000

For the building and equipment, the tax effect at the time of disposal equals (UCC − salvage value) × t

Property	UCC	Salvage Value	Gains (Losses)	Disposal Tax Effect, G
Building	$2,497,079	$2,000,000	$ 497,079	$ 198,831
Equipment	1,224,510	2,500,000	(1,275,490)	(510,196)

The total disposal tax effect, G, is then

$$G = -$100,000 + $198,831 - $510,196 = -$411,365$$

- Cash dividend: 194,311 shares × $2 = $388,622 per year.
- Common stock: When the project terminates and the bonds are retired, the debt ratio is no longer 0.5. If SSI wants to maintain the constant capital structure (0.5), SSI would have to repurchase the common stock in the amount of $5,440,708 at the prevailing market price. In developing Table 12.2, we assumed that this repurchase of common stock had taken place at the end of project years. In practice, a firm may or may not repurchase the common stock. As an alternative means of maintaining the desired capital structure, the firm may use this extra debt capacity released to borrow for other projects.
- Bond: When the bonds mature at the end of year 5, the total face value in the amount of $5,243,948 must be paid to the bondholders.

(b) *Measure of project worth.* The NPW for this project is then

$$PW(20\%) = -$500,000 + $1,654,094(P/F, 20\%, 1) + \ldots$$

$$-$3,074,393(P/F, 20\%, 5)$$

$$= $2,769,647.$$

The investment is nonsimple, and it is also a mixed investment. The RIC at MARR of 20% is 331%. Even though the project requires a significant amount of cash expenditure at the end of its life, it still appears to be a very profitable one.

In Example 12.3, we neither discussed the cost of capital required to finance this project nor explained the relationship between the cost of capital and the MARR. In the remaining sections, these issues will be discussed. As we will see later, in Section 12.3, we can completely ignore the detailed cash flows related to project financing if we adjust our discount rate according to the capital structure, namely, by using the weighted cost of capital.

12.2 Cost of Capital

In most of the capital-budgeting examples in earlier chapters, we assumed that the firms under consideration were financed entirely with equity funds. In those cases, the cost of capital may have represented the firm's required return on equity. However, most firms finance a substantial portion of their capital budget with long-term debt (bonds), and many also use preferred stock as a source of capital. In these cases, a firm's cost of capital must reflect the average cost of the various sources of long-term funds that the firm uses, not only the cost of equity. In this section, we will discuss the ways in which the cost of each individual type of financing (retained earnings, common stock, preferred stock, and debt) can be estimated,[5] given a firm's target capital structure.

12.2.1 Cost of Equity

Whereas debt and preferred stocks are contractual obligations that have easily determined costs, it is not easy to measure the cost of equity. In principle, the cost of equity capital involves an **opportunity cost**. In fact, the firm's after-tax cash flows belong to the stockholders. Management may either pay out these earnings in the form of dividends, or retain the earnings and reinvest them in the business. If management decides to retain the earnings, an opportunity cost is involved: Stockholders could have received the earnings as dividends and invested the money in other financial assets. Therefore, the firm should earn on its retained earnings at least as much as the stockholders themselves could earn in alternative, but comparable, investments.

> **Cost of equity** is the minimum rate of return a firm must offer shareholders to compensate for waiting for their returns, and for bearing some risk.

What rate of return can stockholders expect to earn on retained earnings? This question is difficult to answer, but the value sought is often regarded as the rate of return stockholders require on a firm's common stock. If a firm cannot invest retained earnings so as to earn at least the rate of return on equity, it should pay these funds to the stockholders and let them invest directly in other assets that do provide this return.

When investors are contemplating buying a firm's stock, they have two things in mind: (1) cash dividends and (2) gains (appreciation of shares) at the time of sale. From a conceptual standpoint, investors determine market values of stocks by discounting expected future dividends at a rate that takes into account any future growth. Since investors seek growth companies, a desired growth factor for future dividends is usually included in the calculation.

To illustrate, let's take a simple numerical example. Suppose investors in the common stock of ABC Corporation expect to receive a dividend of $5 by the end of the first year. The future annual dividends will grow at an annual rate of 10%. Investors will hold the stock for two more years and will expect the market price of the stock to rise to $120 by the end of the third year. Given these hypothetical expectations, ABC expects that investors would be willing to pay $100 for this stock in today's market. What is the required

---

[5] Estimating or calculating the cost of capital in any precise fashion is very difficult task.

rate of return k_r on ABC's common stock? We may answer this question by solving the following equation for k_r:

$$\$100 = \frac{\$5}{(1 + k_r)} + \frac{\$5(1 + 0.1)}{(1 + k_r)^2} + \frac{\$5(1 + 0.1)^2 + \$120}{(1 + k_r)^3}.$$

In this case, $k_r = 11.44\%$. This implies that if ABC finances a project by retaining its earnings or by issuing additional common stock at the going market price of $100 per share, it must realize at least 11.44% on new investment just to provide the minimum rate of return required by the investors. Therefore, 11.44% is the specific cost of equity that should be used in calculating the weighted-average cost of capital. Because flotation costs are involved in issuing new stock, the cost of equity will increase. If investors view ABC's stock as risky and therefore are only willing to buy the stock at a price lower than $100 (but with the same expectations), the cost of equity will also increase. There is an increasing tendency for some companies not to pay out dividends, but the principle of requiring a higher cost of equity to account for risk applies nevertheless. The phenomenon is referred to as a risk premium, but the details are beyond the scope of this book. Now we can generalize the preceding result.

Cost of Retained Earnings

Let's assume the same hypothetical situation for ABC. Recall that ABC's retained earnings belong to holders of its common stock. If ABC's current stock is traded for a market price of P_0, with a first-year dividend[6] of D_1, but growing at the annual rate of g thereafter, the specific cost of retained earnings for an infinite period of holding (stocks will change hands over the years, but it does not matter who holds the stock) can be calculated as

$$P_0 = \frac{D_1}{(1 + k_r)} + \frac{D_1(1 + g)}{(1 + k_r)^2} + \frac{D_1(1 + g)^2}{(1 + k_r)^3} + \cdots$$

$$= \frac{D_1}{1 + k_r} \sum_{n=0}^{\infty} \left[\frac{(1 + g)}{(1 + k_r)} \right]^n$$

$$= \frac{D_1}{1 + k_r} \left[\frac{1}{1 - \frac{1 + g}{1 + k_r}} \right], \text{ where } g < k_r.$$

Solving for k_r, we obtain

$$k_r = \frac{D_1}{P_0} + g. \tag{12.1}$$

If we use k_r as the discount rate for evaluating the new project, it will have a positive NPW only if the project's IRR exceeds k_r. Therefore, any project with a positive NPW, calculated at k_r, induces a rise in the market price of the stock. Hence, by definition, k_r is the rate of return required by shareholders and should be used as the cost of the equity component in calculating the weighted average cost of capital.

---

[6] When we check the stock listings in the newspaper, we do not find the expected first-year dividend D_1. Instead, we find the dividend paid out most recently, D_0. So if we expect growth at a rate g, the dividend at the end of one year from now, D_1 may be estimated as $D_1 = D_0(1 + g)$.

Issuing New Common Stock

Again, because flotation costs are involved in issuing new stock, we can modify the cost of retained earnings k_r by

$$k_e = \frac{D_1}{P_0(1 - f_c)} + g, \tag{12.2}$$

where k_e is the cost of common equity and f_c is the flotation cost as a percentage of the stock price.

Either calculation is deceptively simple, because, in fact, several ways are available to determine the cost of equity. In reality, the market price fluctuates constantly, as do a firm's future earnings. Thus, future dividends may not grow at a constant rate, as the model indicates. For a stable corporation with moderate growth, however, the cost of equity as calculated by evaluating either Eq. (12.1) or Eq. (12.2) serves as a good approximation.

Cost of Preferred Stock

A preferred stock is a hybrid security in the sense that it has some of the properties of bonds and other properties that are similar to common stock. Like bondholders, holders of preferred stock receive a fixed annual dividend. In fact, many firms view the payment of the preferred dividend as an obligation just like interest payments to bondholders. It is therefore relatively easy to determine the cost of preferred stock. For the purposes of calculating the weighted average cost of capital, the specific cost of a preferred stock will be defined as

$$k_p = \frac{D^*}{P^*(1 - f_c)}, \tag{12.3}$$

where D^* is the fixed annual dividend, P^* is the issuing price, and f_c is as defined in Eq. (12.2).

Cost of Equity

Once we have determined the specific cost of each equity component, we can determine the weighted-average cost of equity (i_e) for a new project. We have

$$i_e = \left(\frac{c_r}{c_e}\right)k_r + \left(\frac{c_c}{c_e}\right)k_e + \left(\frac{c_p}{c_e}\right)k_p, \tag{12.4}$$

where c_r is the amount of equity financed from retained earnings, c_c is the amount of equity financed from issuing new stock, c_p is the amount of equity financed from issuing preferred stock, and $c_r + c_c + c_p = c_e$. Example 12.4 illustrates how we may determine the cost of equity.

EXAMPLE 12.4 Determining the Cost of Equity

Alpha Corporation needs to raise $10 million for plant modernization. Alpha's target capital structure calls for a debt ratio of 0.4, indicating that $6 million has to be financed from equity.

- Alpha is planning to raise $6 million from the following equity sources:

Source	Amount	Fraction of Total Equity
Retained earnings	$1 million	0.167
New common stock	4 million	0.666
Preferred stock	1 million	0.167

- Alpha's current common stock price is $40, the market price that reflects the firm's future plant modernization. Alpha is planning to pay an annual cash dividend of $5 at the end of the first year, and the annual cash dividend will grow at an annual rate of 8% thereafter.
- Additional common stock can be sold at the same price of $40, but there will be 12.4% flotation costs.
- Alpha can issue $100 par preferred stock with a 9% dividend. (This means that Alpha will calculate the dividend on the basis of the par value, which is $9 per share.) The stock can be sold on the market for $95, and Alpha must pay flotation costs of 6% of the market price.

Determine the cost of equity to finance the plant modernization.

SOLUTION

We will itemize the cost of each component of equity:

- Cost of retained earnings: With $D_1 = \$5$, $g = 8\%$, and $P_0 = \$40$,

$$k_r = \frac{5}{40} + 0.08 = 20.5\%.$$

- Cost of new common stock: With $D_1 = \$5$, $g = 8\%$, and $f_c = 12.4\%$,

$$k_e = \frac{5}{40(1 - 0.124)} + 0.08 = 22.27\%.$$

- Cost of preferred stock: With $D^* = \$9$, $P^* = \$95$, and $f_c = 0.06$,

$$k_p = \frac{9}{95(1 - 0.06)} = 10.08\%.$$

- Cost of equity: With $\dfrac{c_r}{c_e} = 0.167$, $\dfrac{c_c}{c_e} = 0.666$, and $\dfrac{c_p}{c_e} = 0.167$,

$$i_e = (0.167)(0.205) + (0.666)(0.2227) + (0.167)(0.1008)$$
$$= 19.96\%.$$

12.2.2 Cost of Debt

Now let us consider the calculation of the specific cost that is to be assigned to the debt component of the weighted-average cost of capital. The calculation is relatively straightforward and simple. As we said in Section 12.1.2, the two types of debt financing are term loans and bonds. Because the interest payments on both are tax deductible, the effective cost of debt will be reduced.

To determine the after-tax cost of debt (i_d), we evaluate the expression

$$i_d = \left(\frac{c_s}{c_d}\right)k_s(1 - t_m) + \left(\frac{c_b}{c_d}\right)k_b(1 - t_m), \qquad (12.5)$$

where c_s is the amount of the short-term loan, k_s is the before-tax interest rate on the term loan, t_m is the firm's marginal tax rate, k_b is the before-tax interest rate on the bond, c_b is the amount of bond financing, and $c_s + c_b = c_d$.

As for bonds, a new issue of long-term bonds incurs flotation costs. These costs reduce the proceeds to the firm, thereby raising the specific cost of the capital raised. For example, when a firm issues a $1000 par bond, but nets only $940, the flotation cost will be 6%. Therefore, the effective after-tax cost of the bond component will be higher than the nominal interest rate specified on the bond. We will examine this problem with a numerical example.

Cost of debt: The effective rate that a company pays on its current debt.

EXAMPLE 12.5 Determining the Cost of Debt

Consider again Example 12.4, and suppose that Alpha has decided to finance the remaining $4 million by securing a term loan and issuing 20-year $1000 par bonds under the following conditions:

Source	Amount	Fraction	Interest Rate	Flotation Cost
Term loan	$1 million	0.333	12% per year	
Bonds	3 million	0.667	10% per year	6%

If the bond can be sold to net $940 (after deducting the 6% flotation cost), determine the cost of debt to raise $4 million for the plant modernization. Alpha's marginal tax rate is 38%, and it is expected to remain constant in the future.

SOLUTION

First, we need to find the effective after-tax cost of issuing the bond with a flotation cost of 6%. The before-tax specific cost is found by solving the equivalence formula

$$\$940 = \frac{\$100}{(1 + k_b)} + \frac{\$100}{(1 + k_b)^2} + \cdots + \frac{\$100 + \$1000}{(1 + k_b)^{20}}$$

$$= \$100(P/A, k_b, 20) + \$1000(P/F, k_b, 20).$$

Solving for k_b, we obtain $k_b = 10.74\%$. Note that the cost of the bond component increases from 10% to 10.74% after the 6% flotation cost is taken into account.

The after-tax cost of debt is the interest rate on debt, multiplied by $(1 - t_m)$. In effect, the government pays part of the cost of debt because interest is tax deductible. Now we are ready to compute the after-tax cost of debt as follows:

$$i_d = (0.333)(0.12)(1 - 0.38) + (0.667)(0.1074)(1 - 0.38)$$

$$= 6.92\%.$$

12.2.3 Calculating the Cost of Capital

With the specific cost of each financing component determined, we are ready to calculate the tax-adjusted weighted-average cost of capital based on total capital. Then we will define the marginal cost of capital that should be used in project evaluation.

Weighted-Average Cost of Capital

Cost of capital for a firm is a weighted sum of the cost of equity and the cost of debt.

Assuming that a firm raises capital on the basis of the target capital structure and that the target capital structure remains unchanged in the future, we can determine a **tax-adjusted weighted-average cost of capital** (or, simply stated, the **cost of capital**). This cost of capital represents a composite index reflecting the cost of raising funds from different sources. The cost of capital is defined as

$$k = \frac{i_d c_d}{V} + \frac{i_e c_e}{V}, \tag{12.6}$$

where c_d = Total debt capital (such as bonds) in dollars,

$\quad c_e$ = Total equity capital in dollars,

$\quad V = c_d + c_e$,

$\quad i_e$ = Average equity interest rate per period, taking into account all equity sources,

$\quad i_d$ = After-tax average borrowing interest rate per period, taking into account all debt sources, and

$\quad k$ = Tax-adjusted weighted-average cost of capital.

Note that the cost of equity is already expressed in terms of after-tax cost, because any return to holders of either common stock or preferred stock is made after payment of income taxes. Figure 12.3 summarizes the process of determining the cost of capital.

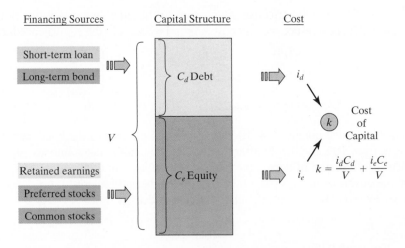

Figure 12.3 Process of calculating the cost of capital.

Marginal Cost of Capital

Now that we know how to calculate the cost of capital, we can ask: "could a typical firm raise unlimited new capital at the same cost?" The answer is no. As a practical matter, as a firm tries to attract more new dollars, the cost of raising each additional dollar will at some point rise. As this occurs, the weighted-average cost of raising each additional new dollar also rises. Thus, the **marginal cost of capital** is defined as the cost of obtaining another dollar of new capital, and the marginal cost rises as more and more capital is raised during a given period. In evaluating an investment project, we are using the concept of the marginal cost of capital. The formula to find the marginal cost of capital is exactly the same as Eq. (12.5); however, the costs of debt and equity in that equation are the interest rates on new debt and equity, not outstanding (or combined) debt or equity. In other words, we are interested in the marginal cost of capital—specifically, to use it in evaluating a new investment project. The rate at which the firm has borrowed in the past is less important for this purpose. Example 12.6 works through the computations for finding the cost of capital (k).

EXAMPLE 12.6 Calculating the Marginal Cost of Capital

Consider again Examples 12.4 and 12.5. The marginal income tax rate (t_m) for Alpha is expected to remain at 38% in the future. Assuming that Alpha's capital structure (debt ratio) also remains unchanged, determine the cost of capital (k) for raising $10 million in addition to existing capital.

SOLUTION

With c_d = $4 million, c_e = $6 million, V = $10 million, i_d = 6.92%, i_e = 19.93%, and Eq. (12.6), we calculate

$$k = \frac{(0.0692)(4)}{10} + \frac{(0.1996)(6)}{10}$$
$$= 14.74\%.$$

This 14.74% would be the marginal cost of capital that a company with the given financial structure would expect to pay to raise $10 million.

12.3 Choice of Minimum Attractive Rate of Return

Thus far, we have said little about what interest rate, or minimum attractive rate of return (MARR), is suitable for use in a particular investment situation. Choosing the MARR is a difficult problem; no single rate is always appropriate. In this section, we will discuss briefly how to select a MARR for project evaluation. Then we will examine the relationship between capital budgeting and the cost of capital.

12.3.1 Choice of MARR When Project Financing Is Known

In Chapter 10, we focused on calculating after-tax cash flows, including situations involving debt financing. When cash flow computations reflect interest, taxes, and debt repayment, what is left is called **net equity flow**. If the goal of a firm is to maximize the wealth of its stockholders, why not focus only on the after-tax cash flow to equity, instead

minimum attractive rate of return (MARR): The required return necessary to make a capital budgeting project—such as building a new factory— worthwhile.

of on the flow to all suppliers of capital? Focusing on only the equity flows will permit us to use the cost of equity as the appropriate discount rate. In fact, we have implicitly assumed that all after-tax cash flow problems, in which financing flows are explicitly stated in earlier chapters, represent net equity flows, so the MARR used represents the **cost of equity** (i_e). Example 12.7 illustrates project evaluation by the net equity flow method.

EXAMPLE 12.7 Project Evaluation by Net Equity Flow

Suppose the Alpha Corporation, which has the capital structure described in Example 12.6, wishes to install a new set of machine tools, which is expected to increase revenues over the next five years. The tools require an investment of $150,000, to be financed with 60% equity and 40% debt. The equity interest rate (i_e), which combines the two sources of common and preferred stocks, is 19.96%. Alpha will use a 12% short-term loan to finance the debt portion of the capital ($60,000), with the loan to be repaid in equal annual installments over five years. The Capital Cost Allowance rate is 30%, subject to the 50% rule, and zero salvage value is expected. Additional revenues and operating costs are expected to be as follows:

n	Revenues ($)	Operating Cost ($)
1	$68,000	$20,500
2	73,000	20,000
3	79,000	20,500
4	84,000	20,000
5	90,000	20,500

The marginal tax rate (combined federal and provincial rate) is 38%. Evaluate this venture by using net equity flows at $i_e = 19.96\%$.

SOLUTION

The calculations are shown in Table 12.3. The NPW and IRR calculations are as follows:

$$\begin{aligned} PW(19.96\%) = &-\$90,000 + \$24,091(P/F, 19.96\%, 1) \\ &+ \$33,055(P/F, 19.96\%, 2) + \$31,623(P/F, 19.96\%, 3) \\ &+ \$31,440(P/F, 19.96\%, 4) + \$43,742(P/F, 19.96\%, 5) \\ = &\ \$4163. \end{aligned}$$

$$IRR = 21.88\% > 19.96\%.$$

The internal rate of return for this cash flow is 21.88%, which exceeds $i_e = 19.96\%$. Thus, the project would be profitable.

TABLE 12.3 After-Tax Cash Flow Analysis When Project Financing Is Known: Net Equity Cash Flow Method (Example 12.7)

End of Year	0	1	2	3	4	5
Income Statement						
Revenue		$68,000	$73,000	$79,000	$84,000	$90,000
Expenses						
Operating cost		20,500	20,000	20,500	20,000	20,500
Interest payment		7,200	6,067	4,797	3,376	1,783
CCA		22,500	38,250	26,775	18,743	13,120
Taxable income		17,800	8,683	26,928	41,882	54,597
Income taxes (38%)		6,764	3,300	10,233	15,915	20,747
Net Income		$11,036	$5,384	$16,695	$25,967	$33,850
Cash Flow Statement						
Operating activities						
Net income		$11,036	$5,384	$16,695	$25,967	$33,850
CCA		22,500	38,250	26,775	18,743	13,120
Investment activities						
Investment and salvage (150,000)						0
Disposal tax effects						11,633
Financing activities						
Borrowed funds	60,000					
Principal repayment		(9,445)	(10,578)	(11,847)	(13,269)	(14,861)
Net cash flow	($90,000)	$24,091	$33,056	$31,623	$31,440	$43,741

$UCC_5 = \$150,000(0.85)(0.7)^4 = \$30,613; G = (\$30,613 - 0)(0.38) = \$11,633$

COMMENT: In this problem, we assumed that the Alpha Corporation would be able to raise the additional equity funds at the same rate of 19.96%, so this 19.96% can be viewed as the marginal cost of capital.

12.3.2 Choice of MARR When Project Financing Is Unknown

You might well ask of what use is the k if we use the cost of equity (i_e) exclusively? The answer to this question is that, by using the value of k, we may evaluate investments without explicitly treating the debt flows (both interest and principal). In this case, we make a tax adjustment to the discount rate by employing the effective after-tax cost of debt. This approach recognizes that the net interest cost is effectively transferred from the tax collector to the creditor in the sense that there is a dollar-for-dollar reduction in taxes up to the debt interest payments. Therefore, debt financing is treated implicitly. The method would be appropriate when debt financing is not identified with individual investments, but rather enables the company to engage in a set of investments. (Except where financing flows are explicitly stated, all previous examples in this book have implicitly assumed the more realistic and appropriate situation wherein debt financing is not identified with individual investment. Therefore, the MARRs represent the weighted cost of capital [k].) Example 12.8 illustrates this concept.

EXAMPLE 12.8 Project Evaluation by Marginal Cost of Capital

In Example 12.7, suppose that Alpha Corporation has not decided how the $150,000 will be financed. However, Alpha believes that the project should be financed according to its target capital structure, with a debt ratio of 40%. Using k, find the NPW and IRR.

SOLUTION

By not accounting for the cash flows related to debt financing, we calculate the after-tax cash flows as shown in Table 12.4. Notice that when we use this procedure, interest and the resulting tax shield are ignored in deriving the net incremental after-tax cash flow. In other words, no cash flow is related to any financing activity. Thus, taxable income is overstated, as are income taxes. To compensate for these overstatements, the discount rate is reduced accordingly. The implicit assumption is that the tax overpayment is exactly equal to the reduction in interest implied by i_d.

TABLE 12.4 After-Tax Cash Flow Analysis When Project Financing Is Unknown: Cost-of-Capital Approach (Example 12.8)

End of Year	0	1	2	3	4	5
Income Statement						
Revenue		$68,000	$73,000	$79,000	$84,000	$90,000
Expenses						
Operating cost		20,500	20,000	20,500	20,000	20,500
CCA		22,500	38,250	26,775	18,743	13,120
Taxable income		25,000	14,750	31,725	45,258	56,380
Income taxes (38%)		9,500	5,605	12,056	17,198	21,424
Net Income		$15,500	$9,145	$19,670	$28,060	$34,956
Cash Flow Statement						
Operating activities						
Net income		$15,500	$9,145	$19,670	$28,060	$34,956
CCA		22,500	38,250	26,775	18,743	13,120
Investment activities						
Investment and salvage	(150,000)					0
Disposal tax effects						11,633
Net cash flow	($150,000)	$38,000	$47,395	$46,445	$46,802	$59,709

The flow at time 0 is simply the total investment, $150,000 in this example. Recall that Alpha's k was calculated to be 14.74%. The internal rate of return for the after-tax flow in the last line of Table 12.4 is calculated as follows:

$$\text{PW}(i) = -\$150,000 + \$38,000(P/F, i, 1)$$
$$+ \$47,395(P/F, i, 2) + \$46,445(P/F, i, 3)$$
$$+ \$46,802(P/F, i, 4) + \$59,709(P/F, i, 5)$$
$$= 0.$$
$$\text{IRR} = 16.54\% > 14.74\%.$$

Since the IRR exceeds the value of k, the investment would be profitable. Here, we evaluated the after-tax flow by using the value of k, and we reached the same conclusion about the desirability of the investment as we did in Example 12.7.

COMMENTS: The net equity flow and the cost-of-capital methods usually lead to the same accept/reject decision for independent projects (assuming the same amortization schedule for debt repayment, such as term loans) and usually rank projects identically for mutually exclusive alternatives. Some differences may be observed as special financing arrangements may increase (or even decrease) the attractiveness of a project by manipulating tax shields and the timing of financing inflows and payments.

In sum, in cases where the exact debt-financing and repayment schedules are known, we recommend the use of the net equity flow method. The appropriate MARR would be the cost of equity, i_e. If no specific assumption is made about the exact instruments that will be used to finance a particular project (but we do assume that the given capital-structure proportions will be maintained), we may determine the after-tax cash flows without incorporating any debt cash flows. Then we use the marginal cost of capital (k) as the appropriate MARR.

12.3.3 Choice of MARR Under Capital Rationing

It is important to distinguish between the cost of capital (k), as calculated in Section 12.2.3, and the MARR (i) used in project evaluation under **capital rationing**—situations in which the funds available for capital investment are not sufficient to cover all potentially acceptable projects. When investment opportunities exceed the available money supply, we must decide which opportunities are preferable. Obviously, we want to ensure that all the selected projects are more profitable than the best rejected project (or the worst accepted project), which is the best opportunity forgone and whose value is called the **opportunity cost**. When a limit is placed on capital, the MARR is assumed to be equal to this opportunity cost, which is usually greater than the marginal cost of capital. In other words, the value of i represents the corporation's time-value trade-offs and partially reflects the available investment opportunities. Thus, there is nothing illogical about borrowing money at k and evaluating cash flows according to the different rate i. Presumably, the money will be invested to earn a rate i or greater. In the next example, we will provide guidelines for selecting a MARR for project evaluation under capital rationing.

A company may borrow funds to invest in profitable projects, or it may return to (invest in) its **investment pool** any unused funds until they are needed for other investment activities. Here, we may view the borrowing rate as a marginal cost of capital (k), as defined in Eq. (12.6). Suppose that all available funds can be placed in investments yielding a return equal to l, the **lending rate**. We view these funds as an investment pool. The firm may withdraw funds from this pool for other investment purposes, but if left in the pool, the funds will earn at the rate l (which is thus the opportunity cost). The MARR is thus related to either the borrowing interest rate or the lending interest rate. To illustrate the relationship among the borrowing interest rate, the lending interest rate, and the MARR, let us define the following variables:

$$k = \text{Borrowing rate (or the cost of capital)},$$

$$l = \text{Lending rate (or the opportunity cost)},$$

$$i = \text{MARR}.$$

Generally (but not always), we might expect k to be greater than or equal to l: We must pay more for the use of someone else's funds than we can receive for "renting out" our own funds (unless we are running a lending institution). Then we will find that the appropriate MARR would be between l and k. The concept for developing a discount rate (MARR) under capital rationing is best understood by a numerical example.

EXAMPLE 12.9 Determining an Appropriate MARR as a Function of the Budget

Sand Hill Corporation has identified six investment opportunities that will last one year. The firm drew up a list of all potentially acceptable projects. The list shows each project's required investment, projected annual net cash flows, life, and IRR. Then it ranks the projects according to their IRR, listing the highest IRR first:

Project	Cash Flow A0	Cash Flow A1	IRR
1	− $10,000	$12,000	20%
2	− 10,000	11,500	15
3	− 10,000	11,000	10
4	− 10,000	10,800	8
5	− 10,000	10,700	7
6	− 10,000	10,400	4

Suppose $k = 10\%$, which remains constant for the budget amount up to $60,000, and $l = 6\%$. Assuming that the firm has available (a) $40,000, (b) $60,000, and (c) $0 for investments, what is the reasonable choice for the MARR in each case?

SOLUTION

We will consider the following steps to determine the appropriate discount rate (MARR) under a capital-rationing environment:

Step 1: Develop the firm's cost-of-capital schedule as a function of the capital budget. For example, the cost of capital can increase as the amount of financing increases. Also, determine the firm's lending rate if any unspent money is lent out or remains invested in the company's investment pool.

Step 2: Plot this investment opportunity schedule by showing how much money the company could invest at different rates of return as shown in Figure 12.4.

   (a) If the firm has $40,000 available for investing, it should invest in Projects 1, 2, 3, and 4. Clearly, it should not borrow at 10% to invest in either Project 5 or Project 6. In these cases, the best rejected project is Project 5. The worst accepted project is Project 4. If you view the opportunity cost as the cost associated with accepting the worst project, the MARR could be 8%.

   (b) If the firm has $60,000 available, it should invest in Projects 1, 2, 3, 4, and 5. It could lend the remaining $10,000 rather than invest these funds in Project 6. For this new situation, we have MARR $= l = 6\%$.

   (c) If the firm has no funds available, it probably would borrow to invest in Projects 1 and 2. The firm might also borrow to invest in Project 3, but it would be indifferent to this alternative, unless some other consideration was involved. In this case, MARR $= k = 10\%$; therefore, we can say that $l \leq$ MARR $\leq k$ when we have complete certainty about future investment opportunities. Figure 12.5 illustrates the concept of selecting a MARR under capital rationing.

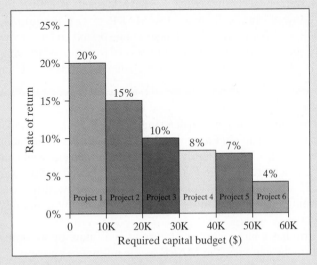

Figure 12.4 An investment opportunity schedule ranking alternatives by the rate of return (Example 12.9).

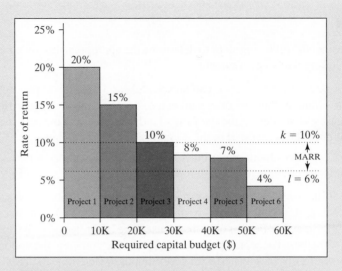

Figure 12.5 A range of MARRs as a function of a budget under capital rationing (Example 12.9): If you have an unlimited budget, MARR = lending rate; if you have no budget, but are allowed to borrow, MARR = 10%.

COMMENTS: In this example, for simplicity, we assumed that the timing of each investment is the same for all competing proposals—say, period 0. If each alternative requires investments over several periods, the analysis will be significantly complicated, as we have to consider both the investment amount and its timing in selecting the appropriate MARR. This is certainly beyond the scope of any introductory engineering economics text, but can be found in C. S. Park and G. P. Sharp-Bette, *Advanced Engineering Economics* (New York: John Wiley, 1990).

Now we can generalize what we have learned. If a firm finances investments through borrowed funds, it should use MARR = k; if the firm is a lender, it should use MARR = l. A firm may be a lender in one period and a borrower in another; consequently, the appropriate rate to use may vary from period to period. In fact, whether the firm is a borrower or a lender may well depend on its investment decisions.

In practice, most firms establish a single MARR for all investment projects. Note the assumption that we made in Example 12.9: **Complete certainty** about investment opportunities was assumed. Under highly uncertain economic environments, the MARR would generally be much greater than k, the firm's cost of capital. For example, if $k = 10\%$, a MARR of 15% would not be considered excessive. Few firms are willing to invest in projects earning only slightly more than their cost of capital, because of elements of *risk* in the project.

If the firm has a large number of current and future opportunities that will yield the desired return, we can view the MARR as the minimum rate at which the firm is willing to invest, and we can also assume that proceeds from current investments can be reinvested to earn at the MARR. Furthermore, *if we choose the "do-nothing" alternative, all available funds are invested at the MARR.* In engineering economics, we also normally separate the risk issue from the concept of MARR. As we will show in Chapter 15, we treat the effects of risk explicitly when we must. Therefore, any reference to the MARR in this book does not account for project risk, although the firm's risk is implicit in the MARR due to the cost of equity estimation.

12.4 Capital Budgeting

In this section, we will examine the process of deciding whether projects should be included in the capital budget. In particular, we will consider decision procedures that should be applied when we have to evaluate a set of multiple investment alternatives for which we have a limited capital budget.

12.4.1 Evaluation of Multiple Investment Alternatives

In Chapter 6, we learned how to compare two or more mutually exclusive projects. Now we shall extend the comparison techniques to a set of multiple decision alternatives that are not necessarily mutually exclusive. Here, we distinguish a **project** from an **investment alternative**, which is a decision option. For a single project, we have two investment alternatives: to accept or reject the project. For two independent projects, we can have four investment alternatives: (1) to accept both projects, (2) to reject both projects, (3) to accept only the first project, and (4) to accept only the second project. As we add interrelated projects, the number of investment alternatives to consider grows exponentially.

To perform a proper capital-budgeting analysis, a firm must group all projects under consideration into decision alternatives. This grouping requires the firm to distinguish between projects that are independent of one another and projects that are dependent on one another in order to formulate alternatives correctly.

Independent Projects

An **independent project** is a project that may be accepted or rejected without influencing the accept–reject decision of another independent project. For example, the purchase of a milling machine, office furniture, and a forklift truck constitutes three independent projects. Only projects that are economically independent of one another can be evaluated separately. (Budget constraints may prevent us from selecting one or more of several independent projects; this external constraint does not alter the fact that the projects are independent.)

Dependent Projects

In many decision problems, several investment projects are related to one another such that the acceptance or rejection of one project influences the acceptance or rejection of others. Two such types of dependencies are mutually exclusive projects and contingent projects. We say that two or more projects are **contingent** if the acceptance of one requires the acceptance of the other. For example, the purchase of a computer printer is dependent upon the purchase of a computer, but the computer may be purchased without purchasing the printer.

12.4.2 Formulation of Mutually Exclusive Alternatives

We can view the selection of investment projects as a problem of selecting a single decision alternative from a set of mutually exclusive alternatives. Note that each investment project is an investment alternative, but that a single investment alternative may entail a whole group of investment projects. The common method of handling various project relationships is to arrange the investment projects so that the selection decision involves only mutually exclusive alternatives. To obtain this set of mutually exclusive alternatives, we need to enumerate all of the feasible combinations of the projects under consideration.

Independent Projects

With a given number of independent investment projects, we can easily enumerate mutually exclusive alternatives. For example, in considering two projects, A and B, we have four decision alternatives, including a do-nothing alternative:

Alternative	Description	X_A	X_B
1	Reject A, Reject B	0	0
2	Accept A, Reject B	1	0
3	Reject A, Accept B	0	1
4	Accept A, Accept B	1	1

In our notation, X_j is a decision variable associated with investment project j. If $X_j = 1$, project j is accepted; if $X_j = 0$, project j is rejected. Since the acceptance of one of these alternatives will exclude any other, the alternatives are mutually exclusive.

Mutually Exclusive Projects

Suppose we are considering two independent sets of projects (A and B). Within each independent set are two mutually exclusive projects (A1, A2) and (B1, B2). The selection of either A1 or A2, however, is also independent of the selection of any project from the set (B1, B2). In other words, you can select A1 and B1 together, but you cannot select A1 and A2 together. For this set of investment projects, the mutually exclusive alternatives are as follows:

Alternative	(X_{A1}, X_{A2})	(X_{B1}, X_{B2})
1	(0, 0)	(0, 0)
2	(1, 0)	(0, 0)
3	(0, 1)	(0, 0)
4	(0, 0)	(1, 0)
5	(0, 0)	(0, 1)
6	(1, 0)	(1, 0)
7	(0, 1)	(1, 0)
8	(1, 0)	(0, 1)
9	(0, 1)	(0, 1)

Note that, with two independent sets of mutually exclusive projects, we have nine different decision alternatives.

Contingent Projects

Suppose the acceptance of C is contingent on the acceptance of both A and B, and the acceptance of B is contingent on the acceptance of A. Then the number of decision alternatives can be formulated as follows:

Alternative	X_A	X_B	X_C
1	0	0	0
2	1	0	0
3	1	1	0
4	1	1	1

Thus, we can easily formulate a set of mutually exclusive investment alternatives with a limited number of projects that are independent, mutually exclusive, or contingent merely by arranging the projects in a logical sequence.

One difficulty with the enumeration approach is that, as the number of projects increases, the number of mutually exclusive alternatives increases exponentially. For example, for 10 independent projects, the number of mutually exclusive alternatives is 2^{10}, or 1024. For 20 independent projects, 2^{20}, or 1,048,576, alternatives exist. As the number of decision alternatives increases, we may have to resort to mathematical programming to find the solution to an investment problem. Fortunately, in real-world business, the number of engineering projects to consider at any one time is usually manageable, so the enumeration approach is a practical one.

12.4.3 Capital-Budgeting Decisions With Limited Budgets

Recall that capital rationing refers to situations in which the funds available for capital investment are not sufficient to cover all the projects. In such situations, we enumerate all investment alternatives as before, but eliminate from consideration any mutually exclusive alternatives that exceed the budget. The most efficient way to proceed in a capital-rationing situation is to select the group of projects that maximizes the total NPW of future cash flows over required investment outlays. Example 12.10 illustrates the concept of an optimal capital budget under a rationing situation.

EXAMPLE 12.10 *Four Energy-Saving Projects Under Budget Constraints*

The facilities department at an electronic instrument firm has four energy-efficiency projects under consideration:

Project 1 (electrical). This project requires replacing the existing standard-efficiency motors in the air-conditioners and exhaust blowers of a particular building with high-efficiency motors.

Project 2 (building envelope). This project involves coating the inside surface of existing fenestration in a building with low-emissivity solar film.

Project 3 (air-conditioning). This project requires the installation of heat exchangers between a building's existing ventilation and relief air ducts.

Project 4 (lighting). This project requires the installation of specular reflectors and the delamping of a building's existing ceiling grid-lighting troffers.

These projects require capital outlays in the $50,000 to $140,000 range and have useful lives of about eight years. The facilities department's first task was to estimate the annual savings that could be realized by these energy-efficiency projects. Currently, the company pays 7.80 cents per kilowatt-hour (kWh) for electricity and $4.85 per thousand cubic metres of air treated. Assuming that the current energy prices would continue for the next eight years, the company has estimated the cash flow and the IRR for each project as follows:

Project	Investment	Annual O&M Cost	Annual Savings (Energy)	Annual Savings (Dollars)	IRR
1	$46,800	$1,200	151,000 kWh	$11,778	15.43%
2	104,850	1,050	513,077 kWh	40,020	33.48%
3	135,480	1,350	6,700,000 m³	32,493	15.95%
4	94,230	942	470,740 kWh	36,718	34.40%

Because each project could be adopted in isolation, at least as many alternatives as projects are possible. For simplicity, assume that all projects are independent, as opposed to being mutually exclusive, that they are equally risky, and that their risks are all equal to those of the firm's average existing assets.

(a) Determine the optimal capital budget for the energy-saving projects.
(b) With $250,000 approved for energy improvement funds during the current fiscal year, the department did not have sufficient capital on hand to undertake all four projects without any additional allocation from headquarters. Enumerate the total number of decison alternatives and select the best alternative.

DISCUSSION: The NPW calculation cannot be shown yet, as we do not know the marginal cost of capital. Therefore, our first task is to develop the **marginal-cost-of-capital** **(MCC)** schedule, a graph that shows how the cost of capital changes as more and more new capital is raised during a given year. The graph in Figure 12.6 is the company's marginal-cost-of-capital schedule. The first $100,000 would be raised at 14%, the next $100,000 at 14.5%, the next $100,000 at 15%, and any amount over $300,000 at 15.5%. We then plot the IRR data for each project as the **investment opportunity schedule** **(IOS)** shown in the graph. The IOS shows how much money the firm could invest at different rates of return.

SOLUTION

(a) Optimal capital budget if projects can be accepted in part:
Consider Project A4. Its IRR is 34.40%, and it can be financed with capital that costs only 14%. Consequently, it should be accepted. Projects A2 and A3 can be analyzed similarly; all are acceptable because the IRR exceeds the marginal cost of capital. Project A1, by contrast, should be rejected because its IRR is less than the marginal cost of capital. Therefore, the firm should accept the three projects A4, A2, and A3, which have rates of return in excess of the cost of capital

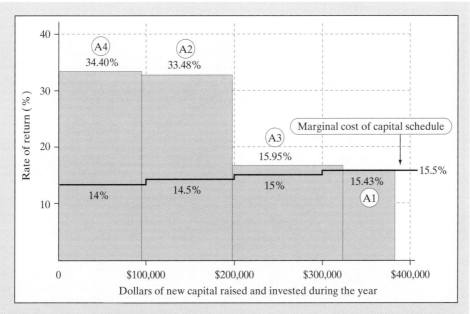

Figure 12.6 Combining the marginal-cost-of-capital schedule and investment opportunity schedule curves to determine a firm's optimal capital budget (Example 12.10).

that would be used to finance them if we end up with a capital budget of $334,560. This should be the amount of the *optimal capital budget*.

In Figure 12.6, even though two rate changes occur in the marginal cost of capital in funding Project A3 (the first change is from 14.5% to 15%, the second from 15% to 15.5%), the accept/reject decision for A3 remains unchanged, as its rate of return exceeds the marginal cost of capital. What would happen if the MCC cut through Project A3? For example, suppose that the marginal cost of capital for any project raised above $300,000 would cost 16% instead of 15.5%, thereby causing the MCC schedule to cut through Project A3. Should we then accept A3? If we can take A3 in part, we would take on only part of it up to 74.49% $((300,000 - 334,560 + 135,480)/135,480 = 0.7449$, or 74.49%).

(b) Optimal capital budget if projects cannot be accepted in part:

If projects cannot be funded partially, we first need to enumerate the number of feasible investment decision alternatives within the budget limit. As shown in Table 12.5, the total number of mutually exclusive decision alternatives that can be obtained from four independent projects is 16, including the do-nothing alternative. However, decision alternatives 13, 14, 15, and 16 are not feasible, because of a $250,000 budget limit. So we need to consider only alternatives 1 through 12.

Now, how do we compare these alternatives as the marginal cost of capital changes for each one? Consider again Figure 12.6. If we take A1 first, it would be acceptable, because its 15.43% return would exceed the 14% cost of capital used to finance it. Why couldn't we do this? The answer is that we are seeking to maximize the excess of returns over costs, or the area above the MCC, but below the IOS. We accomplish that by accepting the most profitable projects first. This logic leads us to conclude that, as long as the budget permits, A4 should be selected first

TABLE 12.5 Mutually Exclusive Decision Alternatives (Example 12.10)

j	Alternative	Required Budget	Combined Annual Savings	
1	0	0	0	
2	A1	$ (46,800)	$ 10,578	
3	A2	(104,850)	38,970	
4	A3	(135,480)	31,145	
5	A4	(94,230)	35,776	
6	A4,A1	(141,030)	46,354	
7	A2,A1	(151,650)	49,548	
8	A3,A1	(182,280)	41,723	
9	A4,A2	(199,080)	74,746	
10	A4,A3	(229,710)	66,921	
11	A2,A3	(240,330)	70,115	
12	A4,A2,A1	(245,880)	85,324	Best alternative
13	A4,A3,A1	(276,510)	77,499	
14	A2,A3,A1	(287,130)	80,693	
15	A4,A2,A3	(334,560)	105,891	Infeasible alternatives
16	A4,A2,A3,A1	(381,360)	116,469	

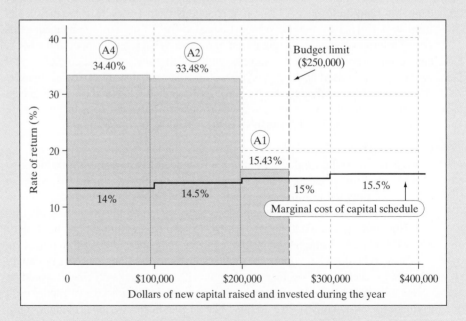

Figure 12.7 The appropriate cost of capital to be used in the capital-budgeting process for decision alternative 12, with a $250,000 budget limit (Example 12.10) is 15%.

and A2 second. This will consume $199,080, which leaves us $50,920 in unspent funds. The question is, What are we going to do with this leftover money? Certainly, it is not enough to take A3 in full, but we can take A1 in full. Full funding for A1 will fit the budget, and the project's rate of return still exceeds the marginal cost of capital $(15.43\% > 15\%)$. Accordingly, unless the leftover funds earn more than 15.43% interest, alternative 12 becomes the best (Figure 12.7).

COMMENTS: In Example 12.9, the MARR was found by applying a capital limit to the investment opportunity schedule. The firm was then allowed to borrow or lend the money as the investment situation dictated. In this example, no such borrowing is explicitly assumed.

SUMMARY

- Methods of financing fall into two broad categories:

    1. **Equity financing** uses retained earnings or funds raised from an issuance of stock to finance a capital investment.

    2. **Debt financing** uses money raised through loans or by an issuance of bonds to finance a capital investment.

- Companies do not simply borrow funds to finance projects. Well-managed firms usually establish a **target capital structure** and strive to maintain the **debt ratio** when individual projects are financed.

- The cost-of-capital formula is a composite index reflecting the cost of funds raised from different sources. The formula is

$$k = \frac{i_d c_d}{V} + \frac{i_e c_e}{V}, V = c_d + c_e.$$

- The selection of an appropriate MARR depends generally upon the **cost of capital**—the rate the firm must pay to various sources for the use of capital:

    1. The **cost of equity** (i_e) is used when debt-financing methods and repayment schedules are known explicitly.

    2. The **cost of capital** (k) is used when exact financing methods are unknown, but a firm keeps its capital structure on target. In this situation, a project's after-tax cash flows contain no debt cash flows, such as principal and interest payment.

- The **marginal cost of capital** is defined as the cost of obtaining another dollar of new capital. The marginal cost rises as more and more capital is raised during a given period.

- Without a capital limit, the choice of MARR is dictated by the availability of financing information:

    1. In cases where the exact debt-financing and repayment schedules are known, we recommend the use of the net equity flow method. The proper MARR would be the cost of equity, i_e.

    2. If no specific assumption is made about the exact instruments that will be used to finance a particular project (but we do assume that the given capital-structure proportions

will be maintained), we may determine the after-tax cash flows without incorporating any debt cash flows. Then we use the marginal cost of capital (k) as the proper MARR.

Conditions	MARR
A firm borrows some capital from lending institutions at the borrowing rate k and some from its investment pool at the lending rate l.	$l <$ MARR $< k$
A firm borrows all capital from lending institutions at the borrowing rate k.	MARR $= k$
A firm borrows all capital from its investment pool at the lending rate l.	MARR $= l$

- Under a highly uncertain economic environment, the MARR generally would be much greater than k, the firm's cost of capital, as the risk increases.

- Under conditions of capital rationing, the selection of the MARR is more difficult, but generally, the following possibilities exist:

  • The cost of capital used in the capital-budgeting process is determined at the intersection of the **IOS** and **MCC** schedules. If the cost of capital at the intersection is used, then the firm will make correct accept/reject decisions, and its level of financing and investment will be optimal. This view assumes that the firm can invest and borrow at the rate where the two curves intersect.

  • If a strict budget is placed in a capital-budgeting problem and no projects can be taken in part, all feasible investment decision scenarios need to be enumerated. Depending upon each such scenario, the cost of capital will also likely change. The task is then to find the best investment scenario in light of a changing-cost-of-capital environment. As the number of projects to consider increases, we may eventually resort to a more advanced technique, such as a mathematical programming procedure.

- **Independent projects** are those for which acceptance or rejection of one project does not affect acceptance or rejection of the other(s). If only independent projects are being considered, the maximum number of investment alternatives is 2^x, where $x =$ the number of independent projects.

- **Dependent projects** are those for which acceptance or rejection of one project affects acceptance or rejection of the other(s). Dependent projects can be:

  1. **Mutually Exclusive**—accepting one project requires rejecting the other(s); or
  2. **Contingent**—accepting one project is contingent upon having first accepted another.

PROBLEMS

Note: Assume all bond coupon payments are annual.

12.1 Optical World Corporation, a manufacturer of peripheral vision storage systems, needs $10 million to market its new robotics-based vision systems. The firm is considering two financing options: common stock and bonds. If the firm decides to raise the capital through issuing common stock, the flotation costs will be 6% and the share price will be $25. If the firm decides to use debt financing, it can sell a 10-year, 12% bond with a par value of $1000. The bond flotation costs will be 1.9%.

(a) For equity financing, determine the flotation costs and the number of shares to be sold to net $10 million.

(b) For debt financing, determine the flotation costs and the number of $1000 par value bonds to be sold to net $10 million. What is the required annual interest payment?

12.2 Consider a project whose initial investment is $300,000, financed at an interest rate of 12% per year. Assuming that the required repayment period is six years, determine the repayment schedule by identifying the principal as well as the interest payments for each of the following methods:

(a) Equal repayment of the principal.

(b) Equal repayment of the interest.

(c) Equal annual installments.

12.3 A chemical plant is considering purchasing a computerized control system. The initial cost is $200,000, and the system will produce net savings of $100,000 per year. If purchased, the system will be written off at a declining balance CCA rate of 30%. The system will be used for four years, at the end of which time the firm expects to sell it for $30,000. The firm's marginal tax rate on this investment is 35%. The firm is considering purchasing the computer control system either through its retained earnings or through borrowing from a local bank. Two commercial banks are willing to lend the $200,000 at an interest rate of 10%, but each requires different repayment plans. Bank A requires four equal annual principal payments, with interest calculated on the basis of the unpaid balance:

Repayment Plan of Bank A		
End of Year	**Principal**	**Interest**
1	$50,000	$20,000
2	50,000	15,000
3	50,000	10,000
4	50,000	5,000

Bank B offers a payment plan that extends over five years, with five equal annual payments:

Repayment Plan of Bank B			
End of Year	**Principal**	**Interest**	**Total**
1	$32,759	$20,000	$52,759
2	36,035	16,724	52,759
3	39,638	13,121	52,759
4	43,602	9,157	52,759
5	47,963	4,796	52,759

(a) Determine the cash flows if the computer control system is to be bought through the company's retained earnings (equity financing).

(b) Determine the cash flows if the asset is financed through either bank A or bank B.

(c) Recommend the best course of financing the project. (Assume that the firm's MARR is known to be 10%.)

12.4 Edison Power Company currently owns and operates a coal-fired combustion turbine plant that was installed 20 years ago. Because of degradation of the system, 65 forced outages occurred during the last year alone and two boiler explosions during the last seven years. Edison is planning to scrap the current plant and install a new, improved gas turbine that produces more energy per unit of fuel than typical coal-fired boilers produce.

The 50-MW gas-turbine plant, which runs on gasified coal, wood, or agricultural wastes, will cost Edison $65 million. Edison wants to raise the capital from three financing sources: 45% common stock, 10% preferred stock (which carries a 6% cash dividend when declared), and 45% borrowed funds. Edison's investment banks quote the following flotation costs:

Financing Source	Flotation Costs	Selling Price	Par Value
Common stock	4.6%	$32/share	$10
Preferred stock	8.1	55/share	15
Bond	1.4	980	1,000

(a) What are the total flotation costs to raise $65 million?

(b) How many shares (both common and preferred) or bonds must be sold to raise $65 million?

(c) If Edison makes annual cash dividends of $2 per common share, and annual bond interest payments are at the rate of 12%, how much cash should Edison have available to meet both the equity and debt obligation? (Note that whenever a firm declares cash dividends to its common stockholders, the preferred stockholders are entitled to receive dividends of 6% of par value.)

12.5 Calculate the after-tax cost of debt under each of the following conditions:

(a) Interest rate, 12%; tax rate, 25%.

(b) Interest rate, 14%; tax rate, 34%.

(c) Interest rate, 15%; tax rate, 40%.

12.6 Sweeney Paper Company is planning to sell $10 million worth of long-term bonds with an 11% interest rate. The company believes that it can sell the $1000 par value bonds at a price that will provide a yield to maturity of 13%. The flotation costs will be 1.9%. If Sweeney's marginal tax rate is 35%, what is its after-tax cost of debt?

12.7 Mobil Appliance Company's earnings, dividends, and stock price are expected to grow at an annual rate of 12%. Mobil's common stock is currently traded at $18 per share. Mobil's last cash dividend was $1.00, and its expected cash dividend for the end of this year is $1.12. Determine the cost of retained earnings (k_r).

12.8 Refer to Problem 12.7, and suppose that Mobil wants to raise capital to finance a new project by issuing new common stock. With the new project, the cash dividend is expected to be $1.10 at the end of the current year, and its growth rate is 10%. The stock now sells for $18, but new common stock can be sold to net Mobil $15 per share.

(a) What is Mobil's flotation cost, expressed as a percentage?

(b) What is Mobil's cost of new common stock (k_e)?

12.9 The Callaway Company's cost of equity is 22%. Its before-tax cost of debt is 13%, and its marginal tax rate is 40%. The firm's capital structure calls for a debt-to-equity ratio of 45%. Calculate Callaway's cost of capital.

12.10 Delta Chemical Corporation is expected to have the following capital structure for the foreseeable future:

Source of After-Tax Financing	Percent of Total Funds	Before-Tax Cost	After-Tax Cost
Debt	30%		
Short term	10	14%	
Long term	20	12	
Equity	70%		
Common stock	55		30%
Prefered stock	15		12

The flotation costs are already included in each cost component. The marginal income tax rate (t_m) for Delta is expected to remain at 40% in the future. Determine the cost of capital (k).

12.11 Maritime Textile Company is considering acquiring a new knitting machine at a cost of $200,000. Because of a rapid change in fashion styles, the need for this particular machine is expected to last only five years, after which the machine is expected to have a salvage value of $50,000. The annual operating cost is estimated at $10,000. The addition of this machine to the current production facility is expected to generate an additional revenue of $90,000 annually and the declining balance CCA rate for this asset is 30%. The income tax rate applicable to Maritime is 36%. The initial investment will be financed with 60% equity and 40% debt. The before-tax debt interest rate, which combines both short-term and long-term financing, is 12%, with the loan to be repaid in equal annual installments. The equity interest rate (i_e), which combines the two sources of common and preferred stocks, is 18%.

(a) Evaluate this investment project by using net equity flows.

(b) Evaluate this investment project by using k.

12.12 The Huron Development Company is considering buying an overhead pulley system. The new system has a purchase price of $100,000, an estimated useful life of five years, and an estimated salvage value of $30,000. It is expected to allow the company to economize on electric power usage, labour, and repair costs, as well as to reduce the number of defective products. A total annual savings of $45,000 will be realized if the new pulley system is installed. The company is in the 30% marginal tax bracket and the system is subject to a 30% CCA rate. The initial investment will be financed with 40% equity and 60% debt. The before-tax debt interest rate, which combines both short-term and long-term financing, is 15%, with the loan to be repaid in equal annual installments over the project life. The equity interest rate (i_e), which combines the two sources of common and preferred stocks, is 20%.

(a) Evaluate this investment project by using net equity flows.

(b) Evaluate this investment project by using k.

12.13 Consider the following two mutually exclusive machines:

	Machine A	Machine B
Initial investment	$40,000	$60,000
Service life	6 years	6 years
Salvage value	$ 4,000	$ 8,000
Annual O&M cost	$ 8,000	$10,000
Annual revenues	$20,000	$28,000
CCA rate	30%	30%

The initial investment will be financed with 70% equity and 30% debt. The before-tax debt interest rate, which combines both short-term and long-term financing, is 10%, with the loan to be repaid in equal annual installments over the project life. The equity interest rate (i_e), which combines the two sources of common and preferred stock, is 15%. The firm's marginal income tax rate is 35%.

(a) Compare the alternatives, using $i_e = 15\%$. Which alternative should be selected?

(b) Compare the alternatives, using k. Which alternative should be selected?

(c) Compare the results obtained in (a) and (b).

12.14 DNA Corporation, a biotech engineering firm, has identified seven R&D projects for funding. Each project is expected to be in the R&D stage for three to five years. The IRR figures shown in the following table represent the royalty income from selling the R&D results to pharmaceutical companies:

Project	Investment Type	Required IRR
1. Vaccines	$15M	22%
2. Carbohydrate chemistry	25M	40
3. Antisense	35M	80
4. Chemical synthesis	5M	15
5. Antibodies	60M	90
6. Peptide chemistry	23M	30
7. Cell transplant/gene therapy	19M	32

DNA Corporation can raise only $100 million. DNA's borrowing rate is 18% and its lending rate is 12%. Which R&D projects should be included in the budget?

12.15 Gene Fowler owns a house that contains 20.2 square metres of windows and 4.0 square metres of doors. Electricity usage totals 46,502 kWh: 7960 kWh for lighting and appliances, 5500 kWh for water heating, 30,181 kWh for space heating to 20°C, and 2861 kWh for space cooling to 25°C. Fowler can borrow money at 12% and lends money at 8%. The following 14 energy-savings alternatives have been suggested by the local power company for Fowler's 162-square-metre home:

Structural Improvement	Annual Savings	Estimated Initial Costs	Payback Period (Years)
1. Add storm windows	$128–156	$455–556	3.5
2. Insulate ceilings to R-30	149–182	408–499	2.7
3. Insulate floors to R-11	158–193	327–399	2.1
4. Caulk windows and doors	25–31	100–122	4.0
5. Weather-strip windows and doors	31–38	224–274	7.2
6. Insulate ducts	184–225	1677–2049	9.1
7. Insulate space-heating water pipes	41–61	152–228	3.7
8. Install heat retardants on E, SE, SW, W windows	37–56	304–456	8.2
9. Install heat-reflecting film on E, SE, SW, W windows	21–31	204–306	9.9
10. Install heat-absorbing film on E, SE, SW, W windows	5–8	204–306	39.5
11. Upgrade 6.5-EER A/C to 9.5-EER unit	21–32	772–1158	36.6
12. Install heat pump water-heating system	115–172	680–1020	5.9
13. Install water heater jacket	26–39	32–48	1.2
14. Install clock thermostat to reduce heat from 20°C to 16°C for 8 hours each night	96–144	88–132	1.1

Note: EER = energy efficiency ratio, R-value indicates the degree of resistance to heat. The higher the number, the greater is the quality of the insulation.

(a) If Fowler stays in the house for the next 10 years, which alternatives would he select with no budget constraint? Assume that his interest rate is 8%. Assume also that all installations would last 10 years. Fowler will be conservative in calculating the net present worth of each alternative (using the minimum annual savings at the maximum cost). Ignore any tax credits available to energy-saving installations.

(b) If Fowler wants to limit his energy-savings investments to $1800, which alternatives should he include in his budget?

Short Case Studies

ST12.1[9] Games Inc. is a publicly traded company that makes computer software and accessories. Games's stock price over the last five years is plotted in Exhibit 12.1, Games's earnings per share for the last five years are shown in Exhibit 12.2, and Games's dividends per share for the last five years are shown in Exhibit 12.3. The company currently has 1,000,000 shares outstanding and a long-term debt of $12,000,000. The company also paid $1,200,000 in interest expenses last year, has other assets of $5,000,000, and had earnings before taxes of $3,500,000 last year.

Exhibit 12.1   Stock price per share.

Exhibit 12.2   Earnings per share.

[9] Contributed by Dr. Luke Miller of Fort Lewis College.

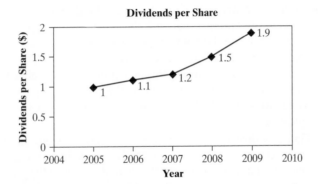

Exhibit 12.3   Dividends per share.

Games has decided to manufacture a new product. In order to make the new product, Games will need to invest in a new piece of equipment that costs $10,000,000. The equipment has a CCA rate of 30%. Equipment installation will require 20 employees working for two weeks and charging $50 per hour. Once the equipment has been installed, the facility is expected to remain operational for two years.

Games intends to maintain its current debt-to-equity ratio and therefore plans on borrowing the appropriate amount today to cover the purchase of the equipment. The interest rate on the loan will be equal to its current cost of debt. The loan will require equal annual interest payments over its life (i.e., the loan rate times the principal borrowed for each year). The principal will be repaid in full at the end of year 2. Games plans on issuing new stock to cover the equity portion of the investment. The underwriter of the new stock issue charges an 11% flotation cost, which will be paid from retained earnings (i.e., rather than from the new stock issue).

Games estimates that its new product will acquire 20% of all North American market share. Even though this product has never been produced before, Games has identified an older product that should have market attributes similar to those of the new product. The older product's unit market sales for North America for the last five years are shown in Exhibit 12.4. It is estimated that the new product will sell for $20 per unit. Costs are estimated at 80% of the selling price for the first year and 60% of the selling price for the second year.

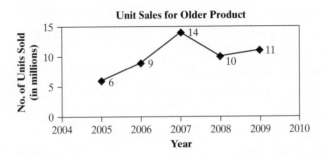

Exhibit 12.4   Total unit sales in North America for the older product.

Due to the large size of the investment required to manufacture the new product, Games's analysts predict that once the decision to accept the project is made, investors will revalue the company's stock price. Because you are the chief engineering economist for Games, you have been requested by upper management to determine how manufacturing this new product will change Games's stock price. Management would like answers to the following questions (*Note:* The first three questions are asked in a sequence that will assist you in answering the last two questions):

(a) Games's investors usually use the corporate value model (CVM) to determine the total market value of the company. In its most basic form, the CVM states that a firm's market value is nothing more than the present value of its expected future net cash flows plus the value of its assets. Using this logic, what must investors currently assess the present value of Games's future net cash flows to be (not including the investment in the new product)?

(b) Determine Games's tax rate.

(c) Determine an appropriate MARR to use in the analysis when the financing source is known. Now do the same when the financing source is unknown.

(d) Assuming that the new-product venture is accepted and the new piece of equipment is disposed of at the end of year 2 for $3,600,000, what is the most likely estimate for Games's stock price?

(e) Now assume that the new piece of equipment is not disposed of in year 2 and that Games has not decided how the $10,000,000 for the equipment will be financed. Instead, assume that the revenues generated from the new product will continue indefinitely. However, any debt incurred will be fully paid off in year 5, with interim payments covering the interest only. Estimate a pessimistic, the most likely, and an optimistic estimate for Games's stock price.

ST12.2 National Food Processing Company is considering investing in plant modernization and plant expansion. These proposed projects would be completed in two years, with varying requirements of money and plant engineering. Management is willing to use the following somewhat uncertain data in selecting the best set of proposals:

No.	Project	Investment Year 1	Investment Year 2	IRR	Engineering Hours
1	Modernize production line	$300,000	0	30%	4,000
2	Build new production line	100,000	$300,000	43%	7,000
3	Provide numerical control for new production line	0	200,000	18%	2,000
4	Modernize maintenance shops	50,000	100,000	25%	6,000
5	Build raw-material processing plant	50,000	300,000	35%	3,000
6	Buy present subcontractor's facilities for raw-material processing	200,000	0	20%	600
7	Buy a new fleet of delivery trucks	70,000	10,000	16%	0

The resource limitations are as follows:

- First-year expenditures: $450,000.
- Second-year expenditures: $420,000.
- Engineering hours: 11,000 hours.

The situation requires that a new or modernized production line be provided (Project 1 or Project 2). The numerical control (Project 3) is applicable only to the new line. The company obviously does not want to both buy (Project 6) and build (Project 5) raw-material processing facilities; it can, if desirable, rely on the present supplier as an independent firm. Neither the maintenance shop project (Project 4) nor the delivery-truck purchase (Project 7) is mandatory.

(a) Enumerate all possible mutually exclusive alternatives without considering the budget and engineering-hour constraints.

(b) Identify all feasible mutually exclusive alternatives.

(c) Suppose that the firm's marginal cost of capital will be 14% for raising the required capital up to $1 million. Which projects would be included in the firm's budget?

ST12.3 Consider the following investment projects:

	Project Cash Flows			
n	A	B	C	D
0	− $2000	− $3000	− $1000	
1	1000	4000	1400	−$1000
2	1000		−100	1090
3	1000			
i^*	23.38%	33.33%	32.45%	9%

Suppose that you have only $3500 available at period 0. Neither additional budgets nor borrowing is allowed in any future budget period. However, you can lend out any remaining funds (or available funds) at 10% interest per period.

(a) If you want to maximize the future worth at period 3, which projects would you select? What is that future worth (the total amount available for lending at the end of period 3)? No partial projects are allowed.

(b) Suppose in (a) that, at period 0, you are allowed to borrow $500 at an interest rate of 13%. The loan has to be repaid at the end of year 1. Which project would you select to maximize your future worth at period 3?

(c) Considering a lending rate of 10% and a borrowing rate of 13%, what would be the most reasonable MARR for project evaluation?

ST12.4 Ally Chemical Corporation (ACC) is a multinational manufacturer of industrial chemical products. ACC has made great progress in reducing energy costs and has implemented several cogeneration projects in the United States and Canada including the completion of a 35-megawatt (MW) unit in Chicago and a 29-MW unit in Vancouver. The division of ACC being considered for one of its more recent cogeneration projects is a chemical plant located in Samia, Ontario. The plant has a power usage of 80 million kilowatt hours (kWh) annually. However, on average, it uses 85% of its 10-MW capacity, which would bring the average power usage to 68 million kWh annually. Ontario Power Generation (OPG) currently charges $0.09 per kWh of electric consumption for the ACC plant, a rate that is considered high throughout the industry. Because ACC's power consumption is so large, the purchase of a cogeneration unit would be desirable. Installation of the unit would allow ACC to generate its own power and to avoid the annual $6,120,000 expense to OPG. The total initial investment cost would be $10,500,000, including $10,000,000 for the purchase of the power unit itself, a gas-fired 10-MW Allison 571, and engineering, design, and site preparation, and $500,000 for the purchase of interconnection equipment, such as poles and distribution lines, that will be used to interface the cogenerator with the existing utility facilities. ACC is considering two financing options:

- ACC could finance $2,000,000 through the manufacturer at 10% for 10 years and the remaining $8,500,000 through issuing common stock. The flotation cost for a common-stock offering is 8.1%, and the stock will be priced at $45 per share.

- Investment bankers have indicated that 10-year 9% bonds could be sold at a price of $900 for each $1000 bond. The flotation costs would be 1.9% to raise $10.5 million.
  (a) Determine the debt-repayment schedule for the term loan from the equipment manufacturer.
  (b) Determine the flotation costs and the number of common stocks to sell to raise the $8,500,000.
  (c) Determine the flotation costs and the number of $1000 par value bonds to be sold to raise $10.5 million.

ST12.5 (Continuation of Problem ST12.4) After ACC management decided to raise the $10.5 million by selling bonds, the company's engineers estimated the operating costs of the cogeneration project. The annual cash flow is composed of many factors: maintenance, standby power, overhaul costs, and miscellaneous expenses. Maintenance costs are projected to be approximately $500,000 per year. The unit must be overhauled every three years at a cost of $1.5 million. Miscellaneous expenses, such as the cost of additional personnel and insurance, are expected to total $1 million. Another annual expense is that for standby power, which is the service provided by the utility in the event of a cogeneration unit trip or scheduled maintenance outage. Unscheduled outages are expected to occur four times annually, each averaging two hours in duration at an annual expense of $6400. Overhauling the unit takes approximately 100 hours and occurs every three years, requiring another triennial power cost of $100,000. Fuel (spot gas) will be consumed at a rate of 8.44 megajoules per kWh, including the heat recovery cycle. At $1.896 per gigajoule, the annual fuel cost will reach $1,280,000. Due to obsolescence, the expected life of the cogeneration project will be 12 years, after which Allison Inc. will pay ACC $1 million for the salvage of all equipment.

Revenue will be realized from the sale of excess electricity to the utility company at a negotiated rate. Since the chemical plant will consume, on average, 85% of the unit's 10-MW output, 15% of the output will be sold at $0.04 per kWh, bringing in an annual revenue of $480,000. ACC's marginal tax rate (combined federal and provincial) is 36%, and the company's minimum required rate of return for any cogeneration project is 27%. The anticipated costs and revenues are summarized as follows:

Initial investment:	
Cogeneration unit, engineering, design, and site preparation (CCA rate = 10%)	$10,000,000
Interconnection equipment (CCA rate = 30%)	500,000
Salvage after 12 years' use	
Cogeneration unit	975,000
Interconnection equipment	25,000
Annual expenses:	
Maintenance	500,000
Misc. (additional personnel and insurance)	1,000,000
Standby power	6,400
Fuel	1,280,000
Other operating expenses:	
Overhaul every 3 years	1,500,000
Standby power during overhaul	100,000
Revenues	
Sale of excess power to OPG	480,000

(a) If the cogeneration unit and other connecting equipment could be financed by issuing corporate bonds at an interest rate of 9% compounded annually, with the flotation expenses as indicated in Problem ST12.4, determine the net cash flow from the cogeneration project.

(b) If the cogeneration unit can be leased, what would be the maximum annual lease amount that ACC is willing to pay?

On the Companion Website that accompanies this text, you will find Excel templates and exercises, as well as the following analysis tools: Cash Flow Analyzer, Depreciation Analysis, Loan Analysis, and Interest Tables.

THIRTEEN

Economic Analysis in the Public Sector

Keeping Ports Safe[1] The Port of Halifax has implemented major upgrades to its marine security arrangements. The centrepiece of this upgrade program is the Halifax Port Authority Command and Control System (HPACCS), which will assist port authorities to deter and respond to both criminal and terrorist activity should either materialize, and to achieve proficiency in response to threat events through scenario-based training.

The HPACCS features the following general system attributes:

- Three-tiered operator interface and control, exercised from numerous geographically dispersed locations.
- An embedded Cruise Vessel Facility (CVF) Command and Control (C2) capability that permits local control and monitoring of CVF system and remote control monitoring of the same when required.
- Command and Control over legacy and planned CCTV surveillance and access control subsystems and hardware.
- Interfaces with the security infrastructure of Port of Halifax marine facilities and other surveillance systems under a progressively expanding program.
- Communication tools that will permit inter-operator communications, as well as operator communications with mobile units.
- Remote, web-based access and collaborative communication tools that will permit limited and/or selected and approved access to HPACCS, which will provide functionality and rapid communications for both training and incident response purposes.
- Planning and decision support tools including two dimensional Halifax Regional Municipality (HRM) land maps, Halifax Harbour charts, and detailed Port of Halifax marine facility models.

[1] www.portofhalifax.ca/english/port-facilities/port-security/security-projects.html.

Source: Illustration based on photo © Dr. Heinz Linke/iStockphoto.

- An independent, secure fibre optic network (SecNet) that hosts the HPACCS and its related surveillance and access control components. This network will permit the transmission of low-level classified material and secure communications with public sector agencies.

The HPA also hosts the following applications:

- Chemical Detection System (CDS)—Integrated with the HPACCS, the CDS sensors provide early warning of the presence or leakage of hazardous chemicals in the vicinity of stowage areas for containerized dangerous goods.
- Video Analytic Software—All existing and planned cameras will be monitored by HPACCS and their imagery feed processed by video analytic in order to support event investigation, intruder identification, object monitoring, licence plate recognition, evidence gathering, and a number of other capabilities.
- Perimeter and Fence Monitoring System—The perimeters and integral fences of several of the Port's marine facilities are monitored by a combination of a microwave intrusion detection system and a cable-based fence monitoring system.

The development of each of these parts of the overall security system involved an assessment in order to choose the best components based on cost, reliability, effectiveness, and other criteria. Now, the HPA is also planning to upgrade systems to improve port access control for a safer environment. The benefits of each new development can be compared against the cost to help prioritize the expenditures to achieve the most "bang for the buck."

Up to this point, we have focused our analysis on the economic issues related to the manufacturing sector of the Canadian economy. The main reason for doing this was that many engineers will be working in that sector. However, an increasing number of engineers are now seeking their careers in the service sector, such as health care, financial institutions, transportation and logistics, and government. In this chapter, we will present some unique features that must be considered in evaluating investment projects in the service sector.

CHAPTER LEARNING OBJECTIVES

After completing this chapter, you should understand the following concepts:

- How to price service.
- How to evaluate investment projects in the health-care industry.
- How to conduct cost benefit analysis and cost effectiveness analysis.

13.1 What Is the Service Sector?[2]

The service sector of the Canadian economy dominates both gross domestic product (GDP) and employment. It is also the fastest-growing part of the economy and the one offering the most fertile opportunities to engineers to improve their productivity. For example, service activities now approach 75% of Canadian employment, far outstripping sectors such as manufacturing (13%) and agriculture (2%), as shown in Figure 13.1.

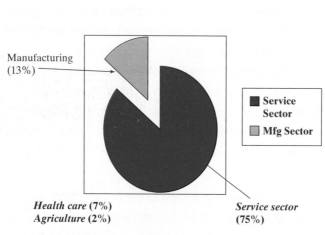

- Real Estate and Rental and Leasing
- Wholesale
- Finance and Insurance
- Retail
- Health Care and Social Assistance
- Public Administration
- Transportation and Warehousing
- Professional, Scientific, and Technical
- Educational
- Information and Cultural
- Administration and Support
- Accommodation and Food
- Arts, Entertainment, and Recreational
- Management of Companies and Enterprises

Manufacturing (13%)

■ Service Sector
▢ Mfg Sector

Health care (7%)
Agriculture (2%)

Service sector (75%)

Figure 13.1 Contribution of the service sector to the Canadian gross domestic product (GDP).

[2] This section is patterned on the National Science Foundation program "Exploratory Research on Engineering the Service Sector (ESS)," NSF 02-029.

The mere scale of the service sector makes it a critical element of the Canadian economy, employing, as it does, many millions of workers producing almost a trillion dollars in economic value. The service sector has expanded far beyond the traditional consumer or institutional service industries and currently includes distributive services, such as transportation, communications, and utilities; the rapidly expanding producer services, such as finance, insurance, real estate, and advertising; wholesale and retail trade and sales; the nonprofit sector, including health, philanthropy, and education; and government. The use of technology in the service sector promotes deskilling (i.e., automation). However, secure and reliable Canadian services provide much of the key infrastructure on which the whole nation depends. Tremendous new challenges have arisen as changes in processes and operations are developed to better assure the safety and reliability of critical services. These changes all lead to significant investment in infrastructures requiring detailed economic analysis.

13.1.1 Characteristics of the Service Sector

What makes the evaluation of a service sector project unique compared with one from the manufacturing sector? Some of the unique features are summarized as follows:

- Services are generally intangible. They have sometimes been defined as "anything of economic value that cannot be held or touched."
- It is usually impossible to build inventories of services. Either the demand for the service must be backlogged, or enough resources need to be provided to meet an acceptable fraction of the demand as it arises.
- Services are more dynamic and responsive to demand than are manufactured products. This means that variability and risk are more central issues in service industries. Indeed, the management of financial risk is an important service in itself.
- Many services (examples are medical treatment and equipment repair) require a diagnostic step to design the service as part of its delivery. Coproduction (i.e., active collaboration between the server and the customer) is also required in many settings.
- Service products are usually less standardized and less subject to design specifications than manufactured goods are, because the outputs are tailored to customer needs as they are delivered. This also makes it harder to distinguish service product design from product manufacture and delivery.
- The dimensions of service quality are more subtle and subjective than those of physical products. Not only are the parameters of services more difficult to express, but customers' perceptions play a much greater role in deciding what is satisfactory or valuable.
- Most service operations are more labour intensive than the production of goods.
- Compared with goods industries, the service economy has a much greater fraction of its operations performed by governments and institutions.
- Information technology is central to service industries. Often, it is the only significant means of multiplying human output.

Certainly, our objective is not to address all aspects of the service sector, but only some of the common economic analysis issues confronted by its engineers.

13.1.2 How to Price Service

Improving service can be many different things. For example, for a delivery business such as Canpar or FedEx, anything that reduces the total time taken from pickup to delivery is considered an improvement of service. For an airline, on-time departures and arrivals, a reduction in the number of mishandled pieces of checked-in baggage, and a speedy check-in at the gate could all be considered important service parameters, because they make airline passengers happy, which in turn will translate into more business volume. Accordingly, one of the critical questions related to improvement in service is "What do service providers gain by improving their own service to suppliers and customers?" If we can quantify improvements in service in terms of dollars, the economic analysis is rather straightforward: If the net increase in revenue due to improvements in service exceeds the investment required, the project can be justified. For this type of decision problem, we can simply use any of the measures of investment worth discussed in Chapter 5.

Improving service is one thing and putting a price on services is another thing. This is far more difficult than pricing products, because the benefits of services are less tangible and service companies often lack well-documented standard unit-production costs to go by. When a company designs an after-service contract to go with equipment sales, customers may be segmented according to their service needs rather than their size, industry, or type of equipment. Companies then develop the pricing, contracting, and monitoring capabilities to support the cost-effective delivery of the service.[3] For example, when customers are segmented according to the service level they need, they tend to fall into one of at least three common categories. The "basic-needs customers" want a standard level of service with basic inspections and periodic maintenance. The "risk avoiders" are looking for coverage to avoid big bills but care less about other elements, such as response times. And the "hand holders" need high levels of service, often with quick and reliable response times, and are willing to pay for the privilege. Therefore, to maximize return, companies need to capture tremendous value from their service businesses by taking a more careful, fact-based approach to designing and pricing services (Figure 13.2).

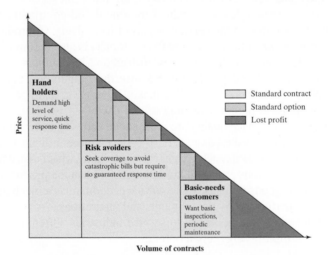

Figure 13.2 How to capture lost profits. By segmenting customers into three categories with appropriate service options, companies can capture much of the lost profit. Options may include guaranteed response time, remote equipment monitoring, or extended warranty.

---

[3] Russell G. Bundschuh and Theodore M. Dezvane, "How to Make After-Sales Services Pay Off," *The McKinsey Quarterly*, 2003, no. 4, pp. 3–13.

   The types of service problems that we are interested in this chapter, however, are
(1) those encountered by nonprofit organizations such as hospitals or universities and
(2) those involving economic decisions by the public sector. In the next section, we will
review some of the common decision tools adopted by the health-care service industry.
The economic issues in the public sector will be discussed in Section 13.3.

13.2 Economic Analysis in Health-Care Service

The health-care service industry alone constitutes about 10% of GDP when all its dimensions are included. Clearly, health care is one of the largest, most research-intensive
service industries in Canada and elsewhere. Accordingly, its medical knowledge base
is expanding at a staggering pace, and many Canadians enjoy unparalleled advance-
ments in medical science and technology. At the same time, the nation's health-care system is not as effective as it could be: Variation in both access to and delivery of care is
considerable, errors are widespread, costs are spiralling, and relatively little attention is
devoted to optimizing system operations and improving the delivery of care. Nevertheless, researchers and economists, such as those at the Institute of Health Economics in
Alberta, continue to study the issues to come up with better ways to use the limited resources more effectively in the health sector.

13.2.1 Economic Evaluation Tools

Economic evaluation can be used to inform decision making and can provide information
to assist in answering the following questions:

- What services do we provide (or improve), when, and at what level?
- How do we provide such services?
- Where do we provide the services?
- What are the costs associated with providing or improving the services?

The following three methods of economic evaluation are related to health service:

- **Cost-effectiveness analysis.** This technique is used in health economics to compare
  the financial costs of therapies whose outcomes can be measured purely in terms of
  their health effect (e.g., years of life saved or ulcers healed). Cost-effectiveness
  analysis (CEA) is the most commonly applied form of economic analysis in health
  economics. However, it does not allow comparisons to be made between courses of
  action that have different health effects.
- **Cost-utility analysis.** This technique is similar to CEA in that there is a defined
  outcome and the cost to achieve that outcome is measured in money. The outcome
  is measured in terms of survival and quality of life (for example, quality-adjusted
  life years, or QALYs). CEA can indicate which one of a number of alternative interventions represents the best value for money, but it is not as useful when comparisons need to be made across different areas of health care, since the outcome
  measures used may be very different. As long as the outcome measure is life years
  saved or gained, a comparison can still be made, but even in such situations CEA
  remains insensitive to the *quality*-of-life dimension. In order to know which areas of
  health care are likely to provide the greatest benefit in improving health status, a

Cost–benefit analysis is the process of weighing the total expected cost vs. the total expected benefits of one or more actions in order to choose the best or most profitable option.

cost-utility analysis[4] needs to be undertaken using a common currency for measuring the outcomes across health-care areas.

- **Cost–benefit analysis.** If information is needed as to which interventions will result in overall resource savings, a cost–benefit analysis (CBA) has to be performed, although, like a cost-utility analysis, a cost-benefit analysis has its own drawbacks. In CBA, the benefit is measured as the associated economic benefit of an intervention; hence, both costs and benefits are expressed in money, and the CBA may ignore many intangible, but very important, benefits that are difficult to measure in monetary terms (e.g., relief of pain and anxiety). Even though the virtue of this analysis is that it enables comparisons to be made between schemes in very different areas of health care, the approach is not widely accepted for use in health economics.

13.2.2 Cost-Effectiveness Analysis

In cost-effectiveness analysis (CEA), outcomes are reported in a single unit of measurement. CEA compares the costs and health effects of an intervention to assess whether it is worth doing from an economic perspective. First of all, CEA is a specific type of economic analysis in which all costs are related to a single, common effect. Decision makers can use it to compare different resource allocation options in like terms. A general misconception is that CEA is merely a means of finding the least expensive alternative or getting the "most bang for the buck." In reality, CEA is a comparison tool; it will not always indicate a clear choice, but it will evaluate options quantitatively and objectively on the basis of a defined model. CEA was designed to evaluate health-care interventions, but the methodology can be used for non-health-economic applications as well. CEA can compare any resource allocation with measurable outcomes.

What Constitutes a Cost?

In CEA, it is common to distinguish between the direct costs and the indirect costs associated with an intervention. Some interventions may also result in intangibles, which are difficult to quantify, but should be included in the cost profile. Examples of these different kinds of costs are as follows:

- **Direct cost.** Drugs, medical staff time, medical equipment, transport, and out-of-pocket expenses by the patients.
- **Indirect costs.** Loss of productive time by the patients during the intervention.
- **Intangibles.** Pain and suffering, and adverse effects from the intervention.

It is essential to specify which costs are included in a CEA and which are not, to ensure that the findings are not subject to misinterpretation.

Cost-Effectiveness Ratio

The cost-effectiveness ratio is simply the sum of all costs, divided by the sum of all health effects:

$$\text{Cost-effectiveness ratio} = \frac{\sum(\text{all costs})}{\sum(\text{all measured health effects})} \tag{13.1}$$

---

[4] We will not discuss any technical details of the cost-utility approach in this chapter, but they can be found in a variety of health economics texts.

The benefits are not measured in terms of just dollars, but as a part of a ratio that incorporates both health outcomes and dollars.

For decisions within a given institution or health agency, cost-effectiveness ratios should be related to the size of relevant budgets to determine the most cost-effective strategies. CEAs compare several program strategies and rank them by cost-effectiveness ratios. An analysis of two screening interventions might show you that one costs $10,000 per life year gained while the other costs $40,000 per life year gained. The first intervention requires monthly screening and the second requires biannual screening. Realizing that compliance is a greater problem with monthly screening, the decision maker would implement the most appropriate coverage strategy for the population in question. Sometimes, the analysis compares an option against a baseline option, such as "Do nothing" or "Give usual care." The last two are valid strategic options.

Discounting

There is often a significant time lag between the investment of health service resources and the arrival of the associated health gain. In general, we prefer to receive benefits now and pay costs in the future. In order to reflect this preference in economic evaluation, costs are discounted.

13.2.3 How to Use a CEA[5]

When we use a CEA, we need to distinguish between those interventions that are completely independent and those that are dependent. Two (or more) interventions are said to be **independent** if the costs and effects of one neither affect nor are affected by the costs and effects of the other. Two (or more) interventions are dependent if the implementation of one results in changes to the costs and effects of the other. The analysis proceeds as follows:

- **Independent interventions.** Using CEA with independent intervention programs requires that cost-effectiveness ratios be calculated for each program and placed in rank order. For example, in Table 13.1, there are three interventions for different patient groups, and each intervention has as an alternative "doing nothing." According to CEA, Program C should be given priority over Program A, since it has a lower cost-effectiveness ratio (CER), but in order to decide which program to implement, the extent of the resources available must be considered. (See Table 13.2.) Clearly, the choice of independent intervention is a function of the budget that is available to

TABLE 13.1 Cost-Effectiveness of Three Independent Intervention Programs

Program	Cost ($)	Health Effect (Life Years Gained)	Cost-Effectiveness Ratio
A	100,000	1,200	83.33
B	120,000	1,350	88.89
C	150,000	1,850	81.08

[5] This section is based on the article "What Is Cost-effectiveness?" by Ceri Phillips and Guy Thompson, vol. 1, no. 3, Hayward Medical Communications, Copyright © 2001; www.evidence-based-medicine.co.uk.

TABLE 13.2 Choices of Program as a Function of Budget

Budget Available ($)	Programs to Be Implemented
Less than $150,000	As much of C as budget allows
$150,000	100% of C
$150,000–$250,000	C and as much of A as budget allows
$250,000	C and A
$250,000–$370,000	C and A, and as much of B as budget allows
$370,000	All three programs, A, B, C

implement. For example, with $200,000, we will go with Program C, and the remaining $50,000 will be available for funding (up to 50%) of Program A.

- **Mutually exclusive interventions.** In reality, the likelihood is that choices will have to be made between different treatment regiments for the same condition and between different dosages or treatments versus prophylaxis (i.e., mutually exclusive interventions). In this case, incremental cost-effectiveness ratios (ΔCERs) should be used:

$$\Delta CER_{2-1} = \frac{\text{Cost of P2} - \text{Cost of P1}}{\text{Effects of P2} - \text{Effects of P1}}$$

The alternative interventions are ranked according to their effectiveness—on the basis of securing the maximum effect rather than considering cost—and CERs are calculated as shown in Table 13.3.

TABLE 13.3 Mutually Exclusive Intervention Programs

Program	Cost	Effects (LifeYears Gained)	Cost-Effectiveness (LifeYears Gained)	Incremental Cost-Effectiveness Ratio
P1	$125,000	1,300	96.15	96.15
P2	$100,000	1,500	66.67	−125
P3	$160,000	2,000	80.00	120
P4	$140,000	2,200	63.63	−100
P5	$170,000	2,600	65.38	75

The analysis proceeds as follows:

- If money is no object, P5 is clearly the best alternative, as the number of life years gained is most significant.
- The least effective intervention (P1) has the same average CER as its incremental cost-effectiveness ratio (ICER), because it is compared with the alternative of "doing nothing":

$$\Delta CER_{1-0} = \frac{125,000 - 0}{1,300 - 0}$$

$$= 96.15.$$

- A comparison between P1 and P2 yields

$$\Delta CER_{2-1} = \frac{\text{Cost of P2} - \text{Cost of P1}}{\text{Effect of P2} - \text{Effect of P1}}$$

$$= \frac{100,000 - 125,000}{1,500 - 1300}$$

$$= -125 < 96.15.$$

The negative ICER for P2 means that adopting P2 rather than P1 results in an improvement in life years gained and a reduction in costs. It also indicates that P2 dominates P1. Hence, we can eliminate P1 at this stage of the analysis.

- A comparison between P2 and P3 yields

$$\Delta CER_{3-2} = \frac{160,000 - 100,000}{2,000 - 1500}$$

$$= 120 > 66.67.$$

The ICER for P3 works out to be 120, which means that it costs $120 to generate each additional life year gained compared with P2. Thus, there is no clear dominance between P2 and P3.

- A comparison between P4 and P3 yields

$$\Delta CER_{4-3} = \frac{140,000 - 160,000}{2200 - 2000}$$

$$= -100 < 80.$$

P4 is more effective than P3 as the incremental cost-effectiveness ratio becomes negative. Also, P4 dominates P3, so we can eliminate P3 from the analysis. Having excluded P1 and P3, we now recalculate for P2, P4, and P5, as shown in Table 13.4.

- A comparison between P2 and P4 yields

$$\Delta CER_{4-2} = \frac{140,000 - 100,000}{2,200 - 1500}$$

$$= \$57.14 < 66.67.$$

Thus, P2 is dominated by P4, since the latter is more effective and costs less to produce an additional unit of effect ($57.14 compared with $66.67). The dominated alternative is then excluded and the ICERs are recalculated again (Table 13.5).

TABLE 13.4 Remaining Mutually Exclusive Alternatives After Eliminating More Costly and Less Effective Programs

Program	Cost	Effects (Life Years Gained)	Cost-Effectiveness (Life Years Gained)	Incremental Cost-Effectiveness Ratio
P2	$100,000	1,500	66.67	66.67
P4	$140,000	2,200	63.63	57.14
P5	$170,000	2,600	65.38	75.00

TABLE 13.5 Remaining Mutually Exclusive Alternatives After Eliminating All Dominated Programs

Program	Cost	Effects (Life Years Gained)	Cost-Effectiveness (Life Years Gained)	Incremental Cost-Effectiveness Ratio
P4	$140,000	2,200	63.63	63.63
P5	$170,000	2,600	65.38	75.00

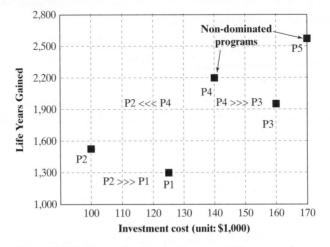

Figure 13.3 Cost-effectiveness diagram—Life years gained.

- Finally, a comparison between P4 and P5 yields

$$\Delta CER_{5-4} = \frac{170,000 - 140,000}{2600 - 2200}$$

$$= 75 > 63.63.$$

No clear dominance exists between P4 and P5. As shown in Figure 13.3, these two programs are therefore the ones that deserve funding consideration. In deciding between them, the size of the available budget must be brought to bear on the matter. If the available budget is $140,000, all patients should receive intervention P4, whereas if the available budget is $170,000, all patients should receive the more effective P5. However, if the budget is, say, $150,000, then, since the cost difference between P4 and P5 is $30,000 and the budget surplus is $10,000, it is possible to switch one-third of the patients to P5 and still remain within budget.

13.3 Economic Analysis in the Public Sector

In earlier chapters, we have focused attention on investment decisions in the private sector; the primary objective of these investments was to increase the wealth of corporations. In the public sector, federal, provincial, and local governments spend hundreds of billions of dollars annually on a wide variety of public activities, such as the port project described in the chapter opener. In addition, governments at all levels regulate the behaviour of individuals and businesses by influencing the use of enormous quantities of productive resources. How can public decision makers determine whether their decisions, which affect the use of these productive resources, are, in fact, in the best public interest?

Many civil engineers work on public-works areas such as highway construction, airport construction, and water projects. In the port security scenario presented at the beginning of the chapter, each option requires a different level of investment and produces a different degree of benefits. One of the most important aspects of airport expansion is to quantify the cost of airport delays in dollar terms. In other words, planners ask, "What is the economic benefit of reducing airport delays?" From the airline's point of view, taxiing and arrival delays mean added fuel costs. For the airport, delays mean lost revenues in landing and departure fees. From the public's point of view, delays mean lost earnings, as people then have to spend more time on transportation. Comparing the investment costs with the potential benefits, an approach known as benefit–cost analysis, is an important feature of economic analysis.

13.3.1 What Is Benefit–Cost Analysis?

Benefit–cost analysis is a decision-making tool that is used to systematically develop useful information about the desirable and undesirable effects of public projects. In a sense, we may view benefit–cost analysis in the public sector as profitability analysis in the private sector. In other words, benefit–cost analysis attempts to determine whether the social benefits of a proposed public activity outweigh the social costs. Usually, public investment decisions involve a great deal of expenditure, and their benefits are expected to occur over an extended period of time. Examples of benefit–cost analysis include studies of (1) public transportation systems, (2) environmental regulations on noise and pollution, (3) public-safety programs, (4) education and training programs, (5) public-health programs, (6) flood control, (7) water resource development, and (8) national defence programs.

> **Benefit–cost analysis:** A technique designed to determine the feasibility of a project or plan by quantifying its costs and benefits.

The three typical aims of benefit–cost analyses are (1) to maximize the benefits for any given set of costs (or budgets), (2) to maximize the net benefits when both benefits and costs vary, and (3) to minimize costs to achieve any given level of benefits (often called "cost-effectiveness" analysis). Three types of decision problems, each having to do with one of these aims, will be considered in this chapter.

13.3.2 Framework of Benefit–Cost Analysis

To evaluate public projects designed to accomplish widely differing tasks, we need to measure the benefits or costs in the same units in all projects so that we have a common perspective by which to judge the different projects. In practice, this means expressing both benefits and costs in monetary units, a process that often must be performed without accurate data. In performing benefit–cost analysis, we define **users** as the public and **sponsors** as the government.

The general framework for benefit–cost analysis can be summarized as follows:

1. Identify all users' benefits expected to arise from the project.
2. Quantify these benefits in dollar terms as much as possible, so that different benefits may be compared against one another and against the costs of attaining them.
3. Identify sponsors' costs.
4. Quantify these costs in dollar terms as much as possible, to allow comparisons.

5. Determine the equivalent benefits and costs during the base period; use an interest rate appropriate for the project.

6. Accept the project if the equivalent users' benefits exceed the equivalent sponsor's costs.

We can use benefit–cost analysis to choose among such alternatives as allocating funds for the construction of a mass-transit system, a dam with irrigation, highways, or an air-traffic control system. If the projects are on the same scale with respect to cost, it is merely a question of choosing the project for which the benefits exceed the costs by the greatest amount. The steps just outlined are for a single (or independent) project evaluation. As in the case of the internal-rate-of-return criterion, in comparing mutually exclusive alternatives, an incremental benefit–cost ratio must be used. Section 13.3.3 illustrates this important issue in detail.

13.3.3 Valuation of Benefits and Costs

In the abstract, the framework we just developed for benefit–cost analysis is no different from the one we have used throughout this text to evaluate private investment projects. The complications, as we shall discover in practice, arise in trying to identify and assign values to all the benefits and costs of a public project.

Users' Benefits

To begin a benefit–cost analysis, we identify all project **benefits** (favourable outcomes) and **disbenefits** (unfavourable outcomes) to the user. We should also consider the indirect consequences resulting from the project—the so-called **secondary effects**. For example, the construction of a new highway will create new businesses such as gas stations, restaurants, and motels (benefits), but it will divert some traffic from the old road, and as a consequence, some businesses would be lost (disbenefits). Once the benefits and disbenefits are quantified, we define the users' benefits as follows:

$$\text{Users' benefits (B)} = \text{Benefits} - \text{Disbenefits}.$$

In identifying user's benefits, we should classify each one as a **primary benefit**—a benefit that is directly attributable to the project—or a **secondary benefit**—a benefit that is indirectly attributable to the project. As an example, the Province of Manitoba and the Canadian government announced in 2009 the establishment of a world-class cold weather testing centre in Thompson, Manitoba. The Canadian Environmental Test Research and Education Centre (CanETREC) will bring many scientists and engineers along with other supporting population to the region. It will focus initially on the aviation industry, with Rolls-Royce Canada Limited and Pratt & Whitney investing in an adjacent large-scale aerospace testing facility. Future plans include extending CanETREC's scope into research on noise and emissions reduction, and alternative fuels and lubricants. Primary national benefits from this facility include the long-term benefits that may accrue as a result of various applications of the research to Canadian businesses. Primary regional benefits may include economic benefits created by the research laboratory activities, which would generate many new supporting businesses. The secondary benefits might include the creation of new economic wealth as a consequence of a possible increase in international trade and any increase in the incomes of various regional producers attributable to a growing population.

The reason for making this distinction is that it may make our analysis more efficient: If primary benefits alone are sufficient to justify project costs, we can save time and effort by *not* quantifying the secondary benefits.

Sponsor's Costs

We determine the cost to the sponsor by identifying and classifying the expenditures required and any savings (or revenues) to be realized by the sponsor. The sponsor's costs should include both capital investment and annual operating costs. Any sales of products or services that take place upon completion of the project will generate some revenues—for example, toll revenues on highways. These revenues reduce the sponsor's costs. Therefore, we calculate the sponsor's costs by combining these cost elements:

Sponsor's cost = Capital cost + Operating and maintenance costs − Revenues.

Social Discount Rate

As we learned in Chapter 12, the selection of an appropriate MARR for evaluating an investment project is a critical issue in the private sector. In public-project analyses, we also need to select an interest rate, called the **social discount rate**, to determine equivalent benefits as well as the equivalent costs. The selection of a social discount rate in public project evaluation is as critical as the selection of a MARR in the private sector.

When present equivalent calculations were initiated to evaluate public water resources and related land-use projects in the 1930s, relatively low discount rates were adopted compared to those existing in markets for private assets. The choice of a suitable **social discount rate** has undergone much debate over the intervening years. There are two prevailing views on the most suitable basis for establishing this rate.

One is referred to as the social rate of time preference (STP), which reflects the trade-off between current and future consumption. There are many reasons why the population may discount future benefits compared with earlier ones. It can be argued that people are nearsighted, preferring immediate satisfaction, or that they discount later benefits since they will not be around (eventually) to enjoy them. Conversely, if people are altruistic they would tend to weight benefits to future generations equally with gains in the near term.

A second perspective, referred to as the social opportunity cost of capital (SOCC), considers what other uses would be made with the money if the current public project investment(s) were not undertaken. One view is that the money would remain in the private sector, in which case the social discount rate is essentially based on expected rates of return averaged across the business sector. However, income taxes complicate this assessment. The government could select an average *before-tax* MARR from the private sector as the benchmark rate for project acceptance, or refer instead to the *after-tax* MARR expected by investors. Conversely, if public capital funds are considered rationed, then the relevant opportunity cost for a public project should be another public project rather than private sector activity. Furthermore, a baseline for public expenditures could be the cost of government debt, as reflected in long-term government bond rates.

In recent years, with the growing interest in performance budgeting and systems analysis in the 1960s, the tendency on the part of government agencies has been to examine the appropriateness of the discount rate in the public sector in relation to the efficient allocation of resources in the economic system as a whole.[6] Two views of the basis for determining the social discount rate prevail:

1. **Projects without private counterparts.** *The social discount rate should reflect only the prevailing government borrowing rate.* Projects such as dams designed

[6] R. F. Mikesell, *The Rate of Discount for Evaluating Public Projects.* American Enterprise Institute for Public Policy Research, 1977.

purely for flood control, access roads for noncommercial uses, and reservoirs for community water supply may not have corresponding private counterparts. In those areas of government activity where benefit–cost analysis has been employed in evaluation, the rate of discount traditionally used has been the cost of government borrowing.

2. **Projects with private counterparts.** *The social discount rate should represent the rate that could have been earned had the funds not been removed from the private sector.* For public projects that are financed by borrowing at the expense of private investment, we may focus on the opportunity cost of capital in alternative investments in the private sector to determine the social discount rate. In the case of public capital projects, similar to some in the private sector that produce a commodity or a service (such as electric power) to be sold on the market, the rate of discount employed would be the average cost of capital as discussed in Chapter 12. The reasons for using the private rate of return as the opportunity cost of capital in projects similar to those in the private sector are (1) to prevent the public sector from transferring capital from higher yielding to lower yielding investments and (2) to force public-project evaluators to employ market standards in justifying projects.

The Treasury Board of Canada provides guidelines for the evaluation of public projects, using 10% per annum after inflation as the base case. Since the acceptability of some projects is very sensitive to the applied social discount rate, a base-case social discount rate is suggested, with a somewhat lower rate and commensurately higher rate suggested to conduct sensitivity evaluations (see Section 15.2.3).

13.3.4 Quantifying Benefits and Costs

Now that we have defined the general framework for benefit–cost analysis and discussed the appropriate discount rate, we will illustrate the process of quantifying the benefits and costs associated with a public project.[7]

Some provinces employ inspection systems for motor vehicles. Critics often claim that these programs lack efficacy and have a poor benefit-to-cost ratio in terms of reducing fatalities, injuries, accidents, and pollution.

Elements of Benefits and Costs

Primary and secondary benefits identified with a motor vehicle inspection program are as follows:

- **Users' Benefits**

  Primary benefits. Deaths and injuries related to motor-vehicle accidents impose financial costs on individuals and society. Preventing such costs through the inspection program has the following primary benefits:

  1. Retention of contributions to society that might be lost due to an individual's death.

  2. Retention of productivity that might be lost while an individual recuperates from an accident.

---

[7] Based on P. D. Loeb and B. Gilad, "The Efficacy and Cost Effectiveness of Vehicle Inspection," *Journal of Transport Economics and Policy*, May 1984: 145–164. The original cost data, which were given in 1981 dollars, were converted to the equivalent cost data in 2000 by using the prevailing consumer price indices during the period.

3. Savings of medical, legal, and insurance services.

4. Savings on property replacement or repair costs.

Secondary benefits. Some secondary benefits are not measurable (e.g., the avoidance of pain and suffering); others can be quantified. Both types of benefits should be considered. A list of secondary benefits is as follows:

1. Savings of income of families and friends of accident victims who might otherwise be tending to the victims.

2. Avoidance of air and noise pollution and savings on fuel costs.

3. Savings on enforcement and administrative costs related to the investigation of accidents.

4. Avoidance of pain and suffering.

- **Users' Disbenefits**

  1. Cost of spending time to have a vehicle inspected (including travel time), as opposed to devoting that time to an alternative endeavour (opportunity cost).

  2. Cost of inspection fees.

  3. Cost of repairs that would not have been made if the inspection had not been performed.

  4. Value of time expended in repairing the vehicle (including travel time).

  5. Cost in time and direct payment for reinspection.

- **Sponsor's Costs**

  1. Capital investments in inspection facilities.

  2. Operating and maintenance costs associated with inspection facilities. (These include all direct and indirect labour, personnel, and administrative costs.)

- **Sponsor's Revenues or Savings**

  1. Inspection fee.

Valuation of Benefits and Costs

The aim of benefit-cost analysis is to maximize the equivalent value of all benefits minus that of all costs (expressed either in present values or in annual values). This objective is in line with promoting the economic welfare of citizens. In general, the benefits of public projects are difficult to measure, whereas the costs are more easily determined. For simplicity, we will attempt only to quantify the primary users' benefits and sponsor's costs on an annual basis. Estimates for a population the size of Ontario are provided as an example.

- **Calculation of Primary Users' Benefits**

  1. **Benefits due to the reduction of deaths.** The equivalent value of the average income stream lost by victims of fatal accidents[8] was estimated at $648,984 per victim

---

[8] These estimates were based on the total average income that these victims could have generated had they lived. The average value on human life was calculated by considering several factors, such as age, sex, and income group.

in 2010. It was also estimated that the inspection program would reduce the number of annual fatal accidents by 344 per year, resulting in a potential savings of

$$(344)(\$648,984) = \$223,250,496.$$

2. **Benefits due to the reduction of damage to property.** The average cost of damage to property per accident was estimated at $6796. This figure includes the cost of repairs for damages to the vehicle, the cost of insurance, the cost of legal and court administration, the cost of police accident investigation, and the cost of traffic delay due to accidents. Accidents are expected to be reduced by 42,910 per year, and about 63% of all accidents result in damage to property only. Therefore, the estimated annual value of benefits due to reduction of property damage is estimated at

$$\$6,796(42,910)(0.63) = \$183,718,306.$$

The overall annual primary benefits are estimated as the following sum:

Value of reduction in fatalities	$223,250,496
Value of reduction in property damage	183,718,307
Total	$406,968,803

- **Calculation of Primary Users' Disbenefits**

  1. **Opportunity cost associated with time spent bringing vehicles for inspection.** This cost is estimated as

     $$C_1 = (\text{Number of cars inspected})$$
     $$\times (\text{Average duration involved in travel})$$
     $$\times (\text{Average wage rate}).$$

     With an estimated average duration of 1.02 travel hours per car, an average wage rate of $13.88 per hour, and 6,940,000 inspected cars per year, we obtain

     $$C_1 = 6,940,000(1.02)(\$13.88)$$
     $$= \$98,253,744.$$

  2. **Cost of inspection fee.** This cost may be calculated as

     $$C_2 = (\text{Inspection fee}) \times (\text{number of cars inspected}).$$

     Assuming an inspection fee of $20 is to be paid for each car, the total annual inspection cost is estimated as

     $$C_2 = (\$20)(6,940,000)$$
     $$= \$138,800,000.$$

3. **Opportunity cost associated with time spent waiting during the inspection process.** This cost may be calculated by the formula

$$C_3 = \text{(Average waiting time in hours)}$$
$$\times \text{(Average wage rate per hour)}$$
$$\times \text{(Number of cars inspected)}.$$

With an average waiting time of 30 minutes (or 0.5 hours),

$$C_3 = 0.5(\$13.88)(6,940,000) = \$48,163,600.$$

4. **Vehicle usage costs for the inspection process.** These costs are estimated as

$$C_4 = \text{(Number of inspected cars)}$$
$$\times \text{(Vehicle operating cost per kilometre)}$$
$$\times \text{(Average round trip distance to inspection station)}.$$

Assuming a $0.40 operating cost per kilometre and 20 kilometres per round trip, we obtain

$$C_4 = 6,940,000(\$0.40)(20) = \$55,520,000.$$

The overall primary annual disbenefits are estimated as follows:

Item	Amount
C_1	$98,253,744
C_2	138,800,000
C_3	48,163,600
C_4	55,520,000
Total disbenefits	$340,737,344, or $49.10 per vehicle

- **Calculation of Primary Sponsor's Costs**

    A provincially run program would incur an expenditure of $110,267,732 for inspection facilities (this value represents the annualized capital expenditure at a 6% discount rate, representative of provincial borrowing rates through bonds) and another annual operating expenditure of $32,544,000 for inspection, adding up to $142,811,732.

- **Calculation of Primary Sponsor's Revenues**

    The sponsor's costs are offset to a large degree by annual inspection revenues, which must be subtracted to avoid double counting. Annual inspection revenues are the same as the direct cost of inspection incurred by the users (C_2), which was calculated as $138,800,000.

Reaching a Final Decision

From the above estimates, the primary benefits of inspection are valued at $406,968,802, as compared to the primary disbenefits of inspection, which total $340,737,344. Therefore, the users' net benefits are

$$\text{User's net annual benefits} = \$406,968,802 - \$340,737,344$$

$$= \$66,231,459.$$

The sponsor's net costs are

$$\text{Sponsor's net annual costs} = \$142,811,732 - \$138,800,000$$

$$= \$4,011,732.$$

Since all benefits and costs are expressed in annual equivalents, we can use these values directly to compute the degree of benefits that exceeds the sponsor's costs annually:

$$\$66,231,459 - \$4,011,732 = \$62,219,727 \text{ per year.}$$

This positive AE amount indicates that the Ontario inspection system would be economically justifiable under the given assumptions. We can assume the AE amount would have been even greater had we also factored in secondary benefits. (For simplicity, we have not explicitly considered vehicle growth in the province of Ontario. For a complete analysis, this growth factor must be considered to account for all related benefits and costs in equivalence calculations.)

13.3.5 Difficulties Inherent in Public-Project Analysis

As we observed in the motor-vehicle inspection program in the previous section, public benefits are very difficult to quantify in a convincing manner. For example, consider the valuation of a saved human life in any category. Conceptually, the total benefit associated with saving a human life may include the avoidance of the insurance administration costs as well as legal and court costs. In addition, the average potential income lost because of premature death (taking into account age and sex) must be included. Obviously, the difficulties associated with any attempt to put precise numbers on human life are insurmountable.

Consider this example: A few years ago, a 50-year-old business executive was killed in a plane accident. The investigation indicated that the plane was not properly maintained according to federal guidelines. The executive's family sued the airline, and the court eventually ordered the airline to pay $5,250,000 to the victim's family. The judge calculated the value of the lost human life assuming that if the executive had lived and worked in the same capacity until his retirement, his remaining lifetime earnings would have been equivalent to $5,250,000 at the time of award. This is an example of how an individual human life was assigned a dollar value, but clearly any attempt to establish an average amount that represents the general population is controversial. We might even take exception to this individual case: Does the executive's salary adequately represent his worth to his family? Should we also assign a dollar value to their emotional attachment to him, and if so, how much?

Now consider a situation in which a local government is planning to widen a typical municipal highway to relieve chronic traffic congestion. Knowing that the project will be financed by local and provincial taxes, but that many out-of-province travellers also are expected to benefit, should the planner justify the project solely on the benefits to local residents? Which point of view should we take in measuring the benefits—the municipal level, the provincial level, or both? It is important that any benefit measure be performed from the appropriate *point of view*.

In addition to valuation and point-of-view issues, many possibilities for tampering with the results of benefit–cost analyses exist. Unlike private projects, many public projects are undertaken because of political pressure rather than on the basis of their economic benefits alone. In particular, whenever the benefit–cost ratio becomes marginal or less than unity, a potential to inflate the benefit figures to make the project look good exists.

13.4 Benefit–Cost Ratios

An alternative way of expressing the worthiness of a public project is to compare the user's benefits (B) with the sponsor's cost (C) by taking the ratio B/C. In this section, we shall define the benefit–cost (B/C) ratio and explain the relationship between it and the conventional NPW criterion.

13.4.1 Definition of Benefit–Cost Ratio

For a given benefit–cost profile, let B and C be the present values of benefits and costs defined respectively by

$$B = \sum_{n=0}^{N} b_n(1 + i)^{-n} \tag{13.2}$$

and

$$C = \sum_{n=0}^{N} c_n(1 + i)^{-n}, \tag{13.3}$$

where b_n = Benefit at the end of period n, $b_n \geq 0$,

c_n = Expense at the end of period n, $c_n \geq 0$,

$A_n = b_n - c_n$,

N = Project life, and

i = Sponsor's interest rate (discount rate).

The sponsor's costs (C) consist of the equivalent capital expenditure (I) and the equivalent annual operating costs (C') accrued in each successive period. (Note the sign convention we use in calculating a benefit–cost ratio. Since we are using a ratio, all benefits and cost flows are expressed in positive units. Recall that in previous equivalent-worth calculations our sign convention was to explicitly assign "+" for cash inflows and "−" for cash outflows.) Let's assume that a series of initial investments is required during the first K periods, while annual operating and maintenance costs accrue in each subsequent period. Then the equivalent present value for each component is

$$I = \sum_{n=0}^{K} c_n(1 + i)^{-n} \tag{13.4}$$

and

$$C' = \sum_{n=K+1}^{N} c_n (1 + i)^{-n}, \tag{13.5}$$

and $C = I + C'$.

The B/C ratio[9] is defined as

$$BC(i) = \frac{B}{C} = \frac{B}{I + C'}, I + C' > 0. \tag{13.6}$$

If we are to accept a project, $BC(i)$ must be greater than unity.

Note that we must express the values of B, C', and I in present-worth equivalents. Alternatively, we can compute these values in terms of annual equivalents and use them in calculating the B/C ratio. The resulting B/C ratio is not affected.

EXAMPLE 13.1 Benefit–Cost Ratio

A public project being considered by a local government has the following estimated benefit–cost profile (Figure 13.4):

n	b_n	c_n	A_n
0		$10	−$10
1		10	−10
2	$20	5	15
3	30	5	25
4	30	8	22
5	20	8	12

---

[9] An alternative measure, called the **net B/C ratio**, $B'C(i)$, considers only the initial capital expenditure as a cash outlay, and annual net benefits are used:

$$B'C(i) = \frac{B - C'}{I} = \frac{B'}{I}, I > 0.$$

The decision rule has not changed—the ratio must still be greater than one. It can be easily shown that a project with $BC(i) > 1$ will always have $B'C(i) > 1$, as long as both C and I are > 0, as they must be for the inequalities in the decision rules to maintain the stated senses. The magnitude of $BC(i)$ will generally be different than that for $B'C(i)$, but the magnitudes are irrelevant for making decisions. All that matters is whether the ratio exceeds the threshold value of one. However, some analysts prefer to use $B'C(i)$ because it indicates the net benefit (B') expected per dollar invested. But why do they care if the choice of ratio does not affect the decision? They may be trying to increase or decrease the magnitude of the reported ratio in order to influence audiences who do not understand the proper decision rule. People unfamiliar with benefit/cost analysis often assume that a project with a higher B/C ratio is better. This is not generally true, as is shown in 13.4.3. An incremental approach must be used to properly compare mutually exclusive alternatives.

Assume that $i = 10\%$, $N = 5$, and $K = 1$. Compute B, C, I, C', and $BC(10\%)$.

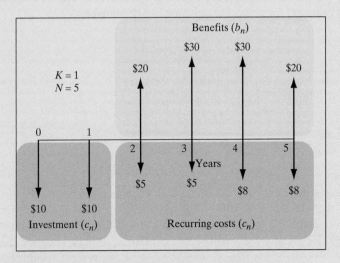

Figure 13.4 Classification of a project's cash flow elements (Example 13.1). Units are in millions of dollars.

SOLUTION

$$B = \$20(P/F, 10\%, 2) + \$30(P/F, 10\%, 3)$$
$$+ \$30(P/F, 10\%, 4) + \$20(P/F, 10\%, 5)$$
$$= \$71.98;$$
$$C = \$10 + \$10(P/F, 10\%, 1) + \$5(P/F, 10\%, 2)$$
$$+ \$5(P/F, 10\%, 3) + \$8(P/F, 10\%, 4) + \$8(P/F, 10\%, 5)$$
$$= \$37.41;$$
$$I = \$10 + \$10(P/F, 10\%, 1)$$
$$= \$19.09;$$
$$C' = C - I$$
$$= \$18.32.$$

Using Eq. (13.6), we can compute the B/C ratio as

$$BC(10\%) = \frac{71.98}{\$19.09 + \$18.32}$$
$$= 1.92 > 1.$$

The B/C ratio exceeds unity, so the users' benefits exceed the sponsor's costs.

13.4.2 Relationship Between B/C Ratio and NPW

The *B/C* ratio yields the same decision for a project as does the NPW criterion. Recall that the BC(*i*) criterion for project acceptance can be stated as

$$\frac{B}{I + C'} > 1. \tag{13.7}$$

If we multiply both sides of the equation by the term $(I + C')$ and transpose $(I + C')$ to the left-hand side, we have

$$B > (I + C'),$$

$$B - (I + C') > 0, \tag{13.8}$$

$$PW(i) = B - C > 0, \tag{13.9}$$

which is the same decision rule[10] as that which accepts a project by the NPW criterion. This implies that we could use the benefit–cost ratio in evaluating private projects instead of using the NPW criterion, or we could use the NPW criterion in evaluating public projects. Either approach will signal consistent project selection. Recall that, in Example 13.1, $PW(10\%) = B - C = \$34.57 > 0$; the project would thus be acceptable under the NPW criterion.

13.4.3 Comparing Mutually Exclusive Alternatives: Incremental Analysis

Let us now consider how we choose among mutually exclusive public projects. As we explained in Chapter 6, we must use the incremental investment approach in comparing alternatives based on any relative measure such as IRR or *B/C*.

Incremental Analysis Based on BC(*i*)

To apply incremental analysis, we compute the incremental differences for each term $(B, I, \text{and } C')$ and take the *B/C* ratio on the basis of these differences. To use BC(*i*) on incremental investment, we may proceed as follows:

1. If one or more alternatives have *B/C* ratios greater than unity, eliminate any alternatives with a *B/C* ratio less than that.
2. Arrange the remaining alternatives in increasing order of the denominator $(I + C')$. Thus, the alternative with the smallest denominator should be the first (*j*), the alternative with the second smallest denominator should be second (*k*), and so forth.
3. Compute the incremental differences for each term $(B, I, \text{and } C')$ for the paired alternatives (j, k) in the list:

$$\Delta B = B_k - B_j,$$

$$\Delta I = I_k - I_j,$$

$$\Delta C' = C'_k - C'_j.$$

---

[10] We can easily verify a similar relationship between the net *B/C* ratio and the NPW criterion.

4. Compute the BC(i) on incremental investment by evaluating

$$BC(i)_{k-j} = \frac{\Delta B}{\Delta I + \Delta C'}.$$

If BC(i)$_{k-j}$ > 1, select alternative k. Otherwise select alternative j.

5. Compare the alternative selected with the next one on the list by computing the incremental benefit–cost ratio.[11] Continue the process until you reach the bottom of the list. The alternative selected during the last pairing is the best one.

We may modify the foregoing decision procedure when we encounter the following situations:

- If $\Delta I + \Delta C' = 0$, we cannot use the benefit–cost ratio, because the equation implies that both alternatives require the same initial investment and operating expenditure. When this happens, we simply select the alternative with the largest B value.
- In situations where public projects with unequal service lives are to be compared, but the projects can be repeated, we may compute all component values (B, C', and I) on an annual basis and use them in incremental analysis.

EXAMPLE 13.2 Incremental Benefit–Cost Ratios

Consider three investment projects: A1, A2, and A3. Each project has the same service life, and the present worth of each component value (B, I, and C') is computed at 10% as follows:

	Projects		
	A1	**A2**	**A3**
I	$5,000	$20,000	$14,000
B	12,000	35,000	21,000
C'	4,000	8,000	1,000
PW(i)	$3,000	$7,000	$6,000

(a) If all three projects are independent, which would be selected on the basis of BC(i)?

(b) If the three projects are mutually exclusive, which would be the best alternative? Show the sequence of calculations that would be required to produce the correct results. Use the B/C ratio on incremental investment.

---

[11] If we use the net B/C ratio as a basis, we need to order the alternatives in increasing order of I and compute the net B/C ratio on the incremental investment.

SOLUTION

(a) Since $PW(i)_1$, $PW(i)_2$, and $PW(i)_3$ are positive, all of the projects are acceptable if they are independent. Also, the $BC(i)$ values for each project are greater than unity, so the use of the benefit–cost ratio criterion leads to the same accept/reject conclusion as does the NPW criterion:

	A1	A2	A3
$BC(i)$	1.33	1.25	1.40

(b) If the projects are mutually exclusive, we must use the principle of incremental analysis. Obviously, if we attempt to rank the projects according to the size of the *B/C* ratio, we will observe a different project preference. For example, if we use the $BC(i)$ on the total investment, we see that A3 appears to be the most desirable and A2 the least desirable, but selecting mutually exclusive projects on the basis of *B/C* ratios is incorrect. Certainly, with $PW(i)_2 > PW(i)_3 > PW(i)_1$, Project A2 would be selected under the NPW criterion. By computing the incremental *B/C* ratios, we will select a project that is consistent with that criterion.

We will first arrange the projects in increasing order of their denominators $(I + C')$ for the $BC(i)$ criterion:[12]

Ranking Base	A1	A3	A2
$I + C'$	\$9,000	\$15,000	\$28,000

- **A1 versus A3.** With the do-nothing alternative, we first drop from consideration any project that has a *B/C* ratio smaller than unity. In our example, the *B/C* ratios of all three projects exceed unity, so the first incremental comparison is between A1 and A3:

$$BC(i)_{3-1} = \frac{\$21,000 - \$12,000}{(\$14,000 - \$5000) + (\$1000 - \$4000)}$$

$$= 1.5 > 1.$$

Since the ratio is greater than unity, we prefer A3 to A1. Therefore, A3 becomes the "current best" alternative.

- **A3 versus A2.** Next, we must determine whether the incremental benefits to be realized from A2 would justify the additional expenditure. Therefore, we need to compare A2 and A3 as follows:

$$BC(i)_{2-3} = \frac{\$35,000 - \$21,000}{(\$20,000 - \$14,000) + (\$8000 - \$1000)}$$

$$= 1.08 > 1.$$

[12] I is used as a ranking base for the $B'C(i)$ criterion. The order still remains unchanged: A1, A3, and A2.

> The incremental *B/C* ratio again exceeds unity; therefore, we prefer A2 over A3. With no further projects to consider, A2 becomes our final choice.[13]

13.5 Analysis of Public Projects Based on Cost-Effectiveness

In evaluating public investment projects, we may encounter situations where competing alternatives have the same goals, but the effectiveness with which those goals can be met may or may not be measurable in dollars. In these situations, we compare decision alternatives directly on the basis of their **cost-effectiveness**. Here, we judge the effectiveness of an alternative in dollars or some nonmonetary measure by the extent to which that alternative, if implemented, will attain the desired objective. The preferred alternative is then either the one that produces the maximum effectiveness for a given level of cost or the one that produces the minimum cost for a fixed level of effectiveness.

13.5.1 Cost-Effectiveness Studies in the Public Sector

A typical cost-effectiveness analysis procedure in the public sector involves the following steps:

Step 1. Establish the goals to be achieved by the analysis.
Step 2. Identify the restrictions imposed on achieving the goals, such as those having to do with the budget or with weight.
Step 3. Identify all the feasible alternatives for achieving the goals.
Step 4. Identify the social interest rate to use in the analysis.
Step 5. Determine the equivalent life-cycle cost of each alternative, including R&D costs, testing costs, capital investment, annual operating and maintenance costs, and salvage value.
Step 6. Determine the basis for developing the cost-effectiveness index. Two approaches may be used: (1) the fixed-cost approach and (2) the fixed-effectiveness approach. If the fixed-cost approach is used, determine the amount of effectiveness obtained at a given cost. If the fixed-effectiveness approach is used, determine the cost required to obtain the predetermined level of effectiveness.

---

[13] Note that if we had to use the net *B/C* ratio on this incremental investment decision, we would obtain the same conclusion. Since all $B'C(i)$ ratios exceed unity, all alternatives are viable. By comparing the first pair of projects on this list, we obtain

$$\mathrm{B'C}(i)_{3-1} = \frac{(\$21{,}000 - \$12{,}000) - (\$1000 - \$4000)}{(\$14{,}000 - \$5000)}$$

$$= 1.33 > 1.$$

Accordingly, Project A3 becomes the "current best." Next, a comparison of A2 and A3 yields

$$\mathrm{B'C}(i)_{2-3} = \frac{(\$35{,}000 - \$21{,}000) - (\$8000 - \$1000)}{(\$20{,}000 - \$14{,}000)}$$

$$= 1.17 > 1.$$

Therefore, A2 becomes the best choice by the net *B/C* ratio criterion.

Step 7. Compute the cost-effectiveness ratio for each alternative, based on the criterion selected in Step 6.

Step 8. Select the alternative with the maximum cost-effective index.

When either the cost or the level of effectiveness is clearly stated in achieving the declared program objective, most cost-effectiveness studies will be relatively straightforward. If that is not the case, however, the decision maker must come up with his or her own program objective by fixing either the cost required in the program or the level of effectiveness to be achieved. The next section shows how such a problem can easily evolve in many public projects or military applications.

13.5.2 A Cost-Effectiveness Case Study[14]

To illustrate the procedures involved in cost-effectiveness analysis, we shall present an example of how the U.S. Army selected the most cost-effective program for providing on-time delivery of time-sensitive, high-priority cargos and key personnel to the battle staging grounds.

Statement of the Problem

The U.S. Army has been engaged in recent conflicts that have transformed from the traditional set-piece battles into operations in distributed locations that require rapid sequences of activities. To support success in these new operations, the Army needs timely delivery of cargos and personnel in the distributed locations. Consequently, the dispersion of forces has created a condition of unsecured land lines of communication (LOC), and aerial delivery is considered as one of the practical solutions to the problem. The time-sensitive nature of these priority cargos makes it most preferable to have direct delivery from intermediate staging bases (ISB) to the forward brigade combat teams (BCT).

The transportation network typically involves a long-haul move (a fixed-wing aerial movement), and a short-haul distribution (by helicopter or truck). The long-haul usually stops at the closest supporting airfield, but with the appropriate asset, a single, direct movement could be made. The long-haul portion has been the bigger problem—trucks take too long and are vulnerable to hostilities; the U.S. Air Force (USAF) assets are not always available or appropriate; and existing helicopters do not have the necessary range to fly. Further, the use of the larger USAF aircraft results in inefficient load, as the high-priority cargos use only a small fraction of the capacity of the aircraft. The Army is investigating how to best correct the problem. They are considering three options to handle the long haul: (1) Use the U.S. Air Force (C-130J) assets; (2) procure a commercially available Future Cargo Aircraft (FCA); or (3) use additional helicopters (CH-47).

Defining the Goals

The solution will be a "best value" selection, providing the best capability for the investment required. The selection process must address three operational perspectives.

- First, a "micro" examination: For a given specific battlefield arrangement (available transportation network, supported organizations and locations), the decision metric is

---

[14] This case study is provided by Dr. George C. Prueitt of CAS, Inc. All numbers used herein do not represent the actual values used by the U.S. Army.

the time required to deliver specific critical cargos and the solutions' comparative operating costs. This metric is a suitability measure, to identify the system that best, most quickly, meets the timely delivery requirement.

- Second, a "macro" examination: Given a broader, more prolonged theatre support requirement, the decision metric becomes what percentage of critical cargos can be delivered, and what is the cost of procuring all the transportation assets needed to create the delivery network combinations (fixed-wing airplanes, helicopters, and trucks) to meet those delivery percentages. This metric is a feasibility measure to determine the number of assets needed to meet the quantities of cargos to be delivered.
- Third, given relatively comparable fixed-wing aircraft fleets (differing quantities for equal capability), what are the 15-year life-cycle costs associated with those alternatives? This is an affordability metric, to ensure that all life-cycle costs are properly identified and can be met within the Army budget.

Description of Alternatives

As mentioned earlier, the Army is considering three alternatives. They are:

- **Alternative 1—Use the USAF C-130J assets.** The USAF C-130J aircraft has been proposed as an alternative for the long-haul task. The C-130J is a new acquisition, and will be considered in two employment concepts: first, in a direct movement pattern, and second, as part of a standard USAF Scheduled Tactical Air Re-supply (STAR) route (analogous to a bus route). Current operations almost exclusively use the STAR route method when C-130 aircraft are employed. It is assumed that transportation planning and theatre priorities would be sufficiently high to permit the C-130J to be employed in the more direct delivery method, and not limited solely to STAR route operations.
- **Alternative 2—Use Future Cargo Aircraft.** Another long-haul alternative is to procure one of two commercially available products capable of delivering these cargos on a timely and efficient basis.
- **Alternative 3—Use additional C-47 helicopters.** Looking first at the long-haul task, most of today's deliveries are provided by either the CH-47 helicopter or the C-130 aircraft in conjunction with another vehicle. The CH-47 is being used to perform this mission because it is the "best available" Army asset. Unfortunately, it is expensive to operate, and its range limitations make it an inefficient method of carrying out the mission. Further, the long distances between the ISB and the forward units are causing the helicopters to accumulate flight hours more rapidly than planned in their service-life projections, and this has generated a significant increase in their maintenance requirements.

These long-haul delivery alternatives are described in Table 13.6.

In addition to the long-haul task, the shorter delivery distribution legs will usually need a complementary ground or helicopter asset to make the final delivery when the fixed-wing assets cannot land close enough to the BCT, because of a lack of improved landing strips or runways of sufficient length. This creates a greater burden on the transportation network to ensure that adequate ground or rotary-wing assets are available to complete the mission. The combinations of long-haul alternatives and final delivery systems create the series of network transportation options described in Table 13.7.

TABLE 13.6 Descriptions of Alternatives for Transportation Network Long-Haul Requirement

Alternative	Description	Comment
Alternative 1—C-130J	Procure additional C-130Js and maximize their use to perform the long-haul task of the intra-theatre delivery of critical, time-sensitive cargo and key personnel. Minimize requirement for additional CH-47s.	C-130Js will use both USAF Scheduled Tactical Air Re-supply (STAR) routes and more direct routes. The C-130J has longer runway requirements than smaller FCA aircraft do.
Alternative 2—Army FCA	Procure a variant of the FCA and maximize its use to perform the long-haul task of the intratheatre delivery of mission-critical, time-sensitive cargo and key personnel, with minimal use of CH-47s.	In some situations, the FCA will be able to land either closer to or at the BCT locations. Two versions of the FCA (Alternative 2A and Alternative 2B) have been included in this examination.
Alternative 3—More CH-47s	Procure additional CH-47s to perform the long-haul task of the intratheatre delivery of critical, time-sensitive cargo and key personnel.	This alternative will include those additional CH-47s needed to carry support systems for the CH-47s that are used to move cargo over extended distances.

Figure 13.5 provides a descriptive portrayal of these transportation network options.

Micro Analysis

To perform a cost-effectiveness analysis, the micro comparisons were made by selecting the delivery time as the performance metric (vertical axis in Figure 13.6), and one-way operating cost as the other metric (horizontal axis). Lower cost and shorter delivery times cause the direction of preference to be in the lower left-hand corner of the figure. The respective operating costs for each system were: C-130J at $3850 per flight hour, FCA 2A at $2800, FCA 2B at $1680, and CH-47 at $4882. Transportation network option 10 (NO_{10}), based on Alternative 2B FCA, and NO_7, based on Alternative 2A FCA, with each flying directly from ISB to BCT, had the shortest times to accomplish the mission. Their respective operating costs pushed them into the most favoured regions on the graph. Not surprisingly, the next quickest option was NO_2, based on Alternative 1 aircraft flying directly to a C-130 Supportable Forward Airfield, with a CH-47 delivering for the last leg; however, the higher operating costs of the C-130J (over the FCA alternatives) and the CH-47 made its operating cost more than double NO_{10}. For the remaining options, NO_3 and NO_4, using the C-130J on a STAR route, required the longest delivery time, although the operating costs were comparable

TABLE 13.7 Descriptions of Transportation Network Options

Network Option (NO_i)	Description
NO_1	ALT 1 (C-130) direct from ISB to C-130 Supportable Forward Airfield,
NO_2	ALT 1 (C-130) direct from ISB to C-130 Supportable Forward Airfield, then CH-47 to BCT
NO_3	ALT 1 (C-130) STAR route from ISB to C-130 Supportable Forward Airfield (stop nearest to BCT), then truck to BCT
NO_4	ALT 1 (C-130) STAR route from ISB to C-130 Supportable Forward Airfield (stop nearest to BCT), then CH-47 to BCT
NO_5	ALT 2A direct from ISB to FCA Supportable Forward Airfield, then truck to BCT
NO_6	ALT 2A direct from ISB to FCA Supportable Forward Airfield, then CH-47 to BCT
NO_7	ALT 2A direct from ISB to BCT
NO_8	ALT 2B direct from ISB to FCA Supportable Forward Airfield, then truck to BCT
NO_9	ALT 2B direct from ISB to FCA Supportable Forward Airfield, then CH-47 to BCT
NO_{10}	ALT 2B direct from ISB to BCT
NO_{11}	ALT 3 (CH-47) direct from ISB to BCT, with multiple refueling stops en route

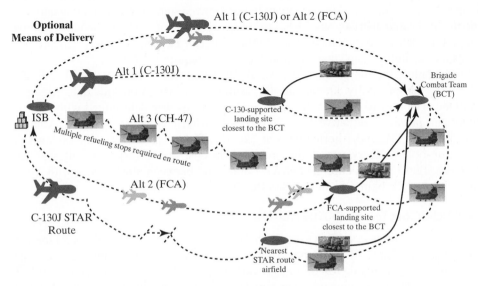

Figure 13.5 Transportation network option schematics.

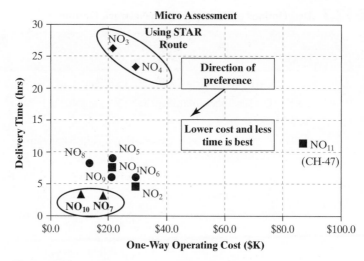

Figure 13.6 Assessment results from micro cost-effectiveness analysis.

to the most favoured solutions. NO_{11}, the CH-47–based option, had the highest operating costs. Figure 13.6 summarizes the results for all 11 transportation network options.

Macro Analysis

The macro analysis was performed with the criterion of capability to deliver high-priority cargos in two simultaneous combat operations over a 30-day period. The minimum acceptable on-time delivery rate is 80% of all required shipments. The requirements were derived from one operation with 10 days of high-intensity combat, followed by 20 days of stability operations, and a second operation consisting of 30 days of low-intensity stability operations. Current transportation asset allocations were used to establish a "baseline" capability. The baseline included 28 CH-47 helicopters and 10 C-130J aircraft for this specific mission, and additional quantities of Army systems (other aircraft and trucks) to do the final cargo distributions. These assets were capable of satisfying about 12% of the total on-time delivery requirement. The shortfall between 12% and the 80% standard represents the capability "gap" that additional assets are required to fill.

To complete the macro analysis, additional quantities of Alternatives 1, 2A, 2B, and 3 were incrementally added, and the resultant increase in on-time deliveries was computed. The cost to procure the additional quantities was determined with the following unit prices: Alternative 1 at $75.6M each, Alternative 2A at $34.5M, Alternative 2B at $29.2M, and Alternative 3 at $26.1M. Figure 13.7 shows the changes in percent-on-time delivery versus the procurement cost of the respective alternatives. There is a significant procurement cost differentially between alternatives, with Alternatives 2A and 2B being the most cost effective.

Life-Cycle Cost for Each Alternative

The third part of the case study is to perform the life-cycle cost (LCC) analysis for the various alternatives. Table 13.8 lists the cost items that are considered in this LCC analysis. In fact, these cost items represent the common decision elements mandated by the U.S. Department of Defense, along with appropriate rules and assumptions, to ensure that each

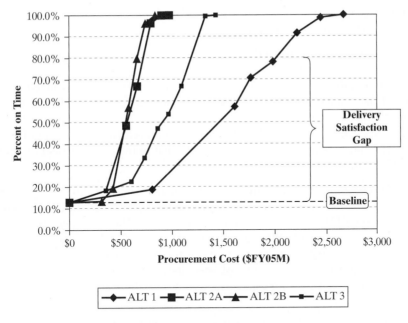

Figure 13.7   Assessment results from macro effectiveness analysis.

TABLE 13.8 Definitions of Life-Cycle Cost Variables

Life-Cycle Cost Element	Cost Variable	Description
1	RDT&E	Government and contractor costs associated with the research, development, test, and evaluation (RDT&E) of the material system.
2	Procurement	Costs resulting from the production and initial fielding of the material system in the Army's operational inventory.
3	Military construction	Cost of all military construction projects associated with a material system. No military construction money is anticipated to be needed for any of the FCA alternatives.
4	Military personnel	Cost of military personnel associated with the development and operation of the material system.
5	O&M	O&M funded cost associated with development, production, fielding, operation, and support of the material system for 20 years.

alternative be evaluated for comparable cost and affordability. First, an LCC estimate was developed for specific quantities of each primary long-haul aircraft in the alternatives. The number of aircraft was chosen to establish equal fleet capabilities (not necessarily fleet size). Second, to determine the number of aircraft required for each alternative, the U.S.

Army used the equal-effectiveness standard of 80% on-time delivery of high-priority cargo and key personnel. Specifically,

- Alternatives 2A and 2B use a fleet size of 56, as derived from the macro analysis and with additional fleet support considerations.
- Alternative 1, the USAF C-130J, uses 40 aircraft, because it was assumed that all additional fleet support requirements would be addressed with the USAF larger program. Similarly,
- Alternative 3 uses 92 CH-47F as its mission requirement, with no additional fleet support requirements, as the 92 would be added to the existing fleet of about 400 CH-47 aircraft.

Cost estimates were developed for the respective alternatives. In a simple form, these estimates provide the future estimated costs for procurement and operations. The procurement costs include both RDT&E and procurement requirements. The operating cost includes operations and maintenance (O&M) and military personnel costs. There were some nominal military construction costs, but they were equal for all alternatives. A 15-year operating cycle (from first procurement through last operation) was assumed in developing the cost estimates. The projected costs for future years are provided in Table 13.9.

TABLE 13.9 Life-Cycle Cost Estimates for Each Alternative

Year	ALT 1		ALT 2A		ALT 2B		ALT 3	
	Proc	Operating	Proc	Operating	Proc	Operating	Proc	Operating
0	$702.80	$ 25.0	$257.5	$ 4.8	$221.1	$ 3.3	$620.2	$ 10.8
1	$860.27	$ 56.1	$529.9	$ 14.7	$450.3	$ 12.5	$654.4	$ 82.0
2	$877.47	$ 88.4	$800.8	$ 43.0	$687.2	$ 35.0	$451.4	$163.2
3	$447.51	$122.0	$431.3	$ 64.0	$376.1	$ 52.6	$145.2	$224.1
4	$ 0.00	$151.6	$206.1	$105.1	$189.2	$ 80.7	$351.0	$251.4
5	$ 0.00	$154.7	$ 0.0	$114.3	$ 0.0	$ 80.5	$262.6	$258.8
6	$ 0.00	$157.8	$ 0.0	$120.6	$ 0.0	$ 81.5	$ 0.0	$262.3
7	$ 0.00	$160.9	$ 0.0	$128.4	$ 0.0	$ 86.8	$ 0.0	$274.7
8	$ 0.00	$164.1	$ 0.0	$132.2	$ 0.0	$ 90.5	$ 0.0	$287.5
9	$ 0.00	$167.4	$ 0.0	$139.9	$ 0.0	$ 95.6	$ 0.0	$301.2
10	$ 0.00	$170.8	$ 0.0	$141.4	$ 0.0	$ 98.3	$ 0.0	$315.5
11	$ 0.00	$174.2	$ 0.0	$145.8	$ 0.0	$102.5	$ 0.0	$324.4
12	$ 0.00	$145.8	$ 0.0	$129.6	$ 0.0	$106.5	$ 0.0	$362.1
13	$ 0.00	$110.0	$ 0.0	$119.8	$ 0.0	$ 89.6	$ 0.0	$248.8
14	$ 0.00	$ 72.6	$ 0.0	$106.2	$ 0.0	$ 77.6	$ 0.0	$ 86.7
15	$ 0.00	$ 33.6	$ 0.0	$100.9	$ 0.0	$ 74.2	$ 0.0	$ 35.4

Notes: All costs are $M.

"Proc" means procurement cost estimate.

"Operating" includes operations, maintenance, and manpower costs.

TABLE 13.10 Present Value of Life-Cycle Cost for Each Alternative

Year	ALT 1 Proc	ALT 1 Operating	ALT 2A Proc	ALT 2A Operating	ALT 2B Proc	ALT 2B Operating	ALT 3 Proc	ALT 3 Operating
0	$ 702.8	$ 25.0	$ 257.5	$ 4.8	$ 221.1	$ 3.3	$ 620.2	$ 10.8
1	$ 796.5	$ 51.9	$ 490.6	$ 13.6	$ 417.0	$ 11.6	$ 606.0	$ 75.9
2	$ 752.3	$ 75.8	$ 686.6	$ 36.8	$ 589.1	$ 30.0	$ 387.0	$ 140.0
3	$ 355.2	$ 96.9	$ 342.4	$ 50.8	$ 298.6	$ 41.8	$ 115.2	$ 177.9
4	$ 0.0	$ 111.5	$ 151.5	$ 77.3	$ 139.1	$ 59.4	$ 258.0	$ 184.8
5	$ 0.0	$ 105.3	$ 0.0	$ 77.8	$ 0.0	$ 54.8	$ 178.7	$ 176.1
6	$ 0.0	$ 99.4	$ 0.0	$ 76.0	$ 0.0	$ 51.4	$ 0.0	$ 165.3
7	$ 0.0	$ 93.9	$ 0.0	$ 74.9	$ 0.0	$ 50.7	$ 0.0	$ 160.3
8	$ 0.0	$ 88.7	$ 0.0	$ 71.4	$ 0.0	$ 48.9	$ 0.0	$ 155.3
9	$ 0.0	$ 83.8	$ 0.0	$ 70.0	$ 0.0	$ 47.8	$ 0.0	$ 150.7
10	$ 0.0	$ 79.1	$ 0.0	$ 65.5	$ 0.0	$ 45.5	$ 0.0	$ 146.1
11	$ 0.0	$ 74.7	$ 0.0	$ 62.6	$ 0.0	$ 43.9	$ 0.0	$ 139.1
12	$ 0.0	$ 57.9	$ 0.0	$ 51.5	$ 0.0	$ 42.3	$ 0.0	$ 143.8
13	$ 0.0	$ 40.4	$ 0.0	$ 44.0	$ 0.0	$ 33.0	$ 0.0	$ 91.5
14	$ 0.0	$ 24.7	$ 0.0	$ 36.2	$ 0.0	$ 26.4	$ 0.0	$ 29.5
15	$ 0.0	$ 10.6	$ 0.0	$ 31.8	$ 0.0	$ 23.4	$ 0.0	$ 11.2
Subtotal	$2,606.9	$1,119.6	$1,928.5	$ 845.0	$1,664.9	$ 614.1	$2,165.1	$1,958.3
ALT Total	$3,726.5		$2,773.5		$2,278.9		$4,123.4	

Note: All figures are in $M. Subtotals and totals are NPW.

For most of the procurements by the U.S. Department of Defense, future inflation factors must be specified in the analysis. For this case study, all costs are assumed to increase by 2% each year. In terms of interest rate, a discount rate of 8% was used to calculate the LCC for each alternative, which is summarized in Table 13.10.

Figure 13.8 illustrates the cost component of each alternative's LCC. The figure shows that Alternatives 2A and 2B, at 56 aircraft, have both lower procurement and operations and maintenance costs than either Alternative 1, the C-130J alternative, for its quantity of 40 aircraft, or Alternative 3, the CH-47F alternative, based on 92 helicopters. As anticipated earlier, the CH-47F alternative turns out to be the most expensive option due to its higher operating costs, even though its procurement of 92 helicopters is less expensive than 40 C-130J aircraft. The total operating costs differ in contribution by type of aircraft. The fixed-wing aircraft estimates are based on 600 hours of flying time per year, while the helicopters are scheduled for less than 200 hours of flying time. However, because the airspeed and flying range for the helicopter are much lower than for the fixed-wing aircraft, more CH-47F helicopters are required to provide the same level of capability. Alternative 1, using the C-130J as the primary aircraft, requires fewer total aircraft than Alternatives 2A and 2B, but

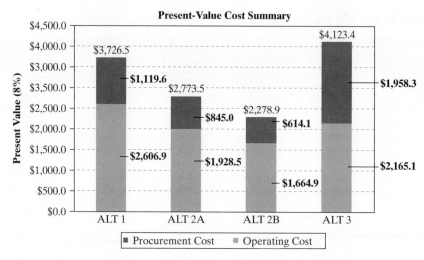

Figure 13.8 Alternatives' present value (8%) for procurement and operations.

with nearly twice the unit cost it is a more expensive option. In conclusion, for the alternatives considered, the commercially available candidates for the FCA (Alternatives 2A and 2B) provide the best opportunity for the Army to satisfy its requirement to deliver mission-critical, time-sensitive cargo and key personnel from an ISB directly to a BCT, with Alternative 2B having a slight edge over 2A, for the metrics used.

SUMMARY

■ The service sector accounts for about 75% of Canada's GDP. Health-care and government services account for almost 35% of the total GDP.

■ In health economics, cost-effectiveness analysis is the most widely used economic evaluation method. The cost-effectiveness ratio is defined as

$$\text{Cost-effectiveness ratio} = \frac{\sum(\text{all costs})}{\sum(\text{all measured health effects})}$$

If independent intervention programs are evaluated, we can rank the programs on the basis of their cost-effectiveness ratios and accept as many programs as the budget permits. When we compare mutually exclusive intervention programs, the incremental cost-effectiveness ratio should be used to determine whether an additional cost can be justified to increase health effects.

■ **Benefit–cost analysis** is commonly used to evaluate public projects; several facets unique to public-project analysis are neatly addressed by benefit–cost analysis:

1. Benefits of a nonmonetary nature can be quantified and factored into the analysis.

2. A broad range of project users distinct from the sponsor can and should be considered; benefits and disbenefits to *all* these users can and should be taken into account.

■ Difficulties involved in public-project analysis include the following:

1. Identifying all the users of the project.

2. Identifying all the benefits and disbenefits of the project.

3. Quantifying all the benefits and disbenefits in dollars or some other unit of measure.

4. Selecting an appropriate **social discount rate** at which to discount benefits and costs to a present value.

■ The *B/C* ratio is defined as

$$\mathrm{BC}(i) = \frac{B}{C} = \frac{B}{I + C'}, I + C' > 0.$$

The decision rule is that if $\mathrm{BC}(i) \geq 1$, then the project is acceptable.

■ The net *B/C* ratio is defined as

$$\mathrm{B'C}(i) = \frac{B - C'}{I} = \frac{B'}{I}, I > 0.$$

The net *B/C* ratio expresses the net benefit expected per dollar invested. The same decision rule applies as for the *B/C* ratio.

■ When selecting among mutually exclusive alternatives based on *B/C* ratio, or net *B/C* ratio, incremental analysis must be used.

■ The **cost-effectiveness method** allows us to compare projects on the basis of cost and nonmonetary effectiveness measures. We may either maximize effectiveness for a given cost criterion or minimize cost for a given effectiveness criterion.

PROBLEMS

13.1 The following table summarizes the costs of treatment of a disease, based on two different antibiotics and associated health benefits (effectiveness):

Which treatment option is the best?

Type of Treatment	Cost of Treating 100 Patients	Effectiveness (Percent Successful Treatment of Infections)
Antibiotic A	$12,000	75%
Antibiotic B	$13,500	80%
Antibiotic C	$14,800	82%

13.2 The Table P13.2 summarizes cervical cancer treatment options and their health effectiveness. Find the best strategy for treating cervical cancer.

TABLE P13.2 A CEA Examining Three Strategies

Strategy	Cost	Marginal Cost	Effectiveness	Marginal Effectiveness	CE Ratio
Nothing	$0	—	0 years	—	—
Simple	$5,000	$5,000	5 years	5 years	$1,000/yr
Complex	$50,000	$45,000	5.5 years	0.5 years	$90,000/yr

13.3 The Province of Quebec is considering a bill that would ban the use of road salt on highways and bridges during icy conditions. Road salt is known to be toxic, costly, corrosive, and caustic. The Canadian Salt Company produces a calcium magnesium acetate (CMA) deicer and sells it for $600 a tonne as Ice-Away Road salts, by contrast, sold for an average of $14 a tonne in 1995. Quebec needs about 600,000 tonnes of road salt each year. (Quebec spent $9.2 million on road salt in 1995.) The Canadian Salt Company estimates that each tonne of salt on the road costs $650 in highway corrosion, $525 in rust on vehicles, $150 in corrosion to utility lines, and $100 in damages to water supplies, for a total of $1425. Unknown salt damage to vegetation and soil surrounding areas of highways has occurred. Quebec would ban road salt (at least on expensive steel bridges or near sensitive lakes) if provincial studies support the Canadian Salt Company's cost claims.

(a) What would be the users' benefits and sponsor's costs if a complete ban on road salt were imposed in Quebec?

(b) How would you go about determining the salt damages (in dollars) to vegetation and soil?

13.4 A public school board in Toronto is considering the adoption of a four-day school week as opposed to the current five-day week. The community is hesitant about the plan, but the superintendent of the board envisions many benefits associated with the four-day system, Wednesday being the "day off." The following pros and cons have been cited:

• Experiments with the four-day system indicate that the "day off" in the middle of the week will cut down on both teacher and pupil absences.

• The longer hours on school days will require increased attention spans, which is not an appropriate expectation for younger children.

• The province bases its expenditures on its local school systems largely on the average number of pupils attending school in the system. Since the number of absences will decrease, provincial expenditures on local systems should increase.

• Older students might want to work on Wednesdays. Unemployment is a problem in this region, however, and any influx of new job seekers could aggravate an existing problem. Community centres, libraries, and other public areas also may experience increased usage on Wednesdays.

• Parents who provide transportation for their children will see a savings in fuel costs. Primarily, only those parents whose children live less than 3 kilometres from the school would be involved. Children living more than 3 kilometres from school are eligible for free transportation provided by the school board.

• Decreases in both public and private transportation should result in fuel conservation, decreased pollution, and less wear on the roads. Traffic congestion should ease on Wednesdays in areas where congestion caused by school traffic is a problem.

- Working parents will be forced to make child-care arrangements (and possibly payments) for one weekday per week.
- Older students will benefit from wasting less time driving to and from school; Wednesdays will be available for study, thus taking the heavy demand off most nights. Bussed students will spend far less time per week waiting for buses.
- The local school board should see some ease in funding problems. The two areas most greatly affected are the transportation system and building operating costs.

(a) For this type of public study, what do you identify as the users' benefits and disbenefits?

(b) What items would be considered as the sponsor's costs?

(c) Discuss any other benefits or costs associated with the four-day school week.

13.5 The Electric Department of the City of Prince George, BC, operates generating and transmission facilities serving approximately 80,000 people in the city. The city has proposed the construction of a $300 million, 235-MW circulating fluidized-bed combustor (CFBC) to power a turbine generator that is currently receiving steam from an existing boiler fuelled by gas or oil. Among the advantages associated with the use of CFBC systems are the following:

- A variety of fuels can be burned, including inexpensive low-grade fuels with high ash and a high sulfur content.
- The relatively low combustion temperatures inhibit the formation of nitrogen oxides. Acid-gas emissions associated with CFBC units would be expected to be significantly lower than emissions from conventional coal-fueled units.
- The sulfur-removal method, low combustion temperatures, and high-combustion efficiency characteristic of CFBC units result in solid wastes, which are physically and chemically more amenable to land disposal than the solid wastes resulting from conventional coal-burning boilers equipped with flue-gas desulfurization equipment.

On the basis of the Ministry of Energy's projections of growth and expected market penetration, the demonstration of a successful 235-MW unit could lead to as much as 41,000 MW of CFBC generation being constructed by the year 2012. The proposed project would reduce the city's dependency on oil and gas fuels by converting its largest generating unit to coal-fuel capability. Consequently, substantial reductions in local acid-gas emissions could be realized in comparison to the permitted emissions associated with oil fuel. The city has requested a $50 million cost share from the province. Cost sharing is considered attractive because the province's cost share would largely offset the risk of using such a new technology. To qualify for cost-sharing money, the city has to address the following questions for the province:

(a) What is the significance of the project at local and provincial levels?

(b) What items would constitute the users' benefits and disbenefits associated with the project?

(c) What items would constitute the sponsor's costs?

Put yourself in the city engineer's position and respond to these questions.

13.6 The Moncton city government is considering two types of town-dump sanitary systems. Design A requires an initial outlay of $400,000, with annual operating and maintenance costs of $50,000 for the next 15 years; design B calls for an investment of $300,000, with annual operating and maintenance costs of $80,000 per year for the next 15 years. Fee collections from the residents would be $85,000 per year. The interest rate is 8%, and no salvage value is associated with either system.

(a) Using the benefit–cost ratio BC(i), which system should be selected?

(b) If a new design (design C), which requires an initial outlay of $350,000 and annual operating and maintenance costs of $65,000, is proposed, would your answer in (a) change?

13.7 The Canadian Government is considering building apartments for government employees working in a foreign country and living in locally owned housing. A comparison of two possible buildings indicates the following:

	Building X	**Building Y**
Original investment by government agencies	$8,000,000	$12,000,000
Estimated annual maintenance costs	$240,000	$180,000
Savings in annual rent now being paid to house employees	$1,960,000	$1,320,000

Assume the salvage or sale value of the apartments to be 60% of the first investment. Use 10% discounting and a 20-year study period to compute the B/C ratio on incremental investment, and make a recommendation. (Assume no do-nothing alternative.)

13.8 Three public-investment alternatives with the same service life are available: A1, A2, and A3. Their respective total benefits, costs, and first costs are given in present worth as follows:

		Proposals	
Present worth	**A1**	**A2**	**A3**
I	$100	$300	$200
B	$400	$700	$500
C'	$100	$200	$150

Assuming no do-nothing alternative, which project would you select on the basis of the benefit–cost ratio BC(i) on incremental investment?

13.9   The city of Calgary operates automobile parking facilities and is evaluating a proposal to erect and operate a structure for parking in the city's downtown area. Three designs for a facility to be built on available sites have been identified. (All figures are in thousands of dollars):

	Design A	Design B	Design C
Cost of site	$240	$180	$200
Cost of building	$2200	$700	$1400
Annual fee collection	$830	$750	$600
Annual maintenance cost	$410	$360	$310
Service life	30 years	30 years	30 years

At the end of the estimated service life, whichever facility had been constructed would be torn down, and the land would be sold. It is estimated that the proceeds from the resale of the land will be equal to the cost of clearing the site. If the city's interest rate is known to be 10%, which design alternative would be selected on the basis of the benefit–cost criterion?

13.10 The Government of Nova Scotia is planning a hydroelectric project for a river basin. In addition to producing electric power, this project will provide flood control, irrigation, and recreation benefits. The estimated benefits and costs expected to be derived from the three alternatives under consideration are listed in the following table:

	Decision Alternatives		
	A	B	C
Initial cost	$8,000,000	$10,000,000	$15,000,000
Annual benefits or costs:			
Power sales	$1,000,000	$1,200,000	$1,800,000
Flood control savings	$250,000	$350,000	$500,000
Irrigation benefits	$350,000	$450,000	$600,000
Recreation benefits	$100,000	$200,000	$350,000
O&M costs	$200,000	$250,000	$350,000

The interest rate is 10%, and the life of each of the projects is estimated to be 50 years.

(a) Find the benefit–cost ratio for each alternative.

(b) Select the best alternative on the basis of BC(i).

13.11 Two different routes are under consideration for a new highway:

	Length of Highway	First Cost	Annual Upkeep
The "long" route	22 km	$21 million	$140,000
Transmountain shortcut	10 km	$45 million	$165,000

For either route, the volume of traffic will be 400,000 cars per year. These cars are assumed to operate at $0.25 per kilometre. Assume a 40-year life for each road and an interest rate of 10%. Determine which route should be selected.

13.12 The Government of PEI is considering undertaking four projects. These projects are mutually exclusive, and the estimated present worth of their costs and the present worth of their benefits are shown in millions of dollars in the following table:

Projects	PW of Benefits	PW of Costs
A1	$40	$85
A2	$150	$110
A3	$70	$25
A4	$120	$73

All of the projects have the same duration.

Assuming there is no do-nothing alternative, which alternative would you select? Justify your choice by using a benefit–cost [BC(i)] analysis on incremental investment.

Short Case Studies

ST13.1 Fast growth in the city of Ottawa-Carleton and surrounding region has resulted in increasing traffic congestion, for both vehicles and pedestrians. Aside from normal road maintenance and minor projects, the city approved the allocation of $24 million to be used for major capital improvements to the road system. This extra funding allows the option of spreading that amount among many smaller projects or concentrating on fewer larger projects. The city engineers and planners were asked to prepare a priority list outlining which roads and facilities could be improved with the extra money. The engineers also computed the possible public benefits associated with each construction project; they accounted for possible reduction in travel time, a reduction in the accident rate, land appreciation, pedestrian safety, and savings in the operating costs of the vehicles.

Region	No.	Project	Type of Improvement	Construction Cost	Annual O&M	Annual Benefits
South (I)	1	Hawthorne Road	4-Lane	$980,000	$9,800	$313,600
	2	Walkley Road	New 4-Lane Extension	$3,500,000	$35,000	$850,000
	3	Hunt Club Road East	4-Lane	$2,800,000	$28,000	$672,000
	4	Conroy Road	4-Lane	$1,400,000	$14,000	$490,000
East (II)	5	St. Joseph Boulevard	4-Lane	$2,380,000	$47,600	$523,600
	6	Place d'Orléans Station	New Transitway	$5,040,000	$100,800	$1,310,400
	7	Blackburn Hamlet	New 4-Lane Bypass	$2,520,000	$50,400	$831,600
	8	Tenth Line Road	4-Lane and Interchange	$4,900,000	$98,000	$1,021,000
West (III)	9	Baseline Road	4-Lane	$1,365,000	$20,475	$245,700
	10	Hunt Club Road West	New 4-Lane Extension	$2,100,000	$31,500	$567,000
	11	Knoxdale Road	Realign	$1,170,000	$17,550	$292,000
	12	March Road	4-Lane	$1,120,000	$16,800	$358,400
Central (IV)	13	Mackenzie King Bridge	Rehabilitate	$2,800,000	$56,000	$980,000
	14	Elgin Street	Reconstruct	$1,690,000	$33,800	$507,000
	15	Wellington Street	Reconstruct	$975,000	$15,900	$273,000
	16	Plaza Bridge	Reconstruct	$1,462,500	$29,250	$424,200

Assuming a 20-year planning horizon, and a municipal interest rate of 10%, which projects would be considered for funding in (a) and (b)?
(a) Due to political pressure, each region will have the same amount of funding, say $6 million.
(b) To compensate for previous imbalances in budget allocations, Regions I and II combined will get $15 million, while Regions III and IV combined will get $9 million. At least two projects from each district must be included.

ST13.2 The City of Winnipeg Sanitation Department is responsible for the collection and disposal of all solid waste within the city limits. The city must collect and dispose of an average of 300 tonnes of garbage each day. The city is considering ways to improve the current solid-waste collection and disposal system.

- The current system uses Dempster Dumpmaster Frontend Loaders for collection, and incineration or landfill for disposal. Each collecting vehicle has a load capacity of 10 tonnes, or 24 cubic metres, and dumping is automatic. The incinerator in use was manufactured in 1942 and was designed to incinerate 150 tonnes per 24 hours. A natural-gas afterburner has been added in an effort to reduce air pollution; however, the incinerator still does not meet provincial air-pollution requirements, and it is operating under a permit from the Manitoba Pollution Control Board. Prison-farm labour is used for the operation of the incinerator. Because the capacity of the incinerator is relatively low, some trash is not incinerated, but is taken to the city landfill. The trash landfill is located approximately 11 kilometres, and the incinerator approximately 5 kilometres, from the centre of the city. The travel distance and costs in person-hours for delivery to the disposal sites are excessive; a high percentage of empty vehicle kilometres and person-hours is required because separate methods of disposal are used and the destination sites are remote from the collection areas. The operating cost for the present system is $905,400, including $624,635 to operate the prison-farm incinerator, $222,928 to operate the existing landfill, and $57,837 to maintain the current incinerator.

- The proposed system locates a number of portable incinerators, each with 100-tonne-per-day capacity for the collection and disposal of waste collected for three designated areas within the city. Collection vehicles will also be staged at these incineration/disposal sites, together with the plant and support facilities that are required for incineration, fuelling and washing of the vehicles, a support building for stores, and shower and locker rooms for collection and site crew personnel. The pick-up and collection procedure remains essentially the same as in the existing system. The disposal/staging sites, however, are located strategically in the city, on the basis of the volume and location of waste collected, thus eliminating long hauls and reducing the number of kilometres the collection vehicles must retravel from pick-up to disposal site.

Four variations of the proposed system are being considered, depending on how the units are distributed among the sites. The type of incinerator is a modular prepackaged unit that can be installed at several sites in the city. Such units exceed all provincial and federal standards for exhaust emissions. The city of Winnipeg needs 24 units, each with a rated capacity of 12.5 tonnes of garbage per 24 hours. The price per unit is $137,600, which means a capital investment of about $3,302,000. The plant facilities, such as housing and foundation, were estimated to cost $200,000 per facility. Each site will have one plant housing eight units and be capable of handling 100 tonnes of garbage per day. Additional plant features, such as landscaping, were estimated to cost $60,000 per plant.

The annual operating cost of the proposed system would vary according to the type of system configuration. It takes about 50 million litres (ML) of fuel to incinerate 1 tonne of garbage. This means that fuel cost $4.25 per tonne of garbage at a cost of $2.50 per ML. Electric requirements at each plant will be 230 kW per day, which means a $0.48-per-tonne cost for electricity if the plant is operating at full capacity. Two people can easily operate one plant, but safety factors dictate three operators at a cost of $7.14 per hour. This translates to a cost of $1.72 per tonne. The maintenance cost of each plant was estimated to be $1.19 per tonne. Since three plants will require fewer transportation kilometres, it is necessary to consider the savings accruing from this operating advantage. Three plant locations will save 6.14 kilometres per truck per day, on the average. At an estimated cost of $0.30 per kilometre, this would mean that an annual savings of $6750 is realized on minimum trips to the landfill disposer, for a total annual savings in transportation of $15,300. Savings in labour are also realized because of the shorter routes, which permit more pick-ups during the day. The annual savings from this source are $103,500. The following table summarizes all costs, in thousands of dollars, associated with the present and proposed systems:

| | | Costs for Proposed Systems | | | |
| | Present | Site Number | | | |
Item	System	1	2	3	4
Capital costs:					
Incinerators		$3302	$3302	$3302	$3302
Plant facilities		$600	$900	$1260	$1920
Annex buildings		$91	$102	$112	$132
Additional features		$60	$80	$90	$100
Total		$4053	$4384	$4764	$5454
Annual O&M costs	$905.4	$342	$480	$414	$408
Annual savings:					
Pick-up transportation		$13.2	$14.7	$15.3	$17.1
Labour		$87.6	$99.3	$103.5	$119.40

A bond will be issued to provide the necessary capital investment at an interest rate of 8% with a maturity date 20 years in the future. The proposed systems are expected to last 20 years, with negligible salvage values. If the current system is to be retained, the annual O&M costs would be expected to increase at an annual rate of 10%. The city will use the bond interest rate as the interest rate for any public-project evaluation.

(a) Determine the operating cost of the current system in terms of dollars per tonne of solid waste.

(b) Determine the economics of each solid-waste disposal alternative in terms of dollars per tonne of solid waste.

ST13.3 Because of a rapid growth in population, a small town in Nova Scotia is considering several options to establish a wastewater treatment facility that can handle a flow of

8 million litres per day. The town has five treatment options available:

- **Option 1: No action.** This option will lead to continued deterioration of the environment. If growth continues and pollution results, fines imposed (as high as $10,000 per day) would soon exceed construction costs.

- **Option 2: Land-treatment facility.** This option will provide a system for land treatment of the wastewater to be generated over the next 20 years and will require the utilization of the most land for treatment of the wastewater. In addition to the need to find a suitable site, pumping of the wastewater for a considerable distance out of town will be required. The land cost in the area is $3000 per hectare. The system will use spray irrigation to distribute wastewater over the site. No more than 1 centimetre of wastewater can be applied in one week per hectare.

- **Option 3: Activated sludge-treatment facility.** This option will provide an activated sludge-treatment facility at a site near the planning area. No pumping will be required, and only 7 hectares of land will be needed for construction of the plant, at a cost of $7000 per hectare.

- **Option 4: Trickling filter-treatment facility.** Provide a trickling filter-treatment facility at the same site selected for the activated sludge plant of Option 3. The land required will be the same as that for Option 3. Both facilities will provide similar levels of treatment, but using different units.

- **Option 5: Lagoon-treatment system.** Utilize a three-cell lagoon system for treatment. The lagoon system requires substantially more land than Options 3 and 4 require, but less than Option 2. Due to the larger land requirement, this treatment system will have to be located some distance outside of the planning area and will require pumping of the wastewater to reach the site.

The following tables summarize, respectively, (1) the land cost and land value for each option, (2) the capital expenditures for each option, and (3) the O&M costs associated with each option:

Land Cost for Each Option			
Option Number	Land Required (hectares)	Land Cost ($)	Land Value (in 20 years)
2	800	$2,400,000	$4,334,600
3	7	$49,000	$88,500
4	7	$49,000	$88,500
5	80	$400,000	$722,400

Option Number	Capital Expenditures			
	Equipment	Structure	Pumping	Total
2	$500,000	$700,000	$100,000	$1,300,000
3	$500,000	$2,100,000	$0	$2,600,000
4	400,000	$2,463,000	$0	$2,863,000
5	$175,000	$1,750,000	$100,000	$2,025,000

Option	Annual O&M costs (Year 1)			
Number	Energy	Labour	Repair	Total
2	$200,000	$95,000	$30,000	$325,000
3	$125,000	$65,000	$20,000	$210,000
4	$100,000	$53,000	$15,000	$168,000
5	$50,000	$37,000	$5,000	$92,000

The price of land is assumed to be appreciating at an annual rate of 3%, and the equipment installed will require a replacement cycle of 15 years. Its replacement cost will increase at an annual rate of 5% (over the initial cost), and its salvage value at the end of the planning horizon will be 50% of the replacement cost. The structure requires replacement after 40 years and will have a salvage value of 60% of the original cost. The cost of energy and repair will increase at an annual rate of 5% and 2%, respectively. The labour cost will increase at an annual rate of 4%.

With the following sets of assumptions, answer (a) and (b).

- Assume an analysis period of 120 years.
- Replacement costs for the equipment, as well as for the pumping facilities, will increase at an annual rate of 5%.
- Replacement cost for the structure will remain constant over the planning period. However, the salvage value of the structure will be 60% of the original cost. (Because it has a 40-year replacement cycle, any increase in the future replacement cost will have very little impact on the solution.)
- The equipment's salvage value at the end of its useful life will be 50% of the original replacement cost. For example, the equipment installed under Option 1 will cost $500,000. Its salvage value at the end of 15 years will be $250,000.
- All O&M cost figures are given for year 1. For example, the annual energy cost of $200,000 for Option 2 means that the actual energy cost during the second operating year will be $200,000(1.05) = $210,000.
- Option 1 is not considered a viable alternative, as its initial annual operating cost exceeds $3,650,000.

  (a) If the interest rate is 10%, which option is the most cost effective?
  (b) Suppose a household discharges about 1600 litres of wastewater per day through the facility selected in (a). What should be the monthly bill assessed for this household?

ST13.4 The Organization for Economic Co-operation and Development (OECD) estimates that Toronto's traffic problems cost the city $3.3 billion in lost productivity annually. Most traffic experts believe that adding and enlarging highway systems will not alleviate the problem. As a result, current research on traffic management is focusing on the development of three areas:

(1) computerized dashboard navigational systems,

(2) roadside sensors and signals that monitor and help manage the flow of traffic,

(3) automated steering and speed controls that might allow cars to drive themselves on certain stretches of highway.

In 1963, the first computerized traffic signal control system was installed in Toronto and by some estimates, this innovation has reduced travel time by 6% to 11%, fuel consumption by 4% to 7%, and pollution emissions by 3% to 6%. In Los Angeles, a city with similar traffic problems to Toronto, there is currently testing of an electronic traffic and navigational system. Costing $40 million, the test would include highway sensors and cars with computerized dashboard maps. To install it throughout Los Angeles could cost upward of $2 billion.

On a national scale, for the entire United States, the estimates for implementing "smart" roads and vehicles would cost $18 billion to build the highways; $4 billion per year to maintain and operate them; $1 billion for research and development of driver-information aids; and $2.5 billion for vehicle-control devices. Advocates say the rewards far outweigh the costs.

(a) How would you identify the users' benefits and disbenefits for this type of public project?

(b) What would be the sponsors' cost?

(c) Suppose that the users' net benefits grow at 3% per year and the sponsor's costs grow at 4% per year. Assuming a social discount rate of 10%, what would be the *B/C* ratio over a 20-year study period?

(d) The U.S. interstate highway system spans approximately 75,000 kilometres, supporting 250 million registered vehicles, whereas the TransCanada Highway spans approximately 11,000 kilometres, supporting 13 million registered vehicles. How would this analysis (for parts a to c) relate to Canada assuming the construction, implementation, and maintenance costs vary proportionately, but the R&D costs are constant?

On the Companion Website that accompanies this text, you will find Excel templates and exercises, as well as the following analysis tools: Cash Flow Analyzer, Depreciation Analysis, Loan Analysis, and Interest Tables.

5

Handling Risk
and Uncertainty

FOURTEEN

Inflation and Its Impact on Project Cash Flows

How Much Will It Cost to Send Your Child to University in the Year 2030? You may have heard your grandparents fondly remembering the "good old days" of penny candy or 35-cents-per-gallon (i.e., 7.7¢/litre) gasoline. But even a university student in his or her early twenties can relate to the phenomenon of escalating costs. Do you remember when a postage stamp cost less than half its current price? When the admission for a movie was under $5?

Item	1979 Price	2006 Price	% Increase
Consumer Price Index (CPI) [2002 = 100]	40.0	108.6	171.5
Annual household operating expense	$3429	$8021	133.9
Annual automobile operating expense	1067	4488	320.6
Store brand bread (675 g)	0.60	1.87	211.7
Large eggs (1 doz, grade A)	1.06	2.30	117.0
Milk (1 L)	0.61	1.83	200.0
Soft drink (per L)	0.68	1.42	108.8
Round steak (kg)	6.42	11.95	86.1
Postage	0.17	0.51	200.0
University tuition	895	4400	391.6

The table above demonstrates price differences between 1979 and 2006 for some commonly bought items. For example, a loaf of bread cost 60 cents in 1979, whereas it cost $1.87 in 2006. In 2006, the same 60 cents bought only a fraction of the bread it would have bought in 1979—specifically, about 32.1% of a loaf. From 1979 to 2006, the 60-cent sum had lost 67.9% of its purchasing power! (See also the Consumer Price Index on the next page, which illustrates the steady increase in the cost of living over the decades.)

Consumer Price Index

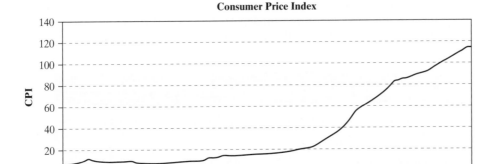

Source: Adapted from Statistics Canada, Consumer Price Index, Historical Summary, www40.statcan. ca/l01/cst01/econ46a-eng.htm (Accessed September 13, 2010).

On a brighter note, we open this chapter with some statistics about engineers' pay rates. In the following table, statistics about pay rates for several engineering disciplines are shown.[1] For example, the average salary of a mechanical engineer employed full-time in Canada was $28,791 in 1980, according to the 1981 Census.[2] By 2006, this had risen to $78,551,[3] which is equivalent to an annual increase of 3.94% over 26 years. During the same period, the annual inflation rate remained at 3.55%, indicating that engineers' pay has more than kept up with the rise in prices in Canada.

Professional Area	1980 Average Pay	2006 Average Pay	% Increase
Chemical	$32,113	$86,669	169.9
Civil	31,181	78,874	153.0
Electrical	29,211	81,836	180.2
Industrial	26,505	74,472	181.0
Mechanical	28,791	78,551	172.8
Metallurgical	31,229	82,851	165.3
Mining	33,669	105,523	213.4
Petroleum	36,519	150,543	312.2

[1] You may be tempted to rate the financial merits of different engineering disciplines based solely on these values. This could be misleading for many reasons. For example, some specialties may have fewer senior engineers in the workforce, either because the demands have changed over time or people get promoted into non-engineering managerial positions. Most provincial engineering associations can provide detailed current salary data broken down by experience, levels of responsibility, and other factors.
[2] Statistics Canada, 1981 Census, Cat. No. 92-930 and Cat. No. 92-919.
[3] Statistics Canada, 2006 Census of Population, Catalogue no. 97-564-XCB2006005.

Up to this point, we have demonstrated how to develop cash flows in a variety of ways and how to compare them under constant conditions in the general economy. In other words, we have assumed that prices remain relatively unchanged over long periods. As you know from personal experience, that is not a realistic assumption. In this chapter, we define and quantify inflation and then go on to apply it in several economic analyses. We will demonstrate inflation's effect on depreciation, borrowed funds, the rate of return of a project, and working capital within the bigger picture of developing projects.

CHAPTER LEARNING OBJECTIVES

After completing this chapter, you should understand the following concepts:

- How to measure inflation.
- Conversion from actual dollars to constant dollars or from constant to actual dollars.
- How to compare the amount of dollars received at different points in time.
- Which inflation rate to use in economic analysis.
- Which interest rate to use in economic analysis (market interest rate versus inflation-free interest rate).
- How to handle multiple inflation rates in project analysis.
- The cost of borrowing and changes in working-capital requirements under inflation.
- How to conduct rate-of-return analysis under inflation.

14.1 Meaning and Measure of Inflation

Inflation is the rate at which the general level of prices of goods and services is rising, and, subsequently, purchasing power is falling.

Historically, the general economy has usually fluctuated in such a way as to exhibit **inflation**, a loss in the purchasing power of money over time. Inflation means that the cost of an item tends to increase over time, or, to put it another way, the same dollar amount buys less of an item over time. **Deflation** is the opposite of inflation, in that prices usually decrease over time; hence, a specified dollar amount gains in purchasing power. Inflation is far more common than deflation in the real world, so our consideration in this chapter will be restricted to accounting for inflation in economic analyses.

CPI is a type of inflationary indicator that measures the change in the cost of a fixed basket of products and services.

14.1.1 Measuring Inflation

Before we can introduce inflation into an engineering economic problem, we need a means of isolating and measuring its effect. Consumers usually have a relative, if not a precise, sense of how their purchasing power is declining. This sense is based on their experience of shopping for food, clothing, transportation, and housing over the years. Economists have developed a measure called the **Consumer Price Index** (**CPI**), which is based on a typical **market basket** of goods and services required by the average consumer. This market basket consists of hundreds of items from eight major groups:

(1) food, (2) shelter, (3) household operations and furnishings, (4) clothing and footwear, (5) transportation, (6) health and personal care, (7) recreation, education, and reading, and (8) alcoholic beverages and tobacco products.

The CPI compares the cost of the typical market basket of goods and services in a current month with its cost 1 month ago, 1 year ago, or 10 years ago. The point in the past with which current prices are compared is called the **base period**. The index value for this base period is set at 100. The CPI for any other year can then be calculated in relation to the base year:

$$\frac{CPI_{Any\ year}}{CPI_{Base\ year}} = \frac{Cost_{Any\ year}}{Cost_{Base\ year}} \text{ or } CPI_{Any\ year} = \frac{Cost_{Any\ year}}{Cost_{Base\ year}} \times 100$$

At the time of writing, the current base period for the CPI in Canada is 2002. That is, the total cost of the prescribed market basket in 2002 is given a price index of 100. The *Consumer Price Index* (Catalogue No 62-001-XWE), a monthly publication prepared by Statistics Canada, lists CPIs for the country, provinces, and select cities. From Table 14.1, the CPI for the year 2008 is 114.1, indicating an increase of $(114.1 - 100)/100 = 14.1\%$ over 2002 aggregate prices. Given the CPI, the consumer product inflation rate can be calculated just as easily between *any* two years, such as 1998–2008:

$$\frac{CPI_{2008} - CPI_{1998}}{CPI_{1998}} = \frac{114.1 - 91.3}{91.3} = 0.2497 = 24.97\%$$

showing a modest increase of almost 25% over the intervening decade. This method of assessing inflation does not imply, however, that consumers actually purchase the same goods and services year after year. Consumers tend to adjust their shopping practices to changes in relative prices, and they tend to substitute other items for those whose prices have increased greatly in relative terms. We must understand that the CPI does not take into account this sort of consumer behavior, because it is predicated on the purchase of a fixed market basket of the same goods and services, in the same proportions, month after month. For this reason, the CPI is called a **price index** rather than a **cost-of-living index**, although the general public often refers to it as a cost-of-living index. Furthermore, the CPI is not a measure of cost-of-living in the traditional sense of "necessities" for survival.

On the other hand, adjustments are made to the basket of goods every four years or so to update expenditure weights as consumers change their spending patterns and to incorporate occasional technical improvements. A significant change in 1995 was to extend the price sampling from cities of 30,000 population or more to *all* areas of Canada, rural and urban. Despite tremendous changes in the consumer world, the CPI has served its function well since its inception in the early 1900s.

The consumer price index is a good measure of the general increase in prices of consumer products. However, it is not a good measure of industrial price increases. In performing engineering economic analysis, the appropriate price indexes must be selected to estimate the price increases of raw materials, finished products, and operating costs. *Industrial Price Indexes*, prepared by Statistics Canada, provide the industrial product price

TABLE 14.1 Selected Price Indexes (Index for Base Year = 100)

Year	CPI	IPPI	RMPI	Capital Equipment	Lumber and wood products	Primary Metal Products
1994	85.7	92.0	89.8	92.7	101.6	89.0
1995	87.6	98.9	97.6	96.0	159.8	104.7
1996	88.9	99.3	101.1	98.0	121.0	94.9
1997	90.4	100.0	100.0	100.0	100.0	100.0
1998	91.3	100.4	86.8	104.9	101.5	90.5
1999	92.9	102.2	93.8	106.6	102.9	91.9
2000	95.4	106.5	114.7	107.3	114.8	100.7
2001	97.8	107.6	113.2	110.7	108.4	92.0
2002	100.0	107.6	112.6	112.6	98.1	94.0
2003	102.8	106.2	114.8	107.1	99.7	96.5
2004	104.7	109.5	128.3	104.5	105.4	113.9
2005	107.0	111.2	145.3	102.5	101.9	117.3
2006	109.1	113.8	161.7	100.2	101.5	153.8
2007	111.5	115.6	174.1	98.3	117.2	165.9
2008	114.1	120.6	196.3	99.2	117.2	141.4

IPPI A type of inflationary indicator published by Statistics Canada to evaluate wholesale price levels in the economy.

index (**IPPI**) and the raw material price index (RMPI) for various industrial goods[4]. Table 14.1 lists the IPPI and RMPI together with select other price indexes over a number of years, with 1997 as the base year. From the table, we can easily calculate the inflation rate of lumber and timber from 2001 to 2002 as follows:

$$\frac{98.1 - 108.4}{98.1} = -0.0950 = -9.50\%$$

Since the inflation rate calculated is negative (deflation), the price of lumber and timber decreased at an annual rate of 9.50% over the year 2001–02. Although the long-term trend is upwards, the prices of some commodities can be quite volatile in the short term, as shown in the table.

Average Inflation Rate (*f*)

To account for the effect of varying yearly inflation rates over a period of several years, we can compute a single rate that represents an **average inflation rate**. Since each individual year's inflation rate is based on the previous year's rate, all these rates have a compounding effect. As an example, suppose we want to calculate the average inflation rate for a two-year period: The first year's inflation rate is 4%, and the second year's rate is 8%, on a base price of $100.

---

[4] Price indexes are available for a nominal fee via the Internet through the Statistics Canada website: www.statcan.ca. Some university libraries may provide access to this data as a service to their members.

Step 1. To find the price at the end of the second year, we use the process of compounding:

First year

$$\overbrace{\$100(1 + 0.04)}(1 + 0.08)=\$112.32.$$

Second year

Step 2. To find the average inflation rate f, we establish the following equivalence equation:

$$\$100(1 + f)^2 = \$112.32 \text{ or } \$100(F/P, f, 2) = \$112.32.$$

Solving for f yields

$$f = 5.98\%.$$

We can say that the price increases in the last two years are equivalent to an average annual percentage rate of 5.98% per year. Note that the average is a geometric, not an arithmetic, average over a several-year period. Our computations are simplified by using a single average rate such as this, rather than a different rate for each year's cash flows.

EXAMPLE 14.1 Average Inflation Rate

Consider again the price increases for the nine items listed in the table at the beginning of this chapter. Determine the average inflation rate for each item over the 27-year period shown.

SOLUTION

Let's take the first item, the annual household operating expense, for a sample calculation. Since we know the prices for 1979 and 2006, we can use the appropriate equivalence formula (single-payment compound amount factor, or growth formula). We state the problem as follows:

Given: $P = \$3429$, $F = \$8021$, and $N = 2006 - 1979 = 27$.

Find: f.

The equation we desire is $F = P(1 + f)^N$, so we have

$$\$8021 = \$3429(1 + f)^{27},$$
$$f = (2.34)^{1/27} - 1$$
$$= 3.20\%.$$

In a similar fashion, we can obtain the average inflation rates for the remaining items. The answers are as follows:

Item	1979 Price	2006 Price	Average Inflation Rate
Annual household operating expense	$3429	$8021	3.20
Annual automobile operating expense	1067	4488	5.46
Store brand bread (675 g)	0.60	1.87	4.30
Large eggs (1 doz, grade A)	1.06	2.30	2.91
Milk (1 L)	0.61	1.83	4.15
Soft drink (per L)	0.68	1.42	2.76
Round steak (kg)	6.42	11.95	2.33
Postage	0.17	0.51	4.15
University tuition	895	4,400	6.08

The cost of university tuition increased most among the items listed in the table, whereas the cost of steak increased relatively slowly.

General Inflation Rate $\bar{f}$ Versus Specific Inflation Rate f_j

When we use the CPI as a base to determine the average inflation rate, we obtain the **general inflation rate**. We need to distinguish carefully between the general inflation rate and the average inflation rate for specific goods:

- **General inflation rate $\bar{f}$.** This average inflation rate is calculated on the basis of the CPI for all items in the market basket. The market interest rate is expected to respond to this general inflation rate.
- **Specific inflation rate f_j.** This rate is based on an index (or the CPI) specific to segment j of the economy. For example, we must often estimate the future cost of a particular item (e.g., labour, material, housing, or gasoline). When we refer to the average inflation rate for just one item, we will drop the subscript j for simplicity.

In terms of the CPI, we define the general inflation rate as

$$CPI_n = CPI_0 (1 + \bar{f})^n, \tag{14.1}$$

or

$$\bar{f} = \left[\frac{CPI_n}{CPI_0} \right]^{1/n} - 1, \tag{14.2}$$

where $\bar{f}$ = The general inflation rate,

   CPI_n = The consumer price index at the end period n, and

   CPI_0 = The consumer price index for the base period.

Knowing the CPI values for two consecutive years, we can calculate the annual general inflation rate as

$$\bar{f}_n = \frac{\text{CPI}_n - \text{CPI}_{n-1}}{\text{CPI}_{n-1}}, \tag{14.3}$$

where $\bar{f}_n$ = the general inflation rate for period n.

As an example, let us calculate the general inflation rate for the year 2008, where $\text{CPI}_{2007} = 111.5$ and $\text{CPI}_{2008} = 114.1$:

$$\frac{114.1 - 111.5}{114.1} = 0.0228 = 2.28\%.$$

This was an unusually good rate for the Canadian economy, compared with the average general inflation rate of 9.62%[5] over the last 50 years.

EXAMPLE 14.2 Yearly and Average Inflation Rates

The following table shows a utility company's cost to supply a fixed amount of power to a new housing development:

Year	Cost
0	$504,000
1	538,400
2	577,000
3	629,500

The indices are specific to the utilities industry. Assume that year 0 is the base period, and determine the inflation rate for each period. Then calculate the average inflation rate over the three years listed in the table.

SOLUTION

Given: Annual cost to supply a fixed amount of power.
Find: Annual inflation rate (f_n) and average inflation rate $(\bar{f})$.

The inflation rate during year 1 (f_1) is

$$\frac{(\$538,400 - \$504,000)}{\$504,000} = 6.83\%.$$

---

[5] To calculate the average general inflation rate from 1958 to 2008, we evaluate

$$\bar{f} = \left[\frac{114.1}{15.2}\right]^{1/50} - 1 = 9.62\%.$$

The inflation rate during year 2 (f_2) is

$$\frac{(\$577,000 - \$538,400)}{\$538,400} = 7.17\%.$$

The inflation rate during year 3 (f_3) is

$$\frac{(\$629,500 - \$577,000)}{\$577,000} = 9.10\%.$$

The average inflation rate over the three years is

$$f = \left(\frac{\$629,500}{\$504,000}\right)^{1/3} - 1 = 0.0769 = 7.69\%.$$

Alternatively $f = [(1.0683)(1.0717)(1.0910)]^{1/3} - 1 = 7.69\%.$

Note that, although the average inflation rate[6] is 7.69% for the period taken as a whole, none of the years within the period had that particular rate.

14.1.2 Actual versus Constant Dollars

To introduce the effect of inflation into our economic analysis, we need to define several inflation-related terms:[7]

Actual (current) dollars: Out-of-pocket dollars paid at the time of purchasing goods and services.

- **Actual (current) dollars (A_n).** Actual dollars are estimates of future cash flows for year n that take into account any anticipated changes in amounts caused by inflationary or deflationary effects. Usually, these amounts are determined by applying an inflation rate to base-year dollar estimates.
- **Constant (real) dollars (A_n').** Constant dollars represent constant purchasing power that is independent of the passage of time. In situations where inflationary effects were assumed when cash flows were estimated, the estimates obtained can be converted to constant dollars (base-year dollars) by adjustment with some readily accepted **general inflation rate**. We will assume that the base year is always time 0, unless we specify otherwise.

Constant (nominal/real) dollars: Dollars as if in some base year, used to adjust for the effects of inflation.

Conversion from Constant to Actual Dollars

Since constant dollars represent dollar amounts expressed in terms of the purchasing power of the base year, we may find the equivalent dollars in year n by using the general inflation rate $\overline{f}$ in the formula

$$A_n = A_n'(1 + \overline{f})^n = A_n'(F/P, \overline{f}, n), \tag{14.4}$$

---

[6] Since we obtained this average rate on the basis of costs that are specific to the utility industry, that rate is not the general inflation rate. Rather, it is a specific inflation rate for the utility in question.

[7] The definitions presented are based on the ANSI Z94 Standards Committee on Industrial Engineering Terminology, *The Engineering Economist*. 1988; 33(2): 145–171.

where A'_n = Constant-dollar expression for the cash flow at the end of year n, and A_n = Actual-dollar expression for the cash flow at the end of year n.

If the future price of a specific cash flow element (j) is not expected to follow the general inflation rate, we will need to use the appropriate average inflation rate f_j applicable to that cash flow element, instead of $\overline{f}$. The conversion process is illustrated in Figure 14.1.

$$A_n = A'_n (1 + \overline{f})^n \quad \leftrightarrow \quad A'_n (F/P, \overline{f}, n)$$

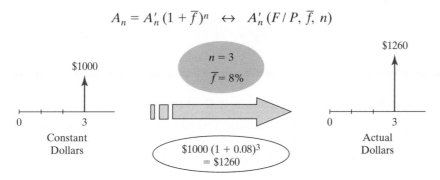

Figure 14.1   Converting a $1000 cash flow at year 3 estimated in constant dollars to its equivalent actual dollars at the same period, assuming an 8% inflation rate.

EXAMPLE 14.3 Conversion From Constant to Actual Dollars

The salary for the current Canadian prime minister, the Right Honourable Stephen Harper, was about $310,800 in 2008. Contrast that with the salary of $25,000 received by Pierre Elliott Trudeau when he first became prime minister in 1968. It seems that Harper's remuneration is much higher. Compare these two salaries in terms of common purchasing power. Whose buying power is greater?

SOLUTION

Given: Two salaries received in 1968 and 2008, respectively.
Find: Compare these two salaries in terms of purchasing power of 2008.

Both salaries are in actual dollars at the time of holding office. So, if we want to compare the salaries on the basis of the purchasing power of the year 2008, then we need to know the consumer price indexes for 1968 as well as 2008.

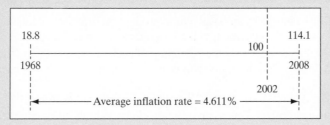

Using the old CPI, we find that the CPI for 1968 is 18.8 and the CPI for 2008 is 114.1. The average inflation rate between these two periods is 4.611%. To find the purchasing power of this $25,000 in 2008, we simply calculate the expression

$$25,000(1 + 0.04611)^{40} = \$151,729.$$

Given inflation and the changing value of money, $25,000 in 1968 is equivalent to $151,729 in terms of 2008 purchasing power. This is clearly much lower than Harper's actual 2008 salary.

Conversion From Actual to Constant Dollars

This conversion, shown in Figure 14.2, is the reverse process of converting from constant to actual dollars. Instead of using the compounding formula, we use a discounting formula (single-payment present-worth factor):

$$A'_n = \frac{A_n}{(1 + \overline{f})^n} = A_n(P/F, \overline{f}, n). \tag{14.5}$$

Once again, we may substitute f_j for $\overline{f}$ if future prices are not expected to follow the general inflation rate.

$$A'_n = A_n (1 + \overline{f})^{-n} \quad \leftrightarrow \quad A_n (P/F, \overline{f}, n)$$

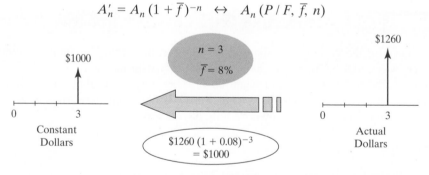

Figure 14.2 Conversion from actual to constant dollars. What it means is that $1260 three years from now will have a purchasing power of $1000 in terms of base dollars (year 0).

EXAMPLE 14.4 Conversion From Actual to Constant Dollars

Jagura Creek Fish Company, an aquacultural production firm, has negotiated a five-year lease on 20 hectares of land, which will be used for fishponds. The annual cost stated in the lease is $20,000, to be paid at the beginning of each of the five years. The general inflation rate $\overline{f} = 5\%$. Find the equivalent cost in constant dollars during each period.

DISCUSSION: Although the $20,000 annual payments are *uniform*, they are not expressed in constant dollars. Unless an inflation clause is built into a contract, any stated amounts refer to *actual dollars*.

SOLUTION

Given: Actual dollars, $\bar{f} = 5\%$.
Find: Constant dollars during each period.

Using Eq. (14.5), we can determine the equivalent lease payments in constant dollars as follows:

End of Period	Cash Flow in Actual $	Conversion at f	Cash Flow in Constant $	Loss of Purchasing Power
0	$20,000	$(1 + 0.05)^0$	$20,000	0%
1	20,000	$(1 + 0.05)^{-1}$	19,048	4.76
2	20,000	$(1 + 0.05)^{-2}$	18,141	9.30
3	20,000	$(1 + 0.05)^{-3}$	17,277	13.62
4	20,000	$(1 + 0.05)^{-4}$	16,454	17.73

Note that, under the inflationary environment, the lender's receipt of the lease payment in year 5 is worth only 82.27% of the first lease payment.

14.2 Equivalence Calculations Under Inflation

In previous chapters, our equivalence analyses took into consideration changes in the **earning power** of money (i.e., interest effects). To factor in changes in **purchasing power** as well—that is, inflation—we may use either (1) constant-dollar analysis or (2) actual-dollar analysis. Either method produces the same solution; however, each method requires the use of a different interest rate and procedure. Before presenting the two procedures for integrating interest and inflation, we will give a precise definition of the two interest rates.

14.2.1 Market and Inflation-Free Interest Rates

Two types of interest rates are used in equivalence calculations: the market interest rate and the inflation-free interest rate. The rate to apply depends on the assumptions used in estimating the cash flow:

- **Market interest rate (i).** This rate takes into account the combined effects of the earning value of capital (earning power) and any anticipated inflation or deflation (purchasing power). Virtually all interest rates stated by financial institutions for loans and savings accounts are market interest rates. Most firms use a market interest rate (also known as **nominal interest rate** or an **inflation-adjusted MARR**) in evaluating their investment projects.

- **Inflation-free interest rate (i').** This rate is an estimate of the true earning power of money when the inflation effects have been removed. Commonly known as the **real interest rate**, it can be computed if the market interest rate and the inflation rate are known. As you will see later in this chapter, in the absence of inflation, the market interest rate is the same as the inflation-free interest rate.

Market interest rate: The interest rate quoted by financial institutions that accounts for both earning and purchasing power.

Real interest rate: The current interest rate minus the current inflation rate.

In calculating any cash flow equivalence, we need to identify the nature of project cash flows. The three common cases are as follows:

Case 1. All cash flow elements are estimated in constant dollars.

Case 2. All cash flow elements are estimated in actual dollars.

Case 3. Some of the cash flow elements are estimated in constant dollars, and others are estimated in actual dollars.

For case 3, we simply convert all cash flow elements into one type—either constant or actual dollars. Then we proceed with either constant-dollar analysis as for case 1 or actual-dollar analysis as for case 2.

14.2.2 Constant-Dollar Analysis

Suppose that all cash flow elements are already given in constant dollars and that you want to compute the equivalent present worth of the constant dollars (A'_n) occurring in year n. In the absence of an inflationary effect, we should use i' to account for only the earning power of the money. To find the present-worth equivalent of this constant-dollar amount at i', we use

$$P_n = \frac{A'_n}{(1 + i')^n}. \qquad (14.6)$$

Constant-dollar analysis is common in the evaluation of many long-term public projects, because governments do not pay income taxes. Typically, income taxes are levied on the basis of taxable incomes in actual dollars.

EXAMPLE 14.5 Equivalence Calculation When Flows Are Stated in Constant Dollars

Transco Company is considering making and supplying computer-controlled traffic-signal switching boxes to be used throughout Western Canada. Transco has estimated the market for its boxes by examining data on new road construction and on the deterioration and replacement of existing units. The current price per unit is $550; the before-tax manufacturing cost is $450. The start-up investment cost is $250,000. The projected sales and net before-tax cash flows in constant dollars are as follows:

Period	Unit Sales	Net Cash Flow in Constant $
0		−$250,000
1	1,000	100,000
2	1,100	110,000
3	1,200	120,000
4	1,300	130,000
5	1,200	120,000

If Transco managers want the company to earn a 12% inflation-free rate of return (i') before tax on any investment, what would be the present worth of this project?

SOLUTION

Given: Cash flow series stated in constant dollars, $i' = 12\%$.
Find: Present worth of the cash flow series.

Since all values are in constant dollars, we can use the inflation-free interest rate. We simply discount the dollar inflows at 12% to obtain the following:

$$PW(12\%) = -\$250{,}000 + \$100{,}000(P/A, 12\%, 5)$$
$$+ \$10{,}000(P/G, 12\%, 4) + \$20{,}000(P/F, 12\%, 5)$$
$$= \$163{,}099 \text{ (in year-0 dollars)}.$$

Since the equivalent net receipts exceed the investment, the project can be justified even before considering any tax effects.

14.2.3 Actual-Dollar Analysis

Now let us assume that all cash flow elements are estimated in actual dollars. To find the equivalent present worth of the actual dollar amount (A_n) in year n, we may use either the **deflation method** or the **adjusted-discount method**.

Deflation Method

The deflation method requires two steps to convert actual dollars into equivalent present-worth dollars. First we convert actual dollars into equivalent constant dollars by discounting by the general inflation rate, a step that removes the inflationary effect. Now we can use i' to find the equivalent present worth.

EXAMPLE 14.6 Equivalence Calculation When Cash Flows Are in Actual Dollars: Deflation Method

Applied Instrumentation, a small manufacturer of custom electronics, is contemplating an investment to produce sensors and control systems that have been requested by a fruit-drying company. The work would be done under a proprietary contract that would terminate in five years. The project is expected to generate the following cash flows in actual dollars:

n	Net Cash Flow in Actual Dollars	n	Net Cash Flow in Actual Dollars
0	$-\$75{,}000$	3	32,800
1	32,000	4	29,000
2	35,700	5	58,000

(a) What are the equivalent year-0 dollars (constant dollars) if the general inflation rate $(\bar{f})$ is 5% per year?

(b) Compute the present worth of these cash flows in constant dollars at $i' = 10\%$.

SOLUTION

Given: Cash flow series stated in actual dollars, $\bar{f} = 5\%$, and $i' = 10\%$.

Find: (a) Cash flow series converted into constant dollars and (b) the present worth of the cash flow series.

The net cash flows in actual dollars can be converted to constant dollars by deflating them, again assuming a 5% yearly deflation factor. The deflated, or constant-dollar, cash flows can then be used to determine the NPW at i'.

(a) We convert the actual dollars into constant dollars as follows:

n	Cash Flows in Actual Dollars	×	Deflation Factor	=	Cash Flows in Constant Dollars
0	−$75,000		1		−$75,000
1	32,000		$(1 + 0.05)^{-1}$		30,476
2	35,700		$(1 + 0.05)^{-2}$		32,381
3	32,800		$(1 + 0.05)^{-3}$		28,334
4	29,000		$(1 + 0.05)^{-4}$		23,858
5	58,000		$(1 + 0.05)^{-5}$		45,445

(b) We compute the equivalent present worth of the constant dollars by using $i' = 10\%$:

n	Cash Flows in Constant Dollars	×	Discounting	=	Equivalent Present Worth
0	−$75,000		1		−$75,000
1	30,476		$(1 + 0.10)^{-1}$		27,706
2	32,381		$(1 + 0.10)^{-2}$		26,761
3	28,334		$(1 + 0.10)^{-3}$		21,288
4	23,858		$(1 + 0.10)^{-4}$		16,295
5	45,445		$(1 + 0.10)^{-5}$		28,218
					$45,268

Adjusted-Discount Method

The two-step process shown in Example 14.6 can be greatly streamlined by the efficiency of the **adjusted-discount method**, which performs deflation and discounting in one step. Mathematically, we can combine this two-step procedure into one with the formula

$$P_n = \frac{\dfrac{A_n}{(1 + \overline{f})^n}}{(1 + i')^n}$$

$$= \frac{A_n}{(1 + \overline{f})^n (1 + i')^n}$$

$$= \frac{A_n}{[(1 + \overline{f})(1 + i')]^n}. \tag{14.7}$$

Since the market interest rate (i) reflects both the earning power and the purchasing power, we have

$$P_n = \frac{A_n}{(1 + i)^n}. \tag{14.8}$$

The equivalent present-worth values in Eqs. (14.7) and (14.8) must be equal at year 0. Therefore,

$$\frac{A_n}{(1 + i)^n} = \frac{A_n}{[(1 + \overline{f})(1 + i')]^n}.$$

This leads to the following relationship among $\overline{f}$, i', and i:

$$(1 + i) = (1 + \overline{f})(1 + i').$$

Simplifying the terms yields

$$i = i' + \overline{f} + i'\overline{f}. \tag{14.9}$$

Eq. (14.9) implies that the market interest rate is a function of two terms: i' and $\overline{f}$. Note that without an inflationary effect, the two interest rates are the same ($\overline{f} = 0 \rightarrow i = i'$). As either i' or $\overline{f}$ increases, i also increases. For example, we can easily observe that, when prices increase due to inflation, bond rates climb, because lenders (i.e., anyone who invests in a money-market fund, a bond, or a certificate of deposit) demand higher rates to protect themselves against erosion in the value of their dollars. If inflation were to remain at 3%, you might be satisfied with an interest rate of 7% on a bond because your return would more than beat inflation. If inflation were running at 10%, however, you would not buy a 7% bond; you might insist instead on a return of at least 14%. By contrast, when prices are coming down, or at least are stable, lenders do not fear any loss of purchasing power with the loans they make, so they are satisfied to lend at lower interest rates.

   In practice, we often approximate the market interest rate (i) simply by adding the inflation rate ($\overline{f}$) to the real interest rate (i') and ignoring the product ($i'\overline{f}$). This practice is fine as long as either i' or $\overline{f}$ is relatively small. With continuous compounding, however, the relationship among i, i', and $\overline{f}$ becomes precisely

$$i' = i - \overline{f}. \tag{14.10}$$

So, if we assume a nominal APR (market interest rate) of 6% per year compounded continuously and an inflation rate of 4% per year compounded continuously, the inflation-free interest rate is exactly 2% per year compounded continuously.

EXAMPLE 14.7 Equivalence Calculation When Flows Are in Actual Dollars: Adjusted-Discounted Method

Consider the cash flows in actual dollars in Example 14.6. Use the adjusted-discount method to compute the equivalent present worth of these cash flows.

SOLUTION

Given: Cash flow series in actual dollars, $\overline{f} = 5\%$, and $i' = 10\%$.
Find: Equivalent present worth.
First, we need to determine the market interest rate i. With $\overline{f} = 5\%$ and $i' = 10\%$, we obtain

$$
\begin{aligned}
i &= i' + \overline{f} + i'\overline{f} \\
&= 0.10 + 0.05 + (0.10)(0.05) \\
&= 15.5\%.
\end{aligned}
$$

Note that the equivalent present worth that we obtain with the adjusted-discount method ($i = 15.5\%$) is exactly the same as the result we obtained in Example 14.6:

n	Cash Flows in Actual Dollars	Multiplied by	=	Equivalent Present Worth
0	−$75,000	1		−$75,000
1	32,000	$(1 + 0.155)^{-1}$		27,706
2	35,700	$(1 + 0.155)^{-2}$		26,761
3	32,800	$(1 + 0.155)^{-3}$		21,288
4	29,000	$(1 + 0.155)^{-4}$		16,296
5	58,000	$(1 + 0.155)^{-5}$		28,217
				$45,268

14.2.4 Mixed-Dollar Analysis

Let us consider the situation in which some cash flow elements are expressed in constant (or today's) dollars and the other elements in actual dollars. In this situation, we can convert all cash flow elements into the same dollar units (either constant or actual). If the cash flow is converted into actual dollars, the market interest rate (i) should be used in calculating the equivalence value. If the cash flow is converted into constant dollars, the inflation-free interest rate (i') should be used.

14.3 Effects of Inflation on Project Cash Flows

We now introduce inflation into some investment projects. We are especially interested in two elements of project cash flows: capital cost allowances and interest expenses. These two elements are essentially immune to the effects of inflation, as they are always given in actual dollars. We will also consider the complication of how to proceed when multiple price indexes have been used to generate various project cash flows.

Because capital cost allowances are calculated on some base-year purchase amount, they do not increase over time to keep pace with inflation. Thus, they lose some of their value to defer taxes, because inflation drives up the general price level and hence taxable income. Similarly, the selling prices of depreciable assets can increase with the general inflation rate and because any gains on salvage values are taxable, they can result in increased taxes. Example 14.8 illustrates how a project's profitability changes under an inflationary economy.

EXAMPLE 14.8 Effects of Inflation on Projects with Depreciable Assets

Consider again the automated machining centre investment project described in Example 10.1. The summary of the financial facts in the absence of inflation is as follows:

Item	Description or Data
Project	Automated machining centre
Required investment	$125,000
Project life	5 years
Salvage value	$50,000
CCA rate	30% declining balance
Annual revenues	$100,000 per year
Annual expenses	
Labour	$20,000 per year
Material	$12,000 per year
Overhead	$8,000 per year
Marginal tax rate	40%
Inflation-free interest rate (i')	15%

The after-tax cash flow for the automated machining centre project was given in Table 10.2, and the net present worth of the project in the absence of inflation was calculated to be $43,443.

What will happen to this investment project if the general inflation rate during the next five years is expected to increase by 5% annually? Sales and operating costs are assumed to increase accordingly. Capital cost allowance will remain unchanged, but taxes, profits, and thus cash flow, will be higher. The firm's inflation-free interest rate (i') is known to be 15%.

(a) Use the deflation method to determine the NPW of the project.

(b) Compare the NPW with that in the inflation-free situation.

DISCUSSION: All cash flow elements, except capital cost allowances, are assumed to be in constant dollars. Since income taxes are levied on actual taxable income, we will use actual-dollar analysis, which requires that all cash flow elements be expressed in actual dollars.

- For the purposes of this illustration, all inflationary calculations are made as of year-end.
- Cash flow elements such as sales, labour, material, overhead, and the selling price of the asset will be inflated at the same rate as the general inflation rate.[8] For example, whereas annual sales had been estimated at $100,000, under conditions of inflation they become 5% greater in year 1, or $105,000; 10.25% greater in year 2; and so forth.

The following table gives the sales conversion from constant to actual dollars for each of the five years of the project life:

Period	Sales in Constant $	Conversion $\bar{f}$	Sales in Actual $
1	$100,000	$(1 + 0.05)^1$	$105,000
2	100,000	$(1 + 0.05)^2$	110,250
3	100,000	$(1 + 0.05)^3$	115,763
4	100,000	$(1 + 0.05)^4$	121,551
5	100,000	$(1 + 0.05)^5$	127,628

Future cash flows in actual dollars for other elements can be obtained in a similar way. Note that

- No change occurs in the investment in year 0 or in capital cost allowances, since these items are unaffected by expected future inflation.
- The selling price of the asset is expected to increase at the general inflation rate. Therefore, the salvage value of the machine, in actual dollars, will be

$$\$50,000(1 + 0.05)^5 = \$63,814.$$

This increase in salvage value will also worsen the tax impact at the time of sale, as the undepreciated capital cost remains unchanged. The calculations for both the capital cost allowance and disposal tax effect are shown in Table 14.2.

---

[8] This is a simplistic assumption. In practice, these elements may have price indexes other than the CPI. Differential price indexes will be treated in Example 14.9.

TABLE 14.2 **Cash Flow Statement for the Automated Machining Center Project Under Inflation**

	Inflation Rate	0	1	2	3	4	5
Income Statement:							
Revenues	5%		$105,000	$110,250	$115,763	$121,551	$127,628
Expenses							
Labour	5%		21,000	22,050	23,153	24,310	25,526
Material	5%		12,600	13,230	13,892	14,586	15,315
Overhead	5%		8,400	8,820	9,261	9,724	10,210
CCA (30%)			18,750	31,875	22,313	15,619	10,933
Taxable Income			$ 44,250	$ 34,275	$ 47,145	$ 57,312	$ 65,644
Income Taxes (40%)			17,700	13,710	18,858	22,925	26,258
Net Income			$ 26,550	$ 20,565	$ 28,287	$ 34,387	$ 39,386
Cash Flow Statement:							
Operating Activities							
Net Income			26,550	20,565	28,287	34,387	39,386
CCA			18,750	31,875	22,313	15,619	10,933
Investment Activities							
Investment		(125,000)					
Salvage	5%						63,814
Disposal Tax Effect							(15,321)
Net Cash Flow							
(in actual dollars)		$(125,000)	$ 45,300	$ 52,440	$ 50,600	$ 50,006	$ 98,812

SOLUTION

Given: Financial data in constant dollars, $\bar{f} = 5\%$, and $i' = 15\%$.
Find: The NPW.

Table 14.2 shows after-tax cash flows in actual dollars. Using the deflation method, we convert the cash flows to constant dollars with the same purchasing power as the dollars used to make the initial investment (year 0), assuming a general inflation rate of 5%. Then we discount these constant-dollar cash flows at i' to determine the NPW. Since NPW = $39,235 > 0$, the investment is still economically attractive.

COMMENTS: Note that the NPW in the absence of inflation was $43,443 in Example 10.1. The $4208 decline just illustrated (known as inflation loss) in the NPW under inflation is due entirely to income tax considerations. The capital cost allowance is a charge against taxable income, which reduces the amount of taxes paid and, as a result, increases the cash flow attributable to an investment by the amount of taxes saved. But the capital cost allowance under existing tax laws is based on historic cost. As time goes by, the capital cost allowance is charged to taxable income in dollars of declining purchasing power; as a result, the "real" cost of the asset is not totally reflected in the capital cost allowance. Depreciation costs are thereby understated, and the taxable income is overstated, resulting in higher taxes. In "real" terms, the amount of this additional income tax is $4208, which is also known as the **inflation tax**. In general, any investment that, for tax purposes, is expensed over time, rather than immediately, is subject to the inflation tax. The following table gives the net cash flow conversion from actual to constant dollars for each of the five years of the project's life:

Year	Net Cash Flow in Actual $	Conversion $\bar{f}$	Net Cash Flow in Constant $	NPW at 15%
0	−$125,000	$(1 + 0.05)^0$	−$125,000	−$125,000
1	45,300	$(1 + 0.05)^{-1}$	43,143	37,516
2	52,440	$(1 + 0.05)^{-2}$	47,565	35,966
3	50,600	$(1 + 0.05)^{-3}$	43,710	28,740
4	50,006	$(1 + 0.05)^{-4}$	41,140	23,522
5	98,812	$(1 + 0.05)^{-5}$	77,422	38,492
				$39,235

14.3.1 Multiple Inflation Rates

As we noted previously, the inflation rate f_j represents a rate applicable to a specific segment j of the economy. For example, if we were estimating the future cost of a piece of machinery, we should use the inflation rate appropriate for that item. Furthermore, we may need to use several rates to accommodate the different costs and revenues in our analysis. The next example introduces the complexity of multiple inflation rates.

EXAMPLE 14.9 Applying Specific Inflation Rates

Let us rework Example 14.8, using different annual indexes (differential inflation rates) in the prices of cash flow components. Suppose that we expect the general rate of inflation, $\bar{f}$, to average 6% per year during the next five years. We also expect that the selling price of the equipment will increase 3% per year, that wages (labour) and overhead will increase 5% per year, and that the cost of materials will increase 4% per year. We expect sales revenue to climb at the general inflation rate. Table 14.3 shows the relevant calculations based on the income statement format. For

TABLE 14.3 **Cash Flow Statement for the Automated Machining Centre Project Under Inflation, With Multiple Price Indexes**

	Inflation Rate	0	1	2	3	4	5
Income Statement:							
Revenues	6%		$ 106,000	$ 112,360	$ 119,102	$ 126,248	$ 133,823
Expenses							
Labour	5%		21,000	22,050	23,153	24,310	25,526
Material	4%		12,480	12,979	13,498	14,038	14,600
Overhead	5%		8,400	8,820	9,261	9,724	10,210
CCA (30%)			18,750	31,875	22,313	15,619	10,933
Taxable Income			$ 45,370	$ 36,636	$ 50,877	$ 62,556	$ 72,554
Income Taxes (40%)			18,148	14,654	20,351	25,023	29,021
Net income			$ 27,222	$ 21,981	$ 30,526	$ 37,534	$ 43,532
Cash Flow Statement:							
Operating Activities							
Net Income			27,222	21,981	30,526	37,534	43,532
CCA			18,750	31,875	22,313	15,619	10,933
Investment Activities							
Investment		(125,000)					
Salvage	3%						57,964
Disposal Tax Effect							(12,981)
Net cash flow (in actual dollars)		$(125,000)	$ 45,972	$ 53,856	$ 52,839	$ 53,153	$ 99,448

simplicity, all cash flows and inflation effects are assumed to occur at year's end. Determine the net present worth of this investment, using the adjusted-discount method.

SOLUTION

Given: Financial data given in Example 14.8, multiple inflation rates.

Find: The NPW, using the adjusted-discount method.

Table 14.3 summarizes the after-tax cash flows in actual dollars. To evaluate the present worth using actual dollars, we must adjust the original discount rate of 15%, which is an inflation-free interest rate i'. The appropriate interest rate to use is the market interest rate:[9]

$$i = i + \bar{f} + i'\bar{f}$$
$$= 0.15 + 0.06 + (0.15)(0.06)$$
$$= 21.90\%.$$

---

[9] In practice, the market interest rate is usually given and the inflation-free interest rate can be calculated when the general inflation rate is known for years in the past or is estimated for time in the future. In our example, we are considering the opposite situation.

The equivalent present worth is obtained as follows:

$$PW(21.90\%) = -\$125,000 + \$45,972(P/F, 21.90\%, 1)$$
$$+ \$53,856(P/F, 21.90\%, 2) + \dots$$
$$+ \$99,448(P/F, 21.90\%, 5)$$
$$= \$39,145.$$

14.3.2 Effects of Borrowed Funds Under Inflation

The repayment of a loan is based on the historical contract amount; the payment size does not change with inflation. Yet inflation greatly affects the value of these future payments, which are computed in year-0 dollars. First, we shall look at how the values of loan payments change under inflation. Interest expenses are usually already stated in the loan contract in actual dollars and need not be adjusted. Under the effect of inflation, the constant-dollar costs of both interest and principal repayments on a debt are reduced. Example 14.10 illustrates the effects of inflation on payments with project financing.

EXAMPLE 14.10 Effects of Inflation on Payments with Financing (Borrowing)

Let us rework Example 14.8 with a debt-to-equity ratio of 0.50, where the debt portion of the initial investment is borrowed at 15.5% annual interest. Assume, for simplicity, that the general inflation rate $\overline{f}$ of 5% during the project period will affect all revenues and expenses (except depreciation and loan payments) and the salvage value of the asset. Determine the NPW of this investment. (Note that the borrowing rate of 15.5% reflects the higher cost of debt financing under an inflationary environment.)

SOLUTION

Given: Cash flow data in Example 14.8, debt-to-equity ratio = 0.50, borrowing interest rate = 15.5%, $\overline{f}$ = 5%, and i' = 15%.
Find: The NPW.

For equal future payments, the actual-dollar cash flows for the financing activity are represented by the circles in Figure 14.3. If inflation were to occur, the cash flow, measured in year-0 dollars, would be represented by the shaded circles in the figure. Table 14.4 summarizes the after-tax cash flows in this situation. For simplicity, assume that all cash flows and inflation effects occur at year's end. To evaluate the present worth with the use of actual dollars, we must adjust the original discount rate

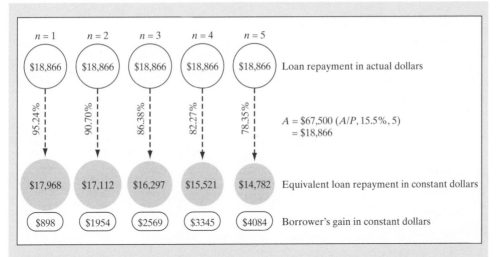

Figure 14.3 Equivalent loan repayment cash flows measured in year-0 dollars and borrower's gain over the life of the loan (Example 14.10).

(MARR) of 15%, which is an inflation-free interest rate i'. The appropriate interest rate to use is thus the market interest rate:

$$i = i' + \overline{f} + i'\overline{f}$$
$$= 0.15 + 0.05 + (0.15)(0.05)$$
$$= 20.75\%.$$

TABLE 14.4 Cash Flow Statement for the Automated Machining Centre Project Under Inflation, With Borrowed Funds

	Inflation Rate	0	1	2	3	4	5
Income Statement:							
Revenues	5%		$105,000	$110,250	$115,763	$121,551	$127,628
Expenses							
Labour	5%		21,000	22,050	23,153	24,310	25,526
Material	5%		12,600	13,230	13,892	14,586	15,315
Overhead	5%		8,400	8,820	9,261	9,724	10,210
CCA (30%)			18,750	31,875	22,313	15,619	10,933
Debt interest			9,688	8,265	6,622	4,724	2,532
Taxable Income			$ 34,563	$ 26,010	$ 40,523	$ 52,588	$ 63,112
Income Taxes (40%)			13,825	10,404	16,209	21,035	25,245
Net Income			$ 20,738	$ 15,606	$ 24,314	$ 31,553	$ 37,867

(Continued)

TABLE 14.4 (Continued)

	Inflation Rate	0	1	2	3	4	5
Cash Flow Statement:							
Operating activities							
Net Income			20,738	15,606	24,314	31,553	37,867
CCA			18,750	31,875	22,313	15,619	10,933
Investment activities							
Investment		(125,000)					
Salvage	5%						63,814
Disposal Tax Effect							(15,321)
Financing activities							
Borrowed funds		62,500					
Principal repayment			(9,178)	(10,601)	(12,244)	(14,142)	(16,334)
Net Cash Flow (in actual dollars)		$ (62,500)	$ 30,309	$ 36,880	$ 34,382	$ 33,029	$ 80,959
Net Cash Flow (in constant dollars)	5%	$ (62,500)	$ 28,866	$ 33,451	$ 29,701	$ 27,173	$ 63,433
Equivalent present worth	15%	$ (62,500)	$ 25,101	$ 25,294	$ 19,529	$ 15,536	$ 31,538
Net present worth	$54,497						

Then, from Table 14.4, we compute the equivalent present worth of the after-tax cash flow as follows:

$$PW(20.75\%) = -\$62,500 + \$30,309(P/F, 20.75\%, 1) + \cdots$$
$$+ \$80,959(P/F, 20.75\%, 5)$$
$$= \$54,497.$$

COMMENTS: In the absence of debt financing, the project would have a net present worth of $39,325, as shown in Example 14.8. Compared with the preceding result of $54,497, the gain in present worth due to debt financing is $15,172. This increase in NPW is due primarily to the debt financing. An inflationary trend decreases the purchasing power of future dollars, which helps long-term borrowers because they can repay a loan with dollars with reduced buying power. That is, the debt-financing cost is reduced in an inflationary environment. In this case, the benefits of financing under inflation have more than offset the *inflation tax* effect on depreciation and the salvage value. The amount of the gain may vary with the interest rate on the loan: The interest rate for borrowing is also generally higher during periods of inflation because it is a market-driven rate.

14.4 Rate-of-Return Analysis Under Inflation

In addition to affecting individual aspects of a project's income statement, inflation can have a profound effect on the overall return—that is, the very acceptability—of an investment project. In this section, we will explore the effects of inflation on the rate of return on an investment and present several examples.

14.4.1 Effects of Inflation on Return on Investment

The effect of inflation on the rate of return on an investment depends on how future revenues respond to the inflation. Under inflation, a company is usually able to compensate for increasing material and labour prices by raising its selling prices. However, even if future revenues increase to match the inflation rate, the allowable depreciation schedule, as we have seen, does not increase. The result is increased taxable income and higher income-tax payments. This increase reduces the available constant-dollar after-tax benefits and, therefore, the inflation-free after-tax rate of return (IRR'), or real IRR. The next example will help us to understand this situation.

EXAMPLE 14.11 IRR Analysis With Inflation

Hartsfield Company, a manufacturer of auto parts, is considering purchasing a set of machine tools at a cost of $30,000. The purchase is expected to generate increased sales of $24,500 per year and increased operating costs of $10,000 per year in each of the next four years. Additional profits will be taxed at a rate of 40%. The asset falls into CCA Class 43 (rate = 30%) for tax purposes. The project has a four-year life with zero salvage value. (All dollar figures represent constant dollars.)

(a) What is the expected internal rate of return?

(b) What is the expected IRR' if the general inflation is 10% during each of the next four years? (Here also, assume that $f_j = \overline{f} = 10\%$.)

(c) If the company requires an inflation-free MARR (or $MARR'$) of 18%, should the company invest in the equipment?

SOLUTION

Given: Financial data in constant dollars, $t_m = 40\%$, and $N = 4$ years.

Find: (a) IRR, (b) IRR' when $\overline{f} = 10\%$, and (c) whether the project can be justified at MARR = 18%.

(a) Rate-of-return analysis without inflation:

We find the rate of return by first computing the after-tax cash flow by the income statement approach, as illustrated in Table 14.5. The first part of the table shows the calculation of additional sales, operating costs, capital cost allowance, and taxes. After four years, there is no expected salvage value. As we emphasized in Chapter 10, depreciation is not a cash expense, although it affects

TABLE 14.5 Rate-of-Return Calculation Without Inflation

	0	1	2	3	4
Income Statement:					
Revenues		$ 24,500	$ 24,500	$ 24,500	$ 24,500
Expenses					
O&M		10,000	10,000	10,000	10,000
CCA (30%)		4,500	7,650	5,355	3,749
Taxable Income		$ 10,000	$ 6,850	$ 9,145	$ 10,752
Income Taxes (40%)		4,000	2,740	3,658	4,301
Net Income		$ 6,000	$ 4,110	$ 5,487	$ 6,451
Cash Flow Statement:					
Operating Activities					
Net Income		6,000	4,110	5,487	6,451
CCA		4,500	7,650	5,355	3,749
Investment Activities					
Investment	($30,000)				
Salvage					0
Disposal Tax Effect					3,499
Net Cash Flow					
(in actual dollars)	($30,000)	$ 10,500	$ 11,760	$ 10,842	$ 13,698

taxable income, and thus cash flow, indirectly. Therefore, we must add CCA to net income to determine the net cash flow.

Thus, if the investment is made, we expect to receive additional annual cash flows of $10,500, $11,760, $10,842, and $13,698. This is a simple investment, so we can calculate the IRR for the project as follows:

$$PW(i') = -\$30,000 + \$10,500(P/F, i', 1) + \$11,760(P/F, i', 2)$$
$$+ \$10,842(P/F, i', 3) + \$13,698(P/F, i', 4)$$
$$= 0$$

Solving for i' yields

$$IRR' = i'^* = 19.66\%.$$

The project has an inflation-free rate of return of 19.66%; that is, the company will recover its original investment ($30,000) plus interest at 19.66% each year for each dollar still invested in the project. Since IRR' > MARR' (18%), the company should buy the equipment.

(b) Rate-of-return analysis under inflation:
With inflation, we assume that sales, operating costs, and the future selling price of the asset will increase. Capital cost allowance will be unchanged, but taxes, profits, and cash flow will be higher. We might think that higher cash flows will

TABLE 14.6 **Rate-of-Return Calculation Under Inflation**

	0	1	2	3	4
Income Statement:					
Revenues		$26,950	$29,645	$32,610	$35,870
Expenses					
O&M		$11,000	$12,100	$13,310	$14,641
CCA (30%)		4,500	7,650	5,355	3,749
Taxable Income		$11,450	$ 9,895	$13,945	$17,481
Income Taxes (40%)		4,580	3,958	5,578	6,992
Net Income		$ 6,870	$ 5,937	$ 8,367	$10,489
Cash Flow Statement:					
Operating Activities					
Net Income		6,870	5,937	8,367	10,489
CCA		4,500	7,650	5,355	3,749
Investment Activities					
Investment	($30,000)				
Salvage					0
Disposal Tax Effect					3,499
Net Cash Flow (in actual dollars)	($30,000)	$11,370	$13,587	$13,722	$17,736
Net Cash Flow (in constant dollars)	($30,000)	$10,336	$11,229	$10,309	$12,114

Note: PW(18%) = $(653); IRR (actual dollars) = 28.59%; IRR' (constant dollars) = 16.90%.

mean an increased rate of return. Unfortunately, this is not the case. We must recognize that cash flows for each year are stated in dollars of declining purchasing power. When the net after-tax cash flows are converted to dollars with the same purchasing power as those used to make the original investment, the resulting rate of return decreases. These calculations, assuming an inflation rate of 10% in sales and operating expenses and a 10% annual decline in the purchasing power of the dollar, are shown in Table 14.6. For example, whereas additional sales had been $24,500 yearly, under conditions of inflation they would be 10% greater in year 1, or $26,950; 21% greater in year 2; and so forth. No change in investment or capital cost allowance will occur, since these items are unaffected by expected future inflation. We have restated the after-tax cash flows (actual dollars) in dollars of a common purchasing power (constant dollars) by deflating them, again assuming an annual deflation factor of 10%. The constant-dollar cash flows are then used to determine the real rate of return. First,

$$PW(i') = -\$30,000 + \$10,336(P/F, i', 1) + \$11,229(P/F, i', 2)$$
$$+ \$10,309(P/F, i', 3) + \$12,114(P/F, i', 4)$$
$$= 0.$$

Then, solving for i' yields

$$i'^* = 16.90\%.$$

The rate of return for the project's cash flows in constant dollars (year-0 dollars) is 16.90%, which is less than the 19.66% return in the inflation-free case. Since IRR$'$ < MARR$'$, the investment is no longer acceptable.

COMMENTS: We could also calculate the rate of return by setting the PW of the actual dollar cash flows to 0. This would give a value of IRR = 28.59%, but that is an inflation-adjusted IRR. We could then convert to an IRR$'$ by deducting the amount caused by inflation:

$$i' = \frac{(1 + i)}{(1 + \bar{f})} - 1$$
$$= \frac{(1 + 0.2859)}{(1 + 0.10)} - 1$$
$$= 16.90\%.$$

This approach gives the same final result of IRR$'$ = 16.90%.

14.4.2 Effects of Inflation on Working Capital

The loss of tax savings from capital cost allowance is not the only way that inflation may distort an investment's rate of return. Another source is working-capital drain. Projects requiring increased levels of working capital suffer from inflation because additional cash must be invested to maintain new price levels. For example, if the cost of inventory increases, additional outflows of cash are required to maintain appropriate inventory levels over time. A similar phenomenon occurs with funds committed to accounts receivable. These additional working-capital requirements can significantly reduce a project's rate of return, as the next example illustrates.

EXAMPLE 14.12 Effect of Inflation on Profits With Working Capital

Suppose that, in Example 14.11, a $1000 investment in working capital is expected and that all the working capital will be recovered at the end of the project's four-year life. Determine the rate of return on this investment.

SOLUTION

Given: Financial data in Example 14.11 and working-capital requirement = $1000. Find: IRR.

Using the data in the upper part of Table 14.7, we can calculate IRR′ = IRR of 18.83% in the absence of inflation. Also, PW(18%) = $524. The lower part of the table includes the effect of inflation on the proposed investment. As illustrated in Figure 14.4, working-capital levels can be maintained only by additional infusions of cash; the working-capital drain also appears in the lower part of Table 14.7. For

TABLE 14.7 Effects of Inflation on Working Capital and After-Tax Rate of Return

Case 1: Without Inflation	0	1	2	3	4
Income Statement:					
Revenues		$ 24,500	$ 24,500	$ 24,500	$ 24,500
Expenses					
O&M		10,000	10,000	10,000	10,000
CCA (30%)		4,500	7,650	5,355	3,749
Taxable Income		$ 10,000	$ 6,850	$ 9,145	$ 10,752
Income Taxes (40%)		4,000	2,740	3,658	4,301
Net Income		$ 6,000	$ 4,110	$ 5,487	$ 6,451
Cash Flow Statement:					
Operating Activities					
Net Income		6,000	4,110	5,487	6,451
CCA		4,500	7,650	5,355	3,749
Investment Activities					
Investment	($30,000)				
Working capital	($ 1,000)				1,000
Salvage					0
Disposal Tax Effect					3,499
Net Cash Flow					
(in actual dollars)	($31,000)	$ 10,500	$ 11,760	$ 10,842	$ 14,698

Note: PW(18%) = $524; IRR′ = 18.83%.

(Continued)

TABLE 14.7 (Continued)

Case 2: With Inflation (10%)	0	1	2	3	4
Income Statement:					
Revenues		$ 26,950	$ 29,645	$ 32,610	$ 35,870
Expenses					
O&M		$ 11,000	$ 12,100	$ 13,310	$ 14,641
CCA (30%)		4,500	7,650	5,355	3,749
Taxable Income		$ 11,450	$ 9,895	$ 13,945	$ 17,481
Income Taxes (40%)		4,580	3,958	5,578	6,992
Net Income		$ 6,870	$ 5,937	$ 8,367	$ 10,489
Cash Flow Statement:					
Operating Activities					
Net Income		6,870	5,937	8,367	10,489
CCA		4,500	7,650	5,355	3,749
Investment Activities					
Investment	($30,000)				
Working capital	($1,000)	(100)	(110)	(121)	$1,331
Salvage					0
Disposal Tax Effect					3,499
Net Cash Flow (in actual dollars)	($31,000)	$ 11,270	$ 13,477	$ 13,601	$ 19,067
Net Cash Flow (in constant dollars)	($31,000)	$ 10,245	$ 11,138	$ 10,218	$ 13,023

Note: PW(18%) = $(1,382); IRR (actual dollars) = 27.34%; IRR' (constant dollars) = 15.76%.

example, the $1000 investment in working capital made in year 0 will be recovered at the end of the first year, assuming a one-year recovery cycle. However, because of 10% inflation, the required working capital for the second year increases to $1100. In addition to reinvesting the $1000 revenues, an additional investment of $100 must be made. This $1100 will be recovered at the end of the second year. However, the project will need a 10% increase, or $1210, for the third year, and so forth.

As Table 14.7 illustrates, the combined effect of CCA and working capital is significant. Given an inflationary economy and investment in working capital, the project's IRR' drops to 15.76%, or PW(18%) = −$1382. By using either IRR analysis or NPW analysis, we end up with the same result (as we must): Alternatives that are attractive when inflation does not exist may not be acceptable when inflation does exist.

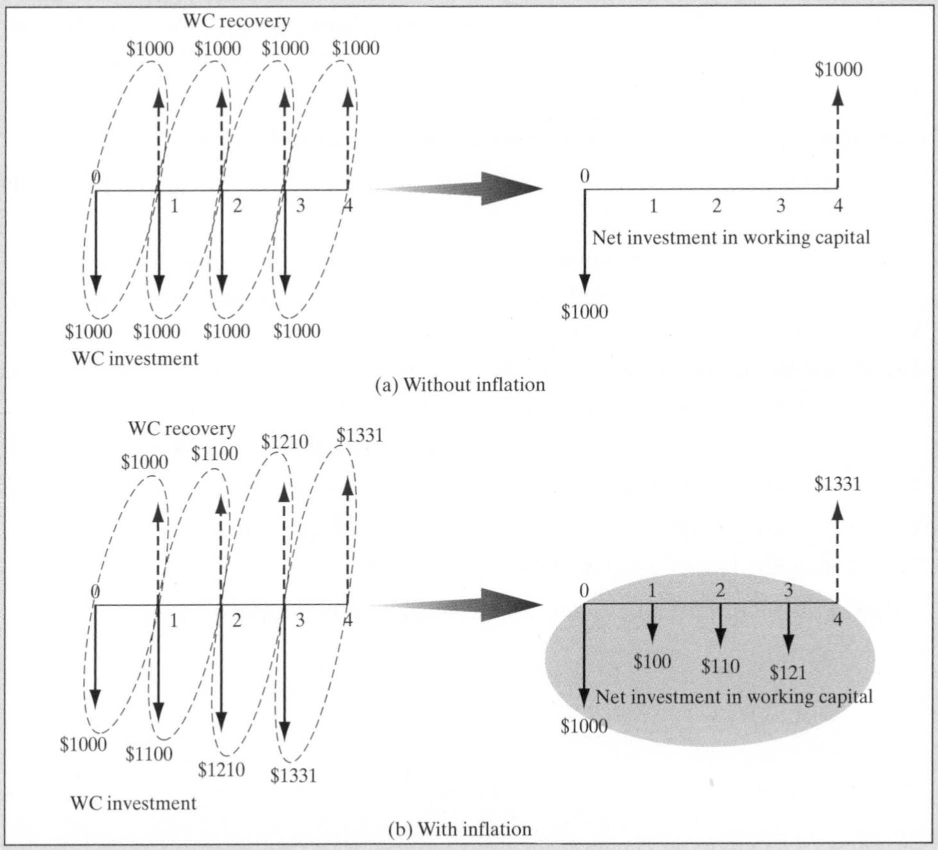

Figure 14.4 Working-capital requirements under inflation: (a) Requirements without inflation and (b) requirements with inflation, assuming a one-year recovery cycle.

SUMMARY

■ The **Consumer Price Index (CPI)** is a statistical measure of change, over time, of the prices of goods and services in major expenditure groups—such as food, housing, clothing, transportation, and health care—typically purchased by urban consumers. Essentially, the CPI compares the cost of a sample "market basket" of goods and services in a specific period with the cost of the same "market basket" in an earlier reference period, designated the **base period**.

■ **Inflation** is the term used to describe a **decline in purchasing power** evidenced in an economic environment of rising prices.

■ **Deflation** is the opposite: an increase in purchasing power evidenced by falling prices.

■ The **general inflation rate** $\bar{f}$ is an average inflation rate based on the CPI. An annual general inflation rate $\bar{f}_j$ can be calculated with the following equation:

$$\bar{f}_n = \frac{\text{CPI}_n - \text{CPI}_{n-1}}{\text{CPI}_{n-1}}.$$

- Specific, individual commodities do not always reflect the general inflation rate in their price changes. We can calculate an **average inflation rate** $\overline{f}_j$ for a specific commodity (j) if we have an index (i.e., a record of historical costs) for that commodity.

- Project cash flows may be stated in one of two forms:

  Actual dollars (A_n): Dollars that reflect the inflation or deflation rate.

  Constant dollars (A'_n): Year 0 (i.e., base year) dollars; dollars with the same purchasing power.

- Interest rates used in evaluating a project may be stated in one of two forms:

  Market interest rate (i) (or real interest rate): A rate that combines the effects of interest and inflation; used with **actual-dollar** analysis.

  Inflation-free interest rate (i'): A rate from which the effects of inflation have been removed; this rate is used with constant-dollar analysis.

- To calculate the present worth of actual dollars, we can use a two-step or a one-step process:

  Deflation method—two steps:

  1. Convert actual dollars by deflating with the general inflation rate $\overline{f}$.

  2. Calculate the PW of constant dollars by discounting at i'.

  Adjusted-discount method—one step (use the market interest rate):

  $$P_n = \frac{A_n}{((1 + \overline{f})(1 + i'))^n}$$

  $$= \frac{A_n}{(1 + i)^n}.$$

- A number of individual elements of project evaluations can be distorted by inflation. These elements are summarized in Table 14.8.

TABLE 14.8 Effects of Inflation on Project Cash Flows and Return

Item	Effects of Inflation
CCA	Capital cost allowance is charged to taxable income in dollars of declining values; taxable income is overstated, resulting in higher taxes.
Salvage values	Inflated salvage values combined with undepreciated capital cost values based on historical costs result in a less beneficial disposal tax effect (i.e., smaller loss or higher gain), hence higher taxes.
Loan repayments	Borrowers repay historical loan amounts with dollars of decreased purchasing power, reducing the cost of financing debt.
Working capital requirement	Known as a *working capital drain*, the cost of working capital increases in an inflationary economy.
Rate of return and NPW	Unless revenues are sufficiently increased to keep pace with inflation, tax effects and a working-capital drain result in a lower rate of return or a lower NPW.

PROBLEMS

Note: *In the problems that follow, the term "market interest rate" represents the "inflation-adjusted MARR" for project evaluation or the "interest rate" quoted by a financial institution for commercial loans.*

14.1 The following data indicate the median average retail prices for regular unleaded gasoline at self-service filling stations, by major urban centres in Ontario:

	Ottawa-Gatineau	Toronto	Thunder Bay
January 2009	74.2	78.9	84.2
February 2009	76.8	81.6	89.2
March 2009	80.8	84.7	87.1
April 2009	80.9	85.2	88.2
May 2009	89.3	94.4	98.2
June 2009	94.4	98.8	107.4
July 2009	89.5	93.9	106.8
August 2009	93.7	98.3	106.2
September 2009	90.5	95.3	104.7
October 2009	91.6	95.6	98.9
November 2009	94.7	98.1	100.0
December 2009	91.3	94.4	96.7
January 2010	95.2	98.9	102.7

(a) Assuming that the base period (price index = 100) is January 2009, compute the average price indexes for January 2010 for all three centres.

(b) Assuming the value for Toronto in January 2009 is the base period index, compute the newly re-indexed values for the entire table.

14.2 The following data indicate the price indexes of primary metal products for the following five-year period (base period 1997 = 100):

Period	Price Index
2001	92.0
2002	94.0
2003	96.5
2004	113.9
2005	117.3
2010	?

(a) Calculate the average annual inflation rate over the 2001–05 period.

(b) If the past trend is expected to continue, how would you estimate the metal product price index in 2010?

(c) Assuming that the base period (price index = 100) is reset to the year 2001, compute the average price indexes for metal for each year between 2001 and 2005.

14.3 For prices that are increasing at an annual rate of 5% the first year and 8% the second year, determine the average inflation rate $(\bar{f})$ over the two years.

14.4 Since the creation of the Bank of Canada, which began operations on March 11, 1935, the Canadian dollar has had an average annual inflation rate of 3.14%. What is the relative value of the dollar in 2010 in relation to its inception in 1935? Assuming the rate is expected to remain the same for the foreseeable future, how many years will it take for the loonie's purchasing power to be one-half of what it is now?

14.5 An annuity provides for 10 consecutive end-of-year payments of $4500. The average general inflation rate is estimated to be 5% annually, and the market interest rate is 12% annually. What is the annuity worth in terms of a single equivalent amount of today's dollars?

14.6 A company is considering buying workstation computers to support its engineering staff. In today's dollars, it is estimated that the maintenance costs for the computers (paid at the end of each year) will be $25,000, $30,000, $32,000, $35,000, and $40,000 for years 1 to 5, respectively. The general inflation rate $(\bar{f})$ is estimated to be 8% per year, and the company will receive 15% per year on its invested funds during the inflationary period. The company wants to pay for maintenance expenses in equivalent equal payments (in actual dollars) at the end of each of the five years. Find the amount of the company's payments.

14.7 The following cash flows are in actual dollars:

n	Cash Flow (in actual $)
0	$1500
4	2500
5	3500
7	4500

Convert to an equivalent cash flow in constant dollars if the base year is time 0. Keep cash flows at the same point in time—that is, years 0, 4, 5, and 7. Assume that the market interest rate is 16% and that the general inflation rate $(\bar{f})$ is 4% per year.

14.8 The purchase of a car requires a $25,000 loan to be repaid in monthly installments for four years at 12% interest compounded monthly. If the general inflation rate is 6% compounded monthly, find the actual- and constant-dollar value of the 20th payment.

14.9 A series of four annual constant-dollar payments beginning with $7000 at the end of the first year is growing at the rate of 8% per year. (Assume that the base year is the current year $(n = 0)$.) If the market interest rate is 13% per year and the general inflation rate $(\bar{f})$ is 7% per year, find the present worth of this series of payments, based on

(a) constant-dollar analysis.

(b) actual-dollar analysis.

14.10 Consider the accompanying cash flow diagrams, where the equal payment cash flow in constant dollars (a) is converted from the equal payment cash flow in actual dollars (b) at an annual general inflation rate of $\bar{f} = 3.8\%$ and $i = 9\%$. What is the amount A in actual dollars equivalent to $A' = \$1000$ in constant dollars?

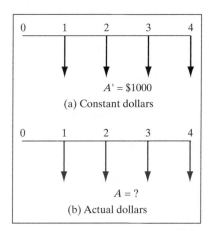

14.11 A 10-year $1000 bond pays a nominal rate of 9% compounded semiannually. If the market interest rate is 12% compounded annually and the general inflation rate is 6% per year, find the actual- and constant-dollar amount (in time-0 dollars) of the 16th interest payment on the bond.

14.12 Suppose that you borrow $20,000 at 12% compounded monthly over five years. Knowing that the 12% represents the market interest rate, you realize that the monthly payment in actual dollars will be $444.90. If the average monthly general inflation rate is expected to be 0.5%, determine the equivalent equal monthly payment series in constant dollars.

14.13 The annual fuel costs required to operate a small solid-waste treatment plant are projected to be $1.5 million, without considering any future inflation. The best estimates indicate that the annual inflation-free interest rate (i') will be 6% and the general inflation rate $(\bar{f})$ will be 5%. If the plant has a remaining useful life of five years, what is the present equivalent of its fuel costs? Use actual-dollar analysis.

14.14 Suppose that you just purchased a used car worth $6000 in today's dollars. Suppose also that you borrowed $5000 from a local bank at 9% compounded monthly over two years. The bank calculated your monthly payment at $228. Assuming that average general inflation will run at 0.5% per month over the next two years,

(a) Determine the annual inflation-free interest rate (i') for the bank.

(b) What equal monthly payments, in terms of constant dollars over the next two years, are equivalent to the series of actual payments to be made over the life of the loan?

14.15 A man is planning to retire in 20 years. Money can be deposited at 6% compounded monthly, and it is also estimated that the future general inflation $(\bar{f})$ rate will be 5% compounded annually. What monthly deposit must be made each month until the man retires so that he can make annual withdrawals of $40,000 in terms of today's dollars over the 15 years following his retirement? (Assume that his first withdrawal occurs at the end of the first six months after his retirement.)

14.16 On her 23rd birthday, a young engineer decides to start saving toward building up a retirement fund that pays 8% interest compounded quarterly (the market interest rate). She feels that $600,000 worth of purchasing power in today's dollars will be adequate to see her through her sunset years after her 63rd birthday. Assume a general inflation rate of 6% per year.

  (a) If she plans to save by making 160 equal quarterly deposits, what should be the amount of her quarterly deposit in actual dollars?

  (b) If she plans to save by making end-of-the-year deposits, increasing by $1000 over each subsequent year, how much would her first deposit be in actual dollars?

14.17 A father wants to save for his eight-year-old son's university expenses. His son will enter university 10 years from now. An annual amount of $20,000 in constant dollars will be required to support the son's university expenses for four years. Assume that these university payments will be made at the beginning of the school year. The future general inflation rate is estimated to be 2% per year, and the market interest rate on the savings account will average 3% compounded annually. Given this information,

  (a) What is the amount of the son's first-year expense in terms of actual dollars?

  (b) What is the equivalent single-sum amount at the present time for these expenses?

  (c) What is the equal amount, in actual dollars, the father must save each year until his son goes to university?

14.18 Consider the following project's after-tax cash flow and the expected annual general inflation rate during the project period:

End of Year	Expected Cash Flow (in actual $)	General Inflation Rate
0	$-$45,000	
1	$26,000	6.5%
2	$26,000	7.7%
3	$26,000	8.1%

  (a) Determine the average annual general inflation rate over the project period.

  (b) Convert the cash flows in actual dollars into equivalent constant dollars with the base year 0.

  (c) If the annual inflation-free interest rate is 5%, what is the present worth of the cash flow? Is this project acceptable?

14.19 Gentry Machines Inc. has just received a special job order from one of its clients. The following financial data have been collected:

- This two-year project requires the purchase of special-purpose production machinery for $55,000. The equipment falls into Class 43 of the CCA, which is a rate of 30%.
- The machine will be sold for $27,000 (today's dollars) at the end of two years.
- The project will bring in additional annual revenue of $114,000 (actual dollars), but it is expected to incur an additional annual operating cost of $53,800 (today's dollars).
- The project requires an investment in working capital in the amount of $12,000 at $n = 0$. In each subsequent year, additional working capital needs to be provided at the general inflation rate. Any investment in working capital will be recovered after the project is terminated.
- To purchase the equipment, the firm expects to borrow $50,000 at 10% over a two-year period. The remaining $5000 will be taken from the firm's retained earnings. The firm will make equal annual payments of $28,810 (actual dollars) to pay off the loan.
- The firm expects a general inflation of 5% per year during the project period. The firm's marginal tax rate is 40%, and its market interest rate is 18%.

 (a) Compute the after-tax cash flows in actual dollars.

 (b) What is the equivalent present value of this amount at time 0?

14.20 Hugh Health Product Corporation is considering purchasing a computer to control plant packaging for a spectrum of health products. The following data have been collected:

- First cost = $120,000, to be borrowed at 9% interest, with only interest paid each year and the principal due in a lump sum at end of year 2.
- Economic service life (project life) = six years.
- Estimated selling price in year-0 dollars = $15,000.
- Depreciation = Class 45 CCA rate = 45%.
- Marginal income tax rate = 40% (remains constant).
- Annual revenue = $145,000 (today's dollars).
- Annual expense (not including depreciation and interest) = $82,000 (today's dollars).
- Market interest rate = 18%.

 (a) With an average general inflation rate of 5% expected during the project period (which will affect all revenues, expenses, and the salvage value of the computer), determine the cash flows in actual dollars.

 (b) Compute the net present value of the project without inflation.

 (c) Compute the net present-value loss (gain) due to inflation.

 (d) In (c), how much is the present-value loss (or gain) due to borrowing?

14.21 Longueil Textile Company is considering automating its piece-goods screen printing system at a cost of $20,000. The firm expects to phase out the new printing system at the end of five years due to changes in style. At that time, the firm could scrap the system for $2000 in today's dollars. The expected net savings due to automation also are in today's dollars (constant dollars) as follows:

End of Year	Cash Flow (in constant $)
1	$15,000
2	$17,000
3–5	$14,000

The system qualifies for a 20% declining balance rate for tax purposes (CCA Class 8). The expected average general inflation rate over the next five years is approximately 5% per year. The firm will finance the entire project by borrowing at 10%. The scheduled repayment of the loan will be as follows:

End of Year	Principal Payment	Interest Payment
1	$6,042	$2,000
2	$6,647	$3,396
3	$7,311	$731

The firm's market interest rate for project evaluation during this inflation-ridden time is 20%. Assume that the net savings and the selling price will be responsive to this average inflation rate. The firm's marginal tax rate is known to be 40%.

(a) Determine the after-tax cash flows of this project, in actual dollars.

(b) Compute the net present value of the project without inflation.

(c) Compute the net present value loss (gain) due to inflation.

14.22 The J. F. Manning Metal Co. is considering the purchase of a new milling machine during year 0. The machine's base price is $135,000, and it will cost another $15,000 to modify it for special use. This results in a $150,000 cost base for a capital cost allowance rate of 30%. The machine will be sold after three years for $80,000 (actual dollars). Use of the machine will require an increase in net working capital (inventory) of $10,000 at the beginning of the project year. The machine will have no effect on revenues, but it is expected to save the firm $80,000 (today's dollars) per year in before-tax operating costs, mainly for labour. The firm's marginal tax rate is 40%, and this rate is expected to remain unchanged over the duration of the project. However, the company expects that the labour cost will increase at an annual rate of 5%, but that the working-capital requirement will grow at an annual rate of 8%, caused by inflation. The selling price of the milling machine is not affected by inflation. The general inflation rate is estimated to be 6% per year over the project period. The firm's market interest rate is 20%.

(a) Determine the project cash flows in actual dollars.

(b) Determine the project cash flows in constant (time − 0) dollars.

(c) Is this project acceptable?

14.23 Magna International is considering purchasing a vertical drill machine. The machine will cost $50,000 and will have an eight-year service life. The selling price of the machine at the end of eight years is expected to be $5000 in today's dollars. The machine will generate annual revenues of $20,000 (today's dollars), but the company expects to have an annual expense (excluding CCA) of $8000 (today's dollars). The asset is classified as Class 43, with a CCA rate of 30%. The project requires a working-capital investment of $10,000 at year 0. The marginal income tax rate for the firm is averaging 35%. The firm's market interest rate is 18%.

(a) Determine the internal rate of return of this investment.

(b) Assume that the firm expects a general inflation rate of 5%, but that it also expects an 8% annual increase in revenue and working capital and a 6% annual increase in expense caused by inflation. Compute the real (inflation-free) internal rate of return. Is this project acceptable?

14.24 Sonja Jensen is considering the purchase of a fast-food franchise. Sonja will be operating on a lot that is to be converted into a parking lot in six years, but that may be rented in the interim for $800 per month. The franchise and necessary equipment will have a total initial cost of $55,000 and a salvage value of $10,000 (in today's dollars) after six years. Sonja is told that the future annual general inflation rate will be 5%. The projected operating revenues and expenses, in actual dollars, other than rent and depreciation for the business, are as follows:

End of Year	Revenue	Expenses
1	$30,000	$15,000
2	35,000	21,000
3	55,000	25,000
4	70,000	30,000
5	70,000	30,000
6	60,000	30,000

The initial investment is classified as Class 14 and Sonja's tax rate will be 30%. Sonja can invest her money at a rate of at least 10% in other investment activities during this inflation-ridden period.

(a) Determine the cash flows associated with the investment over its life.

(b) Compute the projected after-tax rate of return (real) for this investment opportunity.

14.25 Suppose you have $10,000 cash that you want to invest. Normally, you would deposit the money in a savings account that pays an annual interest rate of 6%. However, you are now considering the possibility of investing in a bond. Your alternatives are either a provincial bond paying 9% or a corporate bond paying 12%. Your marginal tax rate is 50% for ordinary income and 30% for capital gains. You expect the general inflation to be 3% during the investment period. Both bonds mature at the end of year 5.

(a) Determine the real (inflation-free) rate of return for each bond.

(b) Without knowing your MARR, can you make a choice between these two bonds?

14.26 Air Nunavut is considering two types of engines for use in its planes. Each engine has the same life, the same maintenance, and the same repair record.

- Engine A costs $100,000 and uses 200,000 litres of fuel per 1000 hours of operation at the average service load encountered in passenger service.
- Engine B costs $200,000 and uses 128,000 litres of fuel per 1000 hours of operation at the same service load.

Both engines are estimated to have 10,000 service hours before any major overhaul of the engines is required. If fuel currently costs 50¢ per litre, and its price is expected to increase at the rate of 8% because of inflation, which engine should the firm install for an expected 2000 hours of operation per year? The firm's marginal income tax rate is 40%, and the engine falls into Class 9 (CCA rate = 25%) for which the 50% rule applies. Assume that the firm's market interest rate is 20%. It is estimated that both engines will retain a market value of 40% of their initial cost (actual dollars) if they are sold on the market after 10,000 hours of operation.

(a) Using the present-worth criterion, which project would you select?

(b) Using the annual-equivalent criterion, which project would you select?

(c) Using the future-worth criterion, which project would you select?

14.27 Yellowknife Chemical Company has just received a special subcontracting job from one of its clients. The two-year project requires the purchase of a special-purpose painting sprayer of $60,000. The equipment has a 30% declining balance rate for CCA calculations. After the subcontracting work is completed, the painting sprayer will be sold at the end of two years for $40,000 (actual dollars). The painting system will require an increase of $5000 in net working capital (for spare-parts inventory, such as spray nozzles). This investment in working capital will be fully recovered after the project is terminated. The project will bring in an additional annual revenue of $120,000 (today's dollars), but it is expected to incur an additional annual operating cost of $60,000 (today's dollars). It is projected that, due to inflation, sales prices will increase at an annual rate of 5%. (This implies that annual revenues will increase at an annual rate of 5%.) An annual increase of 4% for expenses and working-capital requirement is expected. The company has a marginal tax rate of 30%, and it uses a market interest rate of 15% for project evaluation during the inflationary period. If the firm expects a general inflation of 8% during the project period,

(a) Compute the after-tax cash flows in actual dollars.

(b) What is the rate of return on this investment (real earnings)?

(c) Is the special subcontracting project profitable?

14.28 Cadillac Fairview is considering purchasing a bulldozer. The bulldozer will cost $100,000 and will have an estimated salvage value of $30,000 at the end of six years. The asset will generate annual before-tax revenues of $80,000 over the next six years. The asset is depreciated at a CCA rate of 30%. The marginal tax rate is 40%, and the firm's market interest rate is known to be 18%. All dollar figures represent constant dollars at time 0 and are responsive to the general inflation rate $\overline{f}$.

(a) With $\overline{f} = 6\%$, compute the after-tax cash flows in actual dollars.

(b) Determine the real rate of return of this project on an after-tax basis.

(c) Suppose that the initial cost of the project will be financed through a local bank at an interest rate of 12%, with an annual payment of $24,323 over six years. With this additional condition, answer part (a) again.

(d) In part (a), determine the present-value loss due to inflation.

(e) In part (c), determine how much the project has to generate in additional before-tax annual revenue in actual dollars (equal amount) to make up the loss due to inflation.

Short Case Studies

ST14.1 Wilson Machine Tools Inc., a manufacturer of fabricated metal products, is considering purchasing a high-tech computer-controlled milling machine at a cost of $95,000. The cost of installing the machine, preparing the site, wiring, and rearranging other equipment is expected to be $15,000. This installation cost will be added to the cost of the machine to determine the total cost basis for depreciation. Special jigs and tool dies for the particular product will also be required at a cost of $10,000. The milling machine is expected to last 10 years, the jigs and dies only 5 years. Therefore, another set of jigs and dies has to be purchased at the end of five years. The milling machine will have a $10,000 salvage value at the end of its life, and the special jigs and dies are worth only $300 as scrap metal at any time in their lives. The machine has a 30% CCA rate and the special jigs and dies have a 100% CCA rate. With the new milling machine, Wilson expects an additional annual revenue of $80,000 due to increased production. The additional annual production costs are estimated as follows: materials, $9000; labour, $15,000; energy, $4500; and miscellaneous O&M costs, $3000. Wilson's marginal income tax rate is expected to remain at 35% over the project life of 10 years. All dollar figures represent today's dollars. The firm's market interest rate is 18%, and the expected general inflation rate during the project period is estimated at 6%.

(a) Determine the project cash flows in the absence of inflation.

(b) Determine the internal rate of return for the project in (a).

(c) Suppose that Wilson expects price increases during the project period: material at 4% per year, labour at 5% per year, and energy and other O&M costs at 3% per year. To compensate for these increases in prices, Wilson is planning to increase annual revenue at the rate of 7% per year by charging its customers a higher price. No changes in salvage value are expected for the machine or the jigs and dies. Determine the project cash flows in actual dollars.

(d) In (c), determine the real (inflation-free) rate of return of the project.

(e) Determine the economic loss (or gain) in present worth caused by inflation.

ST14.2 Recent biotechnological research has made possible the development of a sensing device that implants living cells on a silicon chip. The chip is capable of detecting physical and chemical changes in cell processes. Proposed uses of the device include researching the mechanisms of disease on a cellular level, developing new therapeutic drugs, and substituting for animals in cosmetic and drug testing. Biotech Device Corporation (BDC) has just perfected a process for mass-producing the chip. The following information has been compiled for the board of directors:

- BDC's marketing department plans to target sales of the device to the larger chemical and drug manufacturers. BDC estimates that annual sales would be 2000 units if the device were priced at $95,000 per unit (in dollars of the first operating year).

- To support this level of sales volume, BDC would need a new manufacturing plant. Once the "go" decision is made, this plant could be built and made ready for production within one year. BDC would need a 10-hectare tract of land that would cost $1.5 million. If the decision were to be made, the land could be purchased on December 31, 2011. The building would cost $5 million and has a 4% CCA rate. The first payment of $1 million would be due to the contractor on December 31, 2012, and the remaining $4 million on December 31, 2013.

- The required manufacturing equipment would be installed late in 2013 and would be paid for on December 31, 2013. BDC would have to purchase the equipment at an estimated cost of $8 million, including transportation, plus a further $500,000 for installation. The equipment has a 30% CCA rate.

- The project would require an initial investment of $1 million in working capital. This investment would be made on December 31, 2013. Then, on December 31 of each subsequent year, net working capital would be increased by an amount equal to 15% of any sales increase expected during the coming year. The investments in working capital would be fully recovered at the end of the project year.

- The project's estimated economic life is six years (excluding the two-year construction period). At that time, the land is expected to have a market value of $2 million, the building a value of $3 million, and the equipment a value of $1.5 million. The estimated variable manufacturing costs would total 60% of the dollar sales. Fixed costs, excluding depreciation, would be $5 million for the first year of operations. Since the plant would begin operations on January 1, 2014, the first operating cash flows would occur on December 31, 2014.

- Sales prices and fixed overhead costs, other than depreciation, are projected to increase with general inflation, which is expected to average 5% per year over the six-year life of the project.

- To date, BDC has spent $5.5 million on research and development (R&D) associated with cell implantation. The company has already expensed $4 million R&D costs. The remaining $1.5 million will be amortized over six years (i.e., the annual amortization expense will be $250,000). If BDC decides not to proceed with the project, the $1.5 million R&D cost could be written off on December 31, 2011.

- BDC's marginal tax rate is 40%, and its market interest rate is 20%. Any capital gains will also be taxed at 40%.

  (a) Determine the after-tax cash flows of the project, in actual dollars.

  (b) Determine the inflation-free (real) IRR of the investment.

  (c) Would you recommend that the firm accept the project?

On the Companion Website that accompanies this text, you will find Excel templates and exercises, as well as the following analysis tools: Cash Flow Analyzer, Depreciation Analysis, Loan Analysis, and Interest Tables.

FIFTEEN

Project Risk
and Uncertainty

Oil Forecasts Are a Roll of the Dice[1] You may know as much as the oil experts. That is, you know that a barrel of oil is pricey and getting pricier. Beyond that, nobody—not even those who get paid to prognosticate—has a real handle on the push and pull that goes into figuring how much oil people need, how much can be pumped, and how much can be refined.

The unreliable data and forecasts have plagued the industry for decades. They became more of a problem once demand and prices

[1] Bhusan Bahree, "Oil Forecasts Are a Roll of the Dice," *The Wall Street Journal*, August 2, 2005, Section C1.

starting climbing in recent years, because the substantial margins of error in these numbers are even larger than the oil industry's shrinking margin of spare pumping capacity.

Put another way, even a small error in predicting oil consumption can cause energy markets to gyrate if demand turns out higher than the market assumed because the industry lacks the ability it had in the 1990s to gin up extra oil on the fly to meet a surge in buying. Yet traders, companies, and consumers have no choice but to rely on the numbers that are out there. One problem anyone who factors oil into an investing decision faces is that accurate oil data come only with a time lag. Oil data on actual demand and supply bounce around because of bad weather, accidents such as pipeline ruptures, or political shocks such as terrorist strikes. Amplifying the fuzziness, meanwhile, is that projections of economic growth, a critical factor in assessing energy needs, are forever changing.

Crude-Oil Price Forecast (August 2005)						
West Texas Intermediate, Spot Price. USD/bbl. Average of Month						
	Dec 2005	Jan 2006	Feb 2006	Mar 2006	Apr 2006	May 2006
Forecast Value	**59.3**	**58.7**	**61.8**	**61.6**	**62.1**	**63.8**
Standard Deviation	0.9	1.2	1.5	1.8	2.2	2.6
Correlation Coefficient	0.9925	0.9912	0.9899	0.9886	0.9873	0.9860

Suppose that your business depends on the price of oil. For example, the airline industry, Canpar, FedEx, and rental companies are all heavily affected by the price of fuel. Now, if your proposed project also depends on the price of crude oil, how would you factor the fluctuation and uncertainty into the analysis?

In previous chapters, cash flows from projects were assumed to be known with complete certainty; our analysis was concerned with measuring the economic worth of projects and selecting the best ones to invest in. Although that type of analysis can provide a reasonable basis for decision making in many investment situations, we should consider the more usual case of *un*certainty. In this type of situation, management rarely has precise expectations about the future cash flows to be derived from a particular project. In fact, the best that a firm can reasonably expect to do is to estimate the range of possible future costs and benefits and the relative chances of achieving a reasonable return on the investment. We use the term **risk** to describe an investment project whose cash flow is not known in advance with absolute certainty, but for which an array of alternative outcomes and their probabilities (odds) are known. We will also use the term **project risk** to refer to variability in a project's NPW. A greater project risk usually means a greater variability in a project's NPW, or simply that the *risk is the potential for loss*. This chapter begins by exploring the origins of project risk.

Risk: The chance that an investment's actual return will be different than expected.

CHAPTER LEARNING OBJECTIVES

After completing this chapter, you should understand the following concepts:

- How to describe the nature of project risk.
- How to conduct a sensitivity analysis of key input variables.
- How to conduct a break-even analysis.
- How to develop a net-present-worth probability distribution.
- How to compare mutually exclusive risky alternatives.
- How to develop a risk simulation model.
- How to make a sequential investment decision with a decision tree.

15.1 Origins of Project Risk

The decision to make a major capital investment such as introducing a new product requires information about cash flow over the life of a project. The profitability estimate of an investment depends on cash flow estimations, which are generally uncertain. The factors to be estimated include the total market for the product; the market share that the firm can attain; the growth in the market; the cost of producing the product, including labour and materials; the selling price; the life of the product; the cost and life of the equipment needed; and the effective tax rates. Many of these factors are subject to substantial uncertainty. A common approach is to make single-number "best estimates" for each of the uncertain factors and then to calculate measures of profitability, such as the NPW or rate of return for the project. This approach, however, has two drawbacks:

1. No guarantee can ever ensure that the "best estimates" will match actual values.
2. No provision is made to measure the risk associated with an investment, or the project risk. In particular, managers have no way of determining either the probability that a project will lose money or the probability that it will generate large profits.

Because cash flows can be so difficult to estimate accurately, project managers frequently consider a range of possible values for cash flow elements. If a range of values for individual cash flows is possible, it follows that a range of values for the NPW of a given project is also possible. Clearly, the analyst will want to gauge the probability and reliability of individual cash flows and, consequently, the level of certainty about the overall project worth.

15.2 Methods of Describing Project Risk

We may begin analyzing project risk by first determining the uncertainty inherent in a project's cash flows. We can do this analysis in a number of ways, which range from making informal judgments to calculating complex economic and statistical quantities. In this section, we will introduce three methods of describing project risk: (1) sensitivity analysis, (2) break-even analysis, and (3) scenario analysis. Each method will be explained with reference to a single example, involving the Windsor Metal Company.

15.2.1 Sensitivity Analysis

One way to glean a sense of the possible outcomes of an investment is to perform a sensitivity analysis. This kind of analysis determines the effect on the NPW of variations in the input variables (such as revenues, operating cost, and salvage value) used to estimate after-tax cash flows. A **sensitivity analysis** reveals how much the NPW will change in response to a given change in an input variable. In calculating cash flows, some items have a greater influence on the final result than others. In some problems, the most significant item may be easily identified. For example, the estimate of sales volume is often a major factor in a problem in which the quantity sold varies with the alternatives. In other problems, we may want to determine the items that have an important influence on the final results so that they can be subjected to special scrutiny.

Sensitivity analysis is sometimes called "what-if" analysis, because it answers questions such as "What if incremental sales are only 1000 units, rather than 2000 units? Then what will the NPW be?" Sensitivity analysis begins with a base-case situation, which is developed by using the most likely values for each input. We then change the specific variable of interest by several specified percentage points above and below the most likely value, while holding other variables constant. Next, we calculate a new NPW for each of the values we obtained. A convenient and useful way to present the results of a sensitivity analysis is to plot **sensitivity graphs**. The slopes of the lines show how sensitive the NPW is to changes in each of the inputs: The steeper the slope, the more sensitive the NPW is to a change in a particular variable. Sensitivity graphs identify the crucial variables that affect the final outcome most. Example 15.1 illustrates the concept of sensitivity analysis.

EXAMPLE 15.1 Sensitivity Analysis

Windsor Metal Company (WMC), a small manufacturer of fabricated metal parts, must decide whether to enter the competition to become the supplier of transmission housings for Gulf Electric, a company that produces the housings in its own in-house manufacturing facility but has almost reached its maximum production capacity. Therefore, Gulf is looking for an outside supplier. To compete, WMC must design a new fixture for the production process and purchase a new forge. The available details for this purchase are as follows:

- The new forge would cost $125,000. This total includes retooling costs for the transmission housings.
- If WMC gets the order, it may be able to sell as many as 2000 units per year to Gulf Electric for $50 each, in which case variable production costs,[2] such as direct labour and direct material costs, will be $15 per unit. The increase in fixed costs,[3] other than capital cost allowance, will amount to $10,000 per year.
- The firm expects that the proposed transmission-housings project will have about a five-year product life. The firm also estimates that the amount ordered by Gulf Electric in the first year will be ordered in each of the subsequent four years. (Due to the nature of contracted production, the annual demand and unit price would remain the same over the project after the contract is signed.)
- The capital investment is eligible for a declining balance CCA rate of 30%, and the marginal income tax rate is expected to remain at 40%. At the end of five years, the forge is expected to retain a market value of about 32% of the original investment.
- On the basis of this information, the engineering and marketing staffs of WMC have prepared the cash flow forecasts shown in Table 15.1. Since the NPW is positive ($40,460) at the 15% opportunity cost of capital (MARR), the project appears to be worth undertaking.

What Makes WMC Managers Worry: WMC's managers are uneasy about this project, because too many uncertain elements have not been considered in the analysis:

- If it decided to take on the project, WMC would have to invest in the forging machine to provide Gulf Electric with some samples as a part of the bidding process. If Gulf Electric were not to like WMC's sample, WMC would stand to lose its entire investment in the forging machine.
- If Gulf were to like WMC's sample, but it was overpriced, WMC would be under pressure to bring the price in line with competing firms. Even the possibility that WMC would get a smaller order must be considered, as Gulf may utilize its overtime capacity to produce some extra units. Finally, WMC is not certain about its projections of variable and fixed costs.

Recognizing these uncertainties, the managers want to assess the various possible future outcomes before making a final decision. Put yourself in WMC's management position, and describe how you may resolve the uncertainty associated with the project. In doing so, perform a sensitivity analysis for each variable and develop a sensitivity graph.

DISCUSSION: Table 15.1 shows WMC's expected cash flows—but that they will indeed materialize cannot be assumed. In particular, WMC is not very confident in its revenue forecasts. The managers think that if competing firms enter the market,

---

[2] These are expenses that change in direct proportion to the change in volume of sales or production, as defined in Section 7.3.

[3] These are expenses that do not vary with the volume of sales or production. For example, property taxes, insurance, depreciation, and rent are usually fixed expenses.

TABLE 15.1 After-Tax Cash Flow for WMC's Transmission-Housings Project (Example 15.1)

	0	1	2	3	4	5
Revenues						
Unit price		$ 50	$ 50	$ 50	$ 50	$ 50
Demand (units)		2,000	2,000	2,000	2,000	2,000
Sales revenue		$100,000	$100,000	$100,000	$100,000	$100,000
Expenses						
Unit variable cost		$ 15	$ 15	$ 15	$ 15	$ 15
Variable cost		$ 30,000	$ 30,000	$ 30,000	$ 30,000	$ 30,000
Fixed cost		$ 10,000	$ 10,000	$ 10,000	$ 10,000	$ 10,000
CCA		$ 18,750	$ 31,875	$ 22,313	$ 15,619	$ 10,933
Taxable income		$ 41,250	$ 28,125	$ 37,688	$ 44,381	$ 49,067
Income taxes (40%)		$ 16,500	$ 11,250	$ 15,075	$ 17,753	$ 19,627
Net income		$ 24,750	$ 16,875	$ 22,613	$ 26,629	$ 29,440
Cash flow statement						
Operating activities						
Net income		$ 24,750	$ 16,875	$ 22,613	$ 26,629	$ 29,440
CCA		$ 18,750	$ 31,875	$ 22,313	$ 15,619	$ 10,933
Investment activities						
Investment	−$125,000					
Salvage						$ 40,000
Disposal tax effect						−$ 5,796
Net cash flow	−$125,500	$ 43,500	$ 48,750	$ 44,925	$ 42,248	$ 74,578

WMC will lose a substantial portion of the projected revenues by not being able to increase its bidding price. Before undertaking the project, the company needs to identify the key variables that will determine whether it will succeed or fail. The marketing department has estimated revenue as follows:

$$\text{Annual revenue} = (\text{Product demand})(\text{unit price})$$
$$= (2,000)(\$50) = \$100,000.$$

The engineering department has estimated variable costs, such as those of labour and materials, at $15 per unit. Since the projected sales volume is 2000 units per year, the total variable cost is $30,000.

After first defining the unit sales, unit price, unit variable cost, fixed cost, and salvage value, we conduct a sensitivity analysis with respect to these key input variables. This is done by varying each of the estimates by a given percentage and determining what effect the variation in that item will have on the final results. If the effect is large, the result is sensitive to that item. Our objective is to locate the most sensitive item(s).

SOLUTION

Sensitivity analysis: We begin the sensitivity analysis with a consideration of the base-case situation, which reflects the best estimate (expected value) for each input variable. In developing Table 15.2, we changed a given variable by 20% in 5% increments, above and below the base-case value, and calculated new NPWs, while other variables were held constant. The values for both sales and operating costs were the expected, or base-case, values, and the resulting $40,460 is the base-case NPW. Now we ask a series of "what-if" questions: What if sales are 20% below the expected level? What if operating costs rise? What if the unit price drops from $50 to $45? Table 15.2 summarizes the results of varying the values of the key input variables.

Sensitivity graph: Next, we construct a sensitivity graph for five of the transmission project's key input variables. (See Figure 15.1.) We plot the base-case NPW of $40,460 on the ordinate of the graph at the value of 0% deviation on the abscissa. Then we reduce the value of product demand to 0.95 of its base-case value and recompute the NPW with all other variables held at their base-case value. We repeat the process by either decreasing or increasing the relative deviation from the base case. The lines for the variable unit price, variable unit cost, fixed cost, and salvage value are obtained in the same manner. In Figure 15.1, we see that the project's NPW is (1) very sensitive to changes in product demand and unit price, (2) fairly sensitive to changes in variable costs, and (3) relatively insensitive to changes in the fixed cost and the salvage value. Note that this analysis does not include any interdependencies between the variables, nor consider the relative likelihood of changes of various magnitudes of these diverse variables, simply what the impact is of a single given change.

TABLE 15.2 Sensitivity Analysis for Five Key Input Variables (Example 15.1)

Deviation	−20%	−15%	−10%	−5%	0%	5%	10%	15%	20%
Unit price	$ 234	$10,291	$20,347	$30,404	$40,460	$50,517	$60,573	$70,630	$80,686
Demand	$12,302	$19,342	$26,381	$33,421	$40,460	$47,500	$54,539	$61,579	$68,618
Unit variable cost	$52,528	$49,511	$46,494	$43,477	$40,460	$37,443	$34,426	$31,410	$28,393
Fixed cost	$44,483	$43,477	$42,472	$41,466	$40,460	$39,455	$38,449	$37,443	$36,438
Salvage value	$38,074	$38,671	$39,267	$39,864	$40,460	$41,057	$41,654	$42,250	$42,847

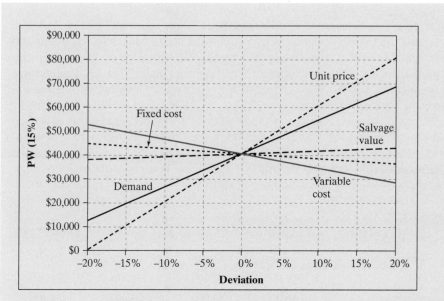

Figure 15.1   Sensitivity graph for WMC's transmission-housings project (Example 15.1).

Graphic displays such as the one in Figure 15.1 provide a useful means to communicate the relative sensitivities of the different variables to the corresponding NPW value. However, sensitivity graphs do not explain any interactions among the variables or the likelihood of realizing any specific deviation from the base case. Certainly, it is conceivable that an answer might not be very sensitive to changes in either of two items, but very sensitive to combined changes in them.

15.2.2 Sensitivity Analysis for Mutually Exclusive Alternatives

In Figure 15.1, each variable is uniformly adjusted by ±20% and all variables are plotted on the same chart. This uniform adjustment assumption can be too simplistic; in many situations, each variable can have a different range of uncertainty. Also, plotting all variables on the same chart could be confusing if there are too many variables to consider. When we perform sensitivity analysis for mutually exclusive alternatives, it may be more effective to plot the NPWs (or any other measures such as AEWs) of all alternatives over the range of each variable, basically one plot for each variable, with units of the variable on the horizontal axis. Example 15.2 illustrates this.

EXAMPLE 15.2 Sensitivity Analysis for Mutually Exclusive Alternatives

A local Canada Post service office is considering purchasing a 2000-kilogram capacity forklift truck, which will be used primarily for processing incoming as well as outgoing postal packages. Forklift trucks traditionally have been fuelled by either gasoline, liquid propane gas (LPG), or diesel fuel. Battery-powered electric forklifts, however, are increasingly popular in many industrial sectors due to the economic and environmental benefits that accrue from their use. Therefore, the postal service is interested in comparing the four different types of fuel. Annual fuel and maintenance costs are measured in terms of number of shifts per year, where one shift is equivalent to eight hours of operation.

The postal service is unsure of the number of shifts per year, but it expects it should be somewhere between 200 and 260 shifts. Canada Post uses 10% as an interest rate for any project evaluation of this nature. This analysis will be conducted on a before-tax basis. Develop a sensitivity graph that shows how the choice of alternatives changes as a function of number of shifts per year.

	Electric	LPG	Gasoline	Diesel Fuel
Life expectancy	7 years	7 years	7 years	7 years
Initial cost	$68,000	$42,400	$40,214	$44,526
Salvage value	$6,000	$4,000	$4,000	$4,400
Fuel consumption/shift	31.25 kWh	26.1 L	29.6 L	18.5 L
Fuel cost/unit	$0.14/kWh	$0.86/L	$0.95/L	$0.78/L
Fuel cost/shift	$4.38	$22.45	$28.12	$14.43
Annual maintenance cost				
Fixed cost	$1,000	$2,000	$2,000	$2,000
Variable cost/shift	$7	$14	$14	$14

SOLUTION

Two annual cost components are pertinent to this problem: (1) Ownership cost (capital cost) and (2) operating cost (fuel and maintenance cost). Since the operating cost is already given on an annual basis, we only need to determine the equivalent annual ownership cost for each alternative.

(a) Ownership cost (capital cost): Using the capital recovery with return formula developed in Eq. (5.5), Section 5.5.3, we compute

Electrical power: $CR(10\%) = (\$68,000 - \$6000)(A/P, 10\%, 7) + (0.10)\$6000$
$= \$13,335.$

LPG: $CR(10\%) = (\$42,400 - \$4000)(A/P, 10\%, 7) + (0.10)\$4000$
$= \$8288.$

Gasoline: $CR(10\%) = (\$40,214 - \$4000)(A/P, 10\%, 7) + (0.10)\$4000$
$= \$7839.$

Diesel fuel: $CR(10\%) = (\$44,526 - \$4400)(A/P, 10\%, 7) + (0.10)\$4400$
$= \$8682.$

(b) Annual operating cost: We can express the annual operating cost as a function of number of shifts per year (M) by combining the variable and fixed cost portions of fuel and maintenance expenditures.

Electrical power: $1000 + (4.38 + 7)M = $1000 + 11.38M.
LPG: $2000 + (22.45 + 14)M = $2000 + 36.45M.
Gasoline: $2000 + (28.12 + 14)M = $2000 + 42.12M.
Diesel fuel: $2000 + (14.43 + 14)M = $2000 + 28.43M.

(c) Total equivalent annual cost: This is the sum of the ownership cost and operating cost.

Electrical power: $AE(10\%) = 14{,}335 + 11.38M.$
LPG: $AE(10\%) = 10{,}288 + 36.45M.$
Gasoline: $AE(10\%) = 9839 + 42.12M.$
Diesel fuel: $AE(10\%) = 10{,}682 + 28.43M.$

In Figure 15.2, these four annual equivalent costs are plotted as a function of number of shifts, M. It appears that the diesel-powered forklift truck is more economic as long as the number of annual shifts does not exceed approximately 215, above which the electric vehicle is cheaper.

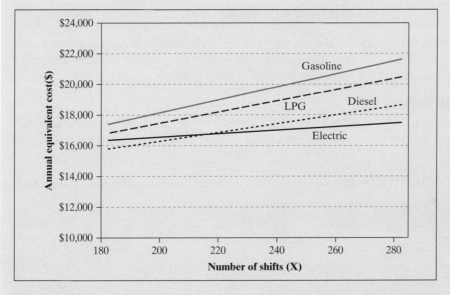

Figure 15.2   Sensitivity analysis for mutually exclusive alternatives (Example 15.2).

15.2.3 Break-Even Analysis

When we perform a sensitivity analysis of a project, we are asking how serious the effect of lower revenues or higher costs will be on the project's profitability. Managers sometimes prefer to ask instead how much sales can decrease below forecasts before the project begins to lose money. This type of analysis is known as **break-even analysis**. In other words, break-even analysis is a technique for studying the effect of variations in output on a firm's NPW (or other measures). We will present an approach to break-even analysis based on the project's cash flows.

To illustrate the procedure of break-even analysis based on NPW, we use the generalized cash flow approach we discussed in Section 10.4. We compute the PW of cash inflows as a function of an unknown variable (say, x), perhaps annual sales. For example,

$$PW \text{ of cash inflows} = f_1(x).$$

Next, we compute the PW of cash outflows as a function of x:

$$PW \text{ of cash outflows} = f_2(x).$$

NPW is, of course, the difference between these two numbers. Accordingly, we look for the break-even value of x that makes

$$f_1(x) = f_2(x).$$

Note that this break-even value is similar to that used to calculate the internal rate of return when we want to find the interest rate that makes the NPW equal zero. The break-even value is also used to calculate many other similar "cutoff values" at which a choice changes.

EXAMPLE 15.3 Break-Even Analysis

Through the sensitivity analysis in Example 15.1, WMC's managers become convinced that the NPW is most sensitive to changes in annual sales volumes. Determine the break-even NPW value as a function of that variable.

SOLUTION

The analysis is shown in Table 15.3, in which the revenues and costs of the WMC transmission-housings project are set out in terms of an unknown amount of annual sales X.

TABLE 15.3 Break-Even Analysis With Unknown Annual Sales (Example 15.3)

	0	1	2	3	4	5
Cash inflow:						
Net salvage						34,204
Revenue:						
$X(1 - 0.4)(\$50)$		$30X$	$30X$	$30X$	$30X$	$30X$
CCA credit						
0.4 (CCA)		7,500	12,750	8,925	6,248	4,373
Cash outflow:						
Investment	−125,000					
Variable cost:						
$-X(1 - 0.4)(\$15)$		$-9X$	$-9X$	$-9X$	$-9X$	$-9X$
Fixed cost:						
$-(1 - 0.4)(\$10,000)$		−6,000	−6,000	−6,000	−6,000	−6,000
Net cash flow	−125,000	$21X+1,500$	$21X+6,750$	$21X+2,925$	$21X+248$	$21X+32,577$

We calculate the PWs of cash inflows and outflows as follows:

- PW of cash inflows:

$$\begin{aligned}
\text{PW}(15\%)_{\text{Inflow}} = {} & (\text{PW of after-tax net revenue}) \\
& + (\text{PW of net salvage value}) \\
& + (\text{PW of tax savings from CCA}). \\
= {} & 30X(P/A, 15\%, 5) + \$34{,}204(P/F, 15\%, 5) \\
& + \$7500(P/F, 15\%, 1) + \$12{,}750(P/F, 15\%, 2) \\
& + \$8925(P/F, 15\%, 3) + \$6248(P/F, 15\%, 4) \\
& + \$4373(P/F, 15\%, 5) \\
= {} & 30X(P/A, 15\%, 5) + \$44{,}782 \\
= {} & 100.5650X + \$44{,}782.
\end{aligned}$$

- PW of cash outflows:

$$\begin{aligned}
\text{PW}(15\%)_{\text{Outflow}} = {} & (\text{PW of capital expenditure}) \\
& + (\text{PW of after-tax expenses}). \\
= {} & \$125{,}000 + (9X + \$6000)(P/A, 15\%, 5) \\
= {} & 30.1694X + \$145{,}113.
\end{aligned}$$

The NPW of cash flows for the WMC is thus

$$\begin{aligned}
\text{PW}(15\%) = {} & 100.5650X + \$44{,}782 \\
& - (30.1694X + \$145{,}113) \\
= {} & 70.3956X - \$100{,}331.
\end{aligned}$$

In Table 15.4, we compute the PW of the inflows and the PW of the outflows as a function of demand (X).

TABLE 15.4 Determination of Break-Even Volume Based on Project's NPW (Example 15.3)

Demand (X)	PW of Inflow ($100.5650X + \$44{,}782$)	PW of Outflow ($30.1694X + \$145{,}113$)	NPW ($70.3956X - \$100{,}331$)
0	$ 44,782	$145,113	$(100,331)
500	95,065	160,198	(65,133)
1000	145,347	175,282	(29,935)
1425	188,087	188,104	(17)
1426	188,188	188,135	53
1500	195,630	190,367	5,262
2000	245,912	205,452	40,460
2500	296,195	220,537	75,658

Break-even volume = 1,426 units.

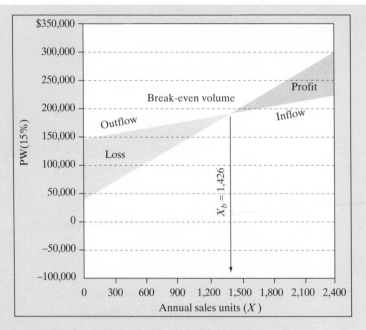

Figure 15.3 Break-even analysis based on net cash flow (Example 15.3).

The NPW will be just slightly positive if the company sells 1426 units. Precisely calculated, the zero-NPW point (break-even volume) is 1425.29 units:

$$PW(15\%) = 70.39356X - \$100,331$$
$$= 0$$
$$X_b = 1426 \text{ units.}$$

In Figure 15.3, we have plotted the PWs of the inflows and outflows under various assumptions about annual sales. The two lines cross when sales are 1426 units, the point at which the project has a zero NPW. Again we see that, as long as sales are greater or equal to 1426, the project has a positive NPW.

15.2.4 Scenario Analysis

Although both sensitivity and break-even analyses are useful, they have limitations. Often, it is difficult to specify precisely the relationship between a particular variable and the NPW. The relationship is further complicated by interdependencies among the variables. Holding operating costs constant while varying unit sales may ease the analysis, but in reality, operating costs do not behave in this manner. Yet, it may complicate the analysis too much to permit movement in more than one variable at a time.

Scenario analysis is a technique that considers the sensitivity of NPW both to changes in key variables and to the range of likely values of those variables. For example, the decision maker may examine two extreme cases: a "worst-case" scenario (low unit sales, low unit price, high variable cost per unit, high fixed cost, and so on) and a "best-case" scenario. The NPWs under the worst and the best conditions are then calculated and compared with the expected, or base-case, NPW. Example 15.4 illustrates a plausible scenario analysis for WMC's transmission-housings project.

EXAMPLE 15.4 Scenario Analysis

Consider again WMC's transmission-housings project first presented in Example 15.1. Assume that the company's managers are fairly confident of their estimates of all the project's cash flow variables, except the estimates of unit sales. Assume further that they regard a drop in unit sales below 1600 or a rise above 2400 as extremely unlikely. Thus, a decremental annual sale of 400 units defines the lower bound, or the worst-case scenario, whereas an incremental annual sale of 400 units defines the upper bound, or the best-case scenario. (Remember that the most likely value was 2000 in annual unit sales.) Discuss the worst- and best-case scenarios, assuming that the unit sales for all five years are equal.

DISCUSSION: To carry out the scenario analysis, we ask the marketing and engineering staffs to give optimistic (best-case) and pessimistic (worst-case) estimates for the key variables. Then we use the worst-case variable values to obtain the worst-case NPW and the best-case variable values to obtain the best-case NPW.

SOLUTION

The results of our analysis are summarized as follows:

Variable Considered	Worst-Case Scenario	Base-Case Scenario	Best-Case Scenario
Unit demand	1,600	2,000	2,400
Unit price ($)	48	50	53
Variable cost ($)	17	15	12
Fixed cost ($)	11,000	10,000	8,000
Salvage value ($)	30,000	40,000	50,000
PW(15%)	−$5,564	$40,460	$104,587

We see that the base (most likely) case produces a positive NPW, the worst case produces a negative NPW, and the best case produces a large positive NPW. Still, by just looking at the results in the table, it is not easy to interpret the scenario analysis or to make a decision based on it. For example, we could say that there is a chance of losing money on the project, but we do not yet have a specific probability for that possibility. Clearly, we need estimates of the probabilities of occurrence of the worst case, the best case, the base case, and all the other possibilities.

The need to estimate probabilities leads us directly to our next step: developing a probability distribution (or, put another way, the probability that the variable in question takes on a certain value). If we can predict the effects on the NPW of variations in the parameters, why should we not assign a probability distribution to the possible outcomes of each parameter and combine these distributions in some way to produce a probability distribution for the possible outcomes of the NPW? We shall consider this issue in the next two sections.

15.3 Probability Concepts for Investment Decisions

In this section, we shall assume that the analyst has available the probabilities (likelihoods) of future events from either previous experience with a similar project or a market survey. The use of probability information can provide management with a range of possible outcomes and the likelihood of achieving different goals under each investment alternative.

15.3.1 Assessment of Probabilities

We begin by defining terms related to probability, such as *random variable*, *probability distribution*, and *cumulative probability distribution*. Quantitative statements about risk are given as numerical probabilities or as likelihoods (odds) of occurrence. Probabilities are given as decimal fractions in the interval from 0.0 to 1.0. An event or outcome that is certain to occur has a probability of 1.0. As the probability of an event approaches 0, the event becomes increasingly less likely to occur. The assignment of probabilities to the various outcomes of an investment project is generally called **risk analysis**.

Risk analysis is a technique to identify and assess factors that may jeopardize the success of a project.

Random Variables

A **random variable** is a parameter or variable that can have more than one possible value (though not simultaneously). The value of a random variable at any one time is unknown until the event occurs, but the probability that the random variable will have a specific value is known in advance. In other words, associated with each possible value of the random variable is a likelihood, or probability, of occurrence. For example, when your school team plays a football game, only two events regarding the outcome of the game are possible: win or lose. The outcome is a random variable, dictated largely by the strength of your opponent.

To indicate random variables, we will adopt the convention of a capital italic letter (for example, X). To denote the situation in which the random variable takes a specific value, we will use a lowercase italic letter (for example, x). Random variables are classified as either discrete or continuous:

- Any random variables that take on only isolated (countable) values are **discrete random variables**.
- **Continuous random variables** may have any value within a certain interval.

For example, the outcome of a game should be a discrete random variable. By contrast, suppose you are interested in the amount of beverage sold on a given day that the game is played. The quantity (or volume) of beverage sold will depend on the weather conditions, the number of people attending the game, and other factors. In this case, the quantity is a continuous random variable—a random variable that takes a value from a continuous set of values.

Probability Distributions

For a discrete random variable, the probability of each random event needs to be assessed. For a continuous random variable, the probability function needs to be assessed, as the event takes place over a continuous domain. In either case, a range of probabilities for each feasible outcome exists. Together, these probabilities make up a **probability distribution**.

Probability assessments may be based on past observations or historical data if the trends that were characteristic of the past are expected to prevail in the future. Forecasting weather or predicting the outcome of a game in many professional sports is done on the basis of compiled statistical data. Any probability assessments based on objective data are called **objective probabilities**. However, in many real investment situations, no objective data are available to consider. In these situations, we assign **subjective probabilities** that we think are appropriate to the possible states of nature. As long as we act consistently with our beliefs about the possible events, we may reasonably account for the economic consequences of those events in our profitability analysis.

For a continuous random variable, we usually try to establish a range of values; that is, we try to determine a **minimum value** (L) and a **maximum value** (H). Next, we determine whether any value within these limits might be *more likely* to occur than the other values; in other words, does the distribution have a **mode** (M_O), or a most frequently occurring value?

- If the distribution does have a mode, we can represent the variable by a **triangular distribution**, such as that shown in Figure 15.4.
- If we have no reason to assume that one value is any more likely to occur than any other, perhaps the best we can do is to describe the variable as a **uniform distribution**, as shown in Figure 15.5.

These two distributions are frequently used to represent the variability of a random variable when the only information we have is its minimum, its maximum, and whether the distribution has a mode. For example, suppose the best judgment of the analyst was that the sales revenue could vary anywhere from $2000 to $5000 per day, and any value within the range is equally likely. This judgment about the variability of sales revenue could be represented by a uniform distribution.

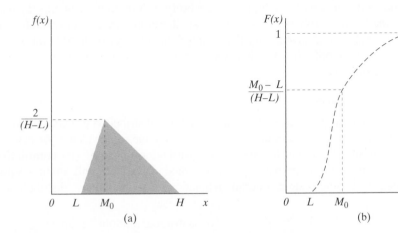

Figure 15.4   A triangular probability distribution: (a) probability function and (b) cumulative probability distribution.

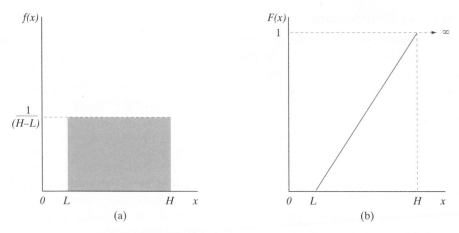

Figure 15.5 A uniform probability distribution: (a) probability function and (b) cumulative probability distribution.

TABLE 15.5 Probability Distributions for Unit Demand (*X*) and Unit Price (*Y*) for WMC's Project

Product Demand (*X*)		Unit Sale Price (*Y*)	
Units (*x*)	$P(X = x)$	Unit Price (*y*)	$P(Y = y)$
1600	0.20	$48	0.30
2000	0.60	50	0.50
2400	0.20	53	0.20

X and *Y* are independent random variables.

For WMC's transmission-housings project, we can think of the discrete random variables (*X* and *Y*) as variables whose values cannot be predicted with certainty at the time of decision making. Let us assume the probability distributions in Table 15.5. We see that the product demand with the highest probability is 2000 units, whereas the unit sale price with the highest probability is $50. These, therefore, are the most likely values. We also see a substantial probability that a unit demand other than 2000 units will be realized. When we use only the most likely values in an economic analysis, we are in fact ignoring these other outcomes.

Cumulative Distribution

As we have observed in the previous section, the probability distribution provides information regarding the probability that a random variable will assume some value *x*. We can use this information, in turn, to define the cumulative distribution function. The **cumulative distribution** function gives the probability that the random variable will attain a value smaller than or equal to some value *x*. A common notation for the cumulative distribution is

$$F(x) = P(X \le x) = \begin{cases} \displaystyle\sum_{j:x_j \le x} p_j & \text{(for a discrete random variable)} \\ \displaystyle\int_{L}^{x} f(x)\, dx & \text{(for a continuous random variable)} \end{cases},$$

where p_j is the probability of occurrence of the x_jth value of the discrete random variable and $f(x)$ is a probability function for a continuous variable. With respect to a continuous random variable, the cumulative distribution rises continuously in a smooth (rather than stairwise) fashion.

Example 15.5 reveals the method by which probabilistic information can be incorporated into our analysis. Again, WMC's transmission-housings project will be used. In the next section, we will show you how to compute some composite statistics with the use of all the data.

EXAMPLE 15.5 Cumulative Probability Distributions

Suppose that the only parameters subject to risk are the number of unit sales (X) to Gulf Electric each year and the unit sales price (Y). From experience in the market, WMC assesses the probabilities of outcomes for each variable as shown in Table 15.5. Determine the cumulative probability distributions for these random variables.

SOLUTION

Consider the demand probability distribution (X) given in Table 15.5 for WMC's transmission-housings project:

Unit Demand (X)	Probability, $P(X = x)$
1,600	0.2
2,000	0.6
2,400	0.2

If we want to know the probability that demand will be less than or equal to any particular value, we can use the following cumulative probability function:

$$F(x) = P(X \leq x) = \begin{cases} 0.2 & x \leq 1{,}600 \\ 0.8 & x \leq 2{,}000. \\ 1.0 & x \leq 2{,}400 \end{cases}$$

For example, if we want to know the probability that the demand will be less than or equal to 2,000, we can examine the appropriate interval $(x \leq 2{,}000)$, and we shall find that the probability is 80%.

We can find the cumulative distribution for Y in a similar fashion. Figures 15.6 and 15.7 illustrate probability and cumulative probability distributions for X and Y, respectively.

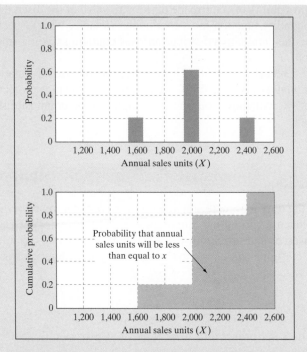

Figure 15.6 Probability and cumulative probability distributions for random variable X (annual sales).

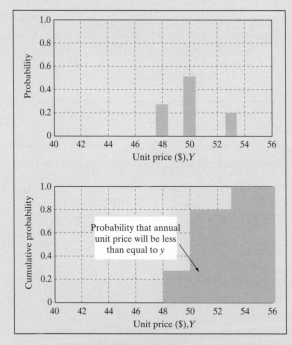

Figure 15.7 Probability and cumulative probability distributions for random variable Y from Table 15.5.

15.3.2 Summary of Probabilistic Information

Although knowledge of the probability distribution of a random variable allows us to make a specific probability statement, a single value that may characterize the random variable and its probability distribution is often desirable. Such a quantity is the **expected value** of the random variable. We also want to know something about how the values of the random variable are dispersed about the expected value (i.e., the **variance**). In investment analysis, this dispersion information is interpreted as the degree of project risk. The expected value indicates the weighted average of the random variable, and the variance captures the variability of the random variable.

Measure of Expectation

The **expected value** (also called the **mean**) is a weighted average value of the random variable, where the weighting factors are the probabilities of occurrence. All distributions (discrete and continuous) have an expected value. We will use $E[X]$ (or μ_x) to denote the expected value of random variable X. For a random variable X that has either discrete or continuous values, we compute the expected value with the formula

$$E[X] = \mu_x = \begin{cases} \displaystyle\sum_{j=1}^{J} (p_j) x_j & \text{(discrete case)} \\ \displaystyle\int_{L}^{H} x f(x)\, dx & \text{(continuous case)} \end{cases}, \qquad (15.1)$$

Expected value represents the average amount one "expects" to win per bet if bets with identical odds are repeated many times.

where J is the number of discrete events and L and H are the lower and upper bounds of the continuous probability distribution.

The expected value of a distribution gives us important information about the "average" value of a random variable, such as the NPW, but it does not tell us anything about the variability on either side of the expected value. Will the range of possible values of the random variable be very small, and will all the values be located at or near the expected value? For example, the following represents the temperatures recorded on a typical April day for two cities in Canada:

Location	Low	High	Average
Calgary	–2°C	11°C	4.5°C
Halifax	1°C	8°C	4.5°C

Even though both cities had identical mean (average) temperatures on that particular day, they had different variations in extreme temperatures. We shall examine this variability issue next.

Measure of Variation

Another measure needed when we are analyzing probabilistic situations is a measure of the risk due to the variability of the outcomes. Among the various measures of the variation of a set of numbers that are used in statistical analysis are the **range**, the **variance**, and the **standard deviation**. The variance and the standard deviation are used most commonly in the analysis of risk. We will use $\text{Var}[X]$ or σ_x^2 to denote the variance, and σ_x to denote the standard deviation, of random variable X. (If there is only one random variable in an analysis, we normally omit the subscript.)

Variance of a random variable is a measure of its statistical dispersion, indicating how far from the expected value its values typically are.

The **variance** tells us the degree of spread, or dispersion, of the distribution on either side of the mean value. As the variance increases, the spread of the distribution increases; the smaller the variance, the narrower is the spread about the expected value.

To determine the variance, we first calculate the deviation of each possible outcome x_j from the expected value $(x_j - \mu)$. Then we raise each result to the second power and multiply it by the probability of x_j occurring (i.e., p_j). The summation of all these products serves as a measure of the distribution's variability. For a random variable that has only discrete values, the equation for computing the variance is[4]

$$\text{Var}[X] = \sigma_x^2 = \sum_{j=1}^{J} (x_j - \mu)^2 (p_j),\tag{15.2}$$

where p_j is the probability of occurrence of the jth value of the random variable (x_j), and μ is as defined by Eq. (15.1). To be most useful, any measure of risk should have a definite value (unit). One such measure is the standard deviation, which we may calculate by taking the positive square root of $\text{Var}[X]$, measured in the same units as X:

$$\sigma_x = \sqrt{\text{Var}[X]}.\tag{15.3}$$

The standard deviation is a probability-weighted deviation (more precisely, the square root of the sum of squared deviations) from the expected value. Thus, it gives us an idea of how far above or below the expected value the actual value is likely to be. For most probability distributions, the actual value will be observed within the $\pm 3\sigma$ range.

In practice, the calculation of the variance is somewhat easier if we use the formula

$$\begin{aligned}\text{Var}[X] &= \Sigma p_j x_j^2 - \left(\Sigma p_j x_j\right)^2\\ &= E[X^2] - (E[X])^2.\end{aligned}\tag{15.4}$$

The term $E[X^2]$ in Eq. (15.4) is interpreted as the mean value of the squares of the random variable (i.e., the actual values squared). The second term is simply the mean value squared. Example 15.6 illustrates how we compute measures of variation.

EXAMPLE 15.6 Calculation of Mean and Variance

Consider once more WMC's transmission-housings project. Unit sales (X) and unit price (Y) are estimated as in Table 15.5. Compute the means, variances, and standard deviations for the random variables X and Y.

SOLUTION

For the product demand variable (X), we have

x_j	p_j	$x_j\,(p_j)$	$(x_j - E\,[X])^2$	$(x_j - E\,[X])^2(p_j)$
1,600	0.20	320	$(-400)^2$	32,000
2,000	0.60	1,200	0	0
2,400	0.20	480	$(400)^2$	32,000
		$E[X] = 2,000$		$\text{Var}[X] = 64,000$
				$\sigma_x = 252.98$

[4] For a continuous random variable, we compute the variance as follows: $\text{Var}[X] = \displaystyle\int_{L}^{H} (x - \mu)^2 f(x)\,dx.$

For the variable unit price (Y), we have

x_j	p_j	$x_j(p_j)$	$(x_j - E[X])^2$	$(x_j - E[X])^2(p_j)$
$48	0.30	$14.40	$(-2)^2$	1.20
50	0.50	25.00	$(0)^2$	0
53	0.20	10.60	$(3)^2$	1.80
		$E[Y] = 50.00$		$\text{Var}[Y] = 3.00$
				$\sigma_y = 1.73$

15.3.3 Joint and Conditional Probabilities

Thus far, we have not looked at how the values of some variables can influence the values of others. It is, however, entirely possible—indeed, it is likely—that the values of some parameters will be dependent on the values of others. We commonly express these dependencies in terms of conditional probabilities. An example is product demand, which will probably be influenced by unit price.

We define a **joint probability** as

$$P(x, y) = P(X = x | Y = y)P(Y = y), \tag{15.5}$$

where $P(X = x | Y = y)$ is the **conditional probability** of observing x, given $Y = y$, and $P(Y = y)$ is the **marginal probability** of observing $Y = y$. Certainly, important cases exist in which a knowledge of the occurrence of event X does *not* change the probability of an event Y. That is, if X and Y are **independent**, then the joint probability is simply

$$P(x, y) = P(x)P(y). \tag{15.6}$$

The concepts of joint, marginal, and conditional distributions are best illustrated by numerical examples.

Suppose that WMC's marketing staff estimates that, for a given unit price of $48, the conditional probability that the company can sell 1600 units is 0.10. The probability of this joint event (unit sales = 1,600 and unit sales price = $48) is

$$P(x, y) = P(1,600, \$48)$$

$$= P(x = 1,600 | y = \$48)P(y = \$48)$$

$$= (0.10)(0.30)$$

$$= 0.03.$$

We can obtain the probabilities of other joint events in a similar fashion, as shown in Table 15.6.

TABLE 15.6 Assessments of Conditional and Joint Probabilities

Unit Price Y	Probability	Conditional Unit Sales X	Joint Probability	Probability
$48	0.30	1,600	0.10	0.03
		2,000	0.40	0.12
		2,400	0.50	0.15
50	0.50	1,600	0.10	0.05
		2,000	0.64	0.32
		2,400	0.26	0.13
53	0.20	1,600	0.50	0.10
		2,000	0.40	0.08
		2,400	0.10	0.02

From Table 15.6, we can see that the unit demand (X) ranges from 1600 to 2400 units, the unit price (Y) ranges from $48 to $53, and nine joint events are possible. These joint probabilities must sum to unity:

Joint Event (xy)	$P(x, y)$
(1,600, $48)	0.03
(2,000, $48)	0.12
(2,400, $48)	0.15
(1,600, $50)	0.05
(2,000, $50)	0.32
(2,400, $50)	0.13
(1,600, $53)	0.10
(2,000, $53)	0.08
(2,400, $53)	0.02
	Sum = 1.00

The marginal distribution for x can be developed from the joint event by fixing x and summing over y:

x_j	$P(x_j) = \sum_y P(x, y)$
1,600	$P(1,600, \$48) + P(1,600, \$50) + P(1,600, \$53) = 0.18$
2,000	$P(2,000, \$48) + P(2,000, \$50) + P(2,000, \$53) = 0.52$
2,400	$P(2,400, \$48) + P(2,400, \$50) + P(2,400, \$53) = 0.30$

This marginal distribution tells us that 52% of the time we can expect to have a demand of 2000 units and 18% and 30% of the time we can expect to have a demand of 1600 and 2400, respectively.

15.3.4 Covariance and Coefficient of Correlation

When two random variables are not independent, we need some measure of their dependence on each other. The parameter that tells the degree to which two variables (X, Y) are related is the covariance $\text{Cov}(X, Y)$, denoted by σ_{xy}. Mathematically, we define

$$
\begin{aligned}
\text{Cov}(X, Y) &= \sigma_{xy} \\
&= E\{(X - E[X])(Y - E[Y])\} \\
&= E(XY) - E(X)E(Y) \\
&= \rho_{xy}\sigma_x\sigma_y,
\end{aligned}
\tag{15.7}
$$

where ρ_{xy} is the coefficient of correlation between X and Y. It is clear that if X tends to exceed its mean whenever Y exceeds its mean, $\text{Cov}(X, Y)$ will be positive. If X tends to fall below its mean whenever Y exceeds its mean, then $\text{Cov}(X, Y)$ will be negative. The sign of $\text{Cov}(X, Y)$, therefore, reveals whether X and Y vary directly or inversely with one another. We can rewrite Eq. (15.7) in terms of ρ_{xy}:

$$
\rho_{xy} = \frac{\text{Cov}(X, Y)}{\sigma_x\sigma_y}.
\tag{15.8}
$$

The value of ρ_{xy} can vary within the range of -1 and $+1$, with $\rho_{xy} = 0$ indicating no correlation between the two random variables. As shown in Table 15.7, the coefficient of correlation between the product demand (X) and the unit price (Y) is negatively correlated, $\rho_{xy} = -0.439$, indicating that as the firm lowers the unit price, it tends to generate a higher demand.

15.4 Probability Distribution of NPW

After we have identified the random variables in a project and assessed the probabilities of the possible events, the next step is to develop the project's NPW distribution.

15.4.1 Procedure for Developing an NPW Distribution

We will consider the situation in which all the random variables used in calculating the NPW are independent. To develop the NPW distribution, we may follow these steps:

- Express the NPW as functions of unknown random variables.
- Determine the probability distribution for each random variable.
- Determine the joint events and their probabilities.
- Evaluate the NPW equation at these joint events.
- Rank the NPW values in increasing order of NPW.

These steps are illustrated in Example 15.7.

TABLE 15.7 Calculating the Correlation Coefficient Between Two Random Variables X and Y

(x, y)	$(x - E[X])$	$(y - E[Y])$	$p(x,y)$	$(x - E[X])(y - E[Y])$	$p(x,y) \times (x - E[X])(y - E[Y])$
(1,600, 48)	(1,600 − 2,000)	(48 − 50)	0.03	800	24
(2,000, 48)	(2,000 − 2,000)	(48 − 50)	0.12	0	0
(2,400, 48)	(2,400 − 2,000)	(48 − 50)	0.15	−800	−120
(1,600, 50)	(1,600 − 2,000)	(50 − 50)	0.05	0	0
(2,000, 50)	(2,000 − 2,000)	(50 − 50)	0.32	0	0
(2,400, 50)	(2,400 − 2,000)	(50 − 50)	0.13	0	0
(1,600, 53)	(1,600 − 2,000)	(53 − 50)	0.10	−1,200	−120
(2,000, 53)	(2,000 − 2,000)	(53 − 50)	0.08	0	0
(2,400, 53)	(2,400 − 2,000)	(53 − 50)	0.02	1,200	24

$$\text{Cov}(X, Y) = -192$$

$$\rho_{xy} = \frac{\text{Cov}(X, Y)}{\sigma_x \sigma_y}$$

$$= \frac{-192}{\left(\sqrt{64{,}000} \times \sqrt{3.00} \right)}$$

$$= -0.439$$

EXAMPLE 15.7 Procedure for Developing an NPW Distribution

Consider again WMC's transmission-housings project first set forth in Example 15.1. Use the unit demand (X) and price (Y) given in Table 15.5, and develop the NPW distribution for the WMC project. Then calculate the mean and variance of the NPW distribution.

SOLUTION

Table 15.8 summarizes the after-tax cash flow for the WMC's transmission-housings project as functions of random variables X and Y. From this table, we can compute the PW of cash inflows as follows:

$$PW(15\%) = 0.6XY(P/A, 15\%, 5) + \$44,782$$
$$= 2.011293XY + \$44,782.$$

The PW of cash outflows is

$$PW(15\%) = \$125,000 + (9X + \$6,000)(P/A, 15\%, 5)$$
$$= 30.1694X + \$145,113.$$

Thus, the NPW is

$$PW(15\%) = 2.011093X(Y - \$15) - \$100,331.$$

If the product demand X and the unit price Y are independent random variables, then PW (15%) will also be a random variable. To determine the NPW distribution, we need to consider all the combinations of possible outcomes.[5] The first possibility is the event in which $x = 1600$ and $y = \$48$. Since X and Y are considered to be independent random variables, the probability of this joint event is

$$P(x = 1600, y = \$48) = P(x = 1600)P(y = \$48)$$
$$= (0.20)(0.30)$$
$$= 0.06.$$

With these values as input, we compute the possible NPW outcome as follows:

$$PW(15\%) = 2.011093X(Y - \$15) - \$100,331$$
$$= 2.011093(1,600)(\$48 - \$15) - \$100,331$$
$$= \$5866.$$

Eight other outcomes are possible: they are summarized with their joint probabilities in Table 15.9 and depicted in Figure 15.8.

The NPW probability distribution in Table 15.9 indicates that the project's NPW varies between $5866 and $83,100, but that no loss occurs under any of the circumstances examined. On the one hand, from the cumulative distribution, we further observe that there is a 0.38 probability that the project would realize an NPW less than that forecast for the base case ($40,460). On the other hand, there is a 0.32 probability that the NPW will be greater than this value. Certainly, the probability distribution provides much more information on the likelihood of each possible event than does the scenario analysis presented in Section 15.2.4.

---

[5] If X and Y are dependent random variables, the joint probabilities developed in Table 15.6 should be used.

TABLE 15.8 After-Tax Cash Flow as a Function of Unknown Unit Demand (X) and Unit Price (Y)

Item	0	1	2	3	4	5
Cash inflow						
Net salvage						34,204
Revenue						
$X(1 - 0.4)Y$		0.6XY	0.6XY	0.6XY	0.6XY	0.6XY
CCA credit						
0.4 (CCA)		7,500	12,750	8,925	6,248	4,373
Cash outflow						
Investment	−125,000					
Variable cost						
$-X(1 - 0.4)(\$15)$		−9X	−9X	−9X	−9X	−9X
Fixed cost						
$-(1 - 0.4)(\$10,000)$		−6,000	−6,000	−6,000	−6,000	−6,000
Net cash flow	−125,000	0.6X(Y − 15) +1,500	0.6X(Y − 15) +6,750	0.6X(Y − 15) +2,925	0.6X(Y − 15) +248	0.6X(Y − 15) +32,577

TABLE 15.9 The NPW Probability Distribution With Independent Random Variables

Event No.	x	y	P(x,y)	Cumulative Joint Probability	NPW
1	1,600	$48.00	0.06	0.06	$ 5,866
2	1,600	50.00	0.10	0.16	12,302
3	1,600	53.00	0.04	0.20	21,956
4	2,000	48.00	0.18	0.38	32,415
5	2,000	50.00	0.30	0.68	40,460
6	2,000	53.00	0.12	0.80	52,528
7	2,400	48.00	0.06	0.86	58,964
8	2,400	50.00	0.10	0.96	68,618
9	2,400	53.00	0.04	1.00	83,100

We have developed a probability distribution for the NPW by considering random cash flows. As we observed, a probability distribution helps us to see what the data imply in terms of project risk. Now we can learn how to summarize the probabilistic information—the mean and the variance:

- For WMC's transmission-housings project, we compute the expected value of the NPW distribution as shown in Table 15.10. Note that this expected value is the same as the most likely value of the NPW distribution. This equality was expected because both X and Y have a symmetrical probability distribution.
- We obtain the variance of the NPW distribution, assuming independence between X and Y and using Eq. (15.2), as shown in Table 15.11. We could obtain the same result more easily by using Eq. (15.4).

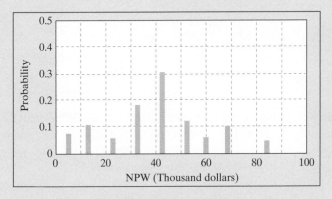

Figure 15.8 The NPW probability distribution when X and Y are independent (Example 15.7).

TABLE 15.10 Calculation of the Mean of the NPW Distribution

Event No.	x	y	P(x,y)	Cumulative Joint Probability	NPW	Weighted NPW
1	1,600	$ 48.00	0.06	0.06	$ 5,866	$ 352
2	1,600	50.00	0.10	0.16	12,302	1,230
3	1,600	53.00	0.04	0.20	21,956	878
4	2,000	48.00	0.18	0.38	32,415	5,835
5	2,000	50.00	0.30	0.68	40,460	12,138
6	2,000	53.00	0.12	0.80	52,528	6,303
7	2,400	48.00	0.06	0.86	58,964	3,538
8	2,400	50.00	0.10	0.96	68,618	6,862
9	2,400	53.00	0.04	1.00	83,100	3,324
					E[NPW] =	$40,460

TABLE 15.11 Calculation of the Variance of the NPW Distribution

Event No.	x	y	P(x,y)	NPW	(NPW − E[NPW])²	Weighted (NPW − E[NPW])²
1	1,600	$48.00	0.06	$ 5,866	1,196,761,483	$71,805,689
2	1,600	50.00	0.10	12,302	792,878,755	79,287,875
3	1,600	53.00	0.04	21,956	342,394,172	13,695,767
4	2,000	48.00	0.18	32,415	64,724,796	11,650,463
5	2,000	50.00	0.30	40,460	0	0
6	2,000	53.00	0.12	52,528	145,630,792	17,475,695
7	2,400	48.00	0.06	58,964	342,394,172	20,543,650
8	2,400	50.00	0.10	68,618	792,878,755	79,287,875
9	2,400	53.00	0.04	83,100	1,818,119,528	72,724,781
					Var[NPW] =	366,471,797
					σ =	$19,143

COMMENTS: We can obtain the mean and variance of the NPW analytically by using the properties of the product of random variables. Let $W = XY$, where X and Y are random variables with known means and variances (μ_x, σ_x^2) and (μ_y, σ_y^2), respectively. If X and Y are independent of each other, then

$$E(W) = E(XY) = \mu_x \mu_y \tag{15.9}$$

and

$$\text{Var}(W) = \text{Var}(XY) = \mu_x^2 \sigma_y^2 + \mu_y^2 \sigma_x^2 + \sigma_x^2 \sigma_y^2. \tag{15.10}$$

The expected net present worth is then

$$E[\text{PW}(15\%)] = 2.011293E(XY) - (2.011293)(15)E(X) - 100{,}331$$
$$= 2.011293(2{,}000)(50) - (2.011293)(15)(2{,}000) - 100{,}331$$
$$= \$40{,}460.$$

Now let $Z = Y - 15$. Then $E(Z) = E(Y) - 15 = \mu_y - 15 = 50 - 15 = 35$ and $\text{Var}[Z] = \text{Var}[Y] = \sigma_y^2 = 3$. So

$$\text{Var}[\text{PW}(15\%)] = \text{Var}[2.011293X(Y - 15) - 100{,}331]$$
$$= \text{Var}[2.011293XZ]$$
$$= (2.011293)^2 \, \text{Var}[XZ]$$
$$= (2.011293)^2 (\mu_x^2 \sigma_y^2 + \mu_y^2 \sigma_x^2 + \sigma_x^2 \sigma_y^2)$$
$$= (2.011293)^2 [(2{,}000^2)(3) + (50^2)(64{,}000) + (64{,}000)(3)]$$
$$= 366{,}471{,}797.$$

For completeness, the standard deviation is calculated as the square root of the variance:

$$\sigma = \sqrt{366{,}471{,}797} = \$19{,}143$$

Note that the mean and variance thus calculated are exactly the same as in Tables 15.10 and 15.11, respectively. Note also that if X and Y are correlated with each other, then Eq. (15.10) cannot be used.[6]

15.4.2 Aggregating Risk Over Time

In the previous section, we developed an NPW equation by aggregating all cash flow components over time. Another approach to estimating the amount of risk present in a particular investment opportunity is to determine the mean and variance of cash flows in each period; then we may be able to aggregate the risk over the project life in terms of net present worth. We have two cases:

- **Independent random variables.** In this case,

$$E[\text{PW}(i)] = \sum_{n=0}^{N} \frac{E(A_n)}{(1 + i)^n} \tag{15.11}$$

and

$$\text{Var}[\text{PW}(i)] = \sum_{n=0}^{N} \frac{\text{Var}(A_n)}{(1 + i)^{2n}} \tag{15.12}$$

where

A_n = cash flow in period n,

$E(A_n)$ = expected cash flow iiod n,

$\text{Var}(A_n)$ = variance of the cash flow in period n.

[6] Analytical treatment for the products of random variables including the correlated case is given by Chan S. Park and Gunter Sharp-Bette, *Advanced Engineering Economics*, New York, John Wiley, 1990 (Chapter 10).

In defining Eq. (15.12), we are also assuming the independence of cash flows, meaning that knowledge of what will happen to one particular period's cash flow will not allow us to predict what will happen to cash flows in other periods.

- **Dependent random variables.** In case we cannot assume a statistical independence among cash flows, we need to consider the degree of dependence among them. The expected-value calculation will not be affected by this dependence, but the project variance will be calculated as

$$\text{Var}[\text{PW}(i)] = \sum_{n=0}^{N} \frac{\text{Var}(A_n)}{(1+i)^{2n}} + 2 \sum_{n=0}^{N-1} \sum_{s=n+1}^{N} \frac{\rho_{ns} \sigma_n \sigma_s}{(1+i)^{n+s}}, \qquad (15.13)$$

where ρ_{ns} = correlation coefficient (degree of dependence) between A_n and A_s. The value of ρ_{ns} can vary within the range from -1 to $+1$, with $\rho_{ns} = -1$ indicating perfect negative correlation and $\rho_{ns} = +1$ indicating perfect positive correlation. The result $\rho_{ns} = 0$ implies that no correlation exists between A_n and A_s. If $\rho_{ns} > 0$, we can say that A_n and A_s are positively correlated, meaning that if the actual realization of cash flow for A_n is higher than its expected value, it is likely that you will also observe a higher cash flow than its expected value for A_s. If $\rho_{ns} < 0$, the opposite relation exists.

Example 15.8 illustrates the mechanics involved in calculating the mean and variance of a project's net present worth.

EXAMPLE 15.8 Aggregation of Risk Over Time

Consider the following financial data for an investment project, where the only random components are the operating expenses in each period:

- Investment required = $10,000
- Project life = three years
- Expected salvage value = $0
- Annual operating revenue = $20,000
- Annual operating expenses are random variables with the following means and variances:

n	X_1	X_2	X_3
Mean	$9,000	$13,000	$15,000
Variance	250,000	490,000	1,000,000

- CCA rate is 10%.
- Tax rate = 40%
- Discount rate (or MARR) = 12%

Determine the expected net present worth and the variance of the NPW assuming the following two situations: (a) X_i are independent random variables and (b) X_i are dependent random variables.

SOLUTION

Step 1: Calculate the net proceeds from disposal of the equipment at the end of the project life:

Salvage value = 0,

$$UCC_3 = 10,000 - \sum_{n=1}^{3} CCA_n = \$7695,$$

Disposal tax effect $G = 0.4(\$7695 - 0) = \$3078$

Net proceeds from sale = Salvage value + G = \$3078.

Step 2: Construct a generalized cash flow table as a function of the random variable X_n:

Description	Cash Flow			
	0	**1**	**2**	**3**
Investment	−10,000			
Revenue × (0.6)		12,000	12,000	12,000
−O&M × (0.6)		−0.6X_1	−0.6X_2	−0.6X_3
CCA$_n$ × (0.4)		200	380	342
Net proceeds from sale				3,078
Net cash flow	−10,000	12,200	12,380	15,420
		−0.6X_1	−0.6X_2	−0.6X_3

$$E[X_1] = 9000, \qquad E[X_2] = 13,000, \qquad E[X_3] = 15,000,$$
$$Var[X_1] = 250,000, \ Var[X_2] = 490,000, \ Var[X_3] = 1,000,000.$$

The net present worth is then

$$PW(12\%) = \left[-10,000 + \frac{12,200}{1.12} + \frac{12,380}{1.12^2} + \frac{15,420}{1.12^3} \right]$$

$$- 0.6 \left[\frac{X_1}{1.12} + \frac{X_2}{1.12^2} + \frac{X_3}{1.12^3} \right]$$

$$= 21,738 - 0.6 \left[\frac{X_1}{1.12} + \frac{X_2}{1.12^2} + \frac{X_3}{1.12^3} \right].$$

- **Case 1.** Independent Cash Flows: Using Eqs. (15.11) and (15.12), we obtain

$$E[PW(12\%)] = 21{,}738 - 0.6\left[\frac{E[X_1]}{1.12} + \frac{E[X_2]}{1.12^2} + \frac{E[X_3]}{1.12^3}\right]$$

$$= 21{,}738 - 17{,}446$$

$$= \$4292,$$

$$Var[PW(12\%)] = \left(\frac{-0.6}{1.12}\right)^2 Var[X_1] + \left(\frac{-0.6}{1.12^2}\right)^2 Var[X_2]$$

$$+ \left(\frac{-0.6}{1.12^3}\right)^2 Var[X_3]$$

$$= 71{,}747.45 + 112{,}105.39 + 182{,}387.20$$

$$= 366{,}240,$$

$$\sigma[PW(12\%)] = \$605.18.$$

- **Case 2.** Dependent Cash Flows: If the random variables X_1, X_2, and X_3 are partially correlated with the correlation coefficients $\rho_{12} = 0.3$, $\rho_{13} = 0.5$, and $\rho_{23} = 0.4$, respectively (see accompanying diagram), then the variance of the NPW can be calculated with Eq. (15.13):

$$Var[PW(12\%)] = (\text{Original variance}) + (\text{Covariance terms})$$

$$= 366{,}240$$

$$+ 2\left(\frac{-0.6}{1.12}\right)\left(\frac{-0.6}{1.12^2}\right)\rho_{12}\sigma_1\sigma_2$$

$$+ 2\left(\frac{-0.6}{1.12}\right)\left(\frac{-0.6}{1.12^3}\right)\rho_{13}\sigma_1\sigma_3$$

$$+ 2\left(\frac{-0.6}{1.12^2}\right)\left(\frac{-0.6}{1.12^3}\right)\rho_{23}\sigma_2\sigma_3$$

$$= 366{,}240$$

$$+ 2\left(\frac{0.6^2}{1.12^3}\right)(0.3)(500)(700)$$

$$+ 2\left(\frac{0.6^2}{1.12^4}\right)(0.5)(500)(1{,}000)$$

$$+ 2\left(\frac{0.6^2}{1.12^5}\right)(0.4)(700)(1{,}000)$$

$$= 366{,}240 + 53{,}810.59 + 114{,}393.25 + 114{,}393.25$$

$$= 648{,}837,$$

$$\sigma[PW(12\%)] = \$805.50.$$

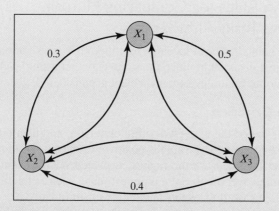

Note that the variance increases significantly when at least some of the random variables are positively correlated.

COMMENTS: How is information such as the preceding used in decision making? Most probability distributions are completely described by six standard deviations—three above, and three below, the mean. As shown in Figure 15.9, the NPW of this project would almost certainly fall between $2477 and $6108 for the independent case and between $1876 and $6709 for the dependent case. In either situation, the NPW below 3σ from the mean is still positive, so we may say that the project is quite safe. If that figure were negative, it would then be up to the decision maker to determine whether it is worth investing in the project, given its mean and standard deviation.

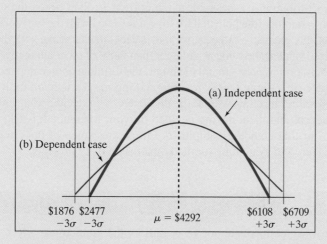

Figure 15.9   NPW distributions with $\pm 3\sigma$: (a) independent case and (b) dependent case.

15.4.3 Decision Rules for Comparing Mutually Exclusive Risky Alternatives

Once the expected value has been located from the NPW distribution, it can be used to make an accept–reject decision in much the same way that a single NPW is used when a single possible outcome for an investment project is considered.

Expected-Value Criterion

The decision rule is called the **expected-value criterion**, and using it, we may accept a single project if its expected NPW value is positive. In the case of mutually exclusive alternatives, we select the one with the highest expected NPW. The use of the expected NPW has an advantage over the use of a point estimate, such as the likely value, because it includes all possible cash flow events and their probabilities.

Law of large numbers implies that the average of a random sample from a large population is likely to be close to the mean of the whole population.

The justification for the use of the expected-value criterion is based on the **law of large numbers**, which states that if many repetitions of an experiment are performed, the average outcome will tend toward the expected value. This justification may seem to negate the usefulness of the criterion, since, in project evaluation, we are most often concerned with a single, nonrepeatable "experiment" (i.e., an investment alternative). However, if a firm adopts the expected-value criterion as a standard decision rule for *all* of its investment alternatives, then, over the long term, the law of large numbers predicts that accepted projects tend to meet their expected values. Individual projects may succeed or fail, but the average project tends to meet the firm's standard for economic success.

Mean-and-Variance Criterion

The expected-value criterion is simple and straightforward to use, but it fails to reflect the variability of the outcome of an investment. Certainly, we can enrich our decision by incorporating information on variability along with the expected value. Since the variance represents the dispersion of the distribution, it is desirable to minimize it. In other words, the smaller the variance, the less the variability (the potential for loss) associated with the NPW. Therefore, when we compare mutually exclusive projects, we may select the alternative with the smaller variance if its expected value is the same as or larger than those of other alternatives.

In cases where preferences are not clear cut, the ultimate choice depends on the decision maker's trade-offs—how far he or she is willing to take the variability to achieve a higher expected value. In other words, the challenge is to decide what level of risk you are willing to accept and then, having decided on your tolerance for risk, to understand the implications of that choice. Example 15.9 illustrates some of the critical issues that need to be considered in evaluating mutually exclusive risky projects.

EXAMPLE 15.9 Comparing Risky Mutually Exclusive Projects

With ever-growing concerns about air pollution, the greenhouse effect, and stricter emission standards in North America, Green Engineering has developed a prototype conversion unit that allows a motorist to switch from gasoline to compressed natural gas (CNG) or vice versa. Driving a car equipped with Green's conversion kit is not much different from driving a conventional model. A small dial switch on the dashboard controls which fuel is to be used. Four different configurations are available, according to the type of vehicle: compact size, midsize, large size, and trucks. In the past, Green has built a few prototype vehicles powered by alternative fuels other than gasoline but has been reluctant to go into higher volume production without more evidence of public demand.

As a result, Green Engineering initially would like to target one market segment (one configuration model) in offering the conversion kit. Green Engineering's marketing group has compiled the potential NPW distribution for each different configuration when marketed independently.

Event (NPW) (unit: thousands)	Probabilities			
	Model 1	Model 2	Model 3	Model 4
$1,000	0.35	0.10	0.40	0.20
1,500	0	0.45	0	0.40
2,000	0.40	0	0.25	0
2,500	0	0.35	0	0.30
3,000	0.20	0	0.20	0
3,500	0	0	0	0
4,000	0.05	0	0.15	0
4,500	0	0.10	0	0.10

Evaluate the expected return and risk for each model configuration, and recommend which one, if any, should be selected.

SOLUTION

For model 1, we calculate the mean and variance of the NPW distribution as follows:

$$E[\text{NPW}]_1 = \$1000(0.35) + \$2000(0.40)$$
$$+ \$3000(0.20) + \$4000(0.05)$$
$$= \$1950;$$
$$\text{Var}[\text{NPW}]_1 = 1000^2(0.35) + 2000^2(0.40)$$
$$+ 3000^2(0.20) + 4000^2(0.05) - (1950)^2$$
$$= 747,500.$$

In a similar manner, we compute the other values as follows:

Configuration	E[NPW]	Var[NPW]
Model 1	$1,950	747,500
Model 2	2,100	915,000
Model 3	2,100	1,190,000
Model 4	2,000	1,000,000

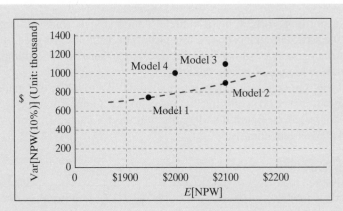

Figure 15.10 Mean–variance chart showing project dominance. Both model 3 and model 4 are dominated by model 2.

The results are plotted in Figure 15.10. If we make a choice solely on the basis of the expected value, we may select either model 2 or model 3, because they have the highest expected NPW.

If we consider the variability along with the expected NPW, however, the correct choice is not obvious. We will first eliminate those alternatives which are clearly inferior to others:

Risk-Return Trade-offs: The greater the amount of risk that an investor is willing to take on, the greater the potential return.

- *Model* 2 *versus Model* 3. We see that models 2 and 3 have the same mean of $2100, but model 2 has a much lower variance. In other words, model 2 dominates model 3, so we eliminate model 3 from further consideration.

- *Model* 2 *versus Model* 4. Similarly, we see that model 2 is preferred over model 4, because model 2 has a higher expected value and a smaller variance; model 2 again dominates model 4, so we eliminate model 4. In other words, model 2 dominates both models 3 and 4 as we consider the variability of NPW.

- *Model* 1 *versus Model* 2. Even though the mean-and-variance rule has enabled us to narrow down our choices to only two models (models 1 and 2), it does not indicate what the ultimate choice should be. In other words, comparing models 1 and 2, we see that $E[\text{NPW}]$ increases from $1950 to $2100 at the price of a higher Var[NPW], which increases from 747,500 to 915,000. The choice, then, will depend on the decision maker's trade-offs between the incremental expected return ($150) and the incremental risk (167,500). We cannot choose between the two simply on the basis of mean and variance, so we must resort to other probabilistic information.[7]

15.5 Risk Simulation

In the previous sections, we examined analytical methods of determining the NPW distributions and computing their means and variances. As we saw in Section 15.4.1, the NPW distribution offers numerous options for graphically presenting to the decision maker probabilistic information, such as the range and likelihoods of occurrence of possible levels

[7] As we seek further refinement in our decision under risk, we may consider the expected-utility theory, or stochastic dominance rules, which is beyond the scope of our text. (See Park, C. S., and Sharp-Bette, G. P., *Advanced Engineering Economics* (New York: John Wiley, 1990), Chapters 10 and 11.)

of NPW. Whenever we can adequately evaluate the risky investment problem by analytical methods, it is generally preferable to do so. However, many investment situations are such that we cannot solve them easily by analytical methods, especially when many random variables are involved. In these situations, we may develop the NPW distribution through computer simulation.

15.5.1 Computer Simulation

Before we examine the details of risk simulation, let us consider a situation in which we wish to train a new astronaut for a future space mission. Several approaches exist for training this astronaut. One (somewhat unlikely) possibility is to place the trainee in an actual space shuttle and to launch him or her into space. This approach certainly would be expensive; it also would be extremely risky, because any human error made by the trainee would have tragic consequences. As an alternative, we can place the trainee in a flight simulator designed to mimic the behaviour of the actual space shuttle in flight. The advantage of this approach is that the astronaut trainee learns all the essential functions of space operation in a simulated space environment. The flight simulator generates test conditions approximating operational conditions, and any human errors made during training cause no harm to the astronaut or to the equipment being used.

The use of computer simulation is not restricted to simulating a physical phenomenon such as the flight simulator. In recent years, techniques for testing the results of some investment decisions before they are actually executed have been developed. As a result, many phases of business investment decisions have been simulated with considerable success. In fact, we can analyze WMC's transmission-housings project by building a simulation model. The general approach is to assign a subjective (or objective) probability distribution to each unknown factor and to combine all these distributions into a probability distribution for the project's profitability as a whole. The essential idea is that if we can simulate the actual state of nature for unknown investment variables on a computer, we may be able to obtain the resulting NPW distribution.

The unit demand (X) in our WMC's transmission-housing project was one of the random variables in the problem. We can know the exact value for this random variable only after the project is implemented. Is there any way to predict the actual value before we make any decision about the project?

The following logical steps are often suggested for a computer program that simulates investment scenarios:

Step 1. Identify all the variables that affect the measure of investment worth (e.g., NPW after taxes).

Step 2. Identify the relationships among all the variables. The relationships of interest here are expressed by the equations or the series of numerical computations by which we compute the NPW of an investment project. These equations make up the model we are trying to analyze.

Step 3. Classify the variables into two groups: the parameters whose values are known with certainty and the random variables for which exact values cannot be specified at the time of decision making.

Step 4. Define distributions for all the random variables.

Step 5. Perform Monte Carlo sampling (see Section 15.5.3) and describe the resulting NPW distribution.

Step 6. Compute the distribution parameters and prepare graphic displays of the results of the simulation.

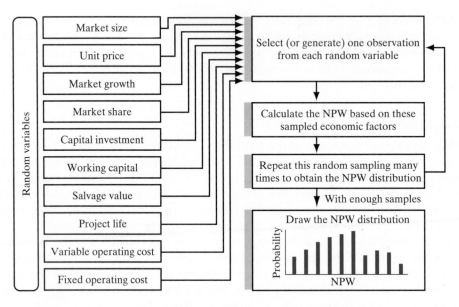

Figure 15.11   Logical steps involved in simulating a risky investment.

Figure 15.11 illustrates the logical steps involved in simulating a risky investment project. The risk simulation process has two important advantages compared with the analytical approach discussed in Section 15.4:

1. The number of variables that can be considered is practically unlimited, and the distributions used to define the possible values of each random variable can be of any type and any shape. The distributions can be based on statistical data if they are available or, more commonly, on subjective judgment.
2. The method lends itself to sensitivity analyses. By defining some factors that have the most significant effect on the resulting NPW values and using different distributions (in terms of either shape or range) for each variable, we can observe the extent to which the NPW distribution is changed.

15.5.2 Model Building

In this section, we shall present some of the procedural details related to the first three steps (model building) outlined in Section 15.5.1. To illustrate the typical procedure involved, we shall work with the investment setting for WMC's transmission-housings project described in Example 15.7.

The initial step is to define the measure of investment worth and the factors which affect that measure. For our presentation, we choose the measure of investment worth as an after-tax NPW computed at a given interest rate i. In fact, we are free to choose any measure of worth, such as annual worth or future worth. In the second step, we must divide into two groups all the variables that we listed in Step 1 as affecting the NPW. One group consists of all the parameters for which values are known. The second group includes all remaining parameters for which we do not know exact values at the time of analysis. The third step is to define the relationships that tie together all the variables. These relationships may take the form of a single equation or several equations.

EXAMPLE 15.10 Developing a Simulation Model

Consider again WMC's transmission-housings project, as set forth in Example 15.7. Identify the input factors related to the project and develop the simulation model for the NPW distribution.

DISCUSSION: For the WMC project, the variables that affect the NPW are the investment required, unit price, demand, variable production cost, fixed production cost, tax rate, and CCA, as well as the firm's interest rate. Some of the parameters that might be included in the known group are the investment cost and interest rate (MARR). If we have already purchased the equipment or have received a price quote, then we also know the CCA amount. In addition, assuming that we are operating in a stable economy, we would probably know the tax rates for computing income taxes due.

The group of parameters with unknown values would usually include all the variables relating to costs and future operating expense and to future demand and sales prices. These are the random variables for which we must assess the probability distributions.

For simplicity, we classify the input parameters or variables for WMC's transmission-housings project as follows:

Assumed to Be Known Parameters	Assumed to Be Unknown Parameters
MARR	Unit price
Tax rate	Demand
CCA amount	Salvage value
Investment amount	
Project life	
Fixed production cost	
Variable production cost	

Note that, unlike the situation in Example 15.7, here we treat the salvage value as a random variable. With these assumptions, we are now ready to build the NPW equation for the WMC project.

SOLUTION

Recall that the basic investment parameters assumed for WMC's five-year project in Example 15.7 were as follows:

- Investment = $125,000
- Marginal tax rate = 0.40
- Annual fixed cost = $10,000
- Variable unit production cost = $15/unit
- MARR(i) = 15%

- Annual CCA amounts:

n	CCA_n
1	$18,750
2	$31,875
3	$22,313
4	$15,619
5	$10,933

The after-tax annual revenue is expressed in terms of functions of product demand (X) and unit price (Y):

$$R_n = XY(1 - t) = 0.6XY.$$

The after-tax annual expenses excluding CCA are also expressed as a function of product demand (X):

$$E_n = (\text{Fixed cost} + \text{variable cost})(1 - t)$$
$$= (\$10,000 + 15X)(0.60)$$
$$= \$6000 + 9X.$$

Then the net after-tax cash revenue is

$$V_n = R_n - E_n$$
$$= 0.6XY - 9X - \$6000.$$

The present worth of the net after-tax cash inflow from revenue is

$$\sum_{n=1}^{5} V_n(P/F, 15\%, n) = [0.6X(Y - 15) - \$6000](P/A, 15\%, 5)$$
$$= 0.6X(Y - 15)(3.3522) - \$20,113.$$

The present worth of the total CCA is

$$\sum_{n=1}^{5} D_n t_m(P/F, i, n) = 0.40[\$18,750(P/F, 15\%, 1) + \$31,875(P/F, 15\%, 2)$$
$$+ \$22,313(P/F, 15\%, 3) + \$15,619(P/F, 15\%, 4)$$
$$+ \$10,933(P/F, 15\%, 5)]$$
$$= \$27,777$$

Since the total CCA amount is $99,490, the UCC at the end of year 5 is $25,510 ($125,000 − $99,490). Any salvage value greater than this UCC is treated as a taxable gain, and this gain is taxed at t. In our example, the salvage value is considered to be a random variable. Thus, the amount of taxable gains (losses) also becomes a random variable. Therefore, the net salvage value after tax adjustment is

$$S - (S - \$25,510)t = S(1 - t) + 25,510t$$
$$= 0.6S + \$10,204.$$

Then the equivalent present worth of this amount is

$$(0.6S + \$10{,}204)(P/F, 15\%, 5) = (0.6S + \$10{,}204)(0.4972).$$

Now the NPW equation can be summarized as

$$
\begin{aligned}
\text{PW}(15\%) &= -\$125{,}000 + 0.6X(Y - 15)(3.3522) - \$20{,}113 + \$27{,}777 \\
&\quad + (0.6S + \$10{,}204)(0.4972) \\
&= -\$112{,}263 + 2.0113X(Y - 15) + 0.2983S.
\end{aligned}
$$

Note that the NPW function is now expressed in terms of the three random variables X, Y, and S.

15.5.3 Monte Carlo Sampling

With some variables, we may base the probability distribution on objective evidence gleaned from the past if the decision maker feels that the same trend will continue to operate in the future. If not, we may use subjective probabilities as discussed in Section 15.3.1. Once we specify a distribution for a random variable, we need to determine how to generate samples from that distribution. **Monte Carlo sampling** is a simulation method in which a random sample of outcomes is generated for a specified probability distribution. In this section, we shall discuss the Monte Carlo sampling procedure for an *independent* random variable.

Monte Carlo sampling is a simulation method using a class of computational algorithms to simulate the behaviour of various physical and mathematical systems.

Random Numbers

The sampling process is the key part of the analysis. It must be done such that the sequence of values sampled will be distributed in the same way as the original distribution. To accomplish this objective, we need a source of independent, identically distributed uniform random numbers between 0 and 1. Toward that end, we can use a table of random numbers, but most digital computers have programs available to generate "equally likely (uniform)" random decimals between 0 and 1. We will use $U(0,1)$ to denote such a statistically reliable uniform random-number generator, and we will use $U_1, U_2, U_3, \dots$ to represent uniform random numbers generated by this routine. (In Microsoft Excel, the RAND function can be used to generate such a sequence of random numbers.)

Sampling Procedure

For any given random numbers, the question is, How are they used to sample a distribution in a simulation analysis? The first task is to convert the distribution into its corresponding cumulative frequency distribution. Then the random number generated is set equal to its numerically equivalent percentile and is used as the entry point on the $F(x)$-axis of the cumulative-frequency graph. The sampled value of the random variable is the x value corresponding to this cumulative percentile entry point.

This method of generating random values works because choosing a random decimal between 0 and 1 is equivalent to choosing a random percentile of the distribution. Then the random value is used to convert the random percentile to a particular value. The method is general and can be used for any cumulative probability distribution, either continuous or discrete.

EXAMPLE 15.11 Monte Carlo Sampling

In Example 15.10, we developed an NPW equation for WMC's transmission-housings project as a function of three random variables—demand (X), unit price (Y), and salvage value (S):

$$PW(15\%) = -\$112,263 + 2.0113X(Y - 15) + 0.2983S.$$

- For random variable X, we will assume the same discrete distribution as defined in Table 15.5.
- For random variable Y, we will assume a triangular distribution with $L = \$48$, $H = \$53$, and $M_O = \$50$.
- For random variable S, we will assume a uniform distribution with $L = \$30,000$ and $H = \$50,000$.

With the random variables (X, Y, and S) distributed as just set forth, and assuming that these random variables are *mutually independent* of each other, we need three uniform random numbers to sample one realization from each random variable. We determine the NPW distribution on the basis of 200 iterations.

DISCUSSION: As outlined in Figure 15.12, a simulation analysis consists of a series of repetitive computations of the NPW. To perform the sequence of repeated simulation trials, we generate a sample observation for each random variable in the model and

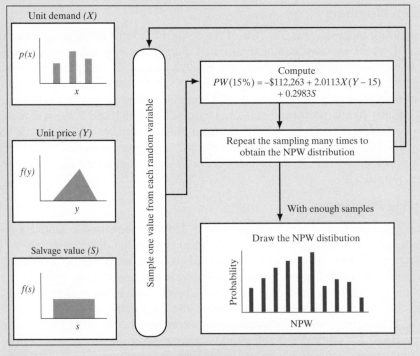

Figure 15.12 A logical sequence of a Monte Carlo simulation to obtain the NPW distribution for WMC's transmission-housings project.

substitute all the values into the NPW equation. Each trial requires that we use a different random number in the sequence to sample each distribution. Thus, if three random variables affect the NPW, we need three random numbers for each trial. After each trial, the computed NPW is stored in the computer. Each value of the NPW computed in this manner represents one state of nature. The trials are continued until a sufficient number of NPW values is available to define the NPW distribution.

SOLUTION

Suppose the following three uniform random numbers are generated for the first iteration: $U_1 = 0.12135$ for X, $U_2 = 0.82592$ for Y, and $U_3 = 0.86886$ for S.

- **Demand (X).** The cumulative distribution for X is already given in Example 15.5. To generate one sample (observation) from this discrete distribution, we first find the cumulative probability function, as depicted in Figure 15.13(a).

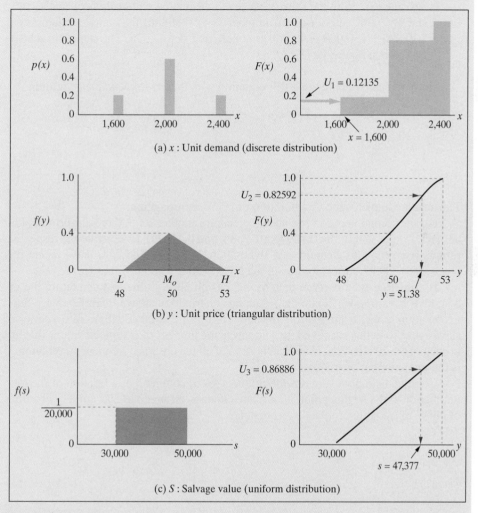

(a) x : Unit demand (discrete distribution)

(b) y : Unit price (triangular distribution)

(c) S : Salvage value (uniform distribution)

Figure 15.13   Illustration of a sampling scheme for discrete and continuous random variables.

In a given trial, suppose the computer gives the random number 0.12135. We then enter the vertical axis at the 12.135th percentile (the percentile numerically equivalent to the random number), read across to the cumulative curve, and then read down to the x-axis to find the corresponding value of the random variable X; this value is 1600, and it is the value of x that we use in the NPW equation. In the next trial, we sample another value of x by obtaining another random number, entering the ordinate at the numerically equivalent percentile and reading the corresponding value of x from the x-axis.

- **Price (Y).** Assuming that the unit-price random variable can be estimated by the three parameters, its probability distribution is shown in Figure 15.13(b). Note that Y takes a continuous value (unlike the assumption of discreteness in Table 15.5). The sampling procedure is again similar to that in the discrete situation. Using the random number $U = 0.82592$, we can approximate $y = \$51.38$ by performing a linear interpolation.

- **Salvage (S).** With the salvage value (S) distributed uniformly between \$30,000 and \$50,000, and a random number $U = 0.86886$, the sample value is $s = \$30,000 + (50,000 - 30,000)0.86886$, or $s = \$47,377$. The sampling scheme is shown in Figure 15.13(c).

Now we can compute the NPW equation with these sample values, yielding

$$PW(15\%) = -\$112,263 + 2.0113(1,600)(\$51.3841 - \$15)$$

$$+ 0.2983(\$47,377)$$

$$= \$18,957.$$

This result completes the first iteration of NPW_1 computation.

For the second iteration, we need to generate another set of three uniform random numbers (suppose they are 0.72976, 0.79885, and 0.41879), to generate the respective sample from each distribution, and to compute $NPW_2 = \$45,044$. If we repeat this process for 200 iterations, we obtain the NPW values listed in Table 15.12.

By ordering the observed data by increasing NPW value and tabulating the ordered NPW values, we obtain the frequency distribution shown in Table 15.13. Such a tabulation results from dividing the entire range of computed NPWs into a series of subranges (20 in this case) and then counting the number of computed values that fall in each of the 20 intervals. Note that the sum of all the observed frequencies (column 3) is the total number of trials.

Column 4 simply expresses the frequencies of column 3 as a fraction of the total number of trials. At this point, all we have done is arrange the 200 numerical values of NPW into a table of relative frequencies.

TABLE 15.12 Observed NPW Values ($) for WMC's Simulation Project (Example 15.11)

No.	U_1	U_2	U_3	X	Y	S	NPW
1	0.12135	0.82592	0.86886	1600	$51.38	$47,377	$18,957
2	0.72976	0.79885	0.41879	2000	$51.26	$38,376	$45,044
3	0.23145	0.58484	0.78720	2000	$50.50	$45,744	$44,202
4	0.67520	0.17786	0.71426	2000	$49.33	$44,285	$39,057
⋮	⋮	⋮	⋮	⋮	⋮	⋮	⋮
200	0.95953	0.70178	0.84848	2400	$50.88	$46,969	$74,969

$18,957	$45,044	$44,202	$39,057	$71,736
36,369	17,840	67,478	40,753	42,239
42,459	35,708	39,624	12,535	9,968
35,024	65,755	42,994	68,907	44,485
60,623	15,053	69,225	36,067	61,402
11,589	32,746	68,015	42,129	47,656
41,946	67,002	34,236	77,236	43,142
39,484	42,678	34,848	47,353	71,315
60,681	71,514	38,540	44,435	8,890
18,071	44,182	65,181	46,359	12,030
73,988	47,975	9,329	41,763	20,554
36,129	42,403	11,597	65,551	40,034
39,728	40,378	8,086	76,318	67,673
51,819	63,063	52,028	14,677	14,616
10,843	37,815	39,093	47,737	34,661
77,240	40,536	43,540	41,768	43,886
41,481	12,312	40,272	43,634	60,644
43,863	39,637	9,035	43,568	68,213
68,937	18,021	12,831	63,691	37,972
41,042	16,140	45,007	43,560	46,653
62,297	50,912	40,781	49,947	42,839
13,007	34,089	39,758	65,079	41,965
69,448	12,223	43,275	17,208	34,167
12,493	42,280	76,754	30,754	75,072
42,833	78,019	72,864	32,782	45,630
12,217	36,548	75,412	37,354	46,015
39,434	45,845	8,127	44,344	38,522
72,550	45,618	39,300	41,816	44,292
45,735	70,083	43,591	66,513	11,656
40,944	42,755	39,983	35,732	14,628
32,043	10,031	46,821	47,448	36,499
43,683	73,704	41,446	40,931	11,638
38,667	65,243	36,958	69,238	38,658
41,642	42,189	38,494	72,450	73,042
48,075	43,216	34,298	14,778	34,031
36,012	67,325	74,614	12,795	17,090
45,000	68,793	39,600	67,268	42,771
18,049	41,051	36,344	46,269	42,956
37,718	71,594	10,198	51,866	34,321
11,207	39,140	36,741	16,672	74,969

TABLE 15.13 Simulated NPW Frequency Distribution for WMC's Transmission-Housings Project (Example 15.11)

Cell No.	Cell Interval	Observed Frequency	Relative Frequency	Cumulative Frequency
1	$8,086 ≤ NPW ≤ $11,583	10	0.05	0.05
2	11,583 < NPW ≤ $15,079	18	0.09	0.14
3	15,079 < NPW ≤ $18,576	8	0.04	0.18
4	18,576 < NPW ≤ $22,073	2	0.01	0.19
5	22,073 < NPW ≤ $25,569	0	0.00	0.19
6	25,569 < NPW ≤ $29,066	0	0.00	0.19
7	29,066 < NPW ≤ $32,563	2	0.01	0.20
8	32,563 < NPW ≤ $36,060	14	0.07	0.27
9	36,060 < NPW ≤ $39,556	23	0.12	0.39
10	39,556 < NPW ≤ $43,053	37	0.19	0.57
11	43,053 < NPW ≤ $46,550	27	0.14	0.71
12	46,550 < NPW ≤ $50,046	9	0.05	0.75
13	50,046 < NPW ≤ $53,543	4	0.02	0.77
14	53,543 < NPW ≤ $57,040	0	0.00	0.77
15	57,040 < NPW ≤ $60,536	0	0.00	0.77
16	60,536 < NPW ≤ $64,033	7	0.04	0.81
17	64,033 < NPW ≤ $67,530	10	0.05	0.86
18	67,530 < NPW ≤ $71,027	10	0.05	0.91
19	71,027 < NPW ≤ $74,523	10	0.05	0.96
20	74,523 < NPW ≤ $78,019	9	0.05	1.00

Cell width = $3,497; mean = $42,436; standard deviation = $18,705; minimum NPW value = $8,086; maximum NPW value = $78,019

15.5.4 Simulation Output Analysis

After a sufficient number of repetitive simulation trials has been run, the analysis is essentially completed. The only remaining tasks are to tabulate the computed NPW values in order to determine the expected value and to make various graphic displays that will be useful to management.

Interpretation of Simulation Results

Once we obtain an NPW frequency distribution (such as that shown in Table 15.13), we need to make the assumption that the actual relative frequencies of column 4 in the table are representative of the probability of having an NPW in each range. That is, we assume that the relative frequencies we observed in the sampling are representative of the proportions we would have obtained had we examined all the possible combinations.

This sampling is analogous to polling the opinions of voters about a candidate for public office. We could speak to every registered voter if we had the time and resources, but a simpler procedure would be to interview a smaller group of persons selected with an unbiased sampling procedure. If 60% of this scientifically selected sample supports the candidate, it probably would be safe to assume that 60% of all registered voters support the candidate. Conceptually, we do the same thing with simulation. As long as we ensure that a sufficient number of representative trials has been made, we can rely on the simulation results.

Once we have obtained the probability distribution of the NPW, we face the crucial question, How do we use this distribution in decision making? Recall that the probability distribution provides information regarding the probability that a random variable will attain some value x. We can use this information, in turn, to define the cumulative distribution, which expresses the probability that the random variable will attain a value smaller than or equal to some x [i.e., $F(x) = P(X \le x)$]. Thus, if the NPW distribution is known, we can also compute the probability that the NPW of a project will be negative. We use this probabilistic information in judging the profitability of the project.

With the assurance that 200 trials was a sufficient number for the WMC project, we may interpret the relative frequencies in column 4 of Table 15.13 as probabilities. The NPW values range between $8086 and $78,019, thereby indicating no loss for any situation. The NPW distribution has an expected value of $42,436 and a standard deviation of $18,705.

Creation of Graphic Displays

Using the output data (the relative frequency) in Table 15.13, we can create the distribution in Figure 15.14(a). A picture such as this can give the decision maker a feel for the

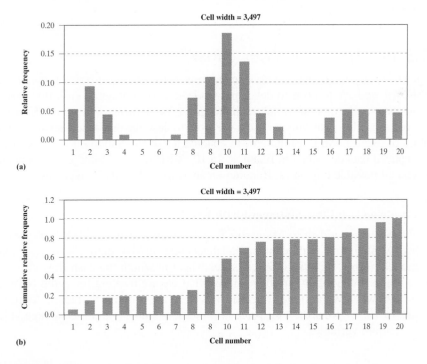

Figure 15.14 Simulation result for WMC's transmission-housings project based on 200 iterations.

ranges of possible NPWs, for the relative likelihoods of loss versus gain, for the range of NPWs that are most probable, and so on.

Another useful display is the conversion of the NPW distribution to the equivalent cumulative frequency, as shown in Figure 15.14(b). Usually, a decision maker is concerned with the likelihood of attaining at least a given level of NPW. Therefore, we construct the cumulative distribution by accumulating the areas under the distribution as the NPW decreases. The decision maker can use Figure 15.14(b) to answer many questions—for example, what is the likelihood of making at least a 15% return on investment (i.e., the likelihood that the NPW will be at least 0)? In our example, this probability is virtually 100%.

Dependent Random Variables

All our simulation examples have involved independent random variables. We must recognize that some of the random variables affecting the NPW may be related to one another. If they are, we need to sample from distributions of the random variables in a manner that accounts for any such dependencies. For example, in the WMC project, both the demand and the unit price are not known with certainty. Each of these parameters would be on our list of variables for which we need to describe distributions, but they could be related inversely. When we describe distributions for these two parameters, we have to account for that dependency. This issue can be critical, as the results obtained from a simulation analysis can be misleading if the analysis does not account for dependencies. The sampling techniques for these dependent random variables are beyond the scope of this text, but can be found in many textbooks on simulation.

15.5.5 Risk Simulation With @RISK[8]

One practical way to conduct a risk simulation is to use an Excel-based program such as @RISK. All we have to do is to develop a worksheet for a project's cash flow statement. Then we identify the random variables in the cash flow elements. With @RISK, we have a variety of probability distributions to choose from to describe our beliefs about the random variables of interest. We illustrate how to conduct a risk simulation on @RISK, using the same financial data described in Example 15.10. @RISK uses Monte Carlo simulation to show you all possible outcomes. Running an analysis with @RISK involves three simple steps.

1. **Create a Cash Flow Statement with Excel**

   @RISK is an add-in to Microsoft Excel. As an add-in, @RISK becomes seamlessly integrated with your spreadsheet, adding risk analysis to your existing models. Table 15.14 is an Excel spreadsheet in which the cash flow entries are a function of the input variables. Here, what we are looking for is the NPW of the project when we assigned specific values to the random variables. As in Example 15.10, we treat the unit price (Y), the demand (X), and the salvage value (S) as random variables.

---

[8] A risk simulation software developed by Palisade Corporation, Ithaca, NY 14850.

TABLE 15.14 An Excel Worksheet of the Transmission-Housing Project Prepared for @RISK

	A	B	C	D	E	F	G	H
1		@Risk Simulation for Example 15.10						
2								
3								
4		Input Data (Base):				Type of Distributions for the random variables:		
5								
6		**Unit Price ($)**	**50.33**			Category	Distribution	
7		**Demand**	**2000**					
8		Var. cost ($/unit)	15			Unit price (Y)	Triangular(48,50,53)	
9		Fixed cost ($)	10000			Demand (X)	Discrete (1600,2000,2400)	
10		**Salvage ($)**	**40000**			Salvage (S)	Uniform(30000,50000)	
11		Tax rate (%)	40%					
12		MARR (%)	15%					
13						Output (NPW)	$41,801	
14								
15			0	1	2	3	4	5
16		**Income Statement**						
17								
18		**Revenues**						
19		Unit price		$ 50.33	$ 50.33	$ 50.33	$ 50.33	$ 50.33
20		Demand (units)		2,000	2,000	2,000	2,000	2,000
21		Sales revenue		$100,667	$100,667	$100,667	$100,667	$100,667
22		**Expenses**						
23		Unit variable cost		$ 15	$ 15	$ 15	$ 15	$ 15
24		Variable cost		$ 30,000	$ 30,000	$ 30,000	$ 30,000	$ 30,000
25		Fixed cost		$ 10,000	$ 10,000	$ 10,000	$ 10,000	$ 10,000
26		CCA		$ 18,750	$ 31,875	$ 22,313	$ 15,619	$ 10,933
27								
28		Taxable income		$ 41,917	$ 28,792	$ 38,354	$ 45,048	$ 49,733
29		Income taxes (40%)		$ 16,767	$ 11,517	$ 15,342	$ 18,019	$ 19,893
30								
31		Net income		$ 25,150	$ 17,275	$ 23,012	$ 27,029	$ 29,840
32								
33		**Cash flow statement**						
34								
35		**Operating activities:**						
36		Net income		$ 25,150	$ 17,275	$ 23,012	$ 27,029	$ 29,840
37		CCA		$ 18,750	$ 31,875	$ 22,313	$ 15,619	$ 10,933
38		**Investment activities**						
39		Investment	($125,000)					
40		Salvage						$ 40,000
41		Disposal tax effect						($ 5,796)
42								
43		**Net Cash Flow**	($125,000)	$ 43,900	$ 49,150	$ 45,325	$ 42,647	$ 74,977

2. Define Uncertainty

We start by replacing uncertain values in our spreadsheet model with @RISK probability distribution functions. These @RISK functions simply represent a range of different possible values that cell could take instead of limiting it to just one value.

Random Variable	Probability Distribution		Probability Functions Provided by @RISK	Cell
Unit price	Triangular (48, 50, 53)		**=RiskTriang (48,50,53)**	C6
Demand	Event	Probability	**=RiskDiscrete ({1600,2000,2400}, {0.2,0.6,0.2})**	C7
	1,600	0.2		
	2,000	0.6		
	2,400	0.2		
Salvage	Uniform (30,000, 50,000)		**=RiskUniform (30000,50000)**	C10

As shown in Figure 15.15, the probability functions are found from the "Define Distribution" menu in the tool bar.

3. Pick Your Bottom Line

We select our output cells, the "bottom line" cells whose values we are interested in. In our case, this is the net present value for the transmission-housing project, located at cell G13. To designate cell G13 as the output cell, we move the cursor to cell G13 and select the "Add Output" command in the tool bar. Then the cell formula in G13 will look like **RiskOutput ("Output (NPW)")** + NPV(C12, D43:H43) + C43.

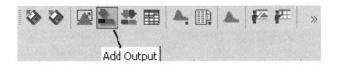

4. Simulate

Now we are ready to simulate. Click the "Start Simulation" button and watch. @RISK recalculates our spreadsheet model hundreds or thousands of times. We can specify the number of iterations through the "Simulation Settings" command. There is no limit to the number of different scenarios we can look at in our simulations. Each time, @RISK samples random values from the @RISK functions we entered and records the resulting outcome. The result: a look at a whole range of possible outcomes, including the probabilities that they will occur! Almost instantly, you will see what critical situations to seek out—or avoid! With @RISK, we can answer questions like, "What is the probability of NPW exceeding $50,000?" or "What are the chances of losing money on this investment?"

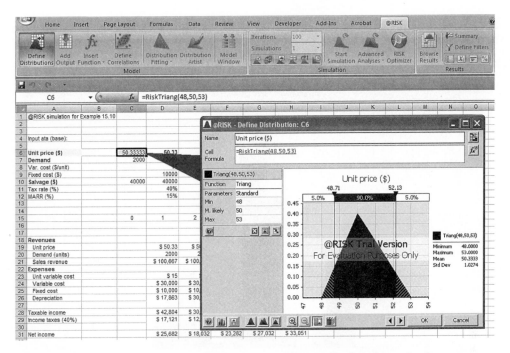

Figure 15.15 @RISK comes with RISKview, a built-in distribution viewer that lets you preview various distributions before selecting them. You can choose distributions from a gallery of thumbnail distribution pictures, and then watch as @RISK builds a graph of the distribution for you while you enter your parameters.

5. **Analyzing the Simulation Result Screen**

@RISK provides a wide range of graphing options for interpreting and presenting the simulation results. @RISK also gives us a full statistical report on our simulations, as well as access to all the data generated. Quick Reports include cumulative graphs, Tornado charts for sensitivity analysis, histograms, and summary statistics. As shown in Figure 15.16, based on 200 iterations, we find that the NPW ranges between $5948 and $81,512, with the mean value of $41,480 and the standard deviation of $18,414. These results are comparable to those obtained in Example 15.10.

15.6 Decision Trees and Sequential Investment Decisions

Most investment problems that we have discussed so far involved only a single decision at the time of investment (i.e., an investment is accepted or rejected), or a single decision that entails a different decision option, such as "make a product in-house" or "farm out," and so on. Once the decision is made, there are no later contingencies or decision options to follow up on. However, certain classes of investment decisions cannot be analyzed in a single step. As an oil driller, for example, you have an opportunity to bid on an offshore lease. Your first decision is whether to bid. But if you decide to bid and eventually win the

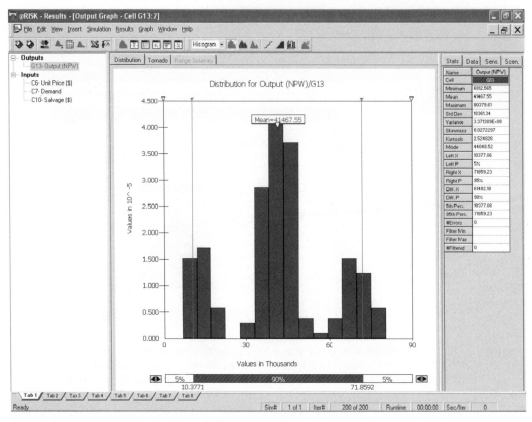

Figure 15.16 The NPW distribution created by @RISK based on 200 iterations. @RISK lets you use the output data to accurately describe what the NPW distribution will look like using the BestFit command, @RISK's built-in distribution fitting tool.

contract, you will have another decision regarding whether to drill immediately or run more seismic (drilling) tests. If you drill a test well that turns out to be dry, you have another decision whether to drill another well or drop the entire drilling project. In a sequential decision problem such as this, in which the actions taken at one stage depend on actions taken in earlier stages, the evaluation of investment alternatives can be quite complex. Certainly, all these future options must be considered when one is evaluating the feasibility of bidding at the initial stage. In this section, we describe a general approach for dealing with more complex decisions that is useful both for structuring the decision problems and for finding a solution to them. The approach utilizes a decision tree—a graphic tool for describing the actions available to the decision maker, the events that can occur, and the relationship between the actions and events.

15.6.1 Structuring a Decision-Tree Diagram

To illustrate the basic decision-tree concept, let's consider an investor, named Bill Henderson, who wants to invest some money in the financial market. He wants to choose between a highly speculative growth stock (d_1) and a relatively safe corporate bond (d_2).

Figure 15.17 illustrates this situation, with the decision point represented by a square box (□) or decision node. The alternatives are represented as branches emanating from the node. Suppose Bill were to select some particular alternative, say, "invest in the stock." There are three chance events that can happen, each representing a potential return on investment. These events are shown in Figure 15.17 as branches emanating from a circular node (○). Note that the branches represent uncertain events over which Bill has no control. However, Bill can assign a probability to each chance event and enter it beside each branch in the decision tree. At the end of each branch is the conditional monetary transaction associated with the selected action and given event.

Relevant Net After-Tax Cash Flow

Once the structure of the decision tree is determined, the next step is to find the relevant cash flow (monetary value) associated with each of the alternatives and the possible chance outcomes. As we have emphasized throughout this book, the decision has to be made on an after-tax basis. Therefore, the relevant monetary value should be on that basis. Since the costs and revenues occur at different points in time over the study period, we also need to convert the various amounts on the tree's branches to their equivalent lump-sum amounts,

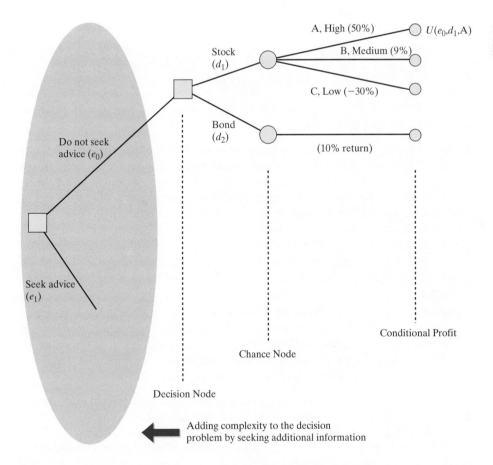

Figure 15.17   Structure of a typical decision tree (Bill's investment problem).

say, net present value. The conditional net present value thus represents the profit associated with the decisions and events along the path from the first part of the tree to the end.

Rollback Procedure

To analyze a decision tree, we begin at the end of the tree and work backwards—the **rollback** procedure. In other words, starting at the tips of the decision tree's branches and working toward the initial node, we use the following two rules:

1. For each chance node, we calculate the expected monetary value (EMV). This is done by multiplying probabilities by conditional profits associated with branches emanating from that node and summing these conditional profits. We then place the EMV in the node to indicate that it is the expected value calculated over all branches emanating from that node.

2. For each decision node, we select the one with the highest EMV (or minimum cost). Then those decision alternatives which are not selected are eliminated from further consideration. On the decision-tree diagram, we draw a mark across the nonoptimal decision branches, indicating that they are not to be followed.

Example 15.12 illustrates how Bill could transform his investment problem into a tree format using numerical values.

EXAMPLE 15.12 Bill's Investment Problem in a Decision-Tree Format

Suppose Bill has $50,000 to invest in the financial market for one year. His choices have been narrowed to two options:

- **Option 1.** Buy 1000 shares of a technology stock at $50 per share that will be held for one year. Since this is a new initial public offering (IPO), there is not much research information available on the stock; hence, there will be a brokerage fee of $100 for this size of transaction (for either buying or selling stocks). For simplicity, assume that the stock is expected to provide a return at any one of three different levels: a high level (A) with a 50% return ($25,000), a medium level (B) with a 9% return ($4500), or a low level (C) with a 30% loss ($-$15,000). Assume also that the probabilities of these occurrences are assessed at 0.25, 0.40, and 0.35, respectively. No stock dividend is anticipated for such a growth-oriented company.

- **Option 2.** Purchase $50,000 worth of bonds issued by Treasure Inc., a large stable financial company. The $1000 bonds cost $909.09 each, and mature in one year. Therefore, Bill can use his $50,000 to buy 55 bonds. These bonds do not pay interest, as they are coupon-free, so their appreciation gets taxed as a capital gain. There is a $150 transaction fee for either buying or selling the bonds.

Bill's dilemma is which alternative to choose to maximize his financial gain. At this point, Bill is not concerned about seeking some professional advice on the stock before making a decision. We will assume that any capital gains will be taxed at 20%. Bill's minimum attractive rate of return is known to be 5% after taxes. Determine the payoff amount at the tip of each branch.

SOLUTION

Figure 15.18(a) shows the before-tax costs and revenues on the branches.

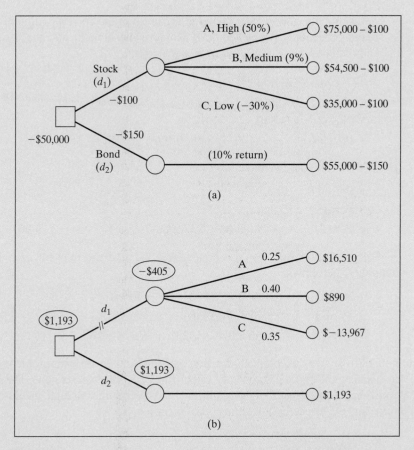

Figure 15.18 Decision tree for Bill's investment problem: (a) Relevant cash flows (before tax) and (b) net present worth for each decision path.

- **Option 1**
  1. With a 50% return (a real winner) over a one-year holding period,
     - The net cash flow associated with this event at Period 0 will include the amount of investment and the brokerage fee. Thus, we have

       Period 0: $(-\$50,000 - \$100) = -\$50,100.$

     - When you sell stock, you need to pay another brokerage fee. However, any investment expenses, such as brokerage commissions, must be included in determining the cost basis for investment. The taxable capital gains will be calculated as $(\$75,000 - \$50,000 - \$100 - \$100) = \$24,800$. Therefore, the net cash flow at Period 1 would be

       Period 1: $(+\$75,000 - \$100) - 0.20(\$24,800) = \$69,940.$

- Then the conditional net present value of this stock transaction is

$$PW(5\%) = -\$50,100 + \$69,940(P/F, 5\%, 1) = \$16,510.$$

This amount of $16,510 is entered at the tip of the corresponding branch. This procedure is repeated for each possible branch, and the resulting amounts are shown in Figure 15.18(b).

2. With a 9% return, we have

- Period 0: −$50,100
- Period 1: $53,540
- $PW(5\%) = -50,100 + 53,540(P/F, 5\%, 1) = \$890$

3. With a 30% loss, we have

- Period 0: −$50,100
- Period 1: $37,940
- $PW(5\%) = -\$50,100 + \$37,940(P/F, 5\%, 1) = -\$13,967$

- **Option 2.** Similarly for the bonds, we find that the relevant cash flows would be as follows:

- Period 0: $(-\$50,000 - \$150) = -\$50,150$
- Period 1: $(+\$55,000 - \$150) - 0.2(\$5,000 - 2(\$150)) = \$53,910$
- $PW(5\%) = -\$50,150 + \$53,910(P/F, 5\%, 1) = \$1,193$

Figure 15.18(b) shows the complete decision tree for Bill's investment problem. Now Bill can calculate the expected monetary value (EMV) at each chance node. The EMV of Option 1 represents the sum of the products of the probabilities of high, medium, and low returns and the respective conditional profits (or losses):

$$EMV = \$16,510(0.25) + \$890(0.40) - \$13,967(0.35) = -\$405.$$

For Option 2, the EMV is simply

$$EMV = \$1193.$$

In Figure 15.18, the expected monetary values are shown in the event nodes. Bill must choose which action to take, and this would be the one with the highest EMV, namely, Option 2, with EMV = $1193. This expected value is indicated in the tree by putting $1,193 in the decision node at the beginning of the tree. Note that the decision tree uses the idea of maximizing expected monetary value developed in the previous section. In addition, the mark ‖ is drawn across the nonoptimal decision branch (Option 1), indicating that it is not to be followed. In this simple example, the benefit of using a decision tree may not be evident. However, as the decision problem becomes more complex, the decision tree becomes more useful in organizing the information flow needed to make the decision. This is true in particular if Bill must make a sequence of decisions, rather than a single decision, as we next illustrate.

15.6.2 Worth of Obtaining Additional Information

In this section, we introduce a general method for evaluating whether it is worthwhile to seek more information. Most of the information that we can obtain is imperfect, in the sense that it will not tell us exactly which event will occur. Such imperfect information may have some value if it improves the chances of making a correct decision—that is, if it improves the expected monetary value. The problem is whether the reduced uncertainty is valuable enough to offset its cost. The gain is in the improved efficiency of decisions that may become possible with better information.

We use the term *experiment* in a broad sense in what follows. An experiment may be a market survey to predict sales volume for a typical consumer product, a statistical sampling of the quality of a product, or a seismic test to give a well-drilling firm some indications of the presence of oil.

The Value of Perfect Information

Let us take Bill's prior decision to "Purchase corporate bonds" as a starting point. How do we determine whether further strategic steps would be profitable? We could do more to obtain additional information about the potential return on an investment in stock, but such steps cost money. Thus, we have to balance the monetary value of reducing uncertainty with the cost of obtaining additional information. In general, we can evaluate the worth of a given experiment only if we can estimate the reliability of the resulting information. In our stock investment problem, an expert's opinion may be helpful in deciding whether to abandon the idea of investing in a stock. This can be of value, however, only if Bill can say beforehand how closely the expert can estimate the performance of the stock.

The best place to start a decision improvement process is with a determination of how much we might improve our incremental profit by removing uncertainty. Although we probably couldn't ever really obtain such uncertainty, the value of perfect information is worth computing as an upper bound to the value of additional information.

We can easily calculate the value of perfect information. First we merely note the mathematical difference of the incremental profit from an optimal decision based on perfect information and the original decision to "Purchase corporate bonds" made without foreknowledge of the actual return on the stock. This difference is called **opportunity loss** and must be computed for each potential level of return on the stock. Then we compute the expected opportunity loss by weighting each potential loss by the assigned probability associated with that event. What turns out in this expected opportunity loss is exactly the **expected value of perfect information** (EVPI). In other words, if we had perfect information about the performance of the stock, we should have made a correct decision about each situation, resulting in no regrets (i.e., no opportunity losses). Example 15.13 illustrates the concept of opportunity loss and how it is related to the value of perfect information.

> **EVPI** is the maximum price that one would be willing to pay in order to gain access to perfect information.

EXAMPLE 15.13 Expected Value of Perfect Information

Reluctant to give up a chance to make a larger profit with the stock option, Bill may wonder whether to obtain further information before acting. As with the decision to "Purchase corporate bonds," Bill needs a single figure to represent the expected value of perfect information (EVPI) about investing in the stock. Calculate the EVPI on the basis of the financial data in Example 15.12 and Figure 15.18(b).

TABLE 15.15 Opportunity Loss Cost Associated with Investing in Bonds

		Decision Option		Optimal Choice with Perfect Information	Opportunity Loss Associated with Investing in Bonds
Potential Return	Probability	Option 1: Invest in Stock	Option 2: Invest in Bonds		
High (A)	0.25	$16,510	$1,193	Stock	$15,317
Medium (B)	0.40	890	1,193	Bond	0
Low (C)	0.35	−13,967	1,193	Bond	0
Expected Value		−$405	$1,193		$ 3,829
			↑		↑
			Prior optimal decision		EVPI

SOLUTION

As regards stock, the decision whether to purchase may hinge on the return on the stock during the next year—the only unknown, subject to a probability distribution. The opportunity loss associated with the decision to purchase bonds is shown in Table 15.15.

- For example, the conditional net present value of $16,510 is the net profit if the return is high. Also, recall that, without Bill's having received any new information, the indicated action (the prior optimal decision) was to select Option 2 ("Purchase corporate bonds"), with an expected net present value of $1,193.

- However, if we know that the stock will be a definite winner (yield a high return), then the prior decision to "Purchase corporate bonds" is inferior to "Invest in stock" to the extent of $16,510 − $1193 = $15,317. With either a medium or low return, however, there is no opportunity loss, as a decision strategy with and without perfect information is the same.

Again, an average weighted by the assigned chances is used, but this time weights are applied to the set of opportunity losses (regrets) in Table 15.15. The result is

$$\text{EVPI} = (0.25)(\$15,317) + (0.40)(\$0) + (0.35)(\$0) = \$3829$$

This EVPI, representing the maximum expected amount that could be gained in incremental profit from having perfect information, places an upper limit on the sum Bill would be willing to pay for additional information.

Updating Conditional Profit (or Loss) After Paying a Fee to the Expert

With a relatively large EVPI in Example 15.13, there is a huge potential that Bill can benefit by obtaining some additional information on the stock. Bill can seek an expert's advice by receiving a detailed research report on the stock under consideration. This report, available at a nominal fee, say, $200, will suggest either of two possible outcomes: (1) The report may come up with a "buy" rating on the stock after concluding that business

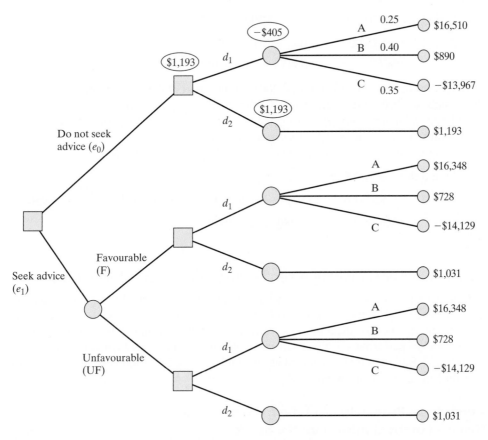

Figure 15.19   Decision tree for Bill's investment problem with an option of getting professional advice.

conditions were favourable, or (2) the report may not recommend any action or may take a neutral stance after concluding that business conditions were unfavourable or that the business model on which the decision is based was not well received in the financial marketplace. Bill's alternatives are either to pay for this service before making any further decision or simply to make his own decision without seeing the report.

If Bill seeks advice before acting, he can base his decision upon the stock report. This problem can be expressed in terms of a decision tree, as shown in Figure 15.19. The upper part of the tree shows the decision process if no advice is sought. This is the same as Figure 15.18, with probabilities of 0.25, 0.40, and 0.35 for high, medium, and low returns, respectively, and an expected loss of $405 for investing in the stock and an expected profit of $1193 from purchasing corporate bonds.

The lower part of the tree, following the branch "Seek advice," displays the results and the subsequent decision possibilities. Using the analyst's research report, Bill can expect two possible events: the expert's recommendation and the actual performance of the stock. After reading either of the two possible report outcomes, Bill must make a decision about whether to invest in the stock. Since Bill has to pay a nominal fee ($200) to get the report, the profit or loss figures at the tips of the branches must also be updated. For the stock investment option, if the actual return proves to be very high (50%), the resulting cash flows would be as follows:

Period 0: $(-\$50,000 - \$100 - \$200) = -\$50,300;$

Period 1: $(+\$75,000 - \$100) - (0.20)(\$25,000 - \$400) = \$69,980;$

$$PW(5\%) = -\$50,300 + \$69,980(P/F, 5\%, 1) = \$16,348.$$

Since the cost of obtaining the report (\$200) will be a part of the total investment expenses, along with the brokerage commissions, the taxable capital gains would be adjusted accordingly. With a medium return, the equivalent net present value would be \$728. With a low return, there would be a net loss in the amount of \$14,129.

Similarly, Bill can compute the net profit for Option 2 as follows:

Period 0: $(-\$50,000 - \$150 - \$200) = -\$50,350;$

Period 1: $(+\$55,000 - \$150) - 0.2(\$5,000 - 2(\$150) - \$200) = \$53,950;$

$$PW(5\%) = -\$50,350 + \$53,950(P/F, 5\%, 1) = \$1,031.$$

15.6.3 Decision Making After Having Imperfect Information

In Figure 15.17, the analyst's recommendation precedes the actual investment decision. This allows Bill to change his prior probabilistic assessments regarding the performance of the stock. In other words, if the analyst's report has any bearing on Bill's reassessments of the stock's performance, then Bill must change his prior probabilistic assessments accordingly. But what types of information should be available and how should the prior probabilistic assessments be updated?

Conditional Probabilities of the Expert's Predictions, Given a Potential Return on the Stock

Suppose that Bill knows an expert (stock analyst) to whom he can provide his research report on the stock described under Option 1. The expert will charge a fee in the amount of \$200 to provide Bill with a report on the stock that is either favourable or unfavourable. This research report is not infallible, but it is known to be pretty reliable. From past experience, Bill estimates that when the stock under consideration is relatively highly profitable (A), there is an 80% chance that the report will be favourable; when the stock return is medium (B), there is a 65% chance that the report will be favourable; and when the stock return is relatively unprofitable (C), there is a 20% chance that the report will be favourable. Bill's estimates are summarized in Table 15.16.

These conditional probabilities can be expressed in a tree format as shown in Figure 15.20. For example, if the stock indeed turned out to be a real winner, the probability

TABLE 15.16 Conditional Probabilities of the Expert's Predictions

What the Report Will Say	Given Level of Stock Performance		
	High (A)	Medium (B)	Low (C)
Favourable (F)	0.80	0.65	0.20
Unfavourable (UF)	0.20	0.35	0.80

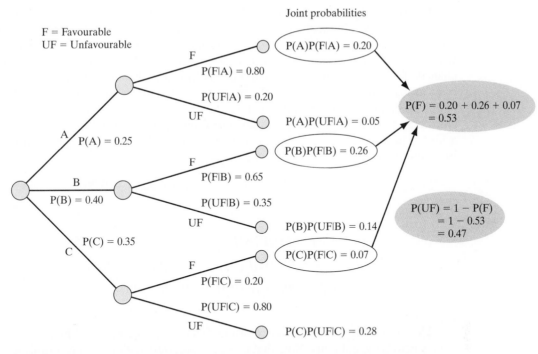

Figure 15.20 Conditional probabilities and joint probabilities expressed in a tree format.

that the report would have predicted the same (with a favourable recommendation) is 0.80, and the probability that the report would have issued an "unfavourable recommendation" is 0.20. Such probabilities would reflect past experience with the analyst's service of this type, modified perhaps by Bill's own judgment. In fact, these probabilities represent the reliability or accuracy of the expert opinion. It is a question of counting the number of times the expert's forecast was "on target" in the past and the number of times the expert "missed the mark," for each of the three actual levels of return on the stock. With this past history of the expert's forecasts, Bill can evaluate the economic worth of the service. Without such a history, no specific value can be attached to seeking the advice.

Joint and Marginal Probabilities

Suppose now that, before seeing the research report, Bill has some critical information on (1) all the probabilities for the three levels of stock performance (prior probabilities) and (2) with what probability the report will say that the return on the stock is favourable or unfavourable, given the three levels of the stock performance in the future. A glance at the sequence of events in Figure 15.19, however, shows that these probabilities are not the ones required to find "expected" values of various strategies in the decision path. What is really needed are the probabilities pertaining to what the report will say about the stock—favourable (a recommendation to buy) or unfavourable (no recommendation or a neutral stance)—and the conditional probabilities of each of the three potential returns after the report is seen. In other words, the probabilities of Table 15.16 are not directly useful in Figure 15.19. Rather, we need the *marginal probabilities* of the "favourable" and "unfavourable" recommendations.

Marginal probability is the probability of one event, regardless of the other event.

To remedy this situation, the probabilities must be put in a different form: We must construct a joint probabilities table. To start with, we have the original (prior) probabilities assessed by Bill before seeing the report: a 0.25 chance that the stock will result in a high return (A), a 0.40 chance of a medium return (B), and a 0.35 chance of a loss (C). From these and the conditional probabilities of Table 15.16, the joint probabilities of Figure 15.20 can be calculated. Thus, the joint probability of both a prediction of a favourable stock market (F) and a high return on the stock (A) is calculated by multiplying the conditional probability of a favourable prediction, given a high return on the stock (which is 0.80 from Table 15.16), by the probability of a high return (A):

$$P(A, F) = P(F|A)P(A) = (0.80)(0.25) = 0.20.$$

Similarly,

$$P(A, UF) = P(UF|A)P(A) = (0.20)(0.25) = 0.05,$$
$$P(B, F) = P(F|B)P(B) = (0.65)(0.40) = 0.26,$$
$$P(B, UF) = P(UF|B)P(B) = (0.35)(0.40) = 0.14,$$

and so on.

The marginal probabilities of return in Table 15.17 are obtained by summing the values across the columns. Note that these are precisely the original probabilities of a high, medium, and low return, and they are designated prior probabilities because they were assessed before any information from the report was obtained.

In understanding Table 15.17, it is useful to think of it as representing the results of 100 past situations identical to the one under consideration. The probabilities then represent the frequency with which the various outcomes occurred. For example, in 25 of the 100 cases, the actual return on the stock turned out to be high; and in these 25 high-return cases, the report predicted a favourable condition in 20 instances [that is $P(A, F) = 0.20$] and predicted an unfavourable one in 5 instances.

TABLE 15.17 Joint as Well as Marginal Probabilities

When Potential Level of Return Is Given	What Report Will Say — Joint Probabilities		Marginal Probabilities of Return
	Favourable (F)	Unfavourable (UF)	
High (A)	0.20	0.05	0.25
Medium (B)	0.26	0.14	0.40
Low (C)	0.07	0.28	0.35
Marginal probabilities	0.53	0.47	1.00

Similarly, the marginal probabilities of prediction in the bottom row of Table 15.17 can be interpreted as the relative frequency with which the report predicted favourable and unfavourable conditions. For example, the survey predicted favourable conditions 53 out of 100 times, 20 of these times when stock returns actually were high, 26 times when returns were medium, and 7 times when returns were low. These marginal probabilities of what the report will say are important to our analysis, for they give us the probabilities associated with the information received by Bill before the decision to invest in the stock is made. The probabilities are entered beside the appropriate branches in Figure 15.19.

Determining Revised Probabilities

We still need to calculate the probabilities for the branches that represent "high return," "medium return," and "low return" in the lower part of Figure 15.19. We cannot use the values 0.25, 0.40, and 0.35 for these events, as we did in the upper part of the tree, because those probabilities were calculated prior to seeking the advice. At this point on the decision tree, Bill will have received information from the analyst, and the probabilities should reflect that information. (If Bill's judgment has not changed even after seeing the report, then there will be no changes in the probabilities.) The required probabilities are the conditional probabilities for the various levels of return on the stock, given the expert's opinion. Thus, using the data from Table 15.17, we can compute the probability $P(A|F)$ of seeing a high return (A), given the expert's buy recommendation (F), directly from the definition of conditional probability:

$$P(A|F) = P(A, F)/P(F) = 0.20/0.53 = 0.38.$$

The probabilities of a medium and a low return, given a buy recommendation, are, respectively,

$$P(B|F) = P(B, F)/P(F) = 0.26/0.53 = 0.49$$

and

$$P(C|F) = P(C, F)/P(F) = 0.07/0.53 = 0.13.$$

These probabilities are called *revised* (or *posterior*) probabilities, since they come after the inclusion of the information received from the report. To understand the meaning of the preceding calculations, think again of Table 15.17 as representing 100 past identical situations. Then, in 53 cases [since $P(F) = 0.53$], 20 actually resulted in a high return. Hence, the posterior probability of a high return is 20/53 = 0.38, as calculated.

The posterior probabilities of other outcomes after seeing an unfavourable recommendation can be calculated similarly:

$$P(A|UF) = P(A, UF)/P(UF) = 0.05/0.47 = 0.11,$$
$$P(B|UF) = P(B, UF)/P(UF) = 0.14/0.47 = 0.30,$$
$$P(C|UF) = P(C, UF)/P(UF) = 0.28/0.47 = 0.59.$$

These values are also shown in Figure 15.21 at the appropriate points in the decision tree.

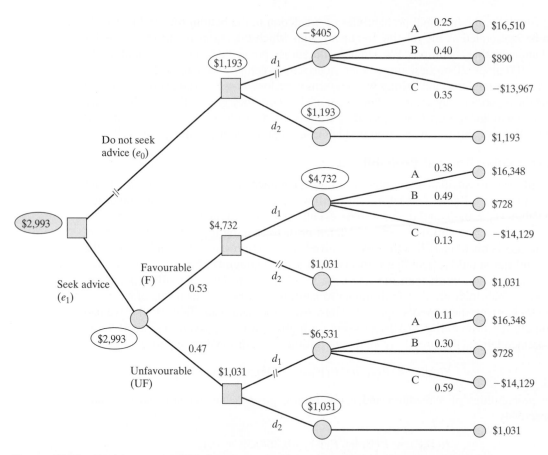

Figure 15.21 Decision tree for Bill's investment problem with an option of having professional advice.

EXAMPLE 15.14 Decision Making After Getting Some Professional Advice

On the basis of Figure 15.21, determine how much the research report is worth to Bill. Should Bill seek an expert's advice (research report)?

SOLUTION

All the necessary information is now available, and Figure 15.21 can be analyzed, starting from the right and working backward. The expected values are shown in the circles. For example, follow the branches "Seek Advice (e_1)," "Favourable (Buy recommendation)," and "Invest in the Stock (d_1)." The expected value of \$4732 shown in the circle at the end of these branches is calculated as

$$0.38(\$16,348) + 0.49(\$728) + 0.13(-\$14,129) = \$4732.$$

Since Bill can expect a profit of \$4732 from investing in the stock after receiving a buy recommendation, the investment in bonds (\$1031) becomes a less profitable alternative. Thus, it can be marked out ($\neq$) of the "Purchase corporate bonds" branch to indicate that it is not optimal.

There is an expected loss of $6531 if the report gives an unfavourable recommendation, but Bill goes ahead with buying the stock. In this situation, the "corporate bonds" option becomes a better choice, and the "Buy stock" branch is marked out. The part of the decision tree relating to seeking the advice is now reduced to two chance events (earning $4732 and earning $1031), and the expected value in the circular node is calculated as

$$0.53(\$4,732) + 0.47(\$1,031) = \$2993.$$

Thus, if the expert's opinion is taken and Bill acts on the basis of the information received, the expected profit is $2993. Since this amount is greater than the $1193 profit that would be obtained in the absence of the expert's advice, we conclude that it is worth spending $200 to receive some additional information from the expert.

In sum, the optimal decision is as follows: (1) Bill should seek professional advice at the outset. (2) If the expert's report indicates a buy recommendation, go ahead and invest in the stock. (3) If the report says otherwise, invest in the corporate bonds.

SUMMARY

- Often, cash flow amounts and other aspects of an investment project analysis are uncertain. Whenever such uncertainty exists, we are faced with the difficulty of **project risk**: the possibility that an investment project will not meet our minimum requirements for acceptability and success.

- Three of the most basic tools for assessing project risk are as follows:
  1. **Sensitivity analysis**—a means of identifying those project variables that, when varied, have the greatest effect on the acceptability of the project.
  2. **Break-even analysis**—a means of identifying the value of a particular project variable that causes the project to exactly break even.
  3. **Scenario analysis**—a means of comparing a "base-case" or expected project measurement (such as the NPW) with one or more additional scenarios, such as the best case and the worst case, to identify the extreme and most likely project outcomes.

- Sensitivity, break-even, and scenario analyses are reasonably simple to apply, but also somewhat simplistic and imprecise in cases where we must deal with multifaceted project uncertainty. **Probability concepts** allow us to further refine the analysis of project risk by assigning numerical values to the likelihood that project variables will have certain values.

- The end goal of a probabilistic analysis of project variables is to produce an NPW distribution from which we can extract such useful information as (1) the **expected NPW value**, (2) the extent to which other NPW values vary from, or are clustered around, the expected value (the **variance**), and (3) the best- and worst-case NPWs.

- Our real task is not to try to find "risk-free" projects; they don't exist in real life. The challenge is to decide what level of risk we are willing to assume and then, having decided on our tolerance for risk, to understand the implications of our choice.

- **Risk simulation**, in general, is the process of modelling reality to observe and weigh the likelihood of the possible outcomes of a risky undertaking.

- **Monte Carlo sampling** is a specific type of randomized sampling method in which a random sample of outcomes is generated for a specified probability distribution. Because Monte Carlo sampling and other simulation techniques often rely on generating a significant number of outcomes, they can be more conveniently performed on the computer than manually.

- The **decision tree** is another technique that can facilitate investment decision making when uncertainty prevails, especially when the problem involves a sequence of decisions. Decision-tree analysis involves the choice of a decision criterion—say, to maximize expected profit. If possible and feasible, an experiment is conducted, and the prior probabilities of the states of nature are revised on the basis of the experimental results. The expected profit associated with each possible decision is computed, and the act with the highest expected profit is chosen as the optimum action.

PROBLEMS

15.1 Ford Construction Company is considering acquiring a new earthmover. The mover's basic price is $70,000, and it will cost another $15,000 to modify it for special use by the company. This earthmover falls into the CCA Class 38 ($d = 30\%$). It will be sold after four years for $30,000. The purchase of the earthmover will have no effect on revenues, but it is expected to save the firm $32,000 per year in before-tax operating costs, mainly labour. The firm's marginal tax rate (federal plus provincial) is 40%, and its MARR is 15%.

(a) Is this project acceptable, based on the most likely estimates given?

(b) Suppose that the project will require an increase in net working capital (spare-parts inventory) of $2000, which will be recovered at the end of year 5. Taking this new requirement into account, would the project still be acceptable?

(c) If the firm's MARR is increased to 20%, what would be the required savings in labour so that the project remains profitable?

15.2 Alberta Metal Forming Company has just invested $500,000 of fixed capital in a manufacturing process that is estimated to generate an after-tax annual cash flow of $200,000 in each of the next five years. At the end of year 5, no further market for the product and no salvage value for the manufacturing process is expected. If a manufacturing problem delays the start-up of the plant for one year (leaving only four years of process life), what additional after-tax cash flow will be needed to maintain the same internal rate of return as would be experienced if no delay occurred?

15.3 A real-estate developer seeks to determine the most economical height for a new office building, which will be sold after five years. The relevant net annual revenues and salvage values are as follows:

| | Height | | | |
	2 Floors	3 Floors	4 Floors	5 Floors
First cost (net after tax)	$500,000	$750,000	$1,250,000	$2,000,000
Lease revenue	199,100	169,200	149,200	378,150
Net resale value (after tax)	600,000	900,000	2,000,000	3,000,000

(a) The developer is uncertain about the interest rate (i) to use, but is certain that it is in the range from 5% to 30%. For each building height, find the range of values of i for which that building height is the most economical.

(b) Suppose that the developer's interest rate is known to be 15%. What would be the cost, in terms of net present value, of a 10% overestimation of the resale value? (In other words, the true value was 10% lower than that of the original estimate.)

15.4 A special-purpose milling machine was purchased four years ago for $40,000. It was estimated at that time that this machine would have a life of 10 years and a salvage value of $1000, with a cost of removal of $1500. These estimates are still good. This machine has annual operating costs of $5000. A new machine that is more efficient will reduce operating costs to $1000, but will require an investment of $25,000. The life of the new machine is estimated to be six years, with a salvage value of $2000. The CCA for both machines is based on a declining balance rate of 30%. An offer of $6000 for the old machine has been made, and the purchaser would pay for removal of the machine. The firm's marginal tax rate is 40%, and its required minimum rate of return is 10%.

(a) What incremental cash flows will occur at the end of years 0 through 6 as a result of replacing the old machine? Should the old machine be replaced now?

(b) Suppose that the annual operating costs for the old milling machine would increase at an annual rate of 9% over the remaining service life of the machine. With this change in future operating costs for the old machine, would the answer in (a) change?

(c) What is the minimum trade-in value for the old machine so that both alternatives are economically equivalent?

15.5 A local telephone company is considering installing a new phone line for a new row of apartment complexes. Two types of cables are being examined: conventional copper wire and fibre optics. Transmission by copper wire cables, although cumbersome, involves much less complicated and less expensive support hardware than does fibre optics. The local company may use five different types of copper wire cables: 100 pairs, 200 pairs, 300 pairs, 600 pairs, and 900 pairs per cable. In calculating the cost per metre of cable, the following equation is used:

Cost per length = [Cost per metre +

cost per pair (number of pairs)](length),

where

22-gauge copper wire = $5.55 per metre

and

Cost per pair = $0.043 per pair.

The annual cost of the cable as a percentage of the initial cost is 18.4%. The life of the system is 30 years.

In fibre optics, a cable is referred to as a ribbon. One ribbon contains 12 fibres, grouped in fours; therefore, one ribbon contains three groups of 4 fibres. Each group can produce 672 lines (equivalent to 672 pairs of wires), and since each ribbon contains three groups, the total capacity of the ribbon is 2016 lines. To transmit signals via fibre optics, many modulators, wave guides, and terminators are needed to convert the signals from electric currents to modulated light waves. Fibre-optic ribbon costs $9321 per kilometre. At each end of the ribbon, three terminators are needed, one for each group of 4 fibres, at a cost of $40,000 per terminator. Twenty-one modulating systems are needed at each end of the ribbon, at a cost of $24,000 for a unit in the central office and $44,000 for a unit in the field. Every 7000 metres, a repeater is required to keep the modulated light waves in the ribbon at an intensity that is intelligible for detection. The unit cost of this repeater is $15,000. The annual cost, including income taxes for the 21 modulating systems, is 12.5% of the initial cost of the units. The annual cost of the ribbon itself is 17.8% initially. The life of the whole system is 30 years. (All figures represent after-tax costs.)

(a) Suppose that the apartments are located 8 kilometres from the phone company's central switching system and that about 2000 telephones will be required. This would require either 2000 pairs of copper wire or one fibre-optic ribbon and related hardware. If the telephone company's interest rate is 15%, which option is more economical?

(b) In (a), suppose that the apartments are located 16 kilometres or 40 kilometres from the phone company's central switching system. Which option is more economically attractive under each scenario?

15.6   A small manufacturing firm is considering purchasing a new boring machine to modernize one of its production lines. Two types of boring machine are available on the market. The lives of machine A and machine B are 8 years and 10 years, respectively. The machines have the following receipts and disbursements:

Item	Machine A	Machine B
First cost	$6000	$8500
Service life	8 years	10 years
Salvage value	$500	$1000
Annual O&M costs	$700	$520
CCA	DB (30%)	DB (30%)

Use a MARR (after tax) of 10% and a marginal tax rate of 30%, and answer the following questions:

(a) Which machine would be most economical to purchase under an infinite planning horizon? Explain any assumption that you need to make about future alternatives.

(b) Determine the break-even annual O&M costs for machine A so that the present worth of machine A is the same as that of machine B.

(c) Suppose that the required service life of the machine is only five years. The salvage values at the end of the required service period are estimated to be $3000 for machine A and $3500 for machine B. Which machine is more economical?

15.7 The management of Langdale Mill is considering replacing a number of old looms in the mill's weave room. The looms to be replaced are two 220-centimetre President looms, 16 135-centimetre President looms, and 22 185-centimetre Draper X-P2 looms. The company may either replace the old looms with new ones of the same kind or buy 21 new shutterless Pignone looms. The first alternative requires purchasing 40 new President and Draper looms and scrapping the old looms. The second alternative involves scrapping the 40 old looms, relocating 12 Picanol looms, and constructing a concrete floor, plus purchasing the 21 Pignone looms and various related equipment.

Description	Alternative 1	Alternative 2
Machinery/related equipment	$2,119,170	$1,071,240
Removal cost of old looms/site preparation	26,866	49,002
Salvage value of old looms	62,000	62,000
Annual sales increase with new looms	7,915,748	7,455,084
Annual labour	261,040	422,080
Annual O&M	1,092,000	1,560,000
CCA	DB (30%)	DB (30%)
Project life	8 years	8 years
Salvage value	169,000	54,000

The undepreciated capital cost of all old looms are negligible. The firm's MARR is 18%, set by corporate executives, who feel that various investment opportunities available for the mills will guarantee a rate of return on investment of at least this much. The mill's marginal tax rate is 40%.

(a) Perform a sensitivity analysis on the project's data, varying the operating revenue, labour cost, annual maintenance cost, and MARR. Assume that each of these variables can deviate from its base-case expected value by ±10%, by ±20%, and by ±30%.

(b) From the results of part (a), prepare sensitivity diagrams and interpret the results.

15.8 Mike Lazenby, an industrial engineer at Energy Conservation Service, has found that the anticipated profitability of a newly developed water-heater temperature control device can be measured by present worth with the formula

$$NPW = 4.028V(2X - \$11) - 77,860,$$

where V is the number of units produced and sold and X is the sales price per unit. Mike also has found that the value of the parameter V could occur anywhere over the range from 1000 to 6000 units and that of the parameter X anywhere between $20 and $45 per unit. Develop a sensitivity graph as a function of the number of units produced and the sales price per unit.

15.9 Susan Campbell is thinking about going into the motel business in Niagara Falls. The cost to build a motel is $2,200,000. The lot costs $600,000. Furniture and furnishings cost $400,000 and have a 20% CCA rate, while the motel building has a 4% CCA rate. The land will appreciate at an annual rate of 5% over the project period, but the building will have a zero salvage value after 25 years. When the motel is full (100% capacity), it takes in (receipts) $4000 per day, 365 days per year. Exclusive of CCA, the motel has fixed operating expenses of $230,000 per year. The variable operating expenses are $170,000 at 100% capacity, and these vary directly with percent capacity down to zero at 0% capacity. If the interest is 10% compounded annually, at what percent capacity over 25 years must the motel operate in order for Susan to break even? (Assume that Susan's tax rate is 31%.)

15.10 A plant engineer wishes to know which of two types of lightbulbs should be used to light a warehouse. The bulbs that are currently used cost $45.90 per bulb and last 14,600 hours before burning out. The new bulb (at $60 per bulb) provides the same amount of light and consumes the same amount of energy, but lasts twice as long. The labour cost to change a bulb is $16.00. The lights are on 19 hours a day, 365 days a year. If the firm's MARR is 15%, what is the maximum price (per bulb) the engineer should be willing to pay to switch to the new bulb? (Assume that the firm's marginal tax rate is 40%.)

15.11 Robert Cooper is considering purchasing a piece of business rental property containing stores and offices at a cost of $250,000. Cooper estimates that annual receipts from rentals will be $35,000 and that annual disbursements, other than income taxes, will be about $12,000. The property is expected to appreciate at the annual rate of 5%. Cooper expects to retain the property for 20 years once it is acquired. The CCA rate for the building is 4%. Cooper's marginal tax rate is 30% and his MARR is 15%. What would be the minimum annual total of rental receipts that would make the investment break even?

15.12 Two different methods of solving a production problem are under consideration. Both methods are expected to be obsolete in six years. Method A would cost $80,000 initially and have annual operating costs of $22,000 a year. Method B would cost $52,000 and costs $17,000 a year to operate. The salvage value realized would be $20,000 with Method A and $15,000 with Method B. Method A would generate $16,000 revenue income a year more than Method B. Investments in both methods have a 30% CCA rate. The firm's marginal income tax rate is 40%. The firm's MARR is 20%. What would be the required additional annual revenue for Method A such that an engineer would be indifferent to choosing one method over the other?

15.13 Rocky Mountain Publishing Company is considering introducing a new morning newspaper in Edmonton. Its direct competitor charges $1.00 at retail, with $0.20 going to the retailer. For the level of news coverage the company desires, it determines the fixed cost of editors, reporters, rent, press-room expenses, and wire service charges to be $1,200,000 per month. The variable cost of ink and paper is $0.40

per copy, but advertising revenues of $0.20 per paper will be generated. To print the morning paper, the publisher has to purchase a new printing press, which will cost $1,200,000. The press machine, with a 30% CCA rate, will be used for 10 years, at which time its salvage value would be about $200,000. Assume 20 weekdays in a month, a 40% tax rate, and a 13% after-tax MARR. How many copies per day must be sold to break even at a retail selling price of $1.00 per paper?

15.14 A corporation is trying to decide whether to buy the patent for a product designed by another company. The decision to buy will mean an investment of $8 million, and the demand for the product is not known. If demand is light, the company expects a return of $1.3 million each year for three years. If demand is moderate, the return will be $2.5 million each year for four years, and high demand means a return of $4 million each year for four years. It is estimated the probability of a high demand is 0.4 and the probability of a light demand is 0.2. The firm's (risk-free) interest rate is 12%. Calculate the expected present worth of the patent. On this basis, should the company make the investment? (All figures represent after-tax values.)

15.15 Juan Carlos is considering two investment projects whose present values are described as follows:

- **Project 1.** $PW(10\%) = 20X + 8XY$, where X and Y are statistically independent discrete random variables with the following distributions:

X		Y	
Event	**Probability**	**Event**	**Probability**
$20	0.6	$10	0.4
40	0.4	20	0.6

- **Project 2.**

PW(10%)	Probability
$0	0.24
400	0.20
1600	0.36
2400	0.20

(*Note*: Cash flows between the two projects are also assumed to be statistically independent.)

(a) Develop the NPW distribution for Project 1.

(b) Compute the mean and variance of the NPW for Project 1.

(c) Compute the mean and variance of the NPW for Project 2.

(d) Suppose that Projects 1 and 2 are mutually exclusive. Which project would you select?

15.16 A financial investor has an investment portfolio worth $350,000. A bond in the portfolio will mature next month and provide him with $25,000 to reinvest. The choices have been narrowed down to the following two options:

- **Option 1.** Reinvest in a foreign bond that will mature in one year. This will entail a brokerage fee of $150. For simplicity, assume that the bond will provide interest of $2450, $2000, or $1675 over the one-year period and that the probabilities of these occurrences are assessed to be 0.25, 0.45, and 0.30, respectively.

- **Option 2.** Reinvest in a $25,000 Guaranteed Investment Certificate (GIC) with a bank. Assume that this GIC has an effective annual rate of 7.5%.

(a) Which form of reinvestment should the investor choose in order to maximize his expected financial gain?

(b) If the investor can obtain professional investment advice from Salomon Brothers, Inc., what would be the maximum amount the investor should pay for this service?

15.17 Kellogg Company is considering the following investment project and has estimated all costs and revenues in constant dollars. The project requires the purchase of a $9000 asset, which will be used for only two years (the project life).
- The salvage value of this asset at the end of two years is expected to be $4000.
- The project requires an investment of $2000 in working capital, and this amount will be fully recovered at the end of the project year.
- The annual revenue, as well as general inflation, are discrete random variables, but can be described by the following probability distributions (both random variables are statistically independent):

Annual Revenue (X)	Probability	General Inflation Rate (Y)	Probability
$10,000	0.30	3%	0.25
20,000	0.40	5%	0.50
30,000	0.30	7%	0.25

- The investment has a CCA Rate of 30%.
- It is assumed that the revenues, salvage value, and working capital are responsive to the general inflation rate.
- The revenue and inflation rate dictated during the first year will prevail over the remainder of the project period.
- The marginal income tax rate for the firm is 40%. The firm's inflation-free interest rate (i') is 10%.

(a) Determine the NPW as a function of X and Y.

(b) In (a), compute the expected NPW of this investment.

(c) In (a), compute the variance of the NPW of the investment.

15.18 A manufacturing firm is considering two mutually exclusive projects, both of which have an economic service life of one year with no salvage value. The initial cost and the net year-end revenue for each project are given in Table P15.18.

We assume that both projects are statistically independent of each other.

(a) If you are an expected-value maximizer, which project would you select?

(b) If you also consider the variance of the project, which project would you select?

TABLE P15.18 Comparison of Mutually Exclusive Projects

First Cost	Project 1 ($1000)		Project 2 ($800)	
	Probability	Revenue	Probability	Revenue
Net revenue, given in PW	0.2	$2000	0.3	$1000
	0.6	3000	0.4	2500
	0.2	3500	0.3	4500

15.19 A business executive is trying to decide whether to undertake one of two contracts or neither one. She has simplified the situation somewhat and feels that it is sufficient to imagine that the contracts provide alternatives as follows:

Contract A		Contract B	
NPW	Probability	NPW	Probability
$100,000	0.2	$40,000	0.3
50,000	0.4	10,000	0.4
0	0.4	−10,000	0.3

(a) Should the executive undertake either one of the contracts? If so, which one? What would she do if she made decisions with an eye toward maximizing her expected NPW?

(b) What would be the probability that Contract A would result in a larger profit than that of Contract B?

15.20 Two alternative machines are being considered for a cost-reduction project.

- Machine A has a first cost of $60,000 and a salvage value (after tax) of $22,000 at the end of six years of service life. The probabilities of annual after-tax operating costs of this machine are estimated as follows:

Annual O&M Costs	Probability
$5,000	0.20
8,000	0.30
10,000	0.30
12,000	0.20

- Machine B has an initial cost of $35,000, and its estimated salvage value (after tax) at the end of four years of service is negligible. The annual after-tax operating costs are estimated to be as follows:

Annual O&M Costs	Probability
$8,000	0.10
10,000	0.30
12,000	0.40
14,000	0.20

The MARR on this project is 10%. The required service period of these machines is estimated to be 12 years, and no technological advance in either machine is expected.

(a) Assuming independence, calculate the mean and variance for the equivalent annual cost of operating each machine.

(b) From the results of part (a), calculate the probability that the annual cost of operating Machine A will exceed the cost of operating Machine B.

15.21 Two mutually exclusive investment projects are under consideration. It is assumed that the cash flows are statistically independent random variables with means and variances estimated as follows:

End of	Project A		Project B	
Year	Mean	Variance	Mean	Variance
0	−$5,000	$1,000^2$	−$10,000	$2,000^2$
1	4,000	$1,000^2$	6,000	$1,500^2$
2	4,000	$1,500^2$	8,000	$2,000^2$

(a) For each project, determine the mean and standard deviation of the NPW, using an interest rate of 15%.

(b) On the basis of the results of part (a), which project would you recommend?

15.22 Delta College's campus police are quite concerned with ever-growing weekend parties taking place at the various dormitories on the campus, where alcohol is commonly served to underage college students. According to reliable information, on a given Saturday night, one may observe a party to take place 60% of the time. Police Lieutenant Shark usually receives a tip regarding student drinking that is to take place in one of the residence halls the next weekend. According to Officer Shark, this tipster has been correct 40% of the time when the party is planned. The tipster has been also correct 80% of the time when the party does not take place. (That is, the tipster says that no party is planned.) If Officer Shark does not raid the residence hall in question at the time of the supposed party, he loses 10 career progress points. (The police chief gets complete information on whether there was a party only after the weekend.) If he leads a raid and the tip is false, he loses 50 career progress points, whereas if the tip is correct, he earns 100 points.

(a) What is the probability that no party is actually planned even though the tipster says that there will be a party?

(b) If the lieutenant wishes to maximize his expected career progress points, what should he do?

(c) What is the EVPI (in terms of career points)?

15.23 As a plant manager of a firm, you are trying to decide whether to open a new factory outlet store, which would cost about $500,000. Success of the outlet store depends on demand in the new region. If demand is high, you expect to gain $1 million per year; if demand is average, the gain is $500,000; and if demand is low, you lose $80,000. From your knowledge of the region and your product, you feel that the chances are 0.4 that sales will be average and equally likely that they will be high or low (0.3, respectively). Assume that the firm's MARR is known to be 15%, and the marginal tax rate will be 40%. Also, assume that the salvage value of the store at the end of 15 years will be about $100,000. The store will be depreciated at a CCA rate of 4%.

(a) If the new outlet store will be in business for 15 years, should you open it? How much would you be willing to pay to know the true state of nature?

(b) Suppose a market survey is available at $1000, with the following reliability (the values shown were obtained from past experience, where actual demand was compared with predictions made by the survey):

Actual	Survey Prediction		
Demand	Low	Medium	High
Low	0.75	0.20	0.05
Medium	0.20	0.60	0.20
High	0.05	0.25	0.70

Determine the strategy that maximizes the expected payoff after taking the market survey. In doing so, compute the EVPI after taking the survey. What is the true worth of the sample information?

Short Case Studies

ST15.1 In Virginia's six-number lottery, or lotto, players pick six numbers from 1 to 44. The winning combination is determined by a machine that looks like a popcorn machine, except that it is filled with numbered table-tennis balls. On February 15, 1992, the Virginia lottery drawing offered the prizes shown in Table ST15.1, assuming that the first prize is not shared.

Common among regular lottery players is this dream: waiting until the jackpot reaches an astronomical sum and then buying every possible number, thereby guaranteeing a winner. Sure, it would cost millions of dollars, but the payoff would be much greater. Is it worth trying? How do the odds of winning the first prize change as you increase the number of tickets purchased?

TABLE ST15.1 Virginia Lottery Prizes

Number of Prizes	Prize Category	Total Amount
1	First prize	$27,007,364
228	Second prizes ($899 each)	204,972
10,552	Third prizes ($51 each)	538,152
168,073	Fourth prizes ($1 each)	168,073
	Total winnings	$27,918,561

ST15.2 Mount Manufacturing Company produces industrial and public safety shirts. As is done in most apparel-manufacturing factories, the cloth must be cut into shirt parts by marking sheets of paper in a particular way. At present, these sheet markings are done manually, and their annual labour cost is running around $103,718. Mount has the option of purchasing one of two automated marking systems: the Lectra System 305 and the Tex Corporation Marking System. The comparative characteristics of the two systems are as follows:

	Lectra System	Tex System
Annual labour cost	$ 51,609	$ 51,609
Annual material savings	$230,000	$274,000
Investment cost	$136,150	$195,500
Estimated life	6 years	6 years
Salvage value	$ 20,000	$ 15,000
CCA rate (DB)	30%	30%

The firm's marginal tax rate is 40%, and the interest rate used for project evaluation is 12% after taxes.

(a) Based on the most likely estimates, which alternative is the best?

(b) Suppose that the company estimates the material savings during the first year for each system on the basis of the following probability distribution:

Lectra System	
Material Savings	Probability
$150,000	0.25
230,000	0.40
270,000	0.35

Tex System	
Material Savings	**Probability**
$200,000	0.30
274,000	0.50
312,000	0.20

Suppose further that the annual material savings for both Lectra and Tex are statistically independent. Compute the mean and variance for the equivalent annual value of operating each system.

(c) In part (b), calculate the probability that the annual benefit of operating Lectra will exceed the annual benefit of operating Tex. (Use @RISK or Excel to answer this question.)

ST15.3 The city of Guelph was having a problem locating land for a new sanitary land-fill when the alternative of burning the solid waste to generate steam was proposed. At the same time, Uniroyal Tire Company seemed to be having a similar problem disposing of solid waste in the form of rubber tires. It was determined that there would be about 200 tonnes per day of waste to be burned, including municipal and industrial waste. The city is considering building a waste-fired steam plant, which would cost $6,688,800. To finance the construction cost, the city will issue resource recovery revenue bonds in the amount of $7,000,000 at an interest rate of 11.5%. Bond interest is payable annually. The differential amount between the actual construction costs and the amount of bond financing ($7,000,000 − $6,688,800 = $311,200) will be used to settle the bond discount and expenses associated with the bond financing. The expected life of the steam plant is 20 years. The expected salvage value is estimated to be about $300,000. The expected labour costs are $335,000 per year. The annual operating and maintenance costs (including fuel, electricity, maintenance, and water) are expected to be $175,000. The plant would generate 4,245 kilograms of waste, along with 3,265 kilograms of waste after incineration, which will have to be disposed of as landfill. At the present rate of $42.88 per kilogram, this will cost the city a total of $322,000 per year. The revenues for the steam plant will come from two sources: (1) sales of steam and (2) tipping fees for disposal. The city expects 20% downtime per year for the waste-fired steam plant. With an input of 200 tonnes per day and 6.637 kilograms of steam per kilogram of refuse, a maximum of 1,327,453 kilograms of steam can be produced per day. However, with 20% downtime, the actual output would be 1,061,962 kilograms of steam per day. The initial steam charge will be approximately $4.00 per tonne, which will bring in $1,550,520 in steam revenue the first year. The tipping fee is used in conjunction with the sale of steam to offset the total plant cost. It is the goal of the Guelph steam plant to phase out the tipping fee as soon as possible. The tipping fee will be $20.85 per tonne in the first year of plant operation and will be phased out at the end of the eighth year. The scheduled tipping fee assessment is as follows:

Year	Tipping Fee
1	$976,114
2	895,723
3	800,275
4	687,153
5	553,301
6	395,161
7	208,585

(a) At an interest rate of 10%, would the steam plant generate sufficient revenue to recover the initial investment?

(b) At an interest rate of 10%, what would be the minimum charge (per tonne) for steam sales to make the project break even?

(c) Perform a sensitivity analysis to determine the input variable of the plant's downtime.

ST15.4 Burlington Motor Carriers, a trucking company, is considering installing a two-way mobile satellite messaging service on its 2000 trucks. On the basis of tests done last year on 120 trucks, the company found that satellite messaging could cut 60% from its $5 million bill for long-distance communication with truck drivers. More important, the drivers reduced the number of "deadhead" kilometres—those driven with nonpaying loads—by 0.5%. Applying that improvement to all 230 million kilometres covered by the Burlington fleet each year would produce an extra $1.25 million savings.

Equipping all 2000 trucks with the satellite hookup will require an investment of $8 million and the construction of a message-relaying system costing $2 million. The equipment and onboard devices will have a service life of eight years and negligible salvage value; they will be depreciated with a 30% CCA rate. Burlington's marginal tax rate is about 38%, and its required minimum attractive rate of return is 18%.

(a) Determine the annual net cash flows from the project.

(b) Perform a sensitivity analysis on the project's data, varying savings in telephone bills and savings in deadhead kilometres. Assume that each of these variables can deviate from its base-case expected value by $\pm 10\%$, $\pm 20\%$, and $\pm 30\%$.

(c) Prepare sensitivity diagrams and interpret the results.

ST15.5 The following is a comparison of the cost structure of conventional manufacturing technology (CMT) with a flexible manufacturing system (FMS) at one Canadian firm.

	Most Likely Estimates	
	CMT	FMS
Number of part types	3,000	3,000
Number of pieces produced/year	544,000	544,000
Variable labour cost/part	$2.15	$1.30
Variable material cost/part	$1.53	$1.10
Total variable cost/part	$3.68	$2.40
Annual overhead	$3.15M	$1.95M
Annual tooling costs	$470,000	$300,000
Annual inventory costs	$141,000	$31,500
Total annual fixed operating costs	$3.76M	$2.28M
Investment	$3.5M	$10M
Salvage value	$0.5M	$1M
Service life	10 years	10 years
CCA rate (DB)	30%	30%

(a) The firm's marginal tax rate and MARR are 40% and 15%, respectively. Determine the incremental cash flow (FMS − CMT), based on the most likely estimates.

(b) Management feels confident about all input estimates in CMT. However, the firm does not have any previous experience operating an FMS. Therefore, many input estimates, except the investment and salvage value, are subject to variation. Perform a sensitivity analysis on the project's data, varying the elements of operating costs. Assume that each of these variables can deviate from its base-case expected value by ±10%, ±20%, and ±30%.

(c) Prepare sensitivity diagrams and interpret the results.

(d) Suppose that probabilities of the variable material cost and the annual inventory cost for the FMS are estimated as follows:

Material Cost	
Cost per Part	Probability
$1.00	0.25
1.10	0.30
1.20	0.20
1.30	0.20
1.40	0.05

Inventory Cost	
Annual Inventory Cost	**Probability**
$25,000	0.10
31,000	0.30
50,000	0.20
80,000	0.20
100,000	0.20

What are the best and the worst cases of incremental NPW?

(e) In part (d), assuming that the random variables of the cost per part and the annual inventory cost are statistically independent, find the mean and variance of the NPW for the incremental cash flows.

(f) In parts (d) and (e), what is the probability that the FMS will be the more expensive of the two investment options? (Use @RISK or Excel to answer this question.)

On the Companion Website that accompanies this text, you will find Excel templates and exercises, as well as the following analysis tools: Cash Flow Analyzer, Depreciation Analysis, Loan Analysis, and Interest Tables.

Personal Investment
Analysis and Strategy

SIXTEEN

Principles of Investing and Personal Income Tax

Jim and Patti Johnston are recent engineering graduates who have good jobs with a large telecommunications company. They have disposable income that they wish to invest wisely to provide some short-term returns. This extra income can help finance their plan for an around-the-world trip in three years' time. After that, they plan to buy a house and perhaps start a family. They know that they will need to save for a down payment, and that a significant portion of their income will go towards monthly mortgage payments. In addition, they are aware that even though retirement is 30 to 40 years away, they should start putting money away for it now. Obviously, Jim and Patti could save up for all these goals by placing money in savings accounts. However, they also believe that they should invest a portion of their savings in stocks and bonds that can provide better returns. How do income taxes affect the returns on their investments? Can Jim and Patti avoid paying income tax on money that they are saving for retirement? In this chapter, we consider personal investments, including income tax effects with respect to bonds, stocks, and Registered Retirement Savings Plans (RRSPs). Registered Education Savings Plans (RESPs) will also be discussed.

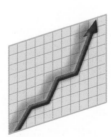

The focus of this book has been on how to perform financial calculations for engineering decisions, but you may have noticed that several of these techniques can also be applied in our personal lives. In fact, many such examples have been peppered throughout the text. How much will your money grow after a specified period when you invest it in a savings account? How large would the monthly mortgage payments be if you buy a house? How do you make money by investing in bonds? If you obtain an "add-on loan" to buy a car, how much does that actually cost you in terms of annual interest charges?

In this chapter we will apply methods presented in earlier chapters to perform financial calculations for personal decisions. We will consider some important choices that many Canadians have to make throughout their lives about what to do with their money. Of course, much of it gets spent on day-to-day living, but it is also important to invest extra funds to save up for large expenses such as your children's university education, unexpected emergencies, large purchases such as a cottage or sailboat, and perhaps the most important one of all: your retirement. The material in this chapter is a cursory overview to introduce the basic concepts and to broaden the use of the tools you have learned earlier in the text. However, it is no substitute for in-depth personal financial planning, which can be done through self-study or with the help of a professional investment planner.

Income taxes also play a big role in personal financial planning. They affect how much of our income we get to keep. The government also uses the tax system to provide benefits to many groups such as students, families, farmers, and people with disabilities. The basic personal tax system will be presented, including the effect it can have on investment decisions. Finally, there is one strong parallel between the personal and corporate tax systems. If you have professional or business income as an individual (sole proprietorship), then the income is treated through the personal income tax system, but you may benefit from some of the deductions available to corporations as described in Chapters 8 and 9, such as depreciation and operating expense deductions.

CHAPTER LEARNING OBJECTIVES

After completing this chapter, you should understand the following concepts:

- How to compare investments in interest-bearing accounts, bonds, and stocks.

- The risk/reward tradeoff in investments.

- How dollar cost averaging smoothes out the volatility in investing.

- The basic elements of personal income taxes.

- How professional activities and sole proprietorship net revenues are taxed.

- How taxes are applied to different types of investment benefits: interest, dividends, and capital gains.

- How the government provides tax shelters to promote savings, such as Registered Retirement Saving Plans (RRSPs) and Registered Education Savings Plans (RESPs).

16.1 Investing in Financial Assets

Most individual investors have three basic investment opportunities in financial assets: stocks, bonds, and cash. Cash investments include money in bank accounts, Guaranteed Investment Certificates (GICs), and Government of Canada Treasury bills (T-bills). You can invest directly in any or all of the three, or indirectly, by buying mutual funds that pool your money with money from other people and then invest it. If you want to invest in financial assets, you have plenty of opportunities. In Canada alone, there are more than 5000 stocks, 11,500 mutual funds, and thousands of corporate and government bonds to choose from. Even though we will discuss investment basics in the framework of financial assets in this chapter, the same economic concepts are applicable to any business assets.

16.1.1 Investment Basics

Selecting the best investment for you depends on your personal circumstances as well as general market conditions. For example, a good investment strategy for a long-term retirement plan may not be a good strategy for a short-term university savings plan. In each case, the right investment is a balance of four things: liquidity, safety, return, and time horizon.

- **Liquidity: How accessible is your money?** If your investment money must be available to cover financial emergencies, you will be concerned about liquidity: how easily you can convert it to cash. Money-market funds and savings accounts are very liquid; so are investments with short maturity dates, such as GICs. However, if you are investing for longer term goals, liquidity is not a critical issue. What you are after in that case is growth, or building your assets. We normally consider certain stocks and stock mutual funds as growth investments.

- **Risk: How safe is your money?** Risk is the chance you take of making or losing money on your investment. For most investors, the biggest risk is losing money, so they look for investments they consider safe. Usually, that means putting money into bank accounts and T-bills, as these investments are either insured or default free. The opposite, but equally important, risk is that your investments will not provide enough growth or income to offset the impact of inflation, the gradual increase in the cost of living. There are additional risks as well, including how the economy is doing. However, *the biggest risk is not investing at all*.

- **Return: How much profit will you be able to expect from your investment?** Safe investments often promise a specific, though limited, return. Those which involve more risk offer the opportunity to make—or lose—a lot of money. Both risk and reward are time dependent. On the one hand, as time progresses, low-yielding investments become more risky because of inflation. On the other hand, the returns associated with higher risk investments could become more stable and predictable over time, thereby reducing the perceived level of risk.

- **Time horizon: When do you expect to use the money that you are investing?** This aspect is intimately linked with the preceding three. If you are saving for a down payment for a house in a few years, the investment strategy may be very different than if you are young and setting funds aside for retirement.

In the following sections, we will look at three of the most common asset classes in more detail: cash, bonds, and stocks.

16.1.2 Cash Investments

Term Deposits (TDs) and Guaranteed Investment Certificates (GICs) are very secure choices for investors who want to preserve their capital and earn a prescribed rate of interest. The term of a GIC is normally from one to five years, but some institutions also offer shorter-term GICs. A wide range of options are available, from fixed rate to variable rate, to escalator rates where the annual interest rate increases each year of the term. A TD typically has a shorter term (one month to one year) than most GICs and has a fixed rate of return but generally has more flexibility for cashing it in before its maturity date. A T-bill is one form of Term Deposit, which is backed by the Canadian government. The prevailing T-bill rate is often used as a benchmark by investment analysts to compare the performance of other classes of investments over the years, such as the stock market.

EXAMPLE 16.1 Rate of Return on Cash Investments

Suppose that you have $3000 in your BMO daily interest savings account, with an effective annual interest rate of 0.25% at the moment. You are considering investing $2000 in one of the following GIC options:

1. A Rate-Riser GIC for a term of three years or five years, with the following annual rates. These investments are not redeemable before they mature.

Years of	Term	
Investment	Three Years	Five Years
1st Year	1.100%	1.100%
2nd Year	1.550%	1.550%
3rd Year	4.300%	2.750%
4th Year	–	3.250%
5th Year	–	6.750%

2. A Term Deposit Receipt that earns interest (payable annually) as shown in the following table for investments from $1000 to $4999 can be cashed before it matures, but with a redemption rate penalty of 0.20%/year.

Term	$1000–$4999
1 year–under 2 years	0.500%
2 years–under 3 years	1.250%
3 years–under 4 years	1.300%
4 years–under 5 years	1.450%
5 years exactly	1.850%

SOLUTION

Given: The annual interest rates for various cash investment options.
Find: The best vehicle to invest $2000.

The equivalent annual rate of return for the three-year GIC is:

$$i_{GIC_{3yr}} = \sqrt[3]{(1 + 0.011)(1 + 0.0155)(1 + 0.043)} - 1 = 2.307\%/\text{year}$$

The equivalent annual rate of return for the five-year GIC is:

$$i_{GIC_{5yr}} = \sqrt[5]{(1.011)(1.0155)(1.0275)(1.0325)(1.0675)} - 1 = 3.061\%/\text{year}$$

Suppose we choose to lock into a three-year TD:

$$i_{TD_{3yr}} = \sqrt[3]{(1 + 0.005)(1 + 0.0125)(1 + 0.013)} - 1 = 1.016\%/\text{year}$$

whereas choosing the five-year TD gives:

$$i_{TD_{5yr}} = \sqrt[5]{(1.005)(1.0125)(1.013)(1.0145)(1.0185)} - 1 = 1.269\%/\text{year}$$

Obviously, the GIC pays better than the TD, and both are a large improvement over the current savings account rate of 0.25%. But the decision hinges on a few things other than strictly the rate of return:

(a) How long do you plan to leave your money invested?
(b) How likely is it that you may need your funds before your planned investment horizon?
(c) What do you think is going to happen to market interest rates over the years during which you have your money invested?

If you plan to use the money in five years with no need to access it before then, then purchase a five-year GIC, unless you think that the market rates are going to rise a lot in the meantime. In that case, perhaps get a one-year TD and hope that the GIC rates available in a year are higher than now. If you think you may need your money before five years, either get the shorter term GIC or if you need more flexibility go with a TD where you can redeem at any time without penalty. If you get a five-year TD and wait for maturity, you'll receive the equivalent of 1.269% per year. However, if you need to withdraw it early after only 3.5 years, you won't receive anything for the partial fourth year, and you will forfeit 0.20%/year off the posted rate:

$$i_{TD_{3yr}} = \sqrt[3]{(1 + 0.003)(1 + 0.0105)(1 + 0.011)} - 1 = 0.816\%/\text{year}$$

DISCUSSION: The bank is offering better rates the longer you lock in your funds because it poses less risk for them that you will remove your money sooner. The daily interest savings account allows you maximum flexibility (liquidity) but with a much lower reward. Note the significant jump in rates the final year of each of the options in this example, as an incentive for customers to choose longer investment timeframes.

16.1.3 Bond Investments

As we saw in Chapter 4, bonds are a means for companies and governments to borrow money. The investor buys the bonds, which provides the issuer with the funds that it needs. In return, the government or company that has issued the bonds promises to pay the investor periodic interest in the form of coupon payments and also to reimburse the face value of the bond to the bondholder on the maturity date.

EXAMPLE 16.2 Investment in Bonds

Continuing Example 16.1, suppose that you are now considering using your $2000 to purchase five-year Saskatchewan provincial bonds at their face value of $1000 with a coupon rate of 2.4% per year payable semiannually. Should you get the bonds instead of a GIC? We will ignore transaction costs for this illustration. Would your response change if you expect interest rates to rise somewhat over the coming years?

SOLUTION

Given: The data on provincial bonds and the GIC options from Example 16.1.

Find: The best investment option under current and expected prevailing interest rates.

Since you would be buying the bond at par, the yield to maturity is the same at the coupon rate, 1.2% semiannually, which is equivalent to $i_a = 2.4144\%$/year. This is marginally higher than the annualized three-year GIC rate but lower than the annualized five-year GIC rate. Although provincial bonds are very secure, they are slightly riskier than GICs, which are insured by the Canadian Deposit Insurance Corporation (CDIC). However, you can also sell your bonds before the maturity date if you need the money. Of course, the amount that you would then earn overall depends on the market price at the time. Suppose the interest rate is expected to rise in the next two years. Then all other things being equal, the value of the bond would drop. If alternative secure investments are paying say 2.0% semiannually starting in two years ($i_a = 4.04\%$/yr), then the value of your bonds at that time would be the present value of the cash flows over the remaining six periods (Figure 16.1). The sensitivity of the NPW to possible discount rate values in two years is shown in Figure 16.2.

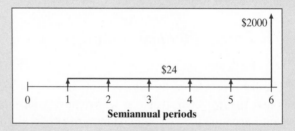

Figure 16.1 Bond cash flows for final three years (Example 16.2).

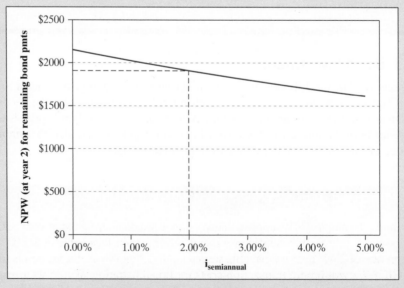

Figure 16.2 Sensitivity of bond price to discount rate (Example 16.2).

Understanding Bond Prices

Corporate bond prices are quoted either by a percentage of the bond's face value or in increments of points and eight fractions of a point, with a par of $1000 as the base. The value of each point is $10 and of each fraction $1.25. For example, a bond quoted at $86\frac{1}{2}$ would be selling for $865, and one quoted at $100\frac{3}{4}$ would be selling for $1007.50. Treasury bonds are measured in thirty-seconds, rather than hundredths, of a point. Each $\frac{1}{32}$ equals 31.25 cents, and we normally drop the fractional part of the cent when stating a price. For example, if a bond is quoted at 100.2, or $100 + \frac{2}{32}$, the price translates to $1000.62.

Are Bonds Safe?

Just because bonds have a reputation as a conservative investment does not mean that they are always safe. There are sources of risk in holding or trading bonds:

- To begin with, not all loans are paid back. Companies, cities, and counties occasionally do go bankrupt. Only Canadian Treasury bonds are considered rock solid.
- Another source of risk for certain bonds is that your loan may be called, or paid back early. While that is certainly better than its not being paid back at all, it forces you to find another, possibly less lucrative, place to put your money.
- The main danger for buy-and-hold investors, however, is a rising inflation rate. Since the dollar amount they earn on a bond investment does not change, the value of that money can be eroded by inflation. Worse yet, with your money locked away in the bond, you will not be able to take advantage of the higher interest rates that are usually available in an inflationary economy. Now you know why bond investors cringe at cheerful headlines about full employment and strong economic growth: These traditional signs of inflation hint that bondholders may soon lose their shirts.

How Do Prices and Yields Works?

Bond quotes follow a few unique conventions. As an example, let's take a look at a corporate bond with a face value of $1000 (issued by Ford Motor Company) and traded on November 18, 2005:

Price:	94.50
Coupon (%):	6.625
Maturity Date:	16-Jun-2008
Yield to Maturity (%):	9.079
Current Yield (%):	7.011
Debt Rating:	BBB
Coupon Payment Frequency:	Semiannual
First Coupon Date:	16-Dec-2005
Type:	Corporate
Industry:	Industrial

- Prices are given as percentages of face value, with the last digits not decimals, but in eighths. The bond on the list, for instance, has just fallen to sell for 94.50, or 94.5% of its $1000 face value. In other words, an investor who bought the bond when it was issued (at 100) could now sell it for a 5.5% discount.

- The discount over face value is explained by examining the bond's coupon rate, 6.625%, and its current yield, 7.011%. The current yield will be higher than the coupon rate whenever the bond is selling for less than its par value.
- The coupon rate is 6.625% paid semiannually, meaning that, for every six-month period, you will receive $66.25/2 = 33.13.
- The bond rating system helps investors determine a company's credit risk. Think of a bond rating as the report card on a company's credit rating. Blue-chip firms, which are safer investments, have a high rating, while risky companies have a low rating. The following chart illustrates the different bond rating scales from Moody's, Standard and Poor's (S&P), and Fitch Ratings.

Bond Rating			
Moody's	S&P/Fitch	Grade	Risk
Aaa	AAA	Investment	Highest quality
Aa	AA	Investment	High quality
A	A	Investment	Strong
Baa	BBB	Investment	Medium grade
Ba, B	BB, B	Junk	Speculative
Caa/Ca/C	CCC/CC/C	Junk	Highly speculative
C	D	Junk	In default

Notice that if the company falls below a certain credit rating, its grade changes from investment quality to junk status. Junk bonds are aptly named: They are the debt of companies that are in some sort of financial difficulty. Because they are so risky, they have to offer much higher yields than any other debt. This brings up an important point: Not all bonds are inherently safer than stocks. Certain types of bonds can be just as risky, if not riskier, than stocks.

- If you buy a bond at face value, its rate of return, or yield, is just the coupon rate. However, a glance at a table of bond quotes (like the preceding one) will tell you that after they are first issued, bonds rarely sell for exactly face value. So how much is the yield then? In our example, if you can purchase the bond for $945, you are getting two bonuses. First, you have effectively bought a bond with a 6.625% coupon, since the $66.25 coupon is 7.011% of your $945 purchase price. (Recall from Chapter 4 that the coupon rate, adjusted for the current price, is the current yield of the bond.) However, there's more: Although you paid $945, in 2008 you will receive the full $1000 face value.

16.1.4 Stock Investments

Stocks (or shares) represent the ownership of part of a publicly traded company. As explained in Chapter 12, most corporations rely on stock offerings as a significant source of capital for their operations. The investor can benefit from investing in stocks through dividends and growth. A company may elect to distribute some of their profits periodically through a payout to shareholders in the form of dividends. The shares may also appreciate in value, and if the shareholders sell them they reap a profit in the form of capital gains. Of course, shares may also fall in value, resulting in a capital loss when sold. There are three main phenomena that drive the price of a stock up or down. One is the underlying value of a company, reflecting how well it is doing in its business. Another is the perception of the company's future performance by investors. If there is a belief that a company is going to improve its performance in the future, then the demand for its shares increases, driving the price up. Conversely, a concern that the company economic health is weakening tends to depress the market for its shares, and the price drops (and perhaps the company's rating drops, as described in the preceding section on bonds). Finally, a general downturn in the economy of a country or of the world tends to push stock prices downwards.

Understanding Stock Prices

There are many measures and means for evaluating a company's current health and prospects, but a detailed discussion is beyond the scope of this text. Publicly traded companies must file annual reports, which provide all the government-mandated financial data as well as descriptions about their operations and goals. These can be accessed through the System for Electronic Document Analysis and Retrieval (SEDAR), a filing system developed for the Canadian Securities Administrators to allow for the public dissemination of Canadian securities information (www.sedar.com). One common measure of a corporation's profitability is the "earnings per share" (EPS), which is the ratio of its net profit at the end of the year divided by the number of outstanding shares. If Widget Inc. generates a net annual profit of $1 million, and there are 2 million shares, then its EPS = $1,000,000/2,000,000 = $0.50 per share. A higher EPS generally indicates that a company is performing well, which tends to push up its share price. Another common indicator is the "price-to-earnings ratio" (P/E ratio). So if the current market price of Widget Inc.'s shares is $7, then its P/E ratio = $7/$0.50 = 14. Although the share price and earnings are not static, this measure gives an indication of how long it would theoretically take the shareholder to recover her investment through earnings, which would be 14 years in this example. Therefore, as a P/E ratio rises over time, interest in the stock wanes and eventually investors are driven out of the market and the share price falls. Another useful calculation is the Dividend Yield, taking the annual dividend per share and dividing by the stock's price. For example, if Widget's annual dividend is $0.63, the Dividend Yield is 9% ($0.63/$7 = 0.09). See Chapter 2 for additional ratios that are useful in both business and personal investment decision making.

Below is an example of TSX stock market listings produced in newspapers and other media:

Stock	Ticker	Close	Net ch	% ch	Vol 000s	Day high	Day low	% yield	P/E	52wk high	52wk low
TD Bank	TD	73.09	−0.07	−0.1	2196	73.15	72.00	3.3	14.9	77.37	61.16
HudBay Mnrls	HBM	13.04	−0.11	−0.8	1126	13.48	12.99	1.5	31.8	17.00	8.05

Each company has a short, unique ticker symbol. Looking first at the TD Bank, the value of its stock at the close of the last trading was $73.09 (the TSX trading hours are business days from 9:30–4:00). This was a drop of $0.07 from its preceding day's close, which was a loss of about 0.1%. The price varied between $73.15 and $72.00 as a total of 2,196,000 shares changed hands. The dividend yield for the past year was 3.3%, and the P/E ratio was 14.9 times earnings. During the past 52 weeks, the stock price went as high as $77.37 and as low as $61.16. Comparing Hudson Bay Minerals to TD, we see that HBM's stock price had declined substantially from its 52-week high, its dividend yield was quite a bit lower, and its P/E ratio notably higher. Its share price also had a wide range over the past year.

EXAMPLE 16.3 Investment in Stocks

Continuing Example 16.2, what if you invested your $2000 in HBM stock instead? Suppose hypothetically that analysts are predicting a compound annual growth rate (CAGR) for the company of 6% per year for the next five years based on expansion of its gold and zinc mining in northern Manitoba, and assume that the annual dividend yield will remain at 1.5% per year. Should you pursue this alternative? Once again, we will ignore transaction costs.

SOLUTION

Given: Predictions for the growth and dividend associated with HBM stock, and alternative investments in bonds or cash from Examples 16.1 and 16.2.

Find: The best investment option.

Over the next five years, your $2000 investment is expected to grow by 6% per year, so you would hope to sell it for $2000(1.06)^5 = 2676.45 at that time. Furthermore, you anticipate earnings at 1.5% of its then current stock price each year, yielding a dividend payout of $2000(1.06)^n \times (0.015)$ in year n as follows:

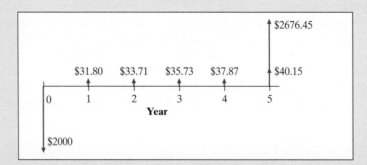

Figure 16.3 Cash flow for stock investment (Example 16.3).

The internal rate of return for this series of cash flows is 7.59% per year. If your goal is to maximize your expected return at the end of five years, with the flexibility to access your money in the interim if needed, then this appears to be the best choice compared with the bond or cash investments.

DISCUSSION: The recommendation to choose the stocks over the bonds or GIC seems compelling, but it ignores some crucial aspects. For one thing, stock prices generally fluctuate a lot, so not only could the predicted share price in five years be wildly wrong, but you could end up losing some of your capital, whether you cash in then or at any other time. Even worse, if a company goes out of business, you may lose all of your investment, or a substantial part of it, as discussed in Chapter 12. Consider the variations in the stock price for HBM in the year preceding this writing, as shown in the Figure 16.4. If you had bought their stock in November 2009 and sold it in May 2010, you would have lost quite a bit of money. On the other hand, if you had invested in August 2009, the share price would have increased by over 50% one year later. This example illustrates that a greater reward from an investment often involves taking a greater risk with your money. This issue will be examined more closely in the following sections.

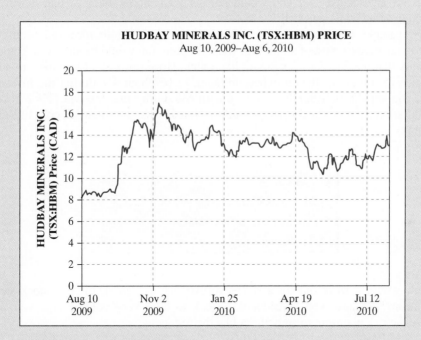

Figure 16.4   Sample stock showing volatility over time.

16.2 Investment Strategies

The optimal strategy seems deceptively simple: achieve our investment goal with the greatest possible return at the lowest possible risk. The following sections explore the tradeoff between risk and return, and how timing and diversification have a major impact on the success of your investment strategy.

16.2.1 How to Determine Your Expected Return

Return is what you get back in relation to the amount you invested. Return is one way to evaluate how your investments in financial assets are doing in relation to each other and to the performance of investments in general. Let us look first at how we may derive rates of return.

Risk-free return:
A theoretical interest rate that would be returned on an investment which was completely free of risk. The three-month Treasury bill is a close approximation, since it is virtually risk-free.

Basic Concepts

Conceptually, the rate of return that we realistically expect to earn on any investment is a function of three components: (1) **risk-free real return**, (2) an inflation factor, and (3) a **risk premium**.

Suppose you want to invest in stock. First, you should expect to be rewarded in some way for not being able to use your money while you are holding the stock. Then, you would be compensated for decreases in purchasing power between the time you invest the money and the time it is returned to you. Finally, you would demand additional rewards for any chance that you would not get your money back or that it will have declined in value while invested.

Risk premium:
The reward for holding a risky investment rather that a risk-free one.

For example, if you were to invest $1000 in risk-free T-bills for a year, you would expect a real rate of return of about 2%. Your risk premium would be also zero. You probably think that the 2% does not sound like much. However, to that you have to add an allowance for inflation. If you expect inflation to be about 4% during the investment period, you should realistically expect to earn 6% during that interval (2% real return +4% inflation factor +0% for risk premium). Here is what the situation looks like in tabular form:

Real return	2%
Inflation (loss of purchasing power)	4%
Risk premium (Treasury bills)	0%
Total expected return	6%

How would it work out for a riskier investment, say, in an Internet stock such as Google.com? Because you consider this stock to be a very volatile one, you would increase the risk premium to something like the following:

Real return	2%
Inflation (loss of purchasing power)	4%
Risk premium (Google.com)	20%
Total expected return	26%

So you will not invest your money in Google.com unless you are reasonably confident of having it grow at an annual rate of 26%. Again, the risk premium of 20% is a perceived value that can vary from one investor to another.

Return on Investment Over Time

Figuring out the actual return on your portfolio investment is not always simple. There are several reasons:

1. The amount of your investment changes. Most investment portfolios are active, with money moving in and out.

2. The method of computing the return can vary. For example, the performance of a stock can be averaged or compounded, which changes the rate of return significantly, as we will demonstrate in Example 16.4.

3. The time you hold specific investments varies. When you buy or sell can have a dramatic effect on the overall return.

EXAMPLE 16.4 Figuring Average versus Compound Return

Consider the following six different cases of the performance of a $1000 investment over a three-year holding period:

| Investment | Annual Investment Yield | | | | | |
	Case 1	Case 2	Case 3	Case 4	Case 5	Case 6
Year 1	9%	5%	0%	0%	−1%	−5%
Year 2	9%	10%	7%	0%	−1%	−8%
Year 3	9%	12%	20%	27%	29%	40%

Compute the average versus compound return for each case.

SOLUTION

Given: Three years' worth of annual investment yield data.

Find: Compound versus average rate of return.

As an illustration, consider case 6 for an investment of $1000. At the end of the first year, the value of the investment decreases to $950; at the end of second year, it decreases again, to $950(1 − 0.08) = $874; at the end of third year, it increases to $874(1 + 0.40) = $1223.60. Therefore, one way you can look at the investment is to ask, "At what annual interest rate would the initial $1000 investment grow to $1223.60 over three years?" This is equivalent to solving the following equivalence problem:

$$\$1223.60 = \$1000(1 + i)^3,$$

$$i = 6.96\%.$$

Average annual return: A figure used when reporting the historical return of a mutual fund.

If someone evaluates the investment on the basis of the average annual rate of return, he or she might proceed as follows:

$$i = \frac{(-5\% - 8\% + 40\%)}{3} = 9\%.$$

If you calculate the remaining cases, you will observe that all six cases have the same average annual rate of return, although their compound rates of return vary from 6.96% to 9%:

	Compound versus Average Rate of Return					
	Case 1	**Case 2**	**Case 3**	**Case 4**	**Case 5**	**Case 6**
Average return	9%	9%	9%	9%	9%	9%
Balance at end of three years	$1,295	$1,294	$1,284	$1,270	$1,264	$1,224
Compound rate of return	9.00%	8.96%	8.69%	8.29%	8.13%	6.96%

Your immediate question is "Are they the same indeed?" Certainly not: You will have the most money with case 1, which also has the highest compound rate of return. The average rate of return is easy to calculate, but it ignores the basic principle of the time value of money. In other words, according to the average-rate-of-return concept, we may view all six cases as indifferent. However, the amount of money available at the end of year 3 would be different in each case. Although the average rate of return is popular for comparing investments in terms of their yearly performance, it is not a correct measure in comparing the performance for investments over a multiyear period.

COMMENTS: You can evaluate the performance of your portfolio by comparing it against standard indexes and averages that are widely reported in the financial press. If you own stocks, you can compare their performance with the performance of the S&P/TSX Composite Index (Standard & Poor's/Toronto Stock Exchange) in Canada or the Dow Jones Industrial Average (DJIA) in the United States. These indexes are calculated from the stock (equity) prices of many companies and therefore give a good indication of how the stock market is doing overall. If you own bonds, you can identify an index that tracks the type you own: corporate, government, or agency bonds. If your investments are in cash, you can follow the movement of interest rates on Treasury bills, GICs, and similar investments. In addition, total-return figures for the performance of mutual funds are reported regularly. You can compare how well your investments are doing against those numbers. Another factor to take into account in evaluating your return is the current inflation rate. Certainly, your return needs to be higher than the inflation rate if your investments are going to have real growth.

Figure 16.5 shows how cash (T-bills), bond, and stock investments perform over a long time horizon, based on a $100 investment in 1950. As expected, the stock indexes do better than bonds and cash investments. The average rate of return across a wide range of companies must exceed the average rates they pay on bond interest. In other words, the wealth of a company must grow faster than its cost of debt, as explained in Chapter 12, or

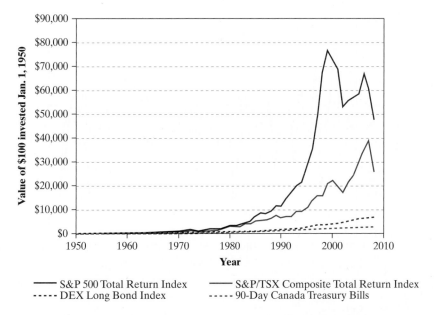

Figure 16.5   Performance of various asset classes over time for a $100 investment. *Source:* Morningstar ANDEX Chart 2010.

they would go bankrupt. So, investing in stocks will likely generate more wealth than bonds investments *when done across many companies and over long periods of time.* These aspects are explored in the following sections. Note in Figure 16.5 that the U.S. stock market (S&P 500) generally does better than the Canadian S&P/TSX.

16.2.2 Investment Risks

As mentioned earlier, pursuing higher returns entails greater risks. However, some types of risk are not as apparent as others. We will now present, more explicitly, five categories of risk that investors face.

Risk That You May Get Back Less Than You Invested

Figure 16.6 represents the same data used in Figure 16.5 in the form of annual returns for each asset class. This provides a very convincing picture that the stock market is relatively volatile, the bond market less so, but in both you could end up losing money on your investment if you "buy high" and "sell low," exactly the opposite of what investors are trying to achieve. In comparison, the T-bills have a positive return every year so you will not lose your capital.

Risk That You Lose All Your Money

The volatility in the stock of an individual company can be much greater than a stock index that averages across a large number of companies. This is because the fluctuations in an index (Figure 16.6) are largely driven by the overall economy (market risk) while an individual company's value (actual or perceived) may jump frequently and suddenly

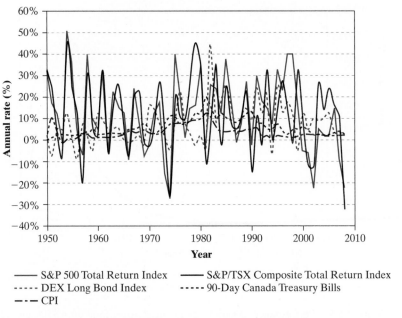

Figure 16.6 Annual returns of various asset classes over time.
Source: Morningstar ANDEX Chart 2010.

(business risk). An associated risk is that a company goes bankrupt, in which case you may lose all of your money. The remaining assets of a bankrupt company are used to repay some of its obligations, and bondholders have priority over shareholders, so bonds are lower risk from this perspective, but both asset classes could result in a complete loss of your invested capital.

Risk That the Buying Power of Your Capital Decreases Over Time

Conservative investors who want to avoid the gyrations of the stock market sometimes opt for safer cash investments, even though the long-run return is much lower. However, these investors are also taking a large risk, although a less apparent one: inflation prevents them from achieving their goals. Figure 16.6 includes a plot of the annual inflation rate based on the Consumer Price Index (CPI). While it is markedly lower in some years than others, it is constantly eating into the earnings from the investments in all asset classes. Note that during periods when even low-risk investments are yielding high returns, such as T-bills in the early 1980s, the inflation rate has risen in tandem, thus negating much of the benefit of the high yield.

EXAMPLE 16.5 Inflation Impact on Investments

Samira Shahrabi has worked hard all her life, putting some money aside throughout her working career to build up a nest egg for emergencies and retirement. Now that she is approaching 60, she has decided to take early retirement. Even though she does not have a company pension plan, she thinks that her investments are sufficient to live on given her modest lifestyle. She has accumulated capital of $600,000. Since the economy is quite strong, she can get a return of 6% in fairly liquid investments, so that she can draw on those funds for her living expenses. The inflation rate is 4% per year. Ignore income taxes for simplicity.

SOLUTION

Given: Samira's savings, and forecast returns on her investments, and the annual inflation rate.
Find: What is the long term impact on her ability to support herself from her savings?

The following table shows that Samira's capital will be seriously eroded by inflation over time. If she withdraws all of the earned interest every year to live off, then the capital does not increase. Furthermore, due to inflation its real value drops by more than one-third by the time she reaches the age of 72, and should she live until 96, the purchasing power of her investments would be less than one-quarter of what it is now. Even worse, her ultimate goal is not the investment itself but its utility for gen-erating income to live off, and it is clear from the last column that her income in today's dollars is insufficient for her stated needs.

Years	Age	Capital @ 4% Inflation	% Drop in Purchasing Power	Income @ 6% RoR
0	60	$600,000		$36,000
12	72	$374,758	37.5%	$22,485
24	84	$234,073	61.0%	$14,044
36	96	$146,201	75.6%	$ 8,772

Risk That You Will Lose Out on Potentially Better Returns

This risk may derive from several behaviours. One might be over-concern about market volatility so that you stick with safe investments thus losing the opportunity to benefit from higher long-term returns from stock markets. This is referred to as being very risk-averse. Of course, this then incurs the risk of neglecting the negative impact of inflation as discussed in the previous section. A related phenomenon is when you already have some investments in the stock market and the market has been dropping considerably. Consider, for example, the large declines in the S&P and TSX indexes in recent years as shown in Figure 16.5. It may be tempting to withdraw your money and put it in GICs, which as we have demonstrated have much lower returns over time than the stock indexes. In general, a low stock market is the time to buy (or leave your money in), not the time to sell. Of course, this assumes that you are leaving your money invested for a long period of time, not wanting to spend it soon. Although these general principles apply to everyone, each investor needs a strategy customized to his or her particular objectives and circumstances as described below. Another consideration for missing out on potentially better returns is how long you lock in your investments. As was shown in Example 16.5, you may choose to invest all your money in a five-year non-redeemable GIC to secure your capital and obtain a known rate of return, but that would prevent you from shifting the money to other instruments should the market heat up.

Risk That You Will Not Achieve Your Objectives

Defining and reaching your objectives is the paramount issue for investment planning, so this risk of failure should be your key consideration. This risk incorporates all of the preceding elements of course, but there are also other aspects to consider. It goes without saying that everyone who invests wants to make money, but how much you want, when you need it, and your total wealth level determine what would be a good investment strategy. So the first thing is to explicitly specify your goals, which a surprising number of investors never do. Then, the goals should be revisited periodically to see if your priorities, your circumstances or your investments' performance require you to revise your objectives. Your attitude toward risk, the effects of inflation and taxes, your other sources of income, and many other factors should be considered to maximize the chances of reaching your objectives.

16.2.3 Investment Timing Issues

One of the most popular fallacies is that the smart way to "beat the market" is to keep a close eye on your stock market investments, and "buy low" and "sell high." Of course, the logic is sound, but the uncertainty is so high that it is virtually impossible to come out ahead with this strategy. Research has consistently shown that even expert investment advisors cannot win at this game. This does not mean that indications that a stock should be sold or bought based on a company's rating and investor interest should be ignored. Sound analysis and trends should be taken into account, but this is very different than micromanaging on a day-to-day basis. Furthermore, there are often transaction costs on every sale or purchase, which eat into profits.

The planning horizon for your investments is perhaps the most important time consideration. The longer you leave your money invested, lower yield asset classes (such as GICs and bonds) become more risky because of inflation as demonstrated earlier, while the relatively volatile stock market becomes less risky because the average fluctuations over a longer period are more damped. This phenomenon is illustrated in Figure 16.7. The behaviour of an investment in four different asset classes from 1950 to 2009 is compared: U.S. stocks, Canadian Stocks, bonds, and GICs. The first group of bars on the left shows what the range of returns was for one-year investments, starting in each month since January 1, 1950, until June 30, 2009. That is, investing for one year on January 1, 1950, is the first period; investing for one-year on February 1, 1950, is the second period; there are 703 such one-year periods. The range of returns for such a short investment period is very high for all asset classes except for the GICs. The 5% of the periods with the best returns and the 5% with the worst returns in each asset class are shown with different shading. The horizontal white line in each bar shows the median return over all 703 periods. At the right side of the chart, the bars represent 30-year investments in each of the asset classes. There are 355 such periods starting in each month from January 1950 onward, until the final 30-year period ends in June 2009.

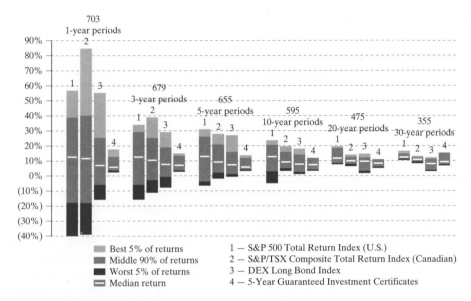

Figure 16.7   Risk/Return analysis (Jan. 1950–June 2009).
Source: Morningstar Risk/Return Analysis.

Although past performance cannot guarantee future outcomes, this chart illustrates many important concepts for investment planning. The longer the investment period, the less is the variability in rates of return. For one-year investment durations, there are some periods where a very high rate of return has occurred in the stock market, reaching almost 87% in one particular period in Canada. However, the chances of losing money are also very high, as in about a quarter of the periods the Canadian stock market lost value (negative return). However, locking in the investment for a three-year period (second set of bars) noticeably reduced the variation in annualized returns. If one had held stocks or bonds for a full 30-year period, starting in any month within the specified timeframe, then no losses would have occurred (last set of bars). In other words, there was not a single 30-year period over which money was lost. This reinforces the principle that investing in stocks over a longer period garners the benefit of higher returns with a reduced risk due to volatility and much less chance of losing part of your capital.

16.2.4 Diversification Reduces Risk

Diversification is the modern investor's equivalent of the well-known proverb: "Don't put all your eggs in one basket." There is too much uncertainty for it to be wise to focus on a single stock, or even a few, to aim at higher returns. The bars shown in Figure 16.7 already represent significant diversification, as each index is a measure of many companies' performance. However, investment advisors generally recommend diversifying your investments across different asset classes as well. The proportion to put in cash, bonds, and stocks depends on several factors. Liquidity is a main issue. If you need to draw on the funds regularly as a retiree would, or keep a certain amount available to withdraw for emergencies as most people do, then you do not want it all locked up in long-term investments that cannot be redeemed. So a certain percentage may be put into lower yielding, but more accessible, instruments. If you are young, a significant portion should be invested in stocks for several years or more to save for big purchases or retirement. Another factor is your personal risk attitude. Some people are naturally more averse to risk than others, so if fluctuations in the stock market are going to cause you to be very stressed, regardless of your newfound awareness of the fundamentals of investing, then having a relatively smaller proportion of your funds in stocks may be the best thing. Investment advisors have standard questionnaires that help you assess your financial goals and your attitude toward risk, so that they can recommend investments that are consistent with your particular circumstances. Another important aspect, discussed later in this chapter, is that the income tax is calculated differently for the benefits from different types of investments, so this may affect the best mix of assets.

While some investors purchase stocks in a broad variety of companies, many invest in mutual funds. A mutual fund is a collection of stocks and/or bonds, where a professional fund manager chooses what to invest in. By investing in a mutual fund, you pool your money with that of other fund investors, allowing the manager to make larger purchases, and it also tends to minimize the relative costs of transaction fees. Mutual funds, however, have to specify the kinds of investments that they will make, so that the investors are aware of the nature of the holdings. Some are "value-based," meaning that they restrict themselves to large, stable companies that generally have a modest expected rate of return but are not too volatile. Other mutual funds concentrate on new or fast-growing companies, or those that work in a less predictable business environment, which is classified as a "growth" fund. Higher returns are possible with such funds, but of course they come with greater risk.

EXAMPLE 16.6 Risk-Reward Tradeoff

Pierre Géroux is a doctor in his mid-30s who has been setting aside some of his earnings every year in GICs. The value of his investments now stands at about $120,000. He recently attended a free seminar put on by a financial advisor, and he realizes that he should be somewhat more aggressive in his investment strategy so that his savings can generate higher returns. Consider the information that was provided in Figure 16.6. What are the implications of using this historical data for approximating Pierre's risk/return options?

SOLUTION

Given: Pierre's interest in diversifying his investments, and historical records for various asset classes from Figure 16.6.

Find: The risk-reward scenarios for asset mix options, for a three-year investment horizon, and for a 10-year horizon.

The average annual rate of return for three-year terms and 10-year terms for each of the asset classes derived from Figure 16.6 are listed in the table below. Within the six-decade timespan, beginning at the start of each year from 1950 there are 57 three-year investment periods. Some of these periods will have an annual RoR greater than the overall average for that asset class, and some will be lower. For example, the TSX index had an annual RoR of 13.32%/year for a three-year investment made in 1967. This is considerably more than the average annual RoR of 10.49% for all three year investments in the TSX index companies. To measure the degree of variability in RoR in each of the asset classes (j) across the six decades of information provided, the standard deviation (SD) is calculated as follows:

$$SD_j = \sum_{i=1}^{N_j} \frac{(x_{i,j} - \overline{x}_j)^2}{N_j - 1}$$

where $\overline{x}_j$ is the average RoR for asset class j, N_j is the number of periods of investment type j, and $x_{i,j}$ is the annual RoR of asset class j in the ith period. This statistic gives a reasonable approximation of the volatility of various investments, although investment analysts actually use more sophisticated computer algorithms to account for additional factors. Assuming that the RoRs in each asset class are normally distributed, there is a 65% chance that the annual return will fall between $\overline{x}_j \pm SD_j$.

	3-Year Investment		10-Year Investment	
	Average	**Std. Dev.**	**Average**	**Std. Dev.**
S&P 500 Total Return Index	11.79%	10.32%	12.01%	5.85%
S&P/TSX Composite Index	10.49%	8.09%	10.19%	2.98%
DEX Long Bond Index	7.80%	6.04%	8.11%	4.22%
90-Day Canada Treasury Bills	5.94%	3.70%	6.41%	3.18%

If Pierre invests all his money in T-bills, his expected return based on past performance will be about 5.94% per year, but there is a 65% chance that the actual return will lie between $5.94 \pm 3.70\%$, that is falling between 2.24% and 9.64%.

Suppose Pierre is considering putting 50% of his funds into Canadian stocks, 30% into bonds, and the remaining 20% into T-bills. Then he can approximate the risk-reward tradeoff by using a weighted average of the mean returns, and of the variances (standard deviation squared) of the three asset classes.

Investment	% of Portfolio	3-Year Investments			
		Return	**Weighted Return**	**Volatility (Variance)**	**Weighted Volatility**
S&P/TSX Comp. Index	50%	10.49%	5.24%	0.00655	0.003276
DEX Long Bond Index	30%	7.80%	2.34%	0.00365	0.001096
90-Day T-Bills	20%	5.94%	1.19%	0.00137	0.000274
		TOTALS:	8.77%		0.004646
					SD = 6.8164%

If Pierre decides to go for this asset mix with three-year terms, the estimation shows how his expected annual return of 8.77% will be less than pure stock investments, but the volatility of $\pm 6.82\%$ for one SD will be somewhat lower as well. Based on the historical data, if he chooses 10-year terms instead, his RoR will be slightly higher, but the volatility will be much lower than the three-year option (SD = 3.44%).

Investment	% of Portfolio	10-year Investments			
		Return	Weighted Return	Volatility (Variance)	Weighted Volatility
S&P/TSX Comp. Index	50%	10.19%	5.09%	0.00089	0.000444
DEX Long Bond Index	30%	8.11%	2.43%	0.00178	0.000534
90-Day T-Bills	20%	6.41%	1.28%	0.00101	0.000203
		TOTALS:	8.81%		0.00118
					SD = 3.4353%

DISCUSSION: This calculation gives a rough estimate of the combined variability but tends to overestimate the risk as it ignores correlations of returns across the assets. Sophisticated software programs produce more accurate assessments. Furthermore, some portfolio analysis methods focus only on the downside risk, as it is the principal concern of more investors.

16.2.5 Dollar Cost Averaging

Since stock prices can be so volatile, investors may be leery about buying a large quantity all at once. Dollar cost averaging (DCA) is a method of gradually acquiring shares by investing a fixed amount of money at regular intervals. This has the effect of acquiring more shares in periods where the price is low and fewer when the price is high.

EXAMPLE 16.7 Dollar Cost Averaging

Consider the problem faced by Pierre Géroux in Example 16.6. Pierre has decided to shift half of his investments to the stock market, having realized the need to pursue higher returns to achieve his goals. However, instead of doing it all at once and investing directly in stocks, he plans to spread out his $60,000 investment over one year by making equal monthly transfers to a stock-based mutual fund. He also wishes to track the fund's performance over the year see the effect of his strategy.

SOLUTION

Given: Pierre's $60,000 investment in a stock-based mutual fund by dollar cost averaging.

Find: The value of his position in the fund after one year.

The table below shows how many units Pierre bought each month for each fixed $5000 investment. At the end of the year, he owns 17,688 units worth $3.87 each, amounting to a total value of $68,451. So, irrespective of any earnings he may have made on his savings not yet transferred to the mutual fund, he earned a profit of $8451 on the $60,000 put into this fund, or 14.085%.

Month	Amount	Fund Unit Price	Units Bought
1	$5,000	$4.00	1,250
2	$5,000	$3.49	1,433
3	$5,000	$3.70	1,351
4	$5,000	$3.21	1,558
5	$5,000	$3.02	1,656
6	$5,000	$3.10	1,613
7	$5,000	$3.57	1,401
8	$5,000	$3.40	1,471
9	$5,000	$2.94	1,701
10	$5,000	$3.13	1,597
11	$5,000	$3.66	1,366
12	$5,000	$3.87	1,292
TOTALS:	$60,000		17,688

DISCUSSION: Even though the fund never regained its initial value of $4.00 per unit over the year, Pierre made a neat profit. With these particular numbers, had he invested all $60,000 at once, he would have acquired 15,000 shares, which would be worth $58,050 at the end of the year, in which case his capital would have lost 3.5% of its value. Despite the appeal of this example, research has shown that dollar cost averaging does not improve the chances for improving returns over a lump sum investment strategy. And if a fund's value is continually dropping, you will lose money whether you choose DCA or lump sum investing (and vice versa for an increasing fund value). However, it provides some peace of mind for the investor as it reduces volatility. In any case, many people benefit from DCA because they are investing part of their income in a fund at regular payday intervals.

16.3 Personal Income Tax

Individuals pay taxes on income from several sources: wages and salaries, investment income, profits from the sale of capital assets, and business income. Typical investment income may include dividends from stocks and interest from bank deposits. Profits from the sale of capital assets such as stocks and bonds are also taxable. The business income from proprietorships and partnerships is taxed in the same manner as individual income.

16.3.1 Calculation of Taxable Income

In calculating taxable income, all taxpayers begin with **total income**, which represents their earnings as determined in accordance with the provisions of the federal income tax laws. To compute income, we add[1]

+ Employment income (wages and salary)

+ Pension income

+ Interest income (savings account, GICs, bond interest, etc.)

+ Dividend income

+ Rental/royalty income

+ Net capital gains (1/2 of net value)

+ Self-employment income (business income; professional income)

= **Total income**

From total income, the Canada Revenue Agency allows you to subtract the following amounts:

− Retirement plan contributions (registered pension plans or registered retirement savings plans)

− Annual union, professional, or like dues

− Child care expenses

− Attendant care expenses

− Business investment losses

− Moving expenses

− Support payments (e.g., alimony)

− Carrying charges and interest expense on money borrowed for investment purposes

− Other deductions

= **Net income before adjustments**

Adjustments subtracted from the **net income before adjustments** include

− Employee home relocation loan deduction

− Other payments deduction (e.g., worker's compensation, social assistance)

− Losses from previous years (includes capital losses)

− Northern resident's deduction

− Additional deductions

= **Taxable income**

---

[1] We are including only the main categories of revenues and expenses.

16.3.2 Income Taxes on Ordinary Income

Canada has four rates for individual taxation. Rates are progressive—that is, the higher the income, the larger the percentage paid in taxes. The federal income tax is computed using the taxable income and the rates in Table 16.1. These rates will change over the years, so you must refer to the most recent tax guides to find the presently applicable rates. For 2009, someone who has taxable income below $40,726 is subjected to a federal tax rate of 15% on ordinary income and interest income. If their taxable income is between $40,726 and $81,452, then the marginal tax rate is 22% in this range, hence they will pay $6109 on the first $40,726 plus 22% of taxable income over $40,726, and so on. The Canada Revenue Agency and the provincial governments periodically add individual surtaxes as a means of collecting additional revenue over the short term. However, some surtaxes last for several years and have become a longer term taxation method. Surtaxes are included in the marginal tax rates shown in Table 16.1

TABLE 16.1 Marginal Personal Income Tax Rates for Canada and by Province/Territory by Income Type for 2009[2]

	Taxable Income	Ordinary Income & Interest	Capital Gains	Canadian Dividends[3]	
				Eligible Dividends[4]	Small Business Dividends
Federal	$10,320–$40,726	15.00%	7.50%	−5.75%	2.08%
	$40,726–$81,452	22.00%	11.00%	4.40%	10.83%
	$81,452–$126,264	26.00%	13.00%	10.20%	15.83%
	$126,264 and over	29.00%	14.50%	14.55%	19.58%
British Columbia	$9,373–$35,716	5.06%	2.53%	−8.61%	1.08%
	$35,716–$71,433	7.70%	3.85%	−4.79%	4.38%
	$71,433–$82,014	10.50%	5.25%	−0.72%	7.88%
	$82,014–$99,588	12.29%	6.15%	1.87%	10.12%
	$99,588 and over	14.70%	7.35%	5.36%	13.13%
Alberta	$16,775 and over	10.00%	5.00%	0.01%	8.13%
Saskatchewan	$13,269–$40,113	11.00%	5.50%	0.00%	6.25%
	$40,113–$114,610	13.00%	6.50%	2.90%	8.75%
	$114,610 and over	15.00%	7.50%	5.80%	11.25%
Manitoba	$8,134–$31,000	10.80%	5.40%	−0.29%	10.38%
	$31,000–$67,000	12.75%	6.38%	2.54%	12.82%
	$67,000 and over	17.40%	8.70%	9.28%	18.63%

[2] Surtaxes included in marginal rates on ordinary income; the marginal tax rates for capital gains and dividends at any income level (say $60,000) are the marginal rates on the next dollar of actual capital gains or actual dividend income, if the taxpayer has $60,000 of taxable income from sources other than dividends.
[3] Marginal tax rate for dividends is a % of actual dividends received (not grossed-up amount).
[4] Negative marginal tax rates on dividends arise for some low tax brackets as the benefit of the tax credit exceeds the tax owing.

(Continued)

TABLE 16.1 Continued

	Taxable Income	Ordinary Income & Interest	Capital Gains	Canadian Dividends Eligible Dividends	Small Business Dividends
Ontario	$8,881–$36,848	6.05%	3.03%	−1.96%	1.80%
	$36,848–$64,882	9.15%	4.58%	2.54%	5.03%
	$64,882–$73,698	10.98%	5.49%	3.04%	6.03%
	$73,698–$76,440	13.39%	6.70%	6.54%	9.05%
	$76,440 and over	17.41%	8.70%	8.51%	11.76%
Quebec[5]	$13,069–$38,385	16.00%	8.00%	5.94%	10.00%
	$38,385–$76,770	20.00%	10.00%	11.75%	15.00%
	$76,770 and over	24.00%	12.00%	17.55%	20.00%
New Brunswick	$8,605–$35,707	9.65%	4.83%	−3.41%	5.44%
	$35,707–$71,415	14.50%	7.25%	3.62%	11.50%
	$71,415–$116,105	16.00%	8.00%	5.80%	13.38%
	$116,105 and over	17.00%	8.50%	7.25%	14.63%
Prince Edward Island	$7,708–$31,984	9.80%	4.90%	−1.02%	8.25%
	$31,984–$63,969	13.80%	6.90%	4.78%	13.25%
	$63,969–$98,143	16.70%	8.35%	8.99%	16.88%
	$98,143 and over	18.37%	9.19%	9.89%	18.57%
Nova Scotia	$7,981–$29,590	8.79%	4.40%	−0.09%	1.37%
	$29,590–$59,180	14.95%	7.48%	8.84%	9.07%
	$59,180–$81,236	16.67%	8.34%	11.34%	11.22%
	$81,236–$93,000	18.34%	9.17%	12.47%	12.34%
	$93,000 and over	19.25%	9.63%	13.80%	13.48%
Newfoundland and Labrador	$7,778–$31,061	7.70%	3.85%	−2.97%	3.38%
	$31,061–$62,121	12.80%	6.40%	4.42%	9.75%
	$62,121 and over	15.50%	7.75%	8.34%	13.13%
Yukon	$10,320–$40,726	7.04%	3.52%	−5.74%	3.24%
	$40,726–$80,595	9.68%	4.84%	−1.91%	6.54%
	$80,595–$81,452	10.16%	5.08%	−2.01%	6.87%
	$81,452–$126,264	12.01%	6.01%	0.67%	9.18%
	$126,264 and over	13.40%	6.70%	2.68%	10.91%
Northwest Territories	$12,664–$36,885	5.90%	2.95%	−8.12%	−0.12%
	$36,885–$73,772	8.60%	4.30%	−4.21%	3.25%
	$73,772–$119,936	12.20%	6.10%	1.01%	7.75%
	$119,936 and over	14.05%	7.03%	3.70%	10.07%
Nunavut	$11,644–$38,832	4.00%	2.00%	−3.20%	0.00%
	$38,832–$77,664	7.00%	3.50%	1.15%	3.75%
	$77,664–$126,264	9.00%	4.50%	4.05%	6.25%
	over $126,264	11.50%	5.75%	7.68%	9.38%

[5] Most Québec taxpayers get a federal tax abatement (reduction) of 16.5% of basic federal tax.

Your federal and provincial income taxes are reduced by **non-refundable tax credits**. The total amount **qualified** for non-refundable tax credits includes the following eligible items. The federal government, and each province/territory, has different threshold amounts for each of these categories.

- Basic personal amount
- Age amount
- Spouse or common-law partner amount
- Amount for dependent children
- Canada or Quebec pension plan contributions
- Employment insurance premiums
- Home buyer's amount
- Pension income amount
- Disability amount
- Tuition, education, and textbook amounts
- Interest on student loans
- Amounts transferred from your spouse or common-law partner
- Medical expenses less adjustment
- Charitable donations[6]

In general, the non-refundable tax credit equals the **total qualified amount** multiplied by the lowest marginal tax rate for the federal or provincial calculations[7] respectively. For example, the **basic federal tax** is calculated as federal income tax, less the non-refundable tax credits times 15%, as well as other tax credits.

In addition to the federal tax payable, provincial/territorial tax on personal income must be considered. Quebec has its own personal income tax that is administered and collected directly by the province. For Quebec residents, the **basic federal tax** is reduced by 16.5% in recognition of a different type of financial arrangement between the federal and Quebec governments. Despite the complexity in the federal and provincial rate structures for personal income tax, the tax system is progressive: the more you earn, the higher the tax rate and the relative amount of taxes. Personal income taxes basically depend upon four factors—where you live, total taxable income, the type of income, and your personal circumstances, which affects deductions and tax credits as discussed above.

EXAMPLE 16.8 **Average and Marginal Tax Rate**

Anya lived in Ontario in 2009 and had a taxable income of $70,000 deriving from wages and investment interest. Estimate her personal marginal tax rate, income tax amount, and average tax rate. The total amount qualified for non-refundable tax credits is $13,000.

SOLUTION

Given: Taxable income of $70,000.
Find: Anya's income tax amount and her marginal and average tax rates.

---

[6] Charitable donations in excess of $200 earn federal non-refundable tax credits at 29% rather than 15%. Similar arrangements apply in some provinces.
[7] Quebec has special rules and rates for tax credits.

(a) With a taxable income of $70,000 the applicable federal rate for that tax bracket is 22.00% from Table 16.1, and the Ontario rate is 10.98%, for a combined marginal tax rate of 32.98%. So, all other things being equal, if Anya earned an additional $100, she would have to pay $32.98 tax on it.

(b) Federal tax $= 15\% \times \$40{,}726 + 22\% \times (\$70{,}000 - \$40{,}726) = \$12{,}549.18$

Minus non-refundable tax credits (15% of $13,000) = $1,950.00

Basic federal tax payable $10,599.18

Ontario tax $= 6.05\% \times \$36{,}848 + 9.15\% \times (\$64{,}882 - \$36{,}848) + 10.98\% \times (\$70{,}000 - \$64{,}882) = \$5{,}356.37$

Minus non-refundable tax credits (6.05% of $13,000) = $786.50

Basic provincial tax payable $4569.87

Therefore, her total tax payable is: $\$10{,}599.18 + \$4{,}569.87 = \$15{,}169.05$

(c) Anya's average tax rate is $(\$15{,}169.05 / \$70{,}000) = 21.67\%$

16.3.3 Timing of Personal Income Tax Payments

In order to include tax effects when evaluating personal investments, a person needs to know when taxes are payable. Depending on personal circumstances, tax payment occurs in various ways and at different times. An employer will automatically withhold a portion of an employee's salary or wage from every paycheque and remit that portion to CRA. However, if a person receives other income that is not taxed at the source like a salary or wage, he or she may be obliged to make monthly or quarterly payments of the estimated taxes on such income. Additionally, by April 30th of each year, an income tax return must be filed for the preceding calendar year. This return represents a formal calculation of the total taxes owed. Any difference between the taxes owed and tax payments already made is either returned as a refund or forwarded as the final payment. The timing of the actual tax payments can be a significant complication in an economic analysis. *For our purposes, we will assume that annual personal income taxes are paid on the last day of each calendar year unless indicated otherwise.*

Based on this assumption on the timing of tax payment, the tax payable on money from an investment in a given calendar year is based on the money or benefit realized in that year and is independent of the point in the year when it was received. For example, $100 of interest received on January 1 has the same tax implication as $100 of interest received on December 31 of the same year. By receiving $100 of interest on January 1, you may choose to spend it or make another investment with it. In the latter case, this second investment may also earn income that results in taxes. These taxes are attributable to the second investment. For GICs and Term Deposits (TDs), the amount of income you report is based on the interest you **earned** during each complete investment year, even if you do not receive the interest until the investment matures. For example, if you invest in a five-year GIC on May 1, 2008, the interest earned up to April 30, 2009, must be reported on your 2009 tax return, even though the interest may be received in 2013. This is called the **accrual method**.

In evaluating personal investments in this book, we will assume that the tax owed as a result of an investment is paid at the end of the calendar year unless specified otherwise.

16.3.4 Professional/Business Income

If you are self-employed, say as an engineering consultant, then you are earning **professional income**. Similarly, a sole proprietor who runs a modest operation is generating **business income**. In both these cases, the individual must prepare a Statement of Business or Professional Activities for income tax purposes (CRA form T2125). The general guidelines for declaring business revenue and claiming operating expenses and capital cost allowance for corporations, as explained in Chapters 8 and 9, also apply in this case, except that the net income from the business is then included in your personal income tax return. So your net income generated through self-employment income is subject to personal taxation rates, not corporate rates.

Your gross business or professional income derive from sales of goods, service fees that you have charged, commissions, and the like. Eligible expenses include the cost of goods sold (e.g. raw materials) if you are producing, operating expenses such as rent, office supplies, utilities, and advertising. Although many self-employed people work out of commercial office space, some use part of their residence as their principal office. A personal car may also be used part-time for the business, such as purchasing or dropping off goods, or visiting clients. In such situations, the individual must prorate the costs of the house or car use, respectively, proportionate to the amount it is used for business. The CRA prescribes the limits on such expenses, and how the calculations may be made. The following example illustrates some of the basic tax elements for a self-employed individual.

EXAMPLE 16.9 Average and Marginal Tax Rate

Phillip had had several years' experience working for a large firm as an industrial engineer. Last New Year's Day his wife got transferred to another city, and since he's always dreamed about being "his own boss," he decided to start up his own consulting business as a quality-control expert. They bought a house for $250,000 (of which the land is worth about $50,000 according to the real estate agent), and he fixed up one of the rooms to use exclusively as his office. The office is 20 m^2, which is one-fifth of the entire floor area of the house. He bought a laptop for $1500 and a multifunction printer for $420 so he could conduct his work effectively. He also bought a car for $25,000, which he used part-time for visiting clients.

He tracked his expenses over the year as follows:

House costs	Heat	$1200
	Electricity	$900
	Water	$300
	Maintenance	$1450
	Insurance	$825
	Mortgage interest	$9700
	Property taxes	$1800
Car costs	Gas	$1900
	Maintenance	$250
Operating costs	Print cartridges	$340
	Stationery	$580
	Advertising	$1400
	Web space	$300

He also kept a driving log and determined that he used the car 40% of the time for business and the rest of the time for personal use.

It was slow getting the word out at first, but through persistent networking and advertising, over the year he succeeded in generating a fair amount of business, charging $73,000 in consulting fees.

SOLUTION

Given: Phillip's self-employment revenue and costs for the past year.
Find: His net income from professional activities.

Proportion of house expenses based on office space: $16,175 \times 1/5 = \$3235$

Proportion of depreciation for the house (Class 1):

$$(\$250,000 - \$50,000) \times 0.04 \times 1/2 * 1/5 = \$800$$

Proportion of car expenses based on business use: $2,150 \times 40\% = \$860$

Proportion of depreciation for car (Class 10): $25,000 \times 0.30 \times 1/2 * 40\% = \$1500$

Operating costs: $2620

Computer hardware depreciation (Class 50): $(\$1,500 + \$420) \times 0.55 \times 1/2 = \$528$

Net income $= \$73,000 - (\$3235 + \$800 + \$860 + \$1500 + \$2620 + \$528)$
$= \$63,457.$

DISCUSSION: This example illustrates how deductions based on operating expenses and use of house and car for business purposes can lower your taxable income. It is also interesting to note that when part of a personal residence is used for business, part of the mortgage interest is tax deductible. However, there may also be income tax implications when the house is sold. Other deductions are commonly used by self-employed individuals, such as health-care premiums, so careful consideration is required to take advantage of all available benefits.

16.4 Effects of Income Taxes on Investment Returns

As noted above, individual income tax rates are progressive and certain types of income receive preferential tax treatment. Such treatment is offered as an incentive for individuals to invest, not only for personal gain, but also as a way of fostering overall economic growth. For example, capital gains are taxed at ½ of their actual value. Table 16.1 shows the effective marginal tax rates for capital gains for federal and provincial taxes. Note that investment income from foreign sources is not taxed as favourably as those from Canadian sources.

Dividends from Canadian corporations also end up being taxed at a lower rate because of a special tax credit. The rates also depend on whether the dividends are eligible for a preferential treatment because they are paid out from Canadian public corporations (or certain qualifying private corporations). Eligible dividends are included in income at 145% of their actual value (referred to as **grossed up amount**) but earn a federal dividend tax credit equal to 18.9655% of the dividend amount included as income. Dividends that are not eligible are included in income at 125% of their actual value but earn a federal dividend tax credit equal to 13.3333% of the dividend amount included as income. Provincial/territorial tax systems also give preferential treatment for dividend income. The marginal tax rates for dividends shown in Table 16.1 already account for the grossing up and the tax credit, so they should be applied to the value of actual dividends received (not the grossed-up amount). At the time of writing, the governments in Canada are actively modifying tax rules concerning dividends to improve their competitiveness.

16.4.1 Bond Investments After Tax

There are two ways for an investor to earn income from a bond—interest and a capital gain on disposal. Most government bonds (Canada Savings Bonds, Alberta Capital Bonds, etc.) can only be purchased and redeemed at their face value. Therefore, there is no possibility of a capital gain (or loss) for these bonds. On the other hand, corporate bonds and certain government bonds are publicly traded and the market price of such a bond is determined by its coupon rate relative to the prevailing money market interest rates and the perceived credibility and financial stability of the issuing company. If the bond coupon rate is greater than the prevailing money market interest rate, investors *may* pay a premium over the face value of the bond to obtain the higher coupon rate. When the bond coupon rate is less than the prevailing money market rate, investors *may* pay less than the face value. Evaluation of bond investments, including tax effects, is based on the usual after-tax cash flow analysis.

EXAMPLE 16.10 **Evaluation of Bond Investment**

A corporate bond with a face value of $1000 was purchased for this amount at its January 1 time of issue. It is a simple interest bond that pays 6% interest on June 30 of each year. Determine the required selling price on December 31 after three years of ownership in order to realize an annual 10% after-tax rate of return. Taxes are paid at the end of each year. The marginal tax rates on interest income and capital gains are 40% and 20% respectively.

SOLUTION

Given: 6% bond, purchased at face value of $1000.

Find: The selling price, S, after three years to provide a 10% after-tax rate of return.

Annual interest payment $= \$1000 \times 0.06 = \$60$/year on June 30.

Income tax on interest $= 0.4 \times \$60 = \$24$/year at the end of year.

Capital gain on sale $= S - \$1000$

Capital gains tax $= 0.2\,(S - \$1000)$ at the end of year.

The after-tax cash flow diagram for this investment is given in Figure 16.8. The required rate of return = 10% on 12-month basis, which is $[(1.10)^{1/2} - 1] \times 100 = 4.88\%$ on a six-month basis.

$$PW = -\$1000 + \$60(P/A, 10\%, 3)(F/P, 4.88\%, 1) - \$24(P/A, 10\%, 3)$$
$$+ [S - 0.2(S - \$1000)](P/F, 10\%, 3)$$
$$= 0$$

Solving the above equation for S yields

$$S = \$1252.68$$

COMMENTS: This means that selling the bond at $1252.68 will yield a 10% after-tax rate of return. If the selling price is higher than $1252.68, the after-tax rate of return will be higher than 10%.

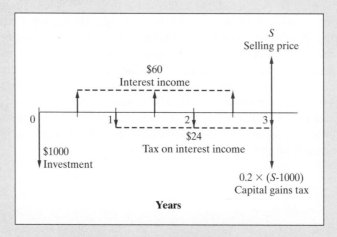

Figure 16.8   Bond after-tax cash flow diagram (Example 16.10).

EXAMPLE 16.11 How Much to Pay for a Bond

A publicly traded bond series has a coupon rate of 8%, a face value of $1000, and a maturity date of December 31, 2015. The interest is paid semiannually. Assume that today's prevailing coupon rate for bonds with similar quality is 7% with interest paid semiannually. How much would an investor be willing to pay to buy this bond on January 1, 2010? Assume that the investor's marginal tax rate is 45% on interest income. Assume that the tax rate on capital gains is 50% of the tax rate on interest income.

SOLUTION

Given: $1000 8% bond, semiannual interest payment.

Find: The purchase price, P, which provides a return equal to that of comparable quality bonds.

The bond under consideration pays a higher interest rate than newly issued bonds of similar quality. Thus, investors would be willing to pay more than its face value for this bond for the higher interest rate. The value of the bond on January 1, 2010, is determined by its earning potential if it is bought and kept to maturity. If the bond is bought for a premium and later redeemed for its face value, there will be a capital loss. We assume that there are other capital gains to offset the capital loss when the bond is redeemed at maturity in year 2015.

$$\text{Interest cash flow} = \$1000 \times 4\%$$
$$= \$40 \text{ every 6 months.}$$
$$\text{Annual tax payment on interest income} = \$40 \times 2 \times 45\%$$
$$= \$36.00 \text{ every year.}$$
$$\text{Tax saving on capital loss at maturity} = 0.5 \times 45\% \times (P - \$1000)$$
$$= 0.225\,P - \$225.$$

The prevailing bond interest rate is 7% per annum compounded semiannually. The corresponding after-tax rate of return is $7\%(1 - 0.45) = 3.85\%$ per annum. The semiannual rate of return is then $3.85\%/2 = 1.925\%$ and the annual effective interest rate is $i_a = (1 + 1.925\%)^2 - 1 = 3.887\%$.

The after-tax cash flow diagram shown in Figure 16.9 is used to find the present value of the bond under consideration. By setting the present equivalent of all cash flows in the after-tax cash flow diagram equal to zero, we obtain the following equation:

$$\text{PW} = -P - \$36.00(P/A, 3.887\%, 6) + \$40(P/A, 1.925\%, 12)$$
$$+(\$1000 + 0.225\,P - \$225)(P/F, 1.925\%, 12)$$
$$= 0.$$

Solving for P, we find $P = \$1037.81$.

COMMENTS: If an investor pays \$1037.81 to buy the bond, he or she would earn the same rate of return as a newly issued bond series on January 1, 2010. If such a bond can be acquired at a price lower than the calculated \$1037.81, the rate of return would be higher than the prevailing bond investments.

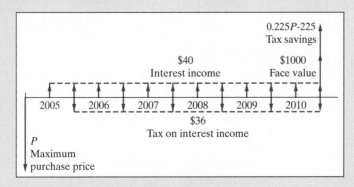

Figure 16.9   Cash flow diagram (Example 16.11).

16.4.2 Stock Investments After Tax

Stock investments with cash flow predictions may be analyzed in a way similar to bond investments, as they also involve dividends and capital gains with their respective tax rules. The following example illustrates the basic principles, noting that the tax implications can be more complicated if shares from a company are actively traded, if foreign stock is owned, or if the investment is sheltered in an RRSP or not (see section 16.4.3).

EXAMPLE 16.12 After-Tax Rate of Return Calculation of Stock Investment

Gerry purchased a newly issued stock on January 1, 2002, for $20 per share. The shares have paid quarterly dividends of 50¢/share since their acquisition. He sold this stock on January 1, 2009, for $29/share. Gerry, who lives in Regina, Saskatchewan, had a taxable income of $73,000 in 2009. Determine the annual after-tax rate of return on this investment, given that the shares were from an eligible Canadian corporation, and taxes are payable on December 31 each year.

SOLUTION

Given: Stock transactions as described above.

Find: The annual after-tax rate of return on the stock.

From Table 16.1, a resident of Saskatchewan in 2009 with a taxable income of $73,000 has a marginal tax rate on capital gains of 17.5% (11% federally and 6.5% provincially) and is taxed on eligible dividends at a combined rate of 7.3% (4.4% federally and 2.9% provincially).

$$\text{Annual taxes on dividends} = 0.073 \times (4 \text{ quarters} \times \$0.50/\text{quarter})$$
$$= \$0.146$$
$$\text{Capital gains} = \$29 - \$20 = \$9 \text{ on January 1, 2009}$$
$$\text{Tax on capital gains} = \$9 \times 0.175 = \$1.575 \text{ on December 31, 2009.}$$

The after-tax cash flow diagram for this investment is shown in Figure 16.10.

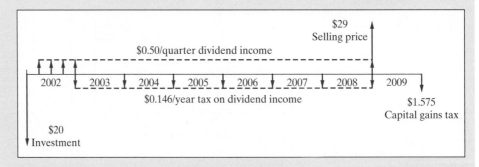

Figure 16.10 Cash flow diagram (Example 16.12).

The effective quarterly after-tax rate of return, i_Q, is found by solving:

$$PW = -\$20 + \$0.50(P/A, i_Q, 28) - \$0.146(A/F, i_Q, 4)\,(P/A, i_Q, 28)$$
$$+ \$29(P/F, i_Q, 28) - \$1.575(P/F, i_Q, 32)$$
$$= 0.$$

We find that PW = 0 at $i_Q = 3.188\%$ per quarter.

The effective annual after-tax rate of return is: $i_a = (1 + i_Q)^4 - 1 = (1 + 2.86\%)^4 - 1 = 13.37\%$.

COMMENTS: As described in the Section 16.4 introduction, taxes on dividends are actually calculated by grossing up the value of the dividends income and then subtracting a percent of the value of the dividends from the total tax owed. However, using a single dividend tax rate as shown in here is often sufficient for evaluating investment alternatives.

The capital gain is easy to calculate if stock is bought all at once, then sold all together. But investors often add to their stock holding, or sell off part of it. Since the capital gain (or loss) at disposal depends on the purchase price, the **adjusted cost base** (ACB) serves as a measure of the average cost of your investment in a stock to date. When you buy more of the same stock, the total number of shares you then have divided by the total spent to buy them, gives the ACB. Any fees you incurred when buying or selling are also included in the ACB.

EXAMPLE 16.13 Adjusted Cost Base

Reconsider Example 16.12. Say Gerry had bought 1000 shares initially (for $20 each on January 1, 2002) and then another 600 shares for $18 each on January 1, 2005. Recalculate his after-tax rate of return when he sells them all on January 1, 2009. Assume all fees are included in the purchase and sale price, and that dividend payouts per share are constant.

SOLUTION

Given: Stock transactions as described above.

Find: The ACB and the new annual after-tax rate of return on the stock.

$$\text{Adjusted Cost Base (Jan. 1, 2005 onward)} = \frac{1000(\$20) + 600(\$18)}{1000 + 600} = \$19.25$$

$$\text{Capital gains} = \$29 - \$19.25 = \$9.75 \text{ on January 1, 2009}$$
$$\text{Tax on capital gains} = 9.75 \times 0.175 = \$1.706 \text{ on December 31, 2009.}$$

The after-tax cash flow diagram for this investment is shown in Figure 16.11.

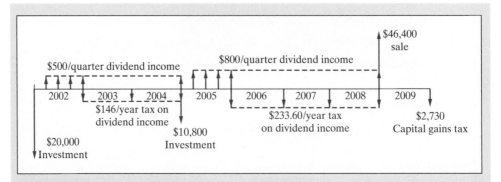

Figure 16.11 Cash flow diagram (Example 16.13).

The effective quarterly after-tax rate of return, i_Q, is found by solving:

$$\text{PW} = -\$20(1000) - \$18(600)\,(P/F, i_Q, 12)$$
$$+ \$29(1600)(P/F, i_Q, 28) - \$1.706(1600)(P/F, i_Q, 32)$$
$$+ \$0.50(1000)(P/A, i_Q, 28) + \$0.50(600)(P/A, i_Q, 16)(P/F, i_Q, 12)$$
$$- \$0.146(1000)(A/F, i_Q, 4)\,(P/A, i_Q, 28)$$
$$- \$0.146(600)(A/F, i_Q, 4)\,(P/A, i_Q, 16)\,(P/F, i_Q, 12)$$
$$= 0.$$

From this, $i_Q = 3.52\%$ per quarter, or $i_a = 14.85\%$ per year.

16.4.3 Registered Retirement Savings Plans (RRSPs)

One of the allowed deductions for tax purposes is RRSP contributions. This tax deductibility means that contributions are made with before-tax dollars. Furthermore, income earned on these contributions is not taxed within the plan. Taxes are payable on any withdrawals from the plan at the time of withdrawal. In general, individuals fall into a lower income tax bracket on retirement so withdrawals after retirement tend to be taxed at a lower rate. In addition, deferring tax payment makes sense even if the individual were in the same tax bracket on retirement because of the time value of the tax savings today. The maximum annual RRSP contribution is regulated by the government. The following example serves to demonstrate the advantage of RRSP investment compared to a savings account.

EXAMPLE 16.14 Comparison of an RRSP Account With a Savings Account

An engineer wishes to save for retirement by making annual contributions over a period of 30 years. These savings would be drawn down over a 10-year period after retirement.

- Marginal tax rate   45%   years 1 to 30
- Average tax rate   35%   years 31 to 40

The savings will earn 10% per year and the investor can afford contributions of $2000 per year provided there is no tax on this money. Compare the annual after-tax cash from these savings after retirement if the savings were in an RRSP versus a savings account.

SOLUTION

Given: Annual contributions to 1) RRSP and 2) savings account, for 30 years and then withdrawals over 10 years.
Find: The annual after-tax cash available after retirement from each of the two options.

1. RRSP contributions are $2000 annually because they can be made with before-tax dollars. The cash flow diagram for this option is given in Figure 16.12.

$$F \text{ at year } 30 = \$2000(F/A, 10\%, 30)$$
$$= \$329\text{K}.$$

Annual withdrawals, A, between years 31 and 40 $= \$329\text{K}(A/P, 10\%, 10)$
$$= \$53.6\text{K}.$$

RRSP withdrawals are taxable at 35%

After-tax cash $= (1 - 0.35) \times \$53.6\text{K}$
$$= \$34.8\text{K/year}.$$

2. Savings account contributions involve after-tax dollars. Annual contributions available are $\$2000 \times (1 - 0.45) = \$1100$.

The cash flow diagram looks the same as the RRSP option except that the annual deposit over the first 30 years is only $1100 per year.

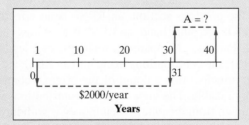

Figure 16.12 Cash flow diagram for RRSP account (Example 16.14).

The interest on a savings account in any year is taxable. Assume year-end interest and tax payments.

$$i_{bt} = \text{before-tax interest rate} = 10\%$$
$$i_{at} = \text{after-tax interest rate}.$$

Over one year, the future equivalent, F, of a present sum, P, after-tax is

$$F = (1 + i_{bt})P - t(i_{bt})P$$
$$= P[1 + (1 - t)i_{bt}]$$
$$= P(1 + i_{at}).$$

The after-tax interest rate is:

$$i_{at} = (1 - t)i_{bt}.$$
$$i_{at} = (1 - 0.45) \times 10\%$$
$$= 5.5\% \text{ for 30 years}$$

$$F \text{ at year } 30 = \$1,100(F/A, 5.5\%, 30)$$
$$= \$79.7\text{K}.$$

Between years 31 and 40

$$i_{at} = (1 - 0.35) \times 10\% = 6.5\%$$
$$A = \$79.7\text{K}(A/P, 6.5\%, 10)$$
$$= \$11.1\text{K}$$

which is the annual after-tax cash available, as the taxes have already been paid! Note that the RRSP provides a significantly larger annual income after retirement.

16.4.4 Registered Education Savings Plans (RESPs)

As we discussed in the previous section, an RRSP is an effective savings vehicle for one's own retirement. On the other hand, an RESP is a savings vehicle for a child's post-secondary education. Parents, grandparents, relatives, or friends can contribute to the RESP. Unlike the RRSP, the contributions made into an RESP are not tax deductible and as a result, when the principal (the amount of the original contributions) is withdrawn from the plan, it is not taxable. The growth of the contributions in the RESP is tax-free until money is withdrawn from the plan. When money is withdrawn from the plan by the beneficiary for post-secondary education, it is taxable in the hands of the beneficiary. As we know, the tax to be paid by a student is often zero or very small. The contribution limits for RESP are $4000 per year for 21 years with a lifetime limit of $50,000 per beneficiary (as of 2009). The money in an RESP can be invested in GICs, stocks, bonds, and/or mutual funds.

Another very attractive feature of RESP is that the contributions to it may qualify for a Canadian Education Saving Grant (CESG). The CESG was introduced in 1998. For the contributions made into each beneficiary's RESP, the federal government will pay a grant equal to 20% of annual contributions to a maximum of $400 per year. For lower income families the grant may reach $500/year. This grant is deposited into the RESP directly, is not a taxable income, and is not included in the calculation of RESP contribution limits. Of course, the growth of this grant in the plan is also tax-free until money is withdrawn from the plan. This grant and its growth are strictly for the beneficiary's post-secondary education. If the child decides not to continue with post-secondary studies, there are many options available to deal with the RESP funds:

- It can be left for up to 26 years, in case later studies are undertaken.
- It can be transferred to a sibling's RESP.
- It can be transferred to an RRSP.
- It can be withdrawn, but the tax and repayment options are complicated.

Interested students may find more information on RESP and CESG on the Internet. The evaluation of the RESP is similar to that of the RRSP.

SUMMARY

- **Personal investments** can provide returns in the form of **interest** (for example, savings accounts, guaranteed investment certificates, and bonds), **dividends** (stocks), and **capital gains** (bonds, stocks, and properties).
- Investing your money involves various types of **risk** such as losing part or all of your capital, not achieving your investment goals, or having inflation erode the value of your investments.
- Pursuing a greater rate of return generally involves taking greater risks, referred to as the **risk/reward tradeoff. Diversification** can reduce the risk.
- **Dollar cost averaging** involves uniform periodic investments in a stock, such that more shares are acquired when the price is low and fewer when the price is high.
- Returns earned on personal investments are taxable at rates appropriate to the **types of income**. Analyses of personal investments must include income tax payments at the appropriate times. The **marginal tax rates** vary by province/territory.
- **Business and professional income** get taxed the same way as personal income, but enjoy some of the same tax benefits as corporations.
- **Registered retirement savings plans**, RRSPs, represent a highly attractive investment for retirement. Contributions to an RRSP account and the income earned by the contributions are not taxable until money is withdrawn from the account.
- **Registered education savings plans**, RESPs, represent a savings vehicle for post-secondary education children. The contributions made into an RESP account are not tax deductible. However, the growth of the money in the account is tax-free until money is withdrawn for post-secondary education purposes. In addition, the federal government offers a grant for contributions in RESPs.

PROBLEMS

Note: Unless stated otherwise, assume that tax rates provided in Table 16.1 are applicable.

16.1 Ajax spends €2 every day at his local café for a double espresso. Fifteen years later he loses his job. How much money would he have saved annually if he didn't purchase his daily espresso? Assume that €1 = $1.30, and neglect inflation.

a) In actual dollars?

b) If he earned $57,000 annually, is a resident of Yukon, what percentage of his after-tax dollars was spent annually on espresso?

c) How much in pre-tax dollars was spent annually on espresso?

16.2 Agamemnon, a resident of Alberta with an annual salary of $82,000, bought a bond for $1800 on January 1, 2004. The bond had a face value of $2000 and a coupon rate of 9% with interest paid semiannually. He sold the bond for $2000 on December 31, 2009. What was the annual after-tax rate of return that he made from this investment?

16.3 Helen bought 1000 Golden Apple's stocks at $45.50 per share 10 years ago. Over the years, the dividend payments had been at 10¢ per share paid at the end of each year. She lost her job, which paid her $10,000/year on Halloween and moved from Quebec to Prince Edward Island. In the New Year, she sold the stocks at $175 per share. What will be her total proceeds after-tax from this investment?

16.4 The Paris Corporation issued a new series of bonds on January 1, 2001. The bonds were sold at par ($1000), have a 12% coupon rate, and mature in 30 years, on December 31, 2030. Coupon interest payments are made semiannually. An investor in Nova Scotia, with a marginal tax-rate of 38.67% has been interested in these bonds.

a) What was the after-tax yield-to-maturity (YTM) of the bond if the investor bought it on January 1, 2001?

b) The interest rate fell to 9% on January 1, 2008. This meant that the coupon rate of newly issued bonds with quality similar to the Paris Corporation's bond was set at 9% with interest paid semiannually. Assume such newly issued bonds could be bought at par and had the same maturity as the Paris Corporation's bond. What was the after-tax YTM of these newly issued bonds?

c) What was the maximum amount that the investor was willing to pay for the Paris Corporation's bond on January 1, 2008?

d) On July 1, 2008, the investor bought some Paris bonds for $922.38 each. What was the after-tax YTM at that date?

e) What was the current after-tax yield on July 1, 2008?

16.5 A $1000, 9.5% semiannual bond is purchased by Manitoban resident Diomedes for $1010. His annual salary is $46,000. If the bond is sold at the end of three years after six interest payments, what should the selling price be to yield a 7% after-tax rate of return?

16.6 Odysseus wishes to sell a bond that has a face value of $1000, bought four years ago for $920. The bond bears an interest rate of 8% payable semiannually.

Odysseus is a resident of Saskatchewan with an annual salary of $51,000. What must be the minimum selling price in order to realize a 10% after-tax rate of return?

16.7 Polyxena is considering purchasing a 6% bond with a face value of $1000 with interest paid semiannually. She desires to earn a 9% annual return on her investment. Polyxena is a resident of Ontario and receives a salary of $100,000. Assume that the bond will mature to its face value five years hence.

a) What is the required purchasing price of the bond?

b) What is her after-tax rate of return?

c) Can you explain why it is much smaller than the required 9%?

16.8 Patroclus had the choice of investing in:

1) A zero coupon bond that costs $513.60 today, pays nothing during its life, and then is redeemed for $1000 after five years

2) A bond that costs $1000 today, pays $113 in interest semiannually, and matures at par ($1000) at the end of five years.

Assuming that Patroclus is from British Columbia and earns $63,000, which bond would provide the higher after-tax yield?

16.9 Suppose Achilles, also from British Columbia, was offered a 12-year, 15% coupon, $1000 par value bond at a price of $1298.68. What after-tax rate of interest (yield-to-maturity) would he earn if he bought the bond and held it to maturity (semiannual interest)? His annual salary is $76,000.

16.10 The Priam Products Company has two bond issues outstanding. Both bonds pay $100 semiannual interest plus $1000 at maturity. Bond Alpha has a remaining maturity of 15 years, and bond Beta a maturity of one year. A wealthy ($88,000 annual income) investor in Alberta is interested in these bonds.

a) What will be the value of each of these bonds now, when the going rate of interest is 9%?

b) Is your answer different from the one you get when tax effects are ignored?

16.11 Hector's bonds have four years remaining to maturity. Interest is paid annually; the bonds have a $1000 par value; the coupon interest rate is 8.75%. Suppose that both you and Hector are residents of Newfoundland, each with an annual income of about $45,000.

a) What is the after-tax yield-to-maturity at a current market price of $1,108?

b) Would you pay $1035 for one of these bonds if you thought that the market rate of interest was 9.5%?

16.12 Suppose Apollo Inc. issued bonds with a 15-year maturity, a $1000 par value, a 12% coupon rate, and semiannual interest payments. Assume that the investor, Jean, is from Moncton, New Brunswick, and has an annual salary of $55,000.

a) Two years after the bonds were issued, the going rate of interest on bonds such as these fell to 9%. At what price would an investor buy the Apollo bonds from Jean?

b) Suppose that, two years after the issue, the going interest rate had risen to 13%. At what price would an investor buy the Apollo bonds?

c) The closing price of the bond at that time was $783.58. What was the after-tax current yield?

16.13 Aeneas bought a stock five years ago for $20.20 per share. Recently he sold it for $42.40 per share. In the past five5 years, he has received dividends of 50¢ per share at the end of each year. What is the after-tax rate of return he has earned from this investment if he is a resident of Ontario with an annual salary of $94,000?

16.14 Cassandra is living in Ontario, has $10,000 available for investment on top of her $120,000 salary, and is comparing the following options:

The Trojan:	A bond series with $1000 face value, 13% coupon rate (semiannual interest payment), selling price of $1020, and maturity in 10 years.
The Greek:	A bond series with $1000 face value, 10% coupon rate (semiannual interest payment), selling price of $980, and maturity in 10 years.
The Hera:	A stock with a price of $10/share, dividends of 10¢/share/year, and estimated price of $30/share in 10 years.
The Zeus:	A stock with price of $15/share, dividends of 2¢/share/year, and estimated price of $50/share in 10 years.

Which one would she prophesize to be the best choice? Why?

16.15 The 2010 Series of Ontario Savings Bonds offers the following rates:

3-Year Fixed-Rate Bond	2.50%
7-Year Fixed-Rate Bond	3.75%
10-Year Fixed-Rate Bond	4.25%

Comparatively, the Step-Up (SU) Bond has the following rate profile:

Period	Annual Interest Rate
6/21/2010–6/20/2011	1.00 %
6/21/2011–6/20/2012	2.00 %
6/21/2012–6/20/2013	3.00 %
6/21/2013–6/20/2014	3.75 %
6/21/2014–6/20/2015	4.25 %

Assume that you are currently a student and thus unemployed; what are the equivalent annual after-tax rates of return for the Step-Up Bond after:
a) three years?
b) five years?
c) How does this compare with the fixed-rate bonds?

16.16 Dilan just made $10,000 working at Tim Hortons over the summer and decided to invest in the stock market. After skimming through the business section, he decided to invest in ENG.TO, a new company specializing in Engineering Economics on the TSX. The stock is priced at $4, and Dilan invests it all. A few weeks go by and the stock has gone down to $3.64. Dilan panics and sells 25% of his shares. A month passes by and now the stock is at $3.28, and he sells 40% of his remaining shares. It is the end of the fiscal quarter and earnings at ENG.TO are not what was expected, the share drops to $2.82. Dilan hears on the news that it is important to buy low and sell high, so he spends all of his cash on hand to buy as many stocks as he can. A few days later the stock continues to slide to $2.72, but Dilan just ignores it and continues to hope for the best. Unexpectedly, another company offers a take-over bid of ENG.TO for $3.72 a share and the board of directors accept it. Dilan gladly sells all of his shares! How much money has he made (or lost)? Assume stocks must be purchased in multiples of 25.

16.17 Onur decides to spend $1000 a month on RIM.TO and the following are the prices that he has bought them at over the last year:

September 1, 2009	$81.21
October 1, 2009	$72.81
November 2, 2009	$60.15
December 1, 2009	$62.61
January 4, 2010	$68.74
February 1, 2010	$68.06
March 1, 2010	$73.47
April 1, 2010	$69.05
May 3, 2010	$72.19
June 1, 2010	$62.27
July 2, 2010	$51.11
August 3, 2010	$56.77

a) How many stocks does he own to date?
b) What price does the stock have to reach before he breaks even?

16.18 Tony decides to buy shares of BP over an eight-month period, starting with 1000 shares on January 4, 2010, and then increasing his purchase quantity by 1000 each month thereafter. Assuming his initial purchase is 1000 shares, how many shares does he own at the end of August, how much has he invested, and at what price does he break-even?

January 4, 2010	$59.15
February 1, 2010	$57.23
March 1, 2010	$53.98
April 1, 2010	$57.74
May 3, 2010	$50.19
June 1, 2010	$36.25
July 1, 2010	$29.39
August 2, 2010	$39.42

16.19 Loki, currently residing in Iqaluit, Nunavut, has decided to invest $100,000 in his PC Financial Interest Plus™ savings account, which provides a rate of 1.50% annually. His regular job also pays him $200,000 annually, but he manages to spend it all each year.
   a) How much tax will he pay in his first year?
   b) How much tax will he pay in his fifth year if he lets the interest compound?
   c) How much tax does he pay in total over the five years?
   d) How much more tax would he pay in the same five-year period if he lived in Yellowknife, Northwest Territories?

16.20 Thor, an investor in Vancouver, British Columbia, is comparing the stocks of the following two companies.

   Muspelheim: Current price = $10/share
                  Dividends = 5¢/share/quarter
                  Expected price in 5 years = $45

   Niflheim: Current price = $5/share
                  Dividends = 2.5¢/share/quarter
                  Expected price in 5 years = $15

   Dividends are paid once per quarter. Thor's current salary is $75,000/year but is expected to rise $5,000/year. After two years, he moves to Fort McMurray, Alberta, for a $30,000 raise with subsequent annual salary increase of $7500/year.
   a) Which stock would provide a higher rate of return?
   b) Which stock would provide a higher after-tax rate of return?
   c) Which stock would provide a higher after-tax rate of return if he stayed in B.C.?

16.21 Freya owns 10,000 shares of Penn West Energy Trust (PWT.UN), which she purchased on September 1, 2009, at $13.85 each. She sells all of her shares on August 9, 2010, at a price of $20.50. Moreover, every month she received a distribution of 15¢/share for a total of 11 periods. Assume distributions are taxed the same way as dividends.
   a) What is her % gain on the share-price?
   b) Using her purchase price, what is the rate of return on her distributions alone?
   c) Using her sale price, what is the rate of return on her distributions alone?
   d) How much tax does she pay if she is from Yukon and her salary is $88,000?

16.22 Odin has decided to set up an RRSP account for his retirement. He wishes to live on $40,000 (constant dollars)/year (withdrawn at the start of each year), when he turns 65 until Ragnarök (his inevitable end), when he turns 85. Assume his investment is expected to earn 10% per year within the RRSP, the average annual inflation rate is 4.2%, and his marginal income tax rate post-retirement will be 25%. How much should he invest on an annual basis if he is turning 30 tomorrow?

Short Case Studies

ST16.1 You have $1500 to invest in a choice of five stocks (A, B, C, D, E). You may put all the money in one stock or spread it evenly between all five. Let stock A=1, B=2, C=3, D=4, E=5. Now, roll a die 10 times. On every odd roll, for the number that comes up that corresponding stock will decrease by 10%, and on every even roll the stock that matches the die will increase by 10%. If a 6 is rolled, the entire portfolio will decrease/increase by 50% depending on whether the roll is an odd one (decrease) or an even one (increase) (e.g., your third roll is an odd one). How well has your portfolio performed when you:
   a) Spread it evenly between all five?
   b) Spread the money evenly in only stocks A, C, E?
   c) Spread the money evenly in only stocks B, D?
   d) Put all the money in one stock? Does the stock you pick make a difference?

ST16.2 Using the data from Table 16.1, create a graph of how much tax an investor would pay in each province as their income rises by $10,000, from $0 to $200,000 in each province, for:
   a) GICs
   b) Capital gains
   c) Dividends
   d) A portfolio of 15% GICs, 55% capital gains, 30% dividends.

On the Companion Website that accompanies this text, you will find Excel templates and exercises, as well as the following analysis tools: Cash Flow Analyzer, Depreciation Analysis, Loan Analysis, and Interest Tables.

ENGLISH/FRENCH CONCORDANCE

English	French
50% rule	règle de 50%
ABC (activity-based costing)	comptabilité par activités
accounting depreciation	amortissement comptable
accrual method	méthode de la comptabilité d'exercice
acquisitions	acquisitions
activity-based costing (ABC)	comptabilité par activités
actual dollars	dollars courants, dollars du moment
add-on interest	intérêt ajouté au montant prêté
add-on loan	prêt avec l'intérêt ajouté
adjusted cost basis	prix de base rajusté
alternatives	alternatives, options
amortization	amortissement
amortized loan	prêt amorti
analysis period	période d'analyse
annual equivalent cost	coût équivalent annuel
annual equivalent worth	valeur annuelle équivalente
annual percentage rate (APR)	taux composé annuel
annuity	annuité
annuity due	annuité de début de période
annuity factor	facteur d'annuité
APR (annual percentage rate)	taux composé annuel
asset	actif
asset class (pool)	catégorie de biens
asset depreciation	dépréciation de l'actif
asset turnover	rotation de l'actif
availability	disponibilité
available-for-use	biens prêts à être mis en service
average cost	coût moyen
average tax rate	taux d'imposition moyen
balance owing	reliquat, solde
balance sheet	bilan
base period	période de base
basic federal tax	impôt fédéral de base
bearer bond	obligation au porteur
benchmark	point de repère
benefit-cost analysis	analyse coûts-avantages
benefit-cost ratio	ratio avantages-coûts
benefits	prestations
betterment	amélioration
bond	obligation
bond financing	financement par obligations
book depreciation	amortissement aux livres
book value	valeur comptable
borrowed funds	fonds empruntés
break-even analysis	analyse du seuil de rentabilité
break-even interest rate	taux d'intérêt au point mort
break-even point	point mort
business	entreprise
business income	revenu d'entreprise
callable bond	obligation remboursable par anticipation
Canada Revenue Agency	Agence de revenu du Canada
Canada Savings Bonds	Obligations d'épargne du Canada

capital	capital
capital assets	immobilisations
capital budgeting	budgétisation des investissements
capital cost	valeur amortissable
capital cost allowance (CCA)	déduction pour amortissement (DPA)
capital cost tax factor	facteur d'impôt du coût en capital
capital expenditure	dépense en capital
capital gains	gains en capital
capital goods	biens d'équipement
capital investment	investissement en capital
capital losses	pertes en capital
capital property	immobilisation
capital rationing	rationnement du capital
capital recovery cost	coût de recouvrement du capital
capital recovery factor	facteur de recouvrement du capital
capital structure	structure du capital
capitalization	capitalisation
capitalized	capitalisé
capitalized cost	coût capitalisé
capitalized equivalent	coût capitalisé équivalent
carrybacks	reports rétrospectifs
cash	valeurs disponibles
cash flow diagram	diagramme de flux monétaires, diagramme de transactions monétaires
cash flow statement	état des flux de trésorerie
cash flows	mouvement de trésorerie
cash inflows	rentrée de fonds, encaissement
cash outflows	sortie de fonds, décaissement
CCA (capital cost allowance)	DPA (déduction pour amortissement)
CCA rate	taux de DPA
challenger	opposant
class number	numéro de catégorie
closed mortgage	prêt hypothécaire fermé
collateral mortgage	hypothèque mobilière
combined corporate tax	taux d'imposition des sociétés combiné
commissioning	mise en service
common service period	période de service commune
common stock	actions ordinaires
common-multiple method	méthode de multiple commun
composite cash flows	flux monétaires composés
composite series	séries monétaires combinées
compound amount factor	facteur de valeur accumulée
compound growth	taux de croissance composé
compound interest	intérêt composé
compounding period	période d'actualisation
compounding process	capitalisation
conditional probability	probabilité conditionnelle
constant dollar	dollar constant
constant-dollar analysis	analyse en dollars constants
Consumer Price Index (CPI)	indice des prix à la consommation
contingent projects	projets conditionnels
contingent, conditional	conditionnel
continuous compounding	capitalisation continue
continuous discounting	actualisation continue
continuous random variables	variables aléatoires continues
continuous-funds flow	flux monétaires de fonds continus

contribution	contribution
conventional mortgage	prêt hypothécaire conventionnel
convertible bond	obligation convertible
corporation	société
cost	coût
cost accounting	comptabilité des coûts
cost basis	coût en capital
cost effectiveness	efficacité des coûts
cost of capital	coût du capital
cost of debt	coût de la dette
cost of equity	coût des fonds propres
cost of goods sold	coût des produits vendus
cost savings/cost reduction	économies sur les coûts/réduction des coûts
cost-benefit analysis	analyse avantages-coûts
cost-effectiveness ratio	ratio coût-efficacité
cost-of-living index	indice du coût de la vie
coupon rate	taux d'intérêt contractuel
CPI (Consumer Price Index)	
creditors	créancier
cumulative distribution	distribution cumulative
current assets	actif à court terme
current liabilities	passif à court terme
current market value	valeur marchande actuelle
current ratio	ratio de liquidité générale
current yield	taux de rendement courant
daily compounding	actualisation quotidienne
debenture	débenture
debenture bond	obligation non garantie
debt	dette
debt financing	financement par emprunt
debt ratio	ratio d'endettement
decision criterion	critère de décision
decision tree	arbre de décision
decisions under risk	décisions dans un contexte aléatoire
decisions under uncertainty	décisions dans un contexte incertain
declining balance depreciation	amortissement avec solde dégressif
defender	défenseur
deferred annuity	pension différée
deferred investment	investissement différé
deflation	déflation
dependent random variables	variables aléatoires dépendantes
depreciable assets	biens amortissables
depreciable life	durée amortissable
depreciation	amortissement
depreciation expense	dotation aux amortissements
depreciation schedule	calendrier d'amortissement
direct cost	coût direct
direct labour	main-d'oeuvre directe
disbenefits	pertes de bénéfices
disbursements	déboursés
discount bond	obligation avec escompte d'émission
discount rate	taux d'actualisation
discounted cash flow	flux de caisse actualisé
discounted payback period	période de récupération actualisée
discounting	actualisation
discounting factor	facteur d'actualisation
discrete	discret
discrete random variables	variables aléatoires discrètes

disposal tax effect	effet de la disposition sur les impôts
disposition	disposition
diversification	diversification
dividend	dividende
dollar-cost averaging	achats périodiques par sommes fixes
do-nothing alternative	option nulle
due date	échéance
earning power	pouvoir de gain
earnings per share	bénéfice par action
economic depreciation	dépréciation économique
economic equivalence	équivalence économique
economic service life	durée de vie économique
effective annual yield	taux annuel effectif
effective interest rate	taux d'intérêt effectif
end-of-period convention	convention de fin de période
engineering economic decisions	décisions économiques en ingénierie
engineering economics	l'économie de l'ingénieur
equal payment series	annuité
equal payment series compound-amount factor	facteur de capitalisation d'une annuité
equal payment series sinking-fund factor	facteur d'amortissement d'une annuité
equipment replacement	remplacement d'équipement
equity	capitaux propres
equity financing	financement par actions
equity ratio	ratio d'endettement
equivalence	équivalence
equivalent annual cost	coût annuel équivalent
EVPI (expected value of perfect information)	valeur espérée de l'information parfaite
expected value	valeur espérée
expected value of perfect information (EVPI)	valeur espérée de l'information parfaite
expenses	dépenses
external interest rate	taux d'intérêt externe
external rate of return	taux externe de rendement
face value (par value)	valeur nominale
fees	frais
financial analysis	analyse financière
financial lease	bail financière
financial risk	risque financière
financial statement	état financier
finished-goods inventory	stock de produits finis
first cost	coût initial
fiscal period	exercice
fixed assets	capital fixe
fixed costs	charges fixes
fixed operating costs	coûts d'exploitation fixes
flotation costs	coûts d'émission
functional depreciation	dépréciation fonctionnelle
future value, future worth	valeur capitalisée, valeur future
general inflation rate	taux d'inflation général
generalized cash flow approach	méthode généralisée de flux monétaire
geometric gradient series	flux monétaires d'un gradient géométrique
geometric growth	croissance géométrique
GIC (Guaranteed Investment Certificate)	certificat de placement garanti
Goal Seek (Excel)	Valeur cible (Excel)
grace period	délai de grâce
gradient	gradient

gross income	revenu brut
gross margin	marge brute
grossed up amount	montant majoré
growth rate	taux de croissance
Guaranteed Investment Certificate (GIC)	certificat de placement garanti
half-year convention	règle de la demi-année
high-ratio mortgage	prêt hypothécaire à rapport prêt-valeur élevé
imputed cost	coûts affectés
income from operations	revenu net provenant des activités
income statement	états des résultats
income tax	impôt sur le revenu
incremental	différentiel
incremental analysis	analyse incrémentielle
incremental costs	coûts additionnels
incremental IRR (internal rate of return)	taux de rendement interne additionnel
incremental tax rate	taux d'impôt additionnel
independent	indépendant(e)
independent random variables	variables aléatoires indépendantes
indirect cost	coût indirect
industry price index	indice des prix industriels
infinite planning horizon	horizon de planification indéfini
inflation	inflation
inflation tax	impôt d'inflation
inflation-adjusted MARR	TRAM indexé
inflation-free interest rate	taux d'intérêt réel
input	intrant
installments	acomptes provisionnels
intangibles	impondérables
interest	intérêt
interest factor	facteur d'intérêt
interest period	période d'actualisation
interest rate	taux d'intérêt
interest tables	tables d'intérêt
internal rate of return (IRR)	taux de rendement interne (TRI)
interpolation	interpolation
inventories	valeurs d'exploitation
inventory turnover	rotation des stocks
investment	investissement
investment alternative	solution de rechange en matière d'investissement
investment in working capital	investissement dans le fonds de roulement
investment income	revenus de placement
investment opportunity schedule	plan des occasions d'investissement
investment pool	fonds d'investissement communs
investment risk profile	profil de risque d'un investissement
investment risks	risques d'investissement
investment strategies	stratégies de placement
investment tax credit	crédit d'impôt à l'investissement
investors	investisseur(euse)
IRR (internal rate of return)	TRI (taux de rendement interne)
irregular series	flux monétaire irrégulier
joint probability	probabilité conjoint
law of large numbers	loi des grands nombres
lease	location
lease financing	crédit-bail
leasehold improvements	améliorations locatives
lease-or-buy decision	décision entre la location ou l'achat
least common multiple	plus petit facteur commun
lending rate	taux prêteur

lessee	locataire
lessor	bailleur
liabilities	passif
life-cycle costing	coût du revient de cycle de vie
linear gradient series	flux monétaire d'un gradient arithmétique
linear interpolation	interpolation monétaire
liquidity	liquidité
loan	prêt
loan balance	reliquat
lowest common multiple	plus petit multiple commun
lump sum	paiement unique
MACRS (Modified Accelerated Cost Recovery System)	méthode modifiée de recouvrement accéléré des coûts
maintenance	entretien
maintenance costs	coûts d'entretien
make-or-buy decision	décision de fabriquer ou d'acheter
manufacturing cost	coût de fabrication
manufacturing overhead	frais généraux de fabrication
marginal	marginal
marginal costs	coûts marginaux
marginal probability	probabilité marginale
marginal tax rate	taux marginal d'imposition
market basket	panier de consommation
market interest rate	taux du marché
market value	valeur marchande
marketable bond	obligation négociable
marketing	commercialisation, marketing
MARR (minimum attractive rate of return)	TRAM (taux de rendement acceptable minimum)
matching principle	principe du rapprochement
mature	venir à l'échéance
maturity date	date d'échéance
minimum attractive rate of return (MARR)	taux de rendement acceptable minimum (TRAM)
mixed financing	base de financement mixte
mode	mode
Modified Accelerated Cost Recovery System (MACRS)	méthode modifiée de recouvrement accéléré des coûts
monetary policy	politique monétaire
Monte Carlo methods	méthodes de Monte Carlo
mortgage	hypothèque
mortgage bond	obligation hypothécaire
mortgage rate	taux d'emprunt hypothécaire
multiple rates of return	taux de rendement multiples
municipal bond	obligation municipale
mutually exclusive	mutuellement exclusif
net cash flow	flux monétaire net
net cash flow rule of signs	règle des signes dans un flux monétaire net
net equity flow	flux monétaire net des capitaux propres
net income	revenu net
net income before adjustments	revenu net avant rajustements
net investment test	test de l'investissement net
net present worth	valeur actualisée nette
net profit	bénéfice net
net revenue	recettes nettes
net worth	capitaux propres
nominal interest rate	taux d'intérêt nominal
non-refundable tax credit	crédits d'impôt non remboursable
nonsimple investment	investissement non simple
objective probability	probabilité objective
open mortgage	prêt hypothécaire ouvert

open-end mortgage	emprunt hypothécaire non plafonné
operating activities	activités d'exploitation
operating costs	coûts d'exploitation
operating expenses	dépenses d'exploitation
operating income	bénéfice d'exploitation
operating lease	bail d'exploitation
operating ratio	ratio d'exploitation
opportunity cost	coût d'opportunité
option	choix, option
output	extrant
overhead costs	frais généraux
par value (face value)	valeur nominale
partnership	société de personnes
payback method	méthode de la période de recouvrement
payback period	délai de récupération
payments	versements
perpetual	perpétuel(le)
personal income tax	impôt sur le revenu des particuliers
physical depreciation	dépréciation physique
planning horizon	horizon de planification
portability	transférabilité
preferred stock	actions privilégiées
premium bond	obligation avec prime d'émission
prepayment privileges	privilèges de remboursement anticipé
present equivalent, present worth, present value	valeur actualisée, valeur présente
present-worth profile	courbe de valeur
present-worth factor	facteur de valeur actuelle
price index	indice des prix
price/earnings ratio	ratio cours-bénéfice
primary benefit	avantage primaire
prime rate	taux préférentiel
principal	principal
probability distribution	distribution des probabilités
proceeds of disposition	produit de disposition
product costs	coûts incorporables
professional income	revenus de profession libérale
profit margin	marge bénéficiaire
project balance	solde du projet
project life	durée du projet
proprietorship	entreprise unipersonnelle
provincial bond	obligation provinciale
public goods	biens publics
purchase price	prix d'achat
purchasing power	pouvoir d'achat
quarterly	trimestrielle
quick (acid-test) ratio	ratio de trésorerie
random number	nombre aléatoire
random variable	variable aléatoire
range	intervalle de variation
rate of return	taux de rendement
raw-materials inventory	stocks de matières premières
real dollars	dollars constants
real interest rate	taux d'intérêt réel
recaptured CCA	DPA récupérée
recaptured depreciation	récupération de l'amortissement
receipts	recettes
recovered capital	capital amorti, capital recouvert
recovery period	période de recouvrement
refundable investment tax credit	crédit d'impôt à l'investissement remboursable

refundable tax credits	crédit d'impôt remboursable
registered bond	obligation nominative
Registered Education Savings Plan (RESP)	régime enregistré d'épargne-études (REEE)
Registered Retirement Savings Plan (RRSP)	régime enregistré d'épargne-retraite (REER)
repair cost	coût de réparation
repayment	remboursement
replacement	remplacement
RESP (Registered Education Savings Plan)	REEE (régime enregistré d'épargne-études)
retained earnings	bénéfices réinvestis
return	rendement
return on equity	rendement des capitaux propres
return on invested capital (RIC)	rendement du capital investi
return on investment	rendement de l'investissement
revenue	revenu
revenue project	projet de revenu
RIC (return on invested capital)	rendement du capital investi
risk	risque
risk analysis	analyse du risque
risk premium	prime de risque
risk-free return	rendement sans risque
risk-return trade-offs	rapport entre les risques et le rendement
RRSP (Registered Retirement Savings Plan)	REER (régime enregistré d'épargne-retraite)
salvage value	valeur de récupération
sampling procedure	méthode d'échantillonnage
scenario analysis	analyse de scénarios
screening	triage
secondary benefit	avantage secondaire
semivariable costs	coûts semivariables
sensitivity analysis	analyse de sensibilité
sensitivity graphs	graphiques de sensibilité
service life	vie de service
service project	projet de service
service sector	secteur des services
shares	actions
simple interest	intérêt simple
simple investment	investissement simple
single payment	paiement unique
single-payment compound-amount factor	facteur de capitalisation d'un flux monétaire unique
single-payment present-worth factor	facteur d'actualisation d'un flux monétaire unique
sinking fund	fonds d'amortissement
SL (straight-line) method	méthode linéaire
small business deduction	déduction accordée aux petites entreprises
social discount rate	taux d'escompte social
sole proprietorship	propriétaire unique
SOYD (sum-of-years'-digits') method	méthode de l'amortissement proportionnel à l'ordre numérique inversé des années
sponsor's costs	coûts du promoteur
standard deviation	écart-type
statement of cash flows	état des flux de trésorerie
stock investment	investissement dans les actions
stock market	bourse de valeurs mobilières
stocks	actions
straight-line depreciation	amortissement linéaire
straight-line (SL) method	méthode linéaire
study period	période d'étude
subjective probability	probabilité subjective
sum-of-years'-digits' (SOYD) method	méthode de l'amortissement proportionnel à l'ordre numérique inversé des années

sunk costs	coûts irrécupérables, coûts perdus
surtax	surtaxe
target capital structure	structure du capital cible
target debt ratio	ratio d'endettement cible
tax credit	crédit d'impôt
tax depreciation	amortissement fiscale
tax incentives	incitations fiscales
tax rate	taux d'imposition
tax savings	économies d'impôt
tax schedule	annexe d'impôt
tax shield	protection fiscale
taxable income	revenus imposables
taxation year	année d'imposition
TD (term deposit)	dépôts à terme
technological change	changement technologique
term deposit (TD)	dépôts à terme
term loans	prêts à terme
terminal loss	perte finale
time horizon	horizon temporel
time value of money	valeur de l'argent dans le temps
total income	revenu total
total investment approach	méthode d'investissement totale
total qualified amount	somme justifiée totale
trade-in value	valeur de reprise
Treasury bills	bons du Trésor
trend analysis	analyse des tendances
trial-and-error method	méthode d'approximations successives
triangular distribution	distribution triangulaire
UCC (undepreciated capital cost)	FNACC (fraction non amortie du coût en capital)
uncertainty	incertitude
undepreciated capital cost (UCC)	fraction non amortie du coût en capital (FNACC)
uniform cash flow series	flux monétaire uniforme
uniform distribution	distribution uniforme
uniform series	annuité
uniform series compound-amount factor	facteur de capitalisation d'une annuité
unit cost	coût unitaire
unit profit	profit unitaire
units-of-production (UP) method	méthode de l'amortissement proportionnel à l'utilisation
unrecovered capital	reliquat
UP (units-of-production) method	méthode de l'amortissement proportionnel à l'utilisation
useful life	durée de vie utile
users' benefits	avantages pour les usagers
users' disbenefits	inconvénients pour les usagers
utility	utilité
valuation	valorisation
value of perfect information	valeur de l'information parfaite
variable cost	coût variable
variable operating costs	coûts d'exploitation variables
variance	variance
volume index	indice du volume
working capital	fonds de roulement
working-capital release	libération
working-capital requirements	fonds de roulement requis
work-in-process inventory	stock de produits en cours
yield	rendement
yield to maturity (YTM)	rendement à l'échéance

Interest Factors for Discrete Compounding

0.25%

	Single Payment		Equal Payment Series				Gradient Series		
N	Compound Amount Factor (F/P,i,N)	Present Worth Factor (P/F,i,N)	Compound Amount Factor (F/A,i,N)	Sinking Fund Factor (A/F,i,N)	Present Worth Factor (P/A,i,N)	Capital Recovery Factor (A/P,i,N)	Gradient Uniform Series (A/G,i,N)	Gradient Present Worth (P/G,i,N)	N
1	1.0025	0.9975	1.0000	1.0000	0.9975	1.0025	0.0000	0.0000	1
2	1.0050	0.9950	2.0025	0.4994	1.9925	0.5019	0.4994	0.9950	2
3	1.0075	0.9925	3.0075	0.3325	2.9851	0.3350	0.9983	2.9801	3
4	1.0100	0.9901	4.0150	0.2491	3.9751	0.2516	1.4969	5.9503	4
5	1.0126	0.9876	5.0251	0.1990	4.9627	0.2015	1.9950	9.9007	5
6	1.0151	0.9851	6.0376	0.1656	5.9478	0.1681	2.4927	14.8263	6
7	1.0176	0.9827	7.0527	0.1418	6.9305	0.1443	2.9900	20.7223	7
8	1.0202	0.9802	8.0704	0.1239	7.9107	0.1264	3.4869	27.5839	8
9	1.0227	0.9778	9.0905	0.1100	8.8885	0.1125	3.9834	35.4061	9
10	1.0253	0.9753	10.1133	0.0989	9.8639	0.1014	4.4794	44.1842	10
11	1.0278	0.9729	11.1385	0.0898	10.8368	0.0923	4.9750	53.9133	11
12	1.0304	0.9705	12.1664	0.0822	11.8073	0.0847	5.4702	64.5886	12
13	1.0330	0.9681	13.1968	0.0758	12.7753	0.0783	5.9650	76.2053	13
14	1.0356	0.9656	14.2298	0.0703	13.7410	0.0728	6.4594	88.7587	14
15	1.0382	0.9632	15.2654	0.0655	14.7042	0.0680	6.9534	102.2441	15
16	1.0408	0.9608	16.3035	0.0613	15.6650	0.0638	7.4469	116.6567	16
17	1.0434	0.9584	17.3443	0.0577	16.6235	0.0602	7.9401	131.9917	17
18	1.0460	0.9561	18.3876	0.0544	17.5795	0.0569	8.4328	148.2446	18
19	1.0486	0.9537	19.4336	0.0515	18.5332	0.0540	8.9251	165.4106	19
20	1.0512	0.9513	20.4822	0.0488	19.4845	0.0513	9.4170	183.4851	20
21	1.0538	0.9489	21.5334	0.0464	20.4334	0.0489	9.9085	202.4634	21
22	1.0565	0.9466	22.5872	0.0443	21.3800	0.0468	10.3995	222.3410	22
23	1.0591	0.9442	23.6437	0.0423	22.3241	0.0448	10.8901	243.1131	23
24	1.0618	0.9418	24.7028	0.0405	23.2660	0.0430	11.3804	264.7753	24
25	1.0644	0.9395	25.7646	0.0388	24.2055	0.0413	11.8702	287.3230	25
26	1.0671	0.9371	26.8290	0.0373	25.1426	0.0398	12.3596	310.7516	26
27	1.0697	0.9348	27.8961	0.0358	26.0774	0.0383	12.8485	335.0566	27
28	1.0724	0.9325	28.9658	0.0345	27.0099	0.0370	13.3371	360.2334	28
29	1.0751	0.9301	30.0382	0.0333	27.9400	0.0358	13.8252	386.2776	29
30	1.0778	0.9278	31.1133	0.0321	28.8679	0.0346	14.3130	413.1847	30
31	1.0805	0.9255	32.1911	0.0311	29.7934	0.0336	14.8003	440.9502	31
32	1.0832	0.9232	33.2716	0.0301	30.7166	0.0326	15.2872	469.5696	32
33	1.0859	0.9209	34.3547	0.0291	31.6375	0.0316	15.7736	499.0386	33
34	1.0886	0.9186	35.4406	0.0282	32.5561	0.0307	16.2597	529.3528	34
35	1.0913	0.9163	36.5292	0.0274	33.4724	0.0299	16.7454	560.5076	35
36	1.0941	0.9140	37.6206	0.0266	34.3865	0.0291	17.2306	592.4988	36
40	1.1050	0.9050	42.0132	0.0238	38.0199	0.0263	19.1673	728.7399	40
48	1.1273	0.8871	50.9312	0.0196	45.1787	0.0221	23.0209	1040.0552	48
50	1.1330	0.8826	53.1887	0.0188	46.9462	0.0213	23.9802	1125.7767	50
60	1.1616	0.8609	64.6467	0.0155	55.6524	0.0180	28.7514	1600.0845	60
72	1.1969	0.8355	78.7794	0.0127	65.8169	0.0152	34.4221	2265.5569	72
80	1.2211	0.8189	88.4392	0.0113	72.4260	0.0138	38.1694	2764.4568	80
84	1.2334	0.8108	93.3419	0.0107	75.6813	0.0132	40.0331	3029.7592	84
90	1.2520	0.7987	100.7885	0.0099	80.5038	0.0124	42.8162	3446.8700	90
96	1.2709	0.7869	108.3474	0.0092	85.2546	0.0117	45.5844	3886.2832	96
100	1.2836	0.7790	113.4500	0.0088	88.3825	0.0113	47.4216	4191.2417	100
108	1.3095	0.7636	123.8093	0.0081	94.5453	0.0106	51.0762	4829.0125	108
120	1.3494	0.7411	139.7414	0.0072	103.5618	0.0097	56.5084	5852.1116	120
240	1.8208	0.5492	328.3020	0.0030	180.3109	0.0055	107.5863	19398.9852	240
360	2.4568	0.4070	582.7369	0.0017	237.1894	0.0042	152.8902	36263.9299	360

0.50%

	Single Payment		Equal Payment Series				Gradient Series		
	Compound Amount Factor	Present Worth Factor	Compound Amount Factor	Sinking Fund Factor	Present Worth Factor	Capital Recovery Factor	Gradient Uniform Series	Gradient Present Worth	
N	(F/P,i,N)	(P/F,i,N)	(F/A,i,N)	(A/F,i,N)	(P/A,i,N)	(A/P,i,N)	(A/G,i,N)	(P/G,i,N)	N
1	1.0050	0.9950	1.0000	1.0000	0.9950	1.0050	0.0000	0.0000	1
2	1.0100	0.9901	2.0050	0.4988	1.9851	0.5038	0.4988	0.9901	2
3	1.0151	0.9851	3.0150	0.3317	2.9702	0.3367	0.9967	2.9604	3
4	1.0202	0.9802	4.0301	0.2481	3.9505	0.2531	1.4938	5.9011	4
5	1.0253	0.9754	5.0503	0.1980	4.9259	0.2030	1.9900	9.8026	5
6	1.0304	0.9705	6.0755	0.1646	5.8964	0.1696	2.4855	14.6552	6
7	1.0355	0.9657	7.1059	0.1407	6.8621	0.1457	2.9801	20.4493	7
8	1.0407	0.9609	8.1414	0.1228	7.8230	0.1278	3.4738	27.1755	8
9	1.0459	0.9561	9.1821	0.1089	8.7791	0.1139	3.9668	34.8244	9
10	1.0511	0.9513	10.2280	0.0978	9.7304	0.1028	4.4589	43.3865	10
11	1.0564	0.9466	11.2792	0.0887	10.6770	0.0937	4.9501	52.8526	11
12	1.0617	0.9419	12.3356	0.0811	11.6189	0.0861	5.4406	63.2136	12
13	1.0670	0.9372	13.3972	0.0746	12.5562	0.0796	5.9302	74.4602	13
14	1.0723	0.9326	14.4642	0.0691	13.4887	0.0741	6.4190	86.5835	14
15	1.0777	0.9279	15.5365	0.0644	14.4166	0.0694	6.9069	99.5743	15
16	1.0831	0.9233	16.6142	0.0602	15.3399	0.0652	7.3940	113.4238	16
17	1.0885	0.9187	17.6973	0.0565	16.2586	0.0615	7.8803	128.1231	17
18	1.0939	0.9141	18.7858	0.0532	17.1728	0.0582	8.3658	143.6634	18
19	1.0994	0.9096	19.8797	0.0503	18.0824	0.0553	8.8504	160.0360	19
20	1.1049	0.9051	20.9791	0.0477	18.9874	0.0527	9.3342	177.2322	20
21	1.1104	0.9006	22.0840	0.0453	19.8880	0.0503	9.8172	195.2434	21
22	1.1160	0.8961	23.1944	0.0431	20.7841	0.0481	10.2993	214.0611	22
23	1.1216	0.8916	24.3104	0.0411	21.6757	0.0461	10.7806	233.6768	23
24	1.1272	0.8872	25.4320	0.0393	22.5629	0.0443	11.2611	254.0820	24
25	1.1328	0.8828	26.5591	0.0377	23.4456	0.0427	11.7407	275.2686	25
26	1.1385	0.8784	27.6919	0.0361	24.3240	0.0411	12.2195	297.2281	26
27	1.1442	0.8740	28.8304	0.0347	25.1980	0.0397	12.6975	319.9523	27
28	1.1499	0.8697	29.9745	0.0334	26.0677	0.0384	13.1747	343.4332	28
29	1.1556	0.8653	31.1244	0.0321	26.9330	0.0371	13.6510	367.6625	29
30	1.1614	0.8610	32.2800	0.0310	27.7941	0.0360	14.1265	392.6324	30
31	1.1672	0.8567	33.4414	0.0299	28.6508	0.0349	14.6012	418.3348	31
32	1.1730	0.8525	34.6086	0.0289	29.5033	0.0339	15.0750	444.7618	32
33	1.1789	0.8482	35.7817	0.0279	30.3515	0.0329	15.5480	471.9055	33
34	1.1848	0.8440	36.9606	0.0271	31.1955	0.0321	16.0202	499.7583	34
35	1.1907	0.8398	38.1454	0.0262	32.0354	0.0312	16.4915	528.3123	35
36	1.1967	0.8356	39.3361	0.0254	32.8710	0.0304	16.9621	557.5598	36
40	1.2208	0.8191	44.1588	0.0226	36.1722	0.0276	18.8359	681.3347	40
48	1.2705	0.7871	54.0978	0.0185	42.5803	0.0235	22.5437	959.9188	48
50	1.2832	0.7793	56.6452	0.0177	44.1428	0.0227	23.4624	1035.6966	50
60	1.3489	0.7414	69.7700	0.0143	51.7256	0.0193	28.0064	1448.6458	60
72	1.4320	0.6983	86.4089	0.0116	60.3395	0.0166	33.3504	2012.3478	72
80	1.4903	0.6710	98.0677	0.0102	65.8023	0.0152	36.8474	2424.6455	80
84	1.5204	0.6577	104.0739	0.0096	68.4530	0.0146	38.5763	2640.6641	84
90	1.5666	0.6383	113.3109	0.0088	72.3313	0.0138	41.1451	2976.0769	90
96	1.6141	0.6195	122.8285	0.0081	76.0952	0.0131	43.6845	3324.1846	96
100	1.6467	0.6073	129.3337	0.0077	78.5426	0.0127	45.3613	3562.7934	100
108	1.7137	0.5835	142.7399	0.0070	83.2934	0.0120	48.6758	4054.3747	108
120	1.8194	0.5496	163.8793	0.0061	90.0735	0.0111	53.5508	4823.5051	120
240	3.3102	0.3021	462.0409	0.0022	139.5808	0.0072	96.1131	13415.5395	240
360	6.0226	0.1660	1004.5150	0.0010	166.7916	0.0060	128.3236	21403.3041	360

0.75%

	Single Payment		Equal Payment Series				Gradient Series		
	Compound Amount Factor	Present Worth Factor	Compound Amount Factor	Sinking Fund Factor	Present Worth Factor	Capital Recovery Factor	Gradient Uniform Series	Gradient Present Worth	
N	(F/P,i,N)	(P/F,i,N)	(F/A,i,N)	(A/F,i,N)	(P/A,i,N)	(A/P,i,N)	(A/G,i,N)	(P/G,i,N)	N
1	1.0075	0.9926	1.0000	1.0000	0.9926	1.0075	0.0000	0.0000	1
2	1.0151	0.9852	2.0075	0.4981	1.9777	0.5056	0.4981	0.9852	2
3	1.0227	0.9778	3.0226	0.3308	2.9556	0.3383	0.9950	2.9408	3
4	1.0303	0.9706	4.0452	0.2472	3.9261	0.2547	1.4907	5.8525	4
5	1.0381	0.9633	5.0756	0.1970	4.8894	0.2045	1.9851	9.7058	5
6	1.0459	0.9562	6.1136	0.1636	5.8456	0.1711	2.4782	14.4866	6
7	1.0537	0.9490	7.1595	0.1397	6.7946	0.1472	2.9701	20.1808	7
8	1.0616	0.9420	8.2132	0.1218	7.7366	0.1293	3.4608	26.7747	8
9	1.0696	0.9350	9.2748	0.1078	8.6716	0.1153	3.9502	34.2544	9
10	1.0776	0.9280	10.3443	0.0967	9.5996	0.1042	4.4384	42.6064	10
11	1.0857	0.9211	11.4219	0.0876	10.5207	0.0951	4.9253	51.8174	11
12	1.0938	0.9142	12.5076	0.0800	11.4349	0.0875	5.4110	61.8740	12
13	1.1020	0.9074	13.6014	0.0735	12.3423	0.0810	5.8954	72.7632	13
14	1.1103	0.9007	14.7034	0.0680	13.2430	0.0755	6.3786	84.4720	14
15	1.1186	0.8940	15.8137	0.0632	14.1370	0.0707	6.8606	96.9876	15
16	1.1270	0.8873	16.9323	0.0591	15.0243	0.0666	7.3413	110.2973	16
17	1.1354	0.8807	18.0593	0.0554	15.9050	0.0629	7.8207	124.3887	17
18	1.1440	0.8742	19.1947	0.0521	16.7792	0.0596	8.2989	139.2494	18
19	1.1525	0.8676	20.3387	0.0492	17.6468	0.0567	8.7759	154.8671	19
20	1.1612	0.8612	21.4912	0.0465	18.5080	0.0540	9.2516	171.2297	20
21	1.1699	0.8548	22.6524	0.0441	19.3628	0.0516	9.7261	188.3253	21
22	1.1787	0.8484	23.8223	0.0420	20.2112	0.0495	10.1994	206.1420	22
23	1.1875	0.8421	25.0010	0.0400	21.0533	0.0475	10.6714	224.6682	23
24	1.1964	0.8358	26.1885	0.0382	21.8891	0.0457	11.1422	243.8923	24
25	1.2054	0.8296	27.3849	0.0365	22.7188	0.0440	11.6117	263.8029	25
26	1.2144	0.8234	28.5903	0.0350	23.5422	0.0425	12.0800	284.3888	26
27	1.2235	0.8173	29.8047	0.0336	24.3595	0.0411	12.5470	305.6387	27
28	1.2327	0.8112	31.0282	0.0322	25.1707	0.0397	13.0128	327.5416	28
29	1.2420	0.8052	32.2609	0.0310	25.9759	0.0385	13.4774	350.0867	29
30	1.2513	0.7992	33.5029	0.0298	26.7751	0.0373	13.9407	373.2631	30
31	1.2607	0.7932	34.7542	0.0288	27.5683	0.0363	14.4028	397.0602	31
32	1.2701	0.7873	36.0148	0.0278	28.3557	0.0353	14.8636	421.4675	32
33	1.2796	0.7815	37.2849	0.0268	29.1371	0.0343	15.3232	446.4746	33
34	1.2892	0.7757	38.5646	0.0259	29.9128	0.0334	15.7816	472.0712	34
35	1.2989	0.7699	39.8538	0.0251	30.6827	0.0326	16.2387	498.2471	35
36	1.3086	0.7641	41.1527	0.0243	31.4468	0.0318	16.6946	524.9924	36
40	1.3483	0.7416	46.4465	0.0215	34.4469	0.0290	18.5058	637.4693	40
48	1.4314	0.6986	57.5207	0.0174	40.1848	0.0249	22.0691	886.8404	48
50	1.4530	0.6883	60.3943	0.0166	41.5664	0.0241	22.9476	953.8486	50
60	1.5657	0.6387	75.4241	0.0133	48.1734	0.0208	27.2665	1313.5189	60
72	1.7126	0.5839	95.0070	0.0105	55.4768	0.0180	32.2882	1791.2463	72
80	1.8180	0.5500	109.0725	0.0092	59.9944	0.0167	35.5391	2132.1472	80
84	1.8732	0.5338	116.4269	0.0086	62.1540	0.0161	37.1357	2308.1283	84
90	1.9591	0.5104	127.8790	0.0078	65.2746	0.0153	39.4946	2577.9961	90
96	2.0489	0.4881	139.8562	0.0072	68.2584	0.0147	41.8107	2853.9352	96
100	2.1111	0.4737	148.1445	0.0068	70.1746	0.0143	43.3311	3040.7453	100
108	2.2411	0.4462	165.4832	0.0060	73.8394	0.0135	46.3154	3419.9041	108
120	2.4514	0.4079	193.5143	0.0052	78.9417	0.0127	50.6521	3998.5621	120
240	6.0092	0.1664	667.8869	0.0015	111.1450	0.0090	85.4210	9494.1162	240
360	14.7306	0.0679	1830.7435	0.0005	124.2819	0.0080	107.1145	13312.3871	360

1.0%

	Single Payment		Equal Payment Series				Gradient Series		
N	Compound Amount Factor (F/P,i,N)	Present Worth Factor (P/F,i,N)	Compound Amount Factor (F/A,i,N)	Sinking Fund Factor (A/F,i,N)	Present Worth Factor (P/A,i,N)	Capital Recovery Factor (A/P,i,N)	Gradient Uniform Series (A/G,i,N)	Gradient Present Worth (P/G,i,N)	N
1	1.0100	0.9901	1.0000	1.0000	0.9901	1.0100	0.0000	0.0000	1
2	1.0201	0.9803	2.0100	0.4975	1.9704	0.5075	0.4975	0.9803	2
3	1.0303	0.9706	3.0301	0.3300	2.9410	0.3400	0.9934	2.9215	3
4	1.0406	0.9610	4.0604	0.2463	3.9020	0.2563	1.4876	5.8044	4
5	1.0510	0.9515	5.1010	0.1960	4.8534	0.2060	1.9801	9.6103	5
6	1.0615	0.9420	6.1520	0.1625	5.7955	0.1725	2.4710	14.3205	6
7	1.0721	0.9327	7.2135	0.1386	6.7282	0.1486	2.9602	19.9168	7
8	1.0829	0.9235	8.2857	0.1207	7.6517	0.1307	3.4478	26.3812	8
9	1.0937	0.9143	9.3685	0.1067	8.5660	0.1167	3.9337	33.6959	9
10	1.1046	0.9053	10.4622	0.0956	9.4713	0.1056	4.4179	41.8435	10
11	1.1157	0.8963	11.5668	0.0865	10.3676	0.0965	4.9005	50.8067	11
12	1.1268	0.8874	12.6825	0.0788	11.2551	0.0888	5.3815	60.5687	12
13	1.1381	0.8787	13.8093	0.0724	12.1337	0.0824	5.8607	71.1126	13
14	1.1495	0.8700	14.9474	0.0669	13.0037	0.0769	6.3384	82.4221	14
15	1.1610	0.8613	16.0969	0.0621	13.8651	0.0721	6.8143	94.4810	15
16	1.1726	0.8528	17.2579	0.0579	14.7179	0.0679	7.2886	107.2734	16
17	1.1843	0.8444	18.4304	0.0543	15.5623	0.0643	7.7613	120.7834	17
18	1.1961	0.8360	19.6147	0.0510	16.3983	0.0610	8.2323	134.9957	18
19	1.2081	0.8277	20.8109	0.0481	17.2260	0.0581	8.7017	149.8950	19
20	1.2202	0.8195	22.0190	0.0454	18.0456	0.0554	9.1694	165.4664	20
21	1.2324	0.8114	23.2392	0.0430	18.8570	0.0530	9.6354	181.6950	21
22	1.2447	0.8034	24.4716	0.0409	19.6604	0.0509	10.0998	198.5663	22
23	1.2572	0.7954	25.7163	0.0389	20.4558	0.0489	10.5626	216.0660	23
24	1.2697	0.7876	26.9735	0.0371	21.2434	0.0471	11.0237	234.1800	24
25	1.2824	0.7798	28.2432	0.0354	22.0232	0.0454	11.4831	252.8945	25
26	1.2953	0.7720	29.5256	0.0339	22.7952	0.0439	11.9409	272.1957	26
27	1.3082	0.7644	30.8209	0.0324	23.5596	0.0424	12.3971	292.0702	27
28	1.3213	0.7568	32.1291	0.0311	24.3164	0.0411	12.8516	312.5047	28
29	1.3345	0.7493	33.4504	0.0299	25.0658	0.0399	13.3044	333.4863	29
30	1.3478	0.7419	34.7849	0.0287	25.8077	0.0387	13.7557	355.0021	30
31	1.3613	0.7346	36.1327	0.0277	26.5423	0.0377	14.2052	377.0394	31
32	1.3749	0.7273	37.4941	0.0267	27.2696	0.0367	14.6532	399.5858	32
33	1.3887	0.7201	38.8690	0.0257	27.9897	0.0357	15.0995	422.6291	33
34	1.4026	0.7130	40.2577	0.0248	28.7027	0.0348	15.5441	446.1572	34
35	1.4166	0.7059	41.6603	0.0240	29.4086	0.0340	15.9871	470.1583	35
36	1.4308	0.6989	43.0769	0.0232	30.1075	0.0332	16.4285	494.6207	36
40	1.4889	0.6717	48.8864	0.0205	32.8347	0.0305	18.1776	596.8561	40
48	1.6122	0.6203	61.2226	0.0163	37.9740	0.0263	21.5976	820.1460	48
50	1.6446	0.6080	64.4632	0.0155	39.1961	0.0255	22.4363	879.4176	50
60	1.8167	0.5504	81.6697	0.0122	44.9550	0.0222	26.5333	1192.8061	60
72	2.0471	0.4885	104.7099	0.0096	51.1504	0.0196	31.2386	1597.8673	72
80	2.2167	0.4511	121.6715	0.0082	54.8882	0.0182	34.2492	1879.8771	80
84	2.3067	0.4335	130.6723	0.0077	56.6485	0.0177	35.7170	2023.3153	84
90	2.4486	0.4084	144.8633	0.0069	59.1609	0.0169	37.8724	2240.5675	90
96	2.5993	0.3847	159.9273	0.0063	61.5277	0.0163	39.9727	2459.4298	96
100	2.7048	0.3697	170.4814	0.0059	63.0289	0.0159	41.3426	2605.7758	100
108	2.9289	0.3414	192.8926	0.0052	65.8578	0.0152	44.0103	2898.4203	108
120	3.3004	0.3030	230.0387	0.0043	69.7005	0.0143	47.8349	3334.1148	120
240	10.8926	0.0918	989.2554	0.0010	90.8194	0.0110	75.7393	6878.6016	240
360	35.9496	0.0278	3494.9641	0.0003	97.2183	0.0103	89.6995	8720.4323	360

1.25%

	Single Payment		Equal Payment Series				Gradient Series		
N	Compound Amount Factor (F/P,i,N)	Present Worth Factor (P/F,i,N)	Compound Amount Factor (F/A,i,N)	Sinking Fund Factor (A/F,i,N)	Present Worth Factor (P/A,i,N)	Capital Recovery Factor (A/P,i,N)	Gradient Uniform Series (A/G,i,N)	Gradient Present Worth (P/G,i,N)	N
1	1.0125	0.9877	1.0000	1.0000	0.9877	1.0125	0.0000	0.0000	1
2	1.0252	0.9755	2.0125	0.4969	1.9631	0.5094	0.4969	0.9755	2
3	1.0380	0.9634	3.0377	0.3292	2.9265	0.3417	0.9917	2.9023	3
4	1.0509	0.9515	4.0756	0.2454	3.8781	0.2579	1.4845	5.7569	4
5	1.0641	0.9398	5.1266	0.1951	4.8178	0.2076	1.9752	9.5160	5
6	1.0774	0.9282	6.1907	0.1615	5.7460	0.1740	2.4638	14.1569	6
7	1.0909	0.9167	7.2680	0.1376	6.6627	0.1501	2.9503	19.6571	7
8	1.1045	0.9054	8.3589	0.1196	7.5681	0.1321	3.4348	25.9949	8
9	1.1183	0.8942	9.4634	0.1057	8.4623	0.1182	3.9172	33.1487	9
10	1.1323	0.8832	10.5817	0.0945	9.3455	0.1070	4.3975	41.0973	10
11	1.1464	0.8723	11.7139	0.0854	10.2178	0.0979	4.8758	49.8201	11
12	1.1608	0.8615	12.8604	0.0778	11.0793	0.0903	5.3520	59.2967	12
13	1.1753	0.8509	14.0211	0.0713	11.9302	0.0838	5.8262	69.5072	13
14	1.1900	0.8404	15.1964	0.0658	12.7706	0.0783	6.2982	80.4320	14
15	1.2048	0.8300	16.3863	0.0610	13.6005	0.0735	6.7682	92.0519	15
16	1.2199	0.8197	17.5912	0.0568	14.4203	0.0693	7.2362	104.3481	16
17	1.2351	0.8096	18.8111	0.0532	15.2299	0.0657	7.7021	117.3021	17
18	1.2506	0.7996	20.0462	0.0499	16.0295	0.0624	8.1659	130.8958	18
19	1.2662	0.7898	21.2968	0.0470	16.8193	0.0595	8.6277	145.1115	19
20	1.2820	0.7800	22.5630	0.0443	17.5993	0.0568	9.0874	159.9316	20
21	1.2981	0.7704	23.8450	0.0419	18.3697	0.0544	9.5450	175.3392	21
22	1.3143	0.7609	25.1431	0.0398	19.1306	0.0523	10.0006	191.3174	22
23	1.3307	0.7515	26.4574	0.0378	19.8820	0.0503	10.4542	207.8499	23
24	1.3474	0.7422	27.7881	0.0360	20.6242	0.0485	10.9056	224.9204	24
25	1.3642	0.7330	29.1354	0.0343	21.3573	0.0468	11.3551	242.5132	25
26	1.3812	0.7240	30.4996	0.0328	22.0813	0.0453	11.8024	260.6128	26
27	1.3985	0.7150	31.8809	0.0314	22.7963	0.0439	12.2478	279.2040	27
28	1.4160	0.7062	33.2794	0.0300	23.5025	0.0425	12.6911	298.2719	28
29	1.4337	0.6975	34.6954	0.0288	24.2000	0.0413	13.1323	317.8019	29
30	1.4516	0.6889	36.1291	0.0277	24.8889	0.0402	13.5715	337.7797	30
31	1.4698	0.6804	37.5807	0.0266	25.5693	0.0391	14.0086	358.1912	31
32	1.4881	0.6720	39.0504	0.0256	26.2413	0.0381	14.4438	379.0227	32
33	1.5067	0.6637	40.5386	0.0247	26.9050	0.0372	14.8768	400.2607	33
34	1.5256	0.6555	42.0453	0.0238	27.5605	0.0363	15.3079	421.8920	34
35	1.5446	0.6474	43.5709	0.0230	28.2079	0.0355	15.7369	443.9037	35
36	1.5639	0.6394	45.1155	0.0222	28.8473	0.0347	16.1639	466.2830	36
40	1.6436	0.6084	51.4896	0.0194	31.3269	0.0319	17.8515	559.2320	40
48	1.8154	0.5509	65.2284	0.0153	35.9315	0.0278	21.1299	759.2296	48
50	1.8610	0.5373	68.8818	0.0145	37.0129	0.0270	21.9295	811.6738	50
60	2.1072	0.4746	88.5745	0.0113	42.0346	0.0238	25.8083	1084.8429	60
72	2.4459	0.4088	115.6736	0.0086	47.2925	0.0211	30.2047	1428.4561	72
80	2.7015	0.3702	136.1188	0.0073	50.3867	0.0198	32.9822	1661.8651	80
84	2.8391	0.3522	147.1290	0.0068	51.8222	0.0193	34.3258	1778.8384	84
90	3.0588	0.3269	164.7050	0.0061	53.8461	0.0186	36.2855	1953.8303	90
96	3.2955	0.3034	183.6411	0.0054	55.7246	0.0179	38.1793	2127.5244	96
100	3.4634	0.2887	197.0723	0.0051	56.9013	0.0176	39.4058	2242.2411	100
108	3.8253	0.2614	226.0226	0.0044	59.0865	0.0169	41.7737	2468.2636	108
120	4.4402	0.2252	275.2171	0.0036	61.9828	0.0161	45.1184	2796.5694	120
240	19.7155	0.0507	1497.2395	0.0007	75.9423	0.0132	67.1764	5101.5288	240
360	87.5410	0.0114	6923.2796	0.0001	79.0861	0.0126	75.8401	5997.9027	360

1.5%

	Single Payment		Equal Payment Series				Gradient Series		
	Compound Amount Factor	Present Worth Factor	Compound Amount Factor	Sinking Fund Factor	Present Worth Factor	Capital Recovery Factor	Gradient Uniform Series	Gradient Present Worth	
N	(F/P,i,N)	(P/F,i,N)	(F/A,i,N)	(A/F,i,N)	(P/A,i,N)	(A/P,i,N)	(A/G,i,N)	(P/G,i,N)	N
1	1.0150	0.9852	1.0000	1.0000	0.9852	1.0150	0.0000	0.0000	1
2	1.0302	0.9707	2.0150	0.4963	1.9559	0.5113	0.4963	0.9707	2
3	1.0457	0.9563	3.0452	0.3284	2.9122	0.3434	0.9901	2.8833	3
4	1.0614	0.9422	4.0909	0.2444	3.8544	0.2594	1.4814	5.7098	4
5	1.0773	0.9283	5.1523	0.1941	4.7826	0.2091	1.9702	9.4229	5
6	1.0934	0.9145	6.2296	0.1605	5.6972	0.1755	2.4566	13.9956	6
7	1.1098	0.9010	7.3230	0.1366	6.5982	0.1516	2.9405	19.4018	7
8	1.1265	0.8877	8.4328	0.1186	7.4859	0.1336	3.4219	25.6157	8
9	1.1434	0.8746	9.5593	0.1046	8.3605	0.1196	3.9008	32.6125	9
10	1.1605	0.8617	10.7027	0.0934	9.2222	0.1084	4.3772	40.3675	10
11	1.1779	0.8489	11.8633	0.0843	10.0711	0.0993	4.8512	48.8568	11
12	1.1956	0.8364	13.0412	0.0767	10.9075	0.0917	5.3227	58.0571	12
13	1.2136	0.8240	14.2368	0.0702	11.7315	0.0852	5.7917	67.9454	13
14	1.2318	0.8118	15.4504	0.0647	12.5434	0.0797	6.2582	78.4994	14
15	1.2502	0.7999	16.6821	0.0599	13.3432	0.0749	6.7223	89.6974	15
16	1.2690	0.7880	17.9324	0.0558	14.1313	0.0708	7.1839	101.5178	16
17	1.2880	0.7764	19.2014	0.0521	14.9076	0.0671	7.6431	113.9400	17
18	1.3073	0.7649	20.4894	0.0488	15.6726	0.0638	8.0997	126.9435	18
19	1.3270	0.7536	21.7967	0.0459	16.4262	0.0609	8.5539	140.5084	19
20	1.3469	0.7425	23.1237	0.0432	17.1686	0.0582	9.0057	154.6154	20
21	1.3671	0.7315	24.4705	0.0409	17.9001	0.0559	9.4550	169.2453	21
22	1.3876	0.7207	25.8376	0.0387	18.6208	0.0537	9.9018	184.3798	22
23	1.4084	0.7100	27.2251	0.0367	19.3309	0.0517	10.3462	200.0006	23
24	1.4295	0.6995	28.6335	0.0349	20.0304	0.0499	10.7881	216.0901	24
25	1.4509	0.6892	30.0630	0.0333	20.7196	0.0483	11.2276	232.6310	25
26	1.4727	0.6790	31.5140	0.0317	21.3986	0.0467	11.6646	249.6065	26
27	1.4948	0.6690	32.9867	0.0303	22.0676	0.0453	12.0992	267.0002	27
28	1.5172	0.6591	34.4815	0.0290	22.7267	0.0440	12.5313	284.7958	28
29	1.5400	0.6494	35.9987	0.0278	23.3761	0.0428	12.9610	302.9779	29
30	1.5631	0.6398	37.5387	0.0266	24.0158	0.0416	13.3883	321.5310	30
31	1.5865	0.6303	39.1018	0.0256	24.6461	0.0406	13.8131	340.4402	31
32	1.6103	0.6210	40.6883	0.0246	25.2671	0.0396	14.2355	359.6910	32
33	1.6345	0.6118	42.2986	0.0236	25.8790	0.0386	14.6555	379.2691	33
34	1.6590	0.6028	43.9331	0.0228	26.4817	0.0378	15.0731	399.1607	34
35	1.6839	0.5939	45.5921	0.0219	27.0756	0.0369	15.4882	419.3521	35
36	1.7091	0.5851	47.2760	0.0212	27.6607	0.0362	15.9009	439.8303	36
40	1.8140	0.5513	54.2679	0.0184	29.9158	0.0334	17.5277	524.3568	40
48	2.0435	0.4894	69.5652	0.0144	34.0426	0.0294	20.6667	703.5462	48
50	2.1052	0.4750	73.6828	0.0136	34.9997	0.0286	21.4277	749.9636	50
60	2.4432	0.4093	96.2147	0.0104	39.3803	0.0254	25.0930	988.1674	60
72	2.9212	0.3423	128.0772	0.0078	43.8447	0.0228	29.1893	1279.7938	72
80	3.2907	0.3039	152.7109	0.0065	46.4073	0.0215	31.7423	1473.0741	80
84	3.4926	0.2863	166.1726	0.0060	47.5786	0.0210	32.9668	1568.5140	84
90	3.8189	0.2619	187.9299	0.0053	49.2099	0.0203	34.7399	1709.5439	90
96	4.1758	0.2395	211.7202	0.0047	50.7017	0.0197	36.4381	1847.4725	96
100	4.4320	0.2256	228.8030	0.0044	51.6247	0.0194	37.5295	1937.4506	100
108	4.9927	0.2003	266.1778	0.0038	53.3137	0.0188	39.6171	2112.1348	108
120	5.9693	0.1675	331.2882	0.0030	55.4985	0.0180	42.5185	2359.7114	120
240	35.6328	0.0281	2308.8544	0.0004	64.7957	0.0154	59.7368	3870.6912	240
360	212.7038	0.0047	14113.5854	0.0001	66.3532	0.0151	64.9662	4310.7165	360

1.75%

	Single Payment		Equal Payment Series				Gradient Series		
	Compound Amount Factor	Present Worth Factor	Compound Amount Factor	Sinking Fund Factor	Present Worth Factor	Capital Recovery Factor	Gradient Uniform Series	Gradient Present Worth	
N	(F/P,i,N)	(P/F,i,N)	(F/A,i,N)	(A/F,i,N)	(P/A,i,N)	(A/P,i,N)	(A/G,i,N)	(P/G,i,N)	N
1	1.0175	0.9828	1.0000	1.0000	0.9828	1.0175	0.0000	0.0000	1
2	1.0353	0.9659	2.0175	0.4957	1.9487	0.5132	0.4957	0.9659	2
3	1.0534	0.9493	3.0528	0.3276	2.8980	0.3451	0.9884	2.8645	3
4	1.0719	0.9330	4.1062	0.2435	3.8309	0.2610	1.4783	5.6633	4
5	1.0906	0.9169	5.1781	0.1931	4.7479	0.2106	1.9653	9.3310	5
6	1.1097	0.9011	6.2687	0.1595	5.6490	0.1770	2.4494	13.8367	6
7	1.1291	0.8856	7.3784	0.1355	6.5346	0.1530	2.9306	19.1506	7
8	1.1489	0.8704	8.5075	0.1175	7.4051	0.1350	3.4089	25.2435	8
9	1.1690	0.8554	9.6564	0.1036	8.2605	0.1211	3.8844	32.0870	9
10	1.1894	0.8407	10.8254	0.0924	9.1012	0.1099	4.3569	39.6535	10
11	1.2103	0.8263	12.0148	0.0832	9.9275	0.1007	4.8266	47.9162	11
12	1.2314	0.8121	13.2251	0.0756	10.7395	0.0931	5.2934	56.8489	12
13	1.2530	0.7981	14.4565	0.0692	11.5376	0.0867	5.7573	66.4260	13
14	1.2749	0.7844	15.7095	0.0637	12.3220	0.0812	6.2184	76.6227	14
15	1.2972	0.7709	16.9844	0.0589	13.0929	0.0764	6.6765	87.4149	15
16	1.3199	0.7576	18.2817	0.0547	13.8505	0.0722	7.1318	98.7792	16
17	1.3430	0.7446	19.6016	0.0510	14.5951	0.0685	7.5842	110.6926	17
18	1.3665	0.7318	20.9446	0.0477	15.3269	0.0652	8.0338	123.1328	18
19	1.3904	0.7192	22.3112	0.0448	16.0461	0.0623	8.4805	136.0783	19
20	1.4148	0.7068	23.7016	0.0422	16.7529	0.0597	8.9243	149.5080	20
21	1.4395	0.6947	25.1164	0.0398	17.4475	0.0573	9.3653	163.4013	21
22	1.4647	0.6827	26.5559	0.0377	18.1303	0.0552	9.8034	177.7385	22
23	1.4904	0.6710	28.0207	0.0357	18.8012	0.0532	10.2387	192.5000	23
24	1.5164	0.6594	29.5110	0.0339	19.4607	0.0514	10.6711	207.6671	24
25	1.5430	0.6481	31.0275	0.0322	20.1088	0.0497	11.1007	223.2214	25
26	1.5700	0.6369	32.5704	0.0307	20.7457	0.0482	11.5274	239.1451	26
27	1.5975	0.6260	34.1404	0.0293	21.3717	0.0468	11.9513	255.4210	27
28	1.6254	0.6152	35.7379	0.0280	21.9870	0.0455	12.3724	272.0321	28
29	1.6539	0.6046	37.3633	0.0268	22.5916	0.0443	12.7907	288.9623	29
30	1.6828	0.5942	39.0172	0.0256	23.1858	0.0431	13.2061	306.1954	30
31	1.7122	0.5840	40.7000	0.0246	23.7699	0.0421	13.6188	323.7163	31
32	1.7422	0.5740	42.4122	0.0236	24.3439	0.0411	14.0286	341.5097	32
33	1.7727	0.5641	44.1544	0.0226	24.9080	0.0401	14.4356	359.5613	33
34	1.8037	0.5544	45.9271	0.0218	25.4624	0.0393	14.8398	377.8567	34
35	1.8353	0.5449	47.7308	0.0210	26.0073	0.0385	15.2412	396.3824	35
36	1.8674	0.5355	49.5661	0.0202	26.5428	0.0377	15.6399	415.1250	36
40	2.0016	0.4996	57.2341	0.0175	28.5942	0.0350	17.2066	492.0109	40
48	2.2996	0.4349	74.2628	0.0135	32.2938	0.0310	20.2084	652.6054	48
50	2.3808	0.4200	78.9022	0.0127	33.1412	0.0302	20.9317	693.7010	50
60	2.8318	0.3531	104.6752	0.0096	36.9640	0.0271	24.3885	901.4954	60
72	3.4872	0.2868	142.1263	0.0070	40.7564	0.0245	28.1948	1149.1181	72
80	4.0064	0.2496	171.7938	0.0058	42.8799	0.0233	30.5329	1309.2482	80
84	4.2943	0.2329	188.2450	0.0053	43.8361	0.0228	31.6442	1387.1584	84
90	4.7654	0.2098	215.1646	0.0046	45.1516	0.0221	33.2409	1500.8798	90
96	5.2882	0.1891	245.0374	0.0041	46.3370	0.0216	34.7556	1610.4716	96
100	5.6682	0.1764	266.7518	0.0037	47.0615	0.0212	35.7211	1681.0886	100
108	6.5120	0.1536	314.9738	0.0032	48.3679	0.0207	37.5494	1816.1852	108
120	8.0192	0.1247	401.0962	0.0025	50.0171	0.0200	40.0469	2003.0269	120
240	64.3073	0.0156	3617.5602	0.0003	56.2543	0.0178	53.3518	3001.2678	240
360	515.6921	0.0019	29410.9747	0.0000	57.0320	0.0175	56.4434	3219.0833	360

2.0%

	Single Payment		Equal Payment Series				Gradient Series		
N	Compound Amount Factor (F/P,i,N)	Present Worth Factor (P/F,i,N)	Compound Amount Factor (F/A,i,N)	Sinking Fund Factor (A/F,i,N)	Present Worth Factor (P/A,i,N)	Capital Recovery Factor (A/P,i,N)	Gradient Uniform Series (A/G,i,N)	Gradient Present Worth (P/G,i,N)	N
1	1.0200	0.9804	1.0000	1.0000	0.9804	1.0200	0.0000	0.0000	1
2	1.0404	0.9612	2.0200	0.4950	1.9416	0.5150	0.4950	0.9612	2
3	1.0612	0.9423	3.0604	0.3268	2.8839	0.3468	0.9868	2.8458	3
4	1.0824	0.9238	4.1216	0.2426	3.8077	0.2626	1.4752	5.6173	4
5	1.1041	0.9057	5.2040	0.1922	4.7135	0.2122	1.9604	9.2403	5
6	1.1262	0.8880	6.3081	0.1585	5.6014	0.1785	2.4423	13.6801	6
7	1.1487	0.8706	7.4343	0.1345	6.4720	0.1545	2.9208	18.9035	7
8	1.1717	0.8535	8.5830	0.1165	7.3255	0.1365	3.3961	24.8779	8
9	1.1951	0.8368	9.7546	0.1025	8.1622	0.1225	3.8681	31.5720	9
10	1.2190	0.8203	10.9497	0.0913	8.9826	0.1113	4.3367	38.9551	10
11	1.2434	0.8043	12.1687	0.0822	9.7868	0.1022	4.8021	46.9977	11
12	1.2682	0.7885	13.4121	0.0746	10.5753	0.0946	5.2642	55.6712	12
13	1.2936	0.7730	14.6803	0.0681	11.3484	0.0881	5.7231	64.9475	13
14	1.3195	0.7579	15.9739	0.0626	12.1062	0.0826	6.1786	74.7999	14
15	1.3459	0.7430	17.2934	0.0578	12.8493	0.0778	6.6309	85.2021	15
16	1.3728	0.7284	18.6393	0.0537	13.5777	0.0737	7.0799	96.1288	16
17	1.4002	0.7142	20.0121	0.0500	14.2919	0.0700	7.5256	107.5554	17
18	1.4282	0.7002	21.4123	0.0467	14.9920	0.0667	7.9681	119.4581	18
19	1.4568	0.6864	22.8406	0.0438	15.6785	0.0638	8.4073	131.8139	19
20	1.4859	0.6730	24.2974	0.0412	16.3514	0.0612	8.8433	144.6003	20
21	1.5157	0.6598	25.7833	0.0388	17.0112	0.0588	9.2760	157.7959	21
22	1.5460	0.6468	27.2990	0.0366	17.6580	0.0566	9.7055	171.3795	22
23	1.5769	0.6342	28.8450	0.0347	18.2922	0.0547	10.1317	185.3309	23
24	1.6084	0.6217	30.4219	0.0329	18.9139	0.0529	10.5547	199.6305	24
25	1.6406	0.6095	32.0303	0.0312	19.5235	0.0512	10.9745	214.2592	25
26	1.6734	0.5976	33.6709	0.0297	20.1210	0.0497	11.3910	229.1987	26
27	1.7069	0.5859	35.3443	0.0283	20.7069	0.0483	11.8043	244.4311	27
28	1.7410	0.5744	37.0512	0.0270	21.2813	0.0470	12.2145	259.9392	28
29	1.7758	0.5631	38.7922	0.0258	21.8444	0.0458	12.6214	275.7064	29
30	1.8114	0.5521	40.5681	0.0246	22.3965	0.0446	13.0251	291.7164	30
31	1.8476	0.5412	42.3794	0.0236	22.9377	0.0436	13.4257	307.9538	31
32	1.8845	0.5306	44.2270	0.0226	23.4683	0.0426	13.8230	324.4035	32
33	1.9222	0.5202	46.1116	0.0217	23.9886	0.0417	14.2172	341.0508	33
34	1.9607	0.5100	48.0338	0.0208	24.4986	0.0408	14.6083	357.8817	34
35	1.9999	0.5000	49.9945	0.0200	24.9986	0.0400	14.9961	374.8826	35
36	2.0399	0.4902	51.9944	0.0192	25.4888	0.0392	15.3809	392.0405	36
40	2.2080	0.4529	60.4020	0.0166	27.3555	0.0366	16.8885	461.9931	40
48	2.5871	0.3865	79.3535	0.0126	30.6731	0.0326	19.7556	605.9657	48
50	2.6916	0.3715	84.5794	0.0118	31.4236	0.0318	20.4420	642.3606	50
60	3.2810	0.3048	114.0515	0.0088	34.7609	0.0288	23.6961	823.6975	60
72	4.1611	0.2403	158.0570	0.0063	37.9841	0.0263	27.2234	1034.0557	72
80	4.8754	0.2051	193.7720	0.0052	39.7445	0.0252	29.3572	1166.7868	80
84	5.2773	0.1895	213.8666	0.0047	40.5255	0.0247	30.3616	1230.4191	84
90	5.9431	0.1683	247.1567	0.0040	41.5869	0.0240	31.7929	1322.1701	90
96	6.6929	0.1494	284.6467	0.0035	42.5294	0.0235	33.1370	1409.2973	96
100	7.2446	0.1380	312.2323	0.0032	43.0984	0.0232	33.9863	1464.7527	100
108	8.4883	0.1178	374.4129	0.0027	44.1095	0.0227	35.5774	1569.3025	108
120	10.7652	0.0929	488.2582	0.0020	45.3554	0.0220	37.7114	1710.4160	120
240	115.8887	0.0086	5744.4368	0.0002	49.5686	0.0202	47.9110	2374.8800	240
360	1247.5611	0.0008	62328.0564	0.0000	49.9599	0.0200	49.7112	2483.5679	360

3.0%

	Single Payment		Equal Payment Series				Gradient Series		
N	Compound Amount Factor (F/P,i,N)	Present Worth Factor (P/F,i,N)	Compound Amount Factor (F/A,i,N)	Sinking Fund Factor (A/F,i,N)	Present Worth Factor (P/A,i,N)	Capital Recovery Factor (A/P,i,N)	Gradient Uniform Series (A/G,i,N)	Gradient Present Worth (P/G,i,N)	N
1	1.0300	0.9709	1.0000	1.0000	0.9709	1.0300	0.0000	0.0000	1
2	1.0609	0.9426	2.0300	0.4926	1.9135	0.5226	0.4926	0.9426	2
3	1.0927	0.9151	3.0909	0.3235	2.8286	0.3535	0.9803	2.7729	3
4	1.1255	0.8885	4.1836	0.2390	3.7171	0.2690	1.4631	5.4383	4
5	1.1593	0.8626	5.3091	0.1884	4.5797	0.2184	1.9409	8.8888	5
6	1.1941	0.8375	6.4684	0.1546	5.4172	0.1846	2.4138	13.0762	6
7	1.2299	0.8131	7.6625	0.1305	6.2303	0.1605	2.8819	17.9547	7
8	1.2668	0.7894	8.8923	0.1125	7.0197	0.1425	3.3450	23.4806	8
9	1.3048	0.7664	10.1591	0.0984	7.7861	0.1284	3.8032	29.6119	9
10	1.3439	0.7441	11.4639	0.0872	8.5302	0.1172	4.2565	36.3088	110
11	1.3842	0.7224	12.8078	0.0781	9.2526	0.1081	4.7049	43.5330	11
12	1.4258	0.7014	14.1920	0.0705	9.9540	0.1005	5.1485	51.2482	12
13	1.4685	0.6810	15.6178	0.0640	10.6350	0.0940	5.5872	59.4196	13
14	1.5126	0.6611	17.0863	0.0585	11.2961	0.0885	6.0210	68.0141	14
15	1.5580	0.6419	18.5989	0.0538	11.9379	0.0838	6.4500	77.0002	15
16	1.6047	0.6232	20.1569	0.0496	12.5611	0.0796	6.8742	86.3477	16
17	1.6528	0.6050	21.7616	0.0460	13.1661	0.0760	7.2936	96.0280	17
18	1.7024	0.5874	23.4144	0.0427	13.7535	0.0727	7.7081	106.0137	18
19	1.7535	0.5703	25.1169	0.0398	14.3238	0.0698	8.1179	116.2788	19
20	1.8061	0.5537	26.8704	0.0372	14.8775	0.0672	8.5229	126.7987	20
21	1.8603	0.5375	28.6765	0.0349	15.4150	0.0649	8.9231	137.5496	21
22	1.9161	0.5219	30.5368	0.0327	15.9396	0.0627	9.3186	148.5094	22
23	1.9736	0.5067	32.4529	0.0308	16.4436	0.0608	9.7093	159.6566	23
24	2.0328	0.4919	34.4265	0.0290	16.9355	0.0590	10.0954	170.9711	24
25	2.0938	0.4776	36.4593	0.0274	17.4131	0.0574	10.4768	182.4336	25
26	2.1566	0.4637	38.5530	0.0259	17.8768	0.0559	10.8535	194.0260	26
27	2.2213	0.4502	40.7096	0.0246	18.3270	0.0546	11.2255	205.7309	27
28	2.2879	0.4371	42.9309	0.0233	18.7641	0.0533	11.5930	217.5320	28
29	2.3566	0.4243	45.2189	0.0221	19.1885	0.0521	11.9558	229.4137	29
30	2.4273	0.4120	47.5754	0.0210	19.6004	0.0510	12.3141	241.3613	30
31	2.5001	0.4000	50.0027	0.0200	20.0004	0.0500	12.6678	253.3609	31
32	2.5751	0.3883	52.5028	0.0190	20.3888	0.0490	13.0169	265.3993	32
33	2.6523	0.3770	55.0778	0.0182	20.7658	0.0482	13.3616	277.4642	33
34	2.7319	0.3660	57.7302	0.0173	21.1318	0.0473	13.7018	289.5437	34
35	2.8139	0.3554	60.4621	0.0165	21.4872	0.0465	14.0375	301.6267	35
40	3.2620	0.3066	75.4013	0.0133	23.1148	0.0433	15.6502	361.7499	40
45	3.7816	0.2644	92.7199	0.0108	24.5187	0.0408	17.1556	420.6325	45
50	4.3839	0.2281	112.7969	0.0089	25.7298	0.0389	18.5575	477.4803	50
55	5.0821	0.1968	136.0716	0.0073	26.7744	0.0373	19.8600	531.7411	55
60	5.8916	0.1697	163.0534	0.0061	27.6756	0.0361	21.0674	583.0526	60
65	6.8300	0.1464	194.3328	0.0051	28.4529	0.0351	22.1841	631.2010	65
70	7.9178	0.1263	230.5941	0.0043	29.1234	0.0343	23.2145	676.0869	70
75	9.1789	0.1089	272.6309	0.0037	29.7018	0.0337	24.1634	717.6978	75
80	10.6409	0.0940	321.3630	0.0031	30.2008	0.0331	25.0353	756.0865	80
85	12.3357	0.0811	377.8570	0.0026	30.6312	0.0326	25.8349	791.3529	85
90	14.3005	0.0699	443.3489	0.0023	31.0024	0.0323	26.5667	823.6302	90
95	16.5782	0.0603	519.2720	0.0019	31.3227	0.0319	27.2351	853.0742	95
100	19.2186	0.0520	607.2877	0.0016	31.5989	0.0316	27.8444	879.8540	100

4.0%

	Single Payment		Equal Payment Series				Gradient Series		
N	Compound Amount Factor (F/P,i,N)	Present Worth Factor (P/F,i,N)	Compound Amount Factor (F/A,i,N)	Sinking Fund Factor (A/F,i,N)	Present Worth Factor (P/A,i,N)	Capital Recovery Factor (A/P,i,N)	Gradient Uniform Series (A/G,i,N)	Gradient Present Worth (P/G,i,N)	N
1	1.0400	0.9615	1.0000	1.0000	0.9615	1.0400	0.0000	0.0000	1
2	1.0816	0.9246	2.0400	0.4902	1.8861	0.5302	0.4902	0.9246	2
3	1.1249	0.8890	3.1216	0.3203	2.7751	0.3603	0.9739	2.7025	3
4	1.1699	0.8548	4.2465	0.2355	3.6299	0.2755	1.4510	5.2670	4
5	1.2167	0.8219	5.4163	0.1846	4.4518	0.2246	1.9216	8.5547	5
6	1.2653	0.7903	6.6330	0.1508	5.2421	0.1908	2.3857	12.5062	6
7	1.3159	0.7599	7.8983	0.1266	6.0021	0.1666	2.8433	17.0657	7
8	1.3686	0.7307	9.2142	0.1085	6.7327	0.1485	3.2944	22.1806	8
9	1.4233	0.7026	10.5828	0.0945	7.4353	0.1345	3.7391	27.8013	9
10	1.4802	0.6756	12.0061	0.0833	8.1109	0.1233	4.1773	33.8814	10
11	1.5395	0.6496	13.4864	0.0741	8.7605	0.1141	4.6090	40.3772	11
12	1.6010	0.6246	15.0258	0.0666	9.3851	0.1066	5.0343	47.2477	12
13	1.6651	0.6006	16.6268	0.0601	9.9856	0.1001	5.4533	54.4546	13
14	1.7317	0.5775	18.2919	0.0547	10.5631	0.0947	5.8659	61.9618	14
15	1.8009	0.5553	20.0236	0.0499	11.1184	0.0899	6.2721	69.7355	15
16	1.8730	0.5339	21.8245	0.0458	11.6523	0.0858	6.6720	77.7441	16
17	1.9479	0.5134	23.6975	0.0422	12.1657	0.0822	7.0656	85.9581	17
18	2.0258	0.4936	25.6454	0.0390	12.6593	0.0790	7.4530	94.3498	18
19	2.1068	0.4746	27.6712	0.0361	13.1339	0.0761	7.8342	102.8933	19
20	2.1911	0.4564	29.7781	0.0336	13.5903	0.0736	8.2091	111.5647	20
21	2.2788	0.4388	31.9692	0.0313	14.0292	0.0713	8.5779	120.3414	21
22	2.3699	0.4220	34.2480	0.0292	14.4511	0.0692	8.9407	129.2024	22
23	2.4647	0.4057	36.6179	0.0273	14.8568	0.0673	9.2973	138.1284	23
24	2.5633	0.3901	39.0826	0.0256	15.2470	0.0656	9.6479	147.1012	24
25	2.6658	0.3751	41.6459	0.0240	15.6221	0.0640	9.9925	156.1040	25
26	2.7725	0.3607	44.3117	0.0226	15.9828	0.0626	10.3312	165.1212	26
27	2.8834	0.3468	47.0842	0.0212	16.3296	0.0612	10.6640	174.1385	27
28	2.9987	0.3335	49.9676	0.0200	16.6631	0.0600	10.9909	183.1424	28
29	3.1187	0.3207	52.9663	0.0189	16.9837	0.0589	11.3120	192.1206	29
30	3.2434	0.3083	56.0849	0.0178	17.2920	0.0578	11.6274	201.0618	30
31	3.3731	0.2965	59.3283	0.0169	17.5885	0.0569	11.9371	209.9556	31
32	3.5081	0.2851	62.7015	0.0159	17.8736	0.0559	12.2411	218.7924	32
33	3.6484	0.2741	66.2095	0.0151	18.1476	0.0551	12.5396	227.5634	33
34	3.7943	0.2636	69.8579	0.0143	18.4112	0.0543	12.8324	236.2607	34
35	3.9461	0.2534	73.6522	0.0136	18.6646	0.0536	13.1198	244.8768	35
40	4.8010	0.2083	95.0255	0.0105	19.7928	0.0505	14.4765	286.5303	40
45	5.8412	0.1712	121.0294	0.0083	20.7200	0.0483	15.7047	325.4028	45
50	7.1067	0.1407	152.6671	0.0066	21.4822	0.0466	16.8122	361.1638	50
55	8.6464	0.1157	191.1592	0.0052	22.1086	0.0452	17.8070	393.6890	55
60	10.5196	0.0951	237.9907	0.0042	22.6235	0.0442	18.6972	422.9966	60
65	12.7987	0.0781	294.9684	0.0034	23.0467	0.0434	19.4909	449.2014	65
70	15.5716	0.0642	364.2905	0.0027	23.3945	0.0427	20.1961	472.4789	70
75	18.9453	0.0528	448.6314	0.0022	23.6804	0.0422	20.8206	493.0408	75
80	23.0498	0.0434	551.2450	0.0018	23.9154	0.0418	21.3718	511.1161	80
85	28.0436	0.0357	676.0901	0.0015	24.1085	0.0415	21.8569	526.9384	85
90	34.1193	0.0293	827.9833	0.0012	24.2673	0.0412	22.2826	540.7369	90
95	41.5114	0.0241	1012.7846	0.0010	24.3978	0.0410	22.6550	552.7307	95
100	50.5049	0.0198	1237.6237	0.0008	24.5050	0.0408	22.9800	563.1249	100

5.0%

	Single Payment		Equal Payment Series				Gradient Series		
N	Compound Amount Factor (F/P,i,N)	Present Worth Factor (P/F,i,N)	Compound Amount Factor (F/A,i,N)	Sinking Fund Factor (A/F,i,N)	Present Worth Factor (P/A,i,N)	Capital Recovery Factor (A/P,i,N)	Gradient Uniform Series (A/G,i,N)	Gradient Present Worth (P/G,i,N)	N
1	1.0500	0.9524	1.0000	1.0000	0.9524	1.0500	0.0000	0.0000	1
2	1.1025	0.9070	2.0500	0.4878	1.8594	0.5378	0.4878	0.9070	2
3	1.1576	0.8638	3.1525	0.3172	2.7232	0.3672	0.9675	2.6347	3
4	1.2155	0.8227	4.3101	0.2320	3.5460	0.2820	1.4391	5.1028	4
5	1.2763	0.7835	5.5256	0.1810	4.3295	0.2310	1.9025	8.2369	5
6	1.3401	0.7462	6.8019	0.1470	5.0757	0.1970	2.3579	11.9680	6
7	1.4071	0.7107	8.1420	0.1228	5.7864	0.1728	2.8052	16.2321	7
8	1.4775	0.6768	9.5491	0.1047	6.4632	0.1547	3.2445	20.9700	8
9	1.5513	0.6446	11.0266	0.0907	7.1078	0.1407	3.6758	26.1268	9
10	1.6289	0.6139	12.5779	0.0795	7.7217	0.1295	4.0991	31.6520	10
11	1.7103	0.5847	14.2068	0.0704	8.3064	0.1204	4.5144	37.4988	11
12	1.7959	0.5568	15.9171	0.0628	8.8633	0.1128	4.9219	43.6241	12
13	1.8856	0.5303	17.7130	0.0565	9.3936	0.1065	5.3215	49.9879	13
14	1.9799	0.5051	19.5986	0.0510	9.8986	0.1010	5.7133	56.5538	14
15	2.0789	0.4810	21.5786	0.0463	10.3797	0.0963	6.0973	63.2880	15
16	2.1829	0.4581	23.6575	0.0423	10.8378	0.0923	6.4736	70.1597	16
17	2.2920	0.4363	25.8404	0.0387	11.2741	0.0887	6.8423	77.1405	17
18	2.4066	0.4155	28.1324	0.0355	11.6896	0.0855	7.2034	84.2043	18
19	2.5270	0.3957	30.5390	0.0327	12.0853	0.0827	7.5569	91.3275	19
20	2.6533	0.3769	33.0660	0.0302	12.4622	0.0802	7.9030	98.4884	20
21	2.7860	0.3589	35.7193	0.0280	12.8212	0.0780	8.2416	105.6673	21
22	2.9253	0.3418	38.5052	0.0260	13.1630	0.0760	8.5730	112.8461	22
23	3.0715	0.3256	41.4305	0.0241	13.4886	0.0741	8.8971	120.0087	23
24	3.2251	0.3101	44.5020	0.0225	13.7986	0.0725	9.2140	127.1402	24
25	3.3864	0.2953	47.7271	0.0210	14.0939	0.0710	9.5238	134.2275	25
26	3.5557	0.2812	51.1135	0.0196	14.3752	0.0696	9.8266	141.2585	26
27	3.7335	0.2678	54.6691	0.0183	14.6430	0.0683	10.1224	148.2226	27
28	3.9201	0.2551	58.4026	0.0171	14.8981	0.0671	10.4114	155.1101	28
29	4.1161	0.2429	62.3227	0.0160	15.1411	0.0660	10.6936	161.9126	29
30	4.3219	0.2314	66.4388	0.0151	15.3725	0.0651	10.9691	168.6226	30
31	4.5380	0.2204	70.7608	0.0141	15.5928	0.0641	11.2381	175.2333	31
32	4.7649	0.2099	75.2988	0.0133	15.8027	0.0633	11.5005	181.7392	32
33	5.0032	0.1999	80.0638	0.0125	16.0025	0.0625	11.7566	188.1351	33
34	5.2533	0.1904	85.0670	0.0118	16.1929	0.0618	12.0063	194.4168	34
35	5.5160	0.1813	90.3203	0.0111	16.3742	0.0611	12.2498	200.5807	35
40	7.0400	0.1420	120.7998	0.0083	17.1591	0.0583	13.3775	229.5452	40
45	8.9850	0.1113	159.7002	0.0063	17.7741	0.0563	14.3644	255.3145	45
50	11.4674	0.0872	209.3480	0.0048	18.2559	0.0548	15.2233	277.9148	50
55	14.6356	0.0683	272.7126	0.0037	18.6335	0.0537	15.9664	297.5104	55
60	18.6792	0.0535	353.5837	0.0028	18.9293	0.0528	16.6062	314.3432	60
65	23.8399	0.0419	456.7980	0.0022	19.1611	0.0522	17.1541	328.6910	65
70	30.4264	0.0329	588.5285	0.0017	19.3427	0.0517	17.6212	340.8409	70
75	38.8327	0.0258	756.6537	0.0013	19.4850	0.0513	18.0176	351.0721	75
80	49.5614	0.0202	971.2288	0.0010	19.5965	0.0510	18.3526	359.6460	80
85	63.2544	0.0158	1245.0871	0.0008	19.6838	0.0508	18.6346	366.8007	85
90	80.7304	0.0124	1594.6073	0.0006	19.7523	0.0506	18.8712	372.7488	90
95	103.0347	0.0097	2040.6935	0.0005	19.8059	0.0505	19.0689	377.6774	95
100	131.5013	0.0076	2610.0252	0.0004	19.8479	0.0504	19.2337	381.7492	100

6.0%

	Single Payment		Equal Payment Series				Gradient Series		
N	Compound Amount Factor (F/P,i,N)	Present Worth Factor (P/F,i,N)	Compound Amount Factor (F/A,i,N)	Sinking Fund Factor (A/F,i,N)	Present Worth Factor (P/A,i,N)	Capital Recovery Factor (A/P,i,N)	Gradient Uniform Series (A/G,i,N)	Gradient Present Worth (P/G,i,N)	N
1	1.0600	0.9434	1.0000	1.0000	0.9434	1.0600	0.0000	0.0000	1
2	1.1236	0.8900	2.0600	0.4854	1.8334	0.5454	0.4854	0.8900	2
3	1.1910	0.8396	3.1836	0.3141	2.6730	0.3741	0.9612	2.5692	3
4	1.2625	0.7921	4.3746	0.2286	3.4651	0.2886	1.4272	4.9455	4
5	1.3382	0.7473	5.6371	0.1774	4.2124	0.2374	1.8836	7.9345	5
6	1.4185	0.7050	6.9753	0.1434	4.9173	0.2034	2.3304	11.4594	6
7	1.5036	0.6651	8.3938	0.1191	5.5824	0.1791	2.7676	15.4497	7
8	1.5938	0.6274	9.8975	0.1010	6.2098	0.1610	3.1952	19.8416	8
9	1.6895	0.5919	11.4913	0.0870	6.8017	0.1470	3.6133	24.5768	9
10	1.7908	0.5584	13.1808	0.0759	7.3601	0.1359	4.0220	29.6023	10
11	1.8983	0.5268	14.9716	0.0668	7.8869	0.1268	4.4213	34.8702	11
12	2.0122	0.4970	16.8699	0.0593	8.3838	0.1193	4.8113	40.3369	12
13	2.1329	0.4688	18.8821	0.0530	8.8527	0.1130	5.1920	45.9629	13
14	2.2609	0.4423	21.0151	0.0476	9.2950	0.1076	5.5635	51.7128	14
15	2.3966	0.4173	23.2760	0.0430	9.7122	0.1030	5.9260	57.5546	15
16	2.5404	0.3936	25.6725	0.0390	10.1059	0.0990	6.2794	63.4592	16
17	2.6928	0.3714	28.2129	0.0354	10.4773	0.0954	6.6240	69.4011	17
18	2.8543	0.3503	30.9057	0.0324	10.8276	0.0924	6.9597	75.3569	18
19	3.0256	0.3305	33.7600	0.0296	11.1581	0.0896	7.2867	81.3062	19
20	3.2071	0.3118	36.7856	0.0272	11.4699	0.0872	7.6051	87.2304	20
21	3.3996	0.2942	39.9927	0.0250	11.7641	0.0850	7.9151	93.1136	21
22	3.6035	0.2775	43.3923	0.0230	12.0416	0.0830	8.2166	98.9412	22
23	3.8197	0.2618	46.9958	0.0213	12.3034	0.0813	8.5099	104.7007	23
24	4.0489	0.2470	50.8156	0.0197	12.5504	0.0797	8.7951	110.3812	24
25	4.2919	0.2330	54.8645	0.0182	12.7834	0.0782	9.0722	115.9732	25
26	4.5494	0.2198	59.1564	0.0169	13.0032	0.0769	9.3414	121.4684	26
27	4.8223	0.2074	63.7058	0.0157	13.2105	0.0757	9.6029	126.8600	27
28	5.1117	0.1956	68.5281	0.0146	13.4062	0.0746	9.8568	132.1420	28
29	5.4184	0.1846	73.6398	0.0136	13.5907	0.0736	10.1032	137.3096	29
30	5.7435	0.1741	79.0582	0.0126	13.7648	0.0726	10.3422	142.3588	30
31	6.0881	0.1643	84.8017	0.0118	13.9291	0.0718	10.5740	147.2864	31
32	6.4534	0.1550	90.8898	0.0110	14.0840	0.0710	10.7988	152.0901	32
33	6.8406	0.1462	97.3432	0.0103	14.2302	0.0703	11.0166	156.7681	33
34	7.2510	0.1379	104.1838	0.0096	14.3681	0.0696	11.2276	161.3192	34
35	7.6861	0.1301	111.4348	0.0090	14.4982	0.0690	11.4319	165.7427	35
40	10.2857	0.0972	154.7620	0.0065	15.0463	0.0665	12.3590	185.9568	40
45	13.7646	0.0727	212.7435	0.0047	15.4558	0.0647	13.1413	203.1096	45
50	18.4202	0.0543	290.3359	0.0034	15.7619	0.0634	13.7964	217.4574	50
55	24.6503	0.0406	394.1720	0.0025	15.9905	0.0625	14.3411	229.3222	55
60	32.9877	0.0303	533.1282	0.0019	16.1614	0.0619	14.7909	239.0428	60
65	44.1450	0.0227	719.0829	0.0014	16.2891	0.0614	15.1601	246.9450	65
70	59.0759	0.0169	967.9322	0.0010	16.3845	0.0610	15.4613	253.3271	70
75	79.0569	0.0126	1300.9487	0.0008	16.4558	0.0608	15.7058	258.4527	75
80	105.7960	0.0095	1746.5999	0.0006	16.5091	0.0606	15.9033	262.5493	80
85	141.5789	0.0071	2342.9817	0.0004	16.5489	0.0604	16.0620	265.8096	85
90	189.4645	0.0053	3141.0752	0.0003	16.5787	0.0603	16.1891	268.3946	90
95	253.5463	0.0039	4209.1042	0.0002	16.6009	0.0602	16.2905	270.4375	95
100	339.3021	0.0029	5638.3681	0.0002	16.6175	0.0602	16.3711	272.0471	100

7.0%

	Single Payment		Equal Payment Series				Gradient Series		
	Compound Amount Factor	Present Worth Factor	Compound Amount Factor	Sinking Fund Factor	Present Worth Factor	Capital Recovery Factor	Gradient Uniform Series	Gradient Present Worth	
N	(F/P,i,N)	(P/F,i,N)	(F/A,i,N)	(A/F,i,N)	(P/A,i,N)	(A/P,i,N)	(A/G,i,N)	(P/G,i,N)	N
1	1.0700	0.9346	1.0000	1.0000	0.9346	1.0700	0.0000	0.0000	1
2	1.1449	0.8734	2.0700	0.4831	1.8080	0.5531	0.4831	0.8734	2
3	1.2250	0.8163	3.2149	0.3111	2.6243	0.3811	0.9549	2.5060	3
4	1.3108	0.7629	4.4399	0.2252	3.3872	0.2952	1.4155	4.7947	4
5	1.4026	0.7130	5.7507	0.1739	4.1002	0.2439	1.8650	7.6467	5
6	1.5007	0.6663	7.1533	0.1398	4.7665	0.2098	2.3032	10.9784	6
7	1.6058	0.6227	8.6540	0.1156	5.3893	0.1856	2.7304	14.7149	7
8	1.7182	0.5820	10.2598	0.0975	5.9713	0.1675	3.1465	18.7889	8
9	1.8385	0.5439	11.9780	0.0835	6.5152	0.1535	3.5517	23.1404	9
10	1.9672	0.5083	13.8164	0.0724	7.0236	0.1424	3.9461	27.7156	10
11	2.1049	0.4751	15.7836	0.0634	7.4987	0.1334	4.3296	32.4665	11
12	2.2522	0.4440	17.8885	0.0559	7.9427	0.1259	4.7025	37.3506	12
13	2.4098	0.4150	20.1406	0.0497	8.3577	0.1197	5.0648	42.3302	13
14	2.5785	0.3878	22.5505	0.0443	8.7455	0.1143	5.4167	47.3718	14
15	2.7590	0.3624	25.1290	0.0398	9.1079	0.1098	5.7583	52.4461	15
16	2.9522	0.3387	27.8881	0.0359	9.4466	0.1059	6.0897	57.5271	16
17	3.1588	0.3166	30.8402	0.0324	9.7632	0.1024	6.4110	62.5923	17
18	3.3799	0.2959	33.9990	0.0294	10.0591	0.0994	6.7225	67.6219	18
19	3.6165	0.2765	37.3790	0.0268	10.3356	0.0968	7.0242	72.5991	19
20	3.8697	0.2584	40.9955	0.0244	10.5940	0.0944	7.3163	77.5091	20
21	4.1406	0.2415	44.8652	0.0223	10.8355	0.0923	7.5990	82.3393	21
22	4.4304	0.2257	49.0057	0.0204	11.0612	0.0904	7.8725	87.0793	22
23	4.7405	0.2109	53.4361	0.0187	11.2722	0.0887	8.1369	91.7201	23
24	5.0724	0.1971	58.1767	0.0172	11.4693	0.0872	8.3923	96.2545	24
25	5.4274	0.1842	63.2490	0.0158	11.6536	0.0858	8.6391	100.6765	25
26	5.8074	0.1722	68.6765	0.0146	11.8258	0.0846	8.8773	104.9814	26
27	6.2139	0.1609	74.4838	0.0134	11.9867	0.0834	9.1072	109.1656	27
28	6.6488	0.1504	80.6977	0.0124	12.1371	0.9824	9.3289	113.2264	28
29	7.1143	0.1406	87.3465	0.0114	12.2777	0.0814	9.5427	117.1622	29
30	7.6123	0.1314	94.4608	0.0106	12.4090	0.0806	9.7487	120.9718	30
31	8.1451	0.1228	102.0730	0.0098	12.5318	0.0798	9.9471	124.6550	31
32	8.7153	0.1147	110.2182	0.0091	12.6466	0.0791	10.1381	128.2120	32
33	9.3253	0.1072	118.9334	0.0084	12.7538	0.0784	10.3219	131.6435	33
34	9.9781	0.1002	128.2588	0.0078	12.8540	0.0778	10.4987	134.9507	34
35	10.6766	0.0937	138.2369	0.0072	12.9477	0.0772	10.6687	138.1353	35
40	14.9745	0.0668	199.6351	0.0050	13.3317	0.0750	11.4233	152.2928	40
45	21.0025	0.0476	285.7493	0.0035	13.6055	0.0735	12.0360	163.7559	45
50	29.4570	0.0339	406.5289	0.0025	13.8007	0.0725	12.5287	172.9051	50
55	41.3150	0.0242	575.9286	0.0017	13.9399	0.0717	12.9215	180.1243	55
60	57.9464	0.0173	813.5204	0.0012	14.0392	0.0712	13.2321	185.7677	60
65	81.2729	0.0123	1146.7552	0.0009	14.1099	0.0709	13.4760	190.1452	65
70	113.9894	0.0088	1614.1342	0.0006	14.1604	0.0706	13.6662	193.5185	70
75	159.8760	0.0063	2269.6574	0.0004	14.1964	0.0704	13.8136	196.1035	75
80	224.2344	0.0045	3189.0627	0.0003	14.2220	0.0703	13.9273	198.0748	80
85	314.5003	0.0032	4478.5761	0.0002	14.2403	0.0702	14.0146	199.5717	85
90	441.1030	0.0023	6287.1854	0.0002	14.2533	0.0702	14.0812	200.7042	90
95	618.6697	0.0016	8823.8535	0.0001	14.2626	0.0701	14.1319	201.5581	95
100	867.7163	0.0012	12381.6618	0.0001	14.2693	0.0701	14.1703	202.2001	100

8.0%

	Single Payment		Equal Payment Series				Gradient Series		
N	Compound Amount Factor (F/P,i,N)	Present Worth Factor (P/F,i,N)	Compound Amount Factor (F/A,i,N)	Sinking Fund Factor (A/F,i,N)	Present Worth Factor (P/A,i,N)	Capital Recovery Factor (A/P,i,N)	Gradient Uniform Series (A/G,i,N)	Gradient Present Worth (P/G,i,N)	N
1	1.0800	0.9259	1.0000	1.0000	0.9259	1.0800	0.0000	0.0000	1
2	1.1664	0.8573	2.0800	0.4808	1.7833	0.5608	0.4808	0.8573	2
3	1.2597	0.7938	3.2464	0.3080	2.5771	0.3880	0.9487	2.4450	3
4	1.3605	0.7350	4.5061	0.2219	3.3121	0.3019	1.4040	4.6501	4
5	1.4693	0.6806	5.8666	0.1705	3.9927	0.2505	1.8465	7.3724	5
6	1.5869	0.6302	7.3359	0.1363	4.6229	0.2163	2.2763	10.5233	6
7	1.7138	0.5835	8.9228	0.1121	5.2064	0.1921	2.6937	14.0242	7
8	1.8509	0.5403	10.6366	0.0940	5.7466	0.1740	3.0985	17.8061	8
9	1.9990	0.5002	12.4876	0.0801	6.2469	0.1601	3.4910	21.8081	9
10	2.1589	0.4632	14.4866	0.0690	6.7101	0.1490	3.8713	25.9768	10
11	2.3316	0.4289	16.6455	0.0601	7.1390	0.1401	4.2395	30.2657	11
12	2.5182	0.3971	18.9771	0.0527	7.5361	0.1327	4.5957	34.6339	12
13	2.7196	0.3677	21.4953	0.0465	7.9038	0.1265	4.9402	39.0463	13
14	2.9372	0.3405	24.2149	0.0413	8.2442	0.1213	5.2731	43.4723	14
15	3.1722	0.3152	27.1521	0.0368	8.5595	0.1168	5.5945	47.8857	15
16	3.4259	0.2919	30.3243	0.0330	8.8514	0.1130	5.9046	52.2640	16
17	3.7000	0.2703	33.7502	0.0296	9.1216	0.1096	6.2037	56.5883	17
18	3.9960	0.2502	37.4502	0.0267	9.3719	0.1067	6.4920	60.8426	18
19	4.3157	0.2317	41.4463	0.0241	9.6036	0.1041	6.7697	65.0134	19
20	4.6610	0.2145	45.7620	0.0219	9.8181	0.1019	7.0369	69.0898	20
21	5.0338	0.1987	50.4229	0.0198	10.0168	0.0998	7.2940	73.0629	21
22	5.4365	0.1839	55.4568	0.0180	10.2007	0.0980	7.5412	76.9257	22
23	5.8715	0.1703	60.8933	0.0164	10.3711	0.0964	7.7786	80.6726	23
24	6.3412	0.1577	66.7648	0.0150	10.5288	0.0950	8.0066	84.2997	24
25	6.8485	0.1460	73.1059	0.0137	10.6748	0.0937	8.2254	87.8041	25
26	7.3964	0.1352	79.9544	0.0125	10.8100	0.0925	8.4352	91.1842	26
27	7.9881	0.1252	87.3508	0.0114	10.9352	0.0914	8.6363	94.4390	27
78	8.6271	0.1159	95.3388	0.0105	11.0511	0.0905	8.8289	97.5687	28
29	9.3173	0.1073	103.9659	0.0096	11.1584	0.0896	9.0133	100.5738	29
30	10.0627	0.0994	113.2832	0.0088	11.2578	0.0888	9.1897	103.4558	30
31	10.8677	0.0920	123.3459	0.0081	11.3498	0.0881	9.3584	106.2163	31
32	11.7371	0.0852	134.2135	0.0075	11.4350	0.0875	9.5197	108.8575	32
33	12.6760	0.0789	145.9506	0.0069	11.5139	0.0869	9.6737	111.3819	33
34	13.6901	0.0730	158.6267	0.0063	11.5869	0.0863	9.8208	113.7924	34
35	14.7853	0.0676	172.3168	0.0058	11.6546	0.0858	9.9611	116.0920	35
40	21.7245	0.0460	259.0565	0.0039	11.9246	0.0839	10.5699	126.0422	40
45	31.9204	0.0313	386.5056	0.0026	12.1084	0.0826	11.0447	133.7331	45
50	46.9016	0.0213	573.7702	0.0017	12.2335	0.0817	11.4107	139.5928	50
55	68.9139	0.0145	848.9232	0.0012	12.3186	0.0812	11.6902	144.0065	55
60	101.2571	0.0099	1253.2133	0.0008	12.3766	0.0808	11.9015	147.3000	60
65	148.7798	0.0067	1847.2481	0.0005	12.4160	0.0805	12.0602	149.7387	65
70	218.6064	0.0046	2720.0801	0.0004	12.4428	0.0804	12.1783	151.5326	70
75	321.2045	0.0031	4002.5566	0.0002	12.4611	0.0802	12.2658	152.8448	75
80	471.9548	0.0021	5886.9354	0.0002	12.4735	0.0802	12.3301	153.8001	80
85	693.4565	0.0014	8655.7061	0.0001	12.4820	0.0801	12.3772	154.4925	85
90	1018.9151	0.0010	12723.9386	0.0001	12.4877	0.0801	12.4116	154.9925	90
95	1497.1205	0.0007	18701.5069	0.0001	12.4917	0.0801	12.4365	155.3524	95
100	2199.7613	0.0005	27484.5157	0.0000	12.4943	0.0800	12.4545	155.6107	100

9.0%

	Single Payment		Equal Payment Series				Gradient Series		
	Compound Amount Factor	Present Worth Factor	Compound Amount Factor	Sinking Fund Factor	Present Worth Factor	Capital Recovery Factor	Gradient Uniform Series	Gradient Present Worth	
N	(F/P,i,N)	(P/F,i,N)	(F/A,i,N)	(A/F,i,N)	(P/A,i,N)	(A/P,i,N)	(A/G,i,N)	(P/G,i,N)	N
1	1.0900	0.9174	1.0000	1.0000	0.9174	1.0900	0.0000	0.0000	1
2	1.1881	0.8417	2.0900	0.4785	1.7591	0.5685	0.4785	0.8417	2
3	1.2950	0.7722	3.2781	0.3051	2.5313	0.3951	0.9426	2.3860	3
4	1.4116	0.7084	4.5731	0.2187	3.2397	0.3087	1.3925	4.5113	4
5	1.5386	0.6499	5.9847	0.1671	3.8897	0.2571	1.8282	7.1110	5
6	1.6771	0.5963	7.5233	0.1329	4.4859	0.2229	2.2498	10.0924	6
7	1.8280	0.5470	9.2004	0.1087	5.0330	0.1987	2.6574	13.3746	7
8	1.9926	0.5019	11.0285	0.0907	5.5348	0.1807	3.0512	16.8877	8
9	2.1719	0.4604	13.0210	0.0768	5.9952	0.1668	3.4312	20.5711	9
10	2.3674	0.4224	15.1929	0.0658	6.4177	0.1558	3.7978	24.3728	10
11	2.5804	0.3875	17.5603	0.0569	6.8052	0.1469	4.1510	28.2481	11
12	2.8127	0.3555	20.1407	0.0497	7.1607	0.1397	4.4910	32.1590	12
13	3.0658	0.3262	22.9534	0.0436	7.4869	0.1336	4.8182	36.0731	13
14	3.3417	0.2992	26.0192	0.0384	7.7862	0.1284	5.1326	39.9633	14
15	3.6425	0.2745	29.3609	0.0341	8.0607	0.1241	5.4346	43.8069	15
16	3.9703	0.2519	33.0034	0.0303	8.3126	0.1203	5.7245	47.5849	16
17	4.3276	0.2311	36.9737	0.0270	8.5436	0.1170	6.0024	51.2821	17
18	4.7171	0.2120	41.3013	0.0242	8.7556	0.1142	6.2687	54.8860	18
19	5.1417	0.1945	46.0185	0.0217	8.9501	0.1117	6.5236	58.3868	19
20	5.6044	0.1784	51.1601	0.0195	9.1285	0.1095	6.7674	61.7770	20
21	6.1088	0.1637	56.7645	0.0176	9.2922	0.1076	7.0006	65.0509	21
22	6.6586	0.1502	62.8733	0.0159	9.4424	0.1059	7.2232	68.2048	22
23	7.2579	0.1378	69.5319	0.0144	9.5802	0.1044	7.4357	71.2359	23
24	7.9111	0.1264	76.7898	0.0130	9.7066	0.1030	7.6384	74.1433	24
25	8.6231	0.1160	84.7009	0.0118	9.8226	0.1018	7.8316	76.9265	25
26	9.3992	0.1064	93.3240	0.0107	9.9290	0.1007	8.0156	79.5863	26
27	10.2451	0.0976	102.7231	0.0097	10.0266	0.0997	8.1906	82.1241	27
28	11.1671	0.0895	112.9682	0.0089	10.1161	0.0989	8.3571	84.5419	28
29	12.1722	0.0822	124.1354	0.0081	10.1983	0.0981	8.5154	86.8422	29
30	13.2677	0.0754	136.3075	0.0073	10.2737	0.0973	8.6657	89.0280	30
31	14.4618	0.0691	149.5752	0.0067	10.3428	0.0967	8.8083	91.1024	31
32	15.7633	0.0634	164.0370	0.0061	10.4062	0.0961	8.9436	93.0690	32
33	17.1820	0.0582	179.8003	0.0056	10.4644	0.0956	9.0718	94.9314	33
34	18.7284	0.0534	196.9823	0.0051	10.5178	0.0951	9.1933	96.6935	34
35	20.4140	0.0490	215.7108	0.0046	10.5668	0.0946	9.3083	98.3590	35
40	31.4094	0.0318	337.8824	0.0030	10.7574	0.0930	9.7957	105.3762	40
45	48.3273	0.0207	525.8587	0.0019	10.8812	0.0919	10.1603	110.5561	45
50	74.3575	0.0134	815.0836	0.0012	10.9617	0.0912	10.4295	114.3251	50
55	114.4083	0.0087	1260.0918	0.0008	11.0140	0.0908	10.6261	117.0362	55
60	176.0313	0.0057	1944.7921	0.0005	11.0480	0.0905	10.7683	118.9683	60
65	270.8460	0.0037	2998.2885	0.0003	11.0701	0.0903	10.8702	120.3344	65
70	416.7301	0.0024	4619.2232	0.0002	11.0844	0.0902	10.9427	121.2942	70
75	641.1909	0.0016	7113.2321	0.0001	11.0938	0.0901	10.9940	121.9646	75
80	986.5517	0.0010	10950.5741	0.0001	11.0998	0.0901	11.0299	122.4306	80
85	1517.9320	0.0007	16854.8003	0.0001	11.1038	0.0901	11.0551	122.7533	85
90	2335.5266	0.0004	25939.1842	0.0000	11.1064	0.0900	11.0726	122.9758	90
95	3593.4971	0.0003	39916.6350	0.0000	11.1080	0.0900	11.0847	123.1287	95
100	5529.0408	0.0002	61422.6755	0.0000	11.1091	0.0900	11.0930	123.2335	100

10.0%

	Single Payment		Equal Payment Series				Gradient Series		
N	Compound Amount Factor (F/P,i,N)	Present Worth Factor (P/F,i,N)	Compound Amount Factor (F/A,i,N)	Sinking Fund Factor (A/F,i,N)	Present Worth Factor (P/A,i,N)	Capital Recovery Factor (A/P,i,N)	Gradient Uniform Series (A/G,i,N)	Gradient Present Worth (P/G,i,N)	N
1	1.1000	0.9091	1.0000	1.0000	0.9091	1.1000	0.0000	0.0000	1
2	1.2100	0.8264	2.1000	0.4762	1.7355	0.5762	0.4762	0.8264	2
3	1.3310	0.7513	3.3100	0.3021	2.4869	0.4021	0.9366	2.3291	3
4	1.4641	0.6830	4.6410	0.2155	3.1699	0.3155	1.3812	4.3781	4
5	1.6105	0.6209	6.1051	0.1638	3.7908	0.2638	1.8101	6.8618	5
6	1.7716	0.5645	7.7156	0.1296	4.3553	0.2296	2.2236	9.6842	6
7	1.9487	0.5132	9.4872	0.1054	4.8684	0.2054	2.6216	12.7631	7
8	2.1436	0.4665	11.4359	0.0874	5.3349	0.1874	3.0045	16.0287	8
9	2.3579	0.4241	13.5795	0.0736	5.7590	0.1736	3.3724	19.4215	9
10	2.5937	0.3855	15.9374	0.0627	6.1446	0.1627	3.7255	22.8913	10
11	2.8531	0.3505	18.5312	0.0540	6.4951	0.1540	4.0641	26.3963	11
12	3.1384	0.3186	21.3843	0.0468	6.8137	0.1468	4.3884	29.9012	12
13	3.4523	0.2897	24.5227	0.0408	7.1034	0.1408	4.6988	33.3772	13
14	3.7975	0.2633	27.9750	0.0357	7.3667	0.1357	4.9955	36.8005	14
15	4.1772	0.2394	31.7725	0.0315	7.6061	0.1315	5.2789	40.1520	15
16	4.5950	0.2176	35.9497	0.0278	7.8237	0.1278	5.5493	43.4164	16
17	5.0545	0.1978	40.5447	0.0247	8.0216	0.1247	5.8071	46.5819	17
18	5.5599	0.1799	45.5992	0.0219	8.2014	0.1219	6.0526	49.6395	18
19	6.1159	0.1635	51.1591	0.0195	8.3649	0.1195	6.2861	52.5827	19
20	6.7275	0.1486	57.2750	0.0175	8.5136	0.1175	6.5081	55.4069	20
21	7.4002	0.1351	64.0025	0.0156	8.6487	0.1156	6.7189	58.1095	21
22	8.1403	0.1228	71.4027	0.0140	8.7715	0.1140	6.9189	60.6893	22
23	8.9543	0.1117	79.5430	0.0126	8.8832	0.1126	7.1085	63.1462	23
24	9.8497	0.1015	88.4973	0.0113	8.9847	0.1113	7.2881	65.4813	24
25	10.8347	0.0923	98.3471	0.0102	9.0770	0.1102	7.4580	67.6964	25
26	11.9182	0.0839	109.1818	0.0092	9.1609	0.1092	7.6186	69.7940	26
27	13.1100	0.0763	121.0999	0.0083	9.2372	0.1083	7.7704	71.7773	27
28	14.4210	0.0693	134.2099	0.0075	9.3066	0.1075	7.9137	73.6495	28
29	15.8631	0.0630	148.6309	0.0067	9.3696	0.1067	8.0489	75.4146	29
30	17.4494	0.0573	164.4940	0.0061	9.4269	0.1061	8.1762	77.0766	30
31	19.1943	0.0521	181.9434	0.0055	9.4790	0.1055	8.2962	78.6395	31
32	21.1138	0.0474	201.1378	0.0050	9.5264	0.1050	8.4091	80.1078	32
33	23.2252	0.0431	222.2515	0.0045	9.5694	0.1045	8.5152	81.4856	33
34	25.5477	0.0391	245.4767	0.0041	9.6086	0.1041	8.6149	82.7773	34
35	28.1024	0.0356	271.0244	0.0037	9.6442	0.1037	8.7086	83.9872	35
40	45.2593	0.0221	442.5926	0.0023	9.7791	0.1023	9.0962	88.9525	40
45	72.8905	0.0137	718.9048	0.0014	9.8628	0.1014	9.3740	92.4544	45
50	117.3909	0.0085	1163.9085	0.0009	9.9148	0.1009	9.5704	94.8889	50
55	189.0591	0.0053	1880.5914	0.0005	9.9471	0.1005	9.7075	96.5619	55
60	304.4816	0.0033	3034.8164	0.0003	9.9672	0.1003	9.8023	97.7010	60
65	490.3707	0.0020	4893.7073	0.0002	9.9796	0.1002	9.8672	98.4705	65
70	789.7470	0.0013	7887.4696	0.0001	9.9873	0.1001	9.9113	98.9870	70
75	1271.8954	0.0008	12708.9537	0.0001	9.9921	0.1001	9.9410	99.3317	75
80	2048.4002	0.0005	20474.0021	0.0000	9.9951	0.1000	9.9609	99.5606	80
85	3298.9690	0.0003	32979.6903	0.0000	9.9970	0.1000	9.9742	99.7120	85
90	5313.0226	0.0002	53120.2261	0.0000	9.9981	0.1000	9.9831	99.8118	90
95	8556.6760	0.0001	85556.7605	0.0000	9.9988	0.1000	9.9889	99.8773	95
100	13780.6123	0.0001	137796.1234	0.0000	9.9993	0.1000	9.9927	99.9202	100

11.0%

	Single Payment		Equal Payment Series				Gradient Series		
	Compound Amount Factor	Present Worth Factor	Compound Amount Factor	Sinking Fund Factor	Present Worth Factor	Capital Recovery Factor	Gradient Uniform Series	Gradient Present Worth	
N	(F/P,i,N)	(P/F,i,N)	(F/A,i,N)	(A/F,i,N)	(P/A,i,N)	(A/P,i,N)	(A/G,i,N)	(P/G,i,N)	N
1	1.1100	0.9009	1.0000	1.0000	0.9009	1.1100	0.0000	0.0000	1
2	1.2321	0.8116	2.1100	0.4739	1.7125	0.5839	0.4739	0.8116	2
3	1.3676	0.7312	3.3421	0.2992	2.4437	0.4092	0.9306	2.2740	3
4	1.5181	0.6587	4.7097	0.2123	3.1024	0.3223	1.3700	4.2502	4
5	1.6851	0.5935	6.2278	0.1606	3.6959	0.2706	1.7923	6.6240	5
6	1.8704	0.5346	7.9129	0.1264	4.2305	0.2364	2.1976	9.2972	6
7	2.0762	0.4817	9.7833	0.1022	4.7122	0.2122	2.5863	12.1872	7
8	2.3045	0.4339	11.8594	0.0843	5.1461	0.1943	2.9585	15.2246	8
9	2.5580	0.3909	14.1640	0.0706	5.5370	0.1806	3.3144	18.3520	9
10	2.8394	0.3522	16.7220	0.0598	5.8892	0.1698	3.6544	21.5217	10
11	3.1518	0.3173	19.5614	0.0511	6.2065	0.1611	3.9788	24.6945	11
12	3.4985	0.2858	22.7132	0.0440	6.4924	0.1540	4.2879	27.8388	12
13	3.8833	0.2575	26.2116	0.0382	6.7499	0.1482	4.5822	30.9290	13
14	4.3104	0.2320	30.0949	0.0332	6.9819	0.1432	4.8619	33.9449	14
15	4.7846	0.2090	34.4054	0.0291	7.1909	0.1391	5.1275	36.8709	15
16	5.3109	0.1883	39.1899	0.0255	7.3792	0.1355	5.3794	39.6953	16
17	5.8951	0.1696	44.5008	0.0225	7.5488	0.1325	5.6180	42.4095	17
18	6.5436	0.1528	50.3959	0.0198	7.7016	0.1298	5.8439	45.0074	18
19	7.2633	0.1377	56.9395	0.0176	7.8393	0.1276	6.0574	47.4856	19
20	8.0623	0.1240	64.2028	0.0156	7.9633	0.1256	6.2590	49.8423	20
21	8.9492	0.1117	72.2651	0.0138	8.0751	0.1238	6.4491	52.0771	21
22	9.9336	0.1007	81.2143	0.0123	8.1757	0.1223	6.6283	54.1912	22
23	11.0263	0.0907	91.1479	0.0110	8.2664	0.1210	6.7969	56.1864	23
24	12.2392	0.0817	102.1742	0.0098	8.3481	0.1198	6.9555	58.0656	24
25	13.5855	0.0736	114.4133	0.0087	8.4217	0.1187	7.1045	59.8322	25
26	15.0799	0.0663	127.9988	0.0078	8.4881	0.1178	7.2443	61.4900	26
27	16.7386	0.0597	143.0786	0.0070	8.5478	0.1170	7.3754	63.0433	27
28	18.5799	0.0538	159.8173	0.0063	8.6016	0.1163	7.4982	64.4965	28
29	20.6237	0.0485	178.3972	0.0056	8.6501	0.1156	7.6131	65.8542	29
30	22.8923	0.0437	199.0209	0.0050	8.6938	0.1150	7.7206	67.1210	30
31	25.4104	0.0394	221.9132	0.0045	8.7331	0.1145	7.8210	68.3016	31
32	28.2056	0.0355	247.3236	0.0040	8.7686	0.1140	7.9147	69.4007	32
33	31.3082	0.0319	275.5292	0.0036	8.8005	0.1136	8.0021	70.4228	33
34	34.7521	0.0288	306.8374	0.0033	8.8293	0.1133	8.0836	71.3724	34
35	38.5749	0.0259	341.5896	0.0029	8.8552	0.1129	8.1594	72.2538	35
40	65.0009	0.0154	581.8261	0.0017	8.9511	0.1117	8.4659	75.7789	40
45	109.5302	0.0091	986.6386	0.0010	9.0079	0.1110	8.6763	78.1551	45
50	184.5648	0.0054	1668.7712	0.0006	9.0417	0.1106	8.8185	79.7341	50
55	311.0025	0.0032	2818.2042	0.0004	9.0617	0.1104	8.9135	80.7712	55
60	524.0572	0.0019	4755.0658	0.0002	9.0736	0.1102	8.9762	81.4461	60

12.0%

	Single Payment		Equal Payment Series				Gradient Series		
N	Compound Amount Factor (F/P,i,N)	Present Worth Factor (P/F,i,N)	Compound Amount Factor (F/A,i,N)	Sinking Fund Factor (A/F,i,N)	Present Worth Factor (P/A,i,N)	Capital Recovery Factor (A/P,i,N)	Gradient Uniform Series (A/G,i,N)	Gradient Present Worth (P/G,i,N)	N
1	1.1200	0.8929	1.0000	1.0000	0.8929	1.1200	0.0000	0.0000	1
2	1.2544	0.7972	2.1200	0.4717	1.6901	0.5917	0.4717	0.7972	2
3	1.4049	0.7118	3.3744	0.2963	2.4018	0.4163	0.9246	2.2208	3
4	1.5735	0.6355	4.7793	0.2092	3.0373	0.3292	1.3589	4.1273	4
5	1.7623	0.5674	6.3528	0.1574	3.6048	0.2774	1.7746	6.3970	5
6	1.9738	0.5066	8.1152	0.1232	4.1114	0.2432	2.1720	8.9302	6
7	2.2107	0.4523	10.0890	0.0991	4.5638	0.2191	2.5515	11.6443	7
8	2.4760	0.4039	12.2997	0.0813	4.9676	0.2013	2.9131	14.4714	8
9	2.7731	0.3606	14.7757	0.0677	5.3282	0.1877	3.2574	17.3563	9
10	3.1058	0.3220	17.5487	0.0570	5.6502	0.1770	3.5847	20.2541	10
11	3.4785	0.2875	20.6546	0.0484	5.9377	0.1684	3.8953	23.1288	11
12	3.8960	0.2567	24.1331	0.0414	6.1944	0.1614	4.1897	25.9523	12
13	4.3635	0.2292	28.0291	0.0357	6.4235	0.1557	4.4683	28.7024	13
14	4.8871	0.2046	32.3926	0.0309	6.6282	0.1509	4.7317	31.3624	14
15	5.4736	0.1827	37.2797	0.0268	6.8109	0.1468	4.9803	33.9202	15
16	6.1304	0.1631	42.7533	0.0234	6.9740	0.1434	5.2147	36.3670	16
17	6.8660	0.1456	48.8837	0.0205	7.1196	0.1405	5.4353	38.6973	17
18	7.6900	0.1300	55.7497	0.0179	7.2497	0.1379	5.6427	40.9080	81
19	8.6128	0.1161	63.4397	0.0158	7.3658	0.1358	5.8375	42.9979	19
20	9.6463	0.1037	72.0524	0.0139	7.4694	0.1339	6.0202	44.9676	20
21	10.8038	0.0926	81.6987	0.0122	7.5620	0.1322	6.1913	46.8188	21
22	12.1003	0.0826	92.5026	0.0108	7.6446	0.1308	6.3514	48.5543	22
23	13.5523	0.0738	104.6029	0.0096	7.7184	0.1296	6.5010	50.1776	23
24	15.1786	0.0659	118.1552	0.0085	7.7843	0.1285	6.6406	51.6929	24
25	17.0001	0.0588	133.3339	0.0075	7.8431	0.1275	6.7708	53.1046	25
26	19.0401	0.0525	150.3339	0.0067	7.8957	0.1267	6.8921	54.4177	26
27	21.3249	0.0469	169.3740	0.0059	7.9426	0.1259	7.0049	55.6369	27
28	23.8839	0.0419	190.6989	0.0052	7.9844	0.1252	7.1098	56.7674	28
29	26.7499	0.0374	214.5828	0.0047	8.0218	0.1247	7.2071	57.8141	29
30	29.9599	0.0334	241.3327	0.0041	8.0552	0.1241	7.2974	58.7821	30
31	33.5551	0.0298	271.2926	0.0037	8.0850	0.1237	7.3811	59.6761	31
32	37.5817	0.0266	304.8477	0.0033	8.1116	0.1233	7.4586	60.5010	32
33	42.0915	0.0238	342.4294	0.0029	8.1354	0.1229	7.5302	61.2612	33
34	47.1425	0.0212	384.5210	0.0026	8.1566	0.1226	7.5965	61.9612	34
35	52.7996	0.0189	431.6635	0.0023	8.1755	0.1223	7.6577	62.6052	35
40	93.0510	0.0107	767.0914	0.0013	8.2438	0.1213	7.8988	65.1159	40
45	163.9876	0.0061	1358.2300	0.0007	8.2825	0.1207	8.0572	66.7342	45
50	289.0022	0.0035	2400.0182	0.0004	8.3045	0.1204	8.1597	67.7624	50
55	509.3206	0.0020	4236.0050	0.0002	8.3170	0.1202	8.2251	68.4082	55
60	897.5969	0.0011	7471.6411	0.0001	8.3240	0.1201	8.2664	68.8100	60

13.0%

	Single Payment		Equal Payment Series				Gradient Series		
N	Compound Amount Factor (F/P,i,N)	Present Worth Factor (P/F,i,N)	Compound Amount Factor (F/A,i,N)	Sinking Fund Factor (A/F,i,N)	Present Worth Factor (P/A,i,N)	Capital Recovery Factor (A/P,i,N)	Gradient Uniform Series (A/G,i,N)	Gradient Present Worth (P/G,i,N)	N
1	1.1300	0.8850	1.0000	1.0000	0.8850	1.1300	0.0000	0.0000	1
2	1.2769	0.7831	2.1300	0.4695	1.6681	0.5995	0.4695	0.7831	2
3	1.4429	0.6931	3.4069	0.2935	2.3612	0.4235	0.9187	2.1692	3
4	1.6305	0.6133	4.8498	0.2062	2.9745	0.3362	1.3479	4.0092	4
5	1.8424	0.5428	6.4803	0.1543	3.5172	0.2843	1.7571	6.1802	5
6	2.0820	0.4803	8.3227	0.1202	3.9975	0.2502	2.1468	8.5818	6
7	2.3526	0.4251	10.4047	0.0961	4.4226	0.2261	2.5171	11.1322	7
8	2.6584	0.3762	12.7573	0.0784	4.7988	0.2084	2.8685	13.7653	8
9	3.0040	0.3329	15.4157	0.0649	5.1317	0.1949	3.2014	16.4284	9
10	3.3946	0.2946	18.4197	0.0543	5.4262	0.1843	3.5162	19.0797	10
11	3.8359	0.2607	21.8143	0.0458	5.6869	0.1758	3.8134	21.6867	11
12	4.3345	0.2307	25.6502	0.0390	5.9176	0.1690	4.0936	24.2244	12
13	4.8980	0.2042	29.9847	0.0334	6.1218	0.1634	4.3573	26.6744	13
14	5.5348	0.1807	34.8827	0.0287	6.3025	0.1587	4.6050	29.0232	14
15	6.2543	0.1599	40.4175	0.0247	6.4624	0.1547	4.8375	31.2617	15
16	7.0673	0.1415	46.6717	0.0214	6.6039	0.1514	5.0552	33.3841	16
17	7.9861	0.1252	53.7391	0.0186	6.7291	0.1486	5.2589	35.3876	17
18	9.0243	0.1108	61.7251	0.0162	6.8399	0.1462	5.4491	37.2714	18
19	10.1974	0.0981	70.7494	0.0141	6.9380	0.1441	5.6265	39.0366	19
20	11.5231	0.0868	80.9468	0.0124	7.0248	0.1424	5.7917	40.6854	20
21	13.0211	0.0768	92.4699	0.0108	7.1016	0.1408	5.9454	42.2214	21
22	14.7138	0.0680	105.4910	0.0095	7.1695	0.1395	6.0881	43.6486	22
23	16.6266	0.0601	120.2048	0.0083	7.2297	0.1383	6.2205	44.9718	23
24	18.7881	0.0532	136.8315	0.0073	7.2829	0.1373	6.3431	46.1960	24
25	21.2305	0.0471	155.6196	0.0064	7.3300	0.1364	6.4566	47.3264	25
26	23.9905	0.0417	176.8501	0.0057	7.3717	0.1357	6.5614	48.3685	26
27	27.1093	0.0369	200.8406	0.0050	7.4086	0.1350	6.6582	49.3276	27
28	30.6335	0.0326	227.9499	0.0044	7.4412	0.1344	6.7474	50.2090	28
29	34.6158	0.0289	258.5834	0.0039	7.4701	0.1339	6.8296	51.0179	29
30	39.1159	0.0256	293.1992	0.0034	7.4957	0.1334	6.9052	51.7592	30
31	44.2010	0.0226	332.3151	0.0030	7.5183	0.1330	6.9747	52.4380	31
32	49.9471	0.0200	376.5161	0.0027	7.5383	0.1327	7.0385	53.0586	32
33	56.4402	0.0177	426.4632	0.0023	7.5560	0.1323	7.0971	53.6256	33
34	63.7774	0.0157	482.9034	0.0021	7.5717	0.1321	7.1507	54.1430	34
35	72.0685	0.0139	546.6808	0.0018	7.5856	0.1318	7.1998	54.6148	35
40	132.7816	0.0075	1013.7042	0.0010	7.6344	0.1310	7.3888	56.4087	40
45	244.6414	0.0041	1874.1646	0.0005	7.6609	0.1305	7.5076	57.5148	45
50	450.7359	0.0022	3459.5071	0.0003	7.6752	0.1303	7.5811	58.1870	50
55	830.4517	0.0012	6380.3979	0.0002	7.6830	0.1302	7.6260	58.5909	55
60	1530.0535	0.0007	11761.9498	0.0001	7.6873	0.1301	7.6531	58.8313	60

14.0%

	Single Payment		Equal Payment Series				Gradient Series		
N	Compound Amount Factor (F/P,i,N)	Present Worth Factor (P/F,i,N)	Compound Amount Factor (F/A,i,N)	Sinking Fund Factor (A/F,i,N)	Present Worth Factor (P/A,i,N)	Capital Recovery Factor (A/P,i,N)	Gradient Uniform Series (A/G,i,N)	Gradient Present Worth (P/G,i,N)	N
1	1.1400	0.8772	1.0000	1.0000	0.8772	1.1400	0.0000	0.0000	1
2	1.2996	0.7695	2.1400	0.4673	1.6467	0.6073	0.4673	0.7695	2
3	1.4815	0.6750	3.4396	0.2907	2.3216	0.4307	0.9129	2.1194	3
4	1.6890	0.5921	4.9211	0.2032	2.9137	0.3432	1.3370	3.8957	4
5	1.9254	0.5194	6.6101	0.1513	3.4331	0.2913	1.7399	5.9731	5
6	2.1950	0.4556	8.5355	0.1172	3.8887	0.2572	2.1218	8.2511	6
7	2.5023	0.3996	10.7305	0.0932	4.2883	0.2332	2.4832	10.6489	7
8	2.8526	0.3506	13.2328	0.0756	4.6389	0.2156	2.8246	13.1028	8
9	3.2519	0.3075	16.0853	0.0622	4.9464	0.2022	3.1463	15.5629	9
10	3.7072	0.2697	19.3373	0.0517	5.2161	0.1917	3.4490	17.9906	10
11	4.2262	0.2366	23.0445	0.0434	5.4527	0.1834	3.7333	20.3567	11
12	4.8179	0.2076	27.2707	0.0367	5.6603	0.1767	3.9998	22.6399	12
13	5.4924	0.1821	32.0887	0.0312	5.8424	0.1712	4.2491	24.8247	13
14	6.2613	0.1597	37.5811	0.0266	6.0021	0.1666	4.4819	26.9009	14
15	7.1379	0.1401	43.8424	0.0228	6.1422	0.1628	4.6990	28.8623	15
16	8.1372	0.1229	50.9804	0.0196	6.2651	0.1596	4.9011	30.7057	16
17	9.2765	0.1078	59.1176	0.0169	6.3729	0.1569	5.0888	32.4305	17
18	10.5752	0.0946	68.3941	0.0146	6.4674	0.1546	5.2630	34.0380	18
19	12.0557	0.0829	78.9692	0.0127	6.5504	0.1527	5.4243	35.5311	19
20	13.7435	0.0728	91.0249	0.0110	6.6231	0.1510	5.5734	36.9135	20
21	15.6676	0.0638	104.7684	0.0095	6.6870	0.1495	5.7111	38.1901	21
22	17.8610	0.0560	120.4360	0.0083	6.7429	0.1483	5.8381	39.3658	22
23	20.3616	0.0491	138.2970	0.0072	6.7921	0.1472	5.9549	40.4463	23
24	23.2122	0.0431	158.6586	0.0063	6.8351	0.1463	6.0624	41.4371	24
25	26.4619	0.0378	181.8708	0.0055	6.8729	0.1455	6.1610	42.3441	25
26	30.1666	0.0331	208.3327	0.0048	6.9061	0.1448	6.2514	43.1728	26
27	34.3899	0.0291	238.4993	0.0042	6.9352	0.1442	6.3342	43.9289	27
28	39.2045	0.0255	272.8892	0.0037	6.9607	0.1437	6.4100	44.6176	28
29	44.6931	0.0224	312.0937	0.0032	6.9830	0.1432	6.4791	45.2441	29
30	50.9502	0.0196	356.7868	0.0028	7.0027	0.1428	6.5423	45.8132	30
31	58.0832	0.0172	407.7370	0.0025	7.0199	0.1425	6.5998	46.3297	31
32	66.2148	0.0151	465.8202	0.0021	7.0350	0.1421	6.6522	46.7979	32
33	75.4849	0.0132	532.0350	0.0019	7.0482	0.1419	6.6998	47.2218	33
34	86.0528	0.0116	607.5199	0.0016	7.0599	0.1416	6.7431	47.6053	34
35	98.1002	0.0102	693.5727	0.0014	7.0700	0.1414	6.7824	47.9519	35
40	188.8835	0.0053	1342.0251	0.0007	7.1050	0.1407	6.9300	49.2376	40
45	363.6791	0.0027	2590.5648	0.0004	7.1232	0.1404	7.0188	49.9963	45
50	700.2330	0.0014	4994.5213	0.0002	7.1327	0.1402	7.0714	50.4375	50

15.0%

	Single Payment		Equal Payment Series				Gradient Series		
N	Compound Amount Factor (F/P,i,N)	Present Worth Factor (P/F,i,N)	Compound Amount Factor (F/A,i,N)	Sinking Fund Factor (A/F,i,N)	Present Worth Factor (P/A,i,N)	Capital Recovery Factor (A/P,i,N)	Gradient Uniform Series (A/G,i,N)	Gradient Present Worth (P/G,i,N)	N
1	1.1500	0.8696	1.0000	1.0000	0.8696	1.1500	0.0000	0.0000	1
2	1.3225	0.7561	2.1500	0.4651	1.6257	0.6151	0.4651	0.7561	2
3	1.5209	0.6575	3.4725	0.2880	2.2832	0.4380	0.9071	2.0712	3
4	1.7490	0.5718	4.9934	0.2003	2.8550	0.3503	1.3263	3.7864	4
5	2.0114	0.4972	6.7424	0.1483	3.3522	0.2983	1.7228	5.7751	5
6	2.3131	0.4323	8.7537	0.1142	3.7845	0.2642	2.0972	7.9368	6
7	2.6600	0.3759	11.0668	0.0904	4.1604	0.2404	2.4498	10.1924	7
8	3.0590	0.3269	13.7268	0.0729	4.4873	0.2229	2.7813	12.4807	8
9	3.5179	0.2843	16.7858	0.0596	4.7716	0.2096	3.0922	14.7548	9
10	4.0456	0.2472	20.3037	0.0493	5.0188	0.1993	3.3832	16.9795	10
11	4.6524	0.2149	24.3493	0.0411	5.2337	0.1911	3.6549	19.1289	11
12	5.3503	0.1869	29.0017	0.0345	5.4206	0.1845	3.9082	21.1849	12
13	6.1528	0.1625	34.3519	0.0291	5.5831	0.1791	4.1438	23.1352	13
14	7.0757	0.1413	40.5047	0.0247	5.7245	0.1747	4.3624	24.9725	14
15	8.1371	0.1229	47.5804	0.0210	5.8474	0.1710	4.5650	26.6930	15
16	9.3576	0.1069	55.7175	0.0179	5.9542	0.1679	4.7522	28.2960	16
17	10.7613	0.0929	65.0751	0.0154	6.0472	0.1654	4.9251	29.7828	17
18	12.3755	0.0808	75.8364	0.0132	6.1280	0.1632	5.0843	31.1565	18
19	14.2318	0.0703	88.2118	0.0113	6.1982	0.1613	5.2307	32.4213	19
20	16.3665	0.0611	102.4436	0.0098	6.2593	0.1598	5.3651	33.5822	20
21	18.8215	0.0531	118.8101	0.0084	6.3125	0.1584	5.4883	34.6448	21
22	21.6447	0.0462	137.6316	0.0073	6.3587	0.1573	5.6010	35.6150	22
23	24.8915	0.0402	159.2764	0.0063	6.3988	0.1563	5.7040	36.4988	23
24	28.6252	0.0349	184.1678	0.0054	6.4338	0.1554	5.7979	37.3023	24
25	32.9190	0.0304	212.7930	0.0047	6.4641	0.1547	5.8834	38.0314	25
26	37.8568	0.0264	245.7120	0.0041	6.4906	0.1541	5.9612	38.6918	26
27	43.5353	0.0230	283.5688	0.0035	6.5135	0.1535	6.0319	39.2890	27
28	50.0656	0.0200	327.1041	0.0031	6.5335	0.1531	6.0960	39.8283	28
29	57.5755	0.0174	377.1697	0.0027	6.5509	0.1527	6.1541	40.3146	29
30	66.2118	0.0151	434.7451	0.0023	6.5660	0.1523	6.2066	40.7526	30
31	76.1435	0.0131	500.9569	0.0020	6.5791	0.1520	6.2541	41.1466	31
32	87.5651	0.0114	577.1005	0.0017	6.5905	0.1517	6.2970	41.5006	32
33	100.6998	0.0099	664.6655	0.0015	6.6005	0.1515	6.3357	41.8184	33
34	115.8048	0.0086	765.3654	0.0013	6.6091	0.1513	6.3705	42.1033	34
35	133.1755	0.0075	881.1702	0.0011	6.6166	0.1511	6.4019	42.3586	35
40	267.8635	0.0037	1779.0903	0.0006	6.6418	0.1506	6.5168	43.2830	40
45	538.7693	0.0019	3585.1285	0.0003	6.6543	0.1503	6.5830	43.8051	45
50	1083.6574	0.0009	7217.7163	0.0001	6.6605	0.1501	6.6205	44.0958	50

16.0%

	Single Payment		Equal Payment Series				Gradient Series		
N	Compound Amount Factor (F/P,i,N)	Present Worth Factor (P/F,i,N)	Compound Amount Factor (F/A,i,N)	Sinking Fund Factor (A/F,i,N)	Present Worth Factor (P/A,i,N)	Capital Recovery Factor (A/P,i,N)	Gradient Uniform Series (A/G,i,N)	Gradient Present Worth (P/G,i,N)	N
1	1.1600	0.8621	1.0000	1.0000	0.8621	1.1600	0.0000	0.0000	1
2	1.3456	0.7432	2.1600	0.4630	1.6052	0.6230	0.4630	0.7432	2
3	1.5609	0.6407	3.5056	0.2853	2.2459	0.4453	0.9014	2.0245	3
4	1.8106	0.5523	5.0665	0.1974	2.7982	0.3574	1.3156	3.6814	4
5	2.1003	0.4761	6.8771	0.1454	3.2743	0.3054	1.7060	· 5.5858	5
6	2.4364	0.4104	8.9775	0.1114	3.6847	0.2714	2.0729	7.6380	6
7	2.8262	0.3538	11.4139	0.0876	4.0386	0.2476	2.4169	9.7610	7
8	3.2784	0.3050	14.2401	0.0702	4.3436	0.2302	2.7388	11.8962	8
9	3.8030	0.2630	17.5185	0.0571	4.6065	0.2171	3.0391	13.9998	9
10	4.4114	0.2267	21.3215	0.0469	4.8332	0.2069	3.3187	16.0399	10
11	5.1173	0.1954	25.7329	0.0389	5.0286	0.1989	3.5783	17.9941	11
12	5.9360	0.1685	30.8502	0.0324	5.1971	0.1924	3.8189	19.8472	12
13	6.8858	0.1452	36.7862	0.0272	5.3423	0.1872	4.0413	21.5899	13
14	7.9875	0.1252	43.6720	0.0229	5.4675	0.1829	4.2464	23.2175	14
15	9.2655	0.1079	51.6595	0.0194	5.5755	0.1794	4.4352	24.7284	15
16	10.7480	0.0930	60.9650	0.0164	5.6685	0.1764	4.6086	26.1241	16
17	12.4677	0.0802	71.6730	0.0140	5.7487	0.1740	4.7676	27.4074	17
18	14.4625	0.0691	84.1407	0.0119	5.8178	0.1719	4.9130	28.5828	18
19	16.7765	0.0596	98.6032	0.0101	5.8775	0.1701	5.0457	29.6557	19
20	19.4608	0.0514	115.3797	0.0087	5.9288	0.1687	5.1666	30.6321	20
21	22.5745	0.0443	134.8405	0.0074	5.9731	0.1674	5.2766	31.5180	21
22	26.1864	0.0382	157.4150	0.0064	6.0113	0.1664	5.3765	32.3200	22
23	30.3762	0.0329	183.6014	0.0054	6.0442	0.1654	5.4671	33.0442	23
24	35.2364	0.0284	213.9776	0.0047	6.0726	0.1647	5.5490	33.6970	24
25	40.8742	0.0245	249.2140	0.0040	6.0971	0.1640	5.6230	34.2841	25
26	47.4141	0.0211	290.0883	0.0034	6.1182	0.1634	5.6898	34.8114	26
27	55.0004	0.0182	337.5024	0.0030	6.1364	0.1630	5.7500	35.2841	27
28	63.8004	0.0157	392.5028	0.0025	6.1520	0.1625	5.8041	35.7073	28
29	74.0085	0.0135	456.3032	0.0022	6.1656	0.1622	5.8528	36.0856	29
30	85.8499	0.0116	530.3117	0.0019	6.1772	0.1619	5.8964	36.4234	30
31	99.5859	0.0100	616.1616	0.0016	6.1872	0.1616	5.9356	36.7247	31
32	115.5196	0.0087	715.7475	0.0014	6.1959	0.1614	5.9706	36.9930	32
33	134.0027	0.0075	831.2671	0.0012	6.2034	0.1612	6.0019	27.2318	33
34	155.4432	0.0064	965.2698	0.0010	6.2098	0.1610	6.0299	37.4441	34
35	180.3141	0.0055	1120.7130	0.0009	6.2153	0.1609	6.0548	37.6327	35
40	378.7212	0.0026	2360.7572	0.0004	6.2335	0.1604	6.1441	38.2992	40
45	795.4438	0.0013	4965.2739	0.0002	6.2421	0.1602	6.1934	38.6598	45
50	1670.7038	0.0006	10435.6488	0.0001	6.2463	0.1601	6.2201	38.8521	50

18.0%

	Single Payment		Equal Payment Series				Gradient Series		
	Compound Amount Factor	Present Worth Factor	Compound Amount Factor	Sinking Fund Factor	Present Worth Factor	Capital Recovery Factor	Gradient Uniform Series	Gradient Present Worth	
N	(F/P,i,N)	(P/F,i,N)	(F/A,i,N)	(A/F,i,N)	(P/A,i,N)	(A/P,i,N)	(A/G,i,N)	(P/G,i,N)	N
1	1.1800	0.8475	1.0000	1.0000	0.8475	1.1800	0.0000	0.0000	1
2	1.3924	0.7182	2.1800	0.4587	1.5656	0.6387	0.4587	0.7182	2
3	1.6430	0.6086	3.5724	0.2799	2.1743	0.4599	0.8902	1.9354	3
4	1.9388	0.5158	5.2154	0.1917	2.6901	0.3717	1.2947	3.4828	4
5	2.2878	0.4371	7.1542	0.1398	3.1272	0.3198	1.6728	5.2312	5
6	2.6996	0.3704	9.4420	0.1059	3.4976	0.2859	2.0252	7.0834	6
7	3.1855	0.3139	12.1415	0.0824	3.8115	0.2624	2.3526	8.9670	7
8	3.7589	0.2660	15.3270	0.0652	4.0776	0.2452	2.6558	10.8292	8
9	4.4355	0.2255	19.0859	0.0524	4.3030	0.2324	2.9358	12.6329	9
10	5.2338	0.1911	23.5213	0.0425	4.4941	0.2225	3.1936	14.3525	10
11	6.1759	0.1619	28.7551	0.0348	4.6560	0.2148	3.4303	15.9716	11
12	7.2876	0.1372	34.9311	0.0286	4.7932	0.2086	3.6470	17.4811	12
13	8.5994	0.1163	42.2187	0.0237	4.9095	0.2037	3.8449	18.8765	13
14	10.1472	0.0985	50.8180	0.0197	5.0081	0.1997	4.0250	20.1576	14
15	11.9737	0.0835	60.9653	0.0164	5.0916	0.1964	4.1887	21.3269	15
16	14.1290	0.0708	72.9390	0.0137	5.1624	0.1937	4.3369	22.3885	16
17	16.6722	0.0600	87.0680	0.0115	5.2223	0.1915	4.4708	23.3482	17
18	19.6733	0.0508	103.7403	0.0096	5.2732	0.1896	4.5916	24.2123	18
19	23.2144	0.0431	123.4135	0.0081	5.3162	0.1881	4.7003	24.9877	19
20	27.3930	0.0365	146.6280	0.0068	5.3527	0.1868	4.7978	25.6813	20
21	32.3238	0.0309	174.0210	0.0057	5.3837	0.1857	4.8851	26.3000	21
22	38.1421	0.0262	206.3448	0.0048	5.4099	0.1848	4.9632	26.8506	22
23	45.0076	0.0222	244.4868	0.0041	5.4321	0.1841	5.0329	27.3394	23
24	53.1090	0.0188	289.4945	0.0035	5.4509	0.1835	5.0950	27.7725	24
25	62.6686	0.0160	342.6035	0.0029	5.4669	0.1829	5.1502	28.1555	25
26	73.9490	0.0135	405.2721	0.0025	5.4804	0.1825	5.1991	28.4935	26
27	87.2598	0.0115	479.2211	0.0021	5.4919	0.1821	5.2425	28.7915	27
28	102.9666	0.0097	566.4809	0.0018	5.5016	0.1818	5.2810	29.0537	28
29	121.5005	0.0082	669.4475	0.0015	5.5098	0.1815	5.3149	29.2842	29
30	143.3706	0.0070	790.9480	0.0013	5.5168	0.1813	5.3448	29.4864	30
31	169.1774	0.0059	934.3186	0.0011	5.5227	0.1811	5.3712	29.6638	31
32	199.6293	0.0050	1103.4960	0.0009	5.5277	0.1809	5.3945	29.8191	32
33	235.5625	0.0042	1303.1253	0.0008	5.5320	0.1808	5.4149	29.9549	33
34	277.9638	0.0036	1538.6878	0.0006	5.5356	0.1806	5.4328	30.0736	34
35	327.9973	0.0030	1816.6516	0.0006	5.5386	0.1806	5.4485	30.1773	35
40	750.3783	0.0013	4163.2130	0.0002	5.5482	0.1802	5.5022	30.5269	40
45	1716.6839	0.0006	9531.5771	0.0001	5.5523	0.1801	5.5293	30.7006	45
50	3927.3569	0.0003	21813.0937	0.0000	5.5541	0.1800	5.5428	30.7856	50

20.0%

	Single Payment		Equal Payment Series				Gradient Series		
N	Compound Amount Factor (F/P,i,N)	Present Worth Factor (P/F,i,N)	Compound Amount Factor (F/A,i,N)	Sinking Fund Factor (A/F,i,N)	Present Worth Factor (P/A,i,N)	Capital Recovery Factor (A/P,i,N)	Gradient Uniform Series (A/G,i,N)	Gradient Present Worth (P/G,i,N)	N
1	1.2000	0.8333	1.0000	1.0000	0.8333	1.2000	0.0000	0.0000	1
2	1.4400	0.6944	2.2000	0.4545	1.5278	0.6545	0.4545	0.6944	2
3	1.7280	0.5787	3.6400	0.2747	2.1065	0.4747	0.8791	1.8519	3
4	2.0736	0.4823	5.3680	0.1863	2.5887	0.3863	1.2742	3.2986	4
5	2.4883	0.4019	7.4416	0.1344	2.9906	0.3344	1.6405	4.9061	5
6	2.9860	0.3349	9.9299	0.1007	3.3255	0.3007	1.9788	6.5806	6
7	3.5832	0.2791	12.9159	0.0774	3.6046	0.2774	2.2902	8.2551	7
8	4.2998	0.2326	16.4991	0.0606	3.8372	0.2606	2.5756	9.8831	8
9	5.1598	0.1938	20.7989	0.0481	4.0310	0.2481	2.8364	11.4335	9
10	6.1917	0.1615	25.9587	0.0385	4.1925	0.2385	3.0739	12.8871	10
11	7.4301	0.1346	32.1504	0.0311	4.3271	0.2311	3.2893	14.2330	11
12	8.9161	0.1122	39.5805	0.0253	4.4392	0.2253	3.4841	15.4667	12
13	10.6993	0.0935	48.4966	0.0206	4.5327	0.2206	3.6597	16.5883	13
14	12.8392	0.0779	59.1959	0.0169	4.6106	0.2169	3.8175	17.6008	14
15	15.4070	0.0649	72.0351	0.0139	4.6755	0.2139	3.9588	18.5095	15
16	18.4884	0.0541	87.4421	0.0114	4.7296	0.2114	4.0851	19.3208	16
17	22.1861	0.0451	105.9306	0.0094	4.7746	0.2094	4.1976	20.0419	17
18	26.6233	0.0376	128.1167	0.0078	4.8122	0.2078	4.2975	20.6805	18
19	31.9480	0.0313	154.7400	0.0065	4.8435	0.2065	4.3861	21.2439	19
20	38.3376	0.0261	186.6880	0.0054	4.8696	0.2054	4.4643	21.7395	20
21	46.0051	0.0217	225.0256	0.0044	4.8913	0.2044	4.5334	22.1742	21
22	55.2061	0.0181	271.0307	0.0037	4.9094	0.2037	4.5941	22.5546	22
23	66.2474	0.0151	326.2369	0.0031	4.9245	0.2031	4.6475	22.8867	23
24	79.4968	0.0126	392.4842	0.0025	4.9371	0.2025	4.6943	23.1760	24
25	95.3962	0.0105	471.9811	0.0021	4.9476	0.2021	4.7352	23.4276	25
26	114.4755	0.0087	567.3773	0.0018	4.9563	0.2018	4.7709	23.6460	26
27	137.3706	0.0073	681.8528	0.0015	4.9636	0.2015	4.8020	23.8353	27
28	164.8447	0.0061	819.2233	0.0012	4.9697	0.2012	4.8291	23.9991	28
29	197.8136	0.0051	984.0680	0.0010	4.9747	0.2010	4.8527	24.1406	29
30	237.3763	0.0042	1181.8816	0.0008	4.9789	0.2008	4.8731	24.2628	30
31	284.8516	0.0035	1419.2579	0.0007	4.9824	0.2007	4.8908	24.3681	31
32	341.8219	0.0029	1704.1095	0.0006	4.9854	0.2006	4.9061	24.4588	32
33	410.1863	0.0024	2045.9314	0.0005	4.9878	0.2005	4.9194	24.5368	33
34	492.2235	0.0020	2456.1176	0.0004	4.9898	0.2004	4.9308	24.6038	34
35	590.6682	0.0017	2948.3411	0.0003	4.9915	0.2003	4.9406	24.6614	35
40	1469.7716	0.0007	7343.8578	0.0001	4.9966	0.2001	4.9728	24.8469	40
45	3657.2620	0.0003	18281.3099	0.0001	4.9986	0.2001	4.9877	24.9316	45

25.0%

	Single Payment		Equal Payment Series				Gradient Series		
N	Compound Amount Factor (F/P,i,N)	Present Worth Factor (P/F,i,N)	Compound Amount Factor (F/A,i,N)	Sinking Fund Factor (A/F,i,N)	Present Worth Factor (P/A,i,N)	Capital Recovery Factor (A/P,i,N)	Gradient Uniform Series (A/G,i,N)	Gradient Present Worth (P/G,i,N)	N
1	1.2500	0.8000	1.0000	1.0000	0.8000	1.2500	0.0000	0.0000	1
2	1.5625	0.6400	2.2500	0.4444	1.4400	0.6944	0.4444	0.6400	2
3	1.9531	0.5120	3.8125	0.2623	1.9520	0.5123	0.8525	1.6640	3
4	2.4414	0.4096	5.7656	0.1734	2.3616	0.4234	1.2249	2.8928	4
5	3.0518	0.3277	8.2070	0.1218	2.6893	0.3718	1.5631	4.2035	5
6	3.8147	0.2621	11.2588	0.0888	2.9514	0.3388	1.8683	5.5142	6
7	4.7684	0.2097	15.0735	0.0663	3.1611	0.3163	2.1424	6.7725	7
8	5.9605	0.1678	19.8419	0.0504	3.3289	0.3004	2.3872	7.9469	8
9	7.4506	0.1342	25.8023	0.0388	3.4631	0.2888	2.6048	9.0207	9
10	9.3132	0.1074	33.2529	0.0301	3.5705	0.2801	2.7971	9.9870	10
11	11.6415	0.0859	42.5661	0.0235	3.6564	0.2735	2.9663	10.8460	11
12	14.5519	0.0687	54.2077	0.0184	3.7251	0.2684	3.1145	11.6020	12
13	18.1899	0.0550	68.7596	0.0145	3.7801	0.2645	3.2437	12.2617	13
14	22.7374	0.0440	86.9495	0.0115	3.8241	0.2615	3.3559	12.8334	14
15	28.4217	0.0352	109.6868	0.0091	3.8593	0.2591	3.4530	13.3260	15
16	35.5271	0.0281	138.1085	0.0072	3.8874	0.2572	3.5366	13.7482	16
17	44.4089	0.0225	173.6357	0.0058	3.9099	0.2558	3.6084	14.1085	17
18	55.5112	0.0180	218.0446	0.0046	3.9279	0.2546	3.6698	14.4147	18
19	69.3889	0.0144	273.5558	0.0037	3.9424	0.2537	3.7222	14.6741	19
20	86.7362	0.0115	342.9447	0.0029	3.9539	0.2529	3.7667	14.8932	20
21	108.4202	0.0092	429.6809	0.0023	3.9631	0.2523	3.8045	15.0777	21
22	135.5253	0.0074	538.1011	0.0019	3.9705	0.2519	3.8365	15.2326	22
23	169.4066	0.0059	673.6264	0.0015	3.9764	0.2515	3.8634	15.3625	23
24	211.7582	0.0047	843.0329	0.0012	3.9811	0.2512	3.8861	15.4711	24
25	264.6978	0.0038	1054.7912	0.0009	3.9849	0.2509	3.9052	15.5618	25
26	330.8722	0.0030	1319.4890	0.0008	3.9879	0.2508	3.9212	15.6373	26
27	413.5903	0.0024	1650.3612	0.0006	3.9903	0.2506	3.9346	15.7002	27
28	516.9879	0.0019	2063.9515	0.0005	3.9923	0.2505	3.9457	15.7524	28
29	646.2349	0.0015	2580.9394	0.0004	3.9938	0.2504	3.9551	15.7957	29
30	807.7936	0.0012	3227.1743	0.0003	3.9950	0.2503	3.9628	15.8316	30
31	1009.7420	0.0010	4034.9678	0.0002	3.9960	0.2502	3.9693	15.8614	31
32	1262.1774	0.0008	5044.7098	0.0002	3.9968	0.2502	3.9746	15.8859	32
33	1577.7218	0.0006	6306.8872	0.0002	3.9975	0.2502	3.9791	15.9062	33
34	1972.1523	0.0005	7884.6091	0.0001	3.9980	0.2501	3.9828	15.9229	34
35	2465.1903	0.0004	9856.7613	0.0001	3.9984	0.2501	3.9858	15.9367	35
40	7523.1638	0.0001	30088.6554	0.0000	3.9995	0.2500	3.9947	15.9766	40

30.0%

	Single Payment		Equal Payment Series				Gradient Series		
N	Compound Amount Factor (F/P,i,N)	Present Worth Factor (P/F,i,N)	Compound Amount Factor (F/A,i,N)	Sinking Fund Factor (A/F,i,N)	Present Worth Factor (P/A,i,N)	Capital Recovery Factor (A/P,i,N)	Gradient Uniform Series (A/G,i,N)	Gradient Present Worth (P/G,i,N)	N
1	1.3000	0.7692	1.0000	1.0000	0.7692	1.3000	0.0000	0.0000	1
2	1.6900	0.5917	2.3000	0.4348	1.3609	0.7348	0.4348	0.5917	2
3	2.1970	0.4552	3.9900	0.2506	1.8161	0.5506	0.8271	1.5020	3
4	2.8561	0.3501	6.1870	0.1616	2.1662	0.4616	1.1783	2.5524	4
5	3.7129	0.2693	9.0431	0.1106	2.4356	0.4106	1.4903	3.6297	5
6	4.8268	0.2072	12.7560	0.0784	2.6427	0.3784	1.7654	4.6656	6
7	6.2749	0.1594	17.5828	0.0569	2.8021	0.3569	2.0063	5.6218	7
8	8.1573	0.1226	23.8577	0.0419	2.9247	0.3419	2.2156	6.4800	8
9	10.6045	0.0943	32.0150	0.0312	3.0190	0.3312	2.3963	7.2343	9
10	13.7858	0.0725	42.6195	0.0235	3.0915	0.3235	2.5512	7.8872	10
11	17.9216	0.0558	56.4053	0.0177	3.1473	0.3177	2.6833	8.4452	11
12	23.2981	0.0429	74.3270	0.0135	3.1903	0.3135	2.7952	8.9173	12
13	30.2875	0.0330	97.6250	0.0102	3.2233	0.3102	2.8895	9.3135	13
14	39.3738	0.0254	127.9125	0.0078	3.2487	0.3078	2.9685	9.6437	14
15	51.1859	0.0195	167.2863	0.0060	3.2682	0.3060	3.0344	9.9172	15
16	66.5417	0.0150	218.4722	0.0046	3.2832	0.3046	3.0892	10.1426	16
17	86.5042	0.0116	285.0139	0.0035	3.2948	0.3035	3.1345	10.3276	17
18	112.4554	0.0089	371.5180	0.0027	3.3037	0.3027	3.1718	10.4788	18
19	146.1920	0.0068	483.9734	0.0021	3.3105	0.3021	3.2025	10.6019	19
20	190.0496	0.0053	630.1655	0.0016	3.3158	0.3016	3.2275	10.7019	20
21	247.0645	0.0040	820.2151	0.0012	3.3198	0.3012	3.2480	10.7828	21
22	321.1839	0.0031	1067.2796	0.0009	3.3230	0.3009	3.2646	10.8482	22
23	417.5391	0.0024	1388.4635	0.0007	3.3254	0.3007	3.2781	10.9009	23
24	542.8008	0.0018	1806.0026	0.0006	3.3272	0.3006	3.2890	10.9433	24
25	705.6410	0.0014	2348.8033	0.0004	3.3286	0.3004	3.2979	10.9773	25
26	917.3333	0.0011	3054.4443	0.0003	3.3297	0.3003	3.3050	11.0045	26
27	1192.5333	0.0008	3971.7776	0.0003	3.3305	0.3003	3.3107	11.0263	27
28	1550.2933	0.0006	5164.3109	0.0002	3.3312	0.3002	3.3153	11.0437	28
29	2015.3813	0.0005	6714.6042	0.0001	3.3317	0.3001	3.3189	11.0576	29
30	2619.9956	0.0004	8729.9855	0.0001	3.3321	0.3001	3.3219	11.0687	30
31	3405.9943	0.0003	11349.9811	0.0001	3.3324	0.3001	3.3242	11.0775	31
32	4427.7926	0.0002	14755.9755	0.0001	3.3326	0.3001	3.3261	11.0845	32
33	5756.1304	0.0002	19183.7681	0.0001	3.3328	0.3001	3.3276	11.0901	33
34	7482.9696	0.0001	24939.8985	0.0000	3.3329	0.3000	3.3288	11.0945	34
35	9727.8604	0.0001	32422.8681	0.0000	3.3330	0.3000	3.3297	11.0980	35

35.0%

	Single Payment		Equal Payment Series				Gradient Series		
	Compound Amount Factor	Present Worth Factor	Compound Amount Factor	Sinking Fund Factor	Present Worth Factor	Capital Recovery Factor	Gradient Uniform Series	Gradient Present Worth	
N	(F/P,i,N)	(P/F,i,N)	(F/A,i,N)	(A/F,i,N)	(P/A,i,N)	(A/P,i,N)	(A/G,i,N)	(P/G,i,N)	N
1	1.3500	0.7407	1.0000	1.0000	0.7407	1.3500	0.0000	0.0000	1
2	1.8225	0.5487	2.3500	0.4255	1.2894	0.7755	0.4255	0.5487	2
3	2.4604	0.4064	4.1725	0.2397	1.6959	0.5897	0.8029	1.3616	3
4	3.3215	0.3011	6.6329	0.1508	1.9969	0.5008	1.1341	2.2648	4
5	4.4840	0.2230	9.9544	0.1005	2.2200	0.4505	1.4220	3.1568	5
6	6.0534	0.1652	14.4384	0.0693	2.3852	0.4193	1.6698	3.9828	6
7	8.1722	0.1224	20.4919	0.0488	2.5075	0.3988	1.8811	4.7170	7
8	11.0324	0.0906	28.6640	0.0349	2.5982	0.3849	2.0597	5.3515	8
9	14.8937	0.0671	39.6964	0.0252	2.6653	0.3752	2.2094	5.8886	9
10	20.1066	0.0497	54.5902	0.0183	2.7150	0.3683	2.3338	6.3363	10
11	27.1439	0.0368	74.6967	0.0134	2.7519	0.3634	2.4364	6.7047	11
12	36.6442	0.0273	101.8406	0.0098	2.7792	0.3598	2.5205	7.0049	12
13	49.4697	0.0202	138.4848	0.0072	2.7994	0.3572	2.5889	7.2474	13
14	66.7841	0.0150	187.9544	0.0053	2.8144	0.3553	2.6443	7.4421	14
15	90.1585	0.0111	254.7385	0.0039	2.8255	0.3539	2.6889	7.5974	15
16	121.7139	0.0082	344.8970	0.0029	2.8337	0.3529	2.7246	7.7206	16
17	164.3138	0.0061	466.6109	0.0021	2.8398	0.3521	2.7530	7.8180	17
18	221.8236	0.0045	630.9247	0.0016	2.8443	0.3516	2.7756	7.8946	18
19	299.4619	0.0033	852.7483	0.0012	2.8476	0.3512	2.7935	2.9547	19
20	404.2736	0.0025	1152.2103	0.0009	2.8501	0.3509	2.8075	8.0017	20
21	545.7693	0.0018	1556.4838	0.0006	2.8519	0.3506	2.8186	8.0384	21
22	736.7886	0.0014	2102.2532	0.0005	2.8533	0.3505	2.8272	8.0669	22
23	994.6646	0.0010	2839.0418	0.0004	2.8543	0.3504	2.8340	8.0890	23
24	1342.7973	0.0007	3833.7064	0.0003	2.8550	0.3503	2.8393	8.1061	24
25	1812.7763	0.0006	5176.5037	0.0002	2.8556	0.3502	2.8433	8.1194	25
26	2447.2480	0.0004	6989.2800	0.0001	2.8560	0.3501	2.8465	8.1296	26
27	3303.7848	0.0003	9436.5280	0.0001	2.8563	0.3501	2.8490	8.1374	27
28	4460.1095	0.0002	12740.3128	0.0001	2.8565	0.3501	2.8509	8.1435	28
29	6021.1478	0.0002	17200.4222	0.0001	2.8567	0.3501	2.8523	8.1481	29
30	8128.5495	0.0001	23221.5700	0.0000	2.8568	0.3500	2.8535	8.1517	30

40.0%

	Single Payment		Equal Payment Series				Gradient Series		
	Compound Amount Factor	Present Worth Factor	Compound Amount Factor	Sinking Fund Factor	Present Worth Factor	Capital Recovery Factor	Gradient Uniform Series	Gradient Present Worth	
N	(F/P,i,N)	(P/F,i,N)	(F/A,i,N)	(A/F,i,N)	(P/A,i,N)	(A/P,i,N)	(A/G,i,N)	(P/G,i,N)	N
1	1.4000	0.7143	1.0000	1.0000	0.7143	1.4000	0.0000	0.0000	1
2	1.9600	0.5102	2.4000	0.4167	1.2245	0.8167	0.4167	0.5102	2
3	2.7440	0.3644	4.3600	0.2294	1.5889	0.6294	0.7798	1.2391	3
4	3.8416	0.2603	7.1040	0.1408	1.8492	0.5408	1.0923	2.0200	4
5	5.3782	0.1859	10.9456	0.0914	2.0352	0.4914	1.3580	2.7637	5
6	7.5295	0.1328	16.3238	0.0613	2.1680	0.4613	1.5811	3.4278	6
7	10.5414	0.0949	23.8534	0.0419	2.2628	0.4419	1.7664	3.9970	7
8	14.7579	0.0678	34.3947	0.0291	2.3306	0.4291	1.9185	4.4713	8
9	20.6610	0.0484	49.1526	0.0203	2.3790	0.4203	2.0422	4.8585	9
10	28.9255	0.0346	69.8137	0.0143	2.4136	0.4143	2.1419	5.1696	10
11	40.4957	0.0247	98.7391	0.0101	2.4383	0.4101	2.2215	5.4166	11
12	56.6939	0.0176	139.2348	0.0072	2.4559	0.4072	2.2845	5.6106	12
13	79.3715	0.0126	195.9287	0.0051	2.4685	0.4051	2.3341	5.7618	13
14	111.1201	0.0090	275.3002	0.0036	2.4775	0.4036	2.3729	5.8788	14
15	155.5681	0.0064	386.4202	0.0026	2.4839	0.4026	2.4030	5.9688	15
16	217.7953	0.0046	541.9883	0.0018	2.4885	0.4018	2.4262	6.0376	16
17	304.9135	0.0033	759.7837	0.0013	2.4918	0.4013	2.4441	6.0901	17
18	426.8789	0.0023	1064.6971	0.0009	2.4941	0.4009	2.4577	6.1299	18
19	597.6304	0.0017	1491.5760	0.0007	2.4958	0.4007	2.4682	6.1601	19
20	836.6826	0.0012	2089.2064	0.0005	2.4970	0.4005	2.4761	6.1828	20
21	1171.3556	0.0009	2925.8889	0.0003	2.4979	0.4003	2.4821	6.1998	21
22	1639.8978	0.0006	4097.2445	0.0002	2.4985	0.4002	2.4866	6.2127	22
23	2295.8569	0.0004	5737.1423	0.0002	2.4989	0.4002	2.4900	6.2222	23
24	3214.1997	0.0003	8032.9993	0.0001	2.4992	0.4001	2.4925	6.2294	24
25	4499.8796	0.0002	11247.1990	0.0001	2.4994	0.4001	2.4944	6.2347	25
26	6299.8314	0.0002	15747.0785	0.0001	2.4996	0.4001	2.4959	6.2387	26
27	8819.7640	0.0001	22046.9099	0.0000	2.4997	0.4000	2.4969	6.2416	27
28	12347.6696	0.0001	30866.6739	0.0000	2.4998	0.4000	2.4977	6.2438	28
29	17286.7374	0.0001	43214.3435	0.0000	2.4999	0.4000	2.4983	6.2454	29
30	24201.4324	0.0000	60501.0809	0.0000	2.4999	0.4000	2.4988	6.2466	30

50.0%

	Single Payment		Equal Payment Series				Gradient Series		
N	Compound Amount Factor (F/P,i,N)	Present Worth Factor (P/F,i,N)	Compound Amount Factor (F/A,i,N)	Sinking Fund Factor (A/F,i,N)	Present Worth Factor (P/A,i,N)	Capital Recovery Factor (A/P,i,N)	Gradient Uniform Series (A/G,i,N)	Gradient Present Worth (P/G,i,N)	N
1	1.5000	0.6667	1.0000	1.0000	0.6667	1.5000	0.0000	0.0000	1
2	2.2500	0.4444	2.5000	0.4000	1.1111	0.9000	0.4000	0.4444	2
3	3.3750	0.2963	4.7500	0.2105	1.4074	0.7105	0.7368	1.0370	3
4	5.0625	0.1975	8.1250	0.1231	1.6049	0.6231	1.0154	1.6296	4
5	7.5938	0.1317	13.1875	0.0758	1.7366	0.5758	1.2417	2.1564	5
6	11.3906	0.0878	20.7813	0.0481	1.8244	0.5481	1.4226	2.5953	6
7	17.0859	0.0585	32.1719	0.0311	1.8829	0.5311	1.5648	2.9465	7
8	25.6289	0.0390	49.2578	0.0203	1.9220	0.5203	1.6752	3.2196	8
9	38.4434	0.0260	74.8867	0.0134	1.9480	0.5134	1.7596	3.4277	9
10	57.6650	0.0173	113.3301	0.0088	1.9653	0.5088	1.8235	3.5838	10
11	86.4976	0.0116	170.9951	0.0058	1.9769	0.5058	1.8713	3.6994	11
12	129.7463	0.0077	257.4927	0.0039	1.9846	0.5039	1.9068	3.7842	12
13	194.6195	0.0051	387.2390	0.0026	1.9897	0.5026	1.9329	3.8459	13
14	291.9293	0.0034	581.8585	0.0017	1.9931	0.5017	1.9519	3.8904	14
15	437.8939	0.0023	873.7878	0.0011	1.9954	0.5011	1.9657	3.9224	15
16	656.8408	0.0015	1311.6817	0.0008	1.9970	0.5008	1.9756	3.9452	16
17	985.2613	0.0010	1968.5225	0.0005	1.9980	0.5005	1.9827	3.9614	17
18	1477.8919	0.0007	2953.7838	0.0003	1.9986	0.5003	1.9878	3.9729	18
19	2216.8378	0.0005	4431.6756	0.0002	1.9991	0.5002	1.9914	3.9811	19
20	3325.2567	0.0003	6648.5135	0.0002	1.9994	0.5002	1.9940	3.9868	20

INDEX

A

absolute (dollar) measures, 312

accelerated methods of depreciation, 436–441

accounting
  accrual method of accounting, 523
  as basis of decision making, 22–24
  depreciation accounting, 428
  matching concept, 388, 430, 523
  negative numbers, 517
  purpose of, 38
  "the language of business," 22
  users of accounting information, 23

accounting depreciation, 430
  see also depreciation

accounting equation, 26

accounting information system, 23

accounting period, 30

accounting profit, 25, 485
  see also net income

accounts payable, 29

accounts receivable, 28

accounts receivable turnover, 44

accrual method of accounting, 523

accrual method (personal income tax), 872

accrued expenses or liabilities, 29

accumulated cash flow sign test, 259, 259f, 288

acid-test ratio, 43

acquisitions cost, 447

activity-based costing (ABC), 413

actual (current) dollars (A_n), 726–729, 750

actual dollars, 62

add-on interest, 171–172

add-on interest loans, 190

additions to depreciable assets, 456–459

Adidas, 137

adjustable-rate mortgage (ARM), 161

adjusted cost base, 879–880

adjusted-discount method, 731, 733–734, 750

administrative costs, 388

administrative expenses budget, 415

AE. *See* annual equivalent worth (AE)

after-tax cash flow diagram approach
  advantage of, 545
  debt financing in form of amortized loan, 547

debt financing in form of bond-type loan, 547–548, 548f
  multiple capital assets, 549
  no debt financing, 546–547, 546f
  other considerations, 549
  single asset and working capital requirement, 549–550

Alberta Centennial Education Savings (ACES) Plan, 203

alterations to depreciable assets, 456–459

amortization
  see also depreciation
  intangible assets, 28
  mortgage, 177

amortized loan, 164, 165–169, 190, 547, 548f

analysis period
  see also planning horizon
  AE analysis for unequal project lives, 331–333
  coinciding with longest project life, 326–329
  common service period, 329
  defined, 320
  implied, 321f
  IRR analysis for unequal project lives, 333–336
  lowest common multiple of project lives, 329–331
  not specified, 329–336
  other common analysis periods, 336
  project life longer than analysis period, 321–323
  project life shorter than analysis period, 324–326
  vs. project lives, 321–329
  required service period, 320

annual compounding, 143

annual effective yield. *See* effective annual interest rate

annual equivalent comparison, 310–311

annual equivalent cost (AEC) method, 238, 582, 587
  see also annual equivalent worth (AE)

annual equivalent worth (AE)
  absolute (dollar) measure, 312
  benefits of AE analysis, 241–242
  capital costs *vs.* operating costs, 238–241
  capital recovery cost, 239, 239f, 240–241

annual equivalent worth (AE) (*continued*)
consistency of report formats, 241
cyclic cash flow pattern, 237–238
described, 234
fundamental decision rule, 234–236
life-cycle cost analysis, 241–242
make-or-buy decision, 337–339
minimum-cost analysis, 346
pricing the use of an asset, 243–244
repeating cash flow cycles, 237–238
replacement analysis, 582–587
single-project evaluation, 235–236
unequal project lives, 331–333
unit costs or profits, 241, 242–243
annual percentage rate (APR), 141, 165,
171, 189
see also nominal interest rate
annual percentage yield (APY), 142
see also effective annual interest rate
annual report, 24, 25
annuity
deferred annuity, 99–100
defined, 98
ANSI Z94 Standards Committee on
Industrial Engineering
Terminology, 726*n*
apartment rental fee, 244
apprenticeship job creation, 491
asset-backed bonds, 190
asset depreciation, 430
see also depreciation
asset management analysis, 43–45
asset turnover, 45, 47
assets
on balance sheet, 26–29
capital assets, 28
capitalized assets, 430, 481
cost basis, 431–432
current assets, 26–28
depreciable assets, 430–431, 495–499,
735–738
depreciable life, 430
depreciation. *See* depreciation
economic service life, 575–580
fixed assets, 6, 28
integral part, 457
liquidity, 26
multiple assets, project requiring,
531–534, 549
other assets, 28–29
pricing the use of an asset, 243–244
real assets, 28*n*

residual value, 172
salvage value. *See* salvage value
trade-in allowance, 432
@RISK, 810–813, 814*f*
attractiveness of investment. *See* investment
attractiveness
automated data collection system, 303–304
automobiles
average cost per km, 395–398
cash purchase, lease or debt financing,
173–176
loan payments, calculation of, 149–150
Autovu Technologies, 206
available-for-use rule, 445–447
average annual return, 857–858
average collection period, 44
average cost, 393, 395–398, 405–406
average inflation rate, 722–724,
725–726, 750
average tax rate, 477, 871–872, 873–874
average unit cost, 393–397

B

Baker, Peter, 8*n*
balance sheet
assets, 26–29
capital assets, 28
common stock, 29
current assets, 26–28
current liabilities, 29
described, 26
fixed assets, 28
liabilities, 29–30
other assets, 28–29
other liabilities, 29
preferred stock, 29
raw-materials inventory, 390
Research In Motion Limited, 27*t*,
33–34, 40*t*
retained earnings, 30
shareholders' equity, 30
work-in-process inventory, 390
Balsillie, Jim, 21
base period, 721
basic EPS, 32*n*
basic federal tax, 871
BCE Inc., 475*t*
bearer bonds, 183
beer bottles, 384
benchmark interest rate, 257
benefit-cost analysis (BCA)

aims of, 681
benefit-cost ratio, 689–695
benefits, 682
defined, 681
described, 340, 681
disbenefits, 682
elements of benefits and costs, 684–685
examples of, 681
final decision, 688
framework, 681–682
primary benefit, 682
quantifying benefits and costs, 684–688
secondary benefit, 682
secondary effects, 682
social discount rate, 683–684
social opportunity cost of capital
   (SOCC), 683
social rate of time preference (STP), 683
sponsors, 681
sponsor's costs, 683
summary, 704–705
users, 681
users' benefits, 682
valuation of benefits and costs, 682–684,
   685–687
when varying benefits among
   alternatives, 340
benefit-cost ratio
   definition of, 689–691
   incremental analysis, 692–695
   incremental benefit-cost ratios, 693–695
   mutually exclusive alternatives,
      comparison of, 692–695
   net *B/C* ratio, 690, 693, 695
   and net present worth, 692
benefits, 682
betterment, 457
BlackBerry, 19–21, 25
Bombardier Inc., 475*t*
bond financing, 631–632
bond investments, 849–852, 875–877
bond-type loan, 547–548, 548*f*
bonds
   asset-backed bonds, 190
   bearer bonds, 183
   bond prices, 185–189, 851
   bond valuation, 185–189
   bond value over time, 188–189
   call feature, 183
   callable bond, 183
   convertible bonds and debentures, 185
   corporate bonds, 184, 185

coupon rate, 183, 190
current yield, 186–188
debenture bond, 190
debentures, 185
defined, 41
discount bond, 183, 190
face value, 183
government bonds, 184–185
market value, 183
marketable bonds, 184
maturity date, 183, 190
mortgage bonds, 185, 190
notes, 185
par value, 183, 190
personal investments in bonds, 849–852,
   875–877
premium bond, 183, 190
rating system, 852
redeemable bond, 183
redemption feature, 183
registered bonds, 183
return on investment, 186
safety of, 851
terminology, 183
types of bonds, 184–185
yield to maturity, 186–188, 851–852
book depreciation methods
   accelerated methods, 436–441
   declining balance (DB) method, 436–440
   described, 433, 434
   revision of book depreciation, 456
   straight-line (SL) method, 434–435,
      438–439
   sum-of-years'-digits (SOYD) method, 436,
      440–441
   vs. tax depreciation, 465*t*
   units-of-production (UP) method, 442
book value, 432
book value per share, 48
borrowed funds
   see also debt financing; loans
   as cash inflow, 514
   and inflation, 740–742
borrowed-funds concept, 223–224
bottom line, 25
break-even analysis, 400–402, 771–774
break-even interest rate, 222, 246
   see also internal rate of return
break-even point, 407
break-even volume, 407
Brin, Sergey, 2–3, 4
British Columbia Lottery Corporation, 135

budget
   administrative expenses budget, 415
   capital budget, 628
   cost-of-goods sold budget, 413–414
   determining an appropriate MARR,
      648–650
   direct labour budget, 412–413
   limited budgets, 653–657
   materials budget, 412
   nonmanufacturing cost budget, 414–415
   optimal capital budget, 655
   overhead budget, 413
   production budget, 410*f,* 411–414
   sales budget, 410–411
   selling expenses budget, 414–415
budget constraints, 653–657
budgeted income statement, 416–417, 417*t*
Build Canada Fund, 570
business environment, 13
business income, 873–874
business limit, 487
business managers, as users of accounting
   information, 23
business organization, types of, 5–6
business taxes, 393

C

C-bonds, 184
CC Media Holdings Inc., 57
call feature, 183
callable bond, 183
Canada
   average selling price of a home, 140
   forms of business organization in, 6
   health-care service industry, 675
   service sector, 672–673, 672*f*
Canada Education Savings Grant
   (CESG), 203
Canada Mortgage and Housing Corporation
   (CMHC), 176
Canada premium bonds, 184
Canada Revenue Agency, 433, 443, 456,
   531*n,* 535*n,* 868
Canada savings bonds, 184
Canadian-controlled private corporations
   (CCPC), 487, 491
Canadian Environmental Test Research and
   Education Centre (CanETREC), 682
Canadian Securities Administrators, 853
capacity cost, 392–393
capital

debt. *See* debt
   equity. *See* equity
   forms of, 39
   marginal efficiency of capital, 245
   need for, 39
   return on invested capital, 247
capital assets, 28
capital budget, 628
capital budgeting
   capital rationing, 647–651
   contingent projects, 652–653
   cost of capital, 637–640
   defined, 628
   dependent projects, 651
   estimation of relevant cash flows, 208
   financing methods, 628–637
   independent project, 651
   independent projects, 652
   investment opportunity schedule (IOS),
      654, 658
   limited budgets, 653–657
   marginal cost of capital, 646–647
   marginal-cost-of-capital (MCC) schedule,
      654, 658
   minimum attractive rate of return
      (MARR), 643–651
   multiple investment alternatives, 651
   mutually exclusive alternatives,
      formulation of, 651–653
   mutually exclusive projects, 652
   net equity flow, 643–645
   optimal capital budget, 655
   project *vs.* investment alternative, 651
   summary, 657–658
capital cost, 238–241, 575–576
   see also cost of capital
capital cost allowance (CCA)
   acquisitions cost, 447
   available-for-use rule, 445–447
   vs. book depreciation, 465*t*
   calculation of, 447–450
   CCA classes, 443–445, 444*t*
   CCA rates, 443–445, 444*t,* 450
   CCA system, 443–445
   class number, 447
   described, 433
   disposal tax effect, 454, 520, 520*n*
   disposal tax effects, 495–499
   enduring benefit, 456
   example, 453*t*
   50% rule, 448, 450–451
   form, 449*f*

for individual projects, 454–456
and inflation, 735
integral part, 457
maintenance *vs.* betterment, 457
maximum allowable CCA, 450
multiple property classes, 451–452
net acquisitions, 450
net adjustments, 447–448
not cash flow, 537
proceeds of dispositions, 448
recaptured CCA, 450, 495
reduced undepreciated capital cost, 450
and replacement analysis, 591–607
revision of, 456–457
schedule, 481
terminal loss, 450
undepreciated capital cost (UCC), 447,
  448, 450
unused CCA, 494
capital expenditure
described, 6
estimation of savings and costs, 25
planning for, 6
tax treatment, 481
capital gains, 478–480, 531*n*
capital gains inclusion rate, 531*n*
capital gains tax, 490, 494
capital investments
anticipated cash inflows, determination
  of, 218
independent projects. *See* independent
  projects
mutually exclusive projects. *See* mutually
  exclusive projects
capital loss, 479
capital rationing, 647–651
capital recovery cost, 239, 239*f,* 240–241,
  576, 577*f*
capital recovery factor, 98
capital stocks, 29
capital structure, 632–637
capital surplus, 29
capitalization, 229–230
capitalized assets, 430, 481
capitalized cost, 230
capitalized equivalent (CE*(i)*), 229, 231
capitalized equivalent-worth analysis, 225,
  229–233
cars. *See* automobiles
case studies
capital budgeting, 664–669
cost concepts, 425

cost-effectiveness case studies, 695–704
economic equivalence, 135–138
financial statements, 56–57
income taxes, 505–506
independent projects, 285–286
inflation, 759–761
money and its management, 63–66
mutually exclusive projects, 374–381
project cash flows, 563–566
project risk, 837–842
public sector, 710–716
replacement analysis, 619–625
time value of money, 135–138
cash
on balance sheet, 28
potential uses of cash, 513
sources of, 35
and time value of money, 484
uses of, 35
cash equivalents, 28
see also cash
Cash Flow Analyzer (CFA), 254*n*
cash flow diagrams, 64, 64*f,* 125
cash flow patterns, 62, 64, 77, 78, 125
cash flow statement
described, 34
financing activities, 35, 38
generic version, 517*f*
for individual projects. *See* cash flow
  statement (individual projects)
investing activities, 35, 38
operating activities, 35, 37
reporting format, 35
Research In Motion Limited, 36*t,*
  37–38, 40*t*
sources and uses of cash, 35
cash flow statement (individual projects)
borrowed funds, financing with, 526–527
cost-only project, 528–531
expansion project, 518–521
generic version, 517*f*
investment tax credits, 534–536
multiple assets required, 531–534
negative numbers, 517
negative taxable income, 528–531
operating and investing activities only,
  518–521
working-capital requirements,
  521–526, 523*f*
cash flows
see also cash flow statement
actual (current) dollars *(A$_n$),* 726–729

cash flows (*continued*)
actual dollars, 62
after-tax cash flow diagram approach,
  545–550
annuity. *See* annuity
cash inflows, 514
cash outflows, 513
classification of cash flow elements,
  514–516, 516*t*
composite cash flows, 115–120
compound growth, 108
constant (real) dollars *(A'_n)*, 726–729
continuous cash flows, 157–161,
  158*f*, 159*t*
under continuous compounding,
  157–161, 159*t*
cyclic cash flow pattern, 237–238
differential cash flows, 512–516
economic effects, determination of, 70
economic equivalence, 70
end-of-period convention, 65
equal payment series (uniform series), 78,
  90–102, 125
from financing activities, 35, 38, 516
funds flow compound amount factor, 161
generalized cash flow approach, 537–545
geometric gradient series, 78,
  108–112, 125
geometric growth, 108
incremental cash flows, 512–516
inflation, effects of, 735–742
interest formulas. *See* interest formulas
from investing activities, 35, 38
irregular (mixed) series, 78, 125
linear gradient series, 78, 102–108, 125
loan cash flow, 209–210, 209*f*
multiple payments, 75–76
net cash flow equations, 537–538
net cash flows, 64
vs. net income, 483–486
from operating activities, 35, 515–516
project cash flows. *See* project cash flows
project evaluation purposes, 485
relevant net after-tax cash flow, 815–816
series of, with changing interest rates,
  163–164
simple borrowing cash flows, 248
single cash flow, 78, 79–86, 125
with subpatterns, 117–118
summing of, to end of compounding
  period, 155–156
types of, 78, 79*f*
uneven payment series, 86–90, 163–164

cash inflows, 514
cash investments, 847–849
cash outflows, 513
challenger, 570–571, 607
  see also replacement analysis
changing interest rates, 161–164
charitable donations, 871*n*
Chevron Overseas Petroleum Inc.,
  619–620
child-care spaces, 491
China, 10
City of Ottawa-Carleton, 710–711
City of Toronto, 715–716
City of Victoria, British Columbia,
  568–569
City of Winnipeg, 712–713
closed mortgage, 177, 178–182
coefficient of correlation, 785
collateral mortgage, 176–177
combined corporate tax rates, 486–488
common base period, 72–74
common service period, 329
common stock
  described, 29
  and dividends, 33
  equity financing, 629–630
  flotation costs, 629, 639
  issuing common stock, 629–630, 639
  vs. preferred stock, 41
complete certainty, 650
complex projects, 511–512
composite cash flows, 115–120, 116*f*
composite series, 102–103
compound-amount factor
  equal payment series, 90–91
  single cash flow, 79
compound growth, 108
compound interest, 65, 66–67, 68–69, 68*f*,
  71, 125, 164
  see also compounding
compound interest bonds, 184
compound return, 857–858
compounding
  see also compound interest; effective
    annual interest rate
  annual compounding, 143
  continuous compounding, 146–148, 152,
    157–161, 159*t*
  decision chart, 156*f*
  discrete compounding, 151, 157
  frequency, determination of, 144–145
  funds deposited after start of compounding
    period, 155–156

funds flow compound amount factor, 161
less frequent than payments, 153–156
more frequent compounding, 143
more frequent than payments, 150–152
payment period equal to compounding period, 149–150
single-payment formulas for continuous compounding, 157
when paid quarterly, 142*t*
compounding formulas. *See* interest formulas
compounding process, 79
computer-aided economic analysis program, 254*n*
computer process control system project (example), 210–211, 213
computer simulation, 799–800
conditional probability, 783–785, 822–823, 823*f*
constant-dollar analysis, 730–731
constant-dollar prices, 331
constant (real) dollars *(A'ₙ)*, 726–729, 750
construction industry, 322
Consumer Price Index (CPI), 719*f*, 720–721, 722*t*, 724–725, 749, 860
contingent projects, 651, 652–653, 658
continuous cash flows, 157–161
continuous compounding, 146–148, 152, 157–161, 159*t*
continuous random variables, 776, 777
conventional manufacturing technology (CMT), 840–842
conventional mortgage, 176
conventional payback method, 212
convertible bonds and debentures, 185
corporate bonds, 184, 185
corporate income tax
  average tax rate, 477
  business limit, 487
  capital cost allowance. *See* capital cost allowance (CCA)
  capital gains, 531*n*
  capital gains inclusion rate, 531*n*
  capital gains (losses), 478–480
  capital gains tax, 490, 494
  combined corporate tax rates, 486–488
  corporate income tax calculation, 486–493
  depreciable property, 430–431
  depreciation for tax purposes (Canada). *See* capital cost allowance (CCA)
  depreciation for tax purposes (U.S), 459–464
  disposal tax effect, 520
  disposal tax effects, 495–499, 520*n*
  effective tax rate, 477
  federal corporate tax structure, 2008, 487*t*
  fundamentals of, 477–480
  incremental tax rate, 478, 493
  investment tax credit (ITC), 491–492, 492*t*, 493, 513, 534–536
  manufacturing and processing profits deduction (MPPD), 487
  marginal tax rate, 477
  maximum deductions, 494
  negative taxable income, 528–531
  net income, 480–486
  non-operating income, 490–491
  non-progressive tax rates, 487–488
  on operating income, 488–490
  operating losses, 490, 528*n*
  payments, 513
  private Canadian corporations, 487
  provincial/territorial corporate tax structure, 2008, 488*t*
  refundable investment tax credit, 492
  and replacement analysis, 591–607
  simple calculation, 477
  small business deduction (SBD), 487
  summary, 499–500
  tax rate definitions, 477–478
  tax savings, 513, 528
  tax shield, 513, 528
  taxable income, 32, 481–482
  and taxable income, 481–482
  timing of corporate tax payments, 494–495
corporation, 6
cost
  acquisitions cost, 447
  activity-based costing (ABC), 413
  administrative costs, 388
  average cost, 393, 395–398
  average unit cost, 393–397
  benefit-cost analysis (BCA). *See* benefit-cost analysis (BCA)
  capacity cost, 392–393
  capital cost, 238–241, 575–576
  capital recovery cost, 239, 239*f*, 240–241, 576
  capitalized cost, 230
  cost classification, 388–397
  in cost-effective analysis (CEA), 676–677
  current market value, 571–572
  of debt, 224
  differential cost, 398–402
  direct cost, 387, 676
  discounted costs, 677
  of equity, 224, 637–640, 644

cost (*continued*)
  fixed costs, 392–393
  flotation costs, 629, 634*n*, 639
  future costs. *See* cost-volume analysis
  historical cost data, 398
  incremental costs, 306, 398–399
  indirect costs, 388, 676
  intangible costs, 676
  inventory costs, 389–390
  lower present-value production cost, 306
  maintenance costs, 513
  manufacturing costs, 386–388, 387*f*, 513
  marginal cost, 404–409
  marketing costs, 388
  mixed cost, 393
  nonmanufacturing costs, 388
  operating costs, 238–241, 392, 513, 573, 575–576
  opportunity cost, 402–404, 647
  overhead, 387–388
  patterns over life of asset, 581
  period costs, 389
  present equivalent incremental cost, 541
  product costs, 389–390
  of retained earnings, 638
  semivariable costs, 393
  sponsor's costs, 683
  summary, 418
  sunk costs, 404, 572–573, 572*f*, 607
  unit costs, 241, 242–243
  unit of use, 243–244
  use of term, 386
  variable cost, 393
cost basis, 431–432
cost behaviour
  average cost, 393, 395–398
  average unit cost, 393–397
  cost-volume diagram, 395–396
  defined, 392
  fixed costs, 392–393
  mixed cost, 393
  prediction of, 392–397
  types of cost behaviour patterns, 392–397
  variable cost, 393
  and volume, 393–398
  volume index, 392
cost-benefit analysis, 676
  see also benefit-cost analysis (BCA)
cost-benefit estimation
  complex projects, 511–512
  described, 510
  simple projects, 511

cost classification
  cost behaviour, prediction of, 392–397
  for financial statements, 388–392
  inventory costs, 389–390
  manufacturing company, 389–390, 390*f*
  period costs, 389
  product costs, 389–390
cost-effective analysis (CEA)
  cost-effectiveness ratio, 676–677
  described, 675
  direct *vs.* indirect costs, 676
  discounting, 677
  how to use, 677–680
  independent interventions, 677–678, 677*t*
  mutually exclusive alternatives, 679*t*, 680*t*
  mutually exclusive interventions, 678, 678*t*
  outcomes, reporting of, 676
  and quality-of-life dimension, 675
cost-effectiveness, 695
cost-effectiveness case studies, 695–704
cost-effectiveness index, 695
cost-effectiveness ratio, 676–677
cost of capital
  calculation of, 642–643, 642*f*
  cost of debt, 637, 640–641
  cost of equity, 224, 637–640, 657
  cost of retained earnings, 638
  defined, 224, 642
  estimate or calculation of, 637*n*
  issuing common stock, 639
  marginal cost of capital, 643, 646–647, 657
  preferred stock, 639
  as rate of return, 224
  tax-adjusted weighted average cost of capital, 642
  use of, 657
cost of debt, 224, 640–641
cost of equity, 224, 637–640, 644, 657
cost of goods sold, 414, 481
  see also cost of sales
cost-of-goods sold budget, 413–414
cost-of-living index, 721
cost of revenue. *See* cost of sales
cost of sales, 30–31
  see also cost of goods sold
cost-only projects, 318–320, 528
cost reduction, 514
cost-reduction projects, 15
cost savings, 514
cost-utility analysis, 675–676
cost-volume analysis

differential cost and revenue, 398–402
future costs *vs.* historical cost data, 398
make-or-buy decision, 402
marginal analysis, 406–409
marginal cost, 404–409
method changes, 399
operations planning, 400–402
opportunity cost, 402–404
summary, 418
sunk costs, 404
cost-volume diagram, 395–396
cost-volume relationships, 399
see also cost-volume analysis
coupon rate, 183, 190
covariance, 785
CPI. *See* Consumer Price Index (CPI)
creditors, as users of accounting information, 23
cumulative distribution, 778–780
current assets
accounts receivables, 28
cash and cash equivalents, 28
described, 26–28
inventories, 28
current assets account, 26
current dollars, 726–729
current liabilities, 29
current market value, 571–572, 607
current ratio, 42–43
current yield, 186–188
cyclic cash flow pattern, 237–238

D

days sales outstanding, 44
debenture bond, 190
debentures, 185
debt, 39
see also debt financing; debt management
debt financing
see also loans
after-tax cash flow diagram approach, 547–548, 548*f*
bond financing, 631–632
borrowed funds, as cash inflow, 514
cash flow statement, project financed with borrowed funds, 526–527
and cost of capital, 637
cost of debt, 224, 631–632
described, 631
and inflation, 740–742

long-term debt financing, 41
pure borrowing, 289
repayment of borrowed funds, 513
short-term debt financing, 39
term loans, 631–632
debt management
commercial loans, 164–172
loan *vs.* lease financing, 172–176
mortgages, 176–182
debt management analysis, 39–42
debt ratio, 41, 526, 632
decision frameworks, 581–582, 582*f*
decision making
accounting information, 38
decision-tree diagram, 814–818, 815*f*
decision trees, 813–818
economic decisions. *See* economic decisions
engineering economic decisions. *See* engineering economic decisions
expected value of perfect information (EVPI), 819–820
expert's advice, 820–822, 822–827
with imperfect information, 822–826
ratios, limitations of, 48–49
ratios, use of, 38–39
rollback procedure, 816
worth of obtaining additional information, 819–822
decision-tree diagram, 814–818, 815*f*
decision trees, 813–818
decline in purchasing power. *See* inflation
declining balance (DB) method, 436–440
declining balance schedule, 440
declining linear gradient, 107–108
decreasing gradient series, 102, 103*f*
defender, 570–571, 607
see also replacement analysis
deferred annuity, 99–100
deferred loan repayment, 99–100
deflation, 720, 722, 749
deflation method, 731–732, 750
delayed consumption, 61*f*
dependent projects, 651, 658
dependent random variables, 792, 810
depreciable assets, 430–431, 495–499, 735–738
depreciable life, 430
depreciable property, 430–431
depreciation
accelerated methods, 436–441
accounting depreciation, 430

depreciation (*continued*)
  additions or alterations to depreciable
      assets, 456–459
  adjustment for overhauled asset, 457–459
  asset depreciation, use of term, 430
  book depreciation methods, 433,
      434–442, 456
  capital cost allowance. *See* capital cost
      allowance (CCA)
  capital expenditures, 481
  classification of, 429*f*
  cost basis, 431–432
  cost basis with trade-in allowance, 432
  declining balance (DB) method, 436–440
  defined, 428
  depreciable asset, 430–431
  described, 28, 428
  economic depreciation, 429–430
  functional depreciation, 429
  inflation tax effect, 742
  inherent factors, 430–434
  intangible assets, 28
  methods, 433–442
  mixed cost, 393
  physical depreciation, 429
  recaptured depreciation, 495
  straight-line (SL) method, 434–435,
      438–439
  sum-of-years'-digits (SOYD) method, 436,
      440–441
  summary, 464–466, 465*t*
  for tax purposes (Canada). *See* capital cost
      allowance (CCA)
  for tax purposes (U.S.), 459–464
  units-of-production (UP) method, 442
  useful life, 433
depreciation accounting, 428
design decisions, 9
design economics
  described, 346
  economical pipe size (example), 350–354
  first-order conditions, 346
  marginal yield, 346
  minimum-cost analysis, 346
  optimal cross-sectional area (example),
      346–349
  second-order conditions, 346
differential cash flows, 512–516
differential cost, 398–402
differential revenue, 398–402
diluted EPS, 32*n*
direct cost, 387, 676

direct labour, 387
direct labour budget, 412–413
direct raw materials, 387
direct solution method, 249–251
disbenefits, 682
discount bond, 183, 190
discount rate, 82, 173
discounted cash flow techniques (DCFs), 217
discounted costs, 677
discounted payback method, 212
discounted payback period, 216–217, 216*t*,
      217*f*, 224*n*
discounting factor, 82
discounting process, 82
discrete compounding, 151, 157
discrete compounding formulas. *See* interest
      formulas
discrete random variable, 776
disposal tax effect, 454, 520, 520*n*
disposal tax effects on depreciable assets,
      495–499
diversification, 863–864
Dividend Yield, 853
dividends, 33, 874–875
"do nothing" decision option, 305–306,
      568, 650
dollar cost averaging, 866–867

E

earning power, 61, 729
earnings per share (EPS)
  basic EPS, 32*n*
  computation of, 32
  diluted EPS, 32*n*
  on income statement, 32
  personal investments, 853
economic decisions
  vs. design decisions, 9
  differences among alternatives, 16
  engineering economic decisions. *See*
      engineering economic decisions
  measurement of investment
      attractiveness, 9
economic depreciation, 429–430
economic equivalence
  actual-dollar analysis, 731–734
  adjusted-discount method, 731, 733–734
  cash flow series, with changing interest
      rates, 163–164
  common base period, 72–74
  composite cash flows, 115–120, 116*f*

constant-dollar analysis, 730–731
with continuous payments, 157–161
defined, 70, 125
deflation method, 731–732
described, 69–70
with effective interest rates, 148–156
equivalence calculations, 71–77
general principles, 72–77
under inflation, 729–734
and interest rate, 74–75
interest rate, determination of, 120–124
lump sum, with changing interest rates,
    161–163
maintained regardless of point of view, 77
mixed-dollar analysis, 734
multiple payments, 75–76
project cash flows, 210–211
simple example, 71–72
summary, 125
unconventional equivalence calculations,
    114–124
economic indifference, 70–71
economic service life, 575–580, 583,
    584–585, 587, 595–601, 608
economical pipe size (example), 350–354
effective annual interest rate
    see also annual percentage rate (APR)
    for add-on interest loan, 171–172
    annual compounding, 143
    compounding less frequent than payments,
        153–156
    compounding more frequent than
        payments, 150–152
    continuous compounding, 146–148, 152
    decision chart, 156*f*
    defined, 189
    described, 142–143
    with different compounding periods, 143*t*
    discrete compounding, 151
    equivalence calculations, 148–156
    payment period equal to compounding
        period, 149–150
    per payment period, 145–146,
        153–154, 156*f*
    with quarterly payment, calculation of,
        147–148
    summing cash flows to end of
        compounding period, 155–156
effective annual yield, 187
effective tax rate, 477
Encana Corp., 475*t,* 476
end-of-period convention, 65

enduring benefit, 456
energy-saving projects under budget
    constraints (examples), 653–657
engineering economic decisions
    see also decision making
    capital budgeting decisions. *See* capital
        budgeting
    classifications of project ideas, 14–15
    common types of, 14–15
    cost reduction, 15
    defined, 6
    difficulties of, 7–9
    equipment or process selection, 14
    equipment replacement, 15
    health care sector. *See* health-care service
        industry
    new product or product expansion, 15
    public projects. *See* public sector
    quality improvement, 15
    replacement decisions. *See* replacement
        analysis
    summary, 17–18
engineering economics
    differences among alternatives, 16
    fundamental principles of, 15–17
    marginal analysis, 16
    risk and return trade-off, 17
    time value of money, 15
engineering projects
    attractiveness of. *See* investment
        attractiveness
    capital cost allowance for individual
        projects, 454–456
    classification, 511–512
    classifications of project ideas, 14–15
    complex projects, 511–512
    contingent projects, 651, 658
    cost-benefit estimation, 510–512
    cost-only projects, 318–320
    dependent projects, 651, 658
    independent project. *See* independent
        projects
    large-scale engineering projects. *See* large-
        scale engineering projects
    mixed investment, 290–291, 292–298
    multiple assets required, 531–534
    mutually exclusive projects. *See* mutually
        exclusive projects
    net investment, 289
    nonsimple investment, 248–249, 259–261,
        287–298, 334–336
    pure investment, 289, 290–292

engineering projects (*continued*)
    revenue projects, 306, 307
    scale of investment, 306–308, 312
    service projects, 306, 307
    simple investments, 248–249,
        256–258, 262
    simple projects, 511
engineers
    capital expenditures, planning for, 6
    and engineering economic decisions, 6
    pay rates, 719, 719*t*
    personal economic decisions, 7
    role of, in business, 5–7
equal payment series
    annuity factor, 98
    capital recovery factor, 98
    combination of uniform series and single
        present and future amount, 93–94
    comparison of three investment plans,
        95–96
    compound-amount factor, 90–91
    deferred annuity, 99–100
    deferred loan repayment, 99–100
    defined, 125
    described, 78
    find *A*, given *P*, *i*, and *N*, 98–99
    find *F*, given *A*, *i*, and *N*, 90–92
    find *P*, given *A*, *i*, and *N*, 100–102
    handling time shifts, 92–93
    net present worth (NPW), 219–220
    present-worth factor, 100–102
    sinking-fund factor, 93
equal payment series capital recovery
        factor, 98
equal payment series compound-amount
        factor, 91
equal payment series sinking-fund factor, 93
equal time span, 320
equipment, purchase of, 513
equipment or process selection, 14
equipment replacement, 15
    see also replacement analysis
equity
    common stock. *See* common stock
    cost of equity, 224, 637–640, 644, 657
    defined, 39
    described, 526
    financing, 628, 629–630, 657
    on mortgage, 176
    net equity flow, 643–645
    preferred stock, 41
    types of, 41

equity financing, 628, 629–630, 657
equivalence. *See* economic equivalence
equivalence calculations. *See* economic
        equivalence
equivalent worth. *See* annual equivalent
        worth (AE)
evolution of typical project idea, 9–12
Excel
    Goal Seek function, 123–124, 123*n*,
        124*f*, 187
    loan repayment schedule, 167*t*
    present-worth profile, 221
    risk simulation with @RISK,
        810–813, 814*f*
expectation, measure of, 781
expected return, 856–859
expected-value criterion, 796
expected value of perfect information
        (EVPI), 819–820
expenses
    cost of goods sold, 414, 481
    leasing expenses, 513
    non-cash expense, 485
    operating expenses, 32, 481
    project expenses, 410, 481
expert's advice, 820–822, 822–827
extremely long service life, 231–233

F

face value, 183
fair market value, 28–29
federal corporate tax structure, 487*t*
fees, 165
50% rule, 448, 450–451
finance charges, 165
financial leases, 541–542
financial leverage, 47
financial ratios. *See* ratios
financial statements
    and accounting equation, 26
    balance sheet, 26–30, 390
    cash flow statement, 34–38
    cost classification, 388–392
    and decision making, 23–24
    financial status for businesses, 24–25
    impact of engineering projects, 13
    income statement, 30–34
    information reported in, 24*f*
    inventory costs, 389–390
    period costs, 389
    product costs, 389–390

Research In Motion Limited, 27t, 31t, 33–34, 34, 36t, 37–38
Research In Motion Ltd. (RIM), 40t
summary, 49
types of, 24
financial status for businesses, 24–25
financing
    and choice of MARR, 643–645
    choices, 628–629
    debt financing, 39–41, 526–527, 628, 631–632
    equity financing, 628, 629–630
    loans *vs.* lease financing, 172–176
    methods of, 628–637
    mixed financing, 633
    optimal capital structure, 633–637
    vehicles, 173–176
financing activities, 35, 38, 516
finished-goods inventory, 390
finite planning horizon, 587–590
finite useful life, 28
first mortgage bonds, 185
first-order conditions, 346
fixed assets
    described, 28
    purchase of, 6
fixed costs, 392–393
    see also operating expenses
fixed-rate mortgage, 140, 177
flexible manufacturing system (FMS), 840–842
flotation costs, 629, 634n, 639
Ford, 376–377
forecasts
    and measurement of investment attractiveness, 9
    uncertainties in forecasted data, 9
forms of business organization
    corporation, 6
    partnership, 5
    proprietorships, 5
Fregin, Doug, 19
functional depreciation, 429
fundamental principles of engineering economics
    differences among alternatives, 16
    marginal analysis, 16
    risk and return trade-off, 17
    time value of money, 15
funds flow compound amount factor, 161
future amount of money, 62
future costs. *See* cost-volume analysis

future value, 72
future worth
    defined, 208
    net future worth (NFW), 224, 226–229
future-worth analysis, 225–229

G

gain
    capital gains, 478–480, 531n
    capital gains tax, 490
    delayed consumption, 61f
    unrecognized gain, 432
general and administrative costs. *See* operating expenses
general inflation rate, 724–725, 726, 749
general mortgage bonds, 185
General Motors (GM), 19
generalized cash flow approach
    compact tabular format, 538
    described, 537
    example, 538–539, 540t
    lease-or-buy decision, 540–545
    net cash flow equations, 537–538
Genetec Inc., 206
geometric gradient, 108
    see also geometric gradient series
geometric gradient series
    at constant rate g, 109, 109f
    defined, 125
    described, 78
    find A_1, given F, g, i, and N, 112
    find P, given A_1, g, i, and N, 109–112
    present-worth factor, 109–112
geometric growth, 108
Goal Seek function, 123–124, 123n, 124f, 187
Goods and Services Tax (GST), 393
goodwill, 28
Google Inc., 2–3, 4, 19, 856
government bonds, 184–185
Government of Canada Treasury bills, 184
grace period, 100
gradient-to-equal-payment series conversion factor, 105–108
graphical method to estimate i^*, 254–256, 255f
gross margin, 32, 416
grossed up amount, 875
Guaranteed Investment Certificates (GICs), 847
guess, 251

H

half-year convention, 461
Halifax Port Authority Command and
    Control System (HPACCS), 670–671
health-care service industry
    cost-benefit analysis, 676
    cost-effective analysis (CEA), 675,
        676–680
    cost-utility analysis, 675–676
    described, 675
    economic evaluation tools, 675–676
    quality-adjusted life years (QALYs), 675
high-ratio mortgage, 176
historical cost data, 398
hybrid electric cars, 9–12, 11*f*, 13–14

I

IMP Aerospace, 426–427
imperfect information, 822–826
Imperial Oil Ltd., 56, 507
income
    see also revenue
    business income, 873–874
    defined, 480
    investment income, 32
    lost rental income, 403–404
    marginal income, 407
    negative taxable income, 528–531
    net income. *See* net income
    non-operating income, 490–491
    operating income, 32, 410, 488–490
    ordinary income, 869–871
    professional income, 873–874
    taxable income, 32, 481–482, 868
    total income, 868
income before income taxes. *See* taxable
    income
income from operations. *See* operating
    income
income statement
    accounting period, 30
    basic income statement equation, 30
    budgeted income statement, 416–417, 417*t*
    cost of sales, 30–31
    described, 30
    earnings per share (EPS), 32
    and finished-goods inventory, 390
    gross margin, 32, 416
    for individual projects, 517, 517*f*
    investment income, 32
    net income, 32
    net profit margin, 417
    net revenue, 30
    operating expenses, 32
    operating income, 32
    operating margin, 416
    reporting format, 30–32
    Research In Motion Limited, 31*t*, 34, 40*t*
    retained earnings, 32–33
    revenue, 30
    taxable income, 32
Income Tax Act, 430, 443, 445
    see also corporate income tax
Income Tax Guides for Corporations and
    Businesses, 445
income taxes. *See* corporate income tax;
    personal income tax
increasing gradient series, 102
incremental analysis
    benefit-cost ratio, 692–695
    cost-only projects, 318–320
    described, 312–320, 355, 401
    flaws in project ranking by IRR, 312
    incremental IRR, 313, 314–318
incremental benefit-cost ratios, 693–695
incremental cash flows, 512–516
incremental costs, 306, 398–399
incremental IRR, 313, 314–318
incremental tax rate, 478, 493
independent interventions, 677–678, 677*t*
independent projects
    annual equivalent-worth analysis, 234–244
    capitalized equivalent-worth analysis, 225,
        229–233
    cost-of-capital methods, 647
    defined, 651, 658
    described, 212
    future-worth analysis, 225–229
    initial project screening method, 212–217
    internal rate of return, 256–261
    investment classification, 248–249
    mutually exclusive alternatives, 652
    vs. mutually exclusive projects, 212
    net equity flow, 647
    nonsimple investment, 259–261, 287–298
    nonsimple investments, 248–249
    present-worth analysis, 217–225
    project cash flows, 209–212
    project with service life of two years of
        return, 249–251
    project with two-flow transaction, 249–251
    rate of return analysis, 245–256

simple investments, 248–249,
      256–258, 262
   summary, 261–262, 263*t*
independent random variables, 783, 791–792
India, 10
indirect costs, 388, 676
indirect labour, 388
indirect materials, 388
individuals, as users of accounting
      information, 23
Industrial Price Indexes, 721–722
industrial product price index (IPPI),
      721–722, 722*t*
infinite planning horizon, 582–587, 601–607
infinite service life, 229–231
inflation
   actual (current) dollars *(A_n),* 750
   actual-dollar analysis, 731–734
   actual *vs.* constant dollars, 726–729
   adjusted-discount method, 731,
      733–734, 750
   average inflation rate, 722–724,
      725–726, 750
   borrowed funds, effects of, 740–742
   and capital cost allowance, 735
   constant-dollar analysis, 730–731
   constant (real) dollars *(A'_n),* 750
   Consumer Price Index (CPI), 720–721,
      724–725, 749
   defined, 720, 749
   deflation method, 731–732, 750
   engineers' pay rates, 719, 719*t*
   equivalence calculations, 729–734
   general inflation rate, 724–725, 726, 749
   industrial product price index (IPPI),
      721–722, 722*t*
   inflation-free interest rate, 729, 750
   inflation-related terms, 726
   internal rate of return, 743–746
   market interest rate, 729, 739*n*, 750
   measurement of, 720–726
   mixed-dollar analysis, 734
   multiple inflation rates, 738–740
   payments with financing (borrowing),
      740–742
   personal investments, effect on, 860–861
   and price changes, 319
   price differences, 1979–2006, 718, 718*t*
   price indexes, 721–722, 722*n*
   project cash flows, effects on, 735–742
   projects with depreciable asset, 735–738
   rate-of-return analysis, 743–749, 750*t*

raw material price index (RMPI), 722, 722*t*
   return on investment, 743–746
   specific inflation rate, 724–725, 738–740
   summary, 749–750, 750*t*
   working capital, 746–749
   yearly inflation rates, 725–726
inflation-adjusted MARR, 729
inflation-free interest rate, 729, 750
inflation loss, 738
inflation tax, 738, 742
information
   expected value of perfect information
      (EVPI), 819–820
   expert's advice, 820–822, 822–827
   imperfect information, 822–826
   perfect information, 819–822
   worth of obtaining additional information,
      819–822
initial project screening method, 212–217
installments, 164–170
intangible assets
   amortization or depreciation, 28
   as other assets, 28
intangible costs, 676
integral part, 457
interest
   see also time value of money
   compound interest, 65, 66–67, 68–69, 68*f*,
      71, 164
   compounding. *See* compounding
   as the cost of money, 60
   defined, 60, 125
   end-of-period convention, 65
   example of interest transaction, 63–64
   future amount of money, 62
   interest period, 62
   methods of calculating interest, 65–67
   number of interest periods, 62
   payments, 513
   plan for receipts or disbursements, 62
   principal, 62
   repayment plans, 63*t*, 70*t*
   simple interest, 65, 66, 68–69, 68*f*, 171
   symbols, 63
   transactions involving interest, 62–65
interest formulas
   described, 125
   equal payment series, 90–102
   factor notation, 81, 91
   geometric gradient series, 108–112
   interest tables, 80–81
   limiting forms, 114, 114*t*

interest formulas (*continued*)
    linear gradient series, 102–108
    single-cash-flow formulas, 79–86
    summary of, 113*t*
    uneven payment series, 86–90
interest period, 62
interest rate
    add-on interest, 171–172
    annual percentage rate (APR), 141, 165,
        171, 189
    annual percentage yield (APY), 142
    benchmark interest rate, 257
    break-even interest rate, 222
    changing interest rates, 161–164
    coupon rate, 183, 190
    defined, 62, 125
    discount rate, 82, 173
    effective annual interest rate, 189
    and equivalence, 74–75
    to establish economic equivalence,
        120–124
    inflation-free interest rate, 729
    market interest rate, 60, 62, 729, 739*n,* 750
    meaning of, 60*f*
    mixed investment, and external interest
        rate, 292–293
    mortgages, 177
    nominal interest rate, 141–142, 189
    periodic interest rate, 165
    prime rate, 139
    real interest rate, 729
    social discount rate, 683–684
    solving for, 83–85
    unknown, with multiple factors, 121–124
    variable interest rates, 161–164
interest rate differential, 177
interest tables, 80–81
internal rate of return
    see also rate of return
    accumulated cash flow sign test, 259,
        259*f,* 288
    benchmark interest rate, 257
    criterion, 256–261, 259*f*
    defined, 247, 262
    described, 221–222
    external interest rate, 292–293
    flaws in project ranking, 312
    incremental IRR, 313, 314–318
    with inflation, 743–746
    minimum attractive rate of return
        (MARR), 293–294
    mixed investment, 290–291, 292–298

net cash flow rule of signs, 259, 259*f,*
    287–288
net investment, 289
net-investment test, 259, 259*f,* 289–292
net present worth (NPW), relationship
    to, 256
nonsimple investments, 259–261, 287–298
predicting multiple *i**s, 287–289
present-worth analysis, relationship to, 256
pure investment, 290–292
real IRR, 743
relative (percentage) measure, 312
return on invested capital (RIC), 293–296
and scale of investment, 312
simple investments, 256–258
trial-and-error method, 297–298
unequal project lives, 333–336
Internal Revenue Service (IRS), 459
interpolation, 123, 251
Interpretation Bulletins (IT), 445
Interpretation Circulars (IC), 445
Interprovincial Lottery Corporation, 58–59
inventory
    described, 28
    finished-goods inventory, 390
    inventory costs, 389–390
    raw-materials inventory, 390
    work-in-process inventory, 390
inventory turnover, 43–44
investing activities, 35, 38, 518–521
investment alternative, 651
investment attractiveness
    annual equivalent-worth analysis, 234–244
    capitalized equivalent-worth analysis, 225,
        229–233
    common measures of, 208
    future-worth analysis, 225–229
    internal rate of return, 256–261
    present-worth analysis, 217–225
    rate of return analysis, 245–256
investment balance. *See* project balance (PB)
investment in working capital, 513, 521–526
investment income, 32
investment opportunity schedule (IOS),
    654, 658
investment pool, 218, 222–223, 223*f,* 648
investment projects. *See* engineering projects
investment risks, 859–862
investment strategies
    diversification, 863–864
    dollar cost averaging, 866–867
    expected return, 856–859

investment risks, 859–862
investment timing issues, 862–863
optimal investment strategy, 855
risk-reward trade-off, 864–866
investment tax credit (ITC), 448, 491–492, 492t, 493, 513, 534–536
investment timing issues, 862–863
investments
    see also engineering projects; personal investments
    bonds. *See* bonds
    classification of, 248–249
    comparison of three different plans, 97
    profitability of. *See* profitability
    return. *See* rate of return
    in working capital, 513, 521–526
investors
    bondholders, 183
    as users of accounting information, 23
irregular (mixed) series, 78, 125

J

Japan, 9–12
Johnson Street Bridge, 568–569
joint probability, 783–785, 823–825, 823f, 824t

K

Kearl Oil Sands Project, 507–509
Kleiner Perkins Caufield & Byers, 3

L

land
    capital gains, 479–480, 494
    and depreciation, 431, 431n
    maintenance of, 431n
    real asset, 28n
large-scale engineering projects
    evaluation, 13–14
    evolution of typical project idea, 9–12
    impact on financial statements, 13
law of large numbers, 796
Lazaridis, Mike, 19–20
lease
    financial leases, 541–542
    lease financing, 540–545, 628n
    lease-or-buy decision, 540–545
    vs. loans, 172–176
    operating leases, 540–541

lease-or-buy decision, 540–545
leasing expenses, 513
lending rate, 648
Lepage, Faye, 58
LetsGoWings, 88
liabilities
    on balance sheet, 29–30
    current liabilities, 29
    long-term liabilities, 29
    other liabilities, 29
life-cycle cost analysis (LCCA), 241–242, 339–345, 340f
lift truck replacement example, 577–580, 595–601
linear gradient series
    as composite series, 102–103
    declining linear gradient, 107–108
    decreasing gradient series, 102, 103f
    defined, 125
    described, 78, 102
    find A, given $G, i,$ and N, 105–107
    find F, given $A_1, G, i,$ and N, 107–108
    find P, given $G, i,$ and N, 103–105
    gradient-to-equal-payment series conversion factor, 105–108
    increasing gradient series, 102
    present-worth factor, 103–105
    strict gradient series, 102, 102f
linear interpolation, 251
liquidity, 846–847, 863
liquidity analysis, 42–43
loan cash flow, 209–210, 209f
loans
    see also debt
    add-on interest loans, 171–172, 190
    after-tax cash flow diagram approach, 547, 548f
    amortized loan, 164, 165–169, 190, 547, 548f
    annual percentage rate (APR), 141, 165, 171
    auto loan payments, calculation of, 149–150
    bond-type loan, 547–548, 548f
    commercial loans, 164–172
    debt management, 164–172
    deferred loan repayment, 99–100
    fees, 165
    finance charges, 165
    grace period, 100
    installments, 164–170
    vs. investment project, 209f

loans (*continued*)
  vs. lease financing, 172–176
  loan cash flow, 209–210, 209*f*
  loan cash flow diagram, 98*f*
  loan repayment schedule, 166–168
  mortgages. *See* mortgage
  periodic interest rate, 165
  present equivalent incremental cost, 541
  principal, 176
  rate of return, 245–246
  remaining-balance method, 169–170
  tabular method, 166–168
  term loans, 631–632
  term of loan, 165
long service life, 231–233
long-term contract, 88–90
long-term liabilities, 29
loss
  capital loss, 479
  defined, 480
  delayed consumption, 61*f*
  operating losses, 490, 528*n*
  opportunity loss, 819
  unrecognized loss, 432
lost rental income, 403–404
lottery, 58–59
lower present-value production cost, 306
lowest common multiple, 329–331

M

MACRS. *See* Modified Accelerated Cost
  Recovery System (MACRS)
maintenance, 457
maintenance costs, 513
make-or-buy decision, 336–339, 402
manufacturing and processing profits
  deduction (MPPD), 487
manufacturing companies
  administrative expenses budget, 415
  budgeted income statement, 416–417, 417*t*
  cost classification, 389–390, 390*f*
  cost flows, 390*f*
  cost-of-goods sold budget, 413–414
  direct labour budget, 412–413
  materials budget, 412
  nonmanufacturing cost budget, 414–415
  nonmanufacturing costs, 388, 513
  overhead budget, 413
  production budget, 410*f*, 411–413
  sales budget, 410–411
  selling expenses budget, 414–415

manufacturing costs, 513
  direct labour, 387
  direct raw materials, 387
  overhead, 387–388
  types of, 386, 387*f*
manufacturing overhead, 387–388
marginal analysis
  defined, 406
  described, 17*f*, 406–407
  as fundamental principle, 16
  profit maximization, 407–409
  replacement decisions, 583
marginal contribution, 407
marginal cost
  vs. average cost, 405–406
  defined, 16, 404
  and marginal revenue, 16
marginal cost of capital, 643, 646–647, 657
marginal-cost-of-capital (MCC) schedule,
  654, 658
marginal efficiency of capital, 245
marginal income, 407
marginal probability, 783, 823–825, 824*t*
marginal revenue
  defined, 16, 404
  and marginal cost, 16
marginal tax rates, 477, 869*t*, 871–872,
  873–874
marginal yield, 346
marginalism, 404
market basket, 720–721
market interest rate, 60, 62, 729, 739*n*, 750
  see also nominal interest rate
market value, current, 571–572, 607
market value analysis, 47–48
market value (of bond), 183
marketable bonds, 184
marketing costs, 388
MARR. *See* minimum attractive rate of
  return (MARR)
Marriott International Inc., 626–627
Marshall, Alfred, 404
matching concept, 388, 430, 523
materials budget, 412
maturity date, 183, 190
maximum allowable CCA, 450
maximum value, 777
mean, 781, 782–783, 790*t*, 796
mean-and-variance criterion, 796
method changes, 399
Microsoft, 4
Millionaire Life, 58–59, 135–136

minimum attractive rate of return (MARR)
  assumed returned at MARR, in PW
    analysis, 223
  basis for selecting MARR, 224–225
  under capital rationing, 647–651
  choice of, 643–651, 657
  and cost of capital, 657
  described, 218
  determining an appropriate MARR,
    648–650
  as established external interest rate, 293
  estimate of MARR, importance of, 222
  financing information, availability of,
    657–658
  funds not invested in project, 307
  inflation-adjusted MARR, 729
  as initial guess, in trial-and-error
    method, 251*n*
  and internal rate of return, 257
  justification of additional investment, 313
  and opportunity cost, 647
  under uncertainty, 650, 658
  when project financing is known,
    643–645
  when project financing is unknown,
    645–647
  when PW value equals zero, 219
minimum-cost analysis, 346
minimum value, 777
Minuit, Peter, 69
mixed cost, 393
mixed-dollar analysis, 734
mixed financing, 633
mixed investment
  described, 290
  external interest rate, 292–293
  internal rate of return for, 294–296
  minimum attractive rate of return
    (MARR), 293–294
  vs. pure investment, 290–291
  return on invested capital (RIC),
    293–296
  trial-and-error method, 297–298
mixed series, 78, 125
Mobil Oil Canada, 507
Mobitex, 19
mode, 777
Modified Accelerated Cost Recovery System
  (MACRS)
  adjustment for overhauled asset,
    463–464
  depreciation rules, 460–464

depreciation schedules, 460*t*
described, 459–460
half-year convention, 461
improvements and repairs, 462
personal property, 461–462
property classification, 459*t*
recovery allowance percentages, 460
recovery period, 459
money
  cost of money, 60
  earning power, 61, 729
  purchasing power, 61
  time, effect of, 96*t*
  time value of money. *See* time value of
    money
Monte Carlo sampling, 799, 803–806
mortgage
  adjustable-rate mortgage (ARM), 161
  amortization, 177
  closed mortgage, 177, 178–182
  collateral mortgage, 176–177
  conventional mortgage, 176
  defined, 178, 190
  equity, 176
  fixed-rate mortgage, 140, 177
  high-ratio mortgage, 176
  interest rate, 177
  interest rate differential, 177
  open mortgage, 177
  origination fee, 178
  payment schedule, 177–178
  penalty charge, 182
  portability, 178
  prepayment privileges, 178–182
  principal, 176
  term of mortgage, 177
  terms and conditions, 177–178
  types of mortgages, 176–177
  variable-rate mortgage (VRM), 139, 177
mortgage bonds, 185, 190
mortgagee, 176
mortgagor, 176
Motorola, 19
multiple assets, 531–534, 549
multiple inflation rates, 738–740
multiple payments, 75–76
municipal bonds, 185
mutually exclusive alternatives
  and benefit-cost ratio, 692–695
  cost-effective analysis (CEA), 679*t*, 680*t*
  formulation of, 651–653
  independent projects, 652

mutually exclusive alternatives (*continued*)
mutually exclusive projects, 652
mutually exclusive risky alternatives,
comparison of, 796–798
sensitivity analysis for, 769–771
mutually exclusive interventions, 678, 678*t*
mutually exclusive projects
alternative, meaning of, 305
analysis period. *See* analysis period
basic elements of mutually exclusive
analysis, 305–308
cost-of-capital methods, 647
cost-only projects, 318–320, 528
defined, 658
described, 212
design economics, 346–354
"do nothing" decision option, 305–306
incremental analysis, 312–320, 355
vs. independent projects, 212
life-cycle cost analysis (LCCA),
339–345, 340*f*
make-or-buy decision, 336–339
mutually exclusive alternatives, 652
net equity flow, 647
project, meaning of, 305
revenue projects, 306, 307
scale of investment, 306–308, 312
service projects, 306, 307, 528
summary, 354–355
total-investment approach, 308–311, 355

N

NASDAQ, 47
National Hockey League, 88–90, 137–138
Naughton, Keith, 10*n*
negative numbers, 517
negative rate of return, 246*n*
negative taxable income, 528–531
net acquisitions, 450
net *B/C* ratio, 690, 693, 695
net cash flow equations, 537–538
net cash flow rule of signs, 259, 259*f*,
287–288
net cash flows, 64
net change in net working capital, 35*n*
net equity flow, 643–645, 647
net future worth (NFW), 224, 226–229, 312
net income
bottom line, 25
calculation of, 480–481
vs. cash flow, 483–486

defined, 480
determination of, 32
matching concept, 523
net income before adjustments, 868
non-cash expense, 485
tabular income statement, 482
within a year, 482–483
net income before adjustments, 868
net investment, 289
net-investment test, 259, 259*f,* 289–292
net operating profit. *See* gross margin
net present value (NPV)
described, 218
internal rate of return, relationship to, 256
NPW profile, 256–257, 256*n*, 259
net-present-value (NPV) method, 217–225
net present worth (NPW)
absolute (dollar) measure, 312
and benefit-cost ratio, 692
borrowed-funds concept, 223–224
criterion, 218–222
defined, 218
and inflation, 738, 742
investment pool, 222–223, 223*f*
mean of NPW distribution, 790*t*
meaning of, 222–224
minimum attractive rate of return
(MARR), 218, 219, 224–225
probability distribution of NPW,
785–798, 809
required rate of return, 218
uneven flows, 220
uniform flows, 219–220
variance of NPW distribution, 790*t*
net profit. *See* net income
net profit margin, 417
net revenue, 30
net salvage value, 496–498, 514
net worth, 22
New Orleans, 8, 8*f*
new product, 15
new production method, adoption of,
399–400
Nokia Corporation, 19, 44, 45, 46, 48*t*
nominal interest rate
see also market interest rate
defined, 189
described, 141–142
with different compounding periods, 143*t*
non-cash expense, 485
non-operating income, 490–491
non-progressive tax rates, 487–488

non-refundable tax credits, 871
nonmanufacturing cost budget, 414–415
nonmanufacturing costs, 388
nonsimple investment, 248–249, 259–261,
    287–298, 334–336
notes, 185
NPW profile, 256*n*, 259
number of interest periods, 62

O

objective probabilities, 777
oil industry, 762–763
open mortgage, 177
operating activities, 35, 37, 515–516,
    518–521
operating costs, 238–241, 392, 513, 573,
    575–576
    see also cost
operating expenses, 32, 481
operating income, 32, 410, 488–490
operating leases, 540–541
operating losses, 490, 528*n*
operating margin, 416
operating revenues, 514
operations planning, 400–402
opportunity cost, 402–404, 647
opportunity cost approach, 573–575,
    592–595, 608
opportunity loss, 819
optimal capital budget, 655
optimal capital structure, 633–637
optimal cross-sectional area (example),
    346–349
optimal investment strategy, 855
ordinary income, 869–871
Organization for Economic Co-operation and
    Development (OECD), 715
origination fee, 178
other assets, 28–29
other liabilities, 29
Ottawa-Carleton, 710–711
outsourcing. *See* make-or-buy decision
overhead, 387–388
overhead budget, 413

P

P/E ratio. *See* price-to-earnings *(P/E)* ratio
Page, Larry, 2–3
paid-in capital, 29
par value, 183, 190

parking meters, 206–207
partnership, 5
pay rates, 719, 719*t*
payback method
    benefits and flaws, 215
    conventional payback method, 212
    defined, 212
    discounted payback method, 212
    discounted payback period, 216–217, 216*t*,
        217*f*, 224*n*
    payback period, 213–215
payback period, 213–215
payback screening, 212–217
Payday, 138
payroll, 393
penalty charge, 182
perfect information, 819–822
period costs, 389
periodic interest rate, 165
perpetual replacements, 595
perpetual service life, 229–231
personal economic decisions, 7
personal income tax
    accrual method, 872
    average tax rate, 871–872, 873–874
    basic federal tax, 871
    bond investments after tax, 875–877
    business income, 873–874
    charitable donations, 871*n*
    and investment returns, 874–883
    marginal tax rates, 869*t*, 871–872,
        873–874
    net income before adjustments, 868
    non-refundable tax credits, 871
    ordinary income, income taxes on,
        869–871
    professional income, 873–874
    progressive tax system, 871
    Quebec, 871, 871*n*
    Registered Education Savings Plan
        (RESP), 203, 882–883
    Registered Retirement Savings Plans
        (RRSPs), 880–882
    stock investments after tax, 878–880
    taxable income, calculation of, 868
    timing of personal tax payments, 872
    total income, 868
    total qualified amount, 871
personal investments
    adjusted cost base, 879–880
    average annual return, 857–858
    "beat the market," 862

personal investments (*continued*)
   bond investments, 849–852
   bond investments after tax, 875–877
   cash investments, 847–849
   compound return, 857–858
   diversification, 863–864
   dividends, 874–875
   dollar cost averaging, 866–867
   expected return, 856–859
   income taxes, effect of, 874–883
   inflation, effect of, 860–861
   investment basics, 846–847
   investment strategies, 855–867
   investment timing issues, 862–863
   liquidity, 846–847, 863
   optimal investment strategy, 855
   planning horizon, 862
   Registered Education Savings Plan
     (RESP), 203, 882–883
   Registered Retirement Savings Plans
     (RRSPs), 880–882
   return, 847
   return on investment over time, 857–859
   risk, 847, 859–862
   risk-free return, 856
   risk-reward trade-off, 864–866
   stock investments, 852–855
   stock investments after tax, 878–880
   time horizon, 847
   university fund, 118–120
PET bottles, 384–385
physical depreciation, 429
pipe size, economical (example), 350–354
plan for receipts or disbursements, 62
planning horizon
   see also analysis period
   described, 581
   finite planning horizon, 587–590
   infinite planning horizon, 582–587, 601–607
   personal investments, 862
   predictions of technological patterns, 581
   replacement analysis, 580–590
point of view, 77
Port of Halifax, 670–671
portability of mortgage, 178
posterior probabilities, 825
potential investment opportunities.
     See profitability
Powell, Donald E., 8
Pratt & Whitney, 682
pre-production mining expenditures, 491
predictions. *See* forecasts

preferred stock
   vs. common stock, 41
   and cost of equity, 639
   described, 29
   and dividends, 33, 41
premium bond, 183
prepaid expenses, 29
prepayment privileges, 178–182
present economic studies, 386
present equivalent cost (PEC), 574, 582, 587,
     594–595, 607
present equivalent incremental cost, 541
present equivalent (PE) method, 217–225
present value
   see also present worth
   choice of point in time, 72
   defined, 63
   lower present-value production cost, 306
   net present value (NPV), 218
present worth
   see also present value; present-worth
     analysis
   defined, 63, 208
   present-worth comparison, 308–310
   of uneven series by decomposition into
     single payments, 87–88
present-worth analysis
   analysis period coinciding with longest
     project life, 327–329
   borrowed-funds concept, 223–224
   discounted cash flow techniques
     (DCFs), 217
   internal rate of return, relationship to, 256
   investment pool, 222–223, 223*f*
   minimum attractive rate of return
     (MARR), 218, 219, 224–225
   net present worth, 218, 222–224
   net-present-worth criterion, 218–222
   present-worth amounts at varying interest
     rates, 221*t*
   present-worth profile, 221, 221*f*
   project life shorter than analysis period,
     324–326
   project lives longer than analysis period,
     322–323
   replacement analysis, 587–590
   unequal lives, lowest-common-multiple
     method, 329–331
   variations of, 225–233
present-worth factor
   equal payment series, 100–102
   geometric gradient series, 109–112

linear gradient series, 103–105
single cash flow, 82
single-payment present-worth factor, 82, 103, 109
price
  see also pricing
  bond prices, 185–189, 851
  inflation. *See* inflation
  and inflation, 319
  oil prices, 762–763
price indexes, 721–722, 722n
price-to-earnings *(P/E)* ratio, 47
pricing
  see also price
  services, 243–244
  the use of an asset, 243–244
primary benefit, 682
prime rate, 139
principal, 62, 176
Prius, 9–12, 11f, 13–14, 15
private Canadian corporations, 487
probabilities
  see also project risk
  assessment of probabilities, 776–780
  coefficient of correlation, 785
  conditional probability, 783–785, 822–823, 823f
  continuous random variables, 776, 777
  covariance, 785
  dependent random variables, 792
  discrete random variable, 776
  expected value, 781
  independent random variables, 783, 791–792
  joint probability, 783–785, 823–825, 823f, 824t
  marginal probability, 783, 823–825, 824t
  maximum value, 777
  mean, 781, 782–783
  minimum value, 777
  mode, 777
  mutually exclusive risky alternatives, comparison of, 796–798
  objective probabilities, 777
  posterior probabilities, 825
  probability distributions. *See* probability distributions
  random variable, 776
  range, 781
  revised probabilities, 825
  risk simulation, 798–813
  standard deviation, 781, 782

  subjective probabilities, 777
  summary of probabilistic information, 781–783
  variance, 781–783
probability distributions
  see also probabilities; project risk
  cumulative distribution, 778–780
  described, 777
  Monte Carlo sampling, 799, 803–806
  NPW distribution, 785–798, 809
  risk simulation, 798–813
  triangular distribution, 777, 777f
  uniform distribution, 777, 778f
  for unit demand *(X)* and unit price *(Y)*, 778t
process selection. *See* equipment or process selection
product costs, 389–390
product expansion, 15
production
  estimate of profit from, 409–417
  new production method, adoption of, 399–400
  production budget, 410f, 411–413
production budget, 410f, 411–413
professional advice, 820–822, 822–827
professional income, 873–874
profit
  accounting profit, 25, 485
  defined, 480
  estimating. *See* profit estimation
  inflation, and profits with working capital, 747–749
  net operating profit. *See* gross margin
  net profit. *See* net income
  net profit margin, 417
  operating profit margin, 416
  unit profits, 241, 242–243
profit estimation
  administrative expenses budget, 415
  budgeted income statement, 416–417, 417t
  cost-of-goods sold budget, 413–414
  direct labour budget, 412–413
  materials budget, 412
  nonmanufacturing cost budget, 414–415
  operating income, calculation of, 410
  overhead budget, 413
  from production, 409–417
  production budget, 410f, 411–413
  sales budget for manufacturing business, 410–411
  selling expenses budget, 414–415

profit margin, 45, 47
profit margin on sales, 45
profit maximization, 407–409
profitability
    annual equivalent-worth analysis, 234–244
    capitalized equivalent-worth analysis, 225, 229–233
    common measures of, 208
    future-worth analysis, 225–229
    in inflationary economy, 735–738
    internal rate of return, 256–261
    present-worth analysis, 217–225
    rate of return analysis, 245–256
profitability analysis, 45–47
project, 651
    see also engineering projects
project balance (PB), 224, 292
project cash flows
    actual (current) dollars (A_n), 726–729
    after-tax cash flow diagram approach, 545–550
    cash flow statements. *See* cash flow statement (individual projects)
    cash inflows, elements of, 514
    cash outflows, elements of, 513
    classification of cash flow elements, 514–516, 516*t*
    constant (real) dollars (A'_n), 726–729
    cost-only project, 528–531
    describing project cash flows, 209–212
    differential cash flows, 512–516
    financing activities, 516
    generalized cash flow approach, 537–545
    identifying, 210–211
    income statement, 517, 517*f*
    incremental cash flows, 512
    inflation, effects of, 735–742
    vs. loan cash flows, 209–212, 209*f*
    operating activities, 515–516
    summary, 550–551
project expenses, 410, 481
project life
    AE analysis for unequal project lives, 331–333
    vs. analysis period, 321–329
    common service period, 329
    construction industry, 322
    economic service life, 575–580, 583, 584–585, 587, 595–601, 608
    extremely long service life, 231–233
    IRR analysis for unequal project lives, 333–336

lowest common multiple, 329–331
    perpetual service life, 229–231
project revenue, 180, 410
project risk
    aggregation of risk over time, 791–795
    break-even analysis, 771–774
    defined, 764
    expected-value criterion, 796
    mean-and-variance criterion, 796
    methods of describing, 765–776
    mutually exclusive risky alternatives, comparison of, 796–798
    origins of, 764–765
    risk simulation, 798–813
    scenario analysis, 774–776
    sensitivity analysis, 765–769
    sensitivity analysis for mutually exclusive alternatives, 769–771
    summary, 827–828
project screening method, 212–217
promissory note, 177
proprietorships, 5, 873
Province of Manitoba, 682
provincial bonds, 184–185
provincial/territorial corporate tax structure, 488*t*
public-project analysis. *See* public sector
public sector
    benefit-cost analysis, 681–688
    benefit-cost ratio, 689–695
    cost-effectiveness, 695
    cost-effectiveness case studies, 695–704
    inherent difficulties, 688–689
pumping system with problem valve (example), 341–345
purchase of new equipment, 513
purchasing power, 61
pure borrowing, 289
pure investment, 289, 290–292

Q

qualified property, 491
quality-adjusted life years (QALYs), 675
quality improvement, 15
Quebec, 871, 871*n*
quick (acid-test) ratio, 43

R

R-bonds, 184
random numbers, 803

random variable
   continuous random variables, 776
   defined, 776
   dependent random variables, 792, 810
   discrete random variable, 776
   expected value, 781
   independent random variables, 783,
      791–792
range, 781
rate of return
   average annual return, 857–858
   cash investments, 847–849
   common definitions of, 245–247
   compound return, 857–858
   defined, 262
   described, 245
   direct solution method, 249–251
   expected return, 856–859
   graphical method, 254–256, 255f
   and inflation, 743–749, 750t
   internal rate of return. See internal rate
      of return
   lending rate, 648
   marginal efficiency of capital, 245
   methods of finding i^*, 249–256
   minimum attractive rate of return
      (MARR). See minimum attractive rate of
      return (MARR)
   negative rate of return, 246n
   personal investments, 847
   required rate of return, 218
   return on invested capital (RIC), 247,
      293–296
   return on investment, 245–246
   risk-free return, 856
   simple vs. nonsimple investments,
      248–249
   trial-and-error method, 251–254
ratios
   asset management analysis, 43–45
   average collection period, 44
   benefit-cost ratio, 689–695
   book value per share, 48
   cost-effectiveness ratio, 676–677
   current ratio, 42–43
   days sales outstanding, 44
   debt management analysis, 39–42
   debt ratio, 41, 526, 632
   earnings per share (EPS), 853
   inventory turnover, 43–44
   limitations of, 48–49
   liquidity analysis, 42–43

   market value analysis, 47–48
   price-to-earnings (P/E) ratio, 47
   profit margin on sales, 45, 47
   profitability analysis, 45–47
   quick (acid-test) ratio, 43
   return on assets (ROA), 46
   return on equity (ROE), 46–47
   RIM vs. Nokia, key financial ratios, 48t
   summary, 49–50
   target debt ratio, 633
   times-interest-earned ratio, 41–42
   total assets turnover ratio, 45, 47
   trend analysis, 49
   types of ratios used, 39f
   use of, 39–40
raw material price index (RMPI), 722, 722t
raw-materials inventory, 390
RBC Capital Trust, 475t
real assets, 28n
real dollars, 726–729
real interest rate, 729
real IRR, 743
recapture of capital cost allowance, 450
recaptured CCA, 450, 495
recaptured depreciation, 495
recovery allowance percentages, 460
recovery period, 459
redeemable bond, 183
redemption feature, 183
refundable investment tax credit, 492
registered bonds, 183
Registered Education Savings Plan (RESP),
    203, 882–883
Registered Retirement Savings Plans
    (RRSPs), 880–882
regular interest bonds, 184
relative (percentage) measure, 312
relevant cash flow, 815–816
relevant range, 392
remaining-balance method, 169–170
rental income, lost, 403–404
replacement analysis
   annual equivalent worth (AE), 582–587
   basic concepts, 570–573
   capital cost allowance, effects of, 591–607
   challenger, 570–571, 607
   current market value, 571, 607
   decision criterion, 582
   decision frameworks, 581–582, 582f
   defender, 570–571, 607
   economic service life, 575–580, 583,
      584–585, 587, 595–601, 608

replacement analysis (*continued*)

   finite planning horizon, 587–590

   fundamentals, 570–573

   infinite planning horizon, 582–587,
      601–607

   net proceeds from disposal of old machine,
      591–592

   operating costs, 573

   opportunity cost approach, 573–575,
      592–595, 608

   perpetual replacements, 595

   planning horizon (study period), 581

   present equivalent cost (PEC), 574, 582,
      587, 594–595, 607

   present-worth approach, 587–590

   relevant information, 571–572

   required assumptions, 581–582

   revenue and cost patterns over asset
      life, 581

   summary, 607–608

   sunk costs, 572–573, 572*f,* 607

   tax considerations, 591–607

   technological change, 581, 590, 608

   terminology, 570–573

   when required service is long, 580–590

replacement problem, 570

   see also replacement analysis

required rate of return, 218

   see also minimum attractive rate of return
      (MARR)

required service period, 320

Research In Motion Ltd. (RIM), 19–21,
   25–38, 27*t,* 31*t,* 36*t,* 40*t,* 41–48, 48*t,* 517

residual value, 172

   see also salvage value

retained earnings

   cost of, 638

   defined, 30

   reporting of, 32–33

   sources of, 629, 629*f*

return on assets (ROA), 46

return on equity (ROE), 46–47

return on invested capital (RIC), 247,
   293–296

return on investment, 186, 245–246, 743–746

return on investment over time, 857–859

revenue

   see also income

   defined, 30

   differential revenue, 398–402

   marginal revenue, 16, 404

   net revenue, 30

   operating revenues, 514

   patterns over life of asset, 581

   project revenue, 180, 410

revenue projects, 306, 307

revised probabilities, 825

RIMgate, 20

risk

   see also project risk

   aggregation of risk over time, 791–795

   decrease in buying power of capital,
      860–861

   defined, 764

   diversification, effect of, 863–864

   failure to reach objectives, 862

   of losing money, 859–860

   missing potentially better returns, 861–862

   personal investments, 847, 859–862

   risk-averse, 861

   risk simulation, 798–813

risk analysis, 776

   see also probabilities

risk-free return, 856

risk premium, 856

risk-return trade-off

   defined, 17, 798

   described, 17

   personal investments, 864–866

risk simulation

   computer simulation, 799–800

   dependent random variables, 810

   graphic displays, 809–810, 809*f*

   model building, 800–803

   Monte Carlo sampling, 799, 803–806

   random numbers, 803

   with @RISK, 810–813, 814*f*

   sampling procedure, 803

   simulation output analysis, 808–810

Rogers Cantel Communications, 19

rollback, 816

Rolls-Royce Canada Limited, 682

Rule of 72, 86

S

sales budget, 410–411

sales taxes, 393

salvage value

   see also residual value

   on cash flow statement, 520, 520*n*

   as cash inflow, 514

   conventional payback period with,
      214–215

defined, 321–322, 433
and depreciation, 433
inflation tax effect, 742
under MACRS, 460
net salvage value, 496–498, 514
sampling procedure, 803
scale of investment, 306–308, 312
scenario analysis, 774–776
scientific research and experimental
  development (SR&ED), 491
scrap value. *See* salvage value
screening capital investments, 212–217
second mortgage bonds, 185
second-order conditions, 346
secondary benefit, 682
secondary effects, 682
SEDAR (System for Electronic Document
  Analysis and Retrieval), 853
self-employment, 873
selling expenses budget, 414–415
semivariable costs, 393
sensitivity analysis, 765–769
sensitivity analysis for mutually exclusive
  alternatives, 769–771
sensitivity graphs, 765
Sequoia Capital, 3
service life. *See* project life
service projects, 306, 307, 528
service sector
  characteristics of, 673
  described, 672–675, 672*f*
  health-care service, 675–680
  improvements, 674
  pricing services, 673–675
  public sector. *See* public sector
  scale of, 673
  summary, 704–705
shareholders' equity, 30
short case studies. *See* case studies
short-run problems
  make-or-buy (outsourcing) decision, 402
  method changes, 399
  operations planning, 400–402
short-term debt financing, 39
simple borrowing cash flows, 248
simple interest, 65, 66, 68–69, 68*f,* 125, 171
simple investments, 248–249, 256–258, 262
simple projects, 511
simulation model, 800–803
  see also risk simulation
single cash flow, 78, 79–86, 125
single-cash-flow formulas

compound amount factor, 79
find *F,* given *i, N,* and *P,* 81–82
find *N,* given *P, F,* and *i,* 85–86
find *P,* given *F, i,* and *N,* 83
interest rates, solving for, 83–85
single-payment compound-amount
  factor, 81
single-payment present-worth factor, 82
time, solving for, 8
single-payment compound-amount factor, 81
single-payment formulas for continuous
  compounding, 157
single-payment present-worth factor, 82,
  103, 109
single projects
  annual equivalent worth (AE), 235–236
  decision rule for simple investments,
    256–257
sinking-fund factor, 93
Sirius XM Radio Inc., 57
small business deduction (SBD), 487
"smart shoe," 137
SNC-Lavalin Group Inc., 475*t,* 476
social discount rate, 683–684
social opportunity cost of capital
  (SOCC), 683
social rate of time preference (STP), 683
specific inflation rate, 724–725, 738–740
sponsors, 681
standard deviation, 781, 782
Statement of Business or Professional
  Activities, 873
statement of cash flows. *See* cash flow
  statement
statement of financial position. *See*
  balance sheet
Statistics Canada, 721, 722*n*
stock
  adjusted cost base, 879–880
  book value per share, 48
  capital stocks, 29
  common stock. *See* common stock
  market value analysis, 47–48
  outstanding stock, 29
  personal investments in stocks, 852–855,
    878–880
  preferred stock. *See* preferred stock
  price-to-earnings *(P/E)* ratio, 47
  prices, 853–854
  treasury stock, 29
stock investments, 852–855, 878–880
stock prices, 25, 25*f*

stockholders' equity. *See* shareholders' equity

straight-line (SL) method, 434–435, 438–439

strict gradient series, 102, 102*f*

study period. *See* analysis period; planning horizon

subjective probabilities, 777

sum-of-years'-digits (SOYD) method, 436, 440–441

Summit International LLC, 384–385

Suncor Energy Inc., 56

sunk costs, 404, 572–573, 572*f*, 607

Syncrude Canada Ltd., 56

T

T2 Corporation - Income Tax Guide, 445, 447

tabular method, 166–168

target capital structure, 633

target debt ratio, 633

tax-adjusted weighted average cost of capital, 642

tax credits

    investment tax credit. *See* investment tax credit (ITC)

    non-refundable tax credits, 871

    total qualified amount, 871

tax depreciation. *See* capital cost allowance (CCA)

tax rates

    average tax rate, 477, 871–872, 873–874

    Canadian-controlled private corporations (CCPC), 487

    combined corporate tax rates, 486–488

    definitions, 477–478

    effective tax rate, 477

    incremental tax rate, 478, 493

    marginal tax rates, 477, 869*t*, 873–874

    non-progressive, 487–488

tax savings, 513, 528

tax shield, 513, 528

taxable income, 32, 481–482, 868

taxation

    corporate income tax. *See* corporate income tax

    personal income tax. *See* personal income tax

    sales taxes, 393

TD Canada Trust, 139

technological change, 581, 590, 608

Term Deposits (TDs), 847

term loans, 631–632

term of loan, 165

term of mortgage, 177

terminal loss, 450

time

    aggregation of risk over time, 791–795

    effect on value of money, 96*t*

    as factor, 18

    Rule of 72, 86

    solving for time, 8

    time shifts, handling, 92–93

time value of money

    see also interest

    and cash, 484

    cash flow diagrams, 64, 64*f*

    defined, 61

    described, 15, 61–62, 62*f*

    and earning power, 61

    illustration of, 16*f*

    and purchasing power, 61

    summary, 125

times-interest-earned ratio, 41–42

Toronto Stock Exchange, 21, 853–854

total assets turnover ratio, 45

total income, 868

total-investment approach

    annual equivalent comparison, 310–311

    described, 308, 355

    present-worth comparison, 308–310

total qualified amount, 871

Toyota Motor Corporation, 9–12, 11*f*, 13–14, 15

trade-in allowance, 432

Treasury bills, 184, 847

Treasury Board of Canada, 684

treasury stock, 29

trend analysis, 49

trial-and-error method, 251–254, 297–298

triangular distribution, 777, 777*f*

U

Ultra Electronics, 303–304

uncertainty, 18, 633*n*, 650, 658

    see also project risk

unconventional equivalence calculations

    cash flow with subpatterns, 117–118

    composite cash flows, 115–120, 116*f*

    described, 114

    interest rate, determination of, 120–124

    university fund, 118–120

    unknown interest rate, with multiple factors, 121–124

undepreciated capital cost (UCC), 447, 448, 450
uneven payment series, 86–90, 163–164, 220
uniform distribution, 777, 778*f*
uniform series. *See* equal payment series
uniform series compound-amount factor, 91
unit costs or profits, 241, 242–243
unit of use, 243–244
United States
    tax depreciation, 459–464
units-of-production (UP) method, 442
university fund, 118–120
unrecognized gain, 432
unrecognized loss, 432
U.S. Army Corps of Engineers, 8
U.S. Army cost-effectiveness case study, 696–704
useful life, 433
    see also project life
users, 681
users of accounting information, 23

V

variable cost, 393
variable interest rates, 161–164
variable-rate mortgage (VRM), 139, 177
variables. *See* random variable
variance, 781–783, 790*t*, 796
variation, measure of, 781–783
volume
    break-even volume, 407
    break-even volume analysis, 400–402
    and cost behaviour, 393–398

cost-volume analysis. *See* cost-volume analysis
    cost-volume diagram, 395–396
    cost-volume relationships, 399
volume index, 392

W

Western Canada Lottery Corporation, 138
"what-if" analysis, 765–769
Winnipeg, 712–713
work-in-process inventory, 390
working capital
    adjustments in, 35
    defined, 42, 513
    inflation, effects of, 746–749
    investments in, 513, 521–526
    net change in net working capital, 35*n*
working-capital release, 514
working-capital requirements, 521–526, 523*f*, 549–550
worth of obtaining additional information, 819–822

Y

Yahoo!, 4
yearly inflation rates, 725–726
yield, 245
    see also rate of return
yield to maturity (YTM), 186–188, 851–852
Young, John, 136

Z

Zetterberg, Henrik, 88–90

Summary of Useful Excel Financial Functions (Part A)

Description		Excel Function	Example	Solution
Single-Payment	Find: F Given: P	$=\text{FV}(i, N, 0, -P)$	Find the future worth of $500 in 5 years at 8%.	$=\text{FV}(8\%, 5, 0, -500)$ $=\$734.66$
Cash Flows	Find: P Given: F	$=\text{PV}(i, N, 0, F)$	Find the present worth of $1300 due in 10 years at a 16% interest rate.	$=\text{PV}(16\%, 10, 0, 1300)$ $=(\$294.69)$
	Find: F Given: A	$=\text{FV}(i, N, A)$	Find the future worth of a payment series of $200 per year for 12 years at 6%.	$=\text{FV}(6\%, 12, -200)$ $=\$3373.99$
Equal-Payment-Series	Find: P Given: A	$=\text{PV}(i, N, A)$	Find the present worth of a payment series of $900 per year for 5 years at 8% interest rate.	$=\text{PV}(8\%, 5, 900)$ $=(\$3593.44)$
	Find: A Given: P	$=\text{PMT}(i, N, -P)$	What equal-annual-payment series is required to repay $25,000 in 5 years at 9% interest rate?	$=\text{PMT}(9\%, 5, -25000)$ $=\$6427.31$
	Find: A Given: F	$=\text{PMT}(i, N, 0, F)$	What is the required annual savings to accumulate $50,000 in 3 years at 7% interest rate?	$=\text{PMT}(7\%, 3, 0, 50000)$ $=(\$15,552.58)$
	Find: NPW Given: Cash flow series	$=\text{NPV}(i, \text{series})$	Consider a project with the following cash flow series at 12% ($n = 0, -\$200$; $n = 1, \$150$; $n = 2, \$300$; $n = 3, 250$)?	$=\text{NPV}(12\%, \text{B3:B5})+\text{B2}$ $=\$351.03$
Measures of Investment Worth	Find: IRR Given: Cash flow series	$=\text{IRR}(\text{values, guess})$	(see table below)	$=\text{IRR}(\text{B2:B5}, 10\%)$ $=89.2\%$
	Find: AW Given: Cash flow series	$=\text{PMT}(i, N, -\text{NPW})$		$=\text{PMT}(12\%, 3, -351.03)$ $=\$146.15$

	A	B
1	Period	Cash Flow
2	0	−200
3	1	150
4	2	300
5	3	250